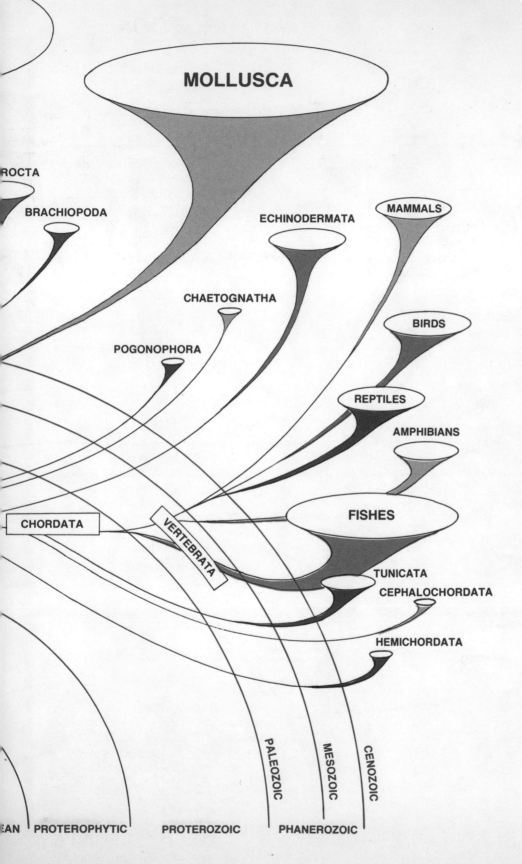

MOLLUSCA

ROCTA

BRACHIOPODA

ECHINODERMATA

MAMMALS

CHAETOGNATHA

BIRDS

POGONOPHORA

REPTILES

AMPHIBIANS

CHORDATA

VERTEBRATA

FISHES

TUNICATA

CEPHALOCHORDATA

HEMICHORDATA

PALEOZOIC

MESOZOIC

CENOZOIC

EAN PROTEROPHYTIC PROTEROZOIC PHANEROZOIC

BIOLOGY OF ANIMALS

CLEVELAND P. HICKMAN
Department of Zoology, DePauw University,
Greencastle, Indiana

CLEVELAND P. HICKMAN, Jr.
Department of Biology, Washington
and Lee University, Lexington, Virginia

FRANCES M. HICKMAN
Department of Zoology, DePauw University,
Greencastle, Indiana

BIOLOGY OF ANIMALS

SECOND EDITION

with 764 illustrations, 149 in color

The C. V. Mosby Company
Saint Louis 1978

SECOND EDITION

Copyright © 1978 by The C. V. Mosby Company

All rights reserved. No part of this book may be reproduced in any manner without written permission of the publisher.

Previous edition copyrighted 1972

Printed in the United States of America

Distributed in Great Britain by Henry Kimpton, London

The C. V. Mosby Company
11830 Westline Industrial Drive, St. Louis, Missouri 63141

Library of Congress Cataloging in Publication Data

Hickman, Cleveland Pendleton, 1895-
    Biology of animals.

    Includes bibliographies and index.
    1. Zoology.   I. Hickman, Cleveland P., joint
author.   II. Hickman, Frances Miller, joint author.
III. Title.   [DNLM: 1. Animals.   QL47.2 H628b]
QL47.2.H53 1978   591   77-10766
ISBN 0-8016-2166-6

TS/VH/VH   9 8 7 6 5 4 3 2 1

To

THE ANIMALS, GREAT AND SMALL

that enrich our lives and
teach us so much about ourselves

PREFACE

Biology of Animals originated as a textbook intended for the beginning college student, regardless of background, seeking a general knowledge of the nature and functioning of animals and of the biosphere in which they live. It was an offspring of the senior author's fourth edition of *Integrated Principles of Zoology,* in part an abridgement of its parent and in part a new approach to animal biology. Despite the success of the first edition of *Biology of Animals,* we felt the need of an extensive revision to accomplish two ends: (1) to seat the entire presentation more firmly into an evolutionary theme and (2) to expand the survey of the animal kingdom, which many readers felt had been trimmed with too sharp a scalpel during its original condensation from the parent book.

We soon discovered that these aims could not be accomplished by simply rearranging and updating the material. We had to rewrite virtually all of the book. Although past readers may be discomfited to find little in the present book that resembles the first edition, we believe this approach is pedagogically more effective and we trust that teachers and students alike will find it stimulating, intellectually lively, and absorbing.

Because the evolutionary process pervades biology, the principles of evolution and the way in which they work form the central theme of the first six chapters of the book. Our treatment begins with an analysis of the earth's remarkable environment and the ecoevolutionary interactions that permitted the evolution of life as we know it. The concepts of ecosystem, biotic community, and population are considered in detail, and examples emphasize the reciprocal relationship between human existence and ecosystem structure. The second chapter traces the evolution of life on earth from its primitive beginning some 3 billion years ago to the appearance of the eukaryotes toward the end of the Precambrian. The chemistry of life's basic molecules—proteins and nucleic acids—is also treated in this chapter. Chapter 3 moves to the organization of the eukaryotic cell, including such topics as cell division, chromosome structure, and membrane function. We progress then to a detailed consideration of the genetic mechanisms that underlie evolution, with special consideration of mendelian genetics and the chromosomal theory of inheritance. The mechanisms of evolutionary change and speciation are then treated in detail, with due consideration given to the historical evidences of evolution. The final chapter in this section deals with the evolution of behavior in animals. Stereotyped and social behavior are discussed with numerous examples to clarify the topics of instinct and learning, aggression, dominance, territoriality, and communication. Although we think Part one can be approached most effectively if all six chapters are considered in order, instructors should find the treatment sufficiently flexible to permit some other order or the use of certain chapters apart from the rest.

Part two deals with animal form and function and shows how the different functional systems are coordinated to serve animals as they interact with their environment. The scope and emphasis of this part remain unchanged from the first edition, in which this material appeared as Part three. However, the text has been updated throughout, and we have made several shifts in placement of material. The most important of these are the expansion of the nutrition chapter to include enzyme physiology and cellular metabolism and the inclusion of a single chapter on reproduction and development, representing the combination of material that appeared in two chapters in Part five of the first edition.

Parts three and four center on the living animal and its activities and on the diversity of animal life. In this systematic consideration of the animal kingdom, all of the invertebrate and vertebrate groups are treated in considerably greater depth than in the first edition of *Biology of Animals*. We have placed greater emphasis on animal diversity and on the evolution, adaptations,

and natural history of each group and less emphasis on the anatomy of representative types. The introductory chapter to Part three, on animal phylogeny, has been rewritten, and a new chapter on chordate ancestry and evolution introduces Part four.

The recent, and we think justified, general acceptance of the five-kingdom system creates a pedagogic problem for the zoologist, since in this scheme the protozoans have been removed from the animal kingdom and included with other single-celled eukaryotic organisms in the kingdom Protista. Yet the protozoans share so many clearly animal-like characteristics that their evolutionary affinity to multicellular animals can hardly be disputed. Thus it is for more than traditional reasons that we retain the protozoans in this treatment of the animal kingdom. The five-kingdom system and the fluid nature of classifications within the kingdoms of life are discussed in Chapter 14.

Perhaps a more relentless problem that we must face with each edition is when to incorporate new taxonomic revisions within the animal phyla. We, of course, always attempt to use the most modern classification available *that has gained general acceptance.* On the other hand, we can think of nothing more provoking to the instructor than to encounter a completely remodeled classification every time a new edition appears. Consequently, in the levels of synthesis through which new concepts and schemes pass—from research report to review article to textbook—we prefer to allow new taxonomic revisions to weather the scrutiny of more than one reviewer before incorporating them in our textbooks.

The instructor and student will find several aids to their study of animal biology. Lists of suggested readings to supplement the text material now follow each chapter. Both the general references and the *Scientific American* articles (which are listed separately) have for the most part been annotated to assist readers. The glossary has been completely revised and expanded. Many new illustrations have been added or replace old ones. The front endpaper bears a complete family tree of the animal kingdom showing supposed group relationships, temporal diversity, and the relative success of the groups in terms of numbers of species as suggested by the size of the ovals. The back endpaper provides a geologic time table and pictorial scheme of the supposed origins and spatial diversity of the kingdoms of life. Both instructor and student should find these highly accessible diagrams useful for repeated reference.

We are especially indebted to a number of people who reviewed principal parts of the manuscript: Ralph Nursall and Royal Ruth of the University of Alberta who read most of Part one; Mary Gardner of the Smithsonian Institution who read all of Part three; Carl Ernst of George Mason University who read all of Part four; Charles Fugler of the University of North Carolina at Wilmington who read part of Part four; W. Preston Adams and Michael D. Johnson of DePauw University who read Chapters 14 and 21, respectively; Diane Nelson of East Tennessee State University who read Chapters 22 and 23; and Joseph Nelson of the University of Alberta who read Chapter 25. We benefited greatly from their helpful suggestions and advice. Rufus Rickenbacher of Maplewood, New Jersey, exercised his exceptionally fine editorial skills on the entire manuscript, correcting syntax and exposition, querying us on logic, and saving us from numerous future embarrassments.

William C. Ober of Crozet, Virginia, prepared numerous excellent new line drawings and paintings for this edition. The fine pencil renderings of animals that introduce many of the chapters were executed by David Hunter of Lexington, Virginia.

Finally, we thank the many teachers who wrote us about the first edition of *Biology of Animals*. Their suggestions and criticisms were of great assistance to us in preparing this edition. We hope they will continue to help us, for there is no one whose guidance we value more.

Cleveland P. Hickman
Cleveland P. Hickman, Jr.
Frances M. Hickman

CONTENTS

Contents

Contents

PART ONE EVOLUTION OF ANIMAL LIFE

"Spaceship earth" viewed from space.

1 FITNESS OF THE ENVIRONMENT

CONDITIONS FOR LIFE ON EARTH

ANALYSIS OF THE ENVIRONMENT

ECOSYSTEMS

Solar radiation and photosynthesis
Production and the food chain
Ecologic pyramids
Nutrient cycles

COMMUNITIES

Ecologic dominance
Ecologic niche concept

POPULATIONS

Natural balance within populations
How population growth is curbed

CONDITIONS FOR LIFE ON EARTH

All life is confined to a thin veneer of the earth called the **biosphere.** Viewers of the first remarkable photographs of earth taken from the Apollo spacecraft, revealing a beautiful blue and white globe lying against the limitless backdrop of space, were struck and perhaps humbled by our isolation and insignificance in the enormity of the universe. The phrase "spaceship earth" became a part of the vocabulary, and the realization evolved that all the resources we will ever have for sustaining life are restricted to a thin layer of land and sea and a narrow veil of atmosphere above it. We could better appreciate just how thin the biosphere is if we could shrink the earth and all of its dimensions to a 5-foot sphere. We would no longer perceive vertical dimensions on the earth's surface. The highest mountains would fail to penetrate a thin coat of paint applied to our shrunken earth; a fingernail's scratch on the surface would exceed the depth of the ocean's deepest trenches.

Our earth is a minor planet circling an ordinary star in one galaxy among billions. More than 10^{20} (100 million million million) stars have been revealed by powerful telescopes. Many of these stars are like our sun and have planetary systems. Among these systems there must be some that closely resemble earth. The astronomer Harlow Shapley has made the conservative estimate that there are 100 million possible places for life in the universe, 100,000 in our galaxy alone. If he is correct, we can no longer maintain that only the planet earth harbors life. Nevertheless, there are so many requirements for life that only a small number of planets can fulfill the special conditions that would permit evolution of life at all similar to that on earth.

First, a planet suitable for life must receive a steady supply of light and heat from its sun for many billions of years. This means that its orbit must be nearly circular.

Second, water must be present on the planet to permit the evolution of complex biochemical systems based on carbon. Fanciful systems of life based on ammonia and silicates rather than water and carbon have been suggested, but such speculations are totally hypothetical and in any case would require environmental conditions vastly different from those on earth.

Third, the temperature must be suitable, meaning practically within the range of $-50°$ to $+100°$ C. Life at temperatures above $100°$ C is impossible because biopolymers based on carbon and water would be rapidly hydrolyzed. Temperatures much below the freezing point of water prevent growth of organisms by slowing chemical processes, although some inactive organisms may survive storage in liquid nitrogen ($-195°$ C) or even in liquid helium ($-269°$ C).

Fourth, all life requires a suitable array of major and minor elements. Oxygen, carbon, hydrogen, and nitrogen form 95% of protoplasm and thus dominate the composition of life on earth. These four are supported by seven other major elements (phosphorus, calcium, potassium, sulfur, sodium, chlorine, and magnesium) and a large number of minor or "trace" elements. Perhaps 46 elements in all are found in protoplasm; many are essential for life, whereas others are present in protoplasm only because they exist in the environment with which the organism interacts. Not all planets possess elements in the same proportions, even in our solar system.

There are many other properties of earth that make it an especially fit environment for life. It is large enough to have a surface density that permits molecules to collect and align properly. Protoplasm is in a colloidal state, an intermediate between conditions too solid to allow change and conditions too fluid or gaseous to permit molecular organization. The earth's gravity is strong enough to hold an extensive gaseous atmosphere but not so strong that more than a trace of free hydrogen remains. Another consideration, especially important for life on earth today, is the oxygen-ozone atmospheric screen that absorbs lethal ultraviolet radiation from the sun. So effective is this absorption that rays with wavelengths shorter than 283 nm (nanometer = 10^{-12} meter) fail to reach the earth's surface.

In his classic book *Fitness of the Environment,* published in 1913, the biochemist L. J. Henderson maintained that earth possesses "the best of all possible environments for life." We may agree; however, we realize that the surface of any body that is the size and age of earth in a similar orbit revolving about a similar sun and having a similar elemental composition should also have an excellent environment for life.

As described in the next chapter, life's origins depended on the formation of a few basic molecules composed of atoms that exist on all stars and planets in the universe. The same laws of physics and chemistry that operated during life's origins on earth must also apply throughout the universe. Therefore the sequence of events that led to life on earth may also have occurred on other planets in our galaxy and on planets in other galaxies as well.

Even so, there is no reason to conclude that all other earthlike planets in the universe have undergone evolutions that produced DNA, chlorophyll, plants, flowers, fungi, molluscs, fishes, frogs, and people. The manner in which living organisms on earth capture and exchange energy, move, respond, reproduce, and grow is not an inevitable pattern of life. There is no objective evidence that the evolutionary process follows a predetermined pattern or is channeled and pointed to some inevitable goal.

When we realize that an advanced mammal such as man possesses an enormous number of inherited characteristics that have appeared in response to changing adaptive pressures during his long evolution, it is obvious that man is only one of an almost infinite variety of structural and intellectual combinations that might have appeared. We are the improbable product of a long ancestry of organisms shaped and molded by their heredity and environment. If we could start all over again at the dawn of life on a primitive earth, it is most unlikely that the result of 3 billion years of evolution would be the same. Life may well exist on other earthlike planets, but the kind of life on each planet is surely unique.

The organism and its environment share a reciprocal relationship. The environment is fit for the organism, and the organism is fitted to the environment and adapts to its changes. As an open system, an animal is forever receiving and giving off materials and energy. The building materials for life are obtained from the physical environment, either directly by producers such as green plants or indirectly by consumers that return inorganic substances to the environment by excretion or by the decay and disintegration of their bodies.

The living form is a transient link that is built up out of environmental materials, which are then returned to the environment to be used again in the re-creation of new life. Life, death, decay, and re-creation have been the cycle of existence since life began.

In this continuous interchange between organism and environment, both are altered in the process and a favorable relationship is preserved. The environment of earth, with its living and nonliving components, is not a static entity but has undergone an evolution in every way as dramatic as the evolution of the animal kingdom. It is still changing today, more rapidly than ever before, under man's heavy-handed influence.

The primitive earth of 3.5 billion years ago, barren, stormy, and volcanic with a reducing atmosphere of ammonia, methane, and water, was wonderfully fit for the prebiotic syntheses that led to life's beginnings. Yet, it was totally unsuited, indeed lethal, for the kinds of living organisms that inhabit the earth today, just as early forms of life could not survive in our present environment. The appearance of free oxygen in the atmosphere, produced largely if not almost entirely by life, is an example of the reciprocity between organism and environment. Although oxygen was at first poisonous to early forms of life, its gradual accumulation over the ages from photosynthesis forced protective biochemical alterations to appear that led eventually to complete dependence on oxygen for life.

Earth's biosphere and the organisms in it have evolved together. As living organisms adapt and evolve, they act on and produce changes in their environment; in so doing they must themselves change.

ANALYSIS OF THE ENVIRONMENT

Life is confined largely to the interfaces between land, air, and water. It does not penetrate very far below the land surface, very high into the atmosphere, or very abundantly into the depths of the ocean. We have already defined where it does occur in the biosphere. Living things within the biosphere may be examined at several different levels of organization.

The highest level of organization is called the **ecosystem,** literally the "environment-system." It includes both living (biotic) and nonliving (abiotic) components of the total environment. Other levels of organization are parts or fractions of the ecosystem. Consequently, the ecosystem may be considered the basic functional unit of ecologic study.

An ecosystem, as defined by the worker studying it, may be large such as a grassland, forest, lake, or cropland, or it may be more restricted as, for example, a river bank or tree hole. Whatever their sizes and differences in physical and biologic structures, all ecosystems function in much the same manner: the sun's energy is fixed by plants and then transferred to consumers and decomposers. The substrate provides nutrients that are cycled and recycled through the various living components of the ecosystem. There is

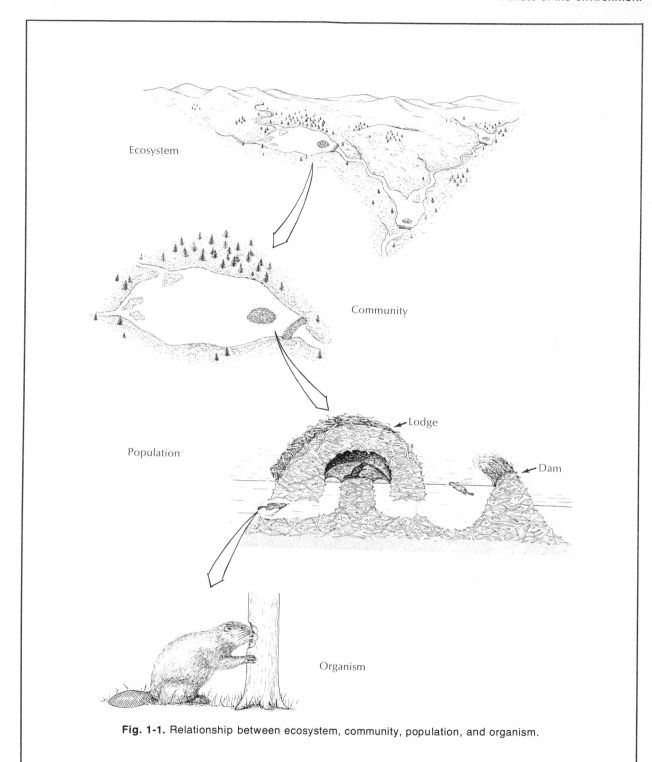

Fig. 1-1. Relationship between ecosystem, community, population, and organism.

always some flow of energy and nutrients between different ecosystems, so that all are interrelated within the earth's biosphere.

The next level of organization is the **community,** an assemblage of living organisms sharing the same environment and having a certain distinctive unity (Fig. 1-1). Communities comprise the living elements of an ecosystem. Like ecosystems, communities may be large or small, ranging from the coniferous forest community that may span a continent to the inhabitants of a rotting log community or the community of microorganisms living in man's large intestine. The elements of a community are closely interdependent.

The **population,** the next lower level of organization, is an interbreeding group of organisms of the *same species* sharing a particular space. Every community is composed of several populations, including those of plants, animals, and microorganisms. Energy and nutrients flow through a population. Its size is regulated by its relationships to other populations in the community and by the abiotic characteristics of the ecosystem in which it is found. Because members of a population are interbreeding, they share a **gene pool** and thus are a distinct genetic unit.

The lowest level of organization within the biosphere, in ecologic terms, is the **organism** itself. It is the living expression of the species. Each organism responds to its environment. The effect of the environment on all members of a species is called **natural selection;** the results of natural selection determine the evolution of the species, that is, its success or failure in response to changes in the environment. Although an organism itself does not evolve, its fate, when taken together with the fates of all other members of its species, is a factor in the evolution of the species.

As the unit upon which natural selection acts, the organism reflects the response of the population to environmental change. If the ecologist is to understand why animals are distributed as they are, he or she must examine the varied mechanisms that animals use to compensate for environmental stresses and alterations.

Ecologists have, in fact, become increasingly interested in the physiologic and behavioral mechanisms of animals. Both are intrinsic to the animal-habitat interrelationship. For example, the success of certain warm-blooded species under extreme temperature conditions such as in the Arctic or in a desert depends on near-perfect balance between heat production and heat loss and between appropriate insulation and special heat exchangers. Other species succeed in these situations by escaping the most extreme conditions by migration, hibernation, or torpidity. Insects, fishes, and other poikilotherms (animals having variable body temperature) compensate for temperature change by altering biochemical and cellular processes involving enzymes, lipid organization, and the neuroendocrine system. Thus the physiologic capacities with which the animal is endowed permit it to live under changing and often adverse environmental conditions. Physiologic studies are necessary to answer the "how" questions of ecology.

The animal's behavioral responses are also part of the animal-habitat interaction and of interest to the ecologist. Behavior is involved in obtaining food, finding shelter, escaping enemies and unfavorable environments, finding a mate, courting, and caring for the young. Genetically determined mechanisms such as behavioral repertoires are acted upon by natural selection. Those that improve adaptability to the environment assist in survival and the evolution of the species.

The term **ecology,** coined in the last century by the German zoologist Ernst Haeckel, is derived from the Greek *oikos,* meaning "house" or "place to live." Haeckel called ecology the "relation of the animal to its organic as well as inorganic environment." Although we no longer restrict ecology to animals alone, Haeckel's definition is still basically sound.

The term **environment** can mean a number of things. It is frequently used in reference to the organism's immediate surroundings, but it is not always clear whether it is meant to include the living, as well as the nonliving, surroundings. To the ecologist it certainly includes both. Ultimately the environment consists of everything in the universe external to the organism. Since ecology is really biology of the environment, it obviously is a broad and far-flung field of study. In a sense, there is very little that is *not* ecology.

The ecologist may choose to focus on any level of organization within the biosphere. **Ecosystem analysis** is largely interdisciplinary, incorporating physics, chemistry, and other sciences to assist in the comprehension of the role of the environment in determining the distribution and abundance of organisms. **Community ecology** is similar but of more restricted scope; it is possible to focus on the interactions of a few species and to study energy transfers in detail. **Population biology** stresses genetics, evolution, seasonal changes, and other phenomena within a single species. Some ecologists study the organism itself **(organismic biology)** to see how it responds to the environment, hour by hour, day by day; such studies have become physiologic and behavioral. All contribute to ecologic understanding. None stands alone.

ECOSYSTEMS

The ecosystem is a complex conceptual unit composed of abiotic and biotic elements. The abiotic component is the nonliving physical and chemical environment. It can be characterized by its physical parameters such as temperature, moisture, light, and altitude and by its chemical features, which include

various essential nutrients. These characteristics determine the basic nature of the ecosystem.

The biotic component—the populations of plants, animals, and microorganisms that form the communities of the ecosystem—may be categorized into producers, consumers, and decomposers. The **producers,** mostly green plants, are **autotrophs** that utilize the energy of the sun to synthesize sugars from carbon dioxide by photosynthesis. This energy is made available to the **consumers** and **decomposers.** They are the **heterotrophs** that exploit the self-nourishing autotrophs by converting organic compounds of plants into compounds required for their own growth and activity. In this section we consider the flow of energy through the ecosystem, which involves the concepts of productivity and the food chain, biogeochemical cycling, and limiting factors within the environment.

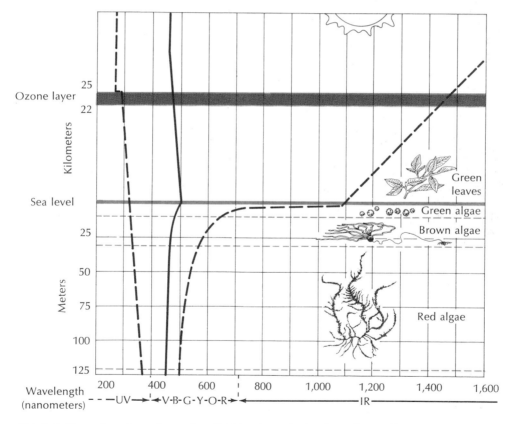

Fig. 1-2. Narrowing of spectrum of sunlight by atmospheric absorption and by absorption in sea water. The solid red line from top to bottom represents wavelengths of maximum intensity. Broken red lines locate the wavelength boundaries within which 90% of the solar energy is concentrated. (After Wald, G., 1959. Life and light. Scientific American, San Francisco, W. H. Freeman & Co.)

Solar radiation and photosynthesis

All life depends upon the energy of the sun. The sun releases electromagnetic energy produced by the nuclear transmutation of hydrogen to helium. Solar radiation received at the earth's surface extends from wavelengths of approximately 280 to 13,500 nm. Ultraviolet radiation with wavelengths of less than 280 nm is cut off sharply by the ozone layer in the upper atmosphere (Fig. 1-2). Radiation with wavelengths greater than 1,050 nm, comprising approximately 45% of the total energy, is long-wave infrared radiation that heats the atmosphere, warms the earth, and produces currents of air and water. The most important part of solar radiation lies between wavelengths of 310 and 1,050 nm; this is the portion that we call **light** because of its effect on man's retina. It is also the range that controls all important photobiologic processes, including photosynthesis, photochemical effects, phototropism (orientation of plants toward light), and animal vision.

The flow of energy through the ecosystem begins with **photosynthesis.** Energy enters the ecosystem as visible light; it is utilized by plants to produce adenosine triphosphate (ATP) and synthesize carbon compounds for themselves and for the animals that feed upon them. Although certain details of photosynthesis remain uncertain, the major events can be outlined.

Light striking a green plant is absorbed by chlorophyll, sending low-energy electrons into a higher energy level. These excited electrons drop back to a ground state in approximately 10^{-7} seconds, but in this brief interval their energy is channeled into a sequence of energy-yielding reactions. Part of the energy is used to synthesize ATP; the remainder causes the reduction of pyridine nucleotides (NADP). Both ATP and reduced NADP are then used to synthesize sugars from carbon dioxide and water. Photosynthesis in the individual leaf begins at low light intensity and increases at first linearly; that is, the rate of photosynthesis increases as light intensity increases until it reaches a maximum. The leaf achieves its highest rate of photosynthesis at only approximately one-tenth the intensity of full sunlight.

The amount of solar energy that reaches our earth's atmosphere is estimated at 15.3×10^8 g-cal/m^2/yr. Much of this energy is dissipated by dust particles or used in the evaporation of water. Only a very small fraction is used in the photosynthetic conversion of carbon dioxide to carbohydrates. Calculated on an annual or growing season basis, the photosynthetic efficiency of land areas is approximately 0.3% and of the ocean approximately 0.13%. These estimates are very low because they are based on the total energy available for the year rather than on the growing season alone. During brief intervals of very active growth, plants may store a maximum of 19% of the available light energy.

Production and the food chain

The energy accumulated by plants in photosynthesis is called **production.** Because it is the first step in the input of energy into the ecosystem, the *rate* of energy storage by plants is known as **primary productivity.** The total rate of energy storage, the **gross productivity,** is not entirely available for growth because plants also use energy for maintenance and reproduction. When this energy consumption, or plant **respiration,** is subtracted from the gross productivity, the **net primary productivity** remains. Plant growth results in the accumulation of plant **biomass.** Biomass is expressed as the *weight* of dry organic matter per unit of area and thus differs from productivity, which is the *rate* at which organic matter is formed by photosynthesis.

The level of productivity of different ecosystems depends upon the availability of nutrients and the limitations of temperature levels and moisture availability. Highly productive ecosystems are flood-plain forests, swamps and marshes, estuaries, coral reefs, and certain crop ecosystems (for example, rice and sugar cane). In such systems net production can exceed 3,000 g/m^2/yr of dry organic matter. Less productive (1,000 to 2,000 g/m^2/yr) are most temperate forests, most agricultural crops, lakes and streams, and grasslands. Least productive (70 to 200 g/m^2/yr) are tundra and alpine regions, deserts, and the open ocean. Extreme desert, rock, and ice regions have virtually zero productivity.

The net productivity of plants is the energy that supports all the rest of life on earth. Plants are eaten by consumers, which are themselves consumed by other consumers, and so on in a series of steps called the **food chain.** Food chains are descriptions of the way energy flows through the ecosystem. A diagram of a food chain shows arrows leading from one species to another, meaning that the first species is food for the second. But the first may be food for several other

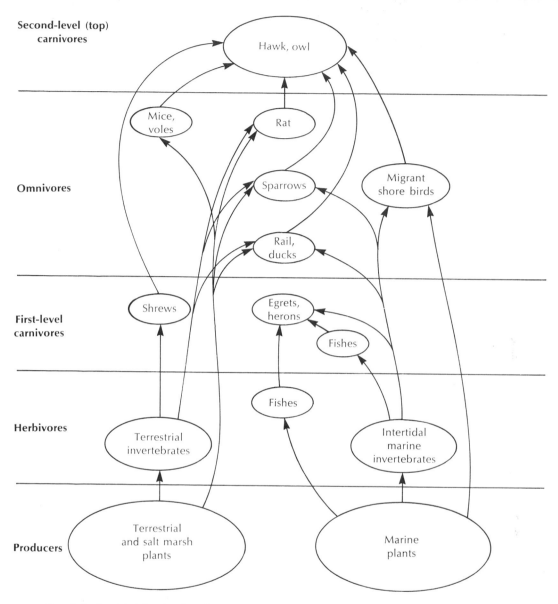

Fig. 1-3. A midwinter food web in a *Salicornia* salt marsh of the San Francisco Bay area. (Adapted from Smith, R. L. 1974. Ecology and field biology, ed. 2, New York, Harper & Row, Publishers; after Johnston, R. F. 1956. Wilson Bull. **68**:91.)

organisms as well. Seldom, in fact, does one organism live exclusively on another. Although some food chains are simple and short, for example, the one in which the whale feeds mainly on plankton, it is more common for several food chains to be interwoven into a complex **food web** (Fig. 1-3).

Despite their complexity, food chains tend to follow

a pattern. Green plants, the base of the food chain, are eaten by grazing **herbivores,** which can convert the energy stored in plants into animal tissue. Herbivores may be eaten by small **carnivores** and these by large carnivores. There may be two or three, sometimes even four, levels of carnivores. At the end of the chain are the **top carnivores** that, lacking predators, die and

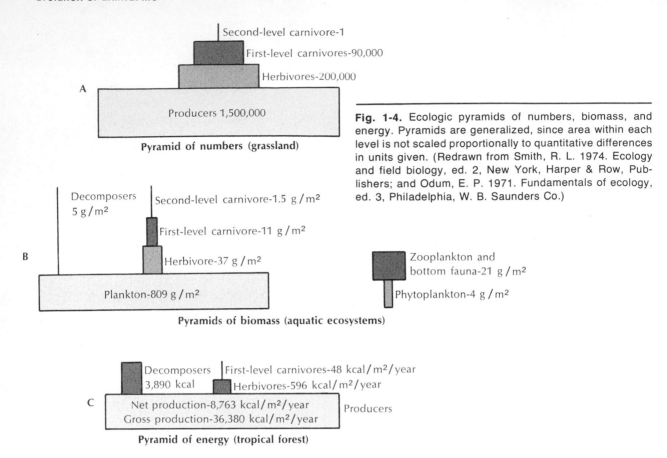

Second-level carnivore-1
First-level carnivores-90,000
Herbivores-200,000
Producers 1,500,000

A

Pyramid of numbers (grassland)

Decomposers 5 g/m²
Second-level carnivore-1.5 g/m²
First-level carnivore-11 g/m²
Herbivore-37 g/m²
Plankton-809 g/m²

B

Zooplankton and bottom fauna-21 g/m²
Phytoplankton-4 g/m²

Pyramids of biomass (aquatic ecosystems)

Decomposers 3,890 kcal
First-level carnivores-48 kcal/m²/year
Herbivores-596 kcal/m²/year
Net production-8,763 kcal/m²/year
Gross production-36,380 kcal/m²/year
Producers

C

Pyramid of energy (tropical forest)

Fig. 1-4. Ecologic pyramids of numbers, biomass, and energy. Pyramids are generalized, since area within each level is not scaled proportionally to quantitative differences in units given. (Redrawn from Smith, R. L. 1974. Ecology and field biology, ed. 2, New York, Harper & Row, Publishers; and Odum, E. P. 1971. Fundamentals of ecology, ed. 3, Philadelphia, W. B. Saunders Co.)

decompose, replenishing the soil with nutrients for plants that start the chain.

There are numerous examples of food chains. In the forest, for instance, there are many small insects (primary consumers) that feed upon plants (producers). A smaller number of spiders and carnivorous insects (secondary consumers) prey on small insects; still fewer birds (tertiary consumers) live on the spiders and carnivorous insects; and finally one or two hawks (quaternary or top consumers) prey on the birds.

The decomposers have traditionally been considered the final step in the herbivore-carnivore food chain since they reduce organic matter into nutrients that become available again to the producers. Ecologists now recognize that the decomposers comprise their own distinct **detritus food chain,** consisting of **detritus feeders,** such as earthworms, mites, millipedes, crabs, aquatic worms, and molluscs, and **microorganisms,** such as bacteria and fungi. Dead organic matter

such as fallen leaves or dead animals is decomposed and utilized by fungi, bacteria, and protozoa. Detritus feeders then eat the microorganisms as well as much of the dead organic matter directly. The detritus feeders are in turn eaten by small carnivores; the detritus food chain thus leads up into herbivore-carnivore food chains.

Ecologic pyramids

At each transfer within the food chain, 80% to 90% of the available energy is lost as heat. This limits the number of steps in the chain to four or five. Therefore the number of top consumers that can be supported by a given biomass of plants depends on the length of the chain.

Man, who occupies a position at the end of the chain, may eat the grain that fixes the sun's energy; this very short chain represents an efficient use of the potential energy. Man also may eat beef from animals that eat grass that fixes the sun's energy; the addition of a step in the chain decreases the available energy by an order of 10. In other words, it requires 10 times as much plant biomass to feed man, the meat eater, as to feed man, the grain eater. Let us consider the man who eats the bass that eats the sunfish that eats the zooplankton that eats the phytoplankton that fixes the sun's energy. The 10-fold loss of energy that occurs at each level in this five-step chain explains why bass do not form a very large part of man's diet. In this particular food chain, for man to gain a pound by eating bass, the pond must produce 5 tons of phytoplankton biomass.

These figures need to be considered as man looks to the sea for food. The productivity of the oceans is, in fact, very low and largely limited to regions of upwelling where nutrients are brought up and made available to the phytoplankton producers. Such areas occupy only approximately 1% of the ocean. The rest is a watery desert.

Ocean fisheries supply 18% of the world's protein, but most of this is used to supplement livestock and poultry feed. If we remember the rule of 10-to-1 loss in energy with each transfer of material, then the use of fish as food for livestock rather than as food for man is a poor use of a valuable resource in a protein-deficient world. Of the fishes that man does eat, the preference is for species such as flounder, tuna, and halibut, which are three or four steps up the food chain. Every 125 g of tuna requires 1 metric ton of phytoplankton food. If man is to derive greater benefit from the oceans as a food source in the future, he must eat more of the less palatable fishes that are lower in the food chain.

When we examine the food chain in terms of biomass at each level, it is apparent that we can construct **ecologic pyramids** of numbers or of biomass. A pyramid of numbers (Fig. 1-4, *A*), also known as an **Eltonian pyramid** (after the British ecologist Charles Elton, who first devised the scheme), depicts the number of individual organisms that is transferred between each level in the food chain. Although providing a vivid impression of the great difference in numbers of organisms involved in each step of the chain, a pyramid of numbers does not indicate the actual weight of organisms at each level. More instructive are pyramids of biomass (Fig. 1-4, *B*), which depict the total bulk of organisms. Such pyramids usually slope upward because mass and

energy are lost at each transfer. However, in some aquatic ecosystems in which the producers are the algae that have short life spans and rapid turnover rates, the pyramid is inverted. This happens because the algae can tolerate heavy exploitation by the zooplankton consumers. Therefore the base of the pyramid is smaller than the biomass it supports.

A third type of pyramid is the pyramid of energy, which shows rate of energy flow between levels (Fig. 1-4, *C*). An energy pyramid is never inverted because less energy is transferred from each level than was put into it. A pyramid of energy gives the best overall picture of community structure because each level reveals its true importance in the community regardless of its biomass.

Nutrient cycles

All of the elements essential for life are derived from the environment, where they are present in the air, soil, rocks, and water. When plants and animals die and their bodies decay or when organic substances are burned or oxidized, the elements and inorganic compounds essential for life processes, which we refer to as nutrients, are released and returned to the environment. Decomposers fulfill an essential role in this process by feeding on plant and animal remains and on fecal material. The result is that nutrients flow in a perpetual cycle between the biotic and abiotic components of the biosphere.

Nutrient cycles are often called **biogeochemical cycles** because they involve exchanges between living organisms (bio-) and the rocks, air, and water of the earth's crust (geo-). Geochemistry is the discipline that deals with the chemical composition of the earth and the exchange of elements therein.

Nutrient and energy cycles are closely interrelated since both influence the abundance of organisms in an ecosystem. However, unlike nutrients, which recirculate, energy flow follows one direction; it does not follow a cycle because it is lost as heat as it is used. The continuous input of energy from the sun keeps nutrients flowing and the ecosystem functioning. This interrelationship is depicted in Fig. 1-5. Among the most important biogeochemical cycles are those of carbon and nitrogen.

Carbon cycle

Carbon is the basic constituent of organic compounds and living tissue and is required by plants for

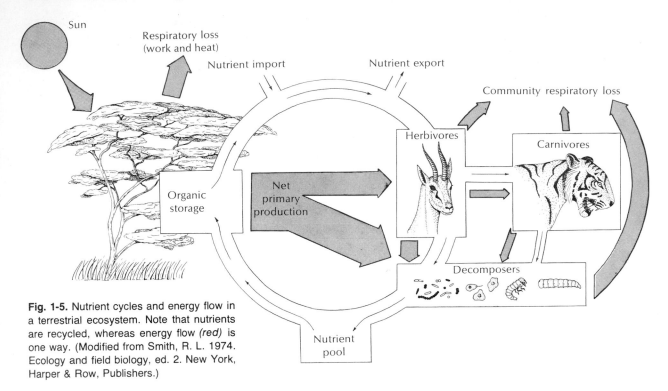

Fig. 1-5. Nutrient cycles and energy flow in a terrestrial ecosystem. Note that nutrients are recycled, whereas energy flow *(red)* is one way. (Modified from Smith, R. L. 1974. Ecology and field biology, ed. 2. New York, Harper & Row, Publishers.)

the fixation of energy by photosynthesis. It is hardly necessary to emphasize the total dependence of life on the availability of this element. Carbon circulates between carbon dioxide (CO_2) gas in atmosphere and living organisms through assimilation and respiration; it is also withdrawn into long-term reserves of fossil fuel deposits (humus and peat and finally coal and oil) (Fig. 1-6).

The cycling of carbon parallels and is linked to the flow of energy that begins with the fixation of energy during photosynthetic production. Plants synthesize glucose, a six-carbon compound, from CO_2 that is withdrawn from the atmosphere; they then use this sugar to build higher carbohydrates, especially cellulose, the structural carbohydrate of plants. Plants require 1.6 kg of CO_2 from the atmosphere for each kilogram of cellulose produced. The concentration of CO_2 in the atmosphere today is only 0.03% of air, in contrast to the relative abundance of atmospheric oxygen, 21% of air.

Two aspects of the low concentration of CO_2 require emphasis. The first is that CO_2 availability limits energy fixation by plants. Physiologists have shown that, if the atmospheric CO_2 is increased by 10%, plant photosynthesis increases 5% to 8%.

The second point is that there has been a tremendous increase in the consumption of fossil fuels by man in the last two decades. More than 10 billion tons of CO_2 enter the atmosphere each year from man's industrial and agricultural activities; of this 75% comes from burning fossil fuels and 25% from the release of CO_2 from soil because of frequent plowing (a surprising and unappreciated effect of present agricultural practice). However, some authorities argue that the input of CO_2 into the atmosphere by man's activities is not really chemical pollution, since the increase is removed by photosynthesis or by precipitation as carbonates in the ocean. There is concern, however, that even small increases in CO_2, which traps radiated heat, may increase the temperature of the earth's biosphere.

Nitrogen cycle

Like carbon, nitrogen (N_2) is also a basic and essential constituent of living material, particularly in the amino acids, which are the building blocks of protein. Despite the high concentration of N_2 in the atmosphere (79% of air), it is almost totally unavailable to life forms in its gaseous state. The most important contribution of the nitrogen cycle is converting N_2 into a chemical form that living organisms can

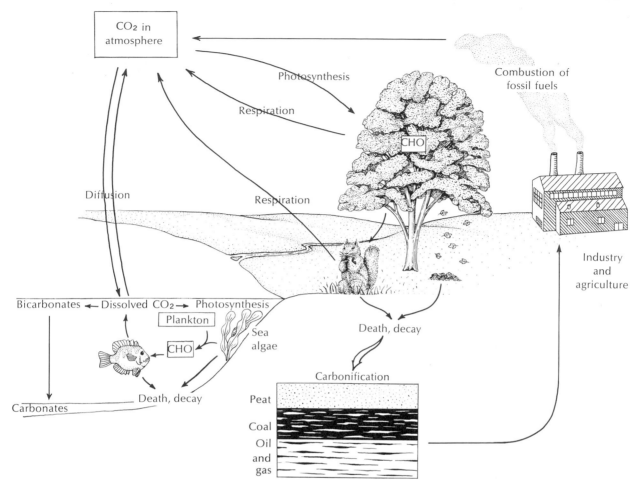

Fig. 1-6. Carbon cycle showing circulation of carbon as CO_2 gas between living components of environment and long-term storage as fossil fuels.

use. This conversion is called **nitrogen fixation.** Some atmospheric N_2 is fixed by lightning, which produces ammonia and nitrates that are carried to earth by rain and snow. But at least 10 times as much N_2 is biologically fixed by bacteria and by blue-green algae (Fig. 1-7).

Most important in terrestrial systems are bacteria associated with legumes (members of the pea family) whose N_2 contribution to soil enrichment is well known. The old agricultural practice of allowing fields to lie fallow (without crops) for a season every few years enabled nitrogen fixers to replenish the available N_2. Aerobic rhizobia bacteria produce nodules on the roots of legumes in which molecular N_2 is converted to ammonia and nitrates, which plants can use to build

protein. Plant proteins are transferred to consumers that build their own proteins from the amino acids supplied. Plants' and animals' waste products (urea and excreta) and their ultimate decomposition provide for the return of organic N_2 to the substrate. Then decomposers break down proteins, freeing ammonia, which is utilized by bacteria in a series of steps as shown in Fig. 1-7. Finally nitrate (NO_3) is produced, which may be utilized once again by plants or may be degraded to inorganic N_2 by denitrifying bacteria and returned to the atmosphere. Nitrates also are carried by runoff to streams and lakes and eventually to the sea. A cycle similar to this terrestrial cycle occurs in aquatic ecosystems except that there is a steady loss of N_2 to deep-sea sediments.

13

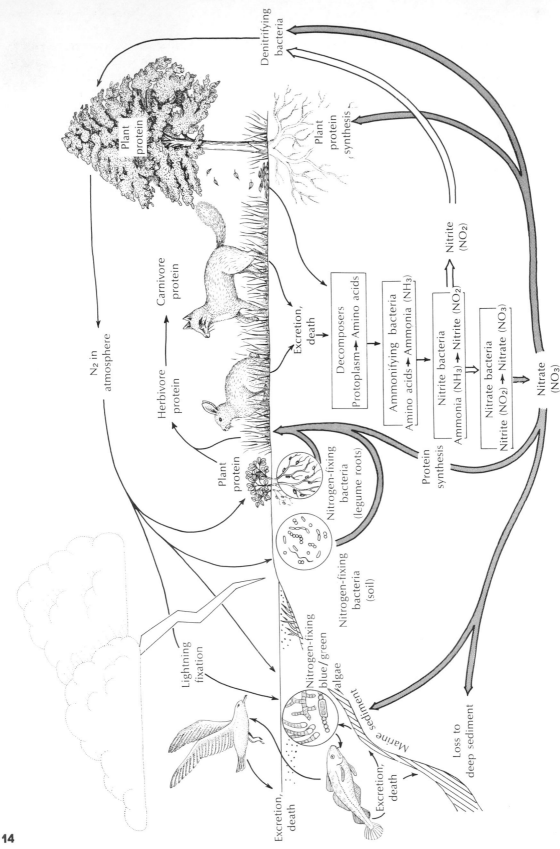

Fig. 1-7. Nitrogen cycle showing circulation of nitrogen between organisms and through environment (*red*). Microorganisms responsible for key conversions are indicated in circles and boxes.

The nitrogen cycle is a near-perfect self-regulating cycle in which N_2 losses in one phase are balanced by N_2 gains in another. There is little absolute change in N_2 in the biosphere as a whole. However, man's activities have caused steady losses of soil N_2 by slowing natural addition of organic N_2 through current agricultural practice and by the harvesting of timber. The latter causes an especially heavy N_2 outflow, from both timber removal and soil disturbance.

COMMUNITIES

Communities represent the most tangible concept in ecology. We are all familiar with the differences between forest, grassland, desert, and salt marsh, and we have little trouble picturing, at least in a general way, the kinds of plant and animal communities associated with each of these ecosystems. Communities comprise the *biotic* portion of the ecosystem; each consists of a certain combination of species that forms a functional unit. Although communities are sometimes difficult to define because the assemblages of species within similar communities are not always the same, communitiies do exist, and they all possess a number of attributes. It is beyond the scope of this

book to discuss all the aspects of community form and function. In this section we examine two important principles that operate in community organization.

Ecologic dominance

Biologic communities are typically dominated by a single species or limited group of species that determines the nature of the local environment. All other species in that community must adapt to conditions created by the dominants. Communities often are named for the dominant species, for example, black spruce forest, beech-maple forest, oyster community, and coral reef. It is not always easy to specify just what constitutes a dominant species. It may be the most numerous, the largest, or the most productive, or it may in some other manner exert the greatest influence on the rest of the community. In a woodland, a tree obviously means more to the community than a poison ivy plant (even though a casual visitor may carry away a more lasting impression of the poison ivy).

Dominant species achieve their status by occupying space that might otherwise be occupied by other species. When a dominant species is eliminated for some reason, the community changes. The American

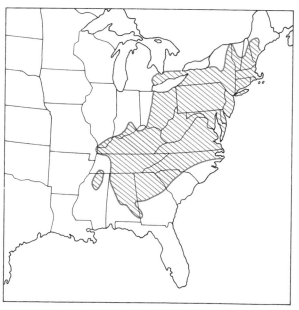

Fig. 1-8. American chestnut tree, *Castanea dentata,* and its original distribution in North America. Photograph taken circa 1900. (Map from Krebs, C. J. 1972. Ecology, New York, Harper & Row, Publishers; after Grimm, W. C. 1967. Familiar trees of America, New York, Harper & Row, Publishers.)

chestnut once dominated large regions of eastern United States where it comprised more than 40% of the overstory trees in climax deciduous forests. After the invasion in 1900 of the chestnut blight (a fungus), which was apparently brought into New York City on nursery stock from Asia, the chestnut was eliminated from its entire range within a few years (Fig. 1-8). The chestnut position was replaced by oak, hickory, beech, and red maple, and chestnut-oak forests became oak and oak-hickory forests. Tree squirrel populations that relied on chestnuts for their major food source decreased to a small fraction of their former abundance.

The appearance of the forests changed. To the subdominant species, whose specialized environmental requirements were attuned to the particular conditions of a chestnut forest, the change required adjustments that some could not meet. The shift in dominance produced changes throughout the community.

Ecologic niche concept

An animal's position in the environment is characterized by more than the habitat in which we expect to find it; it also has a "profession," a special role in life that distinguishes it from all other species. This is its **niche,** which Elton defined as the animal's place in the biotic environment, that is, what it does and its relation to its food and to its enemies. The niche concept has now been broadened to include an animal's position in time and space as well. The expression "every animal has its niche" conveys the idea rather well.

It is an accepted rule that no two species can occupy the same niche at the same time. If they did, they would be in direct competition for exactly the same food and space. Should this happen, one species would have to diverge by natural selection into a different niche or face extinction.

Closely related species that live close together are called **sympatric** (meaning literally "same country") species, as opposed to **allopatric** (different country) species that occupy different geographic regions. Sympatric species might be expected to have similar niches and be in danger of direct competition. This can be avoided in a number of ways, as the following examples illustrate.

In parts of eastern and southern Africa the two species of rhinoceros, the black and the white, are sympatric: they share the same habitat. However, the black rhinoceros is a browser feeding on leaves and woody plants, whereas the white rhinoceros is a grazer

eating grasses and herbs. They do not compete for the same food and consequently occupy distinct ecologic niches.

The three species of boobies of the Galápagos Islands—the white, blue-footed, and red-footed—offer a second example. All three are plunge divers that feed on the same kinds of marine fish frequenting the ocean around the islands. The blue-footed booby, however, always fishes close to the shore; the white booby flies a mile or two from land to fish; and the red-footed booby makes long hunting forays many miles from shore. Again, although sympatric, they do not compete for the same food. The three species have divided up the food resources so that a different portion of the sea becomes the undisputed hunting territory of each species.

The niche concept is a fundamental one to the biologist for it explains why animals avoid endless struggles with other animals. A species and its niche are reflections of the same thing: a unique way of life. A species is master of its own niche and thus is not in direct competition with similar species, even though competition was used to decide the boundaries of the niche in the first place. This concept is again discussed in Chapter 5 when we consider how speciation occurs.

POPULATIONS

As defined earlier, a population is a group of organisms belonging to the same species that share a particular space. Whether the population is grey squirrels or deer in an eastern woods or bluegill sunfish in a farm fishpond, it bears a number of attributes unique to the group. A population shares a common gene pool; it has a certain density, birth rate, death rate, age ratio, and reproductive potential; and it grows and differentiates much like the individual organisms of which it is composed.

Natural balance within populations

One characteristic of populations that we take for granted is their basic stability. We expect to see approximately the same number of robins and starlings on our lawns each year. Fireflies, nighthawks, and whippoorwills are an accepted and appreciated background of summer evenings; they always return each year in approximately the same numbers, and their predictability is comforting. Of course, there are good years and bad years for game animals, and the buzz of cicadas is more pervasive during some summers than during others, but these fluctuations are seldom violent. The populations of plants and animals around us remain in balance, provided that man does not interfere.

Yet with few exceptions animals have reproductive potentials far beyond that required for replacement of their numbers. Insects lay thousands of eggs, field mice can produce as many as 17 litters of 4 to 7 young each year, and a single female codfish may spawn 6 million eggs per season. It has been said that the descendants of a pair of flies, if unchecked, would weigh more than the earth in a few years. The potential for rapid growth of a population, the so-called biotic potential rate, is large but never proceeds unchecked indefinitely because of the limitations of the environment. What keeps populations in check? How is the "balance of nature" explained?

The growth of a population in a limited space and with a limited input of nutrients is described by a **population growth curve** that is sigmoid in shape (Fig. 1-9). In the beginning, with ample space and abundant food, the starting population breeds and grows as fast as its reproductive potential allows. The curve turns steeply upward as more and more animals join the reproductive population; the increase is exponential. But, as space gets crowded and food begins to vanish, the curve straightens out and then flattens as the upper population limit is reached. The reason why the rate of growth of the population becomes zero is, in a word, competition. Growth is suppressed by competition for space and competition for the limited input of food.

The simple sigmoid growth curve is representative of many plant and animal populations, both natural and in the laboratory. It can be easily reproduced in the laboratory with cultures of yeast or paramecium, indeed, with any of a great variety of microorganisms or animals. In nature the curve represents the annual plankton blooms of ponds and lakes, eruptions of insect pests, and growth of weeds in old fields. When populations first approach their growth limit, they tend to overshoot into a temporarily unlimited environment and then decline and oscillate about some lower limit. The level of final stabilization, about which the population fluctuates, is the **carrying capacity** of the environment.

For example, when sheep were introduced on the island of Tasmania during approximately the year 1800, their growth was represented by a sigmoid curve with a small overshoot, followed by mild oscillations around a final population size of 1,700,000 sheep (Fig. 1-10, *A*). A similar pattern but with larger fluctuations was recorded for a population of ring-necked pheasants introduced on an island in Ontario, Canada (Fig. 1-10, *B*). Population overshoots are especially characteristic of populations of higher animals because their life histories are long and complicated and because, after the population already has consumed the resources of the environment, reproduction continues for a time.

The growth of the human population was slow for a very long period of time. For most of his evolutionary history, man was a hunter-gatherer who depended upon and was limited by the natural productivity of the environment. With the development of agriculture, the carrying capacity of the environment increased and the population grew steadily from 5 million around 8000 B.C. when agriculture was introduced to 16 million around 4000 B.C. Despite terrible famines, disease, and war, which took their toll, the population reached 500 million by 1650. With the coming of the Industrial Revolution in Europe and England in the eighteenth century, followed by a medical revolution, discovery of new lands for colonization, and better agriculture practices, the carrying capacity of the earth for man increased dramatically. The population doubled to 1 billion around 1850. It doubled again to 2 billion by 1930 and to 4 billion in 1976. Thus the growth has been exponential and remains high (Fig. 1-10, *C*).

However, recent surveys provide hope that the world population growth is slackening. In just 5 years (1970 to 1975) the annual growth rate decreased from 1.9% to 1.64%. At 1.64%, it will take 42 years for the

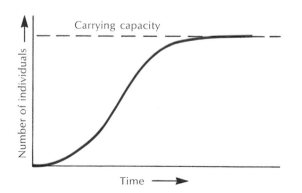

Fig. 1-9. Population growth curve.

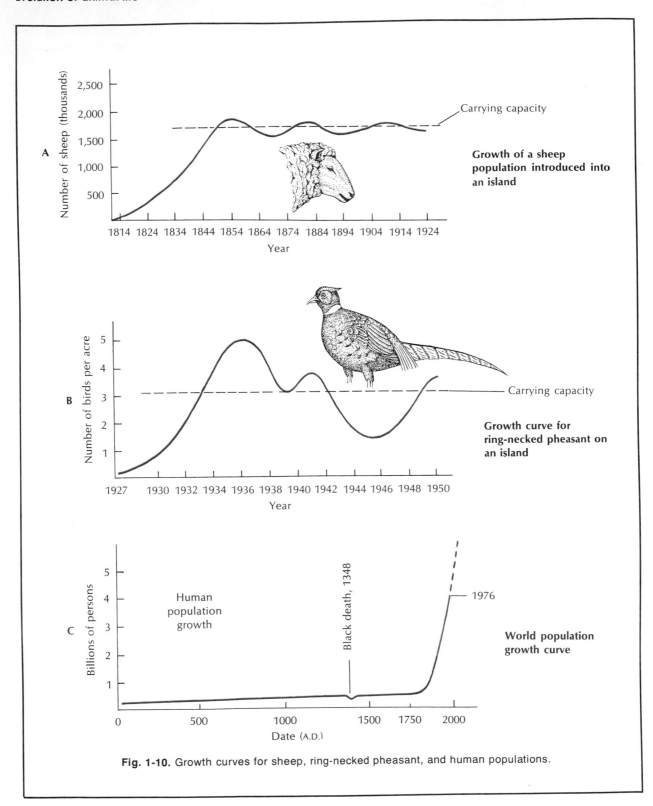

Fig. 1-10. Growth curves for sheep, ring-necked pheasant, and human populations.

world population to double rather than 36 years at the higher annual growth rate figure. The decrease is credited to better family planning programs, but increasing deaths from starvation also are a contributing factor.

Unlike other animals that cannot change and increase the carrying capacity of their environment, man is unique in finding ways. Unfortunately, when the carrying capacity is increased, man responds as would any animal by increasing his population. Because the earth's resources are finite, the time will inevitably come when the carrying capacity can be extended no further. The question is whether man is able to anticipate this limit and check his population growth in time or whether he will overshoot the resources and experience a catastrophic decline as do most other higher animals introduced into a new environment. Indeed, for millions of Third World people time has already run out. The rapid growth of population is not something that thinking people view with equanimity.

How population growth is curbed

What determines the number of animals in a natural population? For a laboratory culture of animals the answer seems fairly clear. They grow until they reach the carrying capacity of the environment, at which point growth is suppressed by competition for the limited resources of space and food. Thus the forces that limit growth arise from within the population; they are **density-dependent** mechanisms caused by crowding. Food and space limitations are obviously density-dependent factors because the intensity of action depends on how large the population is.

It is not difficult to understand how these same limitations, as well as other natural hazards, could act on wild populations in a density-dependent way. As crowding increases, the quest for food becomes increasingly critical. The competition that results may affect the population in either of two ways. All members may scramble for what is left until the resource is subdivided into so many small parts that none of the population gets enough for self-maintenance. Mass starvation follows with only the strongest or luckiest individuals surviving. This **scramble type** of competition, which is typical of certain herbivore populations is wasteful and produces large oscillations in population density.

Or competition may be of the **contest type.** Socially dominant individuals, or those with defended ter-ritories denied to others, claim portions of the food resource for themselves and prevent the unsuccessful from having any part of it. These unfortunates starve or in their weakened condition become victims of disease or predation, but the remainder stay healthy. Population numbers remain high and wild oscillations in density are prevented.

Crowding brings forth other agencies of death that operate to keep populations in check. High density encourages **disease** in both men and animals. Occasionally epidemics sweep through crowded populations as did the terrible bubonic plague (black death) through the crowded, dirty cities of Europe in the fourteenth century. **Predation** increases when prey become abundant and easy to catch. **Shelter** for suitable nest sites and as refuge from bad weather becomes less available with crowding.

The continuous competition for food and space to live brings forth yet another density-dependent force: **stress.** Although the physiology of stress is not thoroughly understood, there is ample evidence that when certain natural populations become crowded, such as populations of lemmings, voles, and snowshoe hares, a neuroendocrine imbalance involving the pituitary and adrenal glands appears. Growth is suppressed. Reproduction fails and individuals become irritable and aggressive. Under such conditions snowshoe hares suffer from a lethal "shock disease" that results in a population crash.

Overcrowding may also cause **emigration** away from the birth area. Overcrowded mice increase their locomotor activity and begin to explore new areas. In the case of lemmings, overpopulation produces the famous mass "marches" recorded at intervals in Scandinavia. When songbird populations become overcrowded the less successful become "vagabond" birds that are forced out into less preferred habitats. In fact, emigration is one major force resulting in the colonization of new habitats. The rapid dispersal of the European starling throughout the United States and southern Canada, following its introduction into New York City in 1890, is an excellent example (see p. 566).

Not all of the forces that limit growth are density dependent. Extreme changes in the weather or unusually cold, hot, wet, or dry weather is a **density-independent** hazard of varying severity to animal populations. Local insect populations are sometimes pushed to the point of extinction by a severe winter. Hurricanes and volcanic eruptions may destroy entire populations. Hailstorms have been known to kill most of the young of wading bird populations. Prairie grass and forest fires kill everything unable to escape.

Vertebrate populations are more resistant than invertebrate populations to catastrophic weather changes but are influenced nonetheless. Populations of Virginia white-tailed deer in the northern parts of its range are much more affected by unusually severe winters than by hunting pressure. One study of deer populations over a period of 70 years in the Adirondack Mountains of New York revealed that during heavy snow years fawn mortality was exceptionally high. Heavy snow covered low-growing food and made travel so difficult that more energy was consumed in searching for scarce food than was gained in eating it. During three successive severe winters (1969 to 1971) five out of six fawns died and the deer population dropped to one third of its former density.

Thus population numbers are influenced by factors generated within the population and by forces from without, and no single mechanism can explain how population growth is curbed. The evidence, however, indicates that most natural populations are commonly controlled by density-dependent forces. Man himself is the greatest force of all. By altering animal habitats he changes the balances that set the old limits to animal numbers. His activities can increase or exterminate whole populations of animals.

SELECTED REFERENCES

Allee, W. C., A. E. Emerson, O. Park, T. Park, and K. P. Schmidt. 1949. Principles of animal ecology. Philadelphia, W. B. Saunders Co. *A valuable comprehensive text and reference, although dated.*

Andrewartha, H. G. 1970. Introduction to the study of animal populations. Chicago, University of Chicago Press. *A summary of the classic 1954 text by Andrewartha and Birch.*

Bates, M. 1960. The forest and the sea. New York, Random House, Inc. *Several editions available. Excellent general approach to ecology. Highly recommended.*

Boughey, A. S. 1971. Fundamental ecology. Scranton, Pa. Intext Educational Publishers. *Elementary text of ecologic principles.*

Boughey, A. S. 1971. Man and the environment. New York, Macmillan, Inc. *Well-written combination of ecology, human evolution, and pollution.*

Clapham, W. B. 1973. Natural ecosystems. New York, Macmillan, Inc. *The ecosystem and how man has altered the environment.*

Colinvaux, P. A. 1973. Introduction to ecology. New York, John Wiley & Sons, Inc. *Highly readable and interesting approach to general ecology. Stresses population ecology.*

Ehrlich, P. R., and A. H. Ehrlich. 1972. Population, resources, environment, 2nd ed. San Francisco, W. H. Freeman & Co., Publishers. *Well-written source book on the subjects in title.*

Elton, C. E. 1942. Voles, mice and lemmings. Oxford, Clarendon Press. *A classic on animal populations.*

Fuller, W. A., and J. C. Holmes. 1972. The life of the Far North. New York, McGraw-Hill, Inc. *Interesting, beautifully illustrated, authoritative treatment of the Arctic. Other excellent books in the "Our Living World of Nature" series published by McGraw-Hill include forest, African plains, and seashore.*

Henderson, L. J. 1913. The fitness of the environment (reprint). Boston, Beacon Press. *A classic that should still be read.*

Krebs, C. J. 1972. Ecology: experimental analysis of distribution and abundance. New York, Harper & Row, Publishers. *Important treatment of population ecology.*

Leopold, A. 1949. A Sand County almanac. New York, Oxford University Press, Inc. *Several editions available. A beautiful account by a great conservationist; now a favorite of the ecology movement.*

MacArthur, R. H. 1972. Geographical ecology. New York, Harper & Row, Publishers. *One of the most important and influential books of recent years.*

Miller, G. T., Jr. 1971. Energy, kinetics and life. Belmont, Calif., Wadsworth Publishing Co. Inc. *Fine general ecology text.*

Odum, E. P. 1971. Fundamentals of ecology, 3rd ed. Philadelphia, W. B. Saunders Co. *One of the most popular general ecology texts; this is the original comprehensive text updated.*

Phillipson, J. 1966. Ecological energetics. New York, St. Martin's Press, Inc. *Brief, clear treatment of the subject.*

Scientific American. 1970. The biosphere. San Francisco, W. H. Freeman & Co., Publishers. *Excellent collection of articles on the ecosystem.*

Smith, R. L. 1974. Ecology and field biology. New York, Harper & Row, Publishers. *Comprehensive, clearly written, well-illustrated general ecology text. An excellent summary follows each chapter.*

SELECTED SCIENTIFIC AMERICAN ARTICLES

Bell, R. H. V. 1971. A grazing ecosystem in the Serengeti. **225:**86-93 (July). *The migration of grazers across Tanzania is synchronized with the availability of specific tissues of grasses.*

Bolin, B. 1970. The carbon cycle. **223:**124-132 (Sept.).

Bormann, F. H., and G. E. Likens. 1970. The nutrient cycles of an ecosystem. **223:**92-101 (Oct.).

Cloud, P., and A. Gibor, 1970. The oxygen cycle. **223:**110-123 (Sept.).

Cole, LaMont C. 1958. The ecosphere. **198:**83-92 (April).

Deevey, E. S., Jr. 1970 Mineral cycles. **223:**148-158 (Sept.).

Delwiche, C. C. 1970. The nitrogen cycle. **223:**136-146 (Sept.).

Gates, D. M. 1971. The flow of energy in the biosphere. **224:**88-100 (Sept.). *Plants capture only a fraction of the solar energy that falls on the earth, but this fraction maintains all life.*

Janick, J., C. H. Noller, and C. L. Rhykerd. 1976. The cycles of plant and animal nutrition. **235:**74-86 (Sept.). *Energy and nutrients for human consumption are processed by chains of organisms.*

Myers, J. H., and C. J. Krebs. 1974. Population cycles in rodents. **230:**38-46 (June). *The cyclic rise and fall in rodent populations are associated with genetic changes in the population.*

Peixoto, J. P., and M. A. Kettoni. 1973. The control of the water cycle. **228:**46-61 (April).

Penman, H. L. 1970. The water cycle. **223:**98-108 (Sept.).

Richards, P. W. 1973. The tropical rain forest. **229:**58-67 (Dec.). *One of the oldest ecosystems is giving way to the activities of man.*

Wynne-Edwards, V. C. 1964. Population control in animals. **211:**68-74 (Aug.).

A green sea turtle labors slowly toward the safety of the ancient sea, having laid approximately 100 eggs in a nest on the beach. Every 2 or 3 years she must briefly leave her adopted marine environment, to which she is so perfectly adapted, to visit her ancestral birthplace, now an alien world, to deposit her leathery eggs. (Drawn from a photo by R. E. Schroeder.)

2 ORIGIN OF LIFE

HISTORICAL PERSPECTIVE

Renewal of inquiry: Haldane-Oparin hypothesis

PRIMITIVE EARTH

Formation of the solar system
Origin of the earth's atmosphere

BASIC MOLECULES OF LIFE

Amino acids and proteins
Nucleic acids
Protein synthesis
Regulation of gene function

CHEMICAL EVOLUTION

Sources of energy
Prebiotic synthesis of small organic molecules
Formation of polymers

ORIGIN OF LIVING SYSTEMS

Origin of metabolism
Appearance of photosynthesis and oxidative metabolism

PRECAMBRIAN LIFE

Prokaryotes and the age of blue-green algae
Appearance of the eukaryotes

Where did life on earth come from? This is an ancient personal question that we must suppose has aroused man's curiosity since the dawn of his cultural development. Although the great religions of man have sought to satisfy this curiosity, they have not provided nor were ever intended to provide a scientific explanation of the detailed sequence of events that culminated in the first appearance of life. To most biologists the question of life's beginnings is one of profound interest. The biologist is struck by the remarkable unity of nature. As more concerning the identifiable components of life is learned, an evolutionary pattern in the structure and function of living things can be seen.

Modern biochemical studies reveal that all organisms, from man to the smallest microbes that transcend the rather arbitrary boundary between life and nonlife, share two kinds of basic biomolecules—nucleic acid and protein. Both molecules are large and complex in form. The nucleic acids DNA and RNA are composed of nitrogenous bases, sugars, and phosphoric acid. They are the basic genetic polymers of all living things and carry the informational blueprint that determines the cell's activities and directs the synthesis of proteins. Proteins are the foundation substance of protoplasm. They are composed of 20 different amino acids joined together with peptide linkages. Thus all organisms use the same simple building blocks: 20 amino acids, 5 bases, 2 sugars, and 1 phosphate.

The remarkable uniformity of life extends also to cell function. The metabolic processes that convert foodstuffs into a usable form of energy consistently occur in the simplest organisms to the most complex. This example, along with other examples of molecular and functional identity, consequently suggests that all life must have had a common beginning.

Even though we acknowledge the kinship of living things, we must admit at the beginning that we do not know how life on earth originated. Until recently the study of life's origins was not considered worthy of serious speculation by biologists because, it was argued, the absence of a geologic record made the course of events resulting in the appearance of life unknowable. This situation has changed.

Since 1950 several laboratories around the world have been devoting full-time research efforts to origin-of-life studies. It is a multidisciplinary effort that requires the contributions of scientists of several specialties—biologists, chemists, physicists, geolo-gists, and astronomers. From such studies it has been possible to reconstruct a scenario of ancient events in which simple single-celled living organisms evolved more than 3 billion years ago from inorganic constituents present on the surface of the earth. These studies are not attempts to prove or disprove any existing belief, religious or philosophic, but rather they are endeavors to solve a great cosmic mystery. Although in some ways a historical reconstruction, origin-of-life studies are buoyed by the recent successful simulation of the "primordial broth" in which life began, the discovery of amino acids in extraterrestrial meteorites, and a vast amount of supportive geochemical information.

HISTORICAL PERSPECTIVE

Man has always been awed and mystified by the question of how life originated on the earth. From ancient times it was commonly believed that life could arise by spontaneous generation from dead material, in addition to arising from parental organisms by reproduction (biogenesis). Frogs appeared to arise from damp earth, mice from putrified matter, insects from dew, maggots from decaying meat, and so on. Warmth, moisture, sunlight, and even starlight were often mentioned as beneficial factors that encouraged spontaneous generation.

These ideas were developed into an elaborate theory by the Greeks and reappeared frequently in the writings of Aristotle (384 to 322 B.C.). They became firmly entrenched in virtually all cultures, including those of the Far East, and were unquestioningly accepted by even relatively recent great figures such as Copernicus, Bacon, Galileo, Harvey, Descartes, Goethe, and Schelling. Spontaneous generation was also supported by Christian philosophers who pointed out that, according to the first chapter of Genesis, God did not create plants and animals directly but bade the waters to bring them forth.

Inevitably the question of spontaneous generation fell under the scrutiny of experimental science (sixteenth and seventeenth centuries). At first such studies were ill-conceived efforts to supplement natural observations of spontaneous generation by artificially producing various organisms in the laboratory. A typical recipe is one for making mice, given by the Belgian plant nutritionist Jean Baptiste van Helmont. "If you press a piece of underwear soiled with sweat together with some wheat in an open jar, after about 21 days the odor changes and the ferment . . . changes the wheat into mice. But what is most remarkable is that the mice which came out of the wheat and underwear were not small mice, not even miniature adults or aborted mice, but adult mice emerge!''

The first attack on the doctrine of spontaneous generation occurred in 1668 when the Italian physician Francesco Redi exposed meat in jars, some of which was uncovered while some was covered with parchment and wire gauze. The meat in all three kinds of vessels spoiled, but only the open vessels had maggots, and he noticed that flies were constantly entering and leaving these vessels. He concluded that, if flies had no access to the meat, no worms would be found.

Although Redi's refutation of spontaneous generation became widely known, the doctrine was too firmly entrenched to be disbelieved. In 1748 the English Jesuit priest John T. Needham boiled mutton broth and put it in corked containers. After a few days the medium was swarming with microscopic organisms. He concluded that spontaneous generation was real because he believed that he had killed all living organisms by boiling the broth and that he had excluded the access of others by the precautions he took in sealing the tubes.

However, an Italian investigator, Abbé Lazzaro Spallanzani (1767), was critical of Needham's experiments and conducted experiments that dealt a blow against the theory of spontaneous generation. He thoroughly boiled extracts of vegetables and meat, placed these extracts in clean vessels, and sealed the necks of the flasks hermetically in flame. He then immersed the sealed flasks in boiling water for several minutes to make sure that all germs were destroyed. As controls, he left some tubes open to the air. At the end

of 2 days he found the open flasks swarming with organisms; the others contained none.

This experiment still did not settle the issue, for the advocates of spontaneous generation maintained either that air, which Spallanzani had excluded, was necessary for the production of new organisms or that the method he used had destroyed the vegetative power of the medium. When oxygen was discovered (1774), the opponents of Spallanzani seized this as the vital principle that he had destroyed in his experiments.

It remained for the great French scientist Louis Pasteur to silence all but the most stubborn proponents of spontaneous generation with an elegant series of experiments with his famous ''swan-neck'' flasks. Pasteur (1861) answered the objection to the lack of air by introducing fermentable material into a flask with a long S-shaped neck that was open to the air (Fig. 2-1). The flask and its contents were then boiled for a long time. Afterwards the flask was cooled and left undisturbed. No fermentation occurred, for all organisms that entered the open end were deposited on the floor of the neck and did not reach the flask contents. When the neck of the flask was cut off, the organisms in the air could fall directly on the fermentable mass and fermentation occurred within it in a short time. Pasteur concluded that, if suitable precautions were taken to keep out the germs and their reproductive elements (eggs, spores, etc.), no fermentation or putrefaction could take place.

Pasteur brought an end to the long and tenacious career of the concept of spontaneous generation. Pasteur's work showed that no living organisms come into existence except as descendants of similar organisms. In announcing his results before the French Academy, Pasteur proclaimed, ''Never will the doctrine of spontaneous generation arise from this mortal

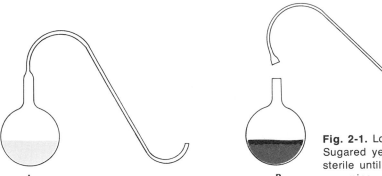

Fig. 2-1. Louis Pasteur's swan-neck flask experiment. **A,** Sugared yeast water boiled in swan-neck flask remains sterile until neck is broken. **B,** Within 48 hours, flask is swarming with life.

A   B

blow.'' Paradoxically, in showing that spontaneous generation did not occur as previously claimed (production of mice, maggots, frogs, etc.), Pasteur also ended for a time further inquiry into the spontaneous origins of life. A lengthy period of philosophic speculation followed, but virtually no experimentation on life's origins was performed for 60 years.

Renewal of inquiry: Haldane-Oparin hypothesis

The rebirth of interest into the origins of life occurred in the 1920s. In this decade the Russian biochemist Alexander I. Oparin and the British biologist J. B. S. Haldane independently proposed that life originated on earth after an inconceivably long period of ''abiogenic molecular evolution.'' Rather than arguing that the first living organisms miraculously originated all at once, a notion that had constrained fresh thinking for so long, Haldane and Oparin suggested that the simplest living units (for example, bacteria) came into being very gradually by the progressive assembly of inorganic molecules into more complex organic molecules. These molecules would react with each other to form living microorganisms.

Haldane proposed that the earth's primitive atmosphere consisted of water, carbon dioxide, and ammonia. When ultraviolet light shines on such a gas mixture, many organic substances such as sugars and amino acids are formed. Ultraviolet light must have been very intense on the primitive earth before the production of oxygen (by plants) that reacted with ultraviolet rays to form ozone, the three-atom form of oxygen. Ozone serves at present as a protective screen to prevent ultraviolet rays from reaching the earth's surface. Haldane believed that the early organic molecules could accumulate in the primitive oceans to form ''a hot dilute soup.'' In this primordial broth carbohydrates, fats, proteins, and nucleic acids might have been assembled to form the earliest microorganisms.

Oparin, too, suggested that the earth's primitive atmosphere lacked oxygen and instead contained hydrogen, methane, ammonia, and other reducing compounds. He proposed that the organic compounds required for life were formed spontaneously in such a reducing atmosphere under the influence of sunlight, lightning, and the intense heat of volcanos.

The Haldane-Oparin hypothesis greatly influenced theoretical speculation on the origins of life during the 1930s and 1940s. Finally in 1953 Stanley Miller, working with Harold Urey in Chicago, made the first attempt to simulate with laboratory apparatus the conditions thought to prevail on the primitive earth. This strikingly successful experiment, which is described in more detail later in this chapter, demonstrated that important biomolecules are formed in surprisingly large amounts when an electric discharge is passed through a reducing atmosphere of the kind proposed by Haldane and Oparin. With the realization that it was possible to simulate successfully a prebiotic milieu in the laboratory, a new era in origin-of-life studies was ushered in. It coincided with the dawn of the space age and a new public interest in the question of life's origins.

PRIMITIVE EARTH
Formation of the solar system

The sun and the planets are believed to have been formed approximately 4.6 billion years ago out of a spheric cloud of cosmic dust and gases that had some angular momentum. The cloud collapsed under the influence of its own gravity into a rotating disc. As the material in the central part of the disc condensed to form the sun, a substantial amount of gravitational energy was released as radiation. The pressure of this outwardly directed radiation prevented the complete collapse of the nebula into the sun. The material left behind began to cool and eventually gave birth to the planets (Fig. 2-2).

The inner planets (Mercury, Venus, Earth, and Mars) are small and composed mostly of nonvolatile elements because the high temperatures of the cooling planets drove off the volatile elements. The outer planets are composed largely of volatile compounds such as methane, ammonia, water, hydrogen, and helium because these materials were blown outward into this region of the collapsing disc by the sun's radiation pressure. In the region occupied by the planets Jupiter and Saturn, massive bodies of hydrogen and helium condensed. The low temperatures and enormous gravitational fields of these bodies prevented their gases from escaping. It is believed that the condensation of the solar system was 95% complete within the relatively short time of 50,000 years.

Origin of the earth's atmosphere

Evolution of the primeval reducing atmosphere. While the earth was still more or less gaseous, the

Fig. 2-2. Solar system showing narrow range of conditions suitable for life.

various atoms were sorted out according to their weight. Heavy elements (silicon, aluminum, nickel, and iron) gravitated toward the center while the lighter elements (hydrogen, oxygen, carbon, and nitrogen) remained in the surface gas. Hydrogen and helium, because of their volatility, continued to escape as the earth condensed, and these elements became severely depleted from the primeval atmosphere. Neon and argon were almost completely lost from the atmosphere. Oxygen, nitrogen, and water, which are the major constituents of our present oxidizing atmosphere, could not escape from the earth because they were present in nonvolatile, chemically combined forms trapped in dust particles. Later as the earth condensed, water, carbon compounds, and nitrogen or ammonia were released into the atmosphere from the earth's interior by volcanic activity, which was much more extensive then than it is now.

It is generally agreed that the earth's primeval atmosphere contained no more than a trace of oxygen. It was a **reducing atmosphere;** that is, it contained a predominance of molecules having less oxygen than hydrogen. Methane (CH_4), ammonia (NH_3), and water (H_2O) are examples of fully reduced compounds; such compounds are believed to have composed the early atmosphere of the earth. Carbon dioxide (CO_2), nitrogen (N_2), and traces of hydrogen (H_2) may also have been present.

The character of the primeval atmosphere is very important in any discussion of the origins of life because the organic compounds of which living organisms are made are not stable in an oxidizing atmosphere. Organic compounds are not synthesized in our oxidizing atmosphere today, and they are not stable if they are introduced into it. Obviously life could not have originated without the basic organic building blocks from which the first organisms were assembled. Consequently modern origin-of-life theories assume that the primeval atmosphere was reducing because the synthesis of compounds of biologic importance occurs only under reducing conditions. Fortunately geochemical evidences tend to support the belief that the early atmosphere was a reducing one that arose by degassing of the earth's interior.

Appearance of oxygen. As water, nitrogen, and carbon compounds continued to enter the atmosphere from volcanos, the atmosphere became saturated with water vapor and rain began to fall. Small lakes formed, enlarging into oceans. The oceans gradually became salted as rocks weathered. During this early period, which lasted 1$\frac{1}{2}$ billion years, the atmosphere was still reducing well after the first living protocells such as algae and bacteria had evolved.

Our atmosphere today is strongly oxidizing. It contains 79% molecular nitrogen, approximately 21% free oxygen, and a small amount of carbon dioxide. Although the time course for its development is much disputed, at some point oxygen began to appear in significant amounts in the atmosphere. In the primitive reducing atmosphere oxygen was formed by the decomposition of water in either of two ways.

The first is the photolytic action of ultraviolet light from the sun on water in the upper atmosphere.

$$\underset{\text{UV light}}{2 H_2O \longrightarrow 2 H_2 + O_2}$$

The hydrogen produced escapes from the earth's gravitational field, leaving free oxygen to accumulate in the atmosphere. The amount produced by the photodissociation of water is now quite small, although it may have been significant over the immense span of time in which it occurred.

The second and probably more important source of oxygen is photosynthesis. Almost all oxygen produced at the present time is produced by algae and land plants. Each day the earth's plants combine approximately 400 million tons of carbon with 70 million tons of hydrogen to set free 1,100 million tons of oxygen. Oceans are the major source of oxygen. Almost all oxygen produced today is consumed by organisms oxidizing their food to carbon dioxide; if this did not occur, the amount of oxygen in the atmosphere would double in approximately 3,000 years. Since Precambrian fossil algae resemble modern algae and since algae today are the greatest producers of oxygen, it seems probable that most of the oxygen in the early atmosphere was produced by photosynthesis.

BASIC MOLECULES OF LIFE

To understand how life on earth originated, we need to know how living things are assembled and how they work. In this section we briefly survey the raw materials and basic molecules that interact to produce life.

Amino acids and proteins

Proteins are large, complex molecules composed of 20 commonly occurring amino acids (Fig. 2-3). The amino acids are linked together by **peptide bonds** to form long, chainlike polymers. The 20 different kinds of amino acids can be arranged in an enormous variety of sequences; therefore it is not difficult to account for practically countless varieties of proteins among living organisms.

A protein is not just a long string of amino acids; it is a highly organized molecule. For convenience, biochemists have recognized four levels of protein organization called primary, secondary, tertiary, and quaternary.

The **primary structure** of a protein is determined by the kind and sequence of amino acids making up the polypeptide chain. The polypeptide chain or chains tend to spiral into a definite helical pattern, like the turns of a screw. This precise coiling, known as the **secondary structure** of the protein, most commonly turns in a clockwise direction to form what is called an **alpha-helix** (Fig. 2-4). The spirals of the chains are stabilized by weak **hydrogen bonds,** usually between a hydrogen atom of one amino acid and the peptide-bond oxygen of another amino acid in an adjacent turn of the helix.

The polypeptide chain (primary structure) not only spirals into helical configurations (secondary structure), but also the helices themselves bend and fold, giving the protein its complex, yet stable, three-dimensional **tertiary structure** (Fig. 2-5). The folded chains are stabilized by the interactions between side groups of amino acids. One of these interactions is the **disulfide bond,** a strong covalent bond between pairs of cysteine (sis'tee-in) molecules that are brought together by folds in the polypeptide chain. Other kinds of bonds that help to stabilize the tertiary structure of proteins are hydrogen bonds, ionic bonds, and hydrophobic bonds.

The term **quaternary structure** describes those proteins that contain more than one polypeptide chain unit. For example, hemoglobin of higher vertebrates is composed of four polypeptide subunits nested together in a single protein molecule.

Proteins as enzymes

Proteins perform many functions in living things. They serve as the structural framework of protoplasm and form hormones, chromosomes, and many other cell components. However, proteins' most important role by far is as **enzymes,** the biologic catalysts required for almost every reaction in the body.

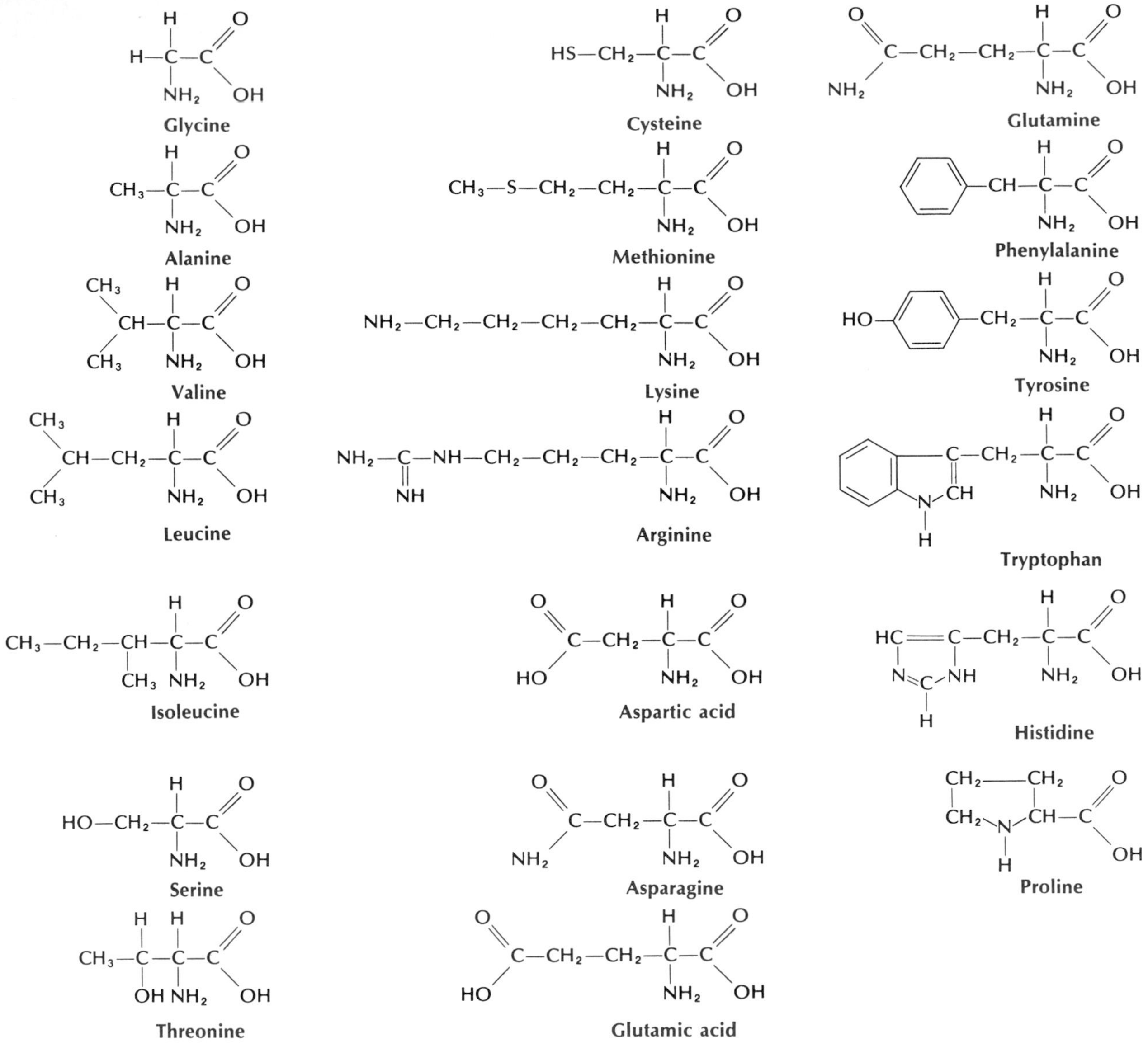

Fig. 2-3. The 20 naturally occurring amino acids.

Enzymes lower the activation energy required for specific reactions and enable life processes to proceed at moderate temperatures. They control the reactions by which food is digested, absorbed, and metabolized. They promote the synthesis of structural materials to replace the wear and tear on the body. They determine the release of energy used in respiration, growth, muscle contraction, physical and mental activities, and many other activities.

Enzymes are the intermediates between the genes (composed of nucleic acids) and the functional elements of cells. Because of their crucial importance, the question of the origin of enzymes takes a central position in origin-of-life theories.

Nucleic acids

Nucleic acids are complex substances of high molecular weight that represent a basic manifestation of

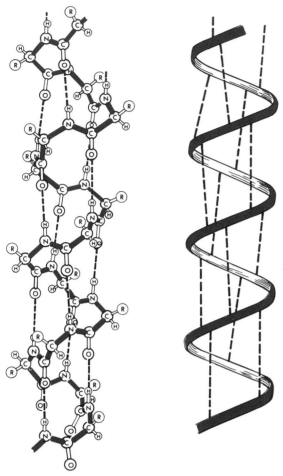

Fig. 2-4. Alpha-helix pattern of a polypeptide chain. *Dashed lines,* Hydrogen bonds that stabilize adjacent turns of the helix. *R,* Amino acid side chains. (Adapted from Green, D. 1956. Currents of biochemical research. New York, Interscience Publishers, Inc.)

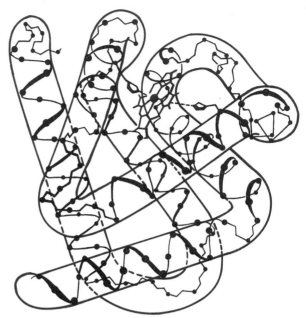

Fig. 2-5. Three-dimensional tertiary structure of the protein myoglobin. Adjacent folds of the polypeptide chain are held together by disulfide bonds that form between pairs of cysteine molecules. In the upper center of the molecule is the heme group, which combines with oxygen. (From Neurath, H. 1964. The proteins, ed. 2, vol. II. New York, Academic Press, Inc.)

life. The genetic information necessary for all aspects of biologic inheritance is encoded in the sequence of these polymeric molecules. Nucleic acids direct the synthesis of proteins, including all of the enzymes that in turn carry out the numerous routine synthetic and functional activities of the cell. Most important, nucleic acids are the only molecules that have the power (with the help of the right enzymes) to replicate themselves. This is the essential process that guarantees the accuracy of reproduction. To understand how nucleic acids perform these roles, we need to look rather closely at their chemical structures.

There are two kinds of nucleic acids in all cells—deoxyribonucleic acid (DNA) and ribonucleic acid (RNA). DNA is a polymer built of repeated units called **nucleotides.** Each nucleotide contains three kinds of organic molecules—a **sugar,** a **phosphate group,** and a **nitrogenous base.**

The sugar in DNA is a pentose (five-carbon) sugar called **deoxyribose.** It has the following structural formula:

The **phosphoric acid** has this structural formula:

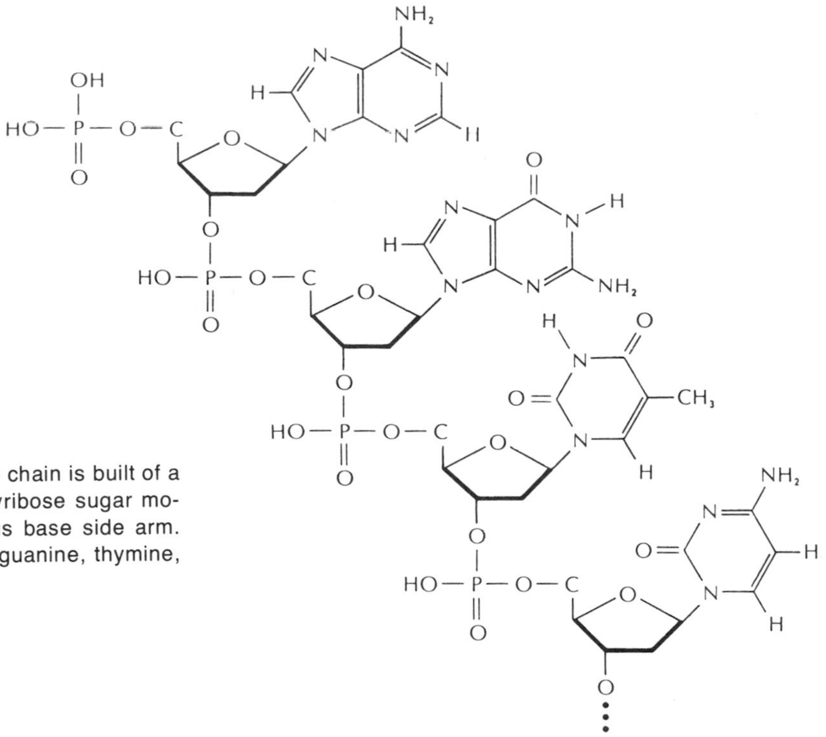

Fig. 2-6. Section of DNA. Polynucleotide chain is built of a backbone of phosphoric acid and deoxyribose sugar molecules. Each sugar holds a nitrogenous base side arm. Shown from top to bottom are adenine, guanine, thymine, and cytosine.

Four **nitrogenous bases** are found in DNA. Two of them are organic compounds composed of nine-membered double rings, classed as **purines.** They are **adenine** and **guanine:**

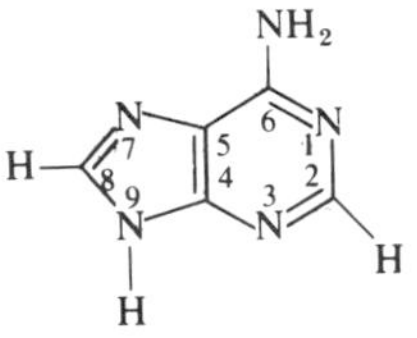

ADENINE 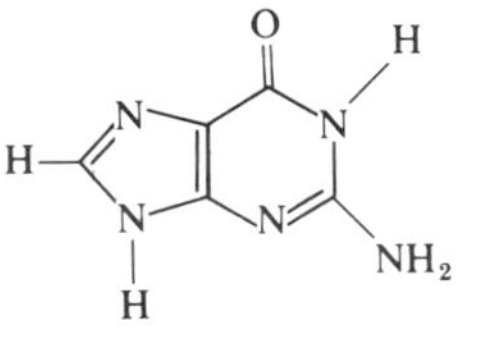   GUANINE

The other two nitrogenous bases belong to a different class of organic compounds called **pyrimidines,** consisting of six-membered rings. The two pyrimidines found in DNA are **thymine** and **cytosine:**

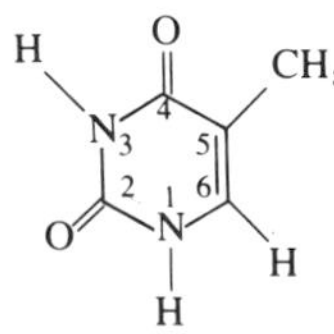

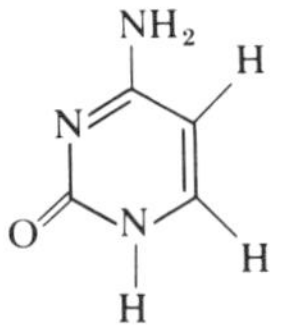

CYTOSINE

The sugar, phosphate group, and nitrogenous base are

linked as shown in this generalized scheme for a **nucleotide:**

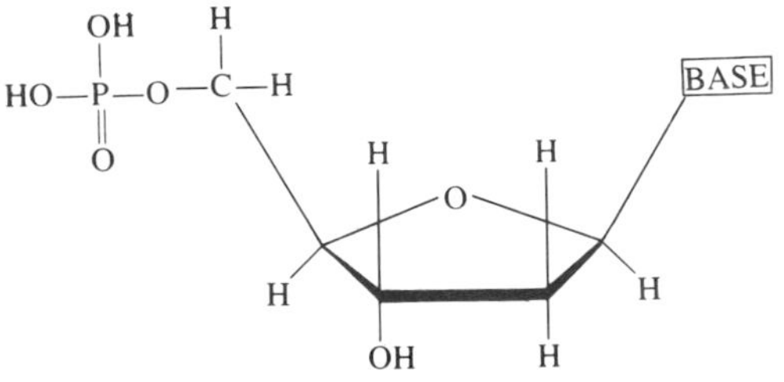

PHOSPHATE   SUGAR   NITROGENOUS BASE

In DNA the backbone of the molecule is built of phosphoric acid and deoxyribose sugar; to this backbone are attached the nitrogenous bases (Fig. 2-6). However, one of the most interesting and important discoveries about the nucleic acids is that DNA is not a single polynucleotide chain; rather it consists of *two* complementary chains that are precisely cross-linked by specific hydrogen bonding between purine and pyrimidine bases. It was found that the number of adenines is equal to the number of thymines, and the number of guanines equals the number of cytosines.

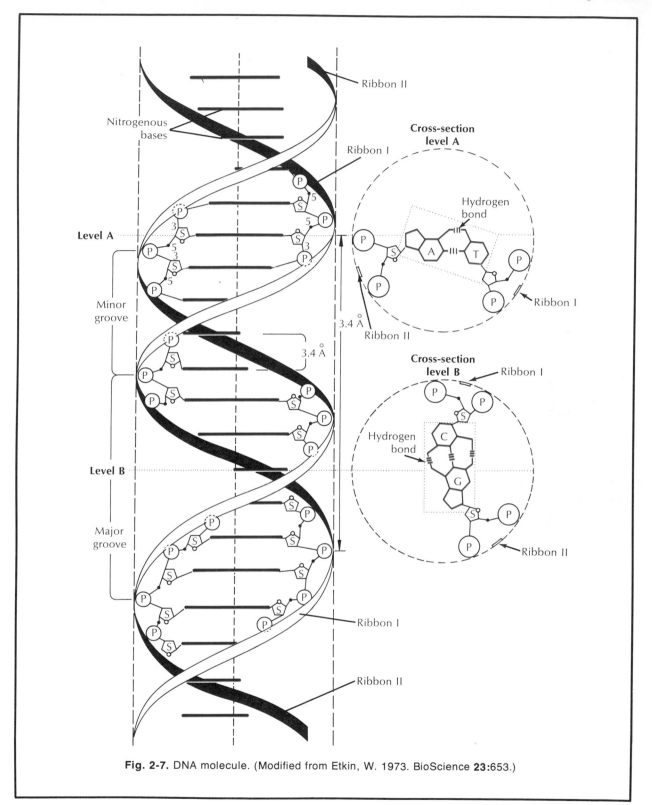

Fig. 2-7. DNA molecule. (Modified from Etkin, W. 1973. BioScience **23:**653.)

This fact suggests a pairing of bases: adenine with thymine (AT) and guanine with cytosine (GC). The larger adenine (a purine) always attaches to the smaller thymine (a pyrimidine) by two hydrogen bonds; the larger guanine (a purine) always attaches to the smaller cytosine (a pyrimidine) by three hydrogen bonds:

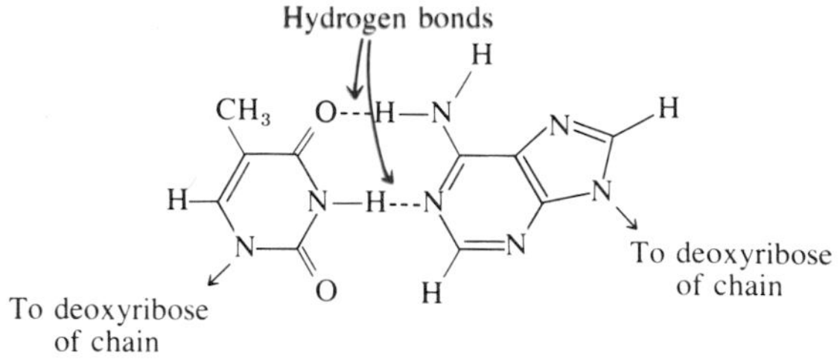

THYMINE-ADENINE

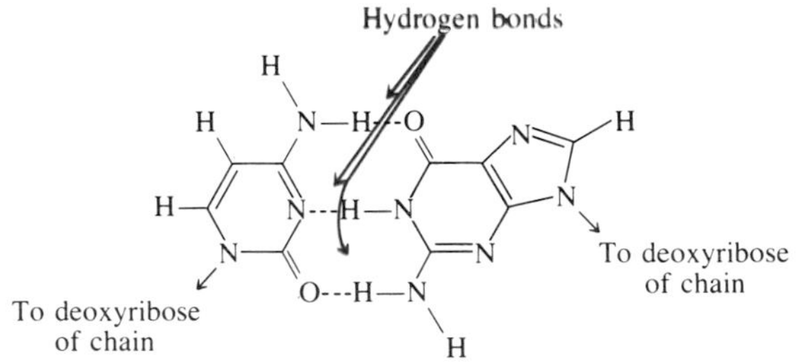

CYTOSINE-GUANINE

The result is a ladder structure. The upright portions are the sugar-phosphate backbones, and the connecting rungs are the paired nitrogenous bases, AT or GC.

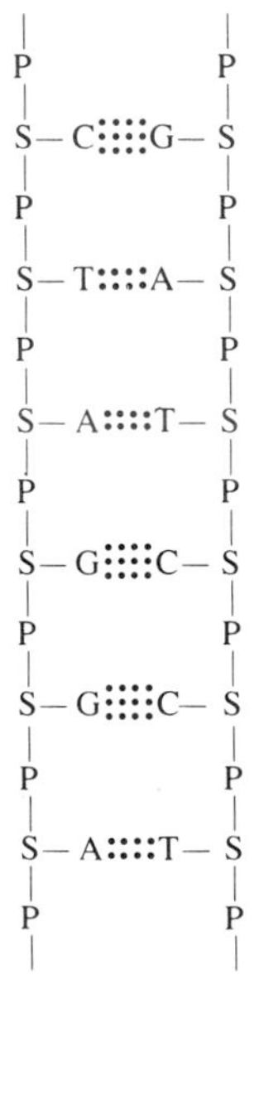

However, the ladder is twisted into a *double helix,* with approximately 10 base pairs for each complete turn of the helix (Fig. 2-7).

The determination of the structure of DNA has been widely acclaimed as the single most important biologic discovery of this century. It was based on the x-ray diffraction studies of Maurice H. F. Wilkins and the ingenious proposals of Francis H. C. Crick and James D. Watson published in 1953. Watson, Crick, and Wilkins were later awarded the Nobel Prize for Medicine and Physiology for their momentous work.

Ribonucleic acid (RNA) is very similar to DNA in structure except that it consists of a *single* polynucleotide chain. In RNA one of the four bases, thymine (T), is replaced by uracil (U), and the sugar deoxyribose is replaced by ribose. In other respects the single RNA chain is joined together like each of the two DNA chains.

Every time a cell divides, the structure of DNA must be precisely copied in the daughter cells. This is called **replication.** During replication, the two strands of the double helix unwind and each separated strand directs the synthesis of a complementary strand using nucleotides from the nucleus or cytoplasm. In this way two new daughter molecules are formed, each identical with the parental double helix. Each daughter molecule is composed of an old and a new strand.

Very rarely mistakes are made when the wrong base pair is inserted into the chain. Such **mutations** are usually disadvantageous, but occasionally they are preserved. This may be one of the important mechanisms of evolution and is discussed in Chapter 4.

Protein synthesis

Proteins are not synthesized directly from DNA. Rather one of the DNA strands behaves as a template that lines up the components of RNA according to base-pairing rules. This is called **transcription.** The **messenger RNA** (mRNA) strand that is transcribed is an intermediary that carries the genetic information stored in the DNA strand to the protein-synthesizing machinery in the cell. The messenger RNA is complementary to the coding DNA strand; that is, A in the coding DNA strand is replaced by U in messenger RNA; C is replaced by G; G is replaced by C; and T is replaced by A.

The process by which messenger RNA determines the sequence of amino acids in a protein is called **translation.** It is a complicated process that involves

Table 2-1. The genetic code—proposed codons (code triplets) between messenger RNA and specific amino acids

Codons	Amino acid
GCU, GCC, GCA, GCG	Alanine
CGU, CGC, CGA, CGG, AGA	Arginine
AAU, AAC	Asparagine
GAU, GAC	Aspartic acid
UGU, UGC	Cysteine
GAA, GAG	Glutamic acid
CAA, CAG	Glutamine
GGU, GGC, GGA, GGG	Glycine
CAU, CAC	Histidine
AUU, AUC, AUA	Isoleucine
CUU, CUC, CUA, CUG, UUA, UUG	Leucine
AAA, AAG	Lysine
AUG	Methionine
UUU, UUC	Phenylalanine
CCU, CCC, CCA, CCG	Proline
AGU, AGC, UCU, UCC, UCA, UCG	Serine
ACU, ACC, ACA, ACG	Threonine
UGG	Tryptophan
UAU, UAC	Tyrosine
GUU, GUC, GUA, GUG	Valine
UAA, UAG, UGA	Termination of code of one gene

more than 100 protein and nucleic acid molecules. This complexity is understandable because messages written in the four-letter alphabet of the nucleic acids (four bases) must specify the exact sequence of the 20 letters (20 amino acids) that appear in proteins.

Actually the principle involved in translation is a simple one. The nucleic acid message lies in the sequence of bases in the messenger RNA. The bases are read three at a time beginning at some fixed point in the chain. The three-base sequence is called a **code triplet** or **codon.** Each code triplet specifies either an amino acid or a signal to stop translation. Since there are 64 different code triplets (that is, the four bases can be arranged 64 different ways in groups of three) and only 20 amino acids, most of the amino acids are specified by more than one code triplet. This is the famous **genetic code** (Table 2-1).

The translation process takes place on **ribosomes,** granular structures composed of protein and nucleic acid. The messenger RNA molecules fix themselves to the ribosomes to form a messenger RNA–ribosome complex. Since only a short section of a messenger RNA molecule is in contact with a single ribosome, the messenger RNA usually fixes itself to several ribosomes at once. This arrangement, called a polyribosome, allows several proteins of the same kind to be synthesized at once, one on each ribosome of the polyribosome (Fig. 2-9).

The assembly of proteins on the messenger RNA–ribosome complex requires the action of another kind of RNA called **transfer RNA.** The transfer RNA molecules collect the free amino acids from the cytoplasm and deliver them to the messenger RNA–ribosome complex, where they are assembled into a protein. There is a special transfer RNA molecule for every amino acid. Furthermore each transfer RNA is accompanied by a special **activating enzyme.** These enzymes are necessary to sort out and attach the correct amino acid to each transfer RNA by a process called **loading.**

The transfer RNAs are surprisingly large molecules that are folded in a complicated way in the form of a cloverleaf (Fig. 2-8). On this cloverleaf a special sequence of three bases (the anticodon) is exposed in just the right way to form base pairs with complementary bases (the codon) in the messenger RNA. The anticodon on the transfer RNA is the key to the correct sequencing of amino acids in the protein being assembled.

For example, alanine is assembled into a protein when it is signaled by the codon GCG in a messenger RNA. This translation is accomplished by alanine transfer RNA in which the anticodon is CGC. The alanine transfer RNA is first loaded with alanine by its activating enzyme. The loaded alanine transfer RNA enters the messenger RNA–ribosome complex where it fits precisely into the right place on the messenger RNA strand. Then the next loaded transfer RNA specified by the messenger RNA code (glycine transfer RNA, for example) enters the messenger RNA–ribosome complex and attaches itself beside the alanine transfer RNA. The two amino acids are united with a peptide bond (with the assistance of a molecule of ATP), and the alanine transfer RNA falls off. The process continues stepwise as the protein chain is built (Fig. 2-9). A protein of 500 amino acids can be assembled in less than 30 seconds.

Regulation of gene function

The previous description explains how the code, carried on the DNA molecules of the nucleus, is

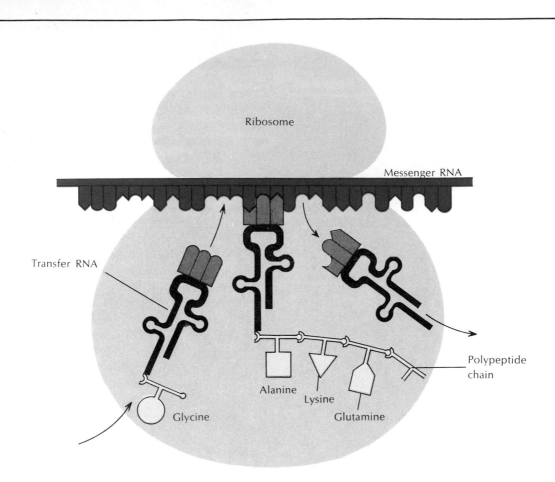

Fig. 2-8. Formation of polypeptide chain on messenger RNA. As ribosome moves down messenger RNA molecule, transfer RNA molecules with attached amino acids enter ribosome *(left)*. Amino acids are joined together into polypeptide chain, and transfer RNA molecules leave ribosome *(right)*.

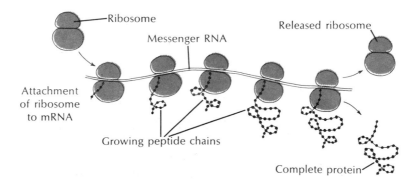

Fig. 2-9. How the protein chain is formed. As ribosomes move along messenger RNA, the amino acids are added stepwise to form the polypeptide chain.

transcribed into a definite protein or enzyme synthesized in the cytoplasm. It does not explain how genes are turned off and on as their products are needed by the cell. Nor does it explain why certain enzymes are not formed when they are not needed. Obviously enzyme-forming systems require control because they have powerful effects, even in minute amounts, on the rate of all cellular metabolic processes. Enzyme synthesis must be responsive to the influences of supply and demand.

Our understanding of how genetic material is utilized to regulate the synthesis of proteins is largely the result of research by two French scientists, J. Monod and F. Jacob (Fig. 2-10). The basic unit carrying the genetic code for protein synthesis is the **structural gene** whose actions are regulated in a complex way. Adjacent to the structural gene is a special segment of DNA called the **operator.** Together, the structural gene and operator comprise a genetic unit called the **operon.** The operon may contain a single structural gene or several structural genes of related function. The operator is in turn negatively controlled by a **regulator gene** that can produce a molecule called a **repressor.** The repressor may bind with and so inhibit the operator. The operator, then, is the receptor site for the repressor. While repressed, the operator prevents the structural gene or genes from forming messenger RNA.

For the operon to resume functioning, that is, form messenger RNA required for protein synthesis, the operator must be **derepressed.** One way that this can be accomplished is for the repressor molecule to bind with and be inactivated by other materials in the cell. This renders the repressor incapable of turning the operator off. The regulatory molecule could be a hormone, or it could be some material synthesized within the cell by the enzymes produced by the operon. For example, if there is a high concentration of a particular enzyme in the cell, this high concentration can act as a feedback through the repressor to block the

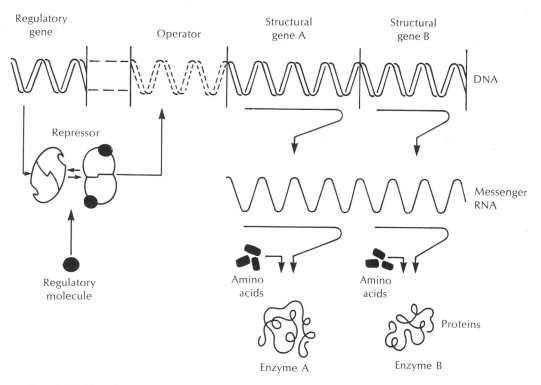

Fig. 2-10. Regulation of gene function. Regulator gene acts by way of repressor on the operator. Repressor, in combination with regulatory molecules (hormones or products of cellular metabolism), binds with and inhibits operator, thus preventing structural genes from producing messenger RNA. (After Monod, J. 1966. Endocrinology **78:**412-425.)

action of the operator so that the structural gene can no longer produce the enzymes. Alternatively, the repressor may be changed to an inactive form by a lower-than-normal concentration of the enzyme.

This important **operon hypothesis,** which has been described in a much-abbreviated form, seeks to explain the mechanism for repressing the synthesis of enzymes when they are not needed and for inducing them when they are needed. Since it is based almost entirely on work with bacteria and viruses, it remains to be seen whether or not it also applies to higher forms of life.

The understanding of protein synthesis by nucleic acids is the great triumph of modern molecular biology. The question we must now try to answer is: How did such an elegant system evolve? This question must really be divided into two reciprocal questions: (1) How could enzymes have arisen without nucleic acids to specify their synthesis? (2) How could nucleic acids have evolved without enzymes to make them? In other words, which came first, enzymes or nucleic acids? It is a chicken-egg question to which we return later.

CHEMICAL EVOLUTION
Sources of energy

According to the Haldane-Oparin hypothesis, a variety of carbon compounds gradually accumulated on the surface of the earth during a lengthy period of prebiotic chemical evolution. The primitive reducing atmosphere contained simple gaseous compounds of

Table 2-2. Present sources of energy averaged over the earth*

Source	Energy (cal/cm²/year)
Total radiation from sun	260,000
Infrared (above 7000 Å)	143,000
Visible (3500-7000 Å)	113,600
Ultraviolet	
2500-3500 Å	2837
2000-2500 Å	522
1500-2000 Å	39
<1500 Å	1.7
Electric discharges	4
Shock waves	1.1
Radioactivity (to 1.0 km depth)	0.8
Volcanos	0.13
Cosmic rays	0.0015

*Modified from Miller, S. L., and L. E. Orgel. 1974. The origins of life on the earth. Englewood Cliffs, N. J. Prentice-Hall, Inc.

carbon, nitrogen, oxygen, and hydrogen, such as carbon dioxide, molecular nitrogen, water vapor, methane, and ammonia. These were the starting materials from which organic compounds were made. However, if these gaseous compounds are mixed together in a closed glass system and allowed to stand at room temperature, they never chemically react with each other. To promote a chemical reaction a continuous source of **free energy** sufficient to overcome reaction-activation barriers must be supplied.

For example, one of the simplest and most important prebiotic chemical reactions is the formation of hydrogen cyanide (HCN) from nitrogen (N_2) and methane (CH_4). When an electric discharge is passed through an atmosphere of nitrogen, some of the molecules absorb enough energy to dissociate into atoms.

$$N_2 \longrightarrow 2\,N$$

Dissociated nitrogen atoms are highly reactive. If methane is present in the atmosphere, nitrogen atoms react to form hydrogen cyanide and hydrogen.

$$N + CH_4 \longrightarrow HCN + \tfrac{3}{2}H_2$$

Other sources of energy such as ultraviolet light or heat could equally well result in the formation of hydrogen cyanide from nitrogen and methane. Once formed, hydrogen cyanide dissolves in rain and is carried into lakes and oceans where it and other reactive molecules can form organic compounds.

The sun is by far the most powerful source of free energy for the earth. Each year each square centimeter of the earth receives an average 260,000 calories of radiant energy. The way in which this energy is distributed over the infrared, visible, and ultraviolet regions of the spectrum is shown in Table 2-2. The gases present in the primitive atmosphere would not have absorbed visible or infrared energy. Since the solar energy available falls off rapidly in the ultraviolet region, only approximately 0.2% of the total energy is at wavelengths shorter than 2,000 Å, where it can be absorbed by molecules such as methane, water, and ammonia. Nevertheless ultraviolet radiation must have been an important source of energy for photochemical reactions in the primitive atmosphere.

Electric discharges must have provided another source of energy for chemical evolution. Although the total amount of electric energy released by lightning is small compared to solar energy, nearly all of the electric energy of lightning is effective in synthesizing organic compounds in a reducing atmosphere. A single flash of lightning through a reducing atmosphere generates a large amount of organic matter. Thunderstorms are thought to have been more prevalent on the earth then than they are today and may have been one of the most important sources of energy for organic synthesis.

Volcanos and hot springs were available on the primitive earth as sources of energy, but it is doubtful that they were a major site of prebiotic organic synthesis. Cosmic rays, radioactivity, and sonic energy generated by ocean waves were also available sources of energy, but their contribution cannot be evaluated and was probably small in any case. Of the other sources of energy, only shock waves generated by meteorites passing through the primitive atmosphere

may have produced a large amount of organic matter. Meteorites generate intense temperatures as high as 20,000° C and pressures exceeding 15,000 atmospheres. A large meteorite could generate millions of tons of organic matter in its wake.

It is not known which energy source was most important for prebiotic synthesis on earth. Most authorities believe both electric discharges and ultraviolet radiation made major contributions, but the quantity and quality of prebiotic energy sources remain a subject of debate and inquiry.

Prebiotic synthesis of small organic molecules

The first steps leading toward life on earth happened when simple organic molecules were generated in the earth's reducing atmosphere by the action of

Fig. 2-11. Dr. S. L. Miller and apparatus used in experiment on the synthesis of amino acids with an electric spark in a reducing atmosphere. (Photo courtesy S. L. Miller.)

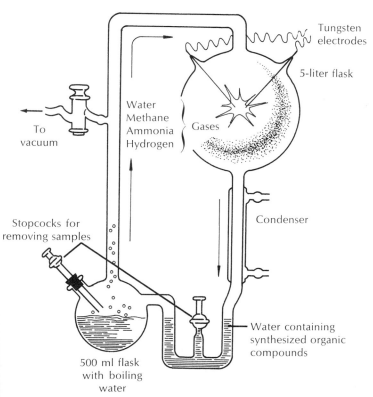

lightning, ultraviolet radiation, and perhaps other kinds of available energy. These simple molecules were washed into lakes and oceans where they formed a "primordial soup." Our understanding of what happened to produce the first reactive organic molecules comes largely from a young branch of chemistry called prebiotic chemistry.

Earlier in this chapter we referred to Stanley Miller's pioneering simulation of primitive earth conditions in the laboratory. Miller built an apparatus designed to circulate a mixture of methane, hydrogen, ammonia, and water past an electric spark (Fig. 2-11). Water in the flask was boiled to produce steam that helped to circulate the gases. The products formed in the electric discharge (representing lightning) were condensed in the condenser and collected in the U-tube and small flask (representing ocean).

After a week of continuous sparking, the water containing the products was analyzed. The results were surprising. Approximately 15% of the carbon that was originally in the reducing "atmosphere" had been converted into organic compounds that collected in the "ocean." The most striking finding was that many compounds related to life were synthesized. These included four amino acids commonly found in proteins, urea, and several simple fatty acids.

We can appreciate the astonishing nature of this synthesis when we consider that there are thousands of known organic compounds with structures that are no more complex than those of the amino acids formed. Yet in Miller's synthesis most of the relatively few substances formed were compounds found in living organisms. This was surely no coincidence, which suggests that prebiotic synthesis on the primitive earth occurred under conditions that were not greatly different from those that Miller chose to simulate.

Miller's work stimulated many other investigators to repeat and extend his experiment. It was soon found that amino acids could be synthesized in many different kinds of gas mixtures that were heated (volcanic heat), irradiated with ultraviolet light (solar radiation), or subjected to electric discharge (lightning). All that was required to produce amino acids was that the gas mixture be reducing and that it be subjected violently to some energy source. It was also confirmed that no amino acids could be produced in an atmosphere containing oxygen.

Miller discovered that amino acids were not formed directly in the spark, but rather were produced by the condensation of certain reactive intermediates, especially hydrogen cyanide and formaldehyde (HCHO) reacting with ammonia. It was found that hydrogen cyanide would react with ammonia under prebiotic conditions to form adenine, one of the four bases found in nucleic acids and a component of ATP, the universal energy intermediate of living systems. Adenine is a purine, a complex molecule chemically (see p. 30). The ease with which it was produced under prebiotic conditions suggests that it came to occupy a central position in biochemistry because it was abundant on the primitive earth. Other nucleic acid bases and sugars have been synthesized under conditions thought to have prevailed on the primitive earth.

Thus the experiments of many scientists have shown that highly reactive intermediate molecules such as hydrogen cyanide, formaldehyde, and cyanoacetylene are formed when a reducing mixture of gases is subjected to a violent energy source. These react with water and ammonia to form more complex organic molecules, including amino acids, fatty acids, urea, aldehydes, sugars, several purine and pyrimidine bases, ATP, porphyrins—indeed all the building blocks required for the synthesis of the most complex organic compounds of living matter.

Formation of polymers

Need for concentration. The next stage in chemical evolution involved the condensation of amino acids, purines, pyrimidines, and sugars to yield larger molecules that resulted in proteins and nucleic acids (see Protein synthesis). Such condensations do not occur easily in dilute solutions because the presence of excess water tends to drive reactions toward decomposition. Although the primitive ocean has been called a primordial soup, it was probably a rather dilute one containing organic material that was approximately one-tenth to one-third as concentrated as chicken bouillon.

Prebiotic synthesis must have occurred in restricted regions where concentrations were higher. The primordial soup might have been concentrated by evaporation in lakes, ponds, or tidepools. Dilute aqueous solutions could also have been concentrated effectively by freezing. As ice freezes, organic solutes are concentrated in the solution that separates from the pure ice. This technique is employed to produce applejack from cider in the northern United States and Canada. When a barrel of cider is allowed to freeze, a liquid residue remains that contains most of the alcohol and flavoring materials.

The Haldane-Oparin hypothesis suggests that the primitive ocean was a *warm* primordial soup. However, there is increasing evidence that prebiotic synthesis may have occurred in a cold rather than in a warm ocean. In a cold ocean, not unlike the ocean today with an average temperature of 4° C, much of the polar water would be frozen. This would be favorable for prebiotic synthesis for two reasons: (1) freezing of the water would concentrate the organic molecules in the residue and (2) the low temperatures would slow the spontaneous decomposition of organic compounds and polymers.

Another way prebiotic molecules might have been concentrated is by adsorption on the surface of clays and other minerals. Clay has the capacity to concentrate and condense large amounts of organic molecules from an aqueous solution.

An attractive related model for prebiotic condensation was proposed by Oparin. He suggested that colloidal droplets assembled from two or more polymers of opposite charge, which he called "coacervates" (from Latin *coacervare,* to assemble or cluster together), provided an important site for the adsorption of soluble organic molecules. Coacervate droplets were formed initially when two or more polymers of opposite charge were mixed. Even though the coacervate droplets were floating in the ocean or in a lake, they would provide a locally nonaqueous environment that was favorable for condensation reactions. Oparin and his Russian associates have prepared coacervate droplets from many different biologic compounds (for example, soaps, serum albumin, gelatin, and gum arabic); however, it has not yet been shown that important prebiotic synthesis occurred within coacervates.

Thermal condensations. Most biologic polymerizations are dehydration reactions; that is, monomers are linked together by the removal of water. The peptide bond is a familiar example. In living systems dehydration reactions always take place in an aqueous (cellular) environment. This apparent paradox—the removal of water with water all around—is possible because in living systems the appropriate dehydrating enzymes are present. Without enzymes and energy supplied by ATP, the macromolecules (proteins and nucleic acids) of living systems soon break down into their constituent monomers.

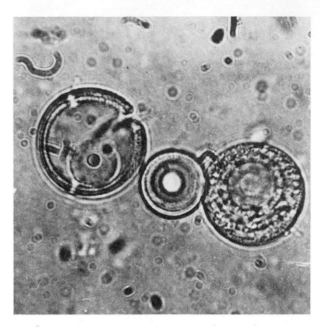

Fig. 2-12. Electron micrograph of proteinoid microspheres. These proteinlike bodies can be produced in the laboratory from polyamino acids and may represent precellular forms. They have definite internal ultrastructure. (×1700.) (Courtesy S. W. Fox, Institute of Molecular Evolution, University of Miami, Coral Gables, Fla.)

One of the most critical problems is explaining how dehydrations occurred in primitive earth conditions before enzymes appeared. The simplest dehydration is accomplished by driving off water from solids by direct heating. For example, if a mixture of all 20 amino acids is heated to 180° C, a good yield of polypeptides is obtained.

The thermal synthesis of polypeptides to form "proteinoids" has been studied extensively by the American scientist Sidney Fox. When proteinoids are boiled in water they form enormous numbers of hard, minute spherules called microspheres. Proteinoid microspheres (Fig. 2-12) possess certain characteristics of living systems. Each is not more than 2 μm in diameter and is comparable in size and shape to spheric bacteria. Their outer walls appear to have a double layer, and they show osmotic and selective diffusion properties. They may grow by accretion or proliferate by budding. There is no way to know whether proteinoids may have been the ancestors of the first cells or whether they are just interesting creations of the chemist's laboratory. They must be formed under conditions that would have been found only in volcanos. Possibly organic polymers might have condensed on or in volcanos and then, wetted by rain or dew, reacted further in solution to form polypeptides or polynucleotides.

ORIGIN OF LIVING SYSTEMS

The first living organisms were cells—autonomous membrane-bound units with a complex functional organization that permitted the essential activities of assimilation, metabolism, excretion, reproduction, irritability, and other characteristic cellular functions. The primitive chemical systems we have described lack all of these properties. The principal problem in understanding the origin of life is explaining how primitive chemical systems could have become organized into living autonomous cells.

As we have seen, a lengthy chemical evolution on the primitive earth produced several molecular components of living forms. In a later stage of evolution, nucleic acids began to behave as simple genetic systems that directed the synthesis of other polymers, especially proteins. Earlier in this chapter we referred to the troublesome chicken-egg question: (1) How could nucleic acids have appeared without enzymes? (2) How could enzymes have evolved without nucleic acids to direct their synthesis? Certainly nucleic acids are necessary for replication, and no great evolutionary progress could have been made without them. But, if nucleic acids (or polymers resembling nucleic acids) evolved before proteins, they must have been able to catalyze their own synthesis, as well as direct the synthesis of other types of compounds.

Proponents of the theory that nucleic acids arose first argue that they behaved as ''naked'' genes that possessed the means for expressing information as well as storing it. Thus far, however, it has not been possible to synthesize nucleic acids with natural linkages under prebiotic conditions. Furthermore no known nucleic acids of living organisms today have any enzymatic capabilities: they are pure repositories of information that cannot be expressed in the absence of proteins. This does not mean that they never did have catalytic powers, but it seems unlikely that nucleic acids would have evolved to code for proteins that did not already exist.

On the other hand, there is some evidence that proteins with rudimentary enzymatic and metabolic functions may have formed primitive cell-like structures with limited reproductive capabilities. The transmission of properties to subsequent generations would be less exact in the absence of nucleic acids, but more exact self-replication would not be an important requirement for early life forms on the primitive earth because lack of environmental competition would encourage, rather than prevent, diverse individuality.

Several authors favor the idea that proteins arose before nucleic acids. This idea is appealing for several reasons. First of all, proteins are the basis of the cell life cycle; the cell is built mainly of and functions principally on proteins. No cell could live on nucleic acids alone. Second, reactant amino acid molecules exhibit a high degree of internal ordering when combined under prebiotic conditions. They link together in nonrandom arrangements to form proteins of very limited diversity. This suggests that nucleic acids would not have been absolutely essential in specifying sequences in primitive proteins. Finally, it is a premise of evolutionary theory that something must exist (for example, a protein) before its existence can be directed (by nucleic acids).

These reasons establish a basis for understanding how the first ''organisms'' may have been self-ordered and self-assembling microsystems of proteins. What did these first creatures look like? They may have consisted of nothing more than grains of clay or mineral onto which were adsorbed amino acids and other micromolecules. These organic molecules, being concentrated in a microenvironment that was segregated from the water around it, were then able to form primitive proteins. The mineral may also have acted as a primitive catalyst to help polymerize amino acids and nucleotides. Only later did the nucleic acids assume the role of administrators that controlled the synthesis of polypeptides and protoenzymes.

Once this stage of organization was reached, natural selection began acting on these primitive self-replicating systems. This was a critical point. Before this stage, biogenesis was shaped by the favorable environmental conditions on the primitive earth and by the nature of the reacting elements themselves. When self-replicating systems became responsive to the forces of natural selection, their subsequent evolution became directed. The more rapidly replicating and more successful systems were favored and they replicated even faster. In short, the most efficient forms survived. From this evolved the genetic code and fully directed protein synthesis. The system was a protocell and it could be called a living organism.

Origin of metabolism

Living cells today are organized systems that possess complex and highly ordered sequences of enzyme-mediated reactions. Some cells trap solar energy and convert it into chemical bond energy, which is stored in glucose, ATP, and other molecules. Other cells are able to utilize these sources of bond energy to grow, divide, and maintain their internal integrity. Indeed, the attributes of life, involving energy conversion, assimilation, secretion, excretion, responsiveness to stimuli, and capacity to reproduce, all depend on the complex metabolic patterns characteristic of contemporary cells. How did such vastly complex metabolic schemes develop?

The earliest microorganisms were most likely anaerobic heterotrophs, probably bacteria that obtained all their nutrients directly from the environment. They would not need and doubtless did not possess any biosynthetic capacity. Chemical evolution had already supplied generous stores of nutrients in the prebiotic soup. There would be neither advantage nor need for the earliest organisms to synthesize their own compounds, as long as they were freely available from the environment. But as soon as deficiencies of certain essential nutrients occurred, alternative sources had to be found. At this point, those microorganisms that could synthesize these essential compounds from other accessible compounds would clearly have a greater advantage for surviving than those that could not.

For example, ATP is the immediate energy coinage of all living organisms. Since it has been formed in simulated prebiotic experiments, there is good reason to believe that it was present in the primitive environment and available to and used by protocells. Thus early organisms would have depended on the environmental supply for their ATP requirements.

Once the supply was exhausted or became precarious, perhaps because of an increase in the numbers of organisms using it, those protocells able to convert a precursor such as glycerate-2-phosphate to ATP would have had a tremendous advantage over those that lacked this capability. They would be selected for survival and would thrive. In the same way, when the supply of glycerate-2-phosphate became limiting it would be necessary to synthesize it from another precursor supplied by the environment. Again, the most successful organisms would have been those that chanced to develop this metabolic capability. Long sequences of reactions could have arisen in this manner.

It is important to realize that an enzyme is required to catalyze each of these reactions. So, when we say that early protocells developed a reaction sequence as we have described (A made from B, B from C, and so on), we are really saying that the appropriate enzymes appeared to catalyze these reactions. The numerous enzymes of cellular metabolism appeared when cells became able to utilize proteins for catalytic functions and thereby gained a selective advantage. No planning was required; the results were achieved through natural selection.

Appearance of photosynthesis and oxidative metabolism

Eventually, almost all utilizable energy-rich nutrients of the prebiotic soup were consumed. This ushered in the next stage of biochemical evolution, the use of readily available solar radiation to provide metabolic energy. Photosynthesis, the production of organic compounds from sunlight and atmospheric carbon dioxide, is the only process that restores free energy to the biosphere. Plant photosynthesis makes possible the richness of life on earth as we know it today. The self-nourishing photoautotrophic organisms, mainly green plants, capture solar energy and use it to convert simple inorganic substances into organic materials. The energy-rich compounds they produce provide not only for their own functioning but for the heterotrophic organisms, mainly animals, that feed upon autotrophs. The heterotrophs in turn release important raw materials for autotrophs. This is the energy cycle of the biosphere that is powered by a steady supply of energy from the sun.

The appearance of photosynthesis was of enormous consequence for evolution, but like other metabolic events it did not appear all at once. In plant photosynthesis, water is the source of the hydrogen that is used to reduce carbon dioxide to sugars. Oxygen is liberated into the atmosphere.

$$6 \, CO_2 + 6 \, H_2O \xrightarrow{\text{light}} C_6H_{12}O_6 + 6 \, O_2$$

However, the first steps in the development of photosynthesis almost certainly did not involve the splitting of water because a large input of energy is required. Hydrogen sulfide is thought to have been abundant in the primitive earth and was probably the first reducing agent used in photosynthesis.

$$6 \text{ CO}_2 + 12 \text{ H}_2\text{S} \xrightarrow{\text{light}} \text{C}_6\text{H}_{12}\text{O}_6 + 12 \text{ S} + 6 \text{ H}_2\text{O}$$

Later, as hydrogen sulfide and other reducing agents except water were used up, oxygen-evolving photosynthesis appeared. This is thought to have occurred approximately 3 billion years ago. Gradually oxygen began to accumulate in the atmosphere. When atmospheric oxygen reached approximately 1% of its present level, it began to interfere with cellular metabolism, which up to this point had evolved under strictly reducing conditions. As the atmosphere slowly changed from a reducing to an oxidizing one, a new and highly efficient kind of energy metabolism appeared: oxidative (aerobic) metabolism. By using the available oxygen to oxidize glucose to carbon dioxide and water, much of the bond energy stored by photosynthesis could be recovered. Most living forms became wholly dependent on oxidative metabolism, and oxygen-evolving photosynthesis became essential for the continuation of life on earth.

The final phase in life's evolution followed. Although a vast span of time—approximately 2 billion years—passed before multicellular organisms appeared, living cells, very much as we know them today, surrounded by semipermeable membranes and supporting an efficient oxygen-consuming form of metabolism, were flourishing on earth.

PRECAMBRIAN LIFE

As depicted on the inside back cover of this book, the Precambrian period spanned the geologic time before the beginning of the Cambrian period 600 million years ago. At the beginning of the Cambrian period most of the major phyla of invertebrate animals made their appearance within a few million years. This has been called the "Cambrian explosion" because before this time fossil deposits were rare and almost devoid of anything more complex than single-celled algae. What forms of life existed on earth before the burst of evolutionary activity in the early Cambrian world?

Prokaryotes and the age of blue-green algae

The earliest microorganisms were probably **anaerobic bacteria.** Their food was organic matter produced abiotically in the primitive oceans. The bacteria proliferated, giving rise to a great variety of bacterial forms, some of which were capable of photosynthesis. From these arose the oxygen-evolving **blue-green algae** (cyanophytes) some 3 billion years ago.

Bacteria and blue-green algae are called **prokaryotes,** meaning literally "before nucleus." They lack not only the nuclei typical of higher unicellular organisms, but other organelles as well; they have no mitochondria, plastids, chromosomes, centrioles, or vacuoles. They contain the genetic material DNA, but it is not complexed with RNA or proteins as it is in higher forms of life. They reproduce primarily by fission or budding, never by true mitotic cell division.

Bacteria and especially blue-green algae ruled the earth's oceans unchallenged for some 1½ to 2 billion years. The blue-green algae reached the zenith of their success approximately 1 billion years ago when filamentous forms produced great floating mats on the ocean surface. This long period of blue-green algae dominance, encompassing approximately two thirds of the history of life, has been called with justification the "age of blue-green algae."* Bacteria and blue-green algae are so completely different from forms of life that evolved later that they are placed in a separate kingdom, Monera (see p. 280).

Appearance of the eukaryotes

Approximately 1 billion years ago or perhaps even earlier organisms with nuclei appeared. These **eukaryotes** (true nucleus), as they are called, have cells with membrane-bound nuclei containing chromosomes in which the DNA is complexed with RNA and proteins. Eukaryotes are larger than prokaryotes, contain at least 1,000 times more DNA, and divide by some form of mitosis. They contain organelles, including mitochondria in which the enzymes for oxidative metabolism are packaged. All of the familiar forms of life (protozoans, fungi, nucleated algae, green plants, and multicellular animals) are composed of eukaryotic cells.

Prokaryotes and eukaryotes are profoundly different from each other and clearly represent a marked dicotomy in the evolution of life. The ascendency of the eukaryotes resulted in a rapid decline in the dominance of blue-green algae since the eukaryotes proliferated and quickly captured rule of the seas.

Why were the eukaryotes immediately so success-

*Schopf, J. W. 1974. Paleobiology of the precambrian: the age of blue-green algae. Evolutionary Biology. **7:**1-43.

ful? Because they developed that most important prerequisite for rapid evolution—sex. Sex promotes almost limitless genetic variability by mixing the genes of two individuals. By preserving favorable genetic variants, natural selection encourages rapid evolutionary change. Prokaryotes, on the other hand, are asexual and can only stamp out carbon copies of parental cells. The prokaryotes propagate themselves effectively and efficiently, but change occurs only when a genetic mutation intervenes.

The first eukaryotes were no doubt unicellular planktonic forms. Some were photosynthetic autotrophs; others were heterotrophs—grazing herbivores that fed on the prokaryotes. As the blue-green algae were cropped, their dense filamentous mats began to thin, providing space for other species. Carnivores appeared to feed on the herbivores. Soon a balanced ecosystem of carnivores, herbivores, and primary producers appeared. This was ideal for evolutionary diversity. By freeing space, cropping herbivores encouraged a greater diversity of producers, which in turn promoted the evolution of new and more specialized croppers. An ecologic pyramid appeared with carnivores at the top.

The burst of evolutionary activity that followed at the end of the Precambrian period and beginning of the Cambrian period was unprecedented; nothing approaching it has occurred since. Nearly all animal and plant phyla appeared and established themselves within a relatively brief period of a few million years. The eukaryotic cell made possible the richness and diversity of life on earth today.

SELECTED REFERENCES

Blum, H. F. 1968. Time's arrow and evolution. Princeton, N. J., Princeton University Press. *Advanced undergraduate level emphasizing the chemical problems of evolution.*

Cairns-Smith, A. G. 1971. The life puzzle. Toronto Ont., University of Toronto Press. *An unorthodox treatment emphasizing the physical forces that lead to order.*

Dose, K., S. W. Fox, G. A. Deborin, and T. E. Pavlovskaya (eds.). 1974. The origin of life and evolutionary biochemistry. New York, Plenum Publishing Corp. *This edited volume of 41 chapters by specialists in the origin of life commemorates the fiftieth anniversary of the appearance of Aleksandr Oparin's first published work in 1924.*

Fox, S. W. and K. Dose. 1972. Molecular evolution and the origin of life. San Francisco, W. H. Freeman & Co., Publishers. *An advanced survey of wide scope. The author's experiments with proteinoids are detailed.*

Kenyon, D. H. and G. Steinman. 1969. Biochemical predestination. New York, McGraw-Hill, Inc. *One of the best books on origin of life. Advanced and detailed but highly readable and with excellent summaries.*

Lemmon, R. M. 1970. Chemical evolution. Chemical Reviews **70:**95-109. *Excellent advanced review of this subject.*

Miller, S. L. 1953. A production of amino acids under possible primitive earth conditions. Science **117:**528. *The first report of the synthesis of amino acids in a reducing atmosphere.*

Miller, S. L. and L. E. Orgel. 1974. The origins of life. Englewood Cliffs, N. J., Prentice-Hall, Inc. *Advanced undergraduate level emphasizing the chemical problems of lifes' origins.*

Oparin, A. I. 1953. The origin of life. New York, Dover Publications, Inc. *This is the paperback edition of Oparin's most important work first published by Macmillan, Inc., in 1938. Later and in some ways less satisfactory versions have been published by Academic Press, Inc.*

Orgel, L. E. 1973. The origins of life: molecules and natural selection. New York, John Wiley & Sons, Inc. *Well-balanced treatment at the intermediate level.*

Ponnamperuma, C. 1972. The origins of life. New York, E. P. Dutton & Co., Inc. *Popularized account. Well-illustrated.*

SELECTED SCIENTIFIC AMERICAN ARTICLES

Barghoorn, E. S. 1971. The oldest fossils. **224:**30-42 (May). *Describes and illustrates fossils of algae and bacteria, some more than three billion years old.*

Cameron, A. G. W. 1975. The origin and evolution of the solar system. **233:**32-41 (Sept.).

Glaessner, M. F. 1961. Precambrian animals. **204:**72-78 (March).

Huang, S. 1960. Life outside the solar system. **202:**55-63 (April).

Lawless, J. G., C. E. Folsome, and K. A. Kvenvolden. 1972. Organic matter in meteorites. **226:**38-46 (June). *Organic compounds found in meteorites appear not to be of biologic origin.*

Sagan, C., and F. Drake. 1975. The search for extraterrestrial life. **232:**80-89 (May). *There is little doubt that civilizations even more advanced than ours exist elsewhere in the universe, but locating them will require great effort.*

Wald, G. 1954. The origin of life. **191:**44-53 (Aug.). *A popular, concise account.*

Developing two-cell frog embryo invested by protective layers of jelly. (Courtesy Carolina Biological Supply Co.)

3 THE CELL AS THE UNIT OF LIFE

CELL CONCEPT

More than 300 years ago the English scientist and inventor Robert Hooke, using a primitive compound microscope, observed boxlike cavities in slices of cork and leaves. He called these compartments "little boxes or cells." In the years that followed Hooke's first demonstration of the remarkable powers of the microscope before the Royal Society of London in 1663, biologists gradually began to realize that cells were far more than simple containers filled with "juices."

Cells are the fabric of life. Even the most primitive cells are enormously complex structures that form the basic units of all living matter. All tissues and organs are composed of cells. In man an estimated 40 trillion cells interact, each performing its specialized role in an organized community. In single-celled organisms, all the functions of life are performed within the confines of one microscopic package. There is no life without cells. The idea that the cell represents the basic structural and functional unit of life is the most important unifying concept of biology.

With the exception of eggs, which are the largest cells (in volume) known, cells are small and mostly invisible to the unaided eye. Consequently, our understanding of cells was paralleled by technical advances in the resolving power of microscopes. The Dutch microscopist A. van Leeuwenhoek, using high-quality single lenses that he had made, sent letters to the Royal Society of London containing detailed descriptions of the numerous organisms he had observed (1673 to 1723). In the early nineteenth century, the improved design of the microscope permitted biologists to separate objects only 1 μm apart. This advance was quickly followed by new discoveries that laid the groundwork for the **modern cell theory**—a theory that states that all living organisms are composed of cells.

In 1838 Matthias Schleiden, a German botanist, announced that all plant tissue was composed of cells. A year later one of his countrymen, Theodor Schwann, described animal cells as being similar to plant cells, which is an understanding that had been long delayed because the animal cell is bounded only by a nearly invisible plasma membrane and because it lacks the distinct cell wall characteristic of the plant cell. Schleiden and Schwann are thus credited with the unifying cell theory that ushered in a new era of productive exploration in cell biology.

In 1840 J. Purkinje introduced the term **protoplasm** to describe the cell contents. Protoplasm was at first thought to be a granular, gel-like mixture with special and elusive life properties of its own; the cell was thus viewed as a bag of thick soup containing a nucleus. Later the interior of the cell became increasingly visible, as microscopes were improved and better tissue-sectioning and staining techniques were introduced. Rather than being a uniform granular soup, the cell interior is composed of numerous **cell organelles,** each performing a specific function in the life of the cell.

How cells are studied

Because cells are submicroscopic structures with complex and delicate internal organization, new technical approaches were required for the visual exploration of cell structure and function. The light microscope, with all its variations and modifications, has contributed more to biologic investigation than any other instrument developed by man. It has been a powerful exploratory tool for 300 years and continues to be so more than 45 years after the invention of the electron microscope. But, until the electron microscope was perfected, our concept of the cell was limited to that which could be seen with magnifications of 1,000 to 2,000 diameters. This is the practical limit of the light microscope.

The electron microscope, invented in Germany in approximately 1931 and developed in the 1940s, employs high voltages to drive a beam of electrons through a vacuum to magnify the object being studied. The wavelength of electrons is approximately .00001 that of ordinary white light, thus permitting far greater magnification (compare *A* and *B* of Fig. 3-1). Electron microscopes can detect objects only .001 the size of objects discernible with the best light microscope. Even large molecules such as DNA and proteins can actually be seen with the electron microscope.

The source of ordinary light for the light microscope is replaced in the electron microscope with a beam of electrons emitted from a tungsten filament; the glass lenses of the light microscope are replaced with magnets for shaping the electron beam; and the man's eye is replaced with a fluorescent screen or a camera (Fig. 3-2). The system must be completely evacuated and samples must be dry, nonvolatile, and cut into extremely thin sections. Image formation depends on differences in electron scattering, which in turn depend upon the density of objects in the electron beam.

45

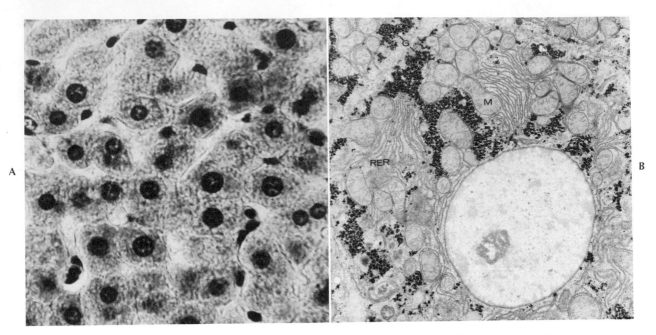

Fig. 3-1. Liver cells of rat. **A,** Magnified approximately 600 times through light microscope. Note prominently stained nucleus in each polyhedral cell. **B,** Portion of single liver cell, magnified approximately 5,000 times through electron microscope. Single large nucleus dominates field; mitochondria *(M)*, rough endoplasmic reticulum *(RER)*, and glycogen granules *(G)* are also seen. (From Morgan, C. R., and R. A. Jersild, Jr. 1970. Anat. Record **166:**575-586.)

Since the atomic nuclei of such elements as oxygen, hydrogen, carbon, and phosphorus, of which the cell is largely composed, have "equivalent mass," the tissue must be treated with "electron stains" containing ions such as osmium, lead, or uranium whose large nuclei block electrons more completely. These ions react with different chemical groups in the cell and are precipitated during these reactions. The electron micrograph shows these locations as dark, electron-opaque parts of the cell.

Recently, the **scanning electron microscope** has been introduced. Because of its great depth of field, this instrument has been used to obtain strikingly beautiful photographs of the surfaces of animals and epithelial tissues. Examples of scanning electron micrographs are shown on pp. 62 and 223.

Advances in the techniques for cell study were not limited to improvements in microscopes. Because most tissues were almost impossible to study in their living state, biologists of the last century found it necessary to kill, preserve, stain, and section tissues to make them optically suitable for examination. Histologists (scientists who specialize in the study of tissue struc-ture) experimented patiently for many years to discover fixatives that arrested life processes almost instantly and preserved the cells and their organelles in the same structural relationships they possessed when alive. The microtome, invented in 1870, allowed controlled sectioning of extremely thin slices of tissues. Organic chemists developed stains and aniline dyes that were applied to provide contrast to cellular structures. The search continues for even more precise physiochemical methods that will reveal specific entities within cells. Special stains, fluorescent antibody techniques, and radioactive tracers have contributed enormously to our present understanding of cell structure and function.

ORGANIZATION OF THE CELL

If we were to restrict our study of cells to fixed and sectioned tissues, we would be left with the erroneous impression that cells are static, quiescent, rigid structures. In fact, the cell interior is in a constant state of upheaval. Most cells are continually changing shape, pulsing and heaving; their organelles twist and regroup in a cytoplasm swimming with starch granules, fat

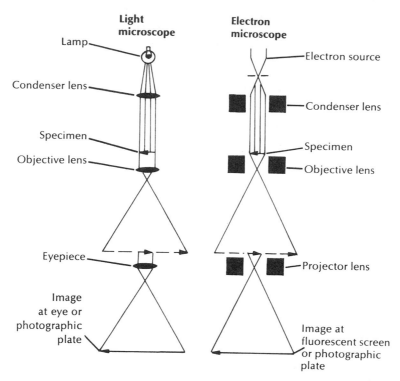

Fig. 3-2. Comparison of optical paths of light and electron microscopes. Note that to facilitate comparison, the schematic of the light microscope has been inverted from its usual orientation with light source below and image above.

globules, and membrane fragments. This description is derived from studies with time-lapse photography of living cell cultures. If we could see the swift shuttling of molecular traffic through gates in the cell membrane and the metabolic energy transformations within cell organelles, we would have an even stronger impression of internal turmoil. But the cell is anything but a bundle of disorganized activity. There is order and harmony in the cell's functioning that represents the elusive phenomenon we call life. Studying this dynamic miracle of evolution through the microscope, we realize that as we gradually comprehend more and more about this unit of life and how it operates, we are gaining a greater understanding of the nature of life itself.

Prokaryotic and eukaryotic cells

The radically different cell plan of prokaryotes and eukaryotes was previously described. A fundamental distinction, expressed in their names, is that prokaryotes lack the membrane-bound nucleus present in all eukaryotic cells. Other major differences are summarized in Table 3-1.

Despite these differences, which are of paramount importance in cell studies, there is much that prokaryotes and eukaryotes have in common. Both have DNA, use the same genetic code, and synthesize proteins. Many specific molecules such as ATP perform similar roles. These fundamental similarities imply common ancestry.

Prokaryotic cells are microbes—bacteria and blue-green algae of the kingdom Monera. The most complex of these are the filamentous forms of blue-green algae and some bacteria. All other organisms are eukaryotes distributed among four kingdoms: the unicellular kingdom Protista (protozoans and nucleated algae) and three multicellular kingdoms, Plantae (green plants), Fungi (true fungi), and Animalia (multicellular animals). The kingdom classifications are discussed in Chapter 14 (p. 280). The following discussion is restricted to eukaryotic cells of which all animals are composed.

47

Table 3-1. Comparison of prokaryotic and eukaryotic cells

Characteristic	Prokaryotic cell	Eukaryotic cell
Cell size	Mostly small (1-10 μm)	Mostly large (10-100 μm)
Genetic system	DNA not associated with proteins; no chromosomes	DNA complexed with proteins in chromosomes
Cell division	Direct by binary fission or budding; no mitosis	Some form of mitosis; centrioles, mitotic spindle present
Sexual system	Absent in most; highly modified if present	Present in most; male and female partners; gametes that fuse
Nutrition	Absorption by most; photosynthesis by some	Absorption, ingestion, photosynthesis
Energy metabolism	Mitochondria absent; oxidative enzymes bound to cell membrane, not packaged separately; great variation in metabolic pattern	Mitochondria present; oxidative enzymes packaged therein; unified pattern of oxidative metabolism throughout
Intracellular movement	None	Cytoplasmic streaming, phagocytosis, pinocytosis

Components of the eukaryotic cell and their function

If the inside of the cheek is gently scraped with a blunt instrument and if the scrapings are put on a slide in a drop of physiologic salt solution and examined unstained with a microscope, a living cell can be seen. The flat circular cells with small nuclei are the squamous epithelial cells that line the mouth region.

Flat epithelial cells are only one variety of many different shapes assumed by cells. Although many cells, because of surface tension forces, assume a spheric shape when freed from restraining influences, others retain their shape under most conditions because of their characteristic cytoskeleton, or framework of microtubules.

A eukaryotic cell includes both its outer wall, or

membranes, and its contents. Typically, it is a semifluid mass of microscopic dimensions, completely enclosed within a thin, differentially permeable **plasma membrane.** It usually contains two distinct regions—the nucleus and cytoplasm. The **nucleus** is enclosed by a **nuclear membrane** and contains **chromatin** and one **nucleolus** or more. Within the cytoplasm are many organelles such as mitochondria, Golgi complex, centrioles, and endoplasmic reticulum. Plant cells may also contain plastids, or chloroplasts.

All structures, or organelles, of the cell have separate, important functions. The **nucleus** (Figs. 3-3 and 3-4) has two important roles: (1) to store and carry hereditary information from generation to generation of cells and individuals and (2) to synthesize messenger RNA, which in cytoplasm translates genetic information into the protein characteristic of the cell and thus determines the cell's specific role in the life process.

One or more **nucleoli** (sing., nucleolus) always occur within the nucleus. These globular bodies are rich in RNA and perform an essential role in synthesizing **ribosomes,** which are required for protein synthesis. In the resting (interphase) nucleus the chromosomes are present as a deeply staining, diffuse material called **chromatin;** individual chromosomes are not visible until the cell begins to divide. The remainder of the nucleus is filled with **nucleoplasm** (Fig. 3-4).

A **plasma membrane,** or plasmalemma, surrounds the cell. It is a sturdy envelope that encloses the cell and behaves as a selective ''gatekeeper'' that determines what can and what cannot enter or leave the cell. Membranes similar to, if not identical to, the cytoplasmic membrane also surround the organelles within the cell. Membranes thus serve as partitions to subdivide the cell space into many self-contained compartments in which biochemical reactions may proceed. With the electron microscope, the cell membrane appears as two dark lines, each approximately 25 to 30 Å thick* at each side of a light zone. The entire membrane is 75 to 100 Å thick.

The **endoplasmic reticulum,** or **ER,** is a complex of membranes that separates some of the products of the cell from the synthetic machinery that produces them. This separation facilitates storage and secretion and ensures that the synthetic machinery is retained

*Units of measurement commonly used in microscopic study are micrometers, nanometers, and angstroms: 1 micrometer (μm) = $\frac{1}{1,000,000}$ meter; 1 nanometer (nm) = $\frac{1}{1,000,000,000}$ meter; 1 angstrom (Å) = $\frac{1}{10,000,000,000}$ meter. Thus 1 m = 10^3 mm = 10^6 μm = 10^9 nm = 10^{10} Å.

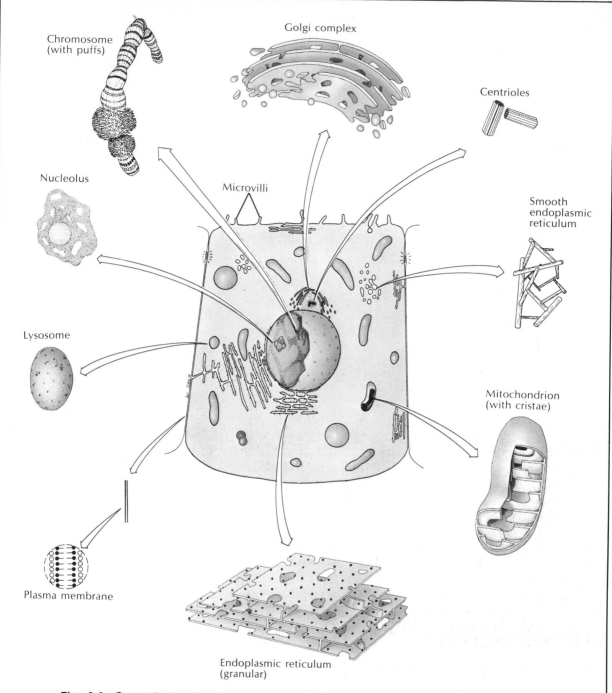

Chromosome
(with puffs)

Golgi complex

Centrioles

Nucleolus

Microvilli

Smooth
endoplasmic
reticulum

Lysosome

Mitochondrion
(with cristae)

Plasma membrane

Endoplasmic reticulum
(granular)

Fig. 3-3. Generalized cell with principal organelles, as might be seen with the electron microscope. Each of the major organelles is shown enlarged. Membranes of organelles are believed to be continuous with, or derived from, the plasma membrane by an infolding process. Structure of other membranes (of nucleus, endoplasmic reticulum, mitochondria, etc.) is probably similar to that of plasma membrane, shown enlarged at lower left.

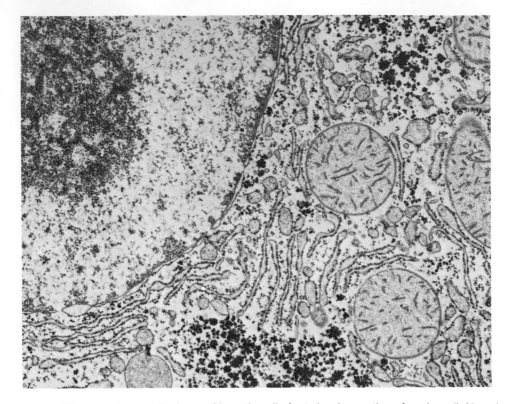

Fig. 3-4. Electron micrograph of part of hepatic cell of rat showing portion of nucleus *(left)* and surrounding cytoplasm. Endoplasmic reticulum and mitochondria are visible in cytoplasm, and pores are seen in nuclear membrane. (×14,000.) (Courtesy G. E. Palade, The Rockefeller University, New York.)

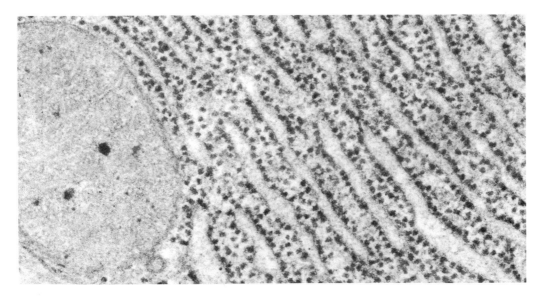

Fig. 3-5. Electron micrograph of portion of pancreatic exocrine cell from guinea pig showing endoplasmic reticulum with ribosomes (small dark granules). Oval body *(left)* is mitochondrion. (×66,000.) (Courtesy G. E. Palade, The Rockefeller University, New York.)

(Figs. 3-4 and 3-5). The nuclear membrane is formed from parts of this membrane system, and it is so arranged that there is direct continuity between nucleus and the synthetic part of the cytoplasm because of openings in the nuclear membrane (Fig. 3-4). There are two types of endoplasmic reticulum, rough-sur-faced and smooth-surfaced. The rough-surfaced type has on its outer surface small granules called ribosomes (Fig. 3-5). The ribosomes are important sites of protein synthesis.

The **Golgi complex** (shown enlarged at the top of Fig. 3-3), one part of the smooth endoplasmic re-

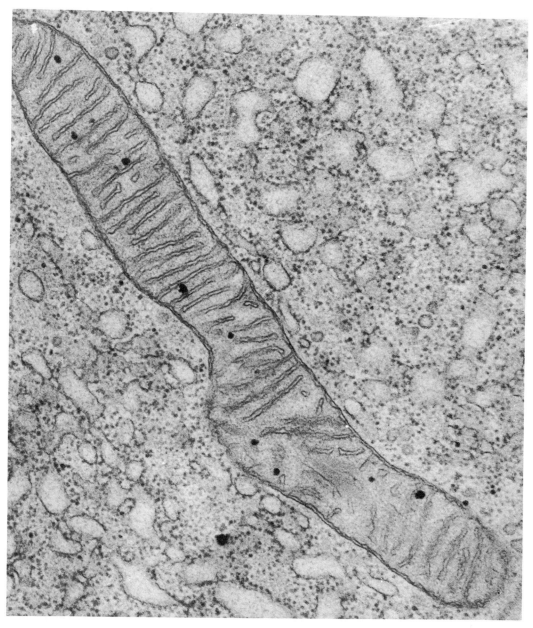

Fig. 3-6. Electron micrograph of elongated mitochondrion in pancreatic exocrine cell of guinea pig. (×50,000.) (Courtesy G. E. Palade, The Rockefeller University, New York.)

ticulum, is the primary site for processing and packaging many secretory products destined to be exported from glandular cells. The Golgi complex also secretes enzyme-rich droplets that remain in the same cell that produces them. These membrane-enclosed droplets are called **lysosomes** (literally "loosening body," a body capable of causing lysis, or disintegration). The enzymes that they contain are involved in the breakdown of foreign material, including bacteria engulfed by the cell. Lysosomes also are capable of breaking down injured or diseased cells, since the enzymes they contain are so powerful that they kill the cell that formed them if the lysosome membrane ruptures. In normal cells the enzymes remain safely enclosed within the protective membrane.

The **mitochondrion** (Fig. 3-6) is a conspicuous organelle present in nearly all eukaryotic cells. They are diverse in shape, size, and number; some are rodlike, and others are more or less spheric in shape. They may be scattered more or less uniformly through the cytoplasm, or they may be localized near cell surfaces and other regions where there is unusual metabolic activity. The mitochondrion is composed of a double membrane. The outer membrane is smooth whereas the inner membrane is folded into numerous platelike projections called **cristae** (Figs. 3-3 and 3-6). These characteristic features serve to make mitochondria easy to identify among the organelles. Mitochondria are often called "powerhouses of the cell" because enzymes located on the cristae carry out the energy-yielding steps of aerobic metabolism. ATP, the most important energy storage molecule of all cells, is produced in this organelle. Mitochondria contain DNA and synthesize some of their own proteins that contribute to their structure, although they cannot synthesize the energy-producing enzymes.

All of these cellular entities (except ribosomes) are composed of or enclosed within membranes. Eukaryotic cells also contain a variety of nonmembranous elements. **Centrioles** (shown in the upper right in Fig. 3-3), determine the orientation of the plane of cell division. An epithelial cell may have several hundred **cilia** on its surface. Each grows out from a **basal body** that is thought to arise by repeated replication of the original pair of centrioles in the cell. **Microtubules** and **microfilaments** are associated with cellular movement phenomena, including cytoplasmic streaming and the movement of chromosomes during cell division. They also form the main supportive and propulsive machinery of cilia and flagella.

Surfaces of cells and their specializations

The free surface of epithelial cells (cells that cover the surface of a structure or line a tube or cavity) frequently bear either **cilia** or **flagella.** These are vibratile, locomotory extensions of the cell surface that serve to sweep materials past the cell. In single-celled animals (many of the protozoans) and some primitive multicellular forms, they propel the entire animal through a liquid medium.

Cilia occur in large numbers on each cell and are relatively short (5 to 10 μm). Flagella typically, though not always, occur singly or in few numbers per cell and are long whip-like structures that may reach 150 μm in length. Both cilia and flagella form from centrioles that are anchored in the cell cytoplasm. Most cilia and flagella have the same internal structure: nine fibrils surrounding a pair of central fibrils.

The surface of continuous cells, or cells packed together, have junction complexes between them.

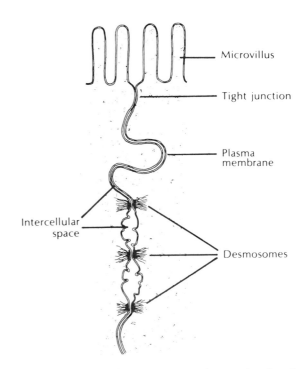

Fig. 3-7. Two opposing plasma membranes forming the boundary between two epithelial cells. Various kinds of junctional complexes are found. Tight junction is a firm, adhesive band completely encircling the cell. Desmosomes are isolated "spot welds" between cells that serve as sites of intercellular communication. Intercellular space may be greatly expanded in epithelial cells of some tissues.

There are several types of these specializations. The adjoining surfaces of cells are sealed only in restricted areas. Nearest the free surface, the two opposing membranes appear to fuse to form a **tight junction** (Fig. 3-7). Below this is a slightly widened **intermediate junction** and then **desmosomes,** small ellipsoid discs scattered between the epithelial cells. Desmosomes act as "spot welds" between apposing plasma membranes. They measure approximately 250 to 410 nm in their greatest diameter. From the cell cytoplasm tufts of fine filaments converge onto the desmosomes. Between the two apposed plates of a desmosome is a narrow intercellular space (20 to 24 nm wide). All of these special junctional complexes produce the **terminal bar,** which is found at the distal junctions of adjacent columnar epithelial cells. They form a complete beltlike junction just beyond the luminal or apical portion of the plasma membrane. It is believed that cell junctions, in addition to serving as points of attachment, serve as avenues of chemical communication between adjacent cells.

Other specializations of the cell surface are the interdigitations of confronted cell surfaces where the plasma membranes of the cells infold and interdigitate very much like a zipper. They are especially common in the epithelium of kidney tubules. The distal or apical boundaries of some epithelial cells, as seen with the electron microscope, show regularly arranged **microvilli.** They are small, fingerlike projections consisting of tubelike evaginations of the plasma membrane with a core of cytoplasm (Fig. 3-7). They are seen clearly in the lining of the intestine where they greatly increase the absorptive or digestive surface. Such specializations appear as brush borders by the light microscope. The spaces between the microvilli are continuous with tubules of the endoplasmic reticulum, which may facilitate the movement of materials into the cells.

CELL DIVISION (MITOSIS)

All cells of the body arise from the division of preexisting cells. All the cells found in most multicellular organisms have originated from the division of a single cell, the **zygote,** which is formed from the union (fertilization) of an **egg** and a **sperm.** Cell division provides the basis for one form of growth, for both sexual and asexual reproduction, and for the transmission of hereditary qualities from one cell generation to another cell generation.

In the formation of **body cells** (somatic cells) the process of nuclear division is referred to as **mitosis.**

The cell divisions that occur in **germ cells** (egg and sperm) are very much like other mitotic divisions, but the last two divisions are called **meiotic divisions** because of specialized differences or changes. Meiosis is described in Chapter 4.

Structure of chromosomes

The nucleus contains the genetic material, deoxyribonucleic acid (DNA). DNA does not occur freely in the nucleus but is combined with special stabilizing proteins called **histones** to form nucleoproteins. These, together with other structural proteins, form **chromatin.** In cells that are not dividing the chromatin is loosely organized into irregular clumps of dispersed material, but in dividing cells the chromatin condenses into the elongate **chromosomes** (color bodies), so named because they stain deeply with biologic dyes.

Chromosomes are of varied lengths and shapes, some bent, some rodlike. Their number is constant for the species, and every body cell (but not the germ cells) has the same number of chromosomes regardless of the cell's function. Man, for example, has 46 chromosomes in each body (somatic) cell.

During mitosis (nuclear division) the chromosomes shorten and become increasingly condensed and distinct, and each assumes a characteristic shape. At some point on the chromosome is a **centromere,** or constriction, to which is attached a spindle fiber that pulls the chromosome toward the pole during mitosis.

With special techniques and the electron microscope, finer details are found such as the presence of a coiled multiple filament **(chromonema)** along which are beadlike, dark-staining enlargements called **chromomeres.** The chromomeres, which perhaps represent superimposed coils or DNA condensations, may contain aggregations of genes, or the genes may be located between them. The DNA is present in the chromosomes as greatly folded threads. The DNA of a human cell exceeds a meter in length, but coiling permits it to be packed into a nucleus whose diameter is no more than .0001 of a meter.

Chromosomes are always arranged in pairs, or two of each kind. Of each pair, one comes from one parent and the other from the other parent. Thus in man there are 23 pairs. Each pair usually has certain characteristics of shape and form that aid in identification. A biparental organism begins with the union of two gametes (sex cells), each of which furnishes a **haploid** set of chromosomes (23 in man) to produce a somatic or **diploid** number of chromosomes (46 in man). The

Interphase

Chromatin material appears granular; each chromosome reaches its maximum length and minimum thickness; duplication of chromosome occurs at this state

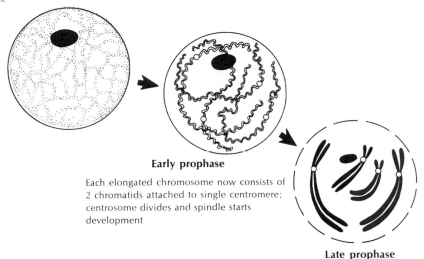

Early prophase

Each elongated chromosome now consists of 2 chromatids attached to single centromere; centrosome divides and spindle starts development

Late prophase

Double nature of short, thick chromosome more apparent; each chromosome made up of 2 half-chromosomes or sister chromatids; nucleolus usually disappears; nuclear envelope disintegrates

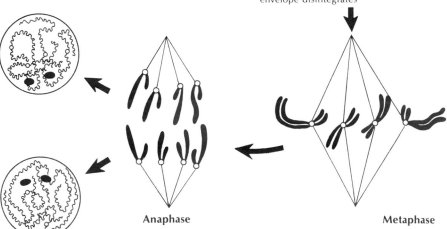

Anaphase

Chromatids, now called daughter chromosomes are in 2 distinct groups; daughter centromeres, which may be attached at various points on different chromosomes, move apart and drag daughter chromosomes toward respective poles

Metaphase

Chromosomes arranged on equatorial plate; centromeres (not yet divided) anchored to equator of spindle

Telophase

Chromosomes become longer and thinner; chromosomes may lose identity; nuclear membrane reappears and spindle-astral fibers fade away; cell body divides into 2 daughter cells, each of which now enters interphase

Fig. 3-8. Stages of mitosis, showing division of a cell with two pairs of chromosomes. One chromosome of each pair is shown in red.

chromosomes of a haploid set are also called a **genome.** Thus a fertilized egg consists of a paternal genome (chromosomes contributed by the father) and a maternal genome (chromosomes contributed by the mother).

Stages in cell division

There are two distinct phases of body cell division, **karyokinesis,** which is the division of the nucleus by mitosis, and **cytokinesis,** which is the division of the cytoplasm. These two phases ordinarily occur at the same time, but there are occasions when the nucleus may divide a number of times without a corresponding division of the cytoplasm. In such a case the resulting mass of protoplasm containing many nuclei is referred to as a **multinucleate** cell. Skeletal muscle is an example. A many-celled mass formed by fusion of cells that have lost their cell membranes is called a **syncytium.**

The process of mitosis is arbitrarily divided for convenience into four successive stages, or phases, although one stage merges into the next without sharp lines of transition. These phases are prophase, meta-phase, anaphase, and telophase (Fig. 3-8). When the cell is not actively dividing, it passes through the "resting" stage, or **interphase.** Before the first visible active phase appears, the chromosome threads and their component genes become chemically duplicated. Genetically the DNA content of the nucleus is the important constituent that is duplicated between divisions. Thus when the cell begins mitosis, it already has a double set of chromosomes.

The chromosomes of a pair are separated and pushed or pulled to opposite poles by a special mitotic apparatus (Fig. 3-9). One requirement for division of animal cells is the presence of **centrioles**—permanent, self-duplicating, rodlike bodies usually found in pairs. Each cell inherits one set of centrioles and produces another set.

At the start of **prophase** the centrioles migrate toward opposite sides of the nucleus. At the same time, fine fibers appear between the centriole complexes to form a **spindle.** Other fibers radiate out from each centriole to form **asters.** The entire structure is the **mitotic apparatus,** and it increases in size as the centrioles move farther apart.

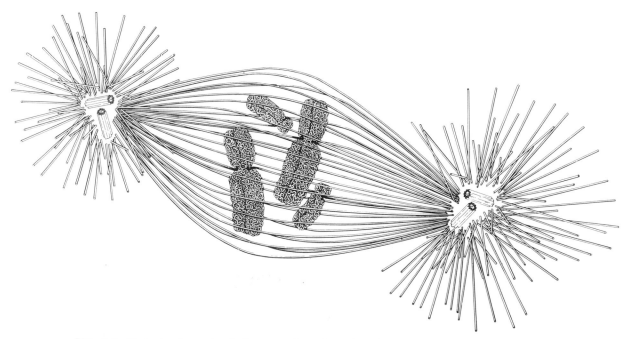

Fig. 3-9. Fine structure of mitotic apparatus. At each pole of the spindles is a clear zone occupied by a pair of centrioles and surrounded by short microtubules. Other microtubules form spindle fibers, some of which extend from pole to pole while others attach to chromosomes. (From DuPraw, E. J. 1968. Cell and molecular biology. New York, Academic Press, Inc.)

The chromosomes move by the control of the spindle. Each chromosome is composed of a double filament (two half chromosomes) that was formed before prophase began. Each half chromosome, or **chromatid,** contains nucleic acids identical to those of the chromosome before duplication. The paired chromatids are joined at a single point by the **centromere.** From each pole of the spindle a fiber makes connection with the centromere of each double chromosome.

During early **metaphase** the chromosomes migrate into the equatorial plane set at right angles to the spindle axis. The two chromatids of each double chromosome separate and are pulled to opposite poles. At this **anaphase** stage the two sets of chromatids (chromosomes) are plainly visible when examined with a microscope; one set moves toward one pole and the other set moves toward the other pole.

When the daughter chromosomes reach their respective poles, **telophase** begins. A **cleavage furrow** encircling the cell surface appears, deepens, and constricts the cell into two daughter cells. The spindle fibers disappear and the chromosomes lose their identity as they are transformed into the diffuse chromatin network.

The result of cell division is the formation of two cells, each with an identical gene set so that each daughter cell is potentially the same as the mother cell. Cell division is important for such processes as growth and replacement and wound healing. Muscle cells rarely divide; nerve cells never divide after birth. The more specialized the cell, the less frequently it divides. However, some tissues continually divide because the body loses a percentage of its cells daily and these must be replaced. Cell reproduction is faster in the embryonic state and slows down with advancing age of the animal, a condition that may be caused by metabolic checks brought on by larger cell populations.

Cell cycle

The complete sequence of events resulting in mitosis, the actual cell division, and the events that follow are called the cell cycle. A cell prepares to divide before the actual division occurs. A man's cell in tissue culture completes a cycle every 18 to 22 hours, yet division occupies only approximately 1 hour of this period.

The most important preparation—replication of DNA—occurs during the interphase, termed the S

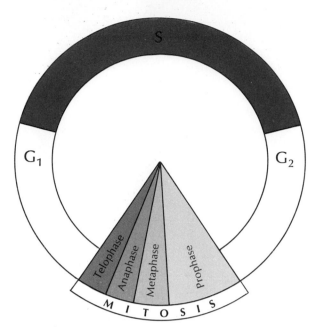

Fig. 3-10. The cell cycle (mitotic cycle) showing relative duration of recognizable stages. *S,* Synthesis of DNA. G_1, Presynthetic phase. G_2, Postsynthetic phase. Actual duration times and relative duration of phases vary considerably in different cells.

period (period of synthesis) (Fig. 3-10). In man, each chromosome contains approximately 175 million nucleotide pairs arranged in a double helix that makes one turn for every 10 nucleotide pairs. Accordingly, there are approximately 17.5 million turns in the DNA in each chromosome, all of which must somehow replicate and untangle during the middle of the interphase (S period).

The S period is preceded and succeeded by G_1 and G_2 periods, respectively (G stands for gap), when no DNA synthesis is occurring. It is believed that enzymes and substrates are being prepared during the G_1 period for the DNA replication that follows. During G_2, spindle and aster proteins are being synthesized. The energy demands of the cell are high during the G_2 period.

MEMBRANE STRUCTURE AND FUNCTION

The incredibly thin, yet sturdy, plasma membrane that encloses every cell is vitally important in maintaining cellular integrity. Once believed to be a rather static entity that defined cell boundaries and kept cell contents from spilling out, the plasma membrane has

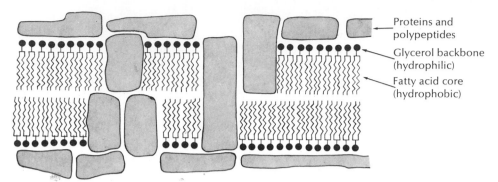

Proteins and polypeptides

Glycerol backbone (hydrophilic)

Fatty acid core (hydrophobic)

Fig. 3-11. Diagram of plasma membrane. Membrane is a lipid bilayer sandwiched between layers of protein molecules that penetrate and may even extend through the membrane in places. The proteins serve as routes for solute penetration.

proved to be a dynamic structure having remarkable activity and selectivity. It is a permeability barrier that separates the internal and external environment of the cell, regulates the vital flow of molecular traffic into and out of the cell, and provides many of the unique functional properties of specialized cells.

Membranes inside the cell—those surrounding the mitochondria, Golgi apparatus, endoplasmic reticulum, lysosomes, and other organelles—also have vital functions and share many of the structural features of the plasma membrane. Internal membranes provide the site for many, perhaps most, of the enzymatic reactions required by the cell. Proteins are formed on the membranes of ribosomes, and DNA is replicated on membrane sites.

Present concept of membrane structure

The plasma membrane is composed almost entirely of proteins and lipids. The lipids that give the membrane its strength and provide other gross structural properties belong to a class of compounds called **phospholipids.** These are **amphipathic** molecules; that is, one end is insoluble in water (hydrophobic), while the opposite end is water soluble (hydrophilic) and polar, carrying an ionic charge. The nonpolar end consists of hydrocarbon chains of **fatty acids,** and the polar (charged) end consists of **glycerol** attached to **phosphate** and other groups (Fig. 3-11).

In 1934 Danielli proposed that the phospholipids were arranged in a bimolecular layer sandwiched between two layers of protein. This remarkably visionary theory was developed from extensive studies of the chemical and biologic properties of the plasma mem-

brane long before it was possible to view the membrane with the electron microscope. Only recently (1971) was it possible to positively confirm the Danielli theory.

High-resolution electron microscopy reveals a very thin double line at the cell surface approximately 75 Å (7.5 nm) thick. The two narrow dark lines are believed to represent the protein layers and the polar ends of the phospholipids; the light middle interspace represents the fatty acid chain bilayer. This three-layered structure—two dark lines separated by a light interspace—is also characteristic of the membranes that surround the organelles within the cell. This observation resulted in the **unit-membrane hypothesis,** which states that all membranes are of similar structure and composed of lipid and protein layers as originally proposed by Danielli.

Although all membranes are built of lipids and proteins, the proteins are not found arranged in an orderly, static array on the lipid bilayer. Some proteins penetrate into the lipid core of the bilayer; others extend all the way through it (Fig. 3-11). Furthermore the membrane appears to be remarkably restless and fluid, with proteins constantly moving and reorganizing their molecular configuration. Some of the proteins that extend through the membrane are thought to behave as "channels" through which small molecules such as sodium, chloride, and potassium are allowed to pass.

Membrane permeability

The plasma membrane acts as a gatekeeper for the entrance and exit of the many substances involved in cell metabolism. Some substances can pass through with ease, others enter slowly and with difficulty, and still others cannot enter at all. This is called the **selective behavior** of the cell membrane. Because conditions outside the cell are different from and more variable than conditions within the cell, it is necessary that the passage of substances across the membrane be rigorously controlled.

We recognize three principal ways that a substance may traverse the cell membrane: (1) by **free diffusion** along a concentration gradient; (2) by a **mediated-transport system,** in which the substance binds to a specific site that in some way assists it across the membrane; and (3) by **endocytosis,** in which the substance is enclosed within a vesicle that forms on and detaches from the membrane surface to enter the cell.

Free diffusion and osmosis. If a living cell surrounded by a membrane is immersed in a solution having more solute molecules than the fluid inside the cell, a **concentration gradient** instantly exists between the two fluids. More solute molecules strike the membrane from the outside than from inside. There is a net movement of solute toward the inside, the side having the lower concentration. If the membrane is **permeable** to the solute, it freely diffuses "downhill" across the membrane until its concentrations on each side are equal.

Most cell membranes are **semipermeable,** that is, permeable to water but selectively permeable or impermeable to solutes. In free diffusion it is this selectiveness that regulates molecular traffic. As a rule, gases (such as O_2 and CO_2), urea, and lipids (such as hydrocarbons and alcohol) are the only solutes that can diffuse through biologic membranes with any degree of freedom. This happens because of the lipid nature of membranes that form a natural barrier to most biologically important molecules that are not lipid soluble. Since many water-soluble molecules readily pass through membranes, such movements cannot be explained by simple diffusion. Instead, sugars, as well as many electrolytes and macromolecules, are moved across membranes by carrier-mediated processes, described in the next section.

If a membrane is placed between two unequal concentrations of solutes to which the membrane is impermeable or only weakly permeable, water flows through the membrane from the more dilute to the more concentrated solution. This is **osmosis.** To understand why this happens we must view the system from the standpoint of the state of the water on each side.

Water exists as a mixture of complexes in equilibrium with free water. The solutes bind water or exaggerate the formation of complexes, leaving less of the water free to diffuse through the membrane. Thus there is a net flow of water across the membrane to the more concentrated solution. Unbound water, like any other material, tends to diffuse "downhill," that is, from an area of higher concentration to an area of lower concentration.

Osmosis differs from unrestricted diffusion in that only the water can diffuse; the solute is restricted by the selectively permeable membrane. Another difference is that the movement of water creates a volume change. This can be demonstrated by a familiar experiment in which a selectively permeable membrane such as collodion membrane is tied over the end of a funnel. The funnel is filled with a sugar solution and placed in a beaker of pure water so that the water levels inside and outside the funnel are equal. In a short time the water level in the glass tube rises,

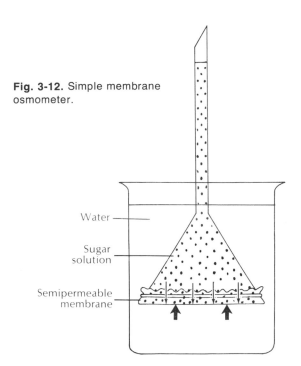

Fig. 3-12. Simple membrane osmometer.

Water

Sugar solution

Semipermeable membrane

indicating that water is passing through the collodion membrane into the sugar solution (Fig. 3-12).

Inside the funnel are sugar molecules, as well as water molecules. In the beaker outside the funnel are only water molecules. Thus the concentration of water is greater on the outside because some of the water on the inside is complexed with sugar molecules. The water therefore goes from the greater concentration (outside) to the lesser concentration (inside).

As the fluid rises in the tube against the force of gravity, it exerts a hydrostatic pressure on the collodion membrane and glass tubing (small arrows in Fig. 3-12). This hydrostatic pressure opposes the movement of water molecules into the funnel. Eventually the hydrostatic pressure becomes so great that there is no further *net* movement of water from the beaker into the bag, and the fluid level in the glass tube stabilizes. We see then that *osmosis can perform work.* A diffusion gradient (large arrows in Fig. 3-12) drives water through the membrane into the solution and creates an opposing **hydrostatic pressure** head. When the hydrostatic pressure (measured by the height of the fluid column) equals the osmotic pressure, no more water enters the osmometer. (Actually water molecules continue to traverse the membrane, but the movement inward is matched by movement outward.) Osmotic pressure can thus be expressed in terms of the height of the fluid column, which in turn depends on the concentration of the sugar solution.

Actually, the *direct* measurement of osmotic pressure in biologic solutions is seldom done today. This is because the osmotic pressures of most biologic solutions are so great that it would be impractical, if not impossible, to measure them with the simple membrane osmometer described. The osmotic pressure of human blood plasma lifts a fluid column more than 250 feet—if we could construct such a long, vertical tube and find a membrane that would not rupture from the pressure.

Indirect methods of measuring osmotic pressure are more practical. By far the most widely used measurement is the **freezing point depression.** This is a much faster and more accurate determination than is the direct measurement of osmotic pressure by the collodion membrane osmometer. Pure water freezes at exactly 0° C. As solutes are added, the freezing point is lowered; the greater the concentration of solutes, the lower the freezing point. Man's blood plasma freezes at approximately $-0.56°$ C; seawater freezes at approximately $-1.80°$ C. Although the lowering of the freezing point of water by the presence of solutes is small, great accuracy of measurement is possible because the instruments used by biologists can detect differences of as little as 0.001° C.

Mediated transport. We have seen that the cell membrane is a very effective barrier to the free diffusion of most molecules of biologic significance. Yet it is essential that such materials enter and leave the cell. Nutrients such as sugars and materials for growth such as amino acids must enter the cell, and the wastes of metabolism must leave. Such molecules are moved across the membrane by special mechanisms built into the structure of the membrane. This is called **mediated** transport, meaning that a specific transport mechanism mediates, or facilitates, transfer across the membrane barrier.

Experimental evidence suggests that mediated transport of sugars, amino acids, and other solutes involves the reversible combination with membrane proteins. Such proteins are called **carriers:** protein molecules positioned within the membrane and capable of shuttling from one membrane surface to the other. It is assumed that the carrier molecule captures a solute molecule to be transported, forming a solute-carrier complex. It moves or rotates to the opposite surface with its fare, where the solute detaches and leaves the membrane. The carrier moves back, again presenting its attachment site for the pickup and transport of another solute. Protein carriers are usually quite specific, recognizing and transporting only a limited group of chemical substances or perhaps even a single substance.

At least two distinctly different kinds of carrier-mediated transport mechanisms are recognized: (1) **facilitated diffusion,** in which the carrier assists a molecule to diffuse through the membrane that it cannot otherwise penetrate and (2) **active transport,** in which energy is supplied to the carrier systems to transport molecules in the uphill direction of the gradient (Fig. 3-13). Facilitated diffusion therefore differs from active transport in that it sponsors movement in a downhill direction only and requires no metabolic energy to drive the carrier system.

In higher animals facilitated diffusion is important for the transport of glucose (blood sugar) into body cells that burn it as a principal energy source for the

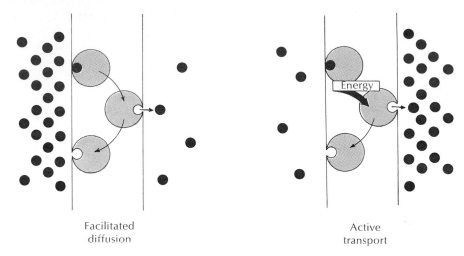

Facilitated
diffusion

Active
transport

Fig. 3-13. Comparison of facilitated diffusion and active transport. Both processes involve protein carrier molecules *(red)* that bind solutes on one surface of the membrane and discharge them on the opposite surface. Facilitated transport is downhill transport requiring no metabolic energy; active transport is uphill and requires the input of energy from the cell.

synthesis of ATP. The concentration of glucose is greater in the blood than in the cells that consume it, favoring inward diffusion, but glucose is a large, polar molecule that does not, by itself, penetrate the membrane rapidly enough to support the metabolism of many cells; the carrier system increases the inward flow of glucose.

In active transport, molecules are moved uphill against the forces of passive diffusion. Active transport always involves the expenditure of energy because materials are pumped against a concentration gradient. In some way not yet understood, cell energy in the form of ATP is coupled to the transport process. Among the most important active transport systems in all animals are those that maintain sodium and potassium gradients between cells and the surrounding extracellular fluid or external environment. Most animal cells require a high internal concentration of potassium, needed for protein synthesis at the ribosome and for certain enzymatic functions. The potassium concentration may be 20 to 50 times greater inside the cell than outside. Sodium, on the other hand, may be 10 times more concentrated outside the cell than inside. Both of these electrolyte gradients are maintained by the active transport of sodium into and potassium out of the cell. It is known that in many cells the outward pumping of sodium is linked to the inward

pumping of potassium; the same carrier molecule is used for both. As much as 10% to 40% of all the energy produced by the cell is used by the **sodium-potassium exchange pump.**

Endocytosis. The ingestion of solid or fluid material by cells was observed by microscopists nearly 100 years before phrases like "active transport" and "protein carrier mechanism" were a part of the biologist's vocabulary. "Endocytosis" is a collective term that describes two similar processes, **phagocytosis** and **pinocytosis.**

Phagocytosis, which literally means "cell eating," is a common method of feeding among the Protozoa and lower Metazoa. It is also the way in which white blood cells (leukocytes) engulf cellular debris and uninvited microbes in the blood. By phagocytosis, the cell membrane forms a pocket that engulfs the solid material. The membrane-enclosed vesicle then detaches from the cell surface and moves into the cytoplasm where its contents are digested by intracellular enzymes.

Pinocytosis, or "cell drinking," is similar to phagocytosis except that drops of fluid are sucked discontinuously through tubular channels into cells to form tiny vesicles. These may combine to form larger vacuoles. Both processes require metabolic energy, and in this respect they are forms of active transport.

SELECTED REFERENCES

Avers, C. J. 1976. Cell biology. New York, D. Van Nostrand Co. *Textbook of cell structure and function; strong on organelle genetics and cytogenetics.*

Jensen, W. A., and R. B. Park. 1967. Cell ultrastructure. Belmont, Calif., Wadsworth Publishing Co., Inc. *A collection of electron micrographs and illustrated cells and their components from both plants and animals.*

Kennedy, D. 1974. Cellular and organismal biology: Readings from Scientific American. San Francisco, W. H. Freeman & Co., Publishers. *An anthology of Scientific American articles on cell structure, regulation, and functional roles.*

Nilsson, L., and J. Linberg. 1973. Behold man. Boston, Little, Brown & Co. *Collection of superb photographs and scanning electron micrographs of organs, tissues, and embryos by an internationally recognized scientific photographer. Excellent explanatory legends accompany illustrations.*

Novikoff, A. B., and E. Holtzman. 1970. Cells and organelles. New York, Holt, Rinehart & Winston, Inc. *Clearly written and well-illustrated description of structure of cells and cell components and their function.*

Pfeiffer, J. 1972. The cell. New York, Time-Life Books. *Structure and function of cells with excellent picture essays.*

Swanson, C. P. 1969. The cell, ed. 3. Englewood Cliffs, N. J., Prentice-Hall, Inc. *One of a series of biologic monographs. A concise, well-illustrated account of the modern concept of the cell.*

SELECTED SCIENTIFIC AMERICAN ARTICLES

Capaldi, R. A. 1974. A dynamic model of cell membranes. **230:**26-33 (March). *Structure and function of the cell membrane are described.*

de Duve, C. 1963. The lysosome. **208:**64-72 (May).

Fox, C. F. 1972. The structure of cell membranes. **226:**30-38 (Feb.).

Koshland, D. E., Jr. 1973. Protein shape and biological control. **229:**52-64 (Oct.). *The ability of enzymes to bend under external influences can explain how they control life processes.*

Margulis, L. 1971. Symbiosis and evolution. **225:**48-53 (Aug.). *The author offers evidence suggesting that mitochondria and chloroplasts were once independent organisms.*

Mazia, D. 1974. The cell cycle. **230:**54-64 (Jan.).

Neutra, M., and C. P. Leblond. 1969. The Golgi apparatus. **220:**100-107 (Feb.).

Racker, E. 1968. The membrane of the mitochondrion. **218:**32-39 (Feb.).

Stent, G. S. 1972. Cellular communication. **227:**42-51 (Sept.). *Cells communicate by means of hormones and nerve impulses.*

Wessells, N. K. 1971. How living cells change shape. **225:**76-82 (Oct.). *Microtubules and microfilaments act as a skeleton and muscle for the cell.*

A fruit fly, *Drosophila melanogaster,* peers at the world through multifaceted compound eyes. This tiny midgelike creature, commonly found in nature hovering near fermenting fruit, has been a favored experimental animal of geneticists for more than half a century. It is easily maintained, has a short (10-day) life cycle, has only four easily distinguishable chromosomes, and exhibits many hundreds of heritable variations. Today more is known of the genetics of this insect than of any other animal. (Photo by P. P. C. Graziadei.)

4 GENETIC BASIS OF EVOLUTION

A basic tenet of modern evolutionary theory is that organisms attain their diversity of form, function, and behavior through hereditary modifications of preexisting lines of ancestors. It means that all known lineages of plants and animals are related by descent from simple common ancestral groups.

The study of genetics is the foundation of modern evolutionary theory. In a very real sense, evolution is change in the relative frequency of genes. Through natural selection, the principal force controlling the course of evolution, certain genes begin to win a higher representation in succeeding generations. Evolution occurs when some individuals within a population, for whatever reason, are more successful at reproduction than others. Thus the inheritable traits—and genes—responsible for their success are preferentially transmitted to the next generation.

The primary function of the organism is to reproduce genes. The organism is a device, a vehicle for the transfer of genes from one generation to the next. It is a repository in which a portion of the gene pool of the population has been temporarily entrusted. Although natural selection acts upon the individual organism, the evolving unit is the population. Therefore to understand how evolution operates, we need to know something about the genetics of populations, and this requires an understanding of the basic principles of genetics.

MENDEL AND THE LAWS OF INHERITANCE
Mendel's investigations

The first man to formulate the cardinal principles of heredity was Gregor Johann Mendel (1822-1884) (Fig. 4-1), who lived with the Augustinian monks at Brünn (Brno), Moravia. At that time Brünn was a part of Austria, but now it is the central part of Czechoslovakia. While conducting experiments in a small monastery garden from 1856 to 1864, he examined with great care and accuracy many thousands of plants. He worked out in elegant simplicity the laws governing the tranmission of characters from parent to offspring. His discoveries, published in 1866, were of great potential significance, coming, as they did, just after Darwin's publication of "Origin of Species." Yet these discoveries remained unappreciated and forgotten until 1900—some 35 years after the completion of the work and 16 years after Mendel's death.

Mendel's classic observations were based on the garden pea because it had been produced in pure strains by gardeners over a long period of time by careful selection. For example, some varieties were definitely dwarf and others were tall. A second reason for selecting peas was that they were self-fertilizing, but also capable of cross-fertilization. To simplify his problem he chose single characters and characters that were sharply contrasted. Mere quantitative and intermediate characters were carefully avoided. Mendel selected seven pairs of these contrasting characters: tall plants, dwarf plants; smooth seeds, wrinkled seeds; green cotyledons, yellow cotyledons; inflated pods, constricted pods; yellow pods, green pods; axial position of flowers, terminal position of flowers; and transparent seed coats, brown seed coats (Fig. 4-1).

Mendel crossed a plant having one of these characters with another having the contrasting character. He did this by removing the stamens from a flower to prevent self-fertilization and then placing on the stigma of this flower the pollen from the flower of the plant that had the contrasting character. He also prevented the experimental flowers from being pollinated from other sources such as wind and insects. When the cross-fertilized flower bore seeds, he noted the kind of plants (hybrids) that were produced from the planted seeds. Subsequently he crossed these hybrids among themselves to see what would happen.

Monohybrid cross

Mendel knew nothing of the cytologic background of heredity, for chromosomes and genes were unknown to him. Instead of using the term "genes" as we do today, he called his inheritance units **factors.** He reasoned that the factors for tallness and dwarfness were units that did not blend when they were together. The F_1 generation (the first generation of hybrids, or first filial generation) contained both these units or factors, but when these plants formed their germ cells the factors separated so that each germ cell had only one factor. In a pure plant both factors were alike; in a hybrid they were different. He concluded that individual germ cells were always pure with respect to a pair of contrasting factors, even though the germ cells were formed from hybrids in which the contrasting characters were mixed.

This idea formed the basis for his first principle, the **law of segregation,** which states that whenever two factors are brought together in a hybrid and whenever that hybrid forms its germ cells, the factors segregate

Experiments on which Mendel based his postulates

Results of monohybrid crosses for first and second generations

Round-wrinkled seeds
F₁ all round
F₂ 5474 round
    1850 wrinkled
Ratio 2.96:1

Colored-white flowers
F₁ all colored
F₂ 705 colored
    224 white
Ratio 3.15:1

Yellow-green cotyledons
F₁ all yellow
F₂ 6022 yellow
    2001 green
Ratio 3.01:1

Green-yellow pods
F₁ all green
F₂ 428 green
    152 yellow
Ratio 2.82:1

Inflated-constricted pods
F₁ all inflated
F₂ 882 inflated
    299 constricted
Ratio 2.95:1

Long-short stems
F₁ all long
F₂ 787 long
    277 short
Ratio 2.84:1

Axial-terminal flowers
F₁ all axial
F₂ 651 axial
    207 terminal
Ratio 3.14:1

Fig. 4-1. Seven experiments of Gregor Mendel. (Portrait courtesy American Museum of Natural History.)

into separate gametes and each germ cell is pure with respect to that character.

In the crosses involving the factors for tallness and dwarfness in which the resulting hybrids were tall, Mendel called the tall factor **dominant** and the short **recessive.** Similarly the other pairs of characters that he studied showed dominance and recessiveness. Thus, when plants with yellow unripe pods were crossed with green unripe pods, the hybrids all contained yellow pods. In the F_2 generation (second filial generation) the expected ratio of 3 yellow to 1 green was obtained. Whenever a dominant factor (gene) was present, the recessive one could not produce an effect. The recessive effect appeared only when both factors were recessive, or, in other words, when the factors were in a pure condition.

In representing his crosses Mendel used letters as symbols. For dominant characters he employed capitals and for recessives, the corresponding small letters. Thus the factors for pure tall plants might be represented by **TT,** the pure recessive by **tt,** and the hybrid of the two plants by **Tt.** In diagram form, one of Mendel's original crosses (tall plant and dwarf plant) could be represented in this manner:

	(tall)		(dwarf)
Parents	**TT**	×	**tt**
Gametes	all **T**		all **t**
F_1		**Tt**	
		(hybrid tall)	

Crossing hybrids	**Tt**		×		**Tt**
Gametes	**T, t**				**T, t**
F_2	**TT**	**Tt**	**tT**	**tt**	
	(3 tall to 1 dwarf)				

It is convenient in most mendelian crosses to use the checkerboard method devised by Punnett for representing the various combinations resulting from a cross. In the F_2 cross the following scheme would apply:

	Eggs	
	T	**t**
T	**TT** (pure tall)	**Tt** (hybrid tall)
t	**Tt** (hybrid tall)	**tt** (pure dwarf)

(left label: Sperm)

Ratio: 3 tall to 1 dwarf.

Genetic terminology

The cross between tall and dwarf plants is called a **monohybrid** cross because it involves only one pair of contrasting characters. When the gametes are united in any cross a zygote is formed. The zygote bears the complete genetic constitution of the organism. In the tall and dwarf plant example, the pure tall plants (**TT**) produce only **T** gametes; the pure dwarf plants (**tt**) produce only **t** gametes. When these gametes unite, the resulting zygote is **Tt** and thus is a mix, or **hybrid,** for this particular characteristic. When the hybrids are crossed (**Tt** × **Tt**), two kinds of plants are produced, tall and dwarf, the tall being **TT** and **Tt** and the dwarf, **tt.** Since the factors **T** and **t** are not alike in the hybrid zygote, it is said to be **heterozygous.** On the other hand, the zygotes **TT** and **tt** are said to be **homozygous,** meaning that the factors, or genes, are alike.

In the cross between tall and dwarf plants there are two types of *visible* characters—tall and dwarf. These are called **phenotypes.** On the basis of genetic formulas there are three hereditary types—**TT, Tt,** and **tt.** These are called **genotypes.** A genotype is a gene combination (**TT, Tt,** or **tt**), and the phenotype is the appearance of the organism (tall or dwarf).

When a gene exists in more than one form, the different forms are called **alleles** (or **allelomorphs**). **T** and **t** are alleles for the gene for plant height. The allele **T** is dominant and the allele **t** is recessive. The position of a gene on a chromosome is called a **locus** (pl., loci).

Testcross

The dominant characters in the offspring of a cross are all of the same phenotypes whether they are homozygous or heterozygous. For instance, in Mendel's experiment of tall and dwarf characters, it is impossible to determine the genetic constitution of the tall plants of the F_2 generation by mere inspection of the tall plants. Three fourths of this generation are tall, but which of them are heterozygous recessive dwarf?

As Mendel reasoned, the test is to cross the F_2 generation (dominant hybrids) with pure recessives. If the tall plant is homozygous, all the plants in such a testcross are tall, thus:

Parents	**TT** (tall) × **tt** (dwarf)
Offspring	**Tt** (hybrid tall)

If, on the other hand, the tall plant is heterozygous, half of the offspring are tall and half dwarf, thus:

Parents   **Tt** × **tt**
Offspring   **Tt** (tall) or **tt** (dwarf)

The testcross is often used in modern genetics for the analysis of the genetic constitution of the offspring, as well as for a quick way to make desirable homozygous stocks of animals and plants.

Intermediate inheritance

A cross that always distinguishes the heterozygotes from the pure dominants is exemplified by the four-o'clock flower *(Mirabilis)* (Fig. 4-2). Whenever a red-flowered variety is crossed with a white-flowered variety, the hybrid (F_1), instead of being red or white according to whichever is dominant, is actually intermediate between the two and is pink. Thus the homozygotes are either red or white, but the heterozygotes are pink. The testcross is therefore unnecessary to determine the nature of the genotype.

In the F_2 generation, when pink flowers are crossed with pink flowers, one fourth are red, one half pink, and one fourth white. This cross may be represented in this fashion:

	(red flower)		(white flower)	
Parents	**RR**	×	**rr**	
Gametes	**R, R**		**r, r**	
F_1		**Rr**		
		(all pink)		
Crossing hybrids	**Rr**	×	**Rr**	
Gametes	**R, r**		**R, r**	
F_2	**RR**	**Rr**	**rR**	**rr**
	(red)	(pink)	(pink)	(white)

Dihybrid and trihybrid crosses

Thus far we have been considering crosses involving alleles of a single gene (monohybrid cross). Mendel also carried out experiments on peas that differed from each other by two or more genes, that is, experiments involving two or more phenotypic characters. When a tall plant with the yellow type of pod was crossed with a dwarf plant bearing green pods, the F_1 generation was all tall and yellow, for these factors were dominant. When the F_1 hybrids were crossed with each other, the result was 9 tall and yellow, 3 tall and green, 3 dwarf and yellow, and 1 dwarf and green.

This dihybrid experiment is demonstrated in the box below.

			(tall, yellow)		(dwarf, green)	
		Parents	**TTYY**	×	**ttyy**	
		Gametes	all **TY**		all **ty**	
		F_1		**TtYy**		
			(hybrid tall, hybrid yellow)			
		Crossing hybrids	**TtYy**	×	**TtYy**	
		Gametes	**TY, Ty, tY, ty**	**TY, Ty, tY, ty**		
		F_2	(see checkerboard)			
	TY	**Ty**		**tY**		**ty**
TY	**TTYY** pure tall pure yellow	**TTYy** pure tall hybrid yellow		**TtYY** hybrid tall pure yellow		**TtYy** hybrid tall hybrid yellow
Ty	**TTYy** pure tall hybrid yellow	**TTyy** pure tall pure green		**TtYy** hybrid tall hybrid yellow		**Ttyy** hybrid tall pure green
tY	**TtYY** hybrid tall pure yellow	**TtYy** hybrid tall hybrid yellow		**ttYY** pure dwarf pure yellow		**ttYy** pure dwarf hybrid yellow
ty	**TtYy** hybrid tall hybrid yellow	**Ttyy** hybrid tall pure green		**ttYy** pure dwarf hybrid yellow		**ttyy** pure dwarf pure green

Ratio: 9 tall yellow to 3 tall green; 3 dwarf yellow to 1 dwarf green.

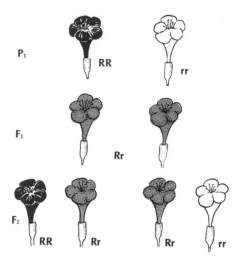

Fig. 4-2. Cross between red and white four-o'clock flowers. Red and white are homozygous; pink is heterozygous.

In the cross involving two pairs of contrasting characters (tall yellow and dwarf green) there are in the F_2 generation four phenotypes: tall yellow, tall green, dwarf yellow, and dwarf green. There are nine genotypes: **TTYY, TTYy, TtYY, TtYy, TTyy, Ttyy, ttYY, ttYy,** and **ttyy.** In this experiment each gene separated independently of the other and showed up in new combinations.

This is Mendel's second law, or the **law of independent assortment,** which states that, whenever two or more pairs of contrasting characters are brought together in a hybrid, the genes of different pairs segregate independently of one another. Rarely do two organisms differ in only one pair of contrasting characters; nearly always they differ in many. The second law of Mendel therefore deals with two or more pairs of contrasting characters.

It happened that all seven pairs of characters that Mendel worked with were on different pairs of chromosomes; but, since his laws have become known, many pairs of characters have been found on the same chromosome and this alters the original mendelian ratios. Since his time, too, the phenomena of linkage and crossing-over have necessitated a modification in his second law. This modification does not detract, however, from the basic significance of his great laws.

The F_2 ratios in any cross involving more than one pair of contrasting pairs can be found by combining the ratios in the cross of one pair of alleles. Thus the number of genotypes are $(3)^n$ and the proportion of phenotypes $(3:1)^n$ when one allele is dominant and the other recessive. For example, let us suppose that in a cross of two pairs of alleles the phenotypes are in the ratio of $(3:1)^2$, or 9:3:3:1. The genotypes in such a cross are $(3)^2$, or 9. If three pairs of characters are involved (trihybrid cross), the proportions, or ratios, of the phenotypes are then $(3:1)^3$, or 27:9:9:9:3:3:3:1. The genotypes are $(3)^3$, or 27. Thus the numerical ratio of the various phenotypes is a power of the binomial $(3 + 1)^n$ whose exponent (n) equals the number of heterozygous genes in F_2. This is true only when one allele of each gene is dominant.

By experience then we can determine the ratios of phenotypes in a cross without using the checkerboard. In a dihybrid (9:3:3:1 ratio), for instance, those phenotypes that make up the dominants of each gene are $9/16$ of the whole F_2 generation; each of the $3/16$ phenotypes consists of one dominant and one recessive; and the $1/16$ phenotype consists of two recessives.

Laws of probability

When Mendel worked out the ratios for his various crosses, they were approximations and not certainties. In his 3 to 1 ratio of tall and short plants, for instance, the resulting phenotypes did not come out exactly 3 tall to 1 short. All genetic experiments are based on probability; that is, the outcome of the events is uncertain and there is an element of chance in the final results. Probabilities are expressed in fractions, or it is always a number between 0 and 1. This probability number (p) is found by dividing the number (m) of favorable cases (for example, a certain event) by the total number (n) of possible outcomes:

$$p = \frac{m}{n}$$

When there are two possible outcomes such as in tossing a coin the chance of getting heads is $p = 1/2$, or 1 chance in 2. The more often a particular event occurs, the more closely the number of favorable cases approaches the number predicted by the p value.

The probability of independent events occurring together involves the **product rule,** which is simply the product of their individual probabilities. When two coins are tossed together, the probability of getting two heads is $1/2 \times 1/2 = 1/4$, or 1 chance in 4. Again, this prediction is most likely to occur if the coins are tossed a sufficient number of times.

The ratios of inheritance in a monohybrid cross of dominant and recessive alleles can be explained by the product rule. In the gametes of the hybrids the sperm may carry either the dominant or the recessive allele: the same applies to the eggs. The probability that the sperm carries the dominant is $\frac{1}{2}$ and the probability of an egg carrying the dominant is also $\frac{1}{2}$. The probability of a zygote obtaining two dominant alleles is $\frac{1}{2} \times \frac{1}{2}$, or $\frac{1}{4}$. Thus 25% of the offspring probably are pure dominants. The same principle applies to the recessive allele, which is pure in 25% of the offspring. The heterozygous gene combinations are found by the sum of the two possible combinations—a sperm with a dominant allele and an egg with a recessive allele, and a sperm with a recessive allele and an egg with a dominant allele—which yields 50% heterozygotes. Thus we have the 1:2:1 ratio.

Multiple alleles

Many dissimilar alleles may occupy the same gene locus on a chromosome but not, of course, all at one time. Thus more than two alternative characters may affect the same character. In the fruit fly *(Drosophila)* there are 18 alleles for eye color alone. Not more than two of these alleles can be in any one individual and only one in a gamete. The remaining 16 alleles are present in other individuals of the population in combinations of many different pairs, each affecting eye color in its own way. In some cases dominance is lacking between two members of a set of multiple alleles; between other members of the same set dominance may be strongly expressed. In *Drosophila* the allele for red eyes (the so-called wild type) is dominant over all other alleles of the eye color series; the allele for white eyes is recessive to all the others.

Multiple alleles arise through mutations at the gene locus over long periods of time. Any gene may mutate in several different ways (see p. 79) if given time, and thus it can give rise to slightly different genes or alleles at the same locus. In this way, many alleles may have arisen at any locus on a chromosome and may have been added to the genetic pool of a population. It is thought that most genes now present in an organism at any locus are mutated alleles of an original gene.

Gene interaction

The types of crosses previously described are simple in that the characters involved result from the action of a single gene, but many cases are known in which the characters are the result of two or more genes. Mendel probably did not appreciate the real significance of the genotype, as contrasted with the visible character—the phenotype. We now know that many different genotypes may be expressed as a single phenotype.

Also many genes have more than a single effect. A gene for eye color, for instance, may be the ultimate cause for eye color, yet at the same time it may be responsible for influencing the development of other characters as well. Some of these special forms of inheritance are described in the following discussions on supplementary and complementary factors, epistasis, and multiple gene inheritance.

Supplementary factors

The variety of comb forms found in chickens illustrates the interaction of supplementary genes (Fig. 4-3). The common forms of comb are rose, pea, walnut, and single. Of these, the pea comb and the rose comb are dominant to the single comb. For example, when a pea comb is crossed with a single comb, all the F_1 offspring are pea and the F_2 offspring show a ratio of 3 pea: 1 single. When the two dominants, pea and rose, are crossed with each other, an entirely new kind of comb, walnut, is found in the F_1 generation. Each of these genes supplements the other in the production of a kind of comb different from each of the dominants. In the F_2 generation the ratio is 9 walnut: 3 rose: 3 pea: 1 single. The walnut

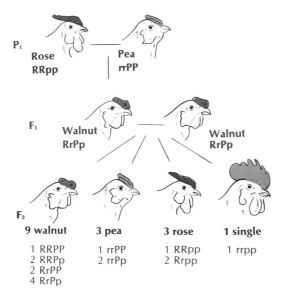

Fig. 4-3. Heredity of comb forms in chickens.

comb cannot thus be considered a unit character but is merely the phenotype's expression of pea and rose when they act together.

Inspection of the ratio reveals that two genes are involved. If **P** represents the gene for the pea comb and **p** represents its recessive allele and if **R** represents the gene for the rose comb and **r** its recessive allele, then the pea comb formula would be **rrPP**; the formula for rose comb would be **RRpp.** Any individual having both dominant alleles has a walnut comb. When no dominant allele is present, the comb is single. The cross may be diagrammed as follows:

Parents	**rrPP**	×	**RRpp**
	(pea comb)		(rose comb)
Gametes	all **rP**		all **Rp**
F_1		**RrPp** × **RrPp**	
		(walnut) (walnut)	
Gametes	**RP, rP, Rp, rp**	×	**RP, rP, Rp, rp**

By the checkerboard method, the F_2 generation shows 9 walnut: 3 pea: 3 rose: 1 single. Genotypes with the combinations of **PR** give walnut phenotypes; those with **P**, pea; those with **R**, rose; and those lacking in both **P** and **R,** single.

Complementary factors

When two genes produce a visible effect together but each alone shows no visible effect, they are referred to as **complementary genes.** Some varieties of sweet peas may be used to illustrate this kind of cross. When two white-flowered varieties of these are crossed, the F_1 generation shows all colored (reddish or purplish) flowers. When these F_1 offspring are self-fertilized, the F_2 offspring show a phenotypic ratio of 9 colored: 7 white flowers. This is really a ratio of 9:3:3:1 because the last three groups cannot be distinguished phenotypically. The explanation is that in one of the white varieties of flowers there is a gene (**C**) for a colorless color base (chromogen) and in the other white variety a gene (**E**) for an enzyme that can change chromogen into a color. Only when chromogen and the enzyme are brought together is a colored flower produced. The cross may be diagrammed in this way:

Parents	**CCee**	×	**ccEE**
	(white)		(white)
Gametes	all **Ce**		all **cE**
F_1		**CcEe** × **CcEe**	
		(colored) (colored)	
Gametes	**CE, Ce, cE, ce**		**CE, Ce, cE, ce**

By the checkerboard method, the F_2 phenotypes are as follows:

9 colored (**CCEE, CCEe, CcEE,** or **CcEe**)—both chromogen and enzyme

7 {
3 white (**CCee,** or **Ccee**)—only chromogen
3 white (**ccEE,** or **ccEe**)—only enzyme
1 white (**ccee**)—no chromogen or enzyme
}

Epistasis

When a gene at one locus affects the expression of a gene at another, the first is said to be epistatic to the second. In rabbits the gene for pigment (of any kind) is epistatic to that for agouti, a barred pigment pattern of the fur. The gene for pigment has two alleles, **C** and **c.** If the allele **C** is present, the rabbit is pigmented and the color depends on other genes. If the individual is homozygous recessive, **cc,** the rabbit is albino, no matter what other genes for pigment may be present.

The expression of the second gene, **A** for agouti, depends on the presence or absence of the **C** gene. This interaction can be summarized as follows:

Parents	**CCAA**	×	**ccaa**
	(agouti)		(albino)
Gametes	all **CA**		all **ca**
F_1		**CcAa** × **CcAa**	
		(agouti) (agouti)	
Gametes	**CA, Ca, cA, cc**		**CA, Cc, cA, cc**
F_2	9 **C-A-**	3 **C-aa** 3 **ccAa**	1 **ccaa**
	(agouti)	(black) (albino)	(albino)

The F_2 phenotypic ratio is 9:3:4. This happens because the last two phenotypic classes are both albino and are combined. The dominant allele **C** must be present for any color to appear. Also if only the recessive gene for agouti is present (homozygous **aa**), the animal's color becomes black, an intermediate between agouti and albino.

Epistasis is not the same as dominance, with which it might be confused. Dominance is the phenotypic expression of one member of a pair of alleles of a *single* gene. In epistasis the alleles of *two* genes on separate loci are involved, with one gene affecting the phenotypic expression of the other gene.

Multiple gene inheritance

Whenever several sets of alleles produce a cumulative effect on the same character, they are called **multiple genes,** or factors. Several characteristics in man are influenced by multiple genes. In such cases the characters, instead of being sharply marked, show continuous variation between two extremes. This is sometimes called **blending,** or **quantitative inheritance.** In this kind of inheritance the children are more or less intermediate between the two parents.

One illustration of such a type is the degree of pigmentation in crosses between the black and white races. The cumulative genes in such crosses have a quantitative expression. A pure Negro has two genes on separate chromosomes for pigmentation (**AABB**). A pure white has the genes (**aabb**) for nonblack. The offspring of a homozygous Negro parent and a homozygous white parent have a skin color intermediate (mulatto) between the black parent and the white. The genes for pigmentation in the cross show incomplete dominance.

The children of two mulatto parents (F_2 generation) show a variety of skin color, depending on the number of genes for pigmentation they inherit. Their skin color ranges from black (**AABB**), through dark brown (**AABb** or **AaBB**), half-colored (**AAbb** or **AaBb** or **aaBB**), light brown (**Aabb** or **aaBb**), to white (**aabb**). In the F_2 generation there is the possibility of a child resembling the skin color of either grandparent, and the others show intermediate grades. It is thus possible for parents heterozygous for skin color to produce children with darker or lighter colors than themselves.

The relationships can be seen in the following diagram:

Parents	**AABB**	×	**aabb**
	(black)		(white)
Gametes	**AB**		**ab**

F_1		**AaBb**	×	**AaBb**
		(mulatto)		(mulatto)
Gametes		**AB, Ab, aB, ab**		**AB, Ab, aB, ab**

By the checkerboard method, the F_2 generation shows this ratio:

1 black (**AABB**)
4 dark brown (**AABb** or **aABB**)
6 half-colored mulattoes (**AaBb, AAbb,** or **aaBB**)
4 light brown (**Aabb, aaBb**)
1 white (**aabb**)

We should realize that when the terms ''white'' and ''black'' are used in a cross involving mulattoes, they refer to skin color and not to other racial characteristics which are inherited independently. Thus in such a cross an individual may have white color (no genes for black) but could have other characteristics of the black parent.

When there are many genes involved in the production of traits, this situation may be represented in graphs as distribution curves, usually bell shaped. A graph for height indicates that the height of most people is near the average height, indicated by the high point of such a bell-shaped curve and that few people are much shorter or taller than average. If one locus is much more important than others, the alleles at this locus might split the graph into two or three bell-shaped curves representing different genotypes. Because we never find this kind of graph for the height of human populations, we know that humans do not have a principle for tall and dwarf lines like that of garden peas.

Penetrance and expressivity

Penetrance refers to the percentage frequency with which a gene manifests a phenotypic effect. If a dominant gene or a recessive gene in a homozygous state always produces a detectable effect, it is said to have **complete penetrance.** If dominant or homozygous recessive genes fail to show phenotypic expression in every case, it is called **incomplete** or **reduced penetrance.** Environmental factors may be responsible for the degree of penetrance because some genes may be more sensitive to such influences than others. The genotype responsible for diabetes mellitus, for instance, may be present, but because of reduced penetrance, the disease does not always occur. All of Mendel's experiments apparently had 100% penetrance.

The phenotypic variation in the expression of a gene is known as **expressivity.** For instance, a heritable allergy may cause more severe symptoms in one person than in another. Environmental factors may cause different degrees in the appearance of a phenotype. Temperature affects the expression of the genes for dark-colored fur in Siamese cats and Himalayan rabbits. Normally the tail, feet, ears, and nose—the areas that have a cooler body temperature—are dark, but under warm conditions there may be less than normal darkening, whereas in colder environmental temperatures there may be some darkening of the entire animal. Other genes in the hereditary constitution may also modify the expression of a trait. What is inherited is a certain genotype, but how it is expressed phenotypically is determined by the environment and other factors.

Genes that have more than one effect are called

pleiotropic. Most genes may have multiple effects. Even those genes that produce visible effects probably have numerous physiologic effects not detected by the geneticist. The recessive gene in fruit flies that produces (in a homozygous condition) vestigial wings also affects other traits such as the halteres (balancers), bristles, reproductive organs, and length of life.

Lethal genes

A lethal gene is one that, when present in a homozygous condition, causes the death of the offspring. It has been known for a long time that the yellow race of the house mouse *(Mus musculus)* is heterozygous. Whenever two yellow mice are bred together, the ratio of the progeny is always 2 yellow: 1 nonyellow rather than the expected ratio of 1 pure yellow: 2 hybrid yellow: 1 pure nonyellow. Examination of pregnant yellow females shows that the homozygous yellow always dies as an embryo; this accounts for the unusual ratio of 2:1.

CHROMOSOMAL THEORY OF INHERITANCE

Heredity is a protoplasmic continuity between parents and offspring. In bisexual animals special **generative cells,** or **gametes** (eggs and sperm), are responsible for establishing this continuity. Scientific explanation of genetic principles required a study of germ cells and their behavior, which meant working backward from certain visible results of inheritance to the mechanism responsible for such results. The nuclei of sex cells were early suspected of furnishing the real answer to the mechanism. This applied especially to the chromosomes, for they appeared to be the only entities passed on in equal quantities from both parents to offspring.

When Mendel's laws were rediscovered in 1900, their parallelism with the cytologic behavior of the chromosomes was obvious. In a series of experiments by Boveri, Sutton, McClung, and Wilson, the mechanism of heredity was definitely assigned to the chromosomes. The next problem was to find out how chromosomes affected the hereditary pattern.

The great American geneticist Thomas Hunt Morgan and his colleagues selected the fruit fly *(Drosophila)* for their studies because it was cheaply and easily maintained in the laboratory. With a new generation produced every 14 days, genetic studies with fruit flies proceeded at least 25 times more rapidly than with organisms that take a year to mature, such as garden peas. Morgan's work led to the mapping of genes on chromosomes and founded the discipline of cytogenetics.

Meiosis: maturation division of germ cells

Every body cell contains many pairs of chromosomes; the two members of a pair have the same size and shape. The members of such a pair are called **homologous chromosomes.** During growth of the individual, the mitotic cell divisions duplicate all the chromosomes of the dividing cells so that each new cell still contains a double set of chromosomes.

In the reproductive organs, the germ cells are formed by a special kind of maturation division, called **meiosis,** which *separates* the double sets of chromosomes. If it were not for this **reductional** division, the union of egg and sperm would produce an individual with twice as many chromosomes as the parents. Continuation of this kind of inheritance in just a few generations would yield body cells with astronomic numbers of chromosomes.

Meiosis consists of *two* nuclear divisions in which the chromosomes divide only once. The result is that mature gametes (eggs and sperm) have only *one* member of each homologous chromosome pair, or a **haploid** (n) number of chromosomes. In man the zygotes and all body cells normally have the **diploid** number (2n), or 46 chromosomes; the gametes have the haploid number (n), or 23.

Most of the unique features of meiosis occur during the prophase of the first meiotic division (Fig. 4-4). The two members of each pair of chromosomes come into side-by-side contact (synapsis). Since each chromosome has already divided into two daughter chromatids, a **tetrad** of two pairs of two chromatids is formed. The most important event in meiosis occurs when the chromatids of one tetrad exchange parts of chromatids. This phenomenon, called **crossing-over,** is clearly shown in Fig. 4-4. Crossing-over is important between a pair of homologous chromosomes (four chromatids, two from each chromosome) because the hereditary material is redistributed among the four chromatids of one tetrad. The chromosomes exchange equivalent sections bearing the same genes, so each chromatid contains a full set of genes. But the genes are in new combinations.

As prophase continues, the tetrads begin to split apart revealing **chiasmata,** the connection points where crossing-over has occurred. There may be one or more chiasmata present in each tetrad, depending on the number of times crossing-over occurs. Toward the end of prophase the chromosomes shorten and thicken and are ready to enter into the first meiotic division.

The two meiotic divisions are similar to mitosis in their morphologic changes. At anaphase of the first meiotic division, the tetrads divide and separate, each forming two pairs of chromatids called **dyads,** which migrate to opposite poles of the cell. Cell division

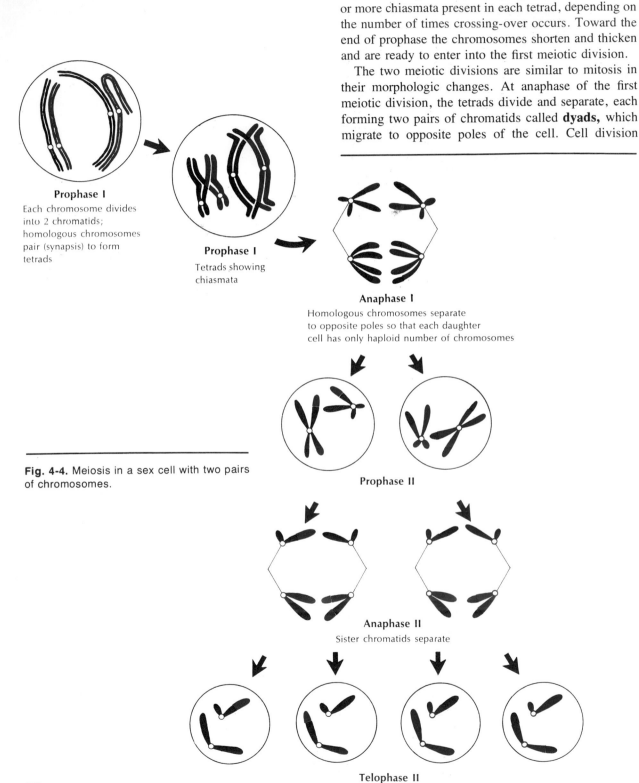

Prophase I

Each chromosome divides into 2 chromatids; homologous chromosomes pair (synapsis) to form tetrads

Prophase I

Tetrads showing chiasmata

Anaphase I

Homologous chromosomes separate to opposite poles so that each daughter cell has only haploid number of chromosomes

Fig. 4-4. Meiosis in a sex cell with two pairs of chromosomes.

Prophase II

Anaphase II

Sister chromatids separate

Telophase II

Four haploid cells (gametes) formed

follows and each resulting daughter cell thus contains half the number of chromosomes (dyads) characteristic of the diploid chromosomes of the organism.

The second meiotic division results in a separation and distribution of the sister chromatids of each chromosome to opposite poles and thus involves no reduction in the number of chromosomes. Each chromatid of the original tetrad exists in a separate nucleus. Four cells (gametes) are formed, each containing one complete, haploid set of chromosomes.

Sex determination

Before the importance of the chromosomes in heredity was realized in the early 1900s, the cause of sex was totally unknown. Speculation produced several incredible theories, for example, that the two testicles of the male contained different types of semen, one begetting males, the other females. It is not difficult to imagine the abuse and mutilation of domestic animals that occurred when attempts were made to alter the sex ratios of herds. Another theory asserted that sex of the offspring was determined by the more heavily sexed parent. An especially masculine father should produce sons, an effeminate father only daughters. Such ideas were not testable and have lingered until recently.

The first really scientific clue to the determination of sex came in 1902 when C. McClung observed that bugs (Hemiptera) produced two kinds of sperm in approximately equal numbers. One kind contained among its regular set of chromosomes a so-called accessory chromosome that was lacking in the other kind of sperm. Since all the eggs of these species had the same number of haploid chromosomes, half the sperm would have the same number of chromosomes as the eggs and half of them would have one chromosome less. When an egg is fertilized by a spermatozoan carrying the accessory (sex) chromosome, the resulting offspring is a female; when fertilized by the spermatozoan without an accessory chromosome, the offspring is a male. There are therefore two kinds of chromosomes in every cell; X chromosomes determine sex (and sex-linked traits), and **autosomes** determine the other bodily traits. The particular type of sex determination just described is often called the XX-XO type, which indicates that the females have 2 X chromosomes and the male only 1 X chromosome (the O represents its absence). The XX-XO method of sex determination can be depicted as follows:

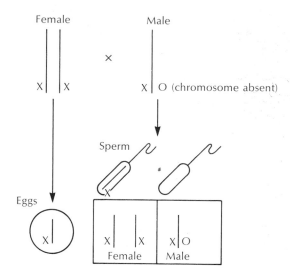

Later, other types of sex determination were discovered. In man and many other forms there are the same number of chromosomes in each sex; however, the sex chromosomes (XX) are alike in the female but unlike (XY) in the male. Hence the human egg contains 22 autosomes + 1 X chromosome; the sperm are of two kinds: half carry 22 autosomes + 1 X and half bear 22 autosomes + 1 Y. The Y chromosome is much smaller than the X. At fertilization, when 2 X chromosomes come together, the offspring are female; when XY, they are male. The XX-XY kind of determination is depicted as follows:

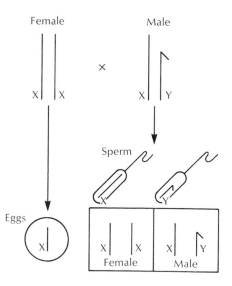

A third type of sex determination is found in birds, moths, and butterflies in which the male has 2 X (or sometimes called ZZ) chromosomes and the female an X and Y (or ZW). In this latter case the male is homozygous for sex and the female is heterozygous.

Sex-linked inheritance

It has long been known that the inheritance of some characters depends on the sex of the parents and the offspring. An example is red-green color blindness in man in which red and green colors are indistinguishable to varying degrees. Color-blind men greatly outnumber color-blind women. When color blindness does appear in women, their fathers are color blind. Furthermore, if a woman with normal vision who is a carrier of color blindness bears sons, half of them are likely to be color blind, regardless of whether the father had normal or affected vision. How are these observations explained?

Sex-linked inheritance refers to the carrying of genes by the X chromosomes for body characters that have nothing to do with sex determination. The sex chromosomes, in addition to determining sex in an organism, also carry genes for other body traits. Because of this, the inheritance of these characters is linked with that of sex. The X chromosome is known to carry many such genes, the Y chromosome only a few because of its small size. Such sex-linked traits are not always limited to one sex but may be transmitted from the mother to her male offspring or from the father to his female offspring.

The color-blindness defect is a recessive trait that is visibly expressed either when both genes are defective in the female or when only one defective gene is present in the male. The inheritance pattern is shown in Fig. 4-5. When the mother is a carrier and the father normal, one half of the sons but none of the daughters are color blind. However, if the father is color blind, one half of the sons *and* one half of the daughters are color blind. It is easy to understand then why the defect is much more prevalent in males: a single sex-linked recessive gene in the male has a visible effect. What would be the outcome of a mating between a homozygous normal woman and a color-blind man?

Another example of a sex-linked character was discovered by Morgan in *Drosophila*. The normal eye color of this fly is red, but mutations for white eyes do occur. The genes for eye color are known to be carried in the X chromosome. If a white-eyed male and a red-eyed female are crossed, all the F_1 offspring are red eyed, because this trait is dominant (Fig. 4-6). If these F_1 offspring are interbred, all the females of F_2 have red eyes, half the males have red eyes, and the other half white eyes. No white-eyed females are found in this generation; only the males have the recessive character (white eyes). The gene for being white eyed is recessive and should appear in a homozygous condition. However, since the male has only one X

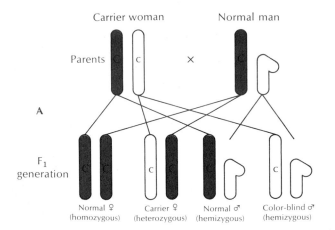
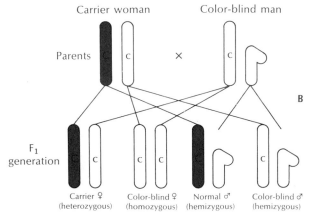

Fig. 4-5. Sex-linked inheritance of red-green color blindness in man. **A,** Carrier mother and normal father produce color blindness in one half of their sons but in none of their daughters. **B,** Half of both sons and daughters of carrier mother and color-blind father are color blind.

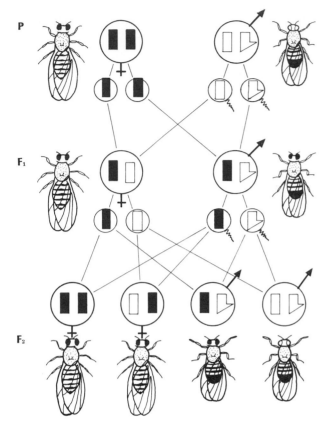

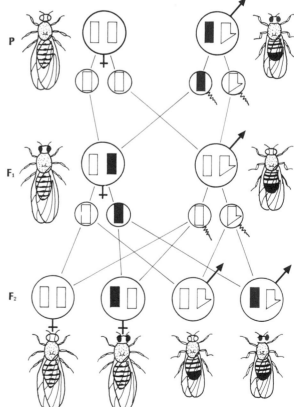

Fig. 4-6. Sex determination and sex-linked inheritance of eye color in fruit fly *(Drosophila)*. Normal red eye color is dominant to white eye color. If a homozygous red-eyed female and a white-eyed male are mated, all F$_1$ flies are red eyed. When F$_1$ flies are intercrossed, F$_2$ yields approximately 1 homozygous red-eyed female and 1 heterozygous red-eyed female to 1 red-eyed male and 1 white-eyed male. Genes for red eyes and white eyes are carried by sex (X) chromosomes; Y carries no genes for eye color.

Fig. 4-7. In cross of a homozygous white-eyed female and a heterozygous red-eyed male (reciprocal cross of Fig. 4-6), F$_1$ consists of white-eyed males and red-eyed females. In the F$_2$, there are equal numbers of red-eyed and white-eyed females and red-eyed and white-eyed males.

chromosome (the Y does not carry a gene for eye color), white eyes appear whenever the X chromosome carries the gene for this trait. If the reciprocal cross is made in which the females are white eyed and the males red eyed, all the F_1 females are red eyed and all the males are white eyed (Fig. 4-7). This is called **crisscross inheritance** because the phenotypes of the parents are exchanged in the offspring. If these F_1 offspring are interbred, the F_2 generation shows equal numbers of red-eyed and white-eyed males and females.

If the allele for red eyed is represented by **R,** white eyed by **r,** and the male sex chromosome lacking a gene for eye color by **y,** the following diagrams show the inheritance of this eye color.

Parents	**RR** (red ♀)		×		**ry** (white ♂)
F_1	**Rr** (red ♀)		×		**Ry** (red ♂)
Gametes	**R, r**				**R, y**
F_2	**RR**	**Ry**		**ry**	**rR**
	red ♀	red ♂		white ♂	red ♀

RECIPROCAL CROSS

Parents	**rr** (white ♀)		×		**Ry** (red ♂)
F_1	**rR** (red ♀)		×		**ry** (white ♂)
Gametes	**r, R**				**r, y**
F_2	**rr**	**ry**		**Rr**	**Ry**
	white ♀	white ♂		red ♀	red ♂

Autosomal linkage and crossing-over

Mendel found that when two or more characters are brought together in the hybrid, the factors of different pairs segregate independently of each other. This is stated in his second law of independent assortment. He found, in effect, that all genes are distributed to the gametes at random. The chance of this happening is $^7/_7 \cdot {}^6/_7 \cdots {}^2/_7 \cdot {}^1/_7 = {}^6/_{1000}$, or approximately 1 chance in 150. It is fortunate that Mendel selected the characters that he did because the results simplified the analysis. Chromosomes were unknown to him, and the confusion of associated inheritance could easily have made his careful study futile.

Linkage

Since Mendel's laws were rediscovered in 1900, it became apparent that, contrary to Mendel's second law, not all factors segregate independently. Indeed, many traits are inherited together. Since the number of chromosomes in any organism is relatively small compared to the number of traits, each chromosome must contain many genes. All genes present in a

chromosome are **linked.** Linkage simply means that the genes are on the same chromosome, and all genes present on homologous chromosomes belong to the same linkage groups. Therefore there should be as many linkage groups as there are chromosome pairs.

In *Drosophila,* in which this principle has been worked out most extensively, there are four linkage groups that correspond to the four pairs of chromosomes found in these fruit flies. Small chromosomes have small linkage groups and large chromosomes have large groups. Five hundred genes have been mapped in the fruit fly and all are distributed among the four pairs.

Let us see how the mendelian ratios can be altered by linkage, as illustrated by one of Morgan's experiments on *Drosophila.* In fruit fly genetics, the normal allele of any gene is called the **wild type,** since that allele is the most widespread in the wild state. It is usually dominant over its sister alleles, which are considered mutations of the normal wild allele. Morgan made a cross between wild-type fruit flies with normal body and normal wings and flies bearing two recessive mutant characters of black body and vestigial wings.

As expected the F_1 generation of this cross was phenotypically the wild type, confirming that the alleles for black body and vestigial wing were recessive. When the F_1 generation hybrids were self-crossed, the F_2 generation was not of the expected 9:3:3:1 ratio, but instead consisted mostly of wild-type flies and flies with black bodies and vestigial wings. Obviously, independent assortment, which would have yielded a 9:3:3:1 ratio, had not occurred in the F_2 generation.

Morgan then made a testcross to learn more about the genotype of the F_1 generation. This was done by breeding back the F_1 hybrid generation to the double recessive flies of the parental generation. With independent assortment we should expect four different phenotypes in approximately equal numbers:

Phenotype	Expected ratio	Numbers obtained
Wild type	1	586 (46%)
Normal body, vestigial wing	1	106 (8%)
Black body, normal wing	1	111 (9%)
Black body, vestigial wing	1	465 (37%)

Instead Morgan obtained an excess of **parental** types (wild type and double-recessive type) and a deficiency of the two gene recombinations (called **recombinant** types). In a test cross such as this, linkage is indicated if the proportion of parental types exceeds 50%. This cross yielded 83% parental types and 17% recombinant types. Morgan concluded that the wild type and black-vestigial type had entered the dihybrid cross together and stayed together, or linked.

Crossing-over

Linkage, however, is usually only partial. In the experiment just described, the parental forms totalled 83% rather than 100% as expected if linkage were complete. The fact that some recombinant types appear means that the linked genes have indeed separated, in this experiment 17% of the time. Separation of genes located on the same chromosome occurs because of **crossing-over.**

As described earlier in this chapter, during the protracted prophase of the first meiotic division, homologous chromosomes break and exchange equivalent portions; genes "cross over" from one chromosome to its homologue, and vice versa (Fig. 4-8). Each chromosome consists of two sister chromatids held

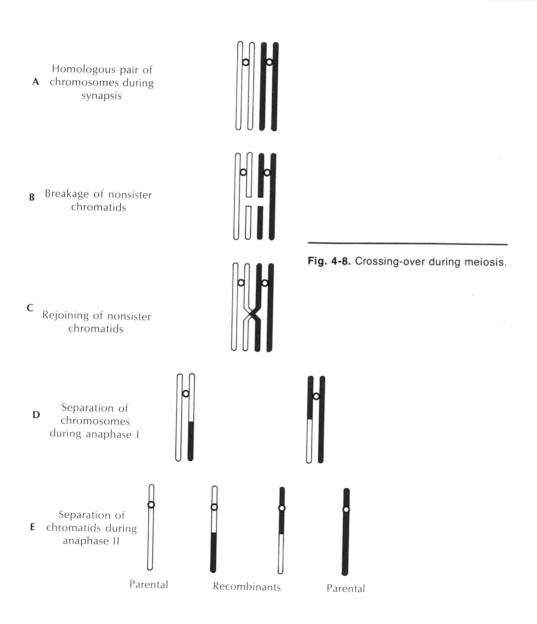

A Homologous pair of chromosomes during synapsis

B Breakage of nonsister chromatids

C Rejoining of nonsister chromatids

D Separation of chromosomes during anaphase I

E Separation of chromatids during anaphase II

Parental   Recombinants   Parental

Fig. 4-8. Crossing-over during meiosis.

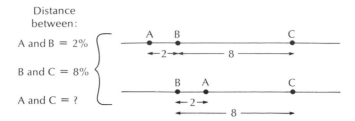

Fig. 4-9. Tentative arrangement of three genes on a chromosome. Distances are given in linkage map units with each unit defined as equal to 1% frequency of recombination. The third cross, *A* and *C*, is necessary to resolve which of the two possible arrangements proposed by the first two crosses is the correct one.

together at the centromere. Breaks and exchanges occur at corresponding points on *nonsister* chromatids. Exchange between sister chromatids apparently does not occur and would accomplish nothing if it did occur since sister chromatids are genetically identical. Crossing-over then is a means for exchanging genes between homologous chromosomes and as such greatly increases the amount of genetic recombination. The frequency of crossing-over varies with the species, but usually at least one and often several crossovers occur each time chromosomes pair.

Gene mapping

Crossing-over makes possible the construction of chromosome maps and provides proof that the genes lie in a linear order on the chromosomes. Crossing-over does not occur randomly throughout the length of the chromosome. The greater the distance is between genes, the greater is the probability that a crossover occurs between them. Two genes located at opposite ends of the chromosome are separated almost every time a break occurs; if they are located close together, they are separated only when the chromosome chances to break between them. Therefore the *frequency of recombination* indicates the relative position of the genes.

In our example of the cross between wild-type and black-vestigial–type flies, the frequency of recombination (percentage of offspring that are recombinants) is 17%. By itself, this value does not indicate the location of the two genes involved. But, if a third gene is added, their arrangement can be determined by making three crosses. Let us take a hypothetical example of three genes (**A**, **B**, **C**) on the same chromosome (Fig. 4-9).

In the determination of their comparative linear position on the chromosome, we first need to find the crossing-over value between any two of these genes. If **A** and **B** have a crossing-over rate of 2% and **B** and **C** of 8%, then the crossing-over percentage between **A** and **C** should be either the sum (2 + 8) or the difference (8 − 2). If it is 10%, **B** lies between **A** and **C**; if 6%, **A** is between **B** and **C**. By laborious genetic experiments for many years, the famed chromosome maps in *Drosophila* were worked out in this manner (Fig. 4-10). More recent cytologic investigations on the giant chromosomes present in the salivary glands of fruit fly larvae tend to prove the correctness of the linear order if not the actual position of the genes on the chromosomes.

SOURCES OF PHENOTYPIC VARIATION

Biologic variation makes evolution possible. Without variability among individuals, there could be no continued adaptation to a changing environment and no evolution. There are actually several sources of variability, some of which we have already described.

The independent assortment (segregation) of chromosomes during meiosis is a completely random process that produces many new recombinations of chromosomes in the gametes formed. In the fruit fly, which has 4 pairs of chromosomes, there are 2^4, or 16, possible kinds of sperm or kinds of eggs. In humans with 23 pairs of chromosomes, the number of possible combinations increases to 2^{23}, or approximately 8.4 million. The number of variations is further increased by chromosomal crossing over during meiosis, which allows recombination of linked genes between homologous chromosomes. Furthermore the random fusion of gametes from *both* parents adds yet another source of variation.

Thus sexual reproduction multiplies variation and

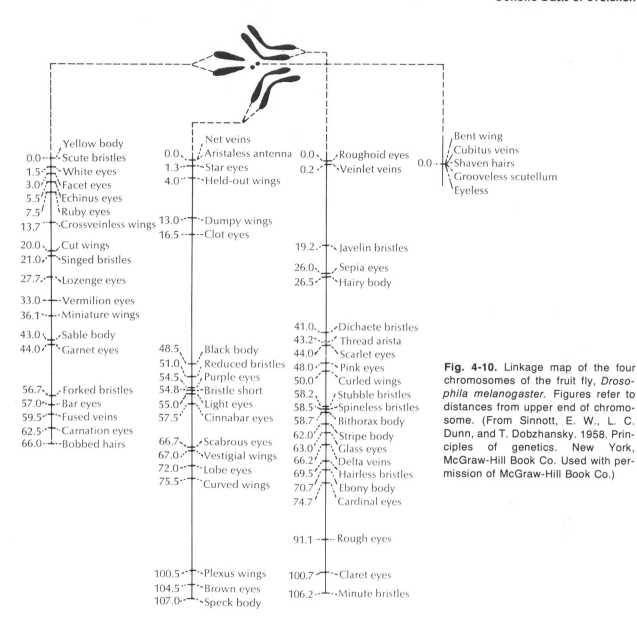

Fig. 4-10. Linkage map of the four chromosomes of the fruit fly, *Drosophila melanogaster*. Figures refer to distances from upper end of chromosome. (From Sinnott, E. W., L. C. Dunn, and T. Dobzhansky. 1958. Principles of genetics. New York, McGraw-Hill Book Co. Used with permission of McGraw-Hill Book Co.)

provides the diversity and plasticity necessary for response to environmental change. Sexual reproduction with its sequence of gene segregation and recombination, generation after generation, is as the geneticist T. Dobzhansky has said, the "master adaptation which makes all other evolutionary adaptations more readily accessible."

Although sexual reproduction reshuffles and amplifies whatever genetic diversity exists in the population, there must be ways to generate *new* genetic material.

Mutation

Mutations are the ultimate source of variability in all living systems. Mutations are random, abrupt changes in gene or chromosome structure, caused by errors in self-copying. They are random because they are unpredictable and unrelated to the needs of the organism. Once a gene is mutated, it faithfully reproduces its new self just as it did before it was mutated. Many, perhaps nearly all, mutant genes are actually harmful because they replace adaptive genes that have evolved and served the organism through its long evolution. Sometimes, however, mutations are advantageous. These are of great significance to evolution because they furnish new possibilities upon which natural selection works. Natural selection determines which new genes have survival merit; the environment imposes a screening process that passes the fit and eliminates the unfit.

When an allele of a gene is mutated into the new allele it tends to be recessive and its effects are normally masked by its partner allele. Only in the homozygous condition can such mutant genes express their phenotypic effect. Thus a population carries a reservoir of mutant recessive genes, some of which are lethal but seldom expressed. Inbreeding encourages the formation of homozygotes and increases the probability of recessive mutants appearing.

Most mutations are destined for a brief existence. There are cases, however, in which mutations may be harmful to one animal under one set of environmental conditions and helpful under a different set. Should the environment change at the same time that a favorable mutation appears, there could be a new adaptation beneficial to the species. The changing environment of earth has provided numerous opportunities for new gene combinations and mutations, as evidenced by the great diversity of animal life today.

Gene mutations

Gene mutations are chemicophysical changes in genes resulting in a visible alteration of the original character. Although the actual mutation cannot be visually detected under the microscope, all gene mutations are believed to be rearrangements of nucleotides in chromosomal DNA. A mutation may involve a codon substitution, the deletion of one or more bases from a codon, or the insertion of additional bases into the DNA chain. Most mutations are "point" mutations involving a single base pair (see p. 32).

Chromosome mutations

Although gene mutation is the fundamental source of variability, there are also heritable variations at the level of the chromosome. Chromosomal aberrations involve structural change affecting many genes at one time or changes in chromosome numbers. A portion of a chromosome may be reversed, placing the linear arrangement of genes in reverse order (inversion); nonhomologous chromosomes may exchange sections (translocation); entire blocks of genes may be lost (deletion); or a section of chromosome may attach to a normal chromosome (duplication). These are all structural changes that often produce phenotypic changes that are inherited in the normal mendelian manner.

Changes in chromosome numbers are another kind of chromosome mutation. A doubling, tripling, or quadrupling of the diploid set is called *polyploidy*. Such mutations are much more common in plants than in animals. Many domestic plant species are polyploid (cotton, wheat, apples, oats, tobacco, etc.) and perhaps one third of flowering plants are believed to have originated in this manner. Horticulturists favor polyploids and often try to develop them because they have more intensely colored flowers and more vigorous vegetative growth.

Frequency of mutations

Although mutation is thought of as a completely random process, there is increasing evidence that some genes are more stable than others. In other words, different mutation rates prevail at different loci. Nevertheless it is possible to estimate average spontaneous mutation rates for different organisms.

All genes are extremely stable. In the well-studied fruit fly *Drosophila* there is approximately 1 detectable mutation per 10,000 loci (rate of 0.01% per locus per generation). The rate for humans is 1 per 10,000 to 1 per 100,000 loci per generation. If we accept the latter, more conservative figure, then a single normal allele is expected to go through 100,000 generations before it is mutated. However, since man's chromosomes contain 100,000 loci, each person carries approximately one new mutation. Similarly, each spermatozoan produced contains, on the average, one mutant allele.

Since most mutations are deleterious, these statistics are anything but cheerful figures. Fortunately, most genes are recessive and are not detected by natural selection until by chance they have increased enough in frequency for homozygotes to be produced. At this point most are eliminated, since the zygote or individual carrying the mutants dies.

All animals carry large numbers of lethal and semilethal alleles in their gene pools. In human populations this has been called the "genetic load" (Muller), or load of hidden mutations. By genetic recombination these continue to surface to produce mutant individuals less "fit" (in the Darwinian sense) than the "wild-type" genotype.

SELECTED REFERENCES

Beadle, G. and M. Beadle. 1966. The language of life. New York, Doubleday & Co., Inc. *Interesting and lucid account of genetics.*

Dobzhansky, T. 1951. Genetics and the origin of species, 3rd ed. New York, Columbia University Press. *Important, advanced treatment of genetics.*

Dunn, L. C. 1965. A short history of genetics. New York, McGraw-Hill, Inc. *Engaging account of the development of the science of genetics.*

Hexter, W., and H. T. Yost, Jr. 1976. The science of genetics. Englewood Cliffs, N.J., Prentice-Hall, Inc. *Introductory text.*

Levine, L. 1973. Biology of the gene, ed. 2. St. Louis, The C. V. Mosby Co. *Good basic treatment of genetics.*

Mayr, E. 1970. Population, species, and evolution. Cambridge, Mass., Harvard University Press. *This excellent reference is an abridgement of the author's "Animal Species and Evolution."*

Mettler, L., and T. G. Gregg. 1969. Population, genetics and evolution. Englewood Cliffs, N.J., Prentice-Hall, Inc. *Sound elementary text.*

Swanson, C. P., T. Merz, and W. J. Young. 1967. Cytogenetics. Englewood Cliffs, N.J., Prentice-Hall, Inc. *Good source of information on the chromosomal theory of inheritance.*

Watson, J. D. 1969. The double helix. New York, Atheneum. *An enlightening and entertaining history of events leading to an understanding of the genetic code.*

Watson, J. D. 1976. Molecular biology of the gene, 3rd ed. New York, The Benjamin Co., Inc. *Concise, well-illustrated popular text.*

SELECTED SCIENTIFIC AMERICAN ARTICLES

Britten, R. J., and D. E. Kohne. 1970. Repeated segments of DNA. **222:**24-31 (April).

Brown, D. D. 1973. The isolation of genes. **229:**20-29 (Aug.).

Clark, B. F. C., and Marcker, K. A. 1968. How proteins start. **218:**36-42 (Jan.).

Crick, F. H. C. 1966. The genetic code: III. **215:**55-63 (Oct.).

Hurwitz, J., and J. J. Furth. 1962. Messenger RNA. **206:**41-49 (Feb.).

Kornberg, A. 1968. The synthesis of DNA. **219:**64-78 (Oct.).

McKusick, V. A. 1971. The mapping of human chromosomes. **224:**104-112 (April).

Miller, O. L. 1973. The visualization of genes in action. **228:**34-42 (March). *Remarkable electron micrographs of the transcription and translation processes in the cell.*

Mirsky, A. E. 1968. The discovery of DNA. **218:**78-88 (June).

Nirenberg, M. W. 1963. The genetic code: II. **208:**80-94 (March).

Nomura, M. 1969. Ribosomes, **221:**28-35 (Oct.).

Stein, G. S., J. S. Stein, and L. J. Kleinsmith. 1975. Chromosomal proteins and gene regulation. **232:**46-57 (Feb.). *The role of histones and nonhistone proteins in turning genes off and on is explained.*

Yanofsky, C. 1967. Gene structure and protein structure. **216:**80-95 (May).

Flightless cormorant (Nannopterum harrisi) of the Galápagos Islands, one of the strangest birds in a strange land. Though completely incapable of flight, it spreads its frail wings to dry after returning from a fishing venture to the sea, just as do flying cormorants from which this species descended. Living on islands unthreatened by predators, and with no need to migrate or fly long distances for food, this species was able to evolve because there was no selection against loss of flight.

5 EVOLUTION OF ANIMAL DIVERSITY

DEVELOPMENT OF THE IDEA OF ORGANIC EVOLUTION

DARWIN'S GREAT VOYAGE OF DISCOVERY

EVIDENCES FOR EVOLUTION

Reconstructing the past
Natural system of classification
Comparative anatomy: homologous resemblances
Comparative embryology
Comparative biochemistry

NATURAL SELECTION

Darwin's theory of natural selection
Appraisal of Darwin's theory
Genetic variation within species

EVOLUTION OF NEW SPECIES

Species concept
Speciation

The singularity of earth is not that it harbors life, for life must surely exist elsewhere in the universe. It is that life on earth is so enormously diverse and pervasive. Even from space the ubiquity of life is evident in the kaleidoscopic patterns of grasslands, forests, lakes, and croplands stretching across the earth's surface, their colors changing with the seasons.

Each ecosystem is biologically and physically distinct. Each is occupied by a great variety of organisms, living interdependently and manifesting every conceivable kind of adaptation to their surroundings. Nutrients are withdrawn and again released; energy captured by plants flows through the system bringing order out of disorder in apparent defiance of the second law of thermodynamics; organisms die and are replaced by offspring, a renewal that threads its descent faithfully through generations of ancestors. It is a drama of incalculable complexity that has been and is now unfolding before our eyes. Incapable at present of fully understanding how such intricacy is perpetuated year after year, we say simply that the balance of nature is preserved.

Yet despite the evident permanence of the natural world, change characterizes all things on earth and in the universe. Countless kinds of animals and plants have flourished and disappeared, leaving behind an imperfect fossil record of their former existence.

The earth itself bears its own record of change, transformed as it is by processes that have occurred over a vast span of time and still are at work today. These changes are irreversible, and for life at least they appear in the broad sense to be directional and progressive, as primitive organisms of the young earth have yielded to the advanced, complex creatures of the present. We call this progression "evolution."

An understanding of what a species is and of the evolutionary process that brought it to the place it is today underlies all biologic knowledge. Charles Darwin's great contribution is that he provided a logical explanation for evolutionary change—natural selection. Natural selection, through its effect on reproductive success, determines the genetic composition of the population and the appearance of individual adaptations.

In our treatment of the principle of organic evolution in this chapter, we consider Darwin's theory of natural selection and the evidences for evolution, the concept of species and how they change with time, and the question of how new species arise.

DEVELOPMENT OF THE IDEA OF ORGANIC EVOLUTION

Evolution is no longer a subject for debate among biologists; as an event it is known and accepted, even though the forces that determine the course of evolution are not fully understood. Of course, it was not always thus. Prior to the eighteenth century, much of the speculation on the origin of species rested on myth and superstition rather than upon observation. Nevertheless, long before this time, there were those who were thinking about and attempting to interpret the order of nature.

Some of the early Greek philosophers, notably Xenophanes, Empedocles, and Aristotle, developed the germ of the idea of change and natural selection within the restrictions of a belief in spontaneous generation. They recognized fossils as evidence of a former life that they thought had been destroyed by some natural catastrophe. Living in a spirit of intellectual inquiry, the Greeks failed to establish an evolutionary concept only because of their limited experience with the natural world.

With the gradual decline of ancient science, beginning well before the rise of Christianity, debate on evolution virtually ended. The opportunity for fresh thinking became even more restricted as the biblical account of earth's creation became accepted as a tenet of faith. The year 4004 B.C. was fixed by Archbishop James Ussher (midseventeenth century) as the time of true creation of life. In this atmosphere evolutionary views were considered rebellious and heretical.

Still, some speculation continued. The French naturalist Buffon (1707-1788) stressed the influence of environment on the modifications of animal type. He also extended the age of earth to 70,000 years from 6,000 years. Another French zoologist, Jean Baptiste de Lamarck (1744-1829), elaborated a theory of the evolutionary ascent of animals on a scale or ladder. He was the first biologist to make a convincing case for the idea that fossils were the remains of extinct animals and not, as some argued, stones molded in chance imitations of life. However, his explanation of the method of evolutionary change was less inspired.

Lamarck supposed that the use and disuse of organs by animals striving to adapt to their environment brought about changes that were passed on by inheritance. In other words, an animal's *need* for a structure encouraged its development. The giraffe, according to Lamarck, developed its long neck as the result of constant stretching for food for many generations. The explanation for the limbless snake was that legs were a handicap in moving through dense vegetation; legs were not used and consequently were lost. Unfortunately, despite the appeal of Lamarck's theory of **inheritance of acquired characteristics,** no critical evidence has been produced to support it.

While eighteenth and early nineteenth century zoologists wrestled with conflicting concepts of organic evolution, geologists were marshalling much sound evidence for the physical evolution of the earth's crust. The geologist Sir Charles Lyell (1797-1875) in his *Principles of Geology* (1830-1833) stated that the forces that produced changes in the earth's surface in the past are the same as those that operate at present. The idea that the same processes act now as in the past came to be called the doctrine of "uniformitarianism." As one writer of the time stated, "no vestige of a beginning—no prospect of an end."

Lyell was able to show that such forces, acting over long periods of time, could account for all observed changes, including the formation of fossil-bearing rocks. His familiarity with fossils, with the natural history of contemporary marine and freshwater animals, and with sedimentary rocks led him to conclude that the earth's age must be reckoned in millions of years rather than in thousands.

Charles Darwin (1809-1882) was thus not the first to propose the idea of evolution nor even to suggest a mechanism for its action. His predecessors had already established an intellectual climate that made a theory of evolution possible, if not inevitable. Darwin himself later acknowledged that the theory of natural selection had been suggested by others before him. But so forcefully did Darwin present his ideas and his array of carefully collected scientific data that no one since has been able to challenge his preeminence in this field. Today the fact of organic evolution can be denied only by abandoning reason. As the noted English biologist Sir Julian Huxley wrote, "Charles Darwin effected the greatest of all revolutions in human thought, greater than Einstein's or Freud's or even Newton's, by simultaneously establishing the fact and discovering the mechanism of organic evolution."*

DARWIN'S GREAT VOYAGE OF DISCOVERY

"After having been twice driven back by heavy south-western gales, Her Majesty's ship *Beagle,* a ten-gun brig, under the command of Captain Fitz Roy, R.N., sailed from Devonport on the 27th of December, 1831." Thus began Charles Darwin's account of the

*Huxley, J. *In* Bowman, R. I. 1966. The Galapagos. Berkeley, University of California Press.

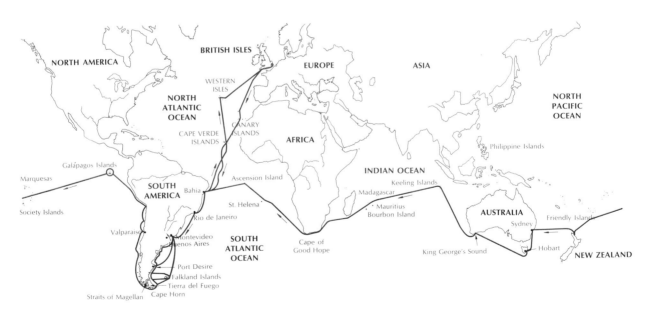

Fig. 5-1. Voyage of the *Beagle.* (Modified from Moorhead, A. 1969. Darwin and the *Beagle.* New York, Harper & Row, Publishers.)

A

B

Fig 5-2. Charles Darwin and H.M.S. *Beagle.* **A,** Darwin in 1840, 4 years after the *Beagle* returned to England. **B,** H.M.S. *Beagle* in the Straits of Magellan. (**A** from a watercolor by George Richmond; **A** and **B** courtesy American Museum of Natural History.)

historic 5-year voyage of the *Beagle* around the world (Fig. 5-1). Darwin, not quite 23 years old, had been invited to serve as naturalist without pay on the *Beagle,* a small vessel only 90 feet in length, which was about to depart on a second extensive surveying voyage to South America and the Pacific. It was the beginning of one of the most important voyages of the nineteenth century.

Charles Darwin (Fig. 5-2) was born on February 12, 1809, on the same day that Abraham Lincoln was born in a Kentucky log cabin. Unlike Lincoln, Darwin was born with position and security. His grandfather, Dr. Erasmus Darwin, naturalist, physician, and poet, had earlier achieved recognition by clearly stating the evolutionary theory of the inheritance of acquired characteristics, later embellished by Lamarck. His grandfather on his mother's side, Josiah Wedgwood, was founder of the famous Wedgwood pottery works. Assured of an inheritance sufficient to support himself comfortably all his life, Charles Darwin's name might well have been lost in obscurity as have numerous others in similar positions. Indeed, his youth promised the possibility of just such a fate.

He was indifferent to schooling. The boyish pursuits of collecting minerals, insects, bird's eggs, and shells later gave way to a passion for hunting and riding. At 16, Charles entered medical school but failed, unable to stand the sight of pain. His concerned father then decreed that Charles should become a clergyman and sent him to Cambridge where, as Charles later said, he largely wasted 3 years. Nevertheless he finished tenth in his class and graduated shortly before joining the *Beagle* expedition. His father, who considered the *Beagle* voyage the final seal to his son's fate as a professional idler, was finally persuaded to give his consent to the venture.

During the 5-year voyage (1831-1836), Darwin endured almost constant seasickness and the erratic companionship of the authoritarian Captain Fitz Roy. But his youthful physical strength and early training as a naturalist equipped him for his work. The *Beagle* made many stops along the harbors and coasts of South America and adjacent regions. Darwin made extensive collections and observations on the fauna and flora of these regions. He unearthed numerous fossils of animals long since extinct and noted the resemblance between fossils of the South American pampas and the known fossils of North America. In the Andes he encountered sea shells embedded in rocks at 13,000

feet. He experienced a severe earthquake and watched the mountain torrents that relentlessly wore away the earth. He read and was impressed by the writings of Charles Lyell. He reflected at length on activities that have shaped the earth over vast periods of time. He began to realize that he was witnessing evolution.

Finally in mid-September of 1835, the *Beagle* arrived at the Galápagos Islands, a volcanic archipelago straddling the equator 600 miles west of the coast of Ecuador (Fig. 5-3). The fame of the islands stems from their infinite strangeness. They are unlike any other islands on earth. Some visitors today are struck with an impression of awe and wonder; others with a sense of depression and dejection. Darwin described them thus:

These islands at a distance have sloping uniform outline, excepting where broken by sundry paps and hillocks; the whole black Lava, completely covered by small leafless brushwood and low trees. The fragments of Lava where most porous, are reddish like cinders; the stunted trees show little signs of life. The black rocks heated by the rays of the Vertical sun, like a stove, give to the air a close and sultry feeling. The plants also smell unpleasantly. The country was compared to what one might imagine the cultivated parts of the Infernal regions to be.

Circled by capricious currents, surrounded by shores of twisted lava, bearing skeletal brushwood baked by the equatorial sun, almost devoid of lush tropical vegetation, inhabited by strange reptiles and by convicts stranded by the Ecuadorian government, the islands indeed had few admirers among mariners. By the middle of the seventeenth century, the islands were already known to the Spaniards as "Las Islas Galápagos"—the tortoise islands. The giant tortoises, used for food first by buccaneers and later by American and British whalers, sealers, and ships of war, were the islands' principal attraction. At the time of Darwin's visit, the tortoises were already being heavily exploited.

During the *Beagle's* 5-week visit to the Galápagos, Darwin began to develop his views of the evolution of life on earth. After returning to England Darwin wrote:

The natural history of these islands is eminently curious, and well deserves attention. Most of the organic productions are aboriginal creations, found nowhere else; there is even a difference between inhabitants of the different islands, yet all show a marked relationship with those of America, though separated from that continent by an open space of ocean, between 500 and 600 miles in width. The archipelago is a little world within itself . . .

Darwin's original observations on the giant tortoises, the marine iguanas that feed on seaweed, and

Fig. 5-3. The Galápagos Islands. (Photo by C. P. Hickman, Jr.)

the family of drab ground finches that had evolved into several species, each with distinct beak and feeding adaptations, all contributed to the turning point in Darwin's thinking.

Darwin was struck by the fact that, although the Galápagos Islands and the Cape Verde Islands (visited earlier in this voyage of the *Beagle*) were similar in climate and topography, their fauna and flora were altogether different. He recognized that Galápagos plants and animals were related to those of the South American mainland, yet differed from them sometimes in curious ways. Each island often contained a unique species of a particular group of animals that was nonetheless related to forms on other islands. In short, Galápagos life must have originated in continental South America and then undergone modification in the various environmental conditions of the separate is-

lands. He concluded that living forms were neither divinely created nor immutable; they were, in fact, the products of evolution. Although Darwin devoted only a few pages to Galápagos animals and plants in his monumental *Origin of Species,* published more than two decades after visiting the islands, his observations on the unique character of the animals and plants were, in his own words, the "origin of all my views."

On October 2, 1836, the *Beagle* landed in England. Most of Darwin's extensive collections had long since preceded him to England, as had most of his notebooks and diaries kept during the 5-year cruise. Three years after the *Beagle*'s return to England, Darwin's journal was published. It was an instant success and required two additional printings within the first year. In later versions, Darwin made extensive changes and abbreviated the original ponderous title typical of nineteenth century books to simply *The Voyage of the Beagle*. The fascinating account of his observations written in a simple, appealing style has made the book one of the most lasting and popular travel books of all time.

Curiously the main product of Darwin's voyage, his theory of evolution, did not appear in print for more than 20 years after the *Beagle*'s return. Darwin began gathering together the evidence for evolution almost as soon as the voyage ended. He carried out an extensive correspondence with botanists and zoologists and followed progress in biology and related fields.

In 1838 he "happened to read for amusement" an essay on populations by T. R. Malthus, who stated that animal and plant populations, including human populations, tend to increase beyond the capacity of the environment to support them. Darwin had already been gathering information on the artificial selection of animals under domestication by man. After reading Malthus' article, Darwin realized that a process of selection in nature, a "struggle for existence" because of overpopulation, could be a powerful force for evolution of wild species.

He allowed the idea to develop in his mind until it was presented in an essay in 1844—still unpublished. Finally in 1856 he began to pull together his voluminous data into a work on the origin of species. He expected to write four volumes, a "very big" book, "as perfect as I can make it." However, his plans were to take an unexpected turn.

In 1858, he received a manuscript from Alfred Russel Wallace (1823-1913), an English naturalist in Malaya with whom he had been corresponding. Darwin was stunned to find that in a few pages, Wallace summarized the main points of the natural selection theory that Darwin had been working on for two

Fig. 5-4. Reconstruction of the appearance of *Corythosaurus*, a dinosaur from the upper Cretaceous of North America, approximately 90 million years ago. **A,** Portion of skeleton as discovered and partially worked out of rocks in New Mexico quarry. **B,** Skeleton, 30 feet in length, on slab. **C,** Drawing prepared from skeleton. **D,** Artist's reconstruction of living animal. (**A** and **B** courtesy American Museum of Natural History; **C** and **D** from Colbert, E. H. 1969. Evolution of the vertebrates, ed. 2. New York, John Wiley & Sons, Inc.)

Fig. 5-4, cont'd. For legend see opposite page.

decades. Rather than withhold his own work in favor of Wallace as he was inclined to do, Darwin was persuaded by two close friends, the geologist Lyell and the botanist Hooker, to publish his views in a brief statement that would appear together with Wallace's paper in the Journal of the Linnaean Society. Portions of both papers were read before an unimpressed audience on July 1, 1858.

For the next year, Darwin worked urgently to prepare an "abstract" of the planned four-volume work. This was published in November, 1859, with the title *On the Origin of Species by Means of Natural Selection or the Preservation of Favoured Races in the Struggle for Life*. The 1,250 copies of the first printing were sold the first day; the book instantly generated a storm that has never completely abated. His views were to have extraordinary consequences upon scientific and religious thought and remain among the greatest intellectual achievements of mankind.

Once Darwin's excessive caution had been swept away by the publication of *Origin of Species,* he entered an incredibly productive period of evolutionary thinking for the next 23 years, producing book after book. He died April 19, 1882, and was buried in Westminster Abbey. The little *Beagle* had already disappeared, having been retired in 1870 and presumably broken up for scrap.*

The biologic fame of the Galápagos Islands that contributed so much to Darwin's thinking had been secured by his first book, *The Voyage of the Beagle.* The unusual fauna and flora of the Galápagos continued to excite biologists, and numerous expeditions visited the islands. Unfortunately the famous tortoises and once-great herds of fur seals were relentlessly exterminated throughout the nineteenth century. At the same time the natural balance of the island communities was seriously damaged by the accidental or purposeful introduction of black rats, dogs, cats, pigs, goats, and donkeys, all of which have gone wild on most islands and continue to menace the native animals today.

Rising concern over these events eventually culminated in the establishment of the Charles Darwin Research Station at Academy Bay, Santa Cruz Island, in 1964. The entire archipelago has now been proclaimed a national park by Ecuador, and legislation protecting the wildlife has been passed. Although the native fauna was seriously disfigured by man's activities and many dangers remain, there is renewed hope that these unique islands are finally receiving the protection they deserve.

EVIDENCES FOR EVOLUTION
Reconstructing the past
Fossils and their formation

The strongest and most direct evidence for evolution is the fossil record of the past. Fossils and the sediments containing them provide dim and imperfect views of an ancient life (Fig. 5-4). The complete record of the past is always beyond our reach since many groups left no fossils. Indeed, were the past by some miracle to be suddenly laid out for our scrutiny, the relationships of organisms would be incalculably more complex than they are now. Yet, incomplete as the record is, biologists rely on the discoveries of new fossils and the continued study of existing fossils to interpret phylogeny and relationships of both plant and animal life. It would be difficult indeed to make sense out of the evolutionary patterns or classification of organisms without the support of the fossil record. The documentary evidence for evolution as a general process, the progressive changes in life from one geologic era to another, the past distribution of lands and seas, and the environmental conditions of the past (paleoecology) are all dependent on what fossils teach us.

A fossil may be defined as the remains of past life uncovered from the crust of the earth. It refers not only to complete remains (mammoths and amber insects), actual hard parts (teeth and bones), and petrified skeletal parts that are infiltrated with silica or other minerals by water seepage (ostracoderms and molluscs), but also to molds, casts, impressions, and fossil excrement (coprolites).

The fossil record is biased because preservation is selective. Vertebrate skeletal parts and invertebrates with shells or other hard structures have left the best record (Fig. 5-5). It is unlikely that jellyfish, worms, caterpillars, and such fossilize, but now and then a rare chance discovery such as the Burgess shale deposits of British Columbia and the Precambrian fossil bed of South Australia reveals an enormous amount of information about soft-bodied organisms. Certain regions have apparently provided ideal conditions for fossil formation, for example, the tar pits of Rancho La Brea in Hancock Park, Los Angeles; the great dinosaur

---

*The history of the famous H.M.S. *Beagle* and its elusive fate are described by K. S. Thomson, 1975. American Scientist **63**(6):664.

Fig. 5-5. This bit of Green River shale from Farson, Wyoming, bears the impression of a "double-armored herring" *(Knightia),* approximately 4 inches long, which swam there during the Eocene Age, approximately 55 million years ago.

beds of Alberta, Canada, and Jensen, Utah; the Olduvai Gorge of South Africa; and many others.

A common method of fossil formation is the quick burial of animals under water-borne sediments. Rapid burial is usually important because it slows or prevents decomposition by oxidation, solution, and bacterial action.

Most fossils are laid down in deposits that become stratified. The numerous layers of rock exposed in riverside cliffs or roadbanks were formed mainly by the accumulation of sand and mud at the bottoms of seas or lakes. If left undisturbed, which is rare, the older strata are the deeper ones; however, the layers are usually tilted or folded or show faults (cracks). Often old deposits exposed by erosion are later covered with new deposits in a different plane. Sedimentary rock such as limestone may be exposed to tremendous pressures or heat during mountain building and may be metamorphosed into rocks such as crystalline quartz-ite, slate, or marble. Fossils are often destroyed during these processes.

Since various fossils are correlated with certain strata, they often serve as a means of identifying the strata of different regions. Certain widespread marine invertebrate fossils such as various foraminifera and echinoderms are such good indicators of specific geologic periods that they are called "index" or "guide" fossils.

Geologic time

Long before the age of the earth was known, geologists began dividing its history into a table of succeeding events, using as a basis the accessible deposits and correlations from sedimentary rock. Time was divided into eons, eras, periods, and epochs. These are shown on the end papers inside the front and back covers of this book. Time during the last eon (Phanerozoic) is expressed in eras (Cenozoic), periods (Tertiary), epochs (Paleocene), and sometimes into smaller divisions of an epoch.

As far as the fossil record is concerned, the recorded history of life begins near the base of the Cambrian period of the Paleozoic era approximately 600 million years B.P. (Before Present). The period before the Cambrian is called the Precambrian era. Although the Precambrian occupies 85% of all geologic time, it is a puzzling era because of the lack of macrofossils. It has received much less attention than later eras partly because of the absence of fossils and partly because oil, which provides the commercial incentive for much geologic work, seldom exists in Precambrian formations. There are, however, evidences for life in the Precambrian: well-preserved bacteria and algae, as well as casts of jellyfish, sponge spicules, soft corals, segmented flatworms, and worm trails. Most, but not all, are microfossils.

There are two explanations for the lack of conventional fossils beyond the Cambrian-Precambrian "barrier": (1) most rocks older than 600 million years have undergone so much metamorphosis or distortion that any fossils that might have been present were destroyed and (2) animal species with hard parts and shells evolved after the early Cambrian. We have already seen (Chapter 2) that life originated perhaps 3 billion years ago and that major phyla of marine invertebrates were well established and had diverged to a considerable extent before the Cambrian. It is probable therefore that toward the end of the Precambrian, the seas contained a wealth of greatly diverse life of which we can never catch more than a glimpse.

How are fossils and rocks dated? The succession of geologic events was well established long before there was any reliable knowledge of the absolute time scale. Paleontologists knew, for example, that dinosaurs and ichthyosaurs were animals of the Mesozoic, that amphibians originated in the Devonian, and that saber-toothed tigers and mastodons roamed the forests of the Pleistocene. At the end of the nineteenth century, the earth was thought to be 100 million years old and the geologic time scale was adjusted accordingly. But with no measure of the absolute passage of time, figures of 1 billion or 10 billion years could just as well have been accepted.

The three principal methods now used for dating geologic formations are based on the radioactive decay of naturally occurring elements into other elements. This proceeds independently of pressure and temperature changes and therefore is not affected by often violent earth-building activities.

One method, potassium-argon dating, depends on the decay of potassium-40 (^{40}K) to argon-40 (^{40}A) (12%) and calcium-40 (^{40}Ca) (88%).

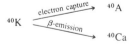

The half-life of ^{40}K is 1.3 billion years. The age of the rock is determined by measuring both the residual potassium content and the ^{40}A content (calcium is unreliable). The method is technically difficult, and certain kinds of rocks cannot be dated because the gaseous argon escapes; however, the procedure is now routinely used in geology laboratories.

The uranium-thorium-lead method uses three isotopes of uranium and thorium that decay to lead at different rates (0.7 to 13.9 billion years) through a complex series of radioactive steps. The age of the rock is determined by measuring the accumulation of stable lead (^{204}Pb).

A third method of dating, the rubidium-strontium method, is complicated and restricted to rubidium-rich minerals.

The well-known ^{14}C method deserves mention, although the short half-life of ^{14}C restricts its use to quite recent events. Since its accuracy is limited to approximately 40,000 years, it is especially useful for archeologic studies. This method is based upon the production of radioactive ^{14}C (half-life of approximately 5,570 years) in the upper atmosphere by bombardment of ^{14}N with cosmic radiation. The radioactive ^{14}C enters the tissue of living animals and plants, and an equilibrium is established between atmospheric ^{14}C and ^{14}C in the organism. At death, ^{14}C exchange with the atmosphere stops. In 5,570 years only half of the original ^{14}C remains in the preserved fossil. Its age is found by comparing the ^{14}C content of the fossil with that of living organisms.

As with all radioactive methods, there are complications. One is that the amount of ^{14}C in the atmosphere is disturbed by fossil-fuel burning (decreases ^{14}C) and thermonuclear explosions (increases ^{14}C). Corrections must be applied for these disturbances.

There are many uncertainties in radioactive dating and the techniques require sophisticated instrumentation and great care in methodology. Not all rocks or fossils can be dated. But, if two or more different methods provide concordant answers, the age can be accepted as reliable since it is most improbable that different isotopes would leach out of the sample at the same rate.

Evolution of the horse

The fossil record provides no more convincing or more complete evolutionary line of descent than that of the horse. The evolution of this form extends back to the Eocene epoch some 60 million years ago, and much of it took place in North America. This record would at first seem to indicate a straight-line evolution, but actually the history of the horse family is made up of many lineages, that is, descent from many lines. The phylogeny of the horse is extensively branched, with most of the branches now extinct. There were millions of years when little change occurred; there

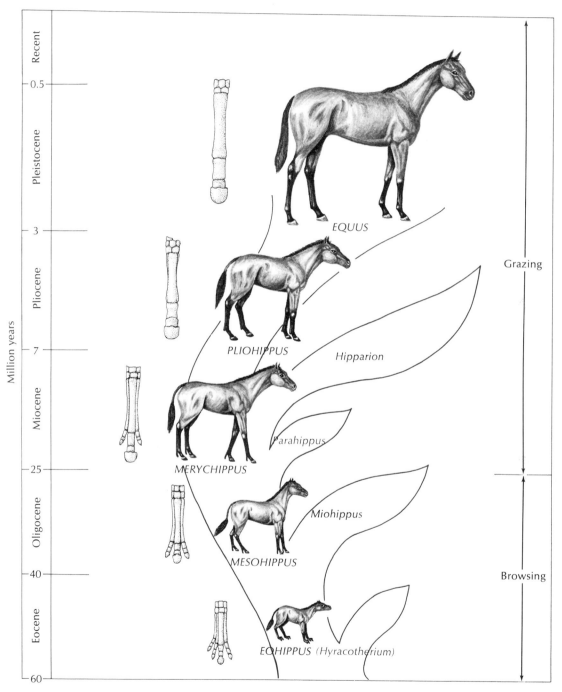

Fig. 5-6. Evolution of the horse family, showing the transition from browsing to grazing and evolution of the hind leg. Only a few of the many lines of horse evolution are shown. See text for further explanation. (Adapted from several sources.)

were other epochs when changes took place relatively rapidly.

No real change in the feet occurred during the Eocene epoch, but at least three types of feet developed later and were found in different groups during the late Cenozoic. Only one of these three types is found today.

There was extensive adaptive radiation throughout the horse's evolutionary history. In the evolution of the horse the morphologic changes of the limbs and teeth were of primary importance, along with a progressive increase in size of most of the types in the direct line of descent.

The earliest member of the horse phylogeny is considered to be *Hyracotherium,* also known as *Eohippus,* the dawn horse (Fig. 5-6). It was rodentlike in appearance and about the size of a small dog. Its forefeet had four digits and the hind legs three. Most of the weight was borne on pads, making the feet well adapted for travel on the soft earth of tropical North America. The teeth had short crowns and long roots and were specialized for browsing on trees and shrubs.

As the tropical conditions of central North America later gave way to open, grassy plains, the horse gradually developed feet better suited for rapid travel on dry land and grinding teeth for grazing. This transition began with *Mesohippus,* which flourished in the Oligocene some 40 million years ago. *Mesohippus,* though still a browser, was larger than *Eohippus* and had three toes on each foot. The middle toe was larger and better developed than the others. *Mesohippus* was a very common and successful animal in North America, as evidenced by the numerous fossil remains discovered.

Mesohippus gave rise to several lineages, one of which, *Merychippus,* was the direct ancestor of modern horses. *Merychippus* lived during the Miocene, some 25 million years ago, when grassy prairies were becoming widespread all over the earth under the influence of an increasingly cooler and drier climate. Its teeth were high-crowned and with complex crests and ridges for grinding; this invention made possible the complete switch to a grazing diet. It was still three-toed, but the lateral toes were high above the ground so that the weight of the body was thrown on the middle toe.

Pliohippus, a descendant of *Merychippus,* persisted into the Pleistocene some 2 million years ago. *Pliohip-*

pus discarded the side toes (although splint bones remained to attest to their former existence) and became a single-toed horse. It gave rise to *Equus,* the modern horse, of which there are some 60 domesticated breeds, and to zebras and asses. During the Pleistocene Ice Age, the *Equus* genus spread from North America into Eurasia, Africa, and South America.

The sequel to the horse evolution story is strange. Having pursued a remarkable evolution on the North American continent for 60 million years, the horse became extinct in both Americas toward the end of the Pleistocene. The cause of its demise is unknown, whether by excessive hunting by early Indians, infectious disease, predation, or some climatic catastrophe. It is certain that no horses were present when white men first came to the Americas. Of course, horses still existed in Europe and Asia and were reintroduced into America from Europe by the early Spanish colonists.

The development of this remarkable animal, which has had an enormous influence on the course of human history, was closely associated with the climate and ecologic changes of its habitat. As the habitat changed, so did the horse. It became a grazer instead of browser and developed the body size and foot structure conducive for fleetness in rapid travel over a firm, prairie surface. It is a beautiful example of evolutionary adaptation to changing ecologic opportunity.

Natural system of classification

Biologists long before Darwin discovered that, despite the great diversity of life, it was possible to arrange living things into a logical system of classification. The "natural system of classification" that originated with Aristotle and later was formalized by Linnaeus is based on the degree of similarity of morphologic characters.

For example, the domestic cat is clearly related to the wild cats and jungle cats of the Old World and all are included in the same genus, *Felis.* This genus shares an obvious relationship with the "big cats" (lions, tigers, leopards, and jaguars of the genus *Panthera*), as well as lynxes and bobcats (genus *Lynx*); so all are combined into the same family Felidae. All members of this family share a number of traits that set them apart from all other families, such as a round head, 30 teeth, and digitigrade feet with retractile claws. At the same time, the cat family is clearly related as a group to a number of other families (for example, the dog family, bear family, raccoon family, weasel and otter family, hyaena family). All of these bear a distinctive anatomy and way of life that characterizes them as flesh-eaters; they are of the order Carnivora.

Thus animals can be grouped together with respect to certain combinations of traits. This fact makes possible the classification of the approximately 2 million species of animals and strongly suggests that groups of animals are interrelated. For centuries it remained a puzzle why the system should work so well. The theory of evolution provided a simple solution: similarities exist because of descent from a common ancestor. True genetic relationship is reflected in the similarity of morphologic traits.

Comparative anatomy: homologous resemblances

Homology (Greek *homologia,* agreement) refers to similarity in structure and development of organs because of common ancestry and similar genetic basis. **Analogy** (Greek *analogia,* proportion) refers to resemblance in function but not in structure or development. The wings of a robin and a moth are analogous because both are used in flight, but they are quite different in structure and origin. The robin's wing is, however, homologous to the arm of a man because they are both pentadactyl (five-toed) structures that share a common embryologic origin; they are homologous despite the different functions they perform.

The similar position and embryologic origin of the pentadactyl limb (Fig. 5-7) is only one of several striking examples of homology among the vertebrates. Mammals characteristically have seven cervical vertebrae in the neck. Mouse, man, elephant, and giraffe—even the porpoise, which has lost the capacity to turn its head—have seven cervical vertebrae, an indication that all have descended from an ancestral mammalian stock that bore this number. Evolutionary theory provides the most rational explanation for the fact that animals differing so much in size and mode of life have seven neck vertebrae, as well as numerous other skeletal similarities.

Vertebrate examples could be extended indefinitely. The skulls of various mammals are unitary structures composed of several bones and obviously constructed from the same pattern. The vertebrate brain presents a common pattern; from fish to philosopher it is con-

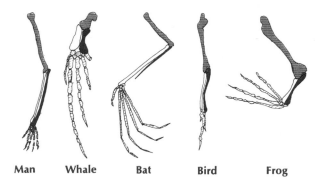

Man   Whale   Bat   Bird   Frog

Fig. 5-7. Forelimbs of five vertebrates to show skeletal homologies. *Red,* humerus; *white,* radius; *black,* ulna; *white striated,* wrist and phalanges. Most generalized or primitive limb is that of man—the feature that has been a primary factor in man's evolution because of its wide adaptability. Various types of limbs have been structurally modified for adaptations to particular functions.

structed of olfactory lobes, cerebral hemispheres, optic lobes, cerebellum, and medulla (see p. 217). All vertebrates share the same array of endocrine glands. Comparative anatomy courses are based on fundamental similarities—homologies—in every vertebrate organ system.

Comparative embryology

If homologies exist in the comparative anatomy of adults, we might expect—and do find—impressive embryologic homologies since adult structures are attained by embryonic development. The early embryos of fish, amphibians, reptiles, birds, and mammals look very similar, and all share several features in common such as gill slits, aortic arches, notochord, neural tube, and postanal tail.

We may examine the vertebrate gill (pharyngeal) arches as a single example (Fig. 5-8). In fish the embryonic gill arches later serve as respiratory organs. In the adults of terrestrial vertebrates the gill arches disappear altogether or become modified beyond recognition. The appearance of gill arches in the embryos of these terrestrial vertebrates and other pronounced similarities between the embryos of fishes and higher vertebrates are a strong indication of a common vertebrate ancestry.

Embryonic development is thus a record, although a considerably modified one, of evolutionary history. Biologists of the last century were so impressed by embryonic similarities between widely separated ver-

95

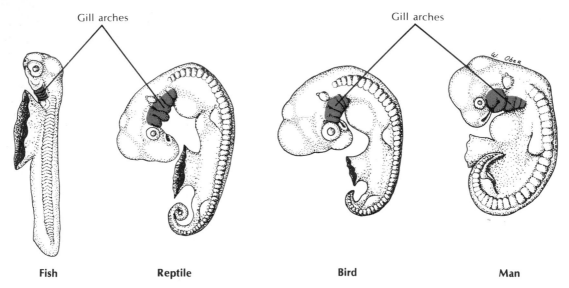

Fig. 5-8. Comparison of gill arches of different vertebrate embryos. All are shown separated from yolk sac.

tebrate groups that they used embryonic development as important evidence in reconstructing lines of evolutionary descent within the animal kingdom.

Comparative biochemistry

Common ancestry can now be demonstrated just as forcefully by homologous biochemical compounds as by homologous anatomic structures. The chemical composition of living things, based as it is on nucleic acids, proteins, carbohydrates, and fats, is itself evidence of the kinship of all life. Despite the diversity of form and function of animals and plants, numerous chemical compounds perform virtually identical metabolic roles in all. The citric acid (Krebs) cycle that releases electrons from carbon compounds; the cytochrome system that passes electrons to molecular oxygen, the ornithine cycle that synthesizes urea from surplus ammonia, the metabolic pathways that oxidize fatty acids; and many more are identical in a wide variety of species.

It has been possible to trace the evolution of at least one protein, cytochrome c, to the origin of eukaryotic cells more than 1 billion years ago. Cytochrome c is a small enzyme composed of 104 amino acid units. As recently noted, the amino acid sequences from various species are different; furthermore, just how much they differ corresponds well with the distance that any two species are separated on the evolutionary tree. Elabo-

rate family trees constructed entirely from sequence diversion agree remarkably well with those obtained by classic morphology and embryology. Such studies are of great importance because they carry evolutionary study much closer to the understanding of the genetic variation that underlies all evolutionary adaptation.

NATURAL SELECTION
Darwin's theory of natural selection

Darwin's concept of natural selection is based on a few simple principles supported by an abundance of facts.

1. *All organisms show variation.* No two individuals are exactly alike. They differ in size, color, physiology, behavior, and other ways. Darwin realized that much of this variation is inherited even though he did not understand how. The genetic basis of heredity was to be discovered many years later. Darwin pointed out that man has taken advantage of inherited variation to produce useful new breeds and races of livestock and plants. He believed that, if selection is possible under human control, it can also be produced by agencies operating in nature. He reasoned that natural selection can have the same effect as artificial selection.

2. *In nature all organisms produce more offspring than can survive.* In every generation the young are more numerous than the parents. Darwin calculated

that, even in a slow-breeding species such as the elephant, a single pair breeding from age 30 to 90 and having only 6 young in this span could give rise to 19 million elephants in 750 years. Why, he asked, are there not more elephants?

3. *Accordingly there is a struggle for survival.* If more individuals are born than can possibly survive, there must be a severe struggle for existence among them. Darwin wrote in *Origin of Species,* "it is the doctrine of Malthus applied with manifold force to the whole animal and vegetable kingdoms." Competition for food, shelter, and space becomes increasingly severe as overpopulation develops.

4. *Some individuals of a species have a better chance for survival than others.* Because individuals of a species differ, some are favored in the struggle for survival; others are handicapped and eliminated.

5. *The result is natural selection.* Out of the struggle for existence there results the survival of the fittest.* Under natural selection, individuals bearing favorable variations survive and have a chance to breed and transmit their characteristics to their offspring. The less fit die without reproducing. Natural selection is simply the differential survival or reproduction of favored variants. The process continues with each succeeding generation so that organisms gradually become better adapted to their environment. Should the environment change, there must also be a change in those characters that have survival value or the species is eliminated. Reproduction is what really counts in natural selection.

6. *Through natural selection new species originate.* The differential reproduction of variants can gradually transform species and result in the long-term improvement of types. According to Darwin, when different parts of an animal or plant population are confronted with slightly different environments, each diverges from the other. In time they differ enough from each other to form separate species. In this way two or more species may arise from a single ancestral species. Through adaptation to a changed environment, a group of animals may also diverge enough from their ancestors to become a separate species. As

divergence continues and species become more and more widely separated, biologists recognize the difference by establishing the higher taxonomic ranks of genera, families, etc.

What happens to characters when they are nonadaptive or indifferent? Variations that are neither useful nor harmful are not affected by natural selection and may be transmitted to succeeding generations as fluctuating variations. This explains many variations that have no significance from an evolutionary viewpoint.

Appraisal of Darwin's theory

Darwin was a pioneer. He convincingly demonstrated evolution and then provided a logical explanation for it. Through the well-chosen body of evidence so lucidly expounded in *Origin of Species,* he made evolution understandable. Nevertheless, despite its elegance it is hardly surprising that parts of his theory have been modernized because of more recent biologic knowledge. When Darwin wrote *Origin of Species,* nothing was known about the processes underlying sexual reproduction, genetics, or variation. It was not even known that sexual reproduction involved the combination of a single sperm with a single egg. Thus we always marvel that Darwin's deductions and conclusions were so sound in view of the scientific ignorance that prevailed at the time.

Two weaknesses in his theory centered around the concepts of variation and inheritance. Darwin made little distinction between variations induced by the environment (physical or chemical) and those that involve alterations of the germ plasm or the chromosomal material. It is now known that many of the types of variations Darwin stressed are noninheritable. Only variations arising from changes in the genes (mutations) are inherited and furnish the material upon which natural selection can act.

In 1866 Gregor Johann Mendel published his enormously important work on inheritance in a journal of the Society of Natural Science of Brünn. Darwin never read the work (although it was found in Darwin's extensive library after his death), and, if there were others who knew of it, they failed to bring it to his attention. It is interesting to speculate on how the genetic implications of Mendel's study would have influenced Darwin's thinking had Mendel written to him of his results.

It is altogether possible that Darwin could not have

---

*The popular phrase "survival of the fittest" was not originated by Darwin but was coined a few years earlier by the British philosopher Herbert Spencer, who anticipated some of Darwin's principles of evolution. Unfortunately the phrase later came to be coupled with unbridled aggression and violence in a bloody, competitive world. In fact, natural selection operates through many tame characteristics of living things; fighting prowess is only one of several means towards successful reproductive advantage.

seen the relationship of the hereditary mechanisms to gradual changes and continuous varieties that represent the hub of Darwinian evolution. It took others years to establish the relationship of genetics to Darwin's natural selection theory. Darwin did not point out the cumulative tradition of man's evolution, the capacity of the human race to transmit experience from generation to generation, or what is now called the "cultural evolution" of man.

The first step in modernizing Darwin's theory was the demonstration that the operative units in inheritance are self-reproducing, nonblending genes. Next there was the discovery of chromosomes in which genes were found to be precisely located in linkage groups. The ultimate source of variation—mutations— then became understandable.

Darwin also did not appreciate the real significance of geographic isolation in the differentiation of new species. Biologists now agree that isolation of some kind is an essential element in the formation of new species. In rare instances it is true that a mutation or chance hybridization may produce a group of organisms that are infertile with the parent stock; accordingly a new species may arise in a single step without isolation. Otherwise geographic or ecologic isolation provides the opportunity for populations to develop their own unique characters independently of the parent stock.

Genetic variation within species

Biologic variation exists within all populations. We can verify this fact simply by measuring a single character in several dozen specimens selected from a local population: tail length in mice, length of a leg segment in grasshoppers, number of gill rakers in sunfish, number of peas in pods, height of adult males of the human species. When the values are graphed with respect to frequency distribution, they tend to approximate a normal, or bell-shaped, probability curve. Most individuals fall near the mode, fewer fall above or below the mode, and the extremes make up the "tails" of the frequency curve with increasing rarity. The larger the population sample, the closer the frequencies resemble a normal curve. Sometimes significant deviations from normality are discovered in populations and such instances may reveal important facts about variations within the population.

As explained in the preceding chapter, biparental reproduction and mendelian assortment furnish a means of shuffling genes into various combinations. Some of those combinations that work best have a better chance to be retained in the population. For instance, biparental reproduction makes it possible for two favorable mutant genes (one from each of two individuals) to come together within the same individual, but within an asexual population the two favorable genes would be together only when both mutations occurred in the same individual. Sexual reproduction thus speeds the rate of evolution and better enables the organisms to fit into a constantly changing environment.

Gene pool. All of the alleles of all genes of a population are referred to collectively as the **gene pool.** The interbreeding population is the visible manifestation of the gene pool, which thus has continuity through successive generations. The gene pool of large populations must be enormous, for at observed mutation rates many mutant alleles can be expected at all gene loci. In some cases more than 40 alleles of the same gene have been demonstrated. Let us suppose that there are two alleles present, **A** and **a.** Among the individuals of the population there are three possible genotypes: **AA, Aa,** and **aa.** When there are three alleles present, **A, a,** and *a,* there are six possible genotypes: **AA, Aa, A***a***, aa, a***a***,** and *aa*. With four alleles present there are 16 possible genotypes. Increasing the number of alleles increases the possible genotypes enormously.

Role of mutations

Changes in uniparental populations occur by the addition or elimination of a mutation; in biparental populations the mutant gene may combine with all existing combinations and thus double the types. With only 10 alleles at each of 100 loci, the number of mating combinations is 10^{100}. When an organism has thousands of pairs of alleles, the amount of diversity is staggering. Even though many genes are found together on a single chromosome and tend to stay together in inheritance, this linkage is often broken by crossing-over.

If no new mutations occurred, the shuffling of the old genes would produce an inconceivably great number of combinations. But this is not the whole story because genes exert different influences in the presence of other genes. Gene A may act differently in the presence of gene B from what it does in the presence of gene C. The diversity produced by this interaction and the addition of new mutations now and then add to the complication of population genetics.

If this diversity is possible in a single population, suppose two different populations with different gene pools should mix by interbreeding. It is easy to see that many more combinations of genes and their phenotypic expression would occur. This means that populations have enormous possibilities for variation.

Variation and natural selection. What does variation signify for evolution? As already stated, genetic variation produced in whatever manner furnishes the material on which natural selection works to produce evolution. Natural selection does this by favoring beneficial variations and eliminating those that are harmful to the organism. Selective advantages of this type represent a very slow process, but on a geologic time scale they can bring about striking evolutionary changes represented by the various taxonomic units (species, genera, etc.), adaptive radiation groups, and the various kinds of adaptations.

The fact must be stressed, however, that natural selection works on the whole animal and not on single hereditary characteristics. The organism that possesses the most beneficial combination of characteristics or "hand of cards" is going to be selected over one not so favored. This concept helps explain some of those puzzling instances in which an animal may have certain characteristics of no advantage or that are actually harmful, but in the overall picture it has a winning combination. Thus, in a population, pools of variations are created on which natural selection can work to produce evolutionary change.

Why recessive genes are not lost from a population. In an interbreeding population why does not the dominant gene gradually supplant the recessive one? It is a common belief that a character dependent on a dominant gene increases in proportion because of its dominance. This is not the case, for there is a tendency in *large* populations for genes to remain in equilibrium generation after generation. A dominant gene does not change in frequency with respect to its allele.

This principle is based on a basic law of population genetics called the **Hardy-Weinberg equilibrium.** According to this law, gene frequencies and genotype ratios in large biparental populations reach an equilibrium in one generation and remain constant thereafter unless disturbed by new mutations, by natural selec-

tion, or by genetic drift (chance). The rule does not operate in small populations.

A rare gene, according to this principle, does not disappear merely because it is rare. That is why certain rare traits, such as albinism, persist for endless generations. Variation is retained even though evolutionary processes are not in active operation. Whatever changes occur in a population—gene flow from other populations, mutations, and natural selection—involve the establishment of a new equilibrium with respect to the gene pool, and this new balance is maintained until upset by disturbing factors.

The Hardy-Weinberg formula is a logical consequence of Mendel's first law of segregation and expresses the tendency toward equilibrium inherent in mendelian heredity. Let us select a pair of alleles such as **T** and **t.** Let p represent the proportion of **T** genes and q the proportion of **t** genes. Therefore, $p + q = 1$, since the genes must be either **T** or **t.** By knowing either p or q, it is possible to calculate the other. Of the male gametes formed, p contains **T** and q contains **t**; the same applies to the female gametes. (See checkerboard in Chapter 4, p. 65.) As we know from Mendel's law, there are three possible genotypic individuals, **TT, Tt,** and **tt,** in the population. By expanding to the second power, the algebraic formula $p + q$ is $(p + q)^2 = p^2 + 2pq + q^2$ (a binomial expansion) in which the proportion of **TT** genotypes is represented by p^2, **Tt** by $2pq$, and **tt** by q^2. The ratio of a mendelian monohybrid is 1:2:1. The homozygotes **TT** and **tt** produce only **T** and **t** gametes, whereas the heterozygotes **Tt** produce equal numbers of **T** and **t** gametes. In the gene pool the frequencies of the **T** and **t** gametes are as follows:

$$\mathbf{T} = p^2 + \tfrac{1}{2}(2pq) = p^2 + pq = p(p + q) = p$$
$$\mathbf{t} = q^2 + \tfrac{1}{2}(2pq) = q^2 + pq = q(q + p) = q$$

In all random mating the gene frequencies of p and q remain constant in sexually reproducing populations (subject to sampling errors). The formula $p^2 + 2pq + q^2$ is the algebraic formula of the checkerboard diagram, and thus the formula can be used for calculating expectations without the aid of the checkerboard.

To illustrate how the Hardy-Weinberg formula applies, let us suppose that a gene pool of a population consisted of 60% **T** genes and 40% **t** genes. Thus:

$$p = \text{frequency of } \mathbf{T} \ (60\% \text{ or } 0.6)$$
$$q = \text{frequency of } \mathbf{t} \ (40\% \text{ or } 0.4)$$

99

Substituting numerical values of gene frequency in the following,

$$p^2 + 2pq + q^2$$
$$(0.36 + 0.48 + 0.16)$$
$$\textbf{TT} \quad \textbf{Tt} \quad \textbf{tt}$$

we find that the proportions of the various genotypes are 36% pure dominants, 48% heterozygotes, and 16% pure recessives. The phenotypes, however, are 84% (36 + 48) dominants and 16% recessives.

On the other hand, if 4% of a population carries a certain recessive trait, then:

$$q^2 = 4\% \text{ or } 0.04$$
$$q = \sqrt{0.04} = 0.2 \text{ or } 20\%$$

Thus 20% of the genes are recessive. Even though a recessive trait may be rare, it is amazing how common a recessive gene may be in a population. Only 1 person in 20,000 is an albino (a recessive trait); yet by formula, it is found that 1 person in every 70 carries the gene or is heterozygous for albinism.

Genetic drift

The Hardy-Weinberg equilibrium can be disturbed, as already stated, by mutation, by natural selection, and by chance or genetic drift. The term **genetic drift** (Wright*) refers to changes in gene frequency resulting from purely random sampling fluctuations. By such means a new mutant gene may be able to spread through a small population until it becomes homozygous in all the organisms of a population (random fixation), or it may be lost altogether from a population (random extinction). Such a condition naturally would upset the gene frequency equilibrium.

How does the principle apply? Let us suppose a few individuals at random became isolated from a large general population. This could happen by some freakish accident of physical conditions, such as a flood carrying a small group of field mice to a remote habitat or a disease epidemic wiping out most of a population. In the general population individuals would be represented by both homozygotes, **TT,** for example, and heterozygotes, **Tt.** It might be possible for the small, isolated group to be made up only of **TT** individuals and the **t** gene would be lost altogether. Also, when only a small number of offspring are produced, certain genes may, by chance, be included in the germ cells

*Wright, S. 1969. Evolution and genetics of populations, vol 2: The theory of gene frequencies. Chicago, University of Chicago Press.

and others not represented. It is possible in this way for heterozygous genes to become homozygous. The new group may in time have gene pools different from the ancestral population.

We should note also that most breeding populations of animals are small. A natural barrier such as a stream may be effective in separating two breeding populations. Thus chance could result in the presence or absence of genes without being directed at first by natural selection.

Genetic drift is the cause of the frequency of certain human traits such as blood groups. Among some American Indian tribes it is known that group B, for instance, is far rarer than among other races, which may be due to small isolated mating units.

There are many who deny the importance of genetic drift, but it is generally agreed that, in bisexually reproducing species, evolution proceeds more rapidly when a population is broken up into isolated or partially isolated breeding communities, and the smaller the population the greater is the importance of genetic drift.

Polymorphism

Many species of plants and animals are represented in nature by two or more clearly distinguishable kinds of individuals, a condition called **polymorphism** (many forms). Polymorphic variation may involve color differences such as blue and white forms of the snow goose or the black and gray forms of the gray squirrel, or some other morphologic or physiologic character may be involved. It does *not* refer to seasonal variations in pelage or plumage (for example, the gray coat in summer of the varying hare changing to white in winter), for such seasonal alterations affect all members of the population alike.

An example of **stable polymorphism** is provided by a European species of ladybird beetle. Individuals show two color phases, red with black spots and black with red spots. Both forms interbreed freely. In summer the black form, behaving as a mendelian dominant, increases and by fall outnumbers the red form. In winter, however, the red form recovers and by spring outnumbers the black form. The recessive red form is in some way favored during the cold winter months.

A classic example of **transient polymorphism** is industrial melanism (dark pigmentation) in the peppered moth of England (Fig. 5-9). Before 1850 the

Fig. 5-9. Light and melanic forms of the peppered moth *(Biston betularia)*, on a lichen-covered tree in an unpolluted countryside (**A**) and on a soot-covered oak near industrial Birmingham, England (**B**). The dark melanic coloration that appeared in 1850 is controlled by a dominant allele. (Photos by H. B. D. Kettlewell.)

peppered moth was always white with black speckling in the wings and body. In 1850 a mutant black form of the species appeared. It became increasingly common, reaching frequencies of 95% in Manchester and other heavily industrialized areas by 1900. The peppered moth, like many similar species, normally rests in exposed places, depending on its cryptic coloration for protection. The mottled pattern of the normal white form blends perfectly with the lichen-covered tree trunks. With increasing industrialization, the soot and grime from thousands of chimneys killed the lichens and darkened the bark of trees for miles around centers such as Manchester. (This part of England is known as the "Black Country.") Against a dark background the white moth is conspicuous to predatory birds, whereas the mutant black form is camouflaged. The result was rapid natural selection: the easier-to-see white form was preferentially selected by birds, whereas the melanic form was subjected to far less predation. Selection pressure thus tended to eliminate the white form while favoring the genes that contributed to black wings and body.

Another way to describe the selective quality of the genes that determine color is in terms of their **fitness.** Fitness describes the way individuals differ in their reproductive success because of differences in the genes that they carry. In this case the genes for the mutant black form increased the fitness of the black moths in England's ''Black Country.'' This same trait, however, confers low fitness in nonpolluted country-side areas. On the other hand, genes for the white form confer high fitness in nonpolluted countryside areas but low fitness near industrial centers.

Of special interest is the rapidity of the change. Rather than requiring thousands of years, the change occurred in less than 100 years. White moths still survived in nonpolluted areas of England, and with the recent institution of pollution-control programs the white forms are beginning to reappear in the woods around cities. Industrial melanism is a dramatic example of shifting gene frequencies as selection pressures change with a changing environment.

Polymorphism clearly has adaptive value to species exposed to different environmental conditions. It is widespread throughout the animal kingdom. Examples are blood types of man and other animals, albinism in many animals, silver foxes in litters of gray foxes, and rufous and gray phases of screech owls in the same brood. Frequently polymorphic forms have been mistaken for separate species. If exposed to a persistent environmental shift that favors reproductive success by one or the other of separate polymorphic forms, one may become dominant over a particular part of the animal's range. This is but a short step to the establishment of a well-defined subspecies.

EVOLUTION OF NEW SPECIES

Previously we have described how populations change over periods of time because of genetic variation. Mutations are the fundamental source of genetic variation, further enhanced by sexual interchange and recombination. Natural selection acting on these variations is the force that produces evolutionary change and maintains the adaptive well-being of populations. In this section we consider how new species arise. The process by which one form becomes genetically isolated from another is called **speciation.**

Species concept

Although a definition of species is needed before we can theorize about species formation, biologists do not agree on a single rigid definition that applies in all cases. Carolus Linnaeus, the great Swedish naturalist who founded our system of classification, saw plant and animal species as fixed entities, static immutable units subject only to minor and unimportant variations. This is the traditional concept of species: namely, a group within which individuals closely resemble one another but that is clearly distinguishable from any other group. The limitation of the fixed entity definition is biologic variation. Many species are made up of individuals that can be arranged into completely intergrading series. Sometimes the gaps between species can be detected only with the greatest difficulty, so that it becomes almost impossible to decide where two species are to be separated. Yet species are realities in nature. We observe and collect them, study them, sample their populations, and describe their behavior.

The difficulty with the traditional system is that classification is based on the animals' appearance alone (size, color, length of various parts, and so on). Many other criteria can be and presently are being used: behavioral differences, reproductive characteristics, genetic composition, and ecologic niche. The single property that maintains the integrity of a species more than any other property, especially among biparental organisms, is *interbreeding*. The members of a species can interbreed freely with each other, produce fertile offspring, and share a common genetic pool. Interbreeding of different species is usually either physically impossible or produces sterile offspring. There are, of course, exceptions, but they are rare. Species are thus usually considered genetically closed systems, whereas **races** within a species are open systems and can exchange genes. **A species therefore may be defined as a group of organisms of interbreeding natural populations that is reproductively isolated from other groups and that shares common gene pools.** (Dobzhansky, 1951, and Mayr, 1970).

Speciation
Geographic isolation

The definition of species states that gene pools of different species are isolated from each other. With the exception of occasional hybridization, gene flow between species does not occur. Isolation, then, is a crucial factor in evolution. Unless some individuals of a population can be segregated for many generations from the parent population, new adaptive variations, which may arise, become lost through interbreeding with the parent population. Geographic isolation permits a unique sample of the population gene pool to be "pinched off" or segregated so that diversification can occur.

Let us imagine a single interbreeding population of mammals that is split into two isolated populations by some geographic change—the uplifting of a mountain barrier, the sinking and flooding of a geologic fault, or a climatic change that creates a hostile ecologic barrier such as a desert. Gene flow between the two isolated populations is no longer possible. The populations are almost certainly different from each other even when first isolated, since just by chance alone one population contains alleles not present in the other.

With the passage of time genetic recombination accentuates the difference, especially if the populations are small. Mutations that appear in the separated populations are certain to be different. Furthermore the climatic conditions of the separated regions are different; one may be warm and moist, the other cool and dry. Natural selection acting on the isolated gene pools favors those mutations and recombinations that best adapt the populations to their respective environments, unique food supplies, and new predator-prey relationships.

The two populations continue to diverge morphologically, physiologically, and behaviorally until distinct geographic races are formed. After some indefinite time span, perhaps only a few thousand years, but more likely a period of hundreds of thousands or even millions of years, evolutionary diversification progresses to the point that two races are reproductively isolated. If the barrier separating them fails (erosion of mountains, shifting of river flow, reestablishment of a hospitable ecologic bridge), the two populations again intermix. However, they may no longer interbreed. They are now distinct species.

Speciation under these conditions is by no means inevitable. It happens only under the most ideal conditions: small population (and gene pool) size, fortunate mutations, and the right combination of selection pressures. The chances are good, in fact, that one or both populations isolated in this way become extinct. But over the vast span of time that organic evolution has had to work, the special conditions required for speciation have been repeated numerous times, indeed millions of times.

Adaptive radiation on islands

Many times in the earth's history, a single parental population has given rise not just to one or two new species, but to an entire family of species. The rapid multiplication of related species, each with their unique specializations that fit them for particular ecologic niches, is called **adaptive radiation.**

Young islands are especially ideal habitats for rapid evolution. If they are formed by volcanos that rose from a platform on the ocean floor, as were the Galápagos Islands, they are at first devoid of life. In time they are colonized by plants and animals from the continent or from other islands. These newcomers arrive to an especially productive situation for evolution because of the abundance of new opportunities.

On the crowded mainland almost every ecologic niche is already occupied. Every animal is specialized and adapted for a particular way of life, and all food sources are exploited by one species or another. Although variations continue to appear within a population, those offspring in each generation most like their parents are most likely to survive, because they are best suited for that particular set of environmental conditions. Competition between different species is too keen on the mainland, where evolution has been proceeding for a very long time, for many new evolutionary experiments to succeed.

On a young island, however, new arrivals find new ecologic opportunities and no competitors. Those that survive the sea voyage and the landfall and are able to become established, multiply, and spread out. They are in a new land that in all probability differs ecologically from their original home. New variations may serve some of the descendants of the colonizers in establishing themselves in new niches. Offspring that differ slightly from their parents may find that these differences enable them to exploit alternate food sources as yet unutilized by other animals. They thus flourish and produce offspring, some of which bear similar characteristics. With each passing generation these animals become increasingly successful at this alternate way of life, at the same time becoming increasingly different from their ancestors. The outcome is a new genetic blueprint, in short, a new species. Equally well, immigrants to an island may fail

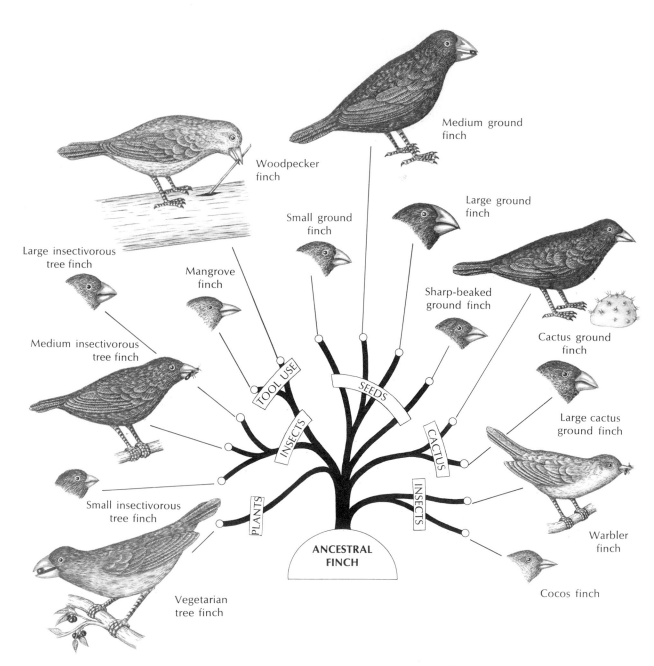

Fig. 5-10. Adaptive radiation in Darwin's finches of the Galápagos Islands.

to become established if the limited variety that is provided by a small group of colonizers does not happen to fit the new environment.

On archipelagos, such as the Galápagos Islands, isolation plays an extremely important evolutionary role. Not only is the entire archipelago isolated from the continent, but each island is geographically isolated from the others by the sea, the most inhospitable environment for land animals. Moreover, each island is different from every other to a greater or lesser extent in its physical, climatic, and biotic characteristics. As colonizing animals find their way to the different islands, they are presented with an environmental challenge that is unique for each island. Moreover, each new arrival carries with it only a small fraction, a biased sample, of the population's gene pool. This further stimulates diversification, already encouraged by isolation, new ecologic opportunities, and lack of competition. Archipelagos, more than any other place on the earth's surface, offer the raw materials and opportunities for rapid evolutionary changes of great magnitude.

Darwin's finches

Let us consider the evolution of the family of 13 famous Galápagos finches (Fig. 5-10). Their fame rests not on their beauty—they are, in fact, inconspicuous, dull colored, and rather unmusical in song—but on the enormous impact they had in molding Darwin's theory of natural selection. Darwin noticed that the Galápagos finches (the name "Darwin's finches" was coined in the 1930s by the British ornithologist David Lack) are clearly related to each other, but that each species differs from the others in several respects, especially in size and shape of the beak and in feeding habits. Darwin reasoned that, if the finches were specially created, it would require the strangest kind of coincidence for 13 similar finches to be created on the Galápagos Islands and nowhere else.

The 13 finches are grouped into several genera according to beak shape and feeding habits. The most clearly finchlike are the six species of **ground finches** *(Geospiza).* They feed mainly on seeds on the ground and are equipped with bills of different strengths and sizes for handling seeds varying in size and hardness. One, the cactus ground finch, feeds mostly on cactus; its slightly curved beak is adapted for probing into the yellow cactus flowers.

The three **tree finches** *(Camarhynchus)* have par-

rotlike beaks, live in trees, and excavate twigs and branches for concealed insects. The single **warbler finch** *(Certhidea olivacea)* is a delicate warblerlike finch that feeds exclusively on insects with its slender bill. The **vegetarian tree finch** *(Platyspiza crassirostris),* the largest Galápagos finch, feeds on fruits and soft seeds with its short, thick beak.

The two most remarkable finches are the **woodpecker finch** and the **mangrove finch** *(Cactospiza).* Lacking the long, protrusive tongue of true woodpeckers, they often use small cactus spines or twigs as tools to probe insect larvae deep in holes in trunks or branches of trees. This is one of the very few recorded cases of tool use among animals below man and apes.

Darwin's finches are an excellent example of how new forms may originate from a single colonizing ancestor. Lacking competitors from other land birds, they underwent adaptive radiation, evolving along various lines to occupy available niches that on the mainland would have been denied to them. They thus assumed the characteristics of mainland families as diverse and unfinchlike as warblers and woodpeckers.

Darwin's fourteenth finch is found on isolated Cocos Island, far to the north of the Galápagos archipelago. It is similar in appearance to the Galápagos finches and almost certainly has descended from the same ancestral stock. Since the Cocos Island is a single small island with no opportunities for isolation of finch populations, there has been no species multiplication.

Allopatric and sympatric speciation

The Cocos Island example emphasizes the fundamental requirement for geographic isolation in speciation. In its absence, there is no other reasonable way for gene pools to become separated long enough for genomic differences to accumulate. This is often called the **allopatric theory of speciation,** meaning that species can originate only in populations that occupy different communities and therefore do not interbreed.

Is there any chance or any mechanism, then, for **sympatric** speciation to occur? That is, can new species originate within a single population without geographic isolation? Most biologists think not, while admitting that it may occur as a highly improbable event.

One clear case of sympatric speciation is the artificial creation of new species by hybridization. In 1924 Karpechenko crossed the radish *(Raphanus sativus)* with the cabbage *(Brassica oleracea)*, each having 18 chromosomes. The 18 chromosome hybrids (9 chromosomes from each parent) were sterile, as hybrids usually are, but by chance some of the hybrids doubled their chromosome number becoming tetraploid. This event is called spontaneous allopolyploidy.

The tetraploid plant with 36 chromosomes, bearing characteristics of both the radish and the cabbage, proved to be fertile. A new sympatric "instant species" was formed called *Raphanobrassica*. It could not be crossed with either of the parents and was reproductively isolated. Unfortunately the new species is of no practical use because it has roots like a cabbage and leaves like a radish.

Spontaneous allopolyploidy has probably occurred as a natural event for the production of some of our food plants such as the bread wheats, potatoes, and raspberries. Although it may be a significant factor in plant evolution, it is an extremely rare process in that of animals.

SELECTED REFERENCES

Darwin, C. 1859. On the origin of species by means of natural selection, or the preservation of favoured races in the struggle for life. London, John Murray. *There were five subsequent editions by the author. The recent Mentor Book edition is introduced by Sir Julian Huxley.*

Darwin, C. 1839. The voyage of the Beagle. New York, Doubleday Co., Inc. (1962). *This Natural History Library edition contains an excellent introduction by Leonard Engel.*

Dobzhansky, T. 1951. Genetics and the origin of species, ed. 3. New York, Columbia University Press. *Important though advanced treatise.*

Dobzhansky, T., et al. 1967. Evolutionary biology, New York, Plenum Publishing Corp. *A series of volumes with chapters by contributors on evolution, appearing annually.*

Eaton, T. H., Jr. 1970. Evolution. New York, W. W. Norton & Co., *Well-written account with an ecologic approach.*

Lack, D. 1947. Darwin's finches: an essay on the general biological theory of evolution. New York, Cambridge University Press. *A classic study of adaptive radiation in the famous Galápagos finches.*

Mayr, E. 1970. Population, species and evolution. Cambridge, Mass., Harvard University Press. *This excellent reference is an abridgement of the author's Animal Species and Evolution.*

Merrill, D. J. 1962. Evolution and genetics. New York, Holt, Rinehart & Winston, Inc. *Elementary introduction, strong on the evidences for evolution.*

Mettler, L., and T. G. Gregg. 1969. Population, genetics and evolution. Englewood Cliffs, N.J., Prentice-Hall, Inc. *Balanced introductory treatment.*

Moody, P. A. 1970. Introduction to evolution, 3rd ed. New York, Harper & Row, Publishers. *A popular introductory text.*

Salthe, S. N. 1972. Evolutionary biology. New York, Holt, Rinehart & Winston, Inc. *Excellent, clearly developed account of evolutionary theory.*

Stebbins, G. L. 1971. Processes of organic evolution. Englewood Cliffs, N.J., Prentice-Hall, Inc. *Solid introductory text.*

Thornton, I. 1971. Darwin's Islands. A natural history of the Galápagos. Garden City, N.Y., Natural History Press. *Best general account of the Galápagos Islands, describing island history, animals, evolution, and future outlook.*

SELECTED SCIENTIFIC AMERICAN ARTICLES

Bishop, J. A., and L. M. Cook. 1975. Moths, melanism and clean air. **232:**90-99 (Jan.). *The evolution of industrial melanism in moths is described.*

Cavalli-Sforza, L. L. 1969. ''Genetic drift'' in an Italian population. **221:**30-37 (Aug.).

Clarke, B. 1975. The causes of biological diversity. **233:**50-60 (Aug.). *Argues that diversity is maintained by, as well as is the basis of, natural selection.*

Cole, Fay-Cooper. 1959. A witness at the Scopes trial. **200:**120-130 (Jan.).

Darlington, C. D. 1959. The origins of Darwinism. **200:**60-66 (May).

Eiseley, L. C., 1959. Alfred Russel Wallace. **200:**70-84 (Feb.).

Evans, H. E., and R. E. Matthews. 1975. The sand wasps of Australia. **233:**108-115 (Dec.). *The genus* Bembix *is remarkably diverse, apparently because they were able to fill available ecologic niches.*

Newell, N. D. 1972. The evolution of reefs. **226:**54-65 (June). *Major events in the earth's history are reflected in changes in tropical reef communities.*

A pair of Galápagos blue-footed boobies (*Sula nebouxii*) displays to each other. The male, at right, is sky pointing; the female, at left, is parading. Such vivid, stereotyped, communicative displays serve to maintain reciprocal stimulation and cooperative behavior during courtship, mating, nesting, and care of the young.

6 EVOLUTION OF BEHAVIOR

For as long as man has walked the earth, his life has been touched by, indeed interwoven with, the lives of other animals. He hunted and fished for them, domesticated them, ate them, and was eaten by them, made pets of them, revered them, hated and feared them, immortalized them in art, song, and verse, fought them, loved them. The very survival of ancient man depended on his knowledge of wild animals. To stalk them he had to know the ways of his quarry. As the hunting society of primitive man gave way to agricultural civilizations, man retained awareness of his interrelationship with other animals.

This is still evident today. Zoos attract more visitors than ever before; wildlife television shows are increasingly popular; game-watching safaris to Africa constitute a thriving enterprise; and millions of pet animals share the cities with man—more than ½ million pet dogs live in New York City alone. Although ethology, the science of animal behavior, is a young discipline of zoology, man has always been a behaviorist. Only recently, however, have we begun to understand how animal societies work.

ETHOLOGY AND THE STUDY OF INSTINCT

In 1973, the Nobel Prize in Physiology and Medicine was awarded to three pioneering zoologists, Karl von Frisch, Konrad Lorenz, and Niko Tinbergen. The citation stated that these three were the principal architects of the new science of ethology. It was the first time any contributor to the behavioral sciences was so honored, and it meant that the discipline of animal behavior, which really takes its roots from the work of Charles Darwin, had arrived.

Ethology, meaning literally "character study," was first used in the late eighteenth century to signify the interpretation of character through the study of gesture. It was an appropriate term for the new field of animal behavior, which had as its objective the study of motor patterns, that is, gestures of animals, with the anticipation that such study would reveal the true characters of animals just as the interpretation of human gestures might reveal the true characters of humans. This decidedly restricted interpretation of ethology as a purely descriptive study of the "habits" of animals was considerably modified during ethology's epoch-making period, 1935 to 1950. Today we may define ethology as a discipline that involves the study of the total repertoire of the innate and learned behavior that animals employ to resolve the problems of survival and reproduction.

The aim of ethologists has always been to describe the behavior of an animal in its *natural habitat*. From the beginning ethologists have been naturalists. Their laboratory has been the out-of-doors and their experiments have been observational ones of animals in their natural surroundings, with nature providing the variables. Ethologists recognized that it makes no more sense to try to study the natural behavior of an animal divorced from its natural surroundings than it does to try to interpret a structural adaptation of an animal apart from the function it serves.

With infinite patience Lorenz, Tinbergen, and their colleagues watched and catalogued the activities and vocalizations of animals during feeding, courtship, and nest building, as well as seemingly insignificant behavioral movements such as head scratching, stretching postures, and turning and shaking movements. Thus these studies concentrated largely on innate motor patterns used for communication within a species.

One of the great contributions of Lorenz and Tinbergen was to demonstrate that behavioral traits are measurable entities like anatomic or physiologic traits. This was to become the central theme of ethology: behavioral traits can be isolated and measured and they have evolutionary histories. They showed that behavior is not the wavering, transient, unpredictable phenomenon often depicted by earlier writers. In short, behavior is genetically mediated. It is apparent that, if behavior is determined by genes in the same way that genes determine morphologic and physiologic characters (ethologists marshaled abundant evidence showing that it is), then behavior evolves and is adaptive. Thus modern behavioral study is founded on the recognition that the Darwinian view of evolution holds for behavioral traits, as well as for anatomic and functional characters.

STEREOTYPED BEHAVIOR

The European ethologists, through step-by-step analysis of the detailed behavior of animals in nature, especially of fishes and birds, focused on the invariant components of behavior. From such detailed studies emerged several concepts that were first made available to large numbers of American readers in Niko Tinbergen's influential book, *The Study of Instinct* (1951).

The basic concepts of ethologic theory can be approached by considering the egg-rolling movement

Fig. 6-1. Egg-rolling movement of the greylag goose *(Anser anser).* (From Lorenz, K., and N. Tinbergen. 1938. Zeit. Tierpsychol. **2:**1-29.)

of the greylag goose, described by Lorenz and Tinbergen in 1938 (Fig. 6-1). If a greylag goose is presented with an egg a short distance from her nest, she reaches out with her bill until her lower beak has contacted the egg. Then, extending her bill just beyond the egg, she contracts her neck pulling the egg carefully toward the nest with the underside of her bill. Since the egg naturally has a tendency to wander off course on the uneven terrain, the goose compensates for this by moving the bill from one side to the other, correcting the course of the egg until she has it in her nest.

If we remove the egg after the goose has begun her retrieval, the goose continues the egg-rolling behavior as though the egg were still there. The side-to-side movements cease, but the tucking movements continue as she slowly pulls her neck back in a straight line toward the nest with the invisible egg. If on another occasion the egg being retrieved slips completely away and rolls down the slope of the nest mound, the goose continues her retrieval movements in vacuo until she is again settled comfortably on her nest. Then, seeing that the egg has not been retrieved, she begins the egg-rolling pattern all over again.

Fixed action pattern

The egg-rolling movement of greylag geese is an example of a **fixed action pattern.** It is a motor act that is highly stereotyped and mostly invariable in its performance. Once the sequence is begun, it is completed whether or not the egg is actually being retrieved. The goose does not have to learn the

movement; it is an innate, or instinctive, skill. The goose behaves as though an internal program had been triggered and must run its course before switching off.

Actually the egg-rolling movement can be separated into two components. The fixed action pattern represents the invarying, straight retrieval movements that continue even in the absence of a stimulus (the egg). However, the lateral corrective movements that the goose makes to steer the egg past obstructions are the taxis components of the retrieval. A **taxis** is an orientation movement that is guided by external stimulation, in this case the changing tactile stimulation on the beak provided by the wandering egg. Fixed action patterns usually occur with orienting movements (taxes) superimposed on them.

Releasers (sign stimuli)

Much of an animal's behavior is triggered by a few key environmental stimuli. Such key signals, or **sign stimuli,** constitute only a small fraction of the total environmental information available. The egg placed outside the nest of the greylag goose is the specific stimulus that releases egg-retrieval behavior. However, at no time do releasers trigger a fixed action pattern in all members of a species. The animal must be in the proper motivational state. Only a *female* greylag goose *in a nest* responds to a wayward egg outside the nest. But, once she is properly motivated and in the "right" surroundings, the greylag goose is not terribly discriminating about what she retrieves. Lorenz and Tinbergen showed that almost any smooth and rounded object placed outside the nest would

Fig. 6-2. An oyster catcher *(Haematopus ostralegus)* attempts to roll a giant egg model into its nest while ignoring its own egg. (From Tinbergen, N. 1951. The study of instinct. Oxford, Oxford University Press.)

trigger the egg-retrieval behavior of this species; even a small toy dog and a large yellow balloon were dutifully retrieved.

Sometimes exaggerated sign stimuli release an exaggerated response. In one famous experiment, an oyster catcher tried mostly unsuccessfully to retrieve and incubate an egg four times the normal size while ignoring its own egg (Fig. 6-2). Such an exceptionally effective sign stimulus is called a **supernormal releaser.**

Because sign stimuli are relevant to the animal's survival and reproduction, they are characteristically simple cues, a fact that reduces the chances of making a mistake or misunderstanding the cue. In every case the response is highly predictable. The alarm call of adult herring gulls releases a crouching freeze response in the chicks. Frogs respond only to the courtship calls of males of their own species. Night-flying moths take evasive maneuvers or drop to the ground when they hear the ultrasonic cries of bats that feed on them; no other sound releases this response. Ethologists have described hundreds of such examples of releasers.

Innate releasing mechanism

The sign stimulus, whether a sound, color, appropriate structure, odor, or movement, releases a highly predictable response. This strict correlation between stimulus and appropriate response led ethologists to conclude that the nervous system is organized into discrete "centers" capable of filtering the sensory input, correctly selecting the releaser stimuli, and releasing an appropriate motor response. Lorenz called

this organization the **innate releasing mechanism (IRM).** An IRM was thought of as an aggregation of nerve cells that, once triggered by the key stimulus, would direct motor neurons to fire in a programmed manner, causing the animal to perform a stereotyped fixed action pattern. Lorenz used an analogy to illustrate how an IRM operates: the sign stimulus is a key that unlocks only one of many possible doors; once the door (the IRM) is opened, the specific response inside is released.

More controversy surrounds the concept of the IRM than perhaps any other ethologic concept. Physiologists are reluctant to believe that "centers" as visualized by Lorenz actually exist. The central nervous system operates as an integrated whole, rather than as a collection of nerve aggregates, each prewired to perform one and only one motor function. Thus the IRM concept has been altered somewhat, but the idea seems to be fundamentally correct; brain "centers" as such probably do not exist, but the brain *works* as though they do. If not interpreted too strictly, the IRM is a useful tool to ethologists for understanding the sequence and coordination of stereotyped behavior.

Concept of drive

Animals are active because of internal motivation. It has been customary to refer to changes in responsiveness as drives. Thus animals show hunger drives, thirst drives, courtship drives, mating drives, migration drives, and so on. Used in this context, a drive refers to a motivational change that results in the performance of certain usually predictable behavior. A hungry animal searches for food in a manner characteristic of its species. The hunting behavior can be thought of as a search for a releasing situation that permits the discharge of prey-capturing behavior.

Ethologists recognize three stages in completing a stereotyped behavior.

1. *Appetitive behavior.* Appetitive behavior is the specific search for a releasing situation; it places an animal in a position to achieve a biologic goal. A hungry fox becomes restless and begins a characteristic hunting behavior that is adapted to the situation; it knows the terrain, the detours to make, the habits of potential quarry, and the location of a farmer's chicken yard. In lower animals, appetitive behavior is more predictable and less subject to variation from previous learning experience.

2. *Consummatory behavior.* The biologic goal of

appetitive behavior is the consummatory act. For the hungry fox, catching and eating a chicken is the climax of the hunger drive. Location of a mate by a songbird following a search (appetitive behavior) releases courtship behavior (consummatory act). A gravid tortoise locates a suitable nest site and lays her eggs. A blowfly's search for food ceases when it has discovered and eaten enough to comfortably fill the foregut. In each case, once the goal is achieved, the goal-directed activity ceases.

3. *Quiescence.* Following the consummatory act, appetitive behavior does not immediately resume. This is the period of quiescence. The sated animal does not hunt. Later, when the animal gets hungry, a new round of appetitive behavior, consummatory act, and quiescence follows.

Although use of the term **drive** is a convenient way to describe a positive motivation toward achieving some biologic goal, the word is increasingly disfavorable. Critics contend that it is vague, nonexplanatory, and is used to embrace a variety of distantly related phenomena, having different neurophysiologic bases. Nevertheless the term persists in the absence of any completely acceptable substitute for expressing in a word the idea of motivation toward some biologic goal.

Instinct and learning

The invariable and predictable nature of stereotyped behavior suggests that it is inherited, or innate, behavior. Many kinds of unpracticed stereotyped behavior appear suddenly in animals and are indistinguishable from similar behavior performed by older, experienced individuals. Newly emerged moths and butterflies fly perfectly as soon as their wings are dry. Orb-weaving spiders "know" how to build their webs without practice and without having watched other spiders build theirs. Newly hatched gull chicks crouch in the nest the first time they hear the alarm call of the parent.

To the observer, these behaviors have a mechanical, programmed appearance. This is even more evident when animals act out behaviors released in inappropriate situations. For example, a male stickleback fish vigorously attacks a plump lump of wax with a red underside placed in its tank by experimenters. The stickleback reacts to this unfishlike model as if it were another male stickleback with a red belly intruding into its territory. A carefully made model that closely resembles a stickleback but lacks the red belly is ignored (Fig. 6-3). Male English robins furiously attack a bundle of red feathers placed in their territory but ignore a stuffed juvenile robin without the red feathers (Fig. 6-4). Male red-winged blackbirds attempt to copulate with a crude model consisting of rump and tail feathers of a female, if the tail is elevated in a precopulatory position.

In these and other examples of stereotyped behavior released by simple sign stimuli, it is difficult to understand how the animal could have learned the response. These "built-in" or innate behavioral patterns are released in complete form and without practice the first time that the animal of certain age and

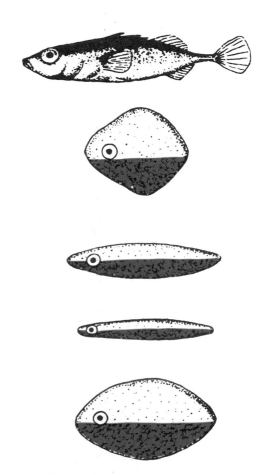

Fig. 6-3. Stickleback models used to study stereotyped behavior. The carefully made model of the stickleback, at top, without a red belly is attacked much less frequently by a territorial male stickleback than the four simple red-bellied models. (From Tinbergen, N. 1951. The study of instinct. Oxford, Oxford University Press.)

motivational state encounters the stimulus. To such behaviors the term **instinct,** meaning "driven from within," is applied. It is easy to understand why instinctive behavior is an important adaptation for survival, especially to lower forms that never know their parents. They must be equipped to respond immediately and correctly to their world as soon as they hatch. It is also evident that more advanced animals with longer lives and with parental care or other opportunities for social interactions have the opportunity to improve or change their behavior by learning.

Learning is the modification of behavior by experience. As a rule, behavioral modifications through learning are adaptive, since they improve the animal's fitness. The crouching behavior of gull chicks previously mentioned provides an excellent illustration of the adaptive value of learning. If the parent calls an alarm, the newly hatched chick crouches down—obviously a useful defensive response to overflying predators. After 2 or 3 days, the chick responds to the alarm call by running away before crouching. Later, it runs and hides in a chosen site before crouching.

As the chick grows stronger and more mobile, it also

Fig. 6-4. Two models of the English robin. The bundle of red feathers, at right, is attacked by male robins, whereas the stuffed juvenile bird without a red breast is ignored. (From Tinbergen, N. 1951. The study of instinct. Oxford, Oxford University Press; after Lack, D. 1943. The life of the robin. London, Cambridge University Press.)

becomes more discriminating in its visual response to moving objects overhead. After several such experiences, it begins to lose its general fear of overflying birds. In other words, its sensitivity to this particular stimulus declines. This very simple kind of learning is called **habituation,** the reduction or elimination of a response in the absence of any reward or punishment. For the gull chick, habituation to overflying objects is clearly appropriate, since running and hiding from every passing shadow consume time and energy better applied to more productive activities.

However, gull chicks do not come to ignore *all* birds flying overhead. Absolute habituation in this instance would be even less appropriate than no habituation at all, since the birds of prey are genuine predators of gull chicks. Chicks that fail to hide from an overflying hawk are less apt to survive than chicks that have retained a healthy fear of this predator.

Lorenz and Tinbergen discovered that chicks discriminate predatory birds from harmless songbirds and ducks on the basis of shape, especially the length of the head and neck. When chicks were presented with a silhouette having a hawklike shape and a short neck such that the head protruded only slightly in front of the wings, they crouched in alarm. Long-necked silhouettes were ignored or aroused only mild interest (Fig. 6-5). In another experiment, a model was built having symmetric wings and with head and tail shaped so that either the front or the rear of the dummy could be regarded as the head or as the tail (Fig. 6-6). When sailed to the left it resembled a goose and caused no alarm; when sailed to the right, however, the model resembled a hawk and did cause alarm. Obviously both the direction of motion and the shape are important in recognition.

Sometimes learning is demonstrated in surprising and amusing ways. Tinbergen describes one such incident. "In order to sail our models, which crossed a meadow where the birds were feeding or resting, at a height of about 10 yards along a wire, running from one tree to another 50 yards away, either Lorenz or I had to climb a tree and mount the dummy we wanted to test out. One family of geese (which also reacted to some of our dummies) very soon associated tree-climbing humans with something dreadful to come, and promptly called the alarm and walked off when one of us went up."*

*Tinbergen, N. 1961. The herring gull's world. New York, Basic Books, Inc.

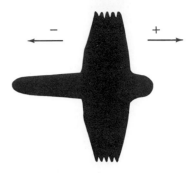

Fig. 6-6. Model that drew positive responses from gull chicks when sailed to the right but none when sailed to the left. (From Tinbergen, N. 1961. The herring gull's world: a study of the social behavior of birds. New York, Basic Books, Inc., Publishers.)

These observations with models might suggest incorrectly that gull chicks (as well as pheasant and turkey chicks that react similarly) have an innate ability to distinguish short-necked predators from harmless birds having longer necks. But, in fact, subsequent experiments demonstrated that newly hatched chicks, which are at first alarmed by anything that passes overhead—even a falling leaf—gradually become habituated to familiar objects. The chick learns that song and shore birds are common and harmless features of its world. But they never become accustomed to short-necked predators because these are seldom seen. It is the unfamiliar that arouses fear.

The alarm response of herring gull chicks is an example of a simple behavior that becomes altered as the result of experience. As they grow, the chicks store information about their world and become increasingly selective in their alarm response. Animals are internally programmed to respond automatically to certain stimuli; however, this response requires adjustment from the environment to become efficient as the animal grows.

What can we say about the role of genes and environment in shaping instinctive and learned behavior? We tread into delicate territory because behaviorists do not all agree on just what constitutes

Fig. 6-5. Models used by Lorenz and Tinbergen for the study of predator reactions in young fowl. Young gull chicks crouch in alarm when hawk silhouettes *(red)* pass overhead but ignore shapes of harmless birds. (From Tinbergen, N. 1951. The study of instinct. Oxford, Oxford University Press.)

Fig. 6-7. Canada goose *(Branta canadensis)* with her imprinted young. (Photo by L. L. Rue III.)

innate behavior. The problem concerns the effects that the environment has on the individual during its development. The idea that behavior must be fixed *either* by heredity *or* by the environment has had a long history, resulting in what is termed the **nature versus nurture controversy.** It has proven to be a useless debate because *any* characteristic of an individual is determined by an interaction between information supplied in the genetic blueprint (nature) and by the environment (nurture). Let us examine once again our definitions of instinct and learning and then try to explain how gene-environment interactions contribute to each.

We defined an instinctive behavior as one that emerges in complete form the first time the animal reacts to the appropriate stimulus. It must have a genetic foundation because without genes there can be no behavior. Specifically the animal's genotype contains instructions that result in the construction of specific neural organization, which permits certain types of behavior. We are not saying that an instinctive behavioral attribute is determined solely by information contained in specific chromosomal loci. The genetic code is an information-generating device that depends on an environment that supplies materials and

provides order for embryonic development. A genotype remains just a genotype if the developing organism cannot obtain the substances required to form tissues, organs, and nervous system.

Learning depends on experience encountered by the organism as it interacts with its environment. It also depends on internal programming because the things an organism learns to do best are determined by the genetic blueprint. The nervous system must be designed to facilitate the acquisition of learning at specific stages in the organism's development. In other words, through its genetically determined development, the brain possesses properties enabling it to "anticipate" its eventual use in the modification of behavior. Learned behavior, like instinctive behavior, contains both genetic and environmental components.

One kind of learned behavior that clearly illustrates the interaction of heredity and environment is **imprinting.** As soon as a newly hatched gosling or duckling is strong enough to walk, it follows its mother away from the nest. After it has followed the mother for some time, it follows no other animal (Fig. 6-7). But, if the eggs are hatched in an incubator or if the mother is separated from the eggs as they hatch, the goslings follow the first large moving object they see.

As they grow, the young geese prefer the artificial "mother" to anything else, including their true mother. The goslings are said to be imprinted on the artificial mother.

Imprinting was observed at least as early as the first century A.D. when the Roman naturalist Pliny the Elder wrote of "a goose which followed Lacydes as faithfully as a dog." Konrad Lorenz first studied the imprinting phenomenon objectively and systematically. When Lorenz hand-reared goslings they formed an immediate and permanent attachment to him and waddled after him wherever he went. They could no longer be induced to follow their own mother or another human being. Lorenz found that the imprinting period is confined to a brief *sensitive* period in the individual's early life and that once established the imprinted bond is retained for life.

What imprinting shows is that the goose or duck brain (or the brain of any of numerous other birds and mammals that show imprinting-like behavior) is designed to accommodate the imprinting experience. The animal's genotype is provided with an internal template that permits the animal to recognize its mother soon after hatching. Natural selection favors the evolution of animals having a brain structure that imprints in this way because following the mother and obeying her commands is important for survival. The fact that a gosling can be made to imprint to a mechanical toy duck or a human being under artificial conditions is a cost to the system that can be tolerated; the disadvantages of the system's simplicity are outweighed by the advantages of its reliability.

Let us cite one final example to complete our consideration of instinct and learning. The males of many species of birds have characteristic territorial songs that identify the singers to other birds and announce territorial rights to other males of that species. Like many other songbirds, the male white-crowned sparrow must learn the song of its species by hearing the song of its father. If the sparrow is hand-reared in acoustic isolation in the laboratory, it develops an abnormal song (Fig. 6-8). But if the isolated bird is allowed to hear recordings of normal white-crowned sparrow songs during a critical period of 10 to 50 days after hatching, it learns to sing normally. It even imitates the local dialect it hears. It might appear

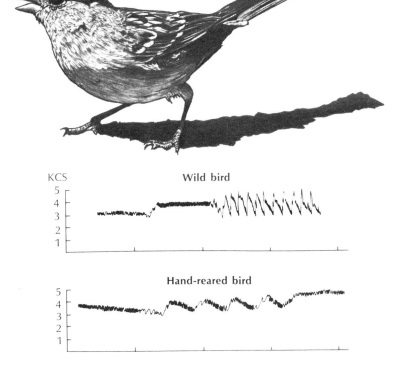

Fig. 6-8. Sound spectrograms of songs of white-crowned sparrows *(Zonotrichia leucophrys). Above,* natural song of wild bird; *below,* abnormal song of isolated bird. (From Alcock, J. 1975. Animal behavior: an evolutionary approach. Sunderland, Mass., Sinauer Associates, Inc.; after M. Konishi, personal communication.)

KCS
Wild bird

Hand-reared bird

from this that song characteristic is determined by learning alone. However, if during the critical learning period, the isolated male white-crowned sparrow is played a recording of another sparrow species, even a closely related one, it does not learn the song. It learns only the song appropriate to its own species. Thus, although the song must be learned, the brain has been programmed in its development to recognize and learn vocalizations produced by males of its species alone. The sparrow *learns* by example but its attention has been *innately* narrowed to focus on the appropriate example. Learning the wrong song would result in behavioral chaos and natural selection quickly eliminates those genotypes that permit such mistakes to occur.

SOCIAL BEHAVIOR

When we think of "social" animals we are apt to think of highly structured honeybee colonies, herds of antelope grazing on the African plains, a school of herring, or a flock of starlings. But social behavior is by no means limited to such obvious examples of animals *of the same species* living together in which individuals influence one another.

In the broad sense, any kind of interaction resulting from the response of one animal to another of the same species represents social behavior. Even a pair of rival males squaring off for a fight over the possession of a female is a social interaction, although our perceptual bias as people might encourage us to label it antisocial. Thus social aggregations are only one kind of social behavior, and indeed not all animal aggregations are social.

Clouds of moths attracted to a light at night, barnacles attracted to a common float, or trout gathering in the coolest pool of a stream are animal groupings responding to *environmental* signals. Social aggregations, on the other hand, depend on *animal* signals. They remain together and do things together by influencing one another.

Of course, not all animals showing sociality are social to the same degree. All sexually reproducing species must at least cooperate enough to achieve fertilization; but among most mammals, breeding is about the only adult sociality to occur. Alternatively, swans, geese, albatrosses, and beavers, to name just a few, form strong monogamous bonds that last a lifetime. Whether or not adult sociality is strongly or weakly developed, the most persistent social bonds usually form between mother and young, and these, for birds and mammals, usually terminate at fledging or weaning.

Advantages of sociality

Living together may be beneficial in many ways. Each species profits in its own particular way; what confers adaptive value to one species may not for another. One obvious benefit for social aggregations is defense, both passive and active, from predators. Musk-oxen that form a passive defensive circle when threatened by a wolf pack are much less vulnerable than an individual facing the wolves alone.

As an example of active defense, a breeding colony of gulls alerted by the alarm calls of a few attack predators *en masse;* this is certain to discourage a predator more effectively than individual attacks. The members of a prairie dog town, though divided into social units called coteries, cooperate by warning each other with a special bark when danger threatens. Thus every individual in a social organization benefits from the eyes, ears, and noses of all other members of the group.

Predators may also be distracted by the confusion effect created by large numbers of prey grouped together. A fish that can chase down and capture a lone crustacean may be unable to concentrate upon a single individual in a large aggregation. Predators may even be frightened by a large aggregate of prey, which they would eat if encountered singly. One ethologist describes the behavior of a captive seal into whose tank a school of anchovies was released. "On previous occasions, when small numbers of the fish were released, the seal rapidly caught and devoured them. The larger number of anchovies, which formed a dense school whose outlines approximated the shape of an animal as large as a seal, not only inhibited pursuit but it also apparently frightened the seal out of its pool. Though anecdotal, the account does illustrate the power of masses."*

Sociality offers several benefits to animal reproduction. Social grouping helps to synchronize reproductive behavior through the mutual stimulation that individuals have on one another. In colonial birds the sounds and displays of courting individuals set in motion prereproductive endocrine changes in other

*Klopfer, P. H. 1974. An introduction to animal behavior, ed. 2, Englewood Cliffs, N.J., Prentice-Hall, Inc.

individuals of the colony. Because there is more social stimulation, large colonies of gulls produce more young per nest than do small colonies. In colonial species, a population size less than a certain minimum results in inadequate social stimulation and reproductive failure. This effect has been suggested as the cause of extinction of the passenger pigeon once heavy hunting had greatly thinned the population (see p. 567).

Another benefit of social grouping to reproductive performance is that it facilitates encounters between males and females. For solitary animals, finding prospective mates may consume much time and energy. Should the population become depleted because of overhunting, habitat destruction, or severe weather, male-female encounters become increasingly rare. For example, on the Galápagos Islands, hunting by man had reduced the giant tortoise population on one large island (Hood Island) to a point at which the few survivors seldom, if ever, met. Lichens grew on the females' backs because there were no males to scrub them off during mating! Research personnel saved the tortoise from inevitable extinction by collecting them together in a pen, where they began to reproduce.

Still another advantage of sociality to reproduction is care of the young. This is especially well developed in the vertebrates and social insects. Parental care, with direct provisioning of the young and food sharing, increases survival of the brood. Social living also provides opportunities for individuals to give aid to and share food with young other than their own. Such interactions in a social network has resulted in some intricate cooperative behavior among parents, their young, and their kin.

Of the many other advantages of social organization noted by ethologists, we may mention only a few in our brief treatment: cooperation in hunting for food; huddling for mutual protection from severe weather; opportunities for division of labor, especially well developed in the social insects with their caste systems; and the potential for learning and transmitting useful information through the society.

Observers of a seminatural colony of macaques in Japan recount an interesting example of passage of tradition in a society. The macaques were provisioned with sweet potatoes and wheat at a feeding station on the beach of an island colony. One day a young female named Imo was observed washing the sand off a sweet potato in sea water. The behavior was quickly imitated by Imo's playmates and later by Imo's mother. Still later when the young members of the troop became mothers they waded into the sea to wash their potatoes; their offspring imitated them without hesitation. The tradition was firmly established in the troop.

Some years later Imo, an adult, discovered that she could separate wheat from sand by tossing a handful of sandy wheat in the water; allowing the sand to sink, she would scoop up the floating wheat to eat. Again, within a few years, wheat-sifting became a tradition in the troop.

Imo's peers and social inferiors copied her innovations most readily. The adult males, her superiors in the social hierarchy, would not adopt the practice but continued laboriously to pick wet sand grains off their sweet potatoes and scour the beach for single grains of wheat.

If social living offers so many benefits, why haven't all animals through natural selection become social? The answer is that a solitary existence offers its own set of advantages. In the diverse array of ecologic situations in nature, species extract their own optimal ways of life. Species that survive by camouflage from potential predators profit by being well spaced out. Large predators benefit from a solitary existence for a different reason, their requirement for a large supply of prey. Thus there is no overriding adaptive advantage to sociality that inevitably selects against the solitary way of life. It depends on the ecologic situation.

Aggression and dominance

Many animals species are social because of the numerous benefits that sociality offers. This requires cooperation. At the same time animals, like governments, tend to look out for their own best interests. In short, they are in competition with one another because of limitations in the common resources that all require for life. Animals may compete for food, water, sexual mates, or shelter, when such requirements are limited in quantity and therefore worth fighting over.

Much of what animals do to resolve competition is called **aggression**, which we may define as an offensive physical action, or threat, to force another to abandon something he owns or might attain. Many ethologists consider aggression to be part of a somewhat more inclusive interaction called **agonistic behavior,** referring to any activity related to fighting, whether aggression, defense, submission, or retreat.

Contrary to the widely held opinion that aggressive

Fig. 6-9. Male Masai giraffes *(Giraffa camelopardalis)* fight for social dominance. Such fights are largely symbolic, seldom resulting in injury. (Photo by L. L. Rue III.)

behavior aims at the destruction or at least defeat of an opponent, most aggressive encounters involve more saber rattling than real combat. Many species possess dangerous weapons such as teeth, beaks, claws, or horns, which could easily inflict serious injury or even kill an opponent; but such well-armed animals seldom use their weapons against members *of their own species*. However, no such inhibiting mechanism prevents their use in killing prey or in defense against attack from another species. A coyote that uses its teeth effectively in killing a prairie dog would never bite a rival coyote in an aggressive encounter over a mate or territory. A giraffe uses its short horns to fight rivals but uses its far more lethal hoofs in defense against predators (Fig. 6-9). Therefore, when biologists speak of aggression, they are usually (though not always) referring to encounters between conspecifics.

Animal aggression within the species seldom results in injury or death because animals have evolved many symbolic aggressive displays that carry mutually understood meanings. Fights over mates, food, or territory become ritualized tournaments rather than bloody, no-holds-barred battles. When fiddler crabs spar for territory, their large claws usually are only slightly opened. Even in the most intense fighting when the claws are used, the crabs grasp each other in a way that prevents reciprocal injury. Rival male poisonous snakes engage in stylized bouts by winding themselves together; each attempts to butt the other's head with its own until one becomes so fatigued that it retreats. The rivals never bite each other. Many species of fish contest territorial boundaries with lateral display threats, the males puffing themselves up to look as threatening as possible. The encounter is usually

Fig. 6-10. Male bighorn sheep *(Ovis canadensis)* fight for social dominance during the breeding season. (Photo by L. L. Rue III.)

settled when either animal perceives itself to be obviously inferior, folds up its fins and swims off.

Thus animals fight as though programmed by rules that prevent serious injury. Fights between rival bighorn rams are spectacular to watch, and the sound of clashing horns may be heard for hundreds of yards (Fig. 6-10). But the skull is so well protected by the massive horns that injury occurs only by accident and death is almost unheard of.

The loser in such encounters may simply run away, or he may signal defeat by a specialized subordination ritual. If it becomes evident to him that he is going to lose anyway, he is better off communicating his submission as quickly as possible and avoiding the cost of a real thrashing. Such submissive displays that signal the end of a fight may be almost the opposite of threat displays (Fig. 6-11). Charles Darwin, whose book *Expression of the Emotions in Man and Animals* (1873) founded the science of ethology, described the seemingly opposite nature of threat and appeasement displays as the ''principle of antithesis.'' The principle remains accepted by ethologists today.

Why doesn't the victor of an aggressive contest kill its opponent? A defeated wolf or dog presents its vulnerable neck to the victor as a sign of complete submission. Although the dominant wolf could easily kill his defeated foe and thus remove a competitor, it never does so. The display of submission has effectively inhibited further aggression by the winner. The best explanation for aggressive restraint is that the winner has little to gain by continuing the fight. His superiority is already assured. By continuing the aggression he merely endangers himself, since a defeated opponent fighting for his life might inflict a wound. It is not difficult to see how natural selection would favor genes that induce aggressive restraint. Aggression that is inappropriate runs counter to the maximization of individual fitness. It is maladaptive and consequently is selected against.

The winner of an aggressive competition is dominant to the loser, the subordinant. For the victor, dominance means enhanced access to all the contested resources that contribute to reproductive success: food, mates, territory, and so on. In a social species,

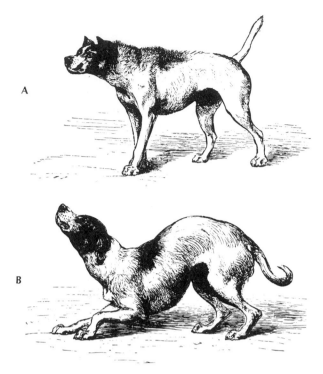

A

B

Fig. 6-11. Darwin's principle of antithesis as exemplified by the postures of dogs. In the upper figure, a dog approaches another dog with hostile, aggressive intentions. In the lower figure, the same dog is in a humble and conciliatory state of mind. The signals of aggressive display have been reversed. (From Darwin, C. 1873. Expression of the emotions in man and animals. New York, D. Appleton & Co.)

During times of food scarcity, the death of the weaker members helps to protect the resource for the stronger members. Rather than sharing food, the population excess is sacrificed. For the species, it is better for the weaker members to starve before the resources dwindle to crisis levels, rather than share the resource, thus denuding the countryside and risking ultimate starvation of the entire population.

Territoriality

Territorial ownership is another facet of sociality in animal populations. Just as an animal requires food for growth and maintenance and a mate with which to reproduce, it must also have space in which to carry out these activities. Since space, which an animal requires to satisfy its biologic needs, is limited, it must somehow be divided among those that require it.

Territorial defense has been observed in numerous animals: insects, crustaceans, other invertebrates, fish, amphibians, lizards, birds, and mammals. A **territory** is a fixed area from which intruders of the same species are excluded. This involves announcing ownership, defending the area from intruders, and spending long periods of time on the site being conspicuous. Like every other competitive endeavor, there are both costs and advantages.

Holding exclusive rights to a piece of property is beneficial when it ensures access to limited resources, *unless* the territory boundaries cannot be maintained with a minimum of effort. The presumed benefits of a territory are, in fact, numerous: uncontested access to a foraging area; enhanced attractiveness to females, thus reducing the problems of pair-bonding, mating, and rearing the young; reduced disease transmission; reduced vulnerability to predators. But the advantages of holding a territory begin to wane if the individual must spend most of his time engaged in boundary disputes with his neighbors.

For territoriality to work, the resources must be defendable, and, in fact, they usually are because natural selection relentlessly disposes of individuals with hopelessly unrealistic territorial appetites. Thus a territory must be large enough to provide the resources needed, but not so large that it cannot be economically defended. In general this is true.

Territory size tends to be flexible. When populations are low, territories are large. When resources, especially food, are abundant, territories tend to shrink, because no benefit is derived from defending large

dominance interactions often take the form of a dominance hierarchy. One animal at the top wins encounters with all other members in the social group; the second in rank wins all but those with the top-ranking individual.

Such a simple, linear hierarchy was first observed in chicken societies by Schjelderup-Ebbe, who called the hierarchy a "peck-order." Once social ranking is established, actual pecking diminishes and is replaced by threats, bluffs, and bows. Top hens and cocks get unquestioned access to feed and water, dusting areas, and the roost. The system works because it reduces the social tensions that would constantly surface if animals had to fight all the time over their social position.

Still, life for the subordinates in any social order is apt to be hard. They are the expendables of the social group. They almost never get a chance to reproduce, and when times get difficult they are the first to die.

ones. On the other hand, in times of population crowding, defended territories may be abandoned altogether because the sheer numbers of individuals make boundary defense impossible.

Sometimes territories move with the individual. This **individual distance,** as it is termed, is characteristic of the species. It can be observed as the spacing between swallows or pigeons on a wire, in gulls lined up on the beach, or in people queued up for a bus.

Most of the time and energy required for territoriality are expended when the territory is first established. Once the boundaries are located they tend to be respected, and aggressive behavior diminishes as territorial neighbors come to recognize each other. Indeed, neighbors may look so peaceful that unless the observer is present when the territories are established he may conclude (incorrectly) that the animals are not territorial. A "beachmaster" sea lion (that is, a dominant male with a harem) seldom quarrels with his neighbors who have their own territories to defend. However, he must be on constant vigilance against bachelor bulls who challenge the beachmaster for harem privileges.

Of all vertebrate classes, birds are the most conspicuously territorial. Most male songbirds establish territories in the early spring and defend these vigorously against all males of the same species during spring and summer when mating and nesting is at its height. A male song sparrow, for example, has a territory of approximately ¾ acre. In any given area, the number of song sparrows remains approximately the same year after year. The population remains stable because the young occupy territories of adults that die or are killed. Any surplus in the song sparrow population is excluded from territories and thus not able to mate or nest.

Sea birds such as gulls, gannets, boobies, and albatrosses occupy colonies that are divided into very small territories just large enough for nesting. These birds' territories cannot include their fishing grounds since they all forage in the sea where the food is always shifting in location and shared by all.

Several species of North American grouse form specialized, small territories called **leks** (Fig. 6-12). Each lek functions exclusively for mating. Females are attracted to the leks where the males engage in gaudy displays. The females tend to mate preferentially with males occupying the central regions of the lekking grounds. Thus the males fight over these preferred mating territories; the most aggressive and ornamental males mate with most of the females, so there is an intense sexual selection for ornamentation and vivid displays. Once the female has mated she leaves the lekking grounds and raises her brood alone.

Fig. 6-12. Sage grouse cock *(Centrocercus urophasianus)* displays before four hens, which he has attracted to his "lek." (Photo by L. L. Rue III.)

Territorial behavior is not so characteristic of mammals as it is of birds. Mammals are less mobile than birds, and this makes it more difficult for them to patrol a territory for trespassers. Instead, many mammals have **home ranges.** A home range is the total area an individual traverses in its activities. It is not an exclusive defended preserve but overlaps with the home ranges of other individuals of the same species.

For example, the home ranges of baboon troops overlap extensively, although a small part of each range becomes the recognized territory of each troop for its exclusive use. Home ranges may shift considerably with the seasons. A baboon troop may have to shift to a new range during the dry season to obtain water and better grass. Elephants, before their movements were restricted by man, made long seasonal migrations across the African savanna to new feeding ranges. However, the home ranges established for each season were remarkably consistent in size.

Not surprisingly, size of the home range increases with body size of mammals, since larger animals require more foraging area to satisfy their energy requirements. Accordingly, the home range may be $1/4$ acre for a field mouse, 100 acres for a deer, 60 square miles for a grizzly bear, and 1500 sq miles for an African hunting dog. In general, carnivores require more foraging space than herbivores, because any given area supports more plant food than animal food.

Communication

Social animals, like people, must be able to communicate with each other. Only through communication can one animal influence the behavior of another. Compared to the enormous communicative potential of human speech, nonhuman communication is severely restricted. Whereas human communication is based mainly, though by no means exclusively, on sounds, animals may communicate by sounds, scents, touch, and movement. Indeed any sensory channel may be used, and in this sense animal communication has richness and variety.

However, unlike man's language, which is composed of words with definite meanings that may be rearranged to generate an almost infinite array of new meanings and images, animal communication consists of a limited repertoire of signals. Typically, each signal conveys one and only one message. These messages cannot be divided or rearranged to construct

new kinds of information. A single message from the sender may, however, contain several bits of relevant information for the receiver.

The song of a cricket announces to an unfertilized female the species of the sender (males of different species have different songs), his sex (only males sing), his location (source of the song), and his social status (only a male able to defend the area around its burrow sings from one location). This is all crucial information to the female and accomplishes a biologic goal. But there is no way for the male to alter his song to provide additional information concerning food, predators, or habitat, which might improve his mate's chances of survival and thus enhance his own fitness.

The limitations of communication are especially evident in the invertebrates and lower vertebrates. Signals are characteristically stereotyped, and the responses highly predictable and constant throughout the species. For example, virgin female silkworm moths have special glands that produce a chemical sex attractant to which the males are sensitive. Adult males smell with their large bushy antennae, covered with thousands of sensory hairs that function as receptors for the attractant. Most of these receptors are sensitive to the chemical attractant (a complex alcohol called bombykol, from the name of the silkworm *Bombyx mori*) and to nothing else.

To attract the male, the female merely sits quietly and emits a minute amount of bombykol, which is carried downwind. When a few molecules reach the male's antennae, he is stimulated to fly upwind in search of the female. His search is at first random, but, when by chance he approaches within a few hundred yards of the female, he encounters a concentration gradient of the attractant. Guided by the gradient, he flies toward the female, finds her, and copulates with her.

In this example of chemical communication the attractant bombykol, which is really a pheromone, serves as a signal to bring the sexes together. Its effectiveness is assured because natural selection favors the evolution of males with antennal receptors sensitive enough to detect the attractant at great distances (several miles). Males with a genotype that produces a less sensitive sensory system fail to locate a female and thus are reproductively eliminated from the population.

Most animals communicate by means of displays. The alarm call of the herring gulls, the release of sex attractant by the female moth, the song of the white-crowned sparrow, and the courtship dance of the sage grouse are all examples of displays. As a rule, the signal is prominent and repeated again and again to reduce the chance of misunderstanding.

The elaborate pair-bonding behavior of the blue-footed boobies, illustrated at the beginning of this chapter (p. 108), exemplifies this point. These displays are performed with maximum intensity when the birds come together after a period of separation. The male at right in the picture is sky pointing: the head and tail are pointed skyward and the wings are swiveled forward in a seemingly impossible position to display their glossy upper surfaces to the female. This is accompanied by a high, piping whistle. The female at left, for her part, is parading. She goose steps with exaggerated slow deliberation, lifting each brilliant blue foot in turn, as if holding it aloft momentarily for the male to admire. Such highly personalized displays, performed with droll solemnity, appear comical, even inane to the observer. Indeed the boobies, whose name is derived from the Spanish word ''bobo'' meaning clown, presumably were so designated for their amusing antics.

Needless to say, for the birds, amusement plays no part in the ceremonies. The exaggerated nature of the displays ensures that the message is not missed or misunderstood. Such displays are essential to establish and maintain a strong pair bond between male and female. This requirement also explains the repetitious nature of the displays that follow one another throughout courtship and until egg laying. Redundancy of displays serves to maintain a state of mutual stimulation between male and female, ensuring the degree of cooperation necessary for copulation and subsequent incubation and care of the young. A sexually aroused male has little success with an uninterested female.

We alluded earlier in this chapter to the stereotyped, mechanical nature of instinctive behavior. Yet in watching the interactions between mates during courtship, nest building, and care of the young, it is difficult for the observer to avoid anthropomorphic interpretations of behavior that resembles man's behavior. It is natural and easy to describe animal behavior in terms of human behavior by using words such as ''love,'' ''deceit,'' ''happiness,'' and ''gentleness.'' This is not necessarily false, especially for the higher primates, but the ethologist must always take care to interpret every animal response by using the simplest mechanism that is known to work.

Animals are capable of highly organized behavior in the absence of any intelligent appreciation of its purpose. Sometimes instinctive behavior misfires, and such incidents often serve to emphasize its stereotyped nature. The following exerpt contains a perfect example of the automatic release of inappropriate behavior in a gannet colony.

A male of an old pair flew into his nest. Normally he would bite his mate on the head with some violence and then go through a long and complicated meeting ceremony, an ecstatic display confined to members of a pair. Unfortunately, the female had caught her lower mandible in a loop of fish netting that was firmly anchored in the structure of the nest. Every time she tried to raise her head to perform the meeting ceremony with her mate she merely succeeded in opening her upper mandible whilst the lower remained fixed in the netting. So she apparently threatened the male with widely gaping beak and he immediately responded by attacking her. With each attack she lowered and turned away her bill (the way in which a female gannet appeases an aggressive male). At once the male stopped biting her and she again turned to greet him but simply repeated the beak-opening and drew another attack. And so it went on despite the fact that these two birds had been mated for years and that the netting, the cause of all the trouble, was clearly visible.*

Despite such limitations to the adaptiveness of instinctive behavior, it obviously functions beautifully most of the time. In recognizing that reasoning and insight are not required for effective, highly organized behavior, we should not conclude that lower animals are little more than nonintelligent machines acting out their lives like so many robots. Although the gannet in this example lacked the ''intelligence'' to free his mate by purposefully disentangling her beak from the netting, he was capable of appropriately analyzing the thousands of strategic choices he must make during his lifetime: how to find and hold a mate, where to build a nest site and how to defend it, how to locate evasive marine food, and what to do when the environment changes. All of this and more requires endless behavioral adjustments to new situations. Conceivably, this might be accomplished by a machine, but only by one of staggering complexity.

*Nelson, B. 1968. Galapagos: islands of birds. London, Longmans, Green & Co., Ltd.

SELECTED REFERENCES

Alcock, J. 1975. Animal behavior: an evolutionary approach. Sunderland, Mass., Sinauer Associates, Inc. *Well-written and illustrated discussion of the genetics, physiology, ecology, and history of behavior in an evolutionary perspective.*

Barash, D. P. 1977. Sociobiology and behavior. New York, Elsevier North-Holland, Inc. *The evolution of social behavior. Clear explanation of how behavior is organized by natural selection.*

Bermant, G. (ed.) 1973. Perspectives on animal behavior. Glenview, Ill., Scott, Foresman & Co. *Collection of 10 original essays by 12 authors. Intermediate level.*

Bramblett, C. A. 1976. Patterns of primate behavior. Palo Alto, Calif., Mayfield Publishing Co. *Behavior of primates with useful field descriptions of 15 best-known primates.*

Brown, J. L. 1975. The evolution of behavior. New York, W. W. Norton Co., Inc. *Advanced text emphasizing social behavior.*

Eibl-Eibesfeldt, I. 1975. Ethology, the biology of behavior, ed. 2. New York, Holt, Rinehart & Winston, Inc. *Excellent general ethology text in the European tradition.*

Eisner, T., and E. O. Wilson. (ed.) 1955-1975. Animal behavior. San Francisco, W. H. Freeman & Co., Publishers. *Collection of articles from Scientific American with introductions by Eisner and Wilson.*

Evans, R. I. 1975. Konrad Lorenz: the man and his ideas. New York, Harcourt Brace Jovanovich, Inc. *Dialogue with one of the world's great ethologists, together with a critique of his work, reprints of four important papers, and bibliography.*

Fox, M. W. 1974. Concepts in ethology. Minneapolis, University of Minnesota Press. *Principles of ethology, with many examples from the canids.*

Hinde, R. A. 1970. Animal behavior: a synthesis of ethology and comparative psychology, ed. 2. New York, McGraw-Hill, Inc. *Advanced text.*

Klopfer, P. H. 1974. An introduction to animal behavior: ethology's first century, ed. 2. Englewood Cliffs, N.J., Prentice-Hall, Inc. *Introductory text.*

Lorenz, K. Z. 1952. King Solomon's ring. New York, Thomas Y. Crowell Co., Inc. *One of the most delightful books ever written about the behavior of animals.*

National Geographic Society. 1972. The marvels of animal behavior. Washington, D.C., National Geographic Society. *Superbly illustrated collection of essays by leading animal behaviorists.*

Scott, J. P. 1972. Animal behavior, ed. 2. Chicago, University of Chicago Press. *Comprehensive introductory text.*

Thorpe, W. H. 1974. Animal nature and human nature. Garden City, Anchor Press. *Comparison of animal and human behavior; especially useful sections on animal communication.*

Tinbergen, N. 1951. The study of instinct. New York, Oxford University Press, Inc. *Classic introduction to ethology.*

Tinbergen, N. 1961. The herring gull's world: a study of the social behaviour of birds. New York, Basic Books, Inc. *One of the most thorough studies of instinctive behavior ever made. Beautifully written.*

Tinbergen, N. 1965. Animal behavior. New York, Time-Life Books. *Popularized but still highly informative account. Fine illustrations.*

Wilson, E. O. 1975. Sociobiology: the new synthesis. Cambridge, Mass., Harvard University Press. *An important synthesis of facts and theories on the biological basis of social behavior.*

SELECTED SCIENTIFIC AMERICAN ARTICLES

Many important articles not listed below are included in the Scientific American anthology edited by Eisner and Wilson.

Burgess, J. W. 1976. Social spiders. **234:**100-106 (March). *Contrary to the usual pattern of solitary behavior among spiders, some species are gregarious and build large communal webs.*

Bertram, B. C. R. 1975. The social system of lions. **232:**54-65 (May). *The social behavior of lions in their natural habitat.*

Eaton, G. G. 1976. The social order of Japanese macaques. **235:**96-106 (Oct.). *Long-term observations of a confined troop of macaques show that the biologic component of their behavior is much modified by the social component.*

Hess, E. H. 1972. "Imprinting" in a natural laboratory. **227:**24-31 (Aug.). *Even before the egg hatches, a duckling and its parent interact in ways that will strengthen the juvenile-to-parent bond.*

Hölldobler, B. 1971. Communication between ants and their guests. **224:**86-93 (March). *Some parasites learn the language of their ant hosts to gain free food and shelter.*

Pooley, A. C., and C. Gans. 1976. The Nile crocodile. **234:**114-124 (April). *The remarkable social behavior of this large reptile includes parental protection of the young.*

Todd, J. H. 1971. The chemical language of fishes. **224:**98-108 (May). *Catfishes produce pheromones to label losers and winners of hierarchical fights.*

Topoff, H. R. 1972. The social behavior of army ants. **227:**70-79 (Nov.). *A complex social organization is described.*

Watts, C. R., and A. W. Stokes. 1971. The social order of turkeys. **224:**112-118 (June). *Describes a rigidly stratified society.*

Wilson, E. O. 1972. Animal communication. **227:**52-60 (Sept.). *The languages of animals and man are compared.*

PART TWO ANIMAL FORM AND FUNCTION

Rearing horse and man. The vertebrate endoskeleton is a flexible framework that provides support, surface for muscle attachment, and protection for brain and lungs. Because it is a living tissue that permits continuous growth, some vertebrate animals have become the most massive in the animal kingdom. (Courtesy American Museum of Natural History.)

7 SUPPORT, PROTECTION, AND MOVEMENT

INTEGUMENT AMONG VARIOUS GROUPS OF ANIMALS

The integument is the outer covering of the body, a protective wrapping that includes the skin and all structures that are derived from or associated with the skin, such as hair, setae, scales, feathers, and horns. In most animals it is tough and pliable, providing mechanical protection against abrasion and puncture and forming an effective barrier against the invasion of bacteria. It provides moisture-proofing against fluid loss or gain. The skin protects the underlying cells against the damaging action of the ultraviolet rays of the sun. But, in addition to being a protective cover, the skin serves a variety of important regulatory functions. For example, in warm-blooded (homeothermic) animals, it is vitally concerned with temperature regulation since most of the body's heat is lost through the skin. The skin contains sensory receptors that provide essential information about the immediate environment. It has excretory functions and in some forms respiratory functions as well. Through skin pigmentation the organism can make itself more or less conspicuous. Skin secretions can make the animal attractive or repugnant or provide olfactory cues that influence behavioral interactions between individuals.

Invertebrate integument

Many protozoans have only the delicate cell or plasma membranes for external coverings; others, such as *Paramecium,* have developed a protective pellicle. Most multicellular invertebrates, however, have more complex tissue coverings. The principal covering is a single-layered **epidermis.** Some invertebrates have added a secreted noncellular **cuticle** over the epidermis for additional protection; some groups, such as many parasitic worms, have only a thick resistant cuticle and lack a cellular epidermis.

The molluscan epidermis is delicate and soft and contains mucous glands, some of which secrete the calcium carbonate of the shell. The cephalopod has developed a more complex integument, consisting of a cuticle, a simple epidermis, a layer of connective tissue, a layer of reflecting cells (iridocytes), and a thicker layer of connective tissue.

Arthropods have the most complex of invertebrate integuments, providing not only protection but also skeletal support. The development of a firm exoskeleton and jointed appendages suitable for the attachment of muscles has been a key feature in the extraordinary evolutionary success of this group, the largest of animal groups. The arthropod integument consists of a single-layered **epidermis** (also called more precisely **hypodermis**), which secretes a complex cuticle of two zones (Fig. 7-1, *A*). The inner zone, the **procuticle,** is composed of protein and chitin. Chitin is a polysaccharide resembling plant cellulose that is laid down in molecular sheets like the veneers of plywood, thus providing great strength to the procuticle layer. The outer zone of cuticle, lying on the external surface above the procuticle, is the thin **epicuticle.** The epicuticle is a nonchitinous complex of proteins and lipids that provides a protective moisture-proofing barrier to the integument.

The arthropod cuticle may remain as a tough but soft and flexible layer, or it may be hardened by one of two ways. In the decapod crustaceans, for example, crabs and lobsters, the cuticle is stiffened by **calcification,** the deposition of calcium carbonate. In insects hardening is achieved by a process called **sclerotization,** in which the protein molecules of the chitin form stabilizing cross-linkages. Arthropod chitin is one of the toughest materials synthesized by animals; it is strongly resistant to pressure and tearing and can withstand boiling in concentrated alkali, yet it is light, having a specific weight of only 1.3.

When arthropods molt, the epidermal cells first divide by mitosis. Enzymes secreted by the epidermis dissolve most of the procuticle; the digested materials are then absorbed and consequently not lost to the body. Then in the space beneath the old cuticle a new epicuticle and procuticle are formed. After the old cuticle is shed, the new cuticle is thickened and calcified or sclerotized.

Vertebrate integument

The basic plan of the vertebrate integument, as exemplified by man's skin (Fig. 7-1, *B*), includes a thin, outer stratified epithelial layer, the **epidermis,** derived from ectoderm and an inner, thicker layer, the **dermis,** or true skin, which is of mesodermal origin and is made up of nerves, blood vessels, connective tissue, pigment, etc.

The epidermis consists usually of several layers of cells. The basal part is made up of columnar cells that continually divide by mitosis to form new cells that are pushed toward the surface. As the cells move upward, they become **keratinized** into scaly plates by the deposition of a fibrous protein, **keratin,** which makes

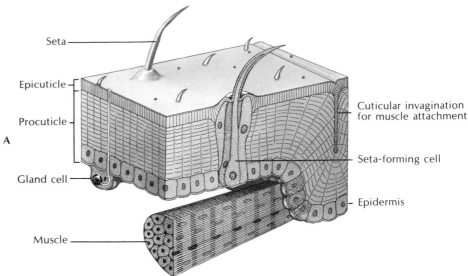

Fig. 7-1. A, Structure of insect integument. This reconstruction shows a block of integument drawn at a point where the cuticle invaginates to provide an exoskeletal muscle attachment. **B,** Structure of human skin.

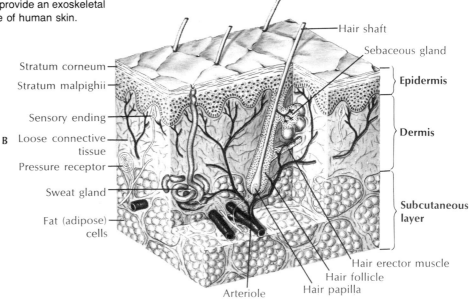

the surface resistant to abrasion, chemical change, and water transfer.

The dermis is comprised of connective tissue, fat cells, muscle, blood vessels, and nerves, as well as epidermally derived structures such as hair follicles, sweat glands, and sebaceous glands. The latter produce a fatty secretion called **sebum** that lubricates the hair and skin.

Although the human species is not very hairy, hair is a distinct mammalian characteristic. The hair follicle from which a hair grows is an epidermal structure even though it lies mostly in the dermis and subdermal (subcutaneous) tissues. The hair grows continuously by rapid proliferation of cells in the base of the follicle. As the hair shaft is pushed upward, new cells are carried away from their source of nourishment and turn

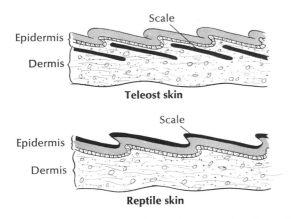

Epidermis {
Dermis {

Scale

Teleost skin

Epidermis {
Dermis {

Scale

Reptile skin

Fig. 7-2. Integument of bony fishes and reptiles. Bony (teleost) fishes have bony scales from dermis, and reptiles have horny scales from epidermis. Dermal scales of fishes are retained throughout life. Since a new growth ring is added to each scale each year, fishery biologists use scales to tell age of fishes. Epidermal scales of reptiles are shed periodically.

into the dense type of keratin **(hard keratin)** that constitutes nails, claws, hooves, and feathers, as well as hair. On a weight basis, hair is by far the strongest material in the body. It has a tensile strength comparable to rolled aluminum, which is nearly twice as strong, weight for weight, as the strongest bone. A small erector muscle is attached to each hair follicle and slants upward toward the epidermis. Contraction of this muscle causes the hair to stand up and the skin to dimple above the muscle attachment. The result is the "gooseflesh" characteristic of certain emotional states (fear and excitement) and of cold stimulation. The color of hair is determined by the amount and quality of pigment in the outer shell, or cortex, of each hair.

The skin is modified in many ways in the different vertebrates. The tough, protective scales of fishes, although appearing to be surface structures, are actually bony plates produced in the dermis (Fig. 7-2). The scales may protrude through the overlying epidermis, but usually they are completely covered by a thin, often transparent layer of epidermis, even though the scales overlap one another. The scales of reptiles, however, are horny, keratinized plates of epidermal origin. In snakes the ventral scales form transverse bands that can be raised and lowered by muscles in a coordinated manner to assist in locomotion. Snakes periodically shed their skins as they grow. The keratinized epidermis of the entire body surface separates

from the living cells beneath. By first loosening the skin around the lips, the snake works the skin back over its head; then, as it crawls along the ground, the old skin is peeled off inside out, revealing the glistening new scales beneath.

Sunburning and tanning. In general, man lacks the special body coverings that protect other land vertebrates from the damaging action of ultraviolet rays of the sun; he must depend on thickening of the outer layer of the epidermis **(stratum corneum)** and on epidermal pigmentation for protection from the sun's spectrum. Sunburn and suntanning are caused largely by exposure to the ultraviolet area of the sun's spectrum (wavelength 300 to 390 nm). This spectral band acts almost entirely on epidermis; very little penetrates to the dermis beneath. The ultraviolet rays photochemically decompose nucleoproteins within the nuclei of cells of the deeper layer of the epidermis. Blood vessels then enlarge and other tissue changes occur, producing the red coloration of sunburn. Light skins suntan through the formation in the deeper epidermis of the pigment **melanin** and by "pigment darkening," that is, the photooxidative blackening of bleached pigment already present in the epidermis.

SKELETAL SYSTEMS

Skeletons are supportive systems that provide rigidity to the body, surfaces for muscle attachment, and protection for vulnerable body organs. The familiar bone of the vertebrate skeleton is only one of several kinds of supportive and connective tissues serving various binding and supportive functions, which we discuss in this section.

Exoskeleton and endoskeleton

Although animal supportive and protective structures take many forms, there are two principal types of skeletons: the **exoskeleton,** typical of molluscs and arthropods, and the **endoskeleton,** characteristic of echinoderms and vertebrates. The invertebrate exoskeleton is mainly protective in function and may take the form of shells, spicules, and calcareous or chitinous plates. It may be rigid, as in molluscs, or jointed and movable as in arthropods. Unlike the endoskeleton, which grows with the animal, the exoskeleton is often a limiting coat of armor that must be periodically shed (molted) to make way for an enlarged replacement. Some invertebrate exoskeletons, such as the shells of snails and bivalves, grow with the animal.

Vertebrates too have traces of exoskeleton that serve to remind us of our invertebrate heritage. These are, for example, scales and plates of fishes, fingernails and claws, hair, feathers, and other keratinized integumentary structures.

The vertebrate endoskeleton is formed inside the body and is composed of bone and cartilage surrounded by soft tissues. It not only supports and protects, but it is also the major body reservoir for calcium and phosphorus. In the higher vertebrates the red blood cells and certain white blood cells are formed in the bone marrow.

Cartilage

Cartilage and bone are the characteristic vertebrate supportive tissues. The **notochord,** the semirigid axial rod of protochordates and vertebrate larvae and embryos, is also a primitive vertebrate supportive tissue. Except in the most primitive chordates, for example, amphioxus and the cyclostomes, the notochord is surrounded or replaced by the backbone during embryonic development. The notochord is composed of large, vacuolated cells and is surrounded by layers of elastic and fibrous sheaths. It is a stiffening device, preserving body shape during locomotion (p. 485).

Vertebrate cartilage is the major skeletal element of primitive vertebrates. Cyclostomes and elasmobranchs have purely cartilaginous skeletons. In contrast, higher vertebrates have principally bony skeletons as adults with some cartilage interspersed. Cartilage is a soft, pliable, characteristically deep-lying tissue. Unlike connective tissue, which is quite variable in form, cartilage is basically the same wherever it is found. The basic form, **hyaline cartilage,** has a clear, glassy appearance. It is composed of cartilage cells **(chondrocytes)** surrounded by firm complex protein gel interlaced with a meshwork of collagenous fibers. Blood vessels are virtually absent. In addition to forming the cartilagenous skeleton of the primitive vertebrates and that of all vertebrate embryos, hyaline cartilage makes up the articulating surfaces of many bone joints of higher adult vertebrates and the supporting tracheal, laryngeal, and bronchial rings.

The basic cartilage has several variants. Among these is **calcified cartilage,** in which calcium salt deposits produce a bonelike structure. **Fibrocartilage,** resembling connective tissue, and **elastic cartilage,** containing many elastic fibers, are other variations of basic hyaline cartilage found among the vertebrates.

Bone

Bone differs from other connective and supportive tissues by having significant deposits of inorganic calcium salts laid down in an extracellular matrix. Its structural organization is such that bone has nearly the tensile strength of cast iron, yet is only one third as heavy. Most bones develop from cartilage **(endochondral bone)** by a complex replacement of embryonic cartilage with bone tissue.

A second type of bone is **membrane bone,** which develops directly from sheets of embryonic cells. In higher vertebrates membrane bone is restricted to bones of the face and cranium; the remainder of the skeleton is endochondral bone. Despite differences in origin, endochondral bone and membrane bone are not distinguishable histologically. In the fishes the dermal scales and plates that may cover most of the body are formed from membrane bone.

Two kinds of bone structure are distinguishable— **spongy** (or **cancellous**) and **compact** (Fig. 7-3). Spongy bone (so named for its appearance) consists of an open, interlacing framework of bony tissue, oriented to give maximum strength under the normal stresses and strains that the bone receives. Compact bone is dense, appearing absolutely solid to the naked eye. Both structural kinds of bone are found in the typical long bones of the body, such as the humerus (upper arm bone) (Fig. 7-3).

Microscopic structure of bone

Compact bone is composed of a calcified bone matrix arranged in concentric rings. The rings contain cavities **(lacunae)** filled with bone cells **(osteocytes)** that are interconnected by many minute passages **(canaliculi).** These serve to distribute nutrients throughout the bone. This entire organization of lacunae and canaliculi is arranged into an elongated cylinder called a **haversian system** (Fig. 7-4). Bone consists of bundles of haversian systems cemented together and interconnected with blood vessels.

Bone growth is a complex restructuring process, involving both its destruction internally by bone-resorbing cells **(osteoclasts)** and its deposition externally by bone-building cells **(osteoblasts).** Both processes occur simultaneously so that the marrow cavity inside grows larger by bone resorption while new bone is laid down outside by bone deposition. Bone growth responds to several hormones, in particular **parathyroid hormone,** which stimulates bone

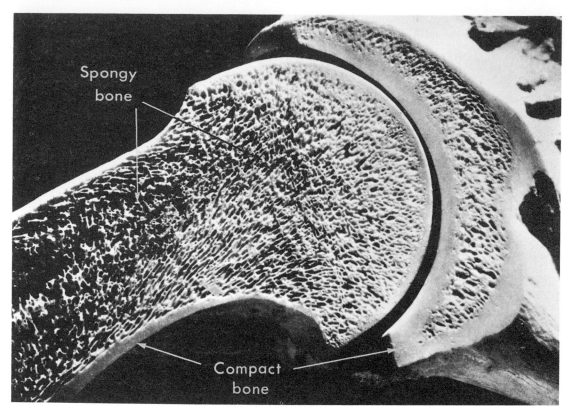

Spongy bone

Compact bone

Fig. 7-3. Section of proximal end of human humerus, showing appearance of spongy and compact bone. (From Feininger: Anatomy of nature, Crown Publishers.)

resorption, and **calcitonin,** which inhibits bone resorption. These two hormones are responsible for maintaining a constant level of calcium in the blood (p. 245).

Plan of the vertebrate skeleton

The vertebrate skeleton is composed of two main divisions: the **axial skeleton,** which includes the skull, vertebral column, sternum, and ribs, and the **appendicular skeleton,** which includes the limbs and the pectoral and pelvic girdles (Fig. 7-5). Not surprisingly the skeleton has undergone extensive remodeling in the course of vertebrate evolution. The move from water to land forced dramatic changes in body form. With cephalization, that is, the concentration of brain, sense organs, and food-gathering and respiratory apparatus in the head, the skull became the most intricate portion of the skeleton. The lower vertebrates have a larger number of skull bones than the more advanced vertebrates. Some fish have 180 skull bones; amphibians

and reptiles, 50 to 95; and mammals, 35 or fewer. Man has 29.

The numerous elements of the head skeleton belong to three structural and functional components: (1) the **neurocranium,** which is the original, primitive skull that encloses the brain and sense organs, (2) the **splanchnocranium,** which supports the respiratory and food-gathering equipment (gill apparatus and jaws), and (3) the **dermocranium,** which arose from ancient, superficial head armor and serves to protect deep-lying delicate tissues. In the lower vertebrates these three skull components are more or less separated from each other; in higher forms they are all fused together or incorporated into a single unit—the vertebrate skull. There is a basic plan of homology in the skull elements of vertebrates from fish to man; evolution has meant reduction in numbers of bones through loss and fusion in accordance with size and functional changes.

The vertebral column is the main stiffening axis of

133

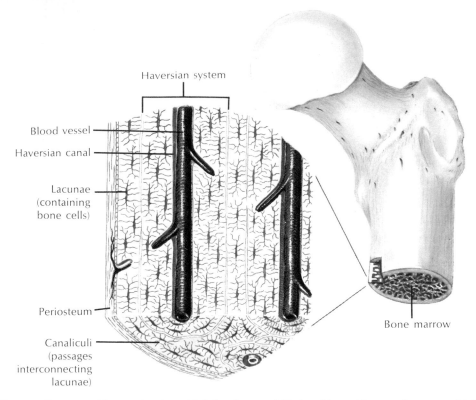

Haversian system

Blood vessel

Haversian canal

Lacunae
(containing
bone cells)

Periosteum

Canaliculi
(passages
interconnecting
lacunae)

Bone marrow

Fig. 7-4. Structure of bone, showing at left the dense calcified matrix and bone cells arranged into haversian systems. Bone cells are entrapped within the cell-like lacunae but receive nutrients from the circulatory system via tiny canaliculi that interlace the calcified matrix. Bone cells were known as osteoblasts when they were building bone, but, in mature bone shown here, they become resting osteocytes. Bone is covered with a compact connective tissue called periosteum.

the postcranial skeleton. In fishes it serves much the same function as the notochord from which it is derived; that is, it provides points for muscle attachment and prevents telescoping of the body during muscle contraction. Since fish musculature is similar throughout the trunk and tail, fish vertebrae are differentiated only into trunk and caudal vertebrae.

With the evolution of tetrapods, the vertebral column becomes structurally adapted to withstand new regional stresses transmitted to the column by the two paired appendages. In the higher tetrapods, the vertebrae are differentiated into **cervical** (neck), **thoracic** (chest), **lumbar** (back), **sacral** (pelvic), and **caudal** (tail) vertebrae. In birds and also in man the caudal vertebrae are reduced in number and size, and the sacral vertebrae are fused. The number of vertebrae varies among the different animals. The python seems to lead the list with 435. In man (Fig. 7-5) there are 33

in the child, but in the adult 5 are fused to form the **sacrum** and 4 to form the **coccyx.** Besides the sacrum and coccyx, man has 7 cervical, 12 thoracic, and 5 lumbar vertebrae. The number of cervical vertebrae (7) is constant in nearly all mammals.

The first two cervical vertebrae, the **atlas** and the **axis,** are modified to support the skull and permit pivotal movements. The atlas bears the globe of the head much as the mythological Atlas bore the earth on his shoulders. The axis, the second vertebra, permits the head to turn from side to side.

Ribs are long or short skeletal structures that articulate medially with vertebrae and extend into the body wall. Primitive forms have a pair of ribs for every vertebra; they serve as stiffening elements in the connective tissue septa that separate the muscle segments and thus improve the effectiveness of muscle contractions. Many fishes have both dorsal and ventral

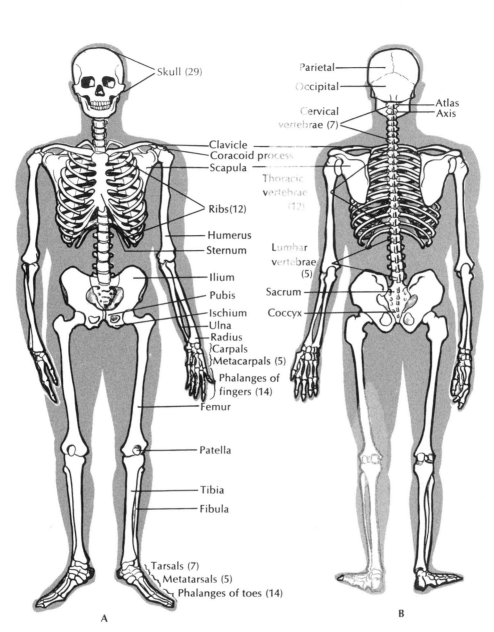

Skull (29)

Parietal
Occipital
Atlas
Axis
Cervical
vertebrae (7)

Clavicle
Coracoid process
Scapula
Thoracic
vertebrae
(12)

Ribs(12)

Humerus
Sternum

Lumbar
vertebrae
(5)

Ilium
Pubis
Sacrum
Ischium
Coccyx
Ulna
Radius
Carpals
Metacarpals (5)
Phalanges of
fingers (14)

Femur

Patella

Tibia
Fibula

Tarsals (7)
Metatarsals (5)
Phalanges of toes (14)

A

B

Fig. 7-5. Human skeleton. **A,** Ventral view. **B,** Dorsal view. Numbers in parentheses indicate number of bones in that unit. In comparison with other mammals, man's skeleton is a patchwork of primitive and specialized parts. Erect posture, brought about by specialized changes in legs and pelvis, enabled primitive arrangement of arms and hands (arboreal adaptation of man's ancestors) to be used for manipulation of tools. Development of skull and brain followed as a consequence of the premium natural selection put upon dexterity, better senses, and ability to appraise environment.

ribs, and some have numerous riblike intermuscular bones as well—all of which increase the difficulty and reduce the pleasure of our eating certain kinds of fish. Higher vertebrates have a reduced number of ribs, and some, such as the familiar leopard frog, have no ribs at all. Others, such as elasmobranchs and some amphibians, have very short ribs. Man has 12 pairs of ribs, but approximately 1 person in 20 has a thirteenth pair. In mammals the ribs together form the thoracic basket, which supports the chest wall and prevents collapse.

Most vertebrates, fishes included, have paired appendages. All fishes except the agnathans have thin pectoral and pelvic fins that are supported by the pectoral and pelvic girdles, respectively. Forms above the fishes (except snakes and limbless lizards) have two pairs of **pentadactyl** (five-toed) limbs, also supported by girdles. The pentadactyl limb is similar in all tetrapods, alive and extinct; even when highly modified for various modes of life, the elements are rather easily homologized.

Modifications of the basic pentadactyl limb for life in different environments involve the distal elements much more frequently than the proximal, and it is far more common for bones to be lost or fused than for new ones to be added. The elongation of the third toe of the horse to facilitate running has already been described (p. 94). The bird wing is a good example of distal modification. The bird embryo bears 13 distinct wrist and hand bones (carpals and metacarpals), which are reduced to three in the adult. Most of the finger bones (phalanges) are lost, leaving four bones in three digits (see p. 553). The proximal bones (humerus, radius, and ulna), however, are little modified in the bird wing.

In nearly all tetrapods the pelvic girdle is firmly attached to the axial skeleton since the greatest locomotory forces transmitted to the body come from the hind limbs. The pectoral girdle, however, is much more loosely attached to the axial skeleton, providing the forelimbs with greater freedom for manipulative movements.

In man the pectoral girdle is made up of 2 scapulae and 2 clavicles; the arm is made up of humerus, ulna, radius, 8 carpals, 5 metacarpals, and 14 phalanges. The pelvic girdle consists of 3 fused bones—ilium, ischium, and pubis; the leg is made up of femur, patella, tibia, fibula, 7 tarsals, 5 metatarsals, and 14 phalanges. Each bone of the leg has its counterpart in the arm with the exception of the patella. This kind of correspondence between anterior and posterior parts is called **serial homology.**

ANIMAL MOVEMENT

Movement is a unique characteristic of animals. Plants may show movement, but this usually results from changes in turgor pressure or growth rather than from specialized contractile proteins as in animals. Movement appears in many forms in animal tissues, ranging from barely discernible streaming of cytoplasm, the swelling of mitochondria, and movement of the mitotic spindle during cell division, to frank movements of powerful striated muscles of vertebrates. Recently it has become evident that virtually all animal movement depends on a single fundamental mechanism: **contractile proteins,** which can change their form to elongate or contract. This contractile machinery is always composed of ultrafine fibrils—fine filaments, striated fibrils, or tubular fibrils (microtubules)—arranged to contract when powered by **ATP.** By far the most important protein contractile system is the **actomyosin system,** composed of two proteins, **actin** and **myosin.** This is an almost universal biomechanical system found in protozoans through vertebrates and performs a long list of diverse functional roles. In this section we examine the three principal kinds of animal movement: ameboid, ciliary, and muscular.

Ameboid movement

Ameboid movement is a form of movement especially characteristic of the freshwater amebae and other sarcodine protozoans; it is also found in many wandering cells of higher animals, such as white blood cells, embryonic mesenchyme, and numerous other mobile cells that move among the tissue spaces. Ameboid cells constantly change their shape by sending out and withdrawing **pseudopodia** (false feet) from any point on the cell surface. Such cells are surrounded by a delicate, highly flexible membrane called **plasmalemma** (Fig. 15-1, p. 289). Beneath this lies a nongranular layer, the **hyaline ectoplasm,** which encloses the **granular ectoplasm.**

Optic studies of an ameba in movement suggest that the outer layer of cytoplasm **(ectoplasm)** actively contracts in the **fountain zone** at the tip of the pseudopod to pull a central core of rather rigid endoplasm forward. The latter is then converted into ectoplasm, which slips posteriorly under the plasma-lemma and joins the endoplasm at the rear to begin another cycle.

There are other theories of ameboid movement, in particular one that favors the posterior end of the animal as the locus of contraction. According to this theory, an ameba is pushed rather than pulled forward. Although no completely satisfactory analysis exists, it is certain that ameboid movement is based on the same fundamental contractile system that powers vertebrate muscles: an actomyosin machinery driven by ATP.

Ciliary movement

Cilia are minute hairlike motile processes that extend from the surfaces of the cells of many animals. They are a particularly distinctive feature of ciliate protozoans, but, except for the nematodes in which cilia are absent and the arthropods in which they are rare, cilia are found in all major groups of animals. Cilia perform many roles either in moving small animals such as protozoans through their aquatic environment or in propelling fluids and materials across the epithelial surfaces of larger animals.

Cilia are of remarkably uniform diameter (0.1 to 0.5 μm) wherever they are found. The electron microscope has shown that each cilium contains a peripheral circle of nine double filaments and an additional two

filaments in the center (Fig. 7-6). (Exceptions to the 9 + 2 arrangement have been noted; certain sperm tails have but one central fibril.) A **flagellum** is a whiplike structure larger than a cilium and usually present singly at one end of a cell. They are found in members of flagellate protozoans, in animal spermatozoa, and in sponges.

According to one currently favored theory of ciliary movement, the fibrils behave as "sliding filaments" that move past one another much like the sliding filaments of vertebrate skeletal muscle that is described in the next section. During contraction, fibrils on the concave side slide outward past fibrils on the convex side to increase curvature of the cilium; during the recovery stroke, fibrils on the opposite side slide outward to bring the cilium back to its starting position. For such a system to work, the fibrils must be interconnected by molecular bridges, which, in fact, cannot be seen with the electron microscope.

Cilia contract in a highly coordinated way, the rhythmic waves of contraction moving across a ciliated epithelium like windwaves across a field of grain. The columns of cilia are coordinated by an interconnected fiber system through which the excitation wave passes with each stroke.

Muscular movement

Contractile tissue is most highly developed in muscle cells called **fibers.** Although muscle fibers themselves can only shorten, they can be arranged in so many different configurations and combinations that almost any movement is possible.

Types of vertebrate muscle

Vertebrate muscle is broadly classified on the basis of the appearance of muscle cells (fibers) when viewed with a light microscope. **Striated muscle** appears transversely striped (striated), with alternating dark and light bands (Fig. 7-7). We can recognize two types of striated muscle: **skeletal** and **cardiac muscle.** A third kind of vertebrate muscle is **smooth** (or visceral) **muscle,** which lacks the characteristic alternating bands of the striated type (Fig. 7-8).

Skeletal muscle is typically organized into sturdy, compact bundles or bands. It is called skeletal muscle because it is attached to skeletal elements and is responsible for movements of the trunk, appendages, respiratory organs, eyes, mouthparts, and so on. Skeletal muscle fibers are extremely long, cylindric,

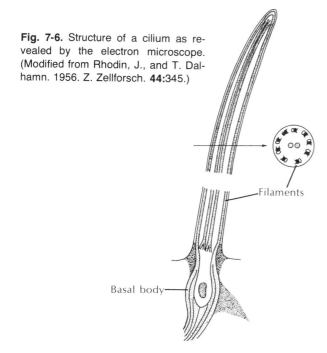

Fig. 7-6. Structure of a cilium as revealed by the electron microscope. (Modified from Rhodin, J., and T. Dalhamn. 1956. Z. Zellforsch. **44:**345.)

Filaments

Basal body

Fig. 7-7. Photomicrograph of skeletal muscle showing several striated fiber lying side by side. (Photo by J. W. Bamberger.)

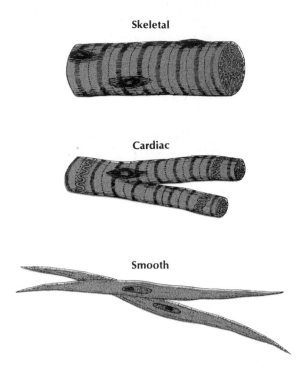

Fig. 7-8. Three kinds of vertebrate muscle fibers.

multinucleate cells, which may reach from one end of the muscle to the other. They are packed into bundles called **fascicles,** which are enclosed by tough connective tissue. The fascicles are in turn grouped into a discrete **muscle** surrounded by a thin connective tissue layer. Most skeletal muscles taper at their ends, where they connect by tendons to bones. Other muscles, such as the ventral abdominal muscles, are flattened sheets.

In most fishes, amphibians, and to some extent reptiles, there is a segmented organization of muscles alternating with the vertebrae. The skeletal muscles of higher vertebrates, by splitting, by fusion, and by shifting, have developed into specialized muscles best suited for manipulating the jointed appendages that have evolved for locomotion on land. Skeletal muscle contracts powerfully and quickly but fatigues more rapidly than does smooth muscle. Skeletal muscle is sometimes called **voluntary muscle** because it is stimulated by motor fibers and is under conscious cerebral control.

Smooth muscle lacks the striations typical of skeletal muscle (Fig. 7-8). The cells are long, tapering strands, each containing a single nucleus. Smooth muscle cells are organized into sheets of muscle circling the walls of the alimentary canal, blood vessels, respiratory passages, and urinary and genital ducts. Smooth muscle is typically slow acting. It is under the control of the autonomic nervous system; thus, unlike skeletal muscle, its contractions are involuntary and unconscious. The principal functions of smooth muscles are to push the contents of a tube, such as the intestine, along its way by active contractions or to regulate the diameter of a tube, such as a blood vessel, by sustained contraction.

Cardiac muscle, the seemingly tireless muscle of the vertebrate heart, combines certain characteristics of both skeletal and smooth muscle. It is fast acting and striated like skeletal muscle, but contraction is under involuntary autonomic control like smooth muscle. Actually the autonomic nerves serving the heart can only speed up or slow down the rate of contraction; the heartbeat originates within specialized cardiac muscle and the heart continues to beat even after all autonomic nerves are severed. Until very recently, cardiac muscle was believed to be one large unseparated mass **(syncytium)** of branching, interconnected fibers. Many histologists, whose understanding was vastly increased by the electron microscope, now consider cardiac muscle to be comprised of closely opposed, but separate, uninucleate cell fibers.

Types of invertebrate muscle

Smooth and striated muscles are also characteristic of invertebrate animals, but there are many variations of both types and even instances in which the structural and functional features of vertebrate smooth and striated muscle are combined in the invertebrates. Striated muscle appears in invertebrate groups as diverse as the primitive coelenterates and the advanced arthropods. The thickest muscle fibers known, approximately 3 mm in diameter and 6 cm long and easily seen with the unaided eye, are those of giant barnacles and Alaska king crabs living along the Pacific coast of North America. These cells are so large that they can be readily cannulated for physiologic studies and are understandably popular with muscle physiologists.

It is not possible in this short space to describe adequately the tremendous diversity of muscle structure and function in the vast assemblage of invertebrates. We will mention only two functional extremes.

Bivalve molluscan muscles contain fibers of two types. One kind can contract rapidly, enabling the bivalve to snap shut its valves when disturbed. Scallops use these "fast" muscle fibers to swim in their awkward manner. The second muscle type is capable of slow, long-lasting contractions. Using these fibers, a bivalve can keep its valves tightly shut for days or even months. Obviously these are no ordinary muscle fibers! It has been discovered that such retractor muscles use very little metabolic energy and receive remarkably few nerve impulses to maintain the activated state. The contracted state has been likened to a "catch mechanism" involving some kind of stable cross-linkage between the contractile proteins within the fiber. However, despite considerable research, no completely satisfactory explanation for this retractor mechanism is known to exist.

Insect flight muscles are virtually the functional antithesis of the slow, holding muscles of bivalves. The wings of some of the small flies operate at frequencies greater than 1,000 per second. The so-called **fibrillar muscle,** which contracts at these incredible frequencies—far greater than even the most active of vertebrate muscles—shows unique characteristics. It has very limited extensibility; that is, the wing leverage system is arranged so that the muscles shorten hardly at all during each downbeat of the wings. Furthermore the muscles and wings operate as a rapidly oscillating system in an elastic thorax (see Fig. 21-21, p. 429). Since the muscles rebound elastically during flight, they receive impulses only periodically rather than one impulse per contraction; 1 reinforcement impulse for every 20 or 30 contractions is enough to keep the system active.

Structure of striated muscle

In recent years the electron microscope and advanced biochemical methods have been focused on the fine structure and function of the striated muscle fiber. These efforts have been so successful that more has been learned of muscle physiology in the last 20 years than in the previous century. The discussion that follows is limited to the striated muscle since its physiology is presently much better understood than is that of smooth muscle.

As we earlier pointed out, striated muscle is so named because of the periodic bands, plainly visible under the light microscope, that pass across the widths of the muscle cells. Each cell, or **fiber,** contains numerous **myofibrils** packed together and invested by the cell membrane, the **sarcolemma** (Fig. 7-9). Also present in each fiber are several hundred nuclei usually located along the edge of the fiber, numerous mitochondria (sometimes called **sarcosomes**), a network of tubules called the **sarcoplasmic reticulum** (to be discussed later), and other cell inclusions typical of any living cell. Most of the fiber, however, is packed with the unique **myofibrils,** each 1 to 2 μm in diameter.

The characteristic banding of the muscle fiber represents the fine structure of the myofibrils that make up the fiber. In the resting fiber are alternating light- and dark-staining bands called the **I bands** and **A bands,** respectively (Fig. 7-9). The functional unit of the myofibril, the **sarcomere,** extends between successive Z lines. The myofibril is actually an aggregate of much smaller parallel units called **myofilaments.** These are of two kinds—thick filaments, 110 Å in diameter composed of the protein **myosin,** and thin filaments, 50 Å in diameter composed of the protein **actin** (Fig. 7-9). These are the actual contractile proteins of muscle. The thick myosin filaments are confined to the A band region. The thin actin filaments are located mainly in the light I bands but extend some distance into the A band as well. In the relaxed muscle, they do not quite meet in the center of the A band. The Z line is a dense protein different from either actin or myosin, which serves as the attachment plane for the thin filaments and keeps them in register. These relationships are diagrammed in Fig. 7-9.

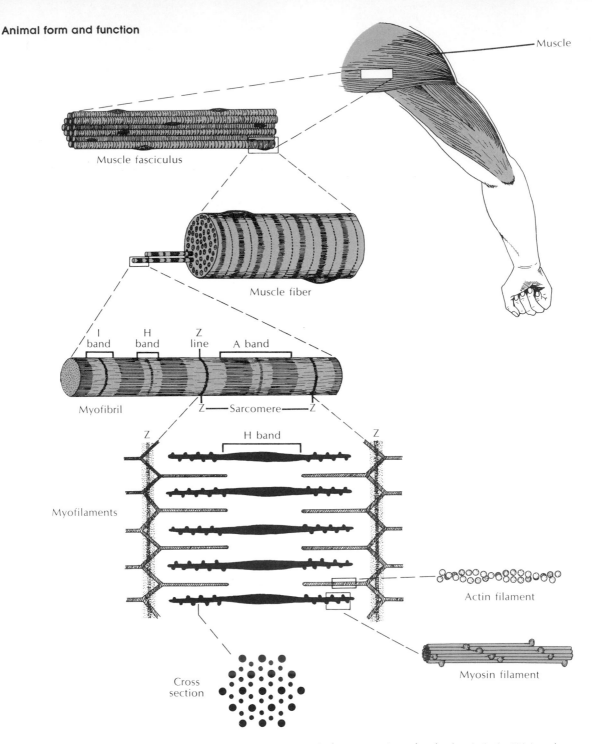

Fig. 7-9. Organization of vertebrate skeletal muscle from gross to molecular level. Actin (thin) and myosin (thick) filaments are enlarged to show supposed shapes of individual molecules and probable positioning of the cross bridges (shown as knobs) on myosin molecules that serve to link thick and thin filaments during contraction. Cross section shows that each thick filament is surrounded by six thin filaments and that each thin filament is surrounded by three thick filaments. (Redrawn with slight modifications from Bloom, W., and D. W. Fawcett. 1968. A textbook of histology. Philadelphia, W. B. Saunders Co.).

Contraction of striated muscle

The thick and thin filaments are spatially arranged in a highly symmetric pattern so that each thick filament is surrounded by six thin filaments; conversely each thin filament lies among three thick filaments (see cross section at bottom of Fig. 7-9). The two kinds of filaments are linked together by molecular bridges, which, it is believed, extend outward from the thick filaments to hook onto active sites on the thin filaments. During contraction the cross bridges swing rapidly back and forth, alternately attaching and releasing the active sites in succession in a kind of ratchet action (Fig. 7-10).

In 1950, the English physiologists A. F. Huxley and H. E. Huxley independently proposed a **sliding filament model** to explain striated muscle contraction. According to this model, the thick and thin filaments

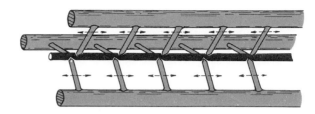

Fig. 7-10. Ratchetlike action of cross bridges between thick and thin filaments of skeletal muscle fibers. Cross bridges swing from site to site, pulling thin filaments past the thick. Each thick filament is actually surrounded by six thin filaments and is linked by six sets of cross bridges. For simplicity, this diagram shows only one set of cross bridges on each thick filament.

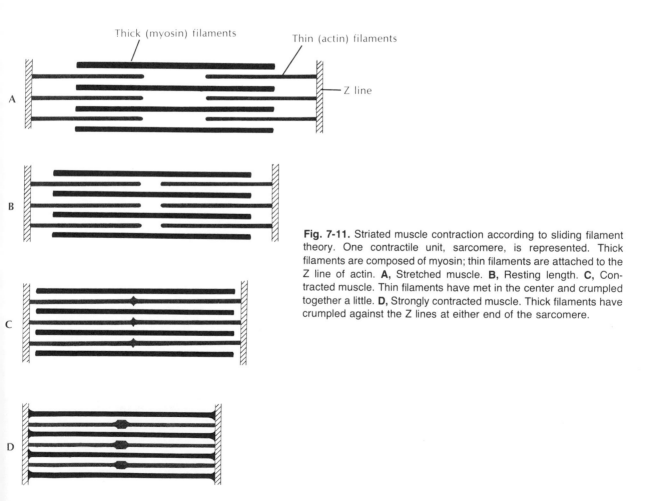

Fig. 7-11. Striated muscle contraction according to sliding filament theory. One contractile unit, sarcomere, is represented. Thick filaments are composed of myosin; thin filaments are attached to the Z line of actin. **A,** Stretched muscle. **B,** Resting length. **C,** Contracted muscle. Thin filaments have met in the center and crumpled together a little. **D,** Strongly contracted muscle. Thick filaments have crumpled against the Z lines at either end of the sarcomere.

slide past one another. Both kinds of filament maintain their original length, but the thin actin filaments extend farther into the A band as shown in Fig. 7-11. As contraction continues, the Z lines are drawn closer together. During very strong contraction the thin filaments touch and crumple in the center of the A band. Striated muscle contracts so rapidly that each cross bridge may attach and release 50 to 100 times per second.

The contractile machinery has been most thoroughly studied in mammals, but recent comparative studies indicate a remarkable uniformity of the sliding-filament mechanism throughout the animal kingdom. Even the contractile proteins myosin and actin are biochemically similar in all animals. The actomyosin contractile system evidently appeared very early in animal evolution and proved so flawless that no significant changes occurred thereafter.

Energy for contraction. Muscles perform work when they contract and, of course, require energy to do so. Resting muscles use little energy but consume large amounts during vigorous exercise. Muscles use only 20% of the energy value of food molecules when contracting; the remainder is released as heat. This is a rapid source of body heat; exercising is the quickest way to warm up.

The immediate source of energy for muscular contraction is ATP. When muscle is stimulated to contract, the energy released by ATP powers the ratchet-like mechanism between actin and myosin, causing the filaments to telescope.

Although the ATP stored in muscle supplies the immediate energy for contraction, the supply is limited and quickly exhausted. However, muscle contains a much larger energy storage form, **creatine phosphate,** which can rapidly transfer energy for the resynthesis of ATP. Eventually even this reserve is used up and must be restored by the breakdown of carbohydrate. Carbohydrate is available from two sources—from **glycogen** stored in the muscle and from **glucose** entering the muscle from the bloodstream. If muscular contraction is not too vigorous or too prolonged, glucose can be completely oxidized to carbon dioxide and water by **aerobic glycolysis.** But, during prolonged or heavy exercise, the blood flow to the muscles, although greatly increased above the resting level, is not sufficient to supply oxygen as rapidly as required for the complete oxidation of glucose. When this happens,

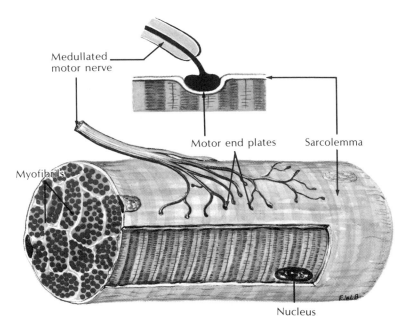

Fig. 7-12. Motor end plates (myoneural junctions) of a motor nerve on a single muscle fiber. (From Schottelius, B. A., and D. D. Schottelius. 1973. Textbook of physiology, ed. 17. St. Louis, The C. V. Mosby Co.)

the contractile machinery receives its energy largely by **anaerobic glycolysis,** a process that does not require oxygen (see p. 206). The presence of this anaerobic pathway, although not nearly as efficient as the aerobic one, is of great importance; without it, all forms of heavy muscular exertion such as running would be impossible.

During anaerobic glycolysis, glucose is degraded to lactic acid with the release of energy. This is used to resynthesize creatine phosphate, which in turn, passes the energy to ADP for the resynthesis of ATP. Lactic acid accumulates in the muscle and diffuses rapidly into the general circulation. If the muscular exertion continues, the buildup of lactic acid causes enzyme inhibition and fatigue. Thus the anaerobic pathway is a self-limiting one since continued heavy exertion leads to exhaustion. The muscles incur an **oxygen debt** because the accumulated lactic acid must be oxidized by extra oxygen. After the period of exertion, oxygen consumption remains elevated until all of the lactic acid has been oxidized or resynthesized to glucose.

In summary, the sequence of chemical sources of energy can be expressed in abridged form as follows:

$$ATP \rightleftharpoons ADP + H_3PO_4 + \text{Energy for contraction}$$

$$\text{Creatine phosphate} \rightleftharpoons \text{Creatine} + H_3PO_4 + $$
$$\text{Energy for resynthesis of ATP (anaerobic)}$$

$$\text{Glucose} \xrightleftharpoons{\text{(anaerobic)}} \text{Lactic acid} + $$
$$\text{Energy for resynthesis of creatine phosphate}$$

$$\text{Glucose} + O_2 \xrightarrow{\text{(aerobic)}} CO_2 + H_2O + $$
$$\text{Energy for resynthesis of creatine phosphate}$$

Stimulation of contraction. To contract, skeletal muscle must, of course, be stimulated. If the nerve supply to a muscle is severed, the muscle **atrophies,** or wastes away. Skeletal muscle fibers are arranged in groups of approximately 100, each group under the control of a single motor nerve fiber. Such a group is called a **motor unit.** As the nerve fiber approaches the muscle fibers, it splays out into many terminal branches. Each branch attaches to a muscle fiber by a special structure, called a **synapse,** or **myoneural junction** (Fig. 7-12). At the synapse is a tiny gap, or cleft, that thinly separates nerve fiber and muscle fiber. In the synapse is stored a chemical, **acetylcholine,** which is released when a nerve impulse reaches the synapse. This substance is a chemical mediator that diffuses across the narrow junction and acts on the muscle fiber membrane to generate an electrical de-polarization. The potential spreads rapidly through the muscle fiber, causing it to contract. Thus the synapse is a special chemical bridge that couples together the electrical potentials of nerve and muscle fibers.

Coupling of excitation and contraction. For a long time physiologists were puzzled as to how the electrical potential at the myoneural junction could spread quickly enough through the fiber to cause simultaneous contraction of all the densely packed filaments within. Recently it was discovered that vertebrate skeletal muscle contains an elaborate communication system that performs just this function. This is the endoplasmic reticulum (called the **sarcoplasmic reticulum** in muscle), a system of fluid-filled channels running parallel to the myofilaments and communicating with the sarcolemma that surrounds the fiber. The system is ideally arranged for speeding the electrical depolarization from the myoneural junction to the myofilament within. It also serves as a distribution network for glucose, oxygen, minerals, and other supplies needed for muscle contraction.

SELECTED REFERENCE

(See also general references for Part two, p. 274.)

Bendall, J. R. 1969. Muscles, molecules and movement. American Elsevier Publishing Co., Inc. *Contains a wealth of well-organized information on muscle structure and physiology, pitched at the advanced undergraduate level.*

SELECTED SCIENTIFIC AMERICAN ARTICLES

Cohen, C. 1975. The protein switch of muscle contraction. **233:**36-45 (Nov.). *The roles of calcium and regulatory proteins in muscle contraction are described.*

Hoyle, G. 1958. The leap of the grasshopper. **198:**30-35 (Jan.). *The powerful muscle system of the grasshopper's hindleg can propel the animal 20 times its body length.*

Hoyle, G. 1970. How is muscle turned on and off? **222:**84-93 (April). *Calcium plays a crucial role in muscle contraction.*

Huxley, H. E. 1965. The mechanism of muscular contraction. **213:**18-27 (Dec.). *The sliding filament theory is described.*

Lester, H. A. 1977. The response to acetylcholine. **236:**106-118 (Feb.). *Action of acetylcholine on cell receptors at myoneural junction is described.*

McLean, F. C. 1955. Bone. **192:**84-91 (Feb.). *Structure and physiology of bone.*

Merton, P. A. 1972. How we control the contraction of our muscles. **226:**30-37 (May). *Voluntary movements of skeletal muscle are controlled by a sensitive feedback mechanism.*

Murray, J. M., and A. Weber. 1974. The cooperative action of muscle proteins. **230:**58-71 (Feb.). *Describes the interactions of muscle proteins in contraction.*

Satir, P. 1974. How cilia move. **231:**44-52 (Oct.).

Male spring peeper *(Hyla crucifer)* sings to attract a mate. One of the great evolutionary contributions of the early amphibians was the transition from gill to lung breathing. This required a buccopharyngeal pump to move air into the lungs and rerouting the circulation to introduce a new blood supply to the lungs. With the addition of vocal cords and a vocal sac–resonating chamber, the first vertebrate sounds became a part of springtime evenings. (Photo by C. P. Hickman, Jr.)

8 INTERNAL FLUIDS
Circulation and respiration

INTERNAL FLUID ENVIRONMENT

    Composition of the body fluids

CIRCULATION

    Plan of the circulatory system

RESPIRATION

    Problems of aquatic and aerial breathing
    Respiration in man

Single-celled organisms live a contact existence with their environment. Nutrients and oxygen are obtained and wastes are released directly across the cell surface. These animals are so small that no special internal transport system, beyond the normal streaming movements of the cytoplasm, is required. Even some primitive multicellular forms, such as sponges, coelenterates, and flatworms, have such a simple internal organization and low rate of metabolism that no circulatory system is needed. Most of the more advanced multicellular organisms, because of their size, activity, and complexity, require a specialized circulatory, or vascular, system to transport nutrients and respiratory gases to and from all tissues of the body. In addition to serving these primary transport needs, circulatory systems have acquired additional functions; hormones are moved about, finding their way to target organs where they assist the nervous system to integrate body function. Water, electrolytes, and the many other constituents of the body fluids are distributed and exchanged between different organs and tissues. An effective response to disease and injury is vastly accelerated by an efficient circulatory system. The warm-blooded birds and mammals depend heavily on the blood circulation to conserve or dissipate heat as required for the maintenance of constant body temperature.

INTERNAL FLUID ENVIRONMENT

The body fluid of a single-celled animal is the cellular cytoplasm, a fluid substance in which the various membrane systems and organelles of the cell are suspended. In multicellular animals the body fluids are divided into two main phases, the **intracellular** and the **extracellular.** The intracellular phase (also called intracellular fluid) is the collective fluid inside all the body's cells. The extracellular phase (or fluid) is the fluid outside and surrounding the cells (Fig. 8-1, *A*).

More than a century ago the great French physiologist Claude Bernard recognized that the body's cells were actually surrounded by two environments. The inner environment, or the extracellular fluid that bathed the cells, he called the **milieu intérieur.** The outer environment, or outside world, he termed the **milieu extérieur.** Thus the cells, the sites of the body's crucial metabolic activities, are bathed by their own aqueous environment, the milieu intérieur, which buffers them from the often harsh physical and chemical changes occurring outside the body. Even today

English-speaking biologists frequently use the French phrase in referring to the extracellular fluid.

In animals having closed circulatory systems (vertebrates, annelids, and a few other invertebrate groups) the extracellular fluid is further subdivided into blood **plasma** and **interstitial** fluid (Fig. 8-1, *A*). The blood plasma is contained within the blood vessels, whereas the interstitial fluid, or tissue fluid as it is sometimes called, occupies the space immediately around the cells. Nutrients and gases passing between the vascular plasma and the cells must traverse this narrow fluid separation. The interstitial fluid is constantly formed from the plasma by filtration through the capillary walls.

Composition of the body fluids

All these fluid spaces—plasma, interstitial, and intracellular—differ from each other in solute composition, but all have one feature in common—they are mostly water. Despite their firm appearance, animals are 70% to 90% water. Man, for example, is approximately 70% water by weight. Of this, 50% is cell water, 15% is interstitial fluid water, and the remaining 5% is in the blood plasma. As Fig. 8-1, *A,* shows, it is the plasma space that serves as the pathway of exchange between the cells of the body and the outside world. This exchange of respiratory gases, nutrients, and wastes is accomplished by specialized organs (kidney, lungs, gill, alimentary canal), as well as by the integument.

The body fluids contain many inorganic and organic substances in solution. Principal among these are the inorganic electrolytes and proteins. Fig. 8-1, *B,* shows that **sodium, chloride,** and **bicarbonate** are the chief extracellular electrolytes, whereas **potassium, magnesium, phosphate,** and **proteins** are the major intracellular electolytes. These differences are dramatic; they are always maintained despite the continuous flow of materials into and out of the cells of the body. The two subdivisions of the extracellular fluid—plasma and interstitial fluid—have similar compositions except that the plasma has more proteins that are too large to filter through the capillary wall into the interstitial fluid.

Composition of blood

Among the lower invertebrates that lack a circulatory system (such as flatworms and coelenterates) it is not possible to distinguish a true "blood." These forms possess a clear, watery tissue fluid containing some primitive phagocytic cells, a little protein, and a mixture of salts similar to sea water. All invertebrates with closed circulatory systems maintain a clear separation between blood contained within blood vessels and tissue (interstitial) fluid surrounding the vessels.

In vertebrates, blood is a complex liquid tissue composed of plasma and formed elements, mostly corpuscles, suspended in the plasma. When the red blood corpuscles and other formed elements are spun down in a centrifuge, the blood is found to be approximately 55% plasma and 45% formed elements.

The composition of mammalian blood is as follows:

Plasma
1. Water 90%
2. Dissolved solids, consisting of the plasma proteins (albumin, globulins, fibrinogen), glucose, amino acids, electrolytes, various enzymes, antibodies, hormones, metabolic wastes, and traces of many other organic and inorganic materials
3. Dissolved gases, especially oxygen, carbon dioxide, and nitrogen

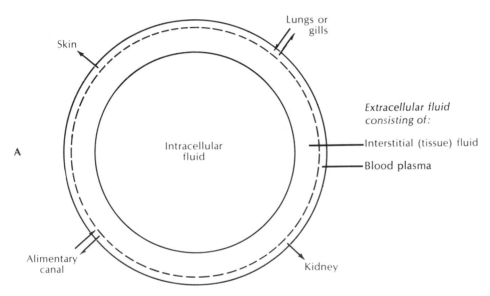

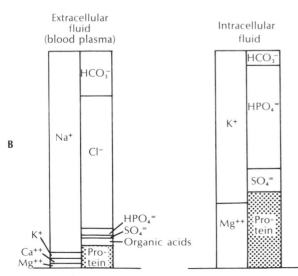

Fig. 8-1. Fluid compartments of body. **A,** All body cells can be represented as belonging to a single large fluid compartment that is completely surrounded and protected by extracellular fluid *(milieu intérieur).* This fluid is further subdivided into plasma and interstitial fluid. All exchanges with the environment occur across the plasma compartment. **B,** Electrolyte composition of extracellular and intracellular fluids. Total equivalent concentration of each major constituent is shown. Equal amounts of anions (negatively charged ions) and cations (positively charged ions) are in each fluid compartment. Note that sodium and chloride, major plasma electrolytes, are virtually absent from intracellular fluid (actually they are present in low concentration). Note the much higher concentration of protein inside the cells.

Formed elements (Fig. 8-2)

1. Red blood corpuscles (erythrocytes), containing hemoglobin for the transport of oxygen and carbon dioxide
2. White blood corpuscles (leukocytes), serving as scavengers and as immunizing agents
3. Platelets (thrombocytes), functioning in blood coagulation

The plasma proteins are a diverse group of large and small proteins that perform numerous functions. The major protein groups are (1) **albumin,** the most abundant plasma protein, which constitutes 60% of the total; (2) the **globulins** (α_1, α_2, β, and γ), a diverse group of high molecular weight proteins (35% of total) that includes immunoglobulins and various metal-binding proteins; and (3) **fibrinogen,** a very large protein that functions in blood coagulation.

Red blood cells, or **erythrocytes,** are present in enormous numbers in the blood, approximately 5.4 billion per milliliter of blood in an adult man and 4.8 billion in women. They are formed continuously from large nucleated **erythroblasts** in the red bone marrow, where hemoglobin is synthesized and the cells divide several times. In mammals the nucleus shrinks during development to a small remnant and eventually disappears altogether. Almost all other characteristics of a typical cell also are lost: ribosomes, mitochondria, and most enzyme systems. What is left is a biconcave disc consisting of a baglike membrane, the **stroma,** packed with 280 million molecules of the blood-transporting pigment **hemoglobin.** Approximately 33% of the erythrocyte by weight is hemoglobin. The biconcave shape (Fig. 8-3) is a mammalian innovation that provides a much larger surface for gas diffusion than would a flat or spheric shape. All other vertebrates have nucleated erythrocytes that are usually ellipsoidal rather than round discs.

The erythrocyte enters the circulation for an average

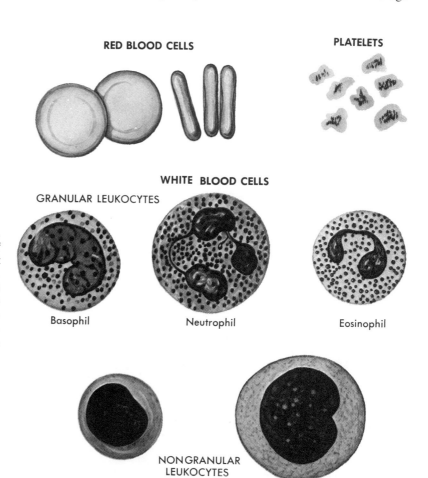

RED BLOOD CELLS

PLATELETS

WHITE BLOOD CELLS

GRANULAR LEUKOCYTES

Basophil

Neutrophil

Eosinophil

NONGRANULAR LEUKOCYTES

Lymphocyte

Monocyte

Fig. 8-2. Formed elements of human blood. Hemoglobin-containing red blood cells of man and other mammals lack nuclei, but those of all lower vertebrates have nuclei. Various leukocytes provide a wandering system of protection for the body. Platelets participate in the body's clotting mechanism. (From Anthony, C. P., and N. J. Kolthoff. 1975. Textbook of anatomy and physiology, ed. 9. St. Louis, The C. V. Mosby Co.)

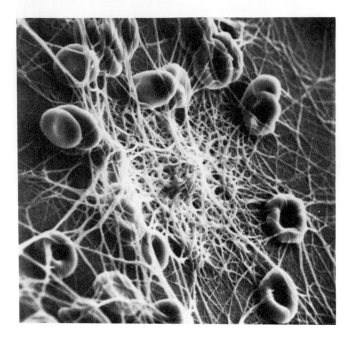

Fig. 8-3. Human red blood cells entrapped in fibrin clot. Clotting is initiated after tissue damage by the disintegration of platelets in the blood, resulting in a complex series of intravascular reactions that end with the conversion of a plasma protein, fibrinogen, into long, tough, insoluble polymers of fibrin. Fibrin and entangled erythrocytes form the blood clot, which arrests bleeding. An aggregation of platelets probably underlies the raised mass of fibrin in center. (Scanning electron micrograph, ×5180; courtesy N. F. Rodman, University of Iowa, Iowa City, Iowa.)

life-span of approximately 4 months. During this time it may journey 700 miles, squeezing repeatedly through the capillaries, which are sometimes so narrow that the erythrocyte must bend to get through. At last it fragments and is quickly engulfed by large scavenger cells called **macrophages** located in the liver, bone marrow, and spleen. The iron from the hemoglobin is salvaged to be used again; the rest of the heme is converted to **bilirubin,** a bile pigment. It is estimated that 10 million erythrocytes are born and another 10 million destroyed every second.

The white blood cells, or **leukocytes,** form a wandering system of protection for the body. In adults they number only approximately 7.5 million per milliliter of blood, a ratio of 1 white cell to 700 red cells. There are several kinds of white blood cells: **granulocytes** (subdivided into neutrophils, basophils, and eosinophils), **lymphocytes,** and **monocytes** (Fig. 8-2).

Defense mechanisms of the body

Of the several kinds of white blood cells, the neutrophils especially have the capacity to pass through the walls of capillaries and wander by ameboid movement through tissues. Neutrophils and, later, monocytes are attracted in huge numbers to any local inflammation caused by the invasion of microorganisms into tissues following an injury. Neutrophils immediately set about engulfing bacteria, viruses, and other foreign particulate matter, a process called **phagocytosis.** Once inside, the bacterium is quickly killed and digested with a variety of powerful hydrolytic enzymes. The battle is not all one sided. If the infection is large or the bacteria are especially virulent, the neutrophils may themselves be killed in large numbers by bacterial toxins and accumulate as pus in the infection site. However, the vast majority of microorganism invasions that are always occurring and of which we are seldom aware are quickly and efficiently disposed of by this **nonspecific defense system.**

Lymphocytes (Fig. 8-2) are responsible for the **specific defense system** of the body. Specific immunity is produced in the body following an exposure to a specific foreign substance. All lymphocytes are formed from precursors in the bone marrow, but they later leave; some take up residence in lymph nodes, whereas others form a wandering system of cells. Lymphocytes elaborate a special group of γ-globulin proteins, called **antibodies,** that are capable of reacting specifically with foreign substances called **antigens.** Almost any foreign cell or molecule having a molecular weight of at least 10,000 can act as an antigen; bacteria, viruses, fungi, parasites, and protein toxins produced by any of these may be antigenic.

Antigens therefore are substances that induce the synthesis of antibodies. Antibodies are highly specific and combine only with the antigen that has stimulated their production. Antibodies appear to behave as bridges between antigens, linking them together into large clumps or chains, which then can be destroyed by neutrophils and other phagocytic cells.

When a microbial antigen first invades the body, it stimulates the development of a specific population of lymphocytes that begin producing antibodies. Once the infection is eliminated, the antibodies may gradually disappear from the blood. However, the lymphocytes retain a ''memory'' of the antigen, and, should a subsequent invasion occur, specific antibodies are immediately poured out in large amounts. This kind of enhanced resistance to an infection, produced as the result of a prior contact with the antigen, is known as **active immunity.** Until recently, suffering a disease or infection was the only way to obtain an active immunity; now it may be achieved safely from the injection of a vaccine or harmless microbial derivative.

Hemostasis: prevention of blood loss

It is essential that animals have ways of preventing the rapid loss of body fluids after an injury. Since blood is flowing and is under considerable hydrostatic pressure, it is especially vulnerable to hemorrhagic loss.

When a vessel is damaged, smooth muscle in the wall contracts, which causes the vessel lumen to narrow, sometimes so strongly that blood flow is completely stopped. This is a primitive but highly effective means of preventing hemorrhage used by invertebrates and vertebrates alike. Beyond this first defense against blood loss, all vertebrates, as well as some of the larger, active invertebrates with high blood pressures, have special cellular elements and proteins in the blood that are capable of forming plugs, or clots, at the injury site.

In higher vertebrates **blood coagulation** is the dominant hemostatic defense. Blood clots form as a tangled network of fibers from one of the plasma proteins, **fibrinogen** (Fig. 8-3). The transformation of fibrinogen into a **fibrin** meshwork that entangles blood cells to form a gel-like clot is catalyzed by the enzyme

thrombin. Thrombin is normally absent from the blood and only appears when vessels are damaged.

In this process, the blood **platelets** (Fig. 8-2) play a dominant role. Platelets are minute, colorless, incomplete cells lacking nuclei that are present in large numbers in the blood. When the surface of a vessel is damaged, platelets rapidly adhere to the injured surface and release a phospholipid substance. This material, in the presence of calcium, initiates a catalytic sequence resulting in the conversion of the plasma protein prothrombin to thrombin. We may summarize the stages in the formation of fibrin as follows:

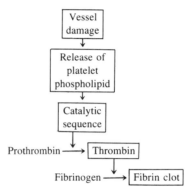

The catalytic sequence in this scheme is unexpectedly complex, involving a series of plasma protein factors, each normally inactive until activated by a previous factor in the sequence. At least 13 different plasma coagulation factors have been recognized. A deficiency of a single factor can delay or prevent the clotting process. Why has such a complex clotting mechanism evolved? Probably it is necessary to provide a fail-safe system capable of responding to any kind of internal or external hemorrhage that might occur and yet a system that cannot be activated into forming dangerous intravascular clots unless injury has occurred.

Several kinds of clotting abnormalities in man are known. Of these, **hemophilia** is perhaps best known. Hemophilia is a condition characterized by the failure of the blood to clot, so that even insignificant wounds can cause continuous severe bleeding. Called the ''disease of kings,'' it once ran through the royal families of Europe, notably those of Queen Victoria of England and Alfonso XIII, the last king of Spain before the Franco dictatorship. Hemophilia is caused by an inherited lack of antihemophilic factor. The disorder is transmitted through females, but almost invariably appears only in males.

Blood groups

In blood transfusions the donor's blood is checked against the blood of the recipient. Blood differs chemically from person to person, and when two different (incompatible) blood types are mixed, **agglutination** (clumping together) results. The basis of these chemical differences is naturally occurring antigens on the membranes of red blood cells. The best known of these inherited immune systems is the ABO blood group. The antigens A and B are inherited as dominant genes. Thus as shown in Table 8-1, an individual with, for example, genes AA or AO develops A antigen (blood type A). The presence of a B gene produces B antigens (blood type B), and for the genotype AB both A and B antigens develop on the erythrocytes (blood type AB).

There is an odd feature about the ABO system. Normally we would expect that a type A individual would develop antibodies against type B blood only if B cells were introduced into the body. In fact, type A persons always have anti-B antibodies in their blood, even without the prior exposure to type B blood. Similarly type B individuals carry anti-A antibodies. Type AB blood has neither anti-A nor anti-B antibodies (since if it did, it would destroy its own blood cells), and type O blood has both anti-A and anti-B antibodies.

We see then that the blood group names identify their *antigen* content. Persons with type O blood are called universal donors because, lacking antigens, their blood can be infused into a person with any blood type. Even though it contains anti-A and anti-B antibodies,

these are so diluted during transfusion that they do not react with A or B antigens in a recipient's blood. In practice, however, clinicians insist on matching blood types to prevent any possibility of incompatibility.

Rh factor. It is difficult to think of another area of physiology as totally linked with the name of a single man as blood grouping is with the name of Karl Landsteiner. This Austrian—later American—physician discovered the ABO blood groups in 1900. The great importance of his work became abundantly clear during World War I when blood transfusion was first attempted on a large scale. In 1927 Landsteiner in collaboration with Philip Levine in the United States discovered the MN blood group, also present in all mankind. This classification is not important in transfusions but may be crucial in determining relationship in paternity cases.

In 1940, 10 years after receiving the Nobel Prize in recognition of contributions that were more than adequate for the lifetime of any scientist, Landsteiner made still another famous discovery. This new group was called the Rh factor, named after the Rhesus monkey, in which it was first found. Approximately 85% of individuals have the factor (positive) and the other 15% do not (negative). He also found that Rh-positive and Rh-negative bloods are incompatible; shock and even death may follow their mixing when Rh-positive blood is introduced into an Rh-negative person who has been sensitized by an earlier transfusion of Rh-positive blood.

The Rh factor is inherited as a dominant gene; this accounts for a peculiar and often fatal form of anemia of newborn infants called **erythroblastosis fetalis.** Although the fetal and maternal bloods are separated by the placenta, this separation is not perfect. Some

Table 8-1. Major blood groups

Blood type	Genotype	Antigens on red corpuscles	Antibodies in serum	Can give blood to	Can receive blood from	Frequency in United States (%)		
						Whites	Blacks	Chinese
O	OO	None	Anti-A and anti-B	All	O	45	38	46
A	AA, AO	A	Anti-B	A, AB	O, A	41	27	28
B	BB, BO	B	Anti-A	B, AB	O, B	10	21	23
AB	AB	AB	None	AB	All	4	4	13

admixture of fetal and maternal blood usually occurs, especially right after birth when the placenta (''after-birth'') separates from the uterine wall. This admixture of blood, normally of no consequence, can be serious *if* the father is Rh positive, the mother Rh negative, and the fetus Rh positive (by inheriting the factor from the father). The fetal blood, containing the Rh antigen, can stimulate the formation of Rh-positive antibodies in the blood of the mother. The mother is permanently immunized against the Rh factor. During the second pregnancy these antibodies may diffuse back into the fetal circulation and produce agglutination and destruction of the fetal red blood cells. Because the mother is usually sensitized at the end of the first pregnancy, subsequent babies are more severely threatened than is the first.

Erythroblastosis fetalis can now be prevented by giving an Rh-negative mother anti-Rh antibodies just after the birth of her first child. These antibodies remain long enough to neutralize any Rh-positive fetal blood cells that may enter her circulation, thus preventing her own antibody machinery from being stimulated to produce the Rh-positive antibodies. Active, permanent immunity is blocked.

CIRCULATION

The circulatory system of vertebrates is made up of a system of tubes, the **blood vessels,** and a propulsive organ, the **heart.** This is a **closed circulation** because the circulating medium, the **blood,** is confined to vessels throughout its journey from the heart to the tissues and back again. Many invertebrates have an **open circulation;** the blood is pumped from the heart into blood vessels that open into tissue spaces. The blood circulates freely in direct contact with the cells and then reenters open blood vessels to be propelled forward again. In invertebrates having open circulatory systems, there is no clear separation of the extracellular fluid into plasma and interstitial fluids, as there is in closed systems. Closed systems are more suitable for large and active animals because the blood can be moved rapidly to the tissues needing it. In addition, flow to various organs can be readjusted to meet changing needs by varying the diameters of the blood vessel.

The closed circulatory system of vertebrates cooperates with the **lymphatic system.** This is a fluid ''pick-up'' system. It re-collects tissue fluid (lymph) that has been squeezed out through the walls of the capillaries and returns it to the blood circulation. In a sense ''closed'' circulatory systems are not absolutely closed because fluid is constantly leaking out into the tissue spaces. However, this leakage is but a small fraction of the total blood flow.

Although it seems obvious to us today that blood flows in a circuit, the first correct description of blood flow by the English physician William Harvey initially received vigorous opposition when published in 1628. Centuries before, Galen had taught that air enters the heart from the windpipe and that blood was able to pass from one ventricle to the other through ''pores'' in the interventricular septum. He also believed that blood first flowed out of the heart into all vessels, arteries, and veins alike and then returned to the heart by these same vessels—an idea of ebb and flow of the blood.

Even though there was almost nothing right about this theory, it was still doggedly trusted at the time of Harvey's publication. Harvey's conclusions were based on sound experimental evidence. He made use of a variety of animals for his experiments, including a little snake found in English meadows. By tying ligatures on arteries, he noticed that the region between the heart and ligature swelled up. When veins were tied off, the swelling occurred beyond the ligature. When blood vessels were cut, blood flowed in arteries from the cut end nearest the heart; the reverse happened in veins. By means of such experiments, Harvey worked out a correct scheme of blood circulation, even though he could not see the capillaries that connected the arterial and venous flows.

Plan of the circulatory system

All vertebrate vascular systems have certain features in common. A **heart** pumps the blood into **arteries** that branch and narrow into **arterioles** and then into a vast system of **capillaries.** Blood leaving the capillaries enters **venules** and then **veins** that return the blood to the heart. Fig. 8-4 compares the circulatory systems of gill-breathing (fish) and lung-breathing (mammal) vertebrates. The principal differences in circulation involve the heart in the transformation from gill to lung breathing.

The fish heart contains two main chambers, the **atrium** (or **auricle**) and the **ventricle.** Although there are also two subsidiary chambers, the **sinus venosus** and **conus arteriosus** (not shown in Fig. 8-4), we still refer to the fish heart as a ''two-chambered'' heart.

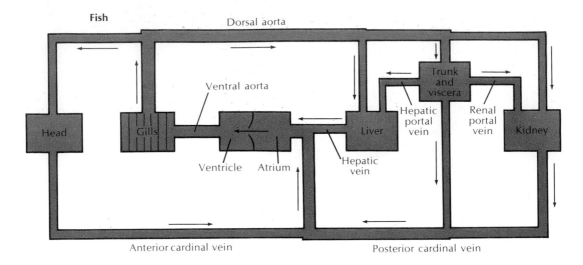

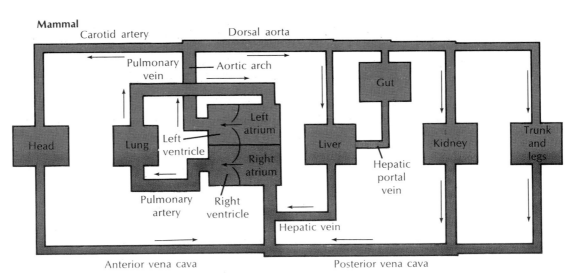

Fig. 8-4. Plan of circulatory system of fish *(above)* and mammal *(below). Red,* Oxygenated blood. *Dark red,* Deoxygenated blood.

Blood makes a single circuit through the fish's vascular system; it is pumped from the heart to the gills, where it is oxygenated, and then flows into the dorsal aorta to be distributed to the body organs. After passing through the capillaries of the body organs and musculature, it returns by veins to the heart. In this circuit the heart must provide sufficient pressure to push the blood through two sequential capillary systems, one in the gills and the other in the organ tissues. The principal disadvantage of the single-circuit system is that the gill capillaries offer so much resistance to blood flow that the pressure drops considerably before

entering the dorsal aorta. This system can never provide high and continuous blood pressure to the body organs.

The evolution of land forms with lungs and their need for highly efficient blood delivery resulted in the introduction of a **double** circulation. One **systemic** circuit with its own pump provides oxygenated blood to the capillary beds of the body organs; another **pulmonary** circuit with its own pump sends deoxygenated blood to the lungs. Rather than actually developing two separate hearts, the existing two-chambered heart was divided down the center into four cham-

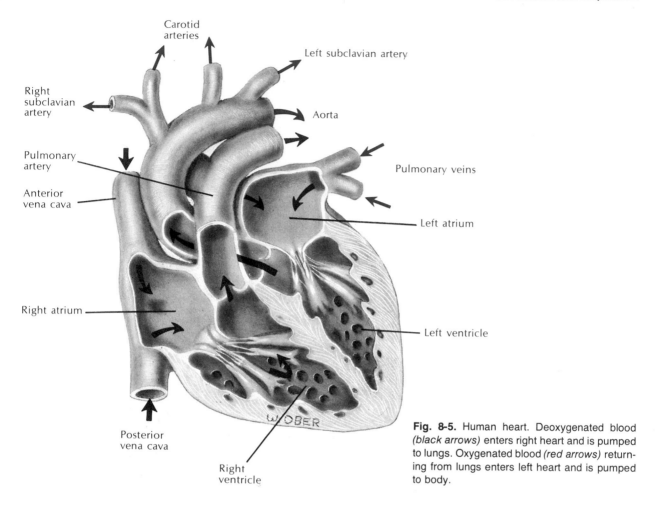

Fig. 8-5. Human heart. Deoxygenated blood *(black arrows)* enters right heart and is pumped to lungs. Oxygenated blood *(red arrows)* returning from lungs enters left heart and is pumped to body.

bers—really two two-chambered hearts lying side-by-side.

Needless to say such a great change in the vertebrate circulatory plan, involving not only the heart but the attendant plumbing as well, took many millions of years to evolve. The partial division of the atrium and ventricle began with the ancestors of present-day lungfishes. Amphibians accomplished the complete separation of the atrium, but the ventricle is still undivided in this group. In some reptiles the ventricle is completely divided, and the four-chambered heart appears for the first time. All birds and mammals have the four-chambered heart and the two separate circuits—one through the lungs (pulmonary) and the other through the body (systemic). The course of the blood through this double circuit is shown in Fig. 8-4.

Heart

The vertebrate heart (Fig. 8-5) is a muscular organ located in the thorax and covered by a tough, fibrous sac, the **pericardium.** As we have seen, the higher vertebrates have a four-chambered heart. Each half consists of a thin-walled atrium and a thick-walled ventricle. Heart (cardiac) muscle is a unique type of muscle found nowhere else in the body. It resembles striated muscle, but the cells are branched, and dense end-to-end attachments between the cells are called intercalated discs.

There are four sets of valves. **Atrioventricular valves** (A-V valves) separate the cavities of the atrium and ventricle in each half of the heart. These permit blood to flow from atrium to ventricle but prevent backflow. Where the great arteries, the **pulmonary**

153

from the right ventricle and the **aorta** from the left ventricle, leave the heart, **semilunar valves** prevent backflow.

The contraction of the heart is called **systole** (sis′to-lee), and the relaxation, **diastole** (dy-as′to-lee). The rate of the heartbeat depends on age, sex, and especially exercise. Exercise may increase the **cardiac output** (volume of blood forced from either ventricle each minute) more than fivefold. Both the heart **rate** and the **stroke volume** increase. Heart rates among vertebrates vary with the general level of metabolism and the body size. The cold-blooded codfish has a heart rate of approximately 30 beats per minute; a warm-blooded rabbit of about the same weight has a rate of 200 beats per minute. Small animals have higher heart rates than do large animals. The heart rate in an elephant is 25 beats per minute, in a man 70 per minute, in a cat 125 per minute, in a mouse 400 per minute, and in the tiny 4-gram shrew, the smallest mammal, the heart rate approaches a prodigious 800 beats per minute. We must marvel that the shrew's heart can sustain this frantic pace throughout this animal's life, brief as it is.

The heart rests only during the short interval between contractions. The mammalian heart does an amazing amount of work during a lifetime. Someone has calculated that the heart of a man approaching the end of his life has beat some 2.5 billion times and pumped 300,000 tons of blood!

Excitation of the heart. The heartbeat originates in a specialized muscle tissue, called the **sinoatrial node,** located in the right atrium near the entrance of the caval veins (Fig. 8-6). This tissue serves as the **pacemaker** of the heart. The contraction spreads across the two atria to the **atrioventricular (A-V) node.** At this point the electrical activity is conducted very rapidly to the apex of the ventricle through specialized fibers (bundle of His and Purkinje fiber system) and then spreads more slowly up the walls of the ventricles. This arrangement allows the contraction to begin at the apex or "tip" of the ventricles and spread upward to squeeze out the blood in the most efficient way; it also ensures that both ventricles contract simultaneously.

Although the vertebrate heart can beat spontaneously—and the excised fish or amphibian heart does beat for hours in a balanced salt solution—the heart rate is normally under nervous control. The control (cardiac) center is located in the medulla and sends out

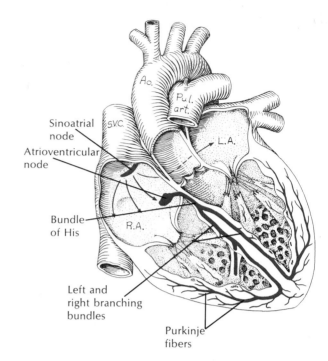

Fig. 8-6. Neuromuscular mechanisms controlling beat of the heart. Arrows indicate spread of excitation from the sinoatrial node (S-A node), across the atria, to the atrioventricular node (A-V node). Wave of excitation is then conducted very rapidly to ventricular muscle over the specialized bundle of His and Purkinje fiber system.

two sets of motor nerves. Impulses sent along one set, the **vagus** (parasympathetic) nerves, apply a brake action to the heart rate, and impulses sent along the other set, the **accelerator** (sympathetic) nerves, speed it up. Both sets of nerves terminate in the sinoatrial node, thus guiding the activity of the pacemaker.

The cardiac center in turn receives sensory information about a variety of stimuli. Pressure receptors (sensitive to blood pressure) and chemical receptors (sensitive to carbon dioxide and pH) are located at strategic points in the vascular system. This information is used by the cardiac center to increase or reduce the heart rate and cardiac output in response to activity or changes in body position. The heart is thus controlled by a series of feedback mechanisms that keep its activity constantly attuned to body needs.

Coronary circulation. It is no surprise that an organ as active as the heart needs a very good blood supply of its own. The heart muscle of the frog and other amphibians is so thoroughly channeled with spaces between the muscle fibers that sufficient oxygenated blood is squeezed through by the heart's own pumping action. In birds and mammals, however, the heart muscle is very thick and has such a high rate of metabolism that it must have its own vascular **(coronary)** circulation. The coronary arteries break up into an extensive capillary network surrounding the muscle fibers and provide them with oxygen and nutrients. Heart muscle has an extremely high oxygen demand, removing 80% of the oxygen from the blood, in contrast to most other body tissues, which remove only approximately 30%.

Arteries

All vessels leaving the heart are called arteries whether they carry oxygenated blood (aorta) or deoxygenated blood (pulmonary artery). To withstand high, pounding pressures, arteries are invested with layers of both elastic and tough, inelastic connective tissue fibers. The elasticity of the arteries allows them to yield to the surge of blood leaving the heart during systole and then to squeeze down on the fluid column during diastole. This smooths out the blood pressure. Thus the arterial pressure in man varies only between a high of 120 mm Hg (systole) and a low of 80 mm Hg (diastole), rather than dropping to zero during diastole as we might expect in a fluid system with an intermittent pump.

As the arteries branch and narrow into **arterioles,** the walls become mostly smooth muscle (Fig. 8-7). Contraction of this muscle narrows the arterioles and reduces the flow of blood. The arterioles thus control the blood flow to body organs, diverting it to where it is needed most. The blood must be given a hydrostatic pressure sufficient to overcome the resistance of the narrow passages through which the blood must flow. Consequently large animals tend to have higher blood pressure than do small animals.

Blood pressure was first measured in 1733 by Stephen Hales, an English clergyman with unusual inventiveness and curiosity. He tied his mare "to have been killed as unfit for service" on her back and exposed the femoral artery. This he cannulated with a brass tube, connecting it to a tall glass tube with the windpipe of a goose. The use of the windpipe was both imaginative and practical; it gave the apparatus flexibility "to avoid inconveniences that might arise if the mare struggled." The blood rose 8 feet in the glass tube and bobbed up and down with the systolic and diastolic beats of the heart. The weight of the 8-foot

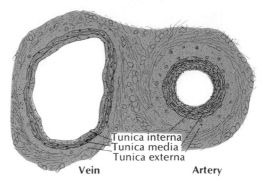

Fig. 8-7. Cross section of vein and corresponding artery.

column of blood was equal to the blood pressure. We now express this as the height of a column of mercury, which is 13.6 times heavier than water. Hales' figures, expressed in millimeters of mercury, indicate that he measured a blood pressure of 180 to 200 mm Hg, about normal for a horse. Today, blood pressure can be measured with great accuracy with a sensitive pressure transducer; the electronic signal from this instrument is displayed on a graphic recorder.

Capillaries

The Italian Marcello Malpighi was the first to describe the capillaries in 1661, thus confirming the existence of the minute links between the arterial and venous systems that Harvey knew must be there but could not see. Malpighi studied the capillaries of the living frog's lung, which incidentally is still one of the simplest and most vivid preparations for demonstrating capillary blood flow.

The capillaries are present in enormous numbers, forming extensive networks in nearly all tissues. In muscle there are more than 2,000 per square millimeter (1,250,000 per square inch), but not all are open at once. Indeed, perhaps less than 1% are open in resting skeletal muscle. But when the muscle is active, all the capillaries may open to bring oxygen and nutrients to the working muscle fibers and to carry away metabolic wastes.

Capillaries are extremely narrow, averaging less than 10 μm in diameter in mammals, which is hardly any wider than the red blood cells that must pass through them. Their walls are formed of a single layer of thin **endothelial** cells, held together by a delicate basement membrane and connective tissue fibers. Capillaries have a built-in leakiness that allows water

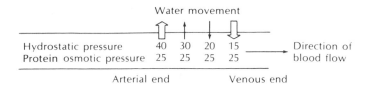

Fig. 8-8. Fluid movement across the wall of a capillary. At arterial end of the capillary, hydrostatic (blood) pressure exceeds protein osmotic pressure contributed by the plasma proteins, and a plasma filtrate (shown as "water movement") is forced out. At venous end, protein osmotic pressure exceeds the hydrostatic pressure, and fluid is drawn back in. In this way plasma nutrients are carried out into the interstitial space where they can enter cells, and metabolic end products from the cells are drawn back into the plasma and carried away.

and most dissolved substances in the blood plasma to filter through into the interstitial space. The capillary wall is **selectively permeable,** however, which means that it filters some dissolved materials and retains others.

In this case the plasma proteins, which are the largest dissolved molecules in the plasma, are held back. These proteins, especially the albumins, contribute an **osmotic pressure** of approximately 25 mm Hg in mammals (Fig. 8-8). Although small, this protein osmotic pressure is of great importance to fluid balance in the tissues. At the arteriole end of the capillaries the blood pressure is approximately 40 mm Hg (in man). This **filtration pressure** forces water and dissolved materials through the capillary endothelium into the tissue space, where they circulate freely around the cells. As the blood proceeds through the narrow capillary, the blood pressure decreases steadily to perhaps 15 mm Hg. At this point the hydrostatic pressure is less than the osmotic pressure of the plasma proteins, still approximately 25 mm Hg. Water is drawn back into the capillaries.

Thus it is the balance between hydrostatic pressure and protein osmotic pressure that determines the direction of capillary fluid shift. Normally water is forced out of the capillary at the arteriole end, where hydrostatic pressure exceeds osmotic pressure, and drawn back into the capillary at the venule end, where osmotic pressure exceeds hydrostatic pressure. Any fluid left behind is picked up and removed by the **lymph capillaries.**

Veins

The venules and veins into which the capillary blood drains for its return journey to the heart are thinner walled, less elastic, and of considerably larger diameter than their corresponding arteries and arterioles (Fig.

8-7). Blood pressure in the venous system is low, from approximately 10 mm Hg, where capillaries drain into venules, to approximately zero in the right atrium. Because pressure is so low, the venous return gets assists from valves in the veins, from muscles surrounding the veins, and from the rhythmic pumping action of the lungs. If it were not for these mechanisms, the blood might pool in the lower extremities of a standing animal—a very real problem for people who must stand for long periods. The veins that lift blood from the extremities to the heart contain valves that serve to divide the long column of blood into segments. When the muscles around the veins contract, as in even slight activity, the blood column is squeezed upward and cannot slip back because of the valves. The well-known risk of fainting while standing at stiff attention in hot weather can usually be prevented by deliberately pumping the leg muscles. The negative pressure created in the thorax by the inspiratory movement of the lungs also speeds the venous return by sucking the blood up the large vena cava into the heart.

Lymphatic system

Gasparo Aselli, an Italian anatomist, first discovered the nature of lacteals in 1627. In a dog that had recently been fed and cut open, he noticed white cordlike bodies in the mesenteries of the intestine that he first mistook for nerves. When he pricked these cords with a scalpel, a milky fluid gushed out. It is now known that this fluid is largely fat that is carried after digestion to the thoracic duct. The thoracic duct and its relations to the lacteals were discovered by the Frenchman Jean Pecquet in 1647. These vessels are part of the complete lymphatic system demonstrated almost simultaneously but independently by O. Rudbeck in Sweden (1651) and T. Bartholin in

Denmark (1653), using dogs and executed criminals.

The lymphatic system is an accessory drainage system for the body. As we have seen, the blood pressure in the arteriole end of the capillaries forces a plasma filtrate through the capillary walls and into the interstitial space. This tissue fluid bathing the cells is **lymph,** a clear, nearly colorless liquid. Lymph and plasma are nearly identical except that lymph contains very little protein, which was screened out as the plasma was squeezed through the capillary walls. Most of the lymph returns to the vascular system at the venous end of the capillaries by the capillary fluid-shift mechanism described earlier. Usually, however, outflow from the capillaries slightly exceeds backflow. This difference is gathered up and returned to the circulatory system by lymphatic vessels. The system begins with tiny, highly permeable lymph capillaries. These lead into larger lymph vessels, which in turn drain into the large **thoracic duct.** This enters the left subclavian vein in the neck region. The rate of lymph flow is very low, a minute fraction of the blood flow.

Located at strategic intervals along the lymph vessels are **lymph nodes** that have several defense-related functions. They are effective filters that remove foreign particles, especially bacteria, that might otherwise enter the general circulation. They are also germinal centers for **lymphocytes** that produce γ-globulin **antibodies**—essential components of the body's defense mechanisms (p. 148).

RESPIRATION

The energy bound up in food must be released by oxidative processes. As oxygen is used by the body cells, carbon dioxide is produced; this process is called **respiration.** Most animals are **aerobic,** meaning that they require and receive the necessary oxygen directly from their environment. A few animals, called **anaerobic,** are able to live in the absence of oxygen. Forms such as worms and arthropods dwelling in the oxygen-depleted mud of lakes and parasites living anaerobically in the intestine derive the necessary oxygen from the metabolism of carbohydrates and fats. However, anaerobic metabolism often occurs in the muscles of basically aerobic animals during vigorous muscle contraction.

Small aquatic animals such as the one-celled protozoans obtain what oxygen they need by direct diffusion from the environment. Carbon dioxide, the gaseous waste of metabolism, is also lost by diffusion to the environment. Such a simple solution to the problem of gas exchange is really only possible for very small animals (less than 1 mm in diameter) or those having very low rates of metabolism.

As animals became larger and evolved a waterproof covering, specialized devices such as lungs and gills developed that greatly increased the effective surface for gas exchange. But, because gases diffuse so slowly through protoplasm, a circulatory system was necessary to distribute the gases to and from the deep tissues of the body. Even these adaptations were inadequate for advanced animals with their high rates of cellular respiration. The solubility of oxygen in the blood plasma is so low that plasma alone could not carry enough to satisfy metabolic demands. With the evolution of special oxygen-transporting blood proteins such as hemoglobin, the oxygen-carrying capacity of the blood was greatly increased. Thus what began as a simple and easily satisfied requirement resulted in the evolution of several complex and essential respiratory and circulatory adaptations.

Problems of aquatic and aerial breathing

How an animal respires is largely determined by the nature of its environment. The two great arenas of animal evolution—water and land—are vastly different in their physical characteristics. The most obvious difference is that air contains far more oxygen—at least 20 times more—than does water. Atmospheric air contains oxygen (approximately 21%), nitrogen (approximately 79%), carbon dioxide (0.03%), a variable amount of water vapor, and very small amounts of inert gases (helium, argon, neon, etc.). These gases are variably soluble in water. The amount of oxygen dissolved depends on the concentration of oxygen in the air and on the water temperature. Water at 5° C fully saturated with air contains approximately 9 ml of oxygen per liter. (Note that by comparison, air contains approximately 210 ml of oxygen per liter.) The solubility of oxygen in water decreases as the temperature rises. For example, water at 15° C contains approximately 7 ml of oxygen per liter, and at 35° C, only 5 ml of oxygen per liter. The relatively low concentration of oxygen dissolved in water is the greatest respiratory problem facing aquatic animals. Unfortunately it is not the only one. Oxygen diffuses much more slowly in water than in air, and water is much denser and more viscous than air. All of this means that successful aquatic animals must have evolved very efficient ways of removing oxygen from water. Yet even the most advanced fishes with highly efficient gills and pumping mechanisms may use as much as 20% of their energy just extracting oxygen from water. By comparison, a mammal uses only 1% to 2% of its resting metabolism to breathe.

It is essential that respiratory surfaces be thin and always kept wet to allow diffusion of gases between the environment and the underlying circulation. This is hardly a problem for aquatic animals, immersed as they are in water, but it is a very real problem for air breathers. To keep the respiratory membranes moist and protected from injury, air breathers have in general developed invaginations of the body surface and then added pumping mechanisms to move air in and out. The lung is the best example of a successful solution to breathing on land. In general, **evaginations** of the body surface, such as gills, are most suitable for aquatic respiration; **invaginations,** such as lungs, are best for air breathing. We can now consider the specific kinds of respiratory organs employed by animals.

Cutaneous respiration

Protozoa, sponges, coelenterates, and many worms respire by direct diffusion of gases between the organism and the environment. We have noted that this kind of **integumentary respiration** is not adequate when the mass of living protoplasm exceeds approximately 1 mm in diameter. But, by greatly increasing the surface of the body relative to the mass, many multicellular animals respire in this way. Integumentary respiration frequently supplements gill or lung breathing in larger animals such as amphibians and fishes. For example, an eel can exchange 60% of its oxygen and carbon dioxide through its highly vascular skin. During their winter hibernation, frogs exchange all their respiratory gases through the skin while submerged in ponds or springs.

Gills

Gills are unquestionably the most effective respiratory device for life in water. Gills may be simple **external** extensions of the body surface, such as the **dermal branchiae** of starfish or the **branchial tufts** of marine worms and aquatic amphibians. Most efficient are the **internal** gills of fishes (described on p. 513) and arthropods. Fish gills are thin filamentous struc-

tures, richly supplied with blood vessels arranged so that blood flow is opposite to the flow of water across the gills. This arrangement, called **countercurrent flow,** provides for the greatest possible extraction of oxygen from water. Water flows over the gills in a steady stream, pushed and pulled by an efficient branchial pump, and often assisted by the fish's forward movement through the water (see Fig. 25-13, p. 513).

Lungs

Gills are unsuitable for life in air because, when removed from the buoying water medium, the gill filaments collapse and stick together; a fish out of water rapidly asphyxiates despite the abundance of oxygen around it. Consequently air-breathing vertebrates possess lungs, highly vascularized internal cavities. Lungs of a sort are found in certain invertebrates (pulmonate snails, scorpions, some spiders, some small crustaceans), but these structures cannot be ventilated and consequently are not very efficient.

Lungs that can be ventilated efficiently are characteristic of the terrestrial vertebrates. The most primitive vertebrate lungs are those of lungfishes (Dipneusti), which use them to supplement, or even replace, gill respiration during periods of drought. Although of simple construction, the lungfish lung is supplied with a capillary network in its largely unfurrowed walls, a tubelike connection to the pharynx, and a primitive ventilating system for moving air in and out of the lung.

Amphibians also have simple baglike lungs, whereas in higher forms the inner surface area is vastly increased by numerous lobulations and folds (Fig. 8-9). This increase is greatest in the mammalian lung, which is complexly divided into many millions of small sacs **(alveoli),** each veiled by a rich vascular network. It has been estimated that man's lungs have a total surface area of from 50 to 90 m^2—50 times the area of the skin surface—and contain 1,000 miles of capillaries.

Moving air into and out of lungs has been an evolutionary design problem that was, of course, solved although we wonder whether an imaginative biologic engineer, if given the proper resources, couldn't come up with a better design. Unlike the efficient one-way flow of water across fish gills, air must enter and exit a lung at the same point. Furthermore, a tube of some length—the bronchi, trachea,

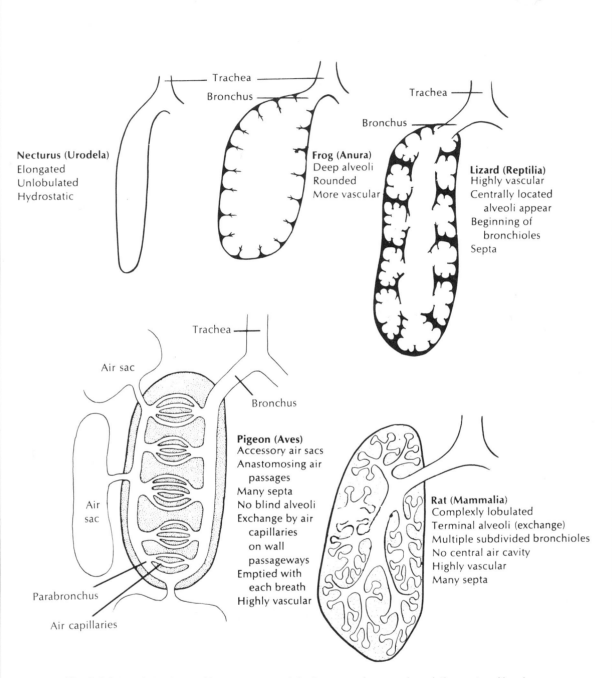

Necturus (Urodela)
Elongated
Unlobulated
Hydrostatic

Frog (Anura)
Deep alveoli
Rounded
More vascular

Lizard (Reptilia)
Highly vascular
Centrally located
      alveoli appear
Beginning of
      bronchioles
Septa

Pigeon (Aves)
Accessory air sacs
Anastomosing air
      passages
Many septa
No blind alveoli
Exchange by air
      capillaries
      on wall
      passageways
Emptied with
      each breath
Highly vascular

Rat (Mammalia)
Complexly lobulated
Terminal alveoli (exchange)
Multiple subdivided bronchioles
No central air cavity
Highly vascular
Many septa

Fig. 8-9. Internal structures of lungs among vertebrate groups. In general, evolutionary trend has been from simple sacs with little exchange surface between blood and air spaces to complex, lobulated structures, each with complex divisions and extensive exchange surfaces.

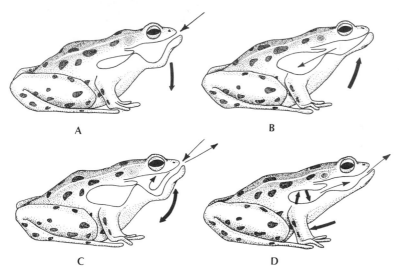

Fig. 8-10. Breathing in frog. The frog, a positive-pressure breather, fills its lungs by forcing air into them. **A,** Floor of mouth is lowered, drawing air in through nostrils. **B,** With nostrils closed and glottis open, the frog forces air into lungs by elevating floor of mouth. **C,** Mouth cavity rhythmically is ventilated for a period. **D,** Lungs are emptied by contraction of body wall musculature and by elastic recoil of lungs. (Modified from Gordon, M. S., et al. 1968. Animal function: principles and adaptations, New York, Macmillan, Inc.)

and mouth cavity—connects the lungs to the outside. This is a "dead-air space" containing a volume of air that shuttles back and forth with each breath, adding to the difficulty of properly ventilating the lungs. In fact, lung ventilation is so inefficient that in normal breathing only approximately one-sixth of the air in the lungs is replenished with each inspiration.

One group of vertebrates, the birds, vastly improved lung efficiency by adding an extensive system of air sacs (Fig. 8-9) that serve as air reservoirs during ventilation. Upon inspiration, some 75% of the incoming air bypasses the lungs to enter the air sacs. At expiration, some of this fresh air passes directly through the lung passages. Thus the air capillaries receive nearly fresh air during both inspiration and expiration (see Fig. 28-6, p. 556). The beautifully designed bird lung is the result of selective pressures during the evolution of flight and its high metabolic demands.

Frogs force air into the lungs by first lowering the floor of the mouth to draw air into the mouth through the external nares (nostrils); then, by closing the nares and raising the floor of the mouth, air is driven into the lungs. Much of the time, however, frogs rhythmically ventilate only the mouth cavity, which serves as a kind of auxiliary "lung" (Fig. 8-10). Amphibians therefore employ a **positive pressure** action to fill their lungs, unlike most reptiles, birds, and mammals, which breathe by sucking air into the lungs (**negative pressure** action).

Tracheae

Insects and certain other terrestrial arthropods (centipedes, millipedes, and some spiders) have a highly specialized type of respiratory system; in many respects it is the simplest, most direct, and most efficient respiratory system found in active animals. It consists of a system of tubes **(tracheae)** that branch repeatedly and extend to all parts of the body. The smallest end channels **(air capillaries),** less than 1 μm in diameter, sink into the plasma membranes of the body cells. Oxygen enters the tracheal system through valvelike openings **(spiracles)** on each side of the body and diffuses directly to all cells of the body. Carbon dioxide diffuses out in the opposite direction. Some insects can ventilate the tracheal system with body movements; the familiar telescoping movement of the bee abdomen is an example. The tracheal system is simple because blood is not needed to transport the respiratory gases; the cells have a direct pipeline to the outside.

Respiration in man

In mammals the respiratory system is made up of the following: the nostrils (external nares); the **nasal chamber,** lined with mucus-secreting epithelium; the **posterior nares,** which connect to the **pharynx** where the pathways of digestion and respiration cross; the **epiglottis,** a flap that folds over the **glottis** (the opening to the larynx) to prevent food from going the wrong way in swallowing; the **larynx,** or voice box;

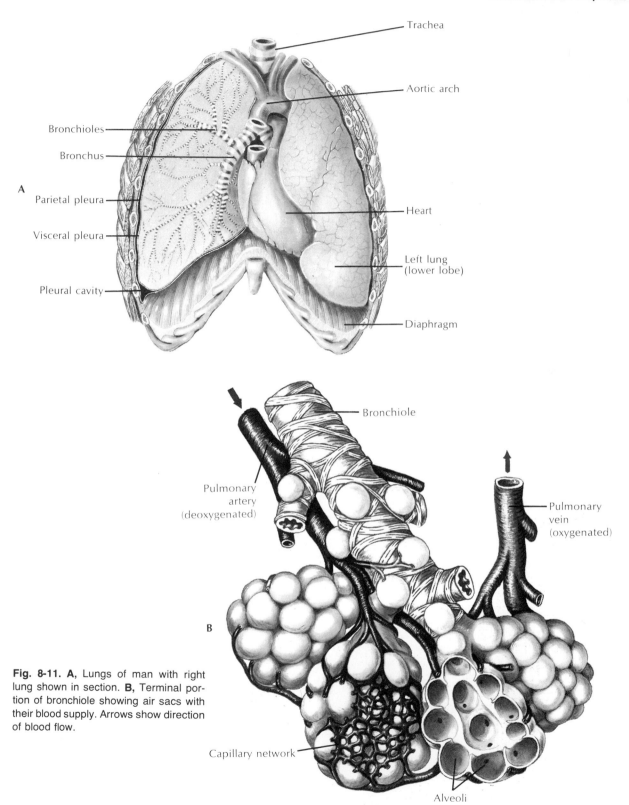

Trachea

Aortic arch

Bronchioles

Bronchus

A

Parietal pleura

Visceral pleura

Heart

Left lung
(lower lobe)

Pleural cavity

Diaphragm

Bronchiole

Pulmonary
artery
(deoxygenated)

Pulmonary
vein
(oxygenated)

B

Fig. 8-11. A, Lungs of man with right
lung shown in section. **B,** Terminal portion of bronchiole showing air sacs with
their blood supply. Arrows show direction
of blood flow.

Capillary network

Alveoli

161

the **trachea,** or windpipe; and the two **bronchi,** one to each lung (Fig. 8-11). Within the lungs each bronchus divides and subdivides into smaller tubes **(bronchioles)** that lead to the air sacs **(alveoli).** The walls of the alveoli are thin and moist to facilitate the exchange of gases between the air sacs and the adjacent blood capillaries. Air passageways are lined with mucus-secreting ciliated epithelium and play an important role in conditioning the air before it reaches the alveoli. There are partial cartilage rings in the walls of the tracheae, bronchi, and even some of the bronchioles to prevent those structures from collapsing.

In its passage to the air sacs the air undergoes three important changes: (1) it is filtered free from most dust and other foreign substances, (2) it is warmed to body temperature, and (3) it is saturated with moisture.

The lungs consist of a great deal of elastic connective tissue and some muscle. They are covered by a thin layer of tough epithelium known as the **visceral pleura.** A similar layer, the **parietal pleura,** lines the inner surface of the walls of the chest (Fig. 8-11). The two layers of the pleura are in contact and slide over one another as the lungs expand and contract. The "space" between the pleura, called the **pleural cavity,** contains a partial vacuum. Actually no real pleural space exists; the two pleura rub together, lubricated by lymph. The chest cavity is bounded by the spine, ribs, and breastbone, and floored by the **diaphragm,** a dome-shaped, muscular partition between chest cavity and abdomen.

Mechanism of breathing

The chest cavity is an air-tight chamber. In **inspiration** the ribs are elevated, the diaphragm is contracted and flattened, and the chest cavity is enlarged. The resultant increase in volume of chest cavity and lungs causes the air pressure in the lungs to fall below atmospheric pressure; air rushes in through the air passageways to equalize the pressure. **Expiration** is a less active process than inspiration. When the muscles relax, the ribs and diaphragm return to their original position and the chest cavity size decreases. The elastic lungs then deflate and force the air out.

Control of breathing

Respiration must adjust itself to the varying needs of the body for oxygen. Breathing is normally involuntary and automatic but may come under voluntary control. The rhythmic inspiratory and expiratory movements are controlled by a nervous mechanism centered in the **medulla oblongata** of the brain (Fig. 8-12). By placing tiny electrodes in various parts of the medulla of experimental animals, neurophysiologists located separate **inspiratory** and **expiratory neurons** that act reciprocally to stimulate the inspiratory and expiratory muscles of the diaphragm and rib cage (intercostal muscles). The rate of breathing is determined by the amount of carbon dioxide in the blood: a slight rise in the blood carbon dioxide stimulates breathing; a fall decreases breathing.

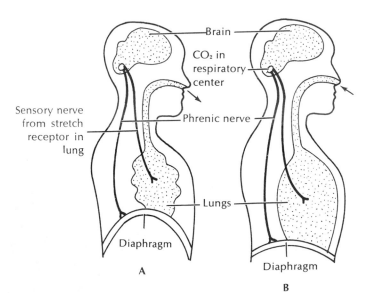

Fig. 8-12. Mechanism of breathing. In inspiration, **B,** carbon dioxide in blood stimulates respiratory center to send impulses by way of phrenic nerves to diaphragm, which, with elevation of ribs, produces inhalation of air. In **A,** impulses from stretch receptors in lungs inhibit respiratory center and exhalation occurs. (See text for explanation.)

Composition of inspired, expired, and alveolar airs

The composition of expired and alveolar airs is not identical. Air in the alveoli contains less oxygen and more carbon dioxide than does the air that leaves the lungs. Inspired air has the composition of atmospheric air. Expired air is really a mixture of alveolar and inspired airs. The variations in the three kinds of air are shown in Table 8-2.

The water given off in expired air depends on the relative humidity of the external air and the activity of the person. At ordinary room temperature and with a relative humidity of approximately 50%, an individual in performing light work loses approximately 350 ml of water from the lungs each day.

Gaseous exchange in lungs

The diffusion of gases takes place in accordance with the laws of physical diffusion; that is, the gases pass from regions of high pressure to those of low pressure. The partial pressure of a gas refers to the pressure that that gas exerts in a mixture of gases. If the atmospheric pressure at sea level is equivalent to 760 mm Hg, the partial pressure of oxygen is 21% (percentage of oxygen in air) of 760, or 159 mm Hg. The partial pressure of oxygen in the lung alveoli is greater (100 mm Hg pressure) than it is in venous blood of lung capillaries (40 mm Hg pressure) (Fig. 8-13). Oxygen then naturally diffuses into the capillaries. In a similar manner the carbon dioxide in the blood of the lung capillaries has a higher concentration (46 mm Hg) than has this same gas in the lung alveoli (40 mm Hg), so that carbon dioxide diffuses from the blood into the alveoli.

In the tissues respiratory gases also move according to their concentration gradients (Fig. 8-13). The concentration of oxygen in the blood (100 mm Hg pressure) is greater than in the tissues (0 to 30 mm Hg pressure), and the carbon dioxide concentration in the tissues (45 to 68 mm Hg pressure) is greater than that in blood (40 mm Hg pressure). The gases in each case go from a high to a low concentration.

Transport of gases in blood

In some invertebrates the respiratory gases are simply carried dissolved in the body fluids. However, the solubility of oxygen is so low in water that it is adequate only for animals with low rates of metabolism. For example, only approximately 1% of man's oxygen requirement can be transported in this way. Consequently in just about all the advanced invertebrates and the vertebrates, nearly all the oxygen and a significant amount of the carbon dioxide are transported by special colored proteins, or **respiratory pigments,** in the blood. In most animals (all vertebrates) these respiratory pigments are packaged into blood corpuscles. This is necessary because, if this amount of respiratory pigment were free in blood, the blood would have the viscosity of syrup and would hardly flow through the blood vessels at all.

Respiratory pigments. The two most widespread respiratory pigments are **hemoglobin,** a red, iron-containing protein present in all vertebrates and many invertebrates, and **hemocyanin,** a blue, copper-containing protein present in the crustaceans and most molluscs. Among other pigments is **chlorocruorin** (klor-a-kroo'o-rin), a green-colored, iron-containing pigment found in four families of polychaete tube worms. Its structure and oxygen-carrying capacity are very similar to those of hemoglobin, but it is carried free in the plasma rather than being enclosed in blood corpuscles. **Hemerythrin** is a red pigment found in some polychaete worms. Although it contains iron, this metal is not present in a heme group (despite its name!) and its oxygen-carrying capacity is poor.

Hemoglobin and oxygen transport. Hemoglobin is a complex protein. Each molecule is made up of 5% **heme,** an iron-containing compound giving the red color to blood, and 95% **globin,** a colorless protein. The heme portion of the hemoglobin has a great affinity for oxygen; each gram of hemoglobin (there are approximately 15 g of hemoglobin in each 100 ml of man's blood) can carry a maximum of approximately 1.3 ml of oxygen; each 100 ml of fully oxygenated blood contains approximately 20 ml of oxygen. Of course, for hemoglobin to be of value to the body it must hold oxygen in a loose, reversible chemical combination so that it can be released to the

Table 8-2. Variation in respired air

	Inspired air (vol. %)	Expired air (vol. %)	Alveolar air (vol. %)
Oxygen	20.96	16	14.0
Carbon dioxide	0.04	4	5.5
Nitrogen	79.00	80	80.5

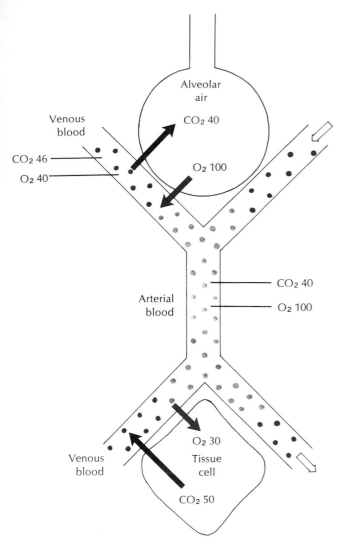

Venous
blood

CO₂ 46
O₂ 40

Alveolar
air

CO₂ 40

O₂ 100

CO₂ 40

O₂ 100

Arterial
blood

O₂ 30

Tissue
cell

Venous
blood

CO₂ 50

Fig. 8-13. Exchange of respiratory gases in lungs and tissue cells. Numbers represent partial pressures in millimeters of mercury.

is continuously consumed by cellular oxidative processes. The oxyhemoglobin releases its bound oxygen, which diffuses into the cells. As the oxygen dissociation curve shows (Fig. 8-14), the lower the surrounding oxygen tension, the greater the quantity of oxygen released. This is an important characteristic because it allows more oxygen to be released to those tissues that need it most (have the lowest oxygen pressure).

Another characteristic facilitating the release of oxygen to the tissues is the sensitivity of oxyhemoglobin to carbon dioxide. Carbon dioxide shifts the oxygen dissociation curve to the right (Fig. 8-14, *B*), a phenomenon that has been called the **Bohr effect** after the Danish scientist who first described it. Therefore as carbon dioxide enters the blood from the respiring tissues, it encourages the release of additional oxygen from the hemoglobin. The opposite event occurs in the lungs; as carbon dioxide diffuses from the venous blood into the alveolar space, the oxygen dissociation curve shifts back to the left, allowing more oxygen to be loaded onto the hemoglobin.

Unfortunately for man and other higher animals, hemoglobin has even a greater affinity for carbon monoxide (CO) than it has for oxygen—in fact, the affinity is approximately 200 times greater for carbon monoxide than for oxygen. Carbon monoxide is becoming an atmospheric contaminant of ever-increasing proportions as the world's population and industrialization continues rapidly upward. This odorless and invisible gas displaces oxygen from hemoglobin to form a stable compound called **carboxyhemoglobin.** Air containing only 0.2% carbon monoxide may be fatal. Children and small animals are poisoned more rapidly than adults because of their higher respiratory rate.

Transport of carbon dioxide by the blood. The same blood that transports oxygen to the tissues from the lungs must carry carbon dioxide back to the lungs on its return trip. However, unlike oxygen that is transported almost exclusively in combination with hemoglobin, carbon dioxide is transported in three major forms.

1. Most of the carbon dioxide, approximately 67%, is converted in the red blood cells into bicarbonate and hydrogen ions by undergoing the following series of reactions:

$$CO_2 + H_2O \rightleftharpoons H_2CO_3$$

CARBONIC
ACID

tissues. The actual amount of oxygen bound to hemoglobin depends on the oxygen partial pressure surrounding the blood corpuscles, a relationship expressed in the oxygen dissociation curve in Fig. 8-14.

When the oxygen tension is high, as it is in the capillaries of the lung alveoli, hemoglobin becomes almost fully saturated to form oxyhemoglobin. When the oxygenated blood leaves the lung and is distributed to the systemic capillaries in the body tissues, it enters regions of low oxygen partial pressure because oxygen

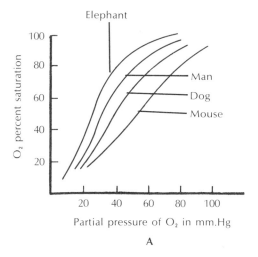

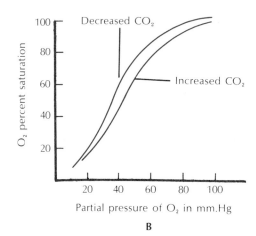

Fig. 8-14. Oxygen dissociation curves. Curves show how the amount of oxygen bound to hemoglobin (oxyhemoglobin) is related to oxygen pressure. **A,** Small animals have blood that gives up oxygen more readily than does the blood of large animals. **B,** Oxyhemoglobin is sensitive to carbon dioxide pressure; as carbon dioxide enters blood from the tissues, it shifts the curve to the right, decreasing affinity of hemoglobin for oxygen.

This reaction would normally proceed very slowly, but an enzyme in the red blood cells, **carbonic anhydrase,** catalyzes the reaction to proceed almost instantly. As soon as carbonic acid forms, it instantly and almost completely ionizes as follows:

$$H_2CO_3 \rightleftharpoons HCO_3^- + H^+$$

| CARBONIC ACID | BICARBONATE ION | HYDROGEN ION |

The hydrogen ion is buffered by several buffer systems in the blood, thus preventing a severe decrease in blood pH. The bicarbonate ion remains in solution in the plasma and red blood cell water since, unlike carbon dioxide, bicarbonate is extremely soluble.

2. Another fraction of the carbon dioxide, approximately 25%, combines reversibly with hemoglobin. It is carried to the lungs where the hemoglobin releases it in exchange for oxygen.

3. A third small fraction of the carbon dioxide, approximately 8%, is carried as the physically dissolved gas in the plasma and red blood cells.

SELECTED REFERENCES

(See also general references for Part two, p. 274.)

Brooks, S. M. 1960. Basic facts of body water and ions. New York, Springer Publishing Co., Inc. *An excellent elementary account of the fluid and electrolyte balance in the body.*

Chapman, G. 1967. The body fluids and their function. Institute of Biology's Studies in Biology no. 8. New York, St. Martin's Press. *Brief, comparative treatment of body fluids, their transport, and regulation.*

Graubard, M. 1964. Circulation and respiration. The evolution of an idea. New York, Harcourt, Brace & World, Inc. *Historic development of blood flow concepts. Selected writings from Aristotle, Galen, Vesalius, Fabricius, Harvey, Malpighi, Boyle, and others.*

Snively, W. D., Jr. 1960. Sea within: the story of our body fluid. Philadelphia, J. B. Lippincott Co. *A popular, interesting treatise on the "interior sea" and its importance in our bodies in health and disease.*

SELECTED SCIENTIFIC AMERICAN ARTICLES

Adolph, E. F. 1967. The heart's pacemaker. **216:**32-37 (March).

Clarke, C. A. 1968. The prevention of "rhesus" babies. **217:**46-52 (Nov.).

Comroe, J. H., Jr. 1966. The lung. **214:**56-68 (Feb.). *Physiology of the human lung.*

Cooper, M. D., and A. R. Lawton III. 1974. The development of the immune system. **231:**58-72 (Nov.).

Laki, K. 1962. The clotting of fibrinogen. **206:**60-66 (March).

Mayerson, H. S. 1963. The lymphatic system. **208:**80-90 (June). *The lymphatics are a crucial "second" circulatory system that picks up fluids leaking from the bloodstream.*

Wood, J. E. 1968. The venous system **218:**86-96 (Jan.).

Zweifach, B. W. 1959. The microcirculation of the blood. **200:**54-60 (Jan.). *Describes the anatomy and function of the capillary bed that serves the body's tissues.*

Many species of flounder live most of their lives in freshwater rivers, only returning to their ancestral home in the sea to spawn. Thus they are able to tolerate and osmotically regulate in both fresh water, which is more dilute than their internal fluids, and seawater, which is more concentrated. This gulf fluke *(Paralichthys albiguttus),* like other flatfishes, is distinguished by its asymmetric body. During larval metamorphosis one eye migrates around the head to join the eye on the opposite side, and the animal assumes a benthic existence, lying and swimming with its blind side down. (Courtesy Shedd Aquarium.)

9 INTERNAL FLUIDS
Excretion and homeostasis

WATER AND OSMOTIC REGULATION

How aquatic animals meet problems of salt and water balance

How terrestrial animals maintain salt and water balance

INVERTEBRATE EXCRETORY STRUCTURES

VERTEBRATE KIDNEY

Vertebrate kidney function

At the beginning of the preceding chapter we described the double-layered environment of the body's cells: the extracellular fluid *(milieu interieur),* which immediately surrounds the cells, and the external environment *(milieu exterieur)* of the outside world. The life-supporting metabolic activities that occur within the body's cells can proceed only as long as they are bathed by a protective extracellular fluid environment of relatively constant composition. Yet there are many activities that threaten to throw the system out of balance. It is apparent that body fluid composition can be altered either by metabolic events occurring within the cells and tissues or by events occurring across the surface of the body. In other words, a living system is "open at both ends."

On the inside, metabolic activities within the cell require a steady supply of materials and these activities turn out a continuous flow of products and wastes. On the outside, materials are constantly being exchanged between the plasma and the external environment. Water, which makes up approximately two thirds of the body weight of animals, is always entering and leaving the body. Water is also formed within the cells as a by-product of oxidative processes. Ionized inorganic and organic salts are continually moving between the cells and the body fluids and also between the animal and its environment. Protein is constantly being formed, transported, and broken down again within the tissues, yielding nitrogenous wastes that must be excreted.

Obviously body composition is a dynamic rather than a static thing. It is often described as operating as a **dynamic steady state.** This means that constancy of composition is maintained despite the continuous shifting of components within the system. This kind of internal regulation is called **homeostasis.**

Homeostasis is maintained by the coordinated activities of numerous body systems, such as the nervous system, endocrine system, and especially the organs that serve as sites of exchange with the external environment, which include the kidneys, lungs or gills, alimentary canal, and skin. Through these organs oxygen, foodstuffs, minerals, and other constituents of the body fluids enter; water is exchanged and metabolic wastes are eliminated.

The kidney is the chief regulator of the body fluids. It is popularly regarded strictly as an organ of excretion that serves to rid the body of assorted metabolic wastes. But, in fact, it is as much a regulatory organ as an excretory organ. It is responsible for individually monitoring and regulating the concentrations of body water, salt ions in the blood, and other major and minor body fluid constituents. In its task of fine-tuning the composition of the internal environment, the kidney is assisted by the other organs of exchange, such as the lungs, skin, and digestive tract, as well as by many internal mechanisms.

Several other specialized structures have evolved among the vertebrates that assist in body fluid regulation in various environments, for example, the salt-secreting cells of fish gills and the salt glands of birds and reptiles.

WATER AND OSMOTIC REGULATION
How aquatic animals meet problems of salt and water balance
Marine invertebrates

Most marine invertebrates are in osmotic equilibrium with their seawater environment. They have body surfaces that are permeable to salts and water so that their body fluid concentration rises or falls in conformity with changes in concentrations of seawater. Because such animals are incapable of regulating their body fluid osmotic pressure, they are referred to as **osmotic conformers.** Invertebrates living in the open sea are seldom exposed to osmotic fluctuations because the ocean is a highly stable environment. Oceanic invertebrates have, in fact, very limited abilities to withstand osmotic change. If they should be exposed to dilute seawater, they die quickly because their body cells cannot tolerate dilution and are helpless to prevent it. These animals are restricted to living in a narrow salinity range, and are said to be **stenohaline** (Gr. *stenos,* narrow; *hals,* salt). An example is the marine spider crab, represented in Fig. 9-1.

Conditions along the coasts and in estuaries and river mouths are much less constant than those of the open ocean. Here animals must be able to withstand large and often abrupt salinity changes as the tides move in and out and mix with fresh water draining from rivers. These animals are referred to as **euryhaline** (Gr. *eurys,* broad; *hals,* salt), meaning that they can survive a wide range of salinity change. Most coastal invertebrates also show varying powers of **osmotic regulation.** For example, the brackish-water shore crab can resist body fluid dilution by dilute (brackish) seawater (Fig. 9-1). Although the body fluid concentration falls, it does so less rapidly than the fall

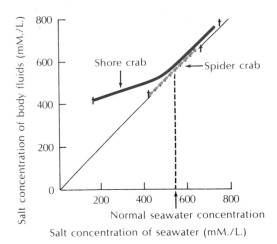

Fig. 9-1. Salt concentration of body fluids of two crabs as affected by variations in the seawater concentration. The 45-degree line represents equal concentration between body fluids and seawater. Since the spider crab cannot regulate its body-fluid salt concentration, it conforms to whatever changes happen in the external seawater environment. The shore crab, however, can regulate osmotic concentration of its body fluids to some degree because in dilute seawater the shore crab can hold its body-fluid concentration above the seawater concentration. For example, when the seawater is 200 mM per liter, the shore crab's body fluids are approximately 430 mM per liter. Crosses at ends of lines indicate tolerance limits of each species.

in seawater concentration. This crab is a **hyperosmotic regulator** because in a dilute environment it can maintain the concentration of its blood above that of the surrounding water.

What is the advantage of hyperosmotic regulation over osmotic conformity, and how is this regulation accomplished? The advantage is that by regulating against excessive dilution, thus protecting the body cells from extreme changes, these crabs can successfully live in the physically unstable but biologically rich coastal environment. Their powers of regulation are limited, however, since if the water is highly diluted, their regulation fails and they die.

To understand how the brackish-water shore crab and other coastal invertebrates achieve hyperosmotic regulation, let us examine the problems they face. First, the salt concentration of the internal fluids is greater than in the dilute seawater outside. This causes a steady osmotic influx of water. As with the membrane osmometer placed in a sugar solution (p. 58), water diffuses inward because it is more concentrated

outside than inside. The shore crab is not nearly as permeable as a membrane osmometer—most of its shelled body surface is, in fact, almost impermeable to water—but the thin respiratory surfaces of the gills are highly permeable. Obviously the crab cannot insulate its gills with an impermeable hide and still breathe. The problem is solved by removing the excess water through the action of the kidney (the antennal gland located in the crab's thorax).

The second problem is salt loss. Again, because the animal is saltier than its environment, it cannot avoid loss of ions by outward diffusion across the gills. Salt is also lost in the urine. This problem is solved by special salt-secreting cells in the gills that can actively remove ions from the dilute seawater and move them into the blood, thus maintaining the internal osmotic concentration. This is an **active transport** process that requires energy because ions must be transported against a concentration gradient, that is, from a lower salt concentration (in the dilute seawater) to an already higher one (in the blood).

Invasion of fresh water

Some 400 million years ago, during the Silurian and Lower Devonian, the major groups of jawed fishes began to penetrate into brackish-water estuaries and then gradually into freshwater rivers. Before them lay a new unexploited habitat already stocked with food in the form of insects and other invertebrates, which had preceded them into fresh water. However, the advantages of this new habitat were traded off for a tough physiologic challenge: the necessity of developing effective osmotic regulation.

Freshwater animals must keep the salt concentration of their body fluids higher than that of the water. Water therefore enters their bodies osmotically and salt is lost by diffusion outward. Their problems are similar to those of the brackish-water shore crab, but more severe and unremitting. Fresh water is much more dilute than are coastal estuaries, and there is no retreat, no salty sanctuary into which the freshwater animal can retire for osmotic relief. It must and has become a permanent and highly efficient hyperosmotic regulator.

The scaled and mucus-covered body surface of a fish is about as waterproof as any flexible surface can be. The water that inevitably enters across the gills is pumped out by the kidney. Even though the kidney is able to make a very dilute urine, some salt is lost; this is replaced by salt in food and by active absorption of

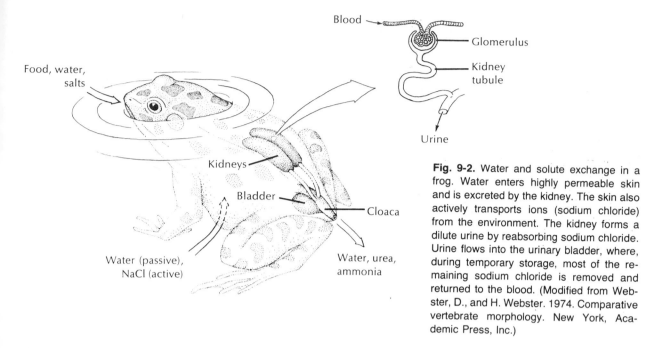

Blood

Glomerulus

Kidney
tubule

Urine

Fig. 9-2. Water and solute exchange in a frog. Water enters highly permeable skin and is excreted by the kidney. The skin also actively transports ions (sodium chloride) from the environment. The kidney forms a dilute urine by reabsorbing sodium chloride. Urine flows into the urinary bladder, where, during temporary storage, most of the remaining sodium chloride is removed and returned to the blood. (Modified from Webster, D., and H. Webster. 1974. Comparative vertebrate morphology. New York, Academic Press, Inc.)

Food, water, salts

Kidneys

Bladder

Cloaca

Water (passive), NaCl (active)

Water, urea, ammonia

salt (primarily sodium and chloride) across the gills. The bony fishes that inhabit our lakes and streams today are so well adapted to their dilute surroundings that they need expend very little energy to regulate themselves osmotically. Osmotic regulation in fishes is described on p. 514 and illustrated in Fig. 25-14, p. 515.

Crayfishes, aquatic insect larvae, mussels, and other freshwater animals are also hyperosmotic regulators and face the same hazards as freshwater fishes; they tend to gain too much water and lose too much salt. Not surprisingly, all of these forms solved these problems in the same direct way that fishes did. They excrete the excess water as urine, and they actively absorb salt from the water by some salt-transporting mechanism on the body surface.

Amphibians, when they are living in water, also must compensate for salt loss by absorbing salt from the water (Fig. 9-2). They use their skin for this purpose. Physiologists learned some years ago that pieces of frog skin continue to actively transport sodium and chloride for hours when removed and placed in a specially balanced salt solution. Fortunately for biologists, but unfortunately for frogs, these animals are so easily collected and maintained in the laboratory that frog skin has become a favorite membrane system for studies of ion-transport phenomena.

Marine fishes

The great families of bony fishes that inhabit the seas today maintain the salt concentration of their body fluids at approximately one third that of seawater (body fluids = 0.3 to 0.4 M; seawater = 1 M). Obviously they are osmotic regulators. Bony fishes living in the oceans today are descendants of earlier freshwater bony fishes that moved back into the sea during the Triassic approximately 200 million years ago. The return to their ancestral sea was probably prompted by unfavorable climatic conditions on land and the deterioration of freshwater habitats, but we can only guess at the reasons. During the many millions of years that the freshwater fishes were adapting themselves so well to their environment, they established a body fluid concentration equivalent to approximately one-third that of seawater, thus setting the pattern for all the vertebrates that were to evolve later, whether aquatic, terrestrial, or aerial. The ionic composition of vertebrate body fluid is remarkably similar to dilute seawater too, a fact that is undoubtedly related to their marine heritage.

When some of the freshwater bony fishes of the Triassic ventured back to the sea, they encountered a new set of problems. Having a much lower internal osmotic concentration than the seawater around them, they lost water and gained salt. Indeed the marine bony fish literally risks drying out, much like a desert mammal deprived of water. The way the marine bony fishes regulate osmotically is described on p. 514.

In brief, to compensate for water loss, the marine teleost drinks seawater. This is absorbed from the intestine, and the major sea salt, sodium chloride, is carried by the blood to the gills, where specialized salt-secreting cells transport it back into the surrounding sea. The ions remaining in the intestinal residue, especially magnesium, sulfate, and calcium, are voided with the feces or excreted by the kidney. In this roundabout way, marine fishes rid themselves of the excess sea salts they have drunk, resulting in a net gain of water, which replaces the water lost by osmosis. Samuel Taylor Coleridge's ancient mariner, surrounded by "water, water, everywhere, nor any drop to drink." would undoubtedly have been tormented even more had he known of the marine fishes' simple solution for thirst. A marine fish carefully regulates the amount of seawater it drinks, consuming only enough to replace water loss and no more.

The cartilaginous sharks and rays (elasmobranchs) solve their water balance problems in a completely different way. This primitive group is almost totally marine. The salt composition of shark's blood is similar to that of the bony fishes, but the blood also contains a large amount of organic compounds, especially urea and trimethylamine oxide. Urea is, of course, a metabolic waste that most animals quickly excrete in the urine. The shark kidney, however, conserves urea, causing it to accumulate in the blood. The blood urea, added to the usual blood electrolytes, raises the blood osmotic pressure to slightly exceed that of seawater. In this way the sharks and their kin turn an otherwise useless waste material into an asset, eliminating the osmotic problem encountered by the marine bony fishes.

How terrestrial animals maintain salt and water balance

The problems of living in an aquatic environment seem small indeed compared to the problems of life on land. Since our bodies are mostly water, all metabolic activities proceed in water, and the origins of life itself were conceived in water, it seems obvious that animals were meant to stay in water. Yet many animals, like the plants preceding them, inevitably moved onto land, carrying their watery composition with them. Once on land, the terrestrial animals continued their adaptive

radiation, undaunted by the threat of desiccation, until they became abundant even in some of the most arid parts of the earth.

Terrestrial animals lose water by evaporation from the lungs and body surface, by excretion in the urine, and by elimination in the feces. Such losses are replaced by water in the food, by drinking water if it is available, and by forming **metabolic water** in the cells by the oxidation of foodstuffs, especially carbohydrates. In some desert rodents, metabolic water may constitute most of the animals' water gain.

Particularly revealing is a comparison (Table 9-1) of water balance in man, a nondesert mammal that drinks water, with that of the kangaroo rat, a desert rodent that may drink no water at all.

The kangaroo rat gains all of its water from its food (90% as metabolic water derived from the oxidation of foodstuffs, 10% as free moisture in the food). Even though man eats foods with a much higher water content than the dry seeds that comprise much of the kangaroo rat's diet, he must still drink half his total water requirement.

The excretion of wastes presents a special problem in water conservation. The primary end-product of protein catabolism is ammonia, a highly toxic material. Fishes can easily excrete ammonia across their gills since there is an abundance of water to wash it away. The terrestrial insects, reptiles, and birds have no convenient way to rid themselves of toxic ammonia; instead they convert it into uric acid, a nontoxic, almost insoluble compound. This enables them to excrete a semisolid urine with little water loss. The use

Table 9-1. Water balance in man and kangaroo rat, a desert rodent*

	Man (%)	Kangaroo rat (%)
Gains		
Drinking	48	0
Free water in food	40	10
Metabolic water	12	90
Losses		
Urine	60	25
Evaporation (lungs and skin)	34	70
Feces	6	5

*Partly from Schmidt-Nielsen, K. 1972. How animals work, New York, Cambridge University Press.

of uric acid has another important benefit. All of these animals lay eggs enclosing the embryos, their stores of food and water, and whatever wastes that accumulate during development. By converting ammonia to uric acid, the developing embryo's waste can be precipitated into solid crystals, which are stored harmlessly within the egg until hatching.

Marine birds and turtles have evolved a unique solution for excreting the large loads of salt eaten with their food. Located above each eye is a special **salt gland** capable of excreting a highly concentrated solution of sodium chloride—up to twice the concentration of seawater. In birds the salt solution runs out the nares (see p. 557). Marine turtles and lizards shed their salt gland secretion as salty tears. Salt glands are important accessory organs of salt excretion to these animals since their kidney cannot produce a concentrated urine, as can the mammalian kidney.

INVERTEBRATE EXCRETORY STRUCTURES

In such a large and varied group as the invertebrates it is hardly surprising that there is a great variety of morphologic structures serving as excretory organs. Many Protozoa and some freshwater sponges have special excretory organelles called contractile vacuoles. The more advanced invertebrates have excretory organs that are basically tubular structures that form urine by first producing an ultrafiltrate or fluid secretion of the blood. This enters the proximal end of the tubule and is modified continuously as it flows down the tubule. The final product is urine.

Contractile vacuole

This tiny spheric intracellular vacuole of protozoans and freshwater sponges is not a true excretory organ since ammonia and other nitrogenous wastes of metabolism readily leave the cell by direct diffusion across the cell membrane into the surrounding water. The contractile vacuole is really an organ of water balance. Because the cytoplasm of freshwater Protozoa is considerably saltier than their freshwater environment, they tend to draw water into themselves by osmosis. In *Paramecium* this excess water is collected by minute canals within the cytoplasm and conveyed to the contractile vacuole (Fig. 15-12, p. 301). This grows larger as water accumulates within it. Finally the vacuole is emptied through a pore on the surface, and the cycle is rhythmically repeated. Although the contractile vacuole has been carefully studied, it is not

yet known how this system is able to pump out pure water while retaining valuable salts within the animal. Contractile vacuoles are common in freshwater Protozoa but rare or absent from marine Protozoa, which are isosmotic with seawater and consequently neither lose nor gain too much water.

Nephridia

The nephridium is the most common type of invertebrate excretory organ. All nephridia are tubular structures, but there are large differences in degree of complexity. One of the simplest arrangements is the flame cell system (or **protonephridia**) of the flatworm.

In *Planaria* and other flatworms this takes the form of two highly branched systems of tubules distributed throughout the body (Fig. 9-3). Fluid enters the system through specialized "flame" cells, moves slowly into and down the tubules, and is excreted through pores that open at intervals on the body surface. It is believed that fluid containing wastes enters the flame bulb from the surrounding tissues by **pinocytosis** (cell drinking). In this process fluid-filled vesicles are formed just under the outer cell surface. The vesicles are transported across the cell and set free into the ciliated lumen of the flame cell. Here, the rhythmic beat of the ciliary tuft (the "flame") creates a negative fluid pressure that drives the fluid into the tubular portion of the system. It is probable that, as the fluid passes down the tubules, the tubular epithelial cells add certain waste materials to the tubular fluid (secretion) and withdraw valuable materials from it (reabsorption) to complete the formation of urine.

The flame cell system, like the contractile vacuole of protozoans, is primarily a water balance system, since it is best developed in free-living freshwater forms. Branched flame cell systems are typical of primitive invertebrates that lack circulatory systems. Since there is no circulation to carry wastes to a compact excretory organ such as the kidney of higher invertebrates and vertebrates, the flame cell system must be distributed to reach the cells directly.

The protonephridium just described is a **"closed"** system, that is, the urine is formed from a fluid that must first enter the tubule by being transported across the flame cells. Another type of nephridium, typical of many annelids (segmented worms), is the **open,** or "true," nephridium. In the earthworm *Lumbricus* there are paired nephridia in every segment of the

171

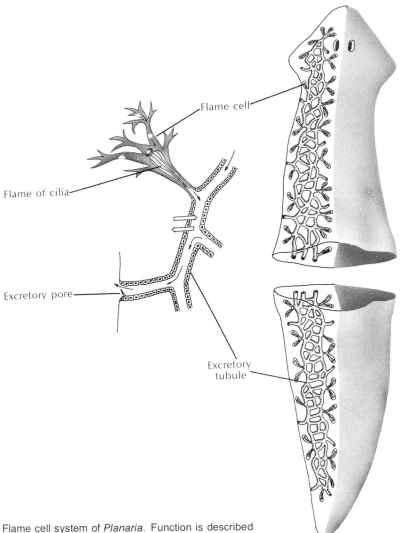

Flame cell

Flame of cilia

Excretory pore

Excretory tubule

Fig. 9-3. Flame cell system of *Planaria*. Function is described in text.

body, except the first three and the last one (Fig. 20-2, p. 389). Each nephridium occupies parts of two successive segments. In the earthworm each nephridium is a tiny, self-contained "kidney" that independently drains to the outside through pores **(nephridiopores)** in the body wall.

Coelomic fluid containing wastes to be excreted is swept into a ciliated, funnellike opening **(nephrostome)** of the nephridium and carried through a long, twisted tubule of increasing diameter. It then enters a bladder and is finally expelled through a nephridiopore to the outside. The nephridial tubule is surrounded by an extensive network of blood vessels. Solutes, es-

pecially sodium and chloride, are reabsorbed from the formative urine during its travel through the tubule. The addition of the blood vascular network to the annelid nephridium makes it a much more versatile and effective system than the flame cell system. However, the basic process of urine formation is the same: fluid flows continuously through a tubule while materials are added here and taken away there, until urine is formed.

Arthropod kidneys

The **antennal glands** of crustaceans form a single, paired tubular structure located in the ventral part of

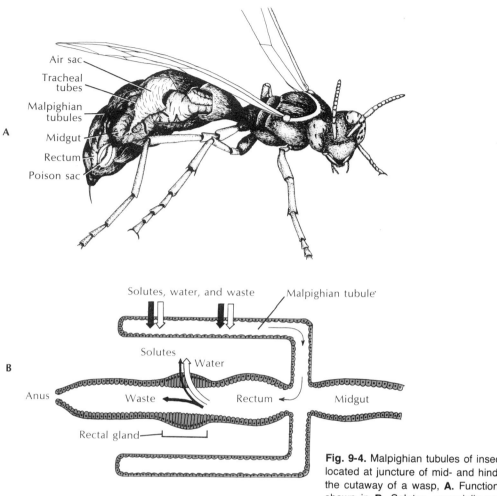

Fig. 9-4. Malpighian tubules of insect. Malpighian tubules are located at juncture of mid- and hindgut (rectum) as shown in the cutaway of a wasp, **A**. Function of malpighian tubules is shown in **B**. Solutes, especially potassium, are actively secreted into the tubules. Water and wastes follow. This fluid moves into the rectum where solutes and water are actively reabsorbed, leaving wastes to be excreted.

the head. Their structure and function are described on p. 423. These excretory devices are an advanced design of the basic nephridial organ. However, they lack open nephrostomes. Instead, a protein-free filtrate of the blood (ultrafiltrate) is formed in the end sac by the hydrostatic pressure of the blood. In the tubular portion of the gland, the filtrate is modified by the selective reabsorption of certain salts and the active secretion of others. Thus crustaceans have excretory organs that are basically vertebrate-like in the functional sequence of urine formation.

Insects and spiders have a unique excretory system consisting of **malpighian tubules** that operate in conjunction with specialized glands in the wall of the rectum (Fig. 9-4). The thin, elastic, blind malpighian tubules are closed and lack an arterial supply. Consequently urine formation cannot be initiated by blood ultrafiltration as in the crustaceans and vertebrates. Instead salts, largely potassium, are actively secreted into the tubules. This primary secretion of ions creates an osmotic drag that pulls water, solutes, and waste materials into the tubule. The fluid, or "urine," then drains from the tubules into the intestine, where specialized rectal glands actively reabsorb most of the potassium and water, leaving behind wastes such as uric acid. This unique excretory system is ideally suited for life in dry environments. We must assume that it has contributed to the great success of this, the most abundant and widespread group of land animals.

VERTEBRATE KIDNEY

The ancestral vertebrate kidney is believed to have extended the length of the coelomic cavity and to have been made up of segmentally arranged uriniferous tubules. Each tubule opened at one end into the coelom by a nephrostome and at the other end into a common archinephric duct. Such a kidney has been called an **archinephros,** or **holonephros,** and is found in the embryos of hagfishes (Fig. 9-5).

Kidneys of higher vertebrates developed from this primitive plan. Embryologic evidence indicates that there are three generations of kidneys: **pronephros, mesonephros,** and **metanephros.** In all vertebrate embryos, the pronephros is the first and most primitive kidney to appear. As its name implies, it is located anteriorly in the body. It becomes the persistent kidney of adult hagfishes. In all other vertebrates it degenerates during development and is replaced by a more centrally located and more structurally advanced kidney, the mesonephros. The mesonephros becomes the persistent kidney of adult fishes and amphibians. But in the developing embryos of amniotes (reptiles, birds, and mammals) the mesonephros is replaced in turn by the metanephros. The metanephros develops behind the mesonephros and is structurally and functionally the most advanced of the three kidney types. Thus three kidneys are formed in succession, each more

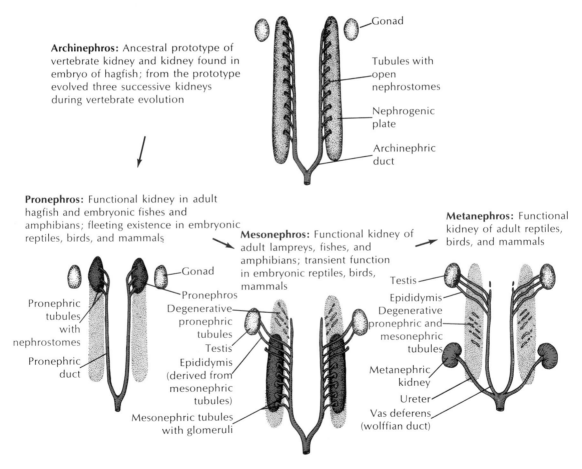

Fig. 9-5. Evolution of male vertebrate kidney from archinephric prototype. *Red,* Functional structures. *Light red,* Degenerative or undeveloped parts.

advanced and each located more caudally than its predecessor.

Vertebrate kidney function

The kidneys of man and other vertebrates play a critical role in the body's economy. As vital organs their failure means death; in this respect they are neither more nor less important than are the heart, lungs, and liver. The kidney is part of many interlocking mechanisms that maintain **homeostasis**—constancy of the internal environment. However, the kidney's share in this regulatory council is an especially large one. It must and does individually monitor and regulate most of the major constituents of the blood and several minor constituents as well. In addition it silently labors to remove a variety of potentially harmful substances that animals deliberately or unconsciously eat, drink, or inhale.

Perhaps even more remarkable is the way in which the kidney does its job. These small organs, which in man weigh less than 0.5% of the body weight, receive nearly 25% of the total cardiac output, amounting to approximately 2,000 liters of blood per day. This vast blood flow is channeled to approximately 2 million **nephrons,** which comprise the bulk of the two human kidneys. Each nephron is a tiny excretory unit consisting of a pressure filter **(glomerulus)** and a long **nephric tubule.** Urine formation begins in the glomerulus where an ultrafiltrate of the blood is squeezed into the nephric tubule by the hydrostatic blood pressure. The ultrafiltrate then flows steadily down the twisted tubule. During its travel some substances are added to

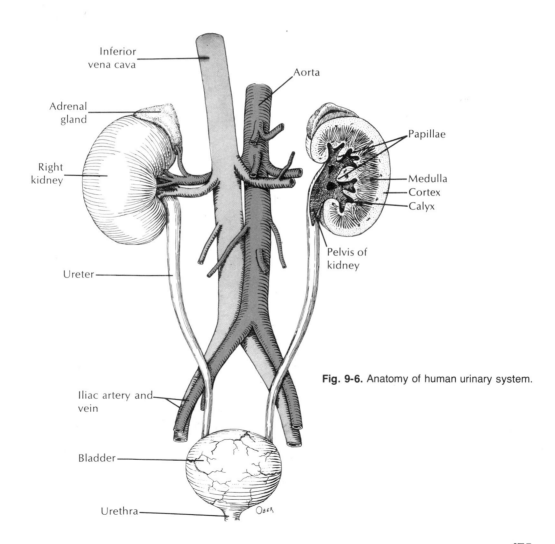

Fig. 9-6. Anatomy of human urinary system.

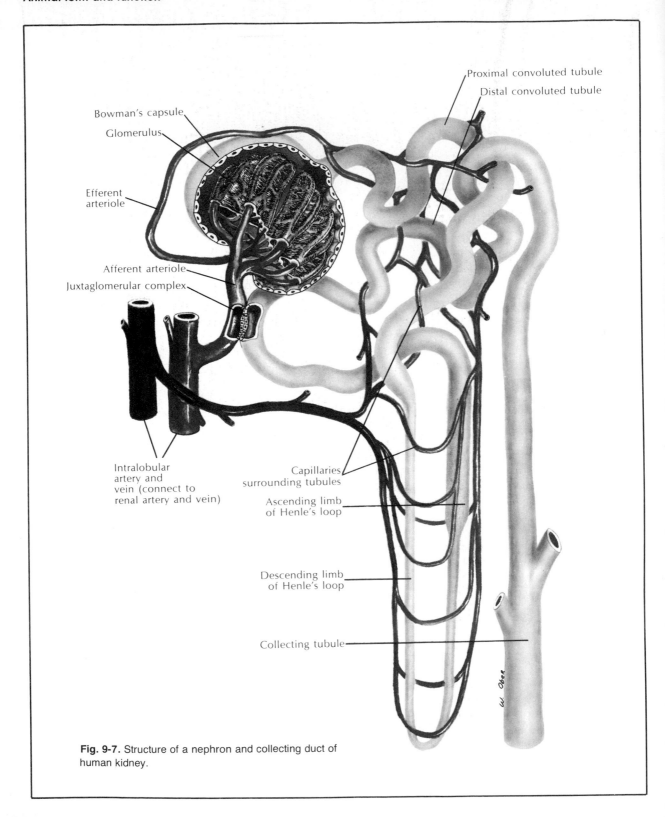

Proximal convoluted tubule

Distal convoluted tubule

Bowman's capsule

Glomerulus

Efferent arteriole

Afferent arteriole

Juxtaglomerular complex

Intralobular artery and vein (connect to renal artery and vein)

Capillaries surrounding tubules

Ascending limb of Henle's loop

Descending limb of Henle's loop

Collecting tubule

Fig. 9-7. Structure of a nephron and collecting duct of human kidney.

and others are subtracted from the ultrafiltrate. The final product of this process is urine.

All mammalian kidneys are paired structures that lie embedded in fat, anchored against the dorsal abdominal wall. Each of the two **ureters** 25 to 30 cm (10 to 12 in) long in man, extend from the **renal pelvis** to the dorsal surface of the **urinary bladder.** Urine is discharged from the bladder by way of the single **urethra** (Fig. 9-6). In the male the urethra is the terminal portion of the reproductive system as well as of the excretory system. In the female the urethra is solely excretory in function, opening to the outside just anterior to the vagina.

Since each of the thousands of nephrons in the kidney forms urine independently, each is in a way a tiny, self-contained kidney that produces a miniscule amount of urine—perhaps only a few nanoliters per hour. This amount, multiplied by the number of nephrons in the kidney, produces the total urine flow. The kidney is an "in parallel" system of independent units. However, these "independent" nephrons actually work together to create large osmotic gradients in the kidney medulla. This makes it possible for the

mammalian kidney to concentrate urine well above the salt concentration of the blood.

As previously indicated, the nephron, with its pressure filter and tubule, is intimately associated with the blood circulation (Fig. 9-7). Blood from the aorta is delivered to the kidney by way of the large **renal artery,** which breaks up into a branching system of smaller arteries. The arterial blood flows to each nephron through an **afferent arteriole** to the **glomerulus** (glo-mer'yoo-lus), which is a tuft of blood capillaries enclosed within a thin, cuplike **Bowman's capsule.** Blood leaves the glomerulus via the **efferent arteriole.** This vessel immediately breaks up again into an extensive system of capillaries, the **peritubular capillaries,** which completely surround the nephric tubules. Finally the blood from these many capillaries is collected by veins that unite to form the **renal vein.** This vein returns the blood to the vena cava.

Glomerular filtration

Let us now return to the glomerulus, where the process of urine formation begins. The glomerulus acts as a specialized mechanical filter in which a protein-

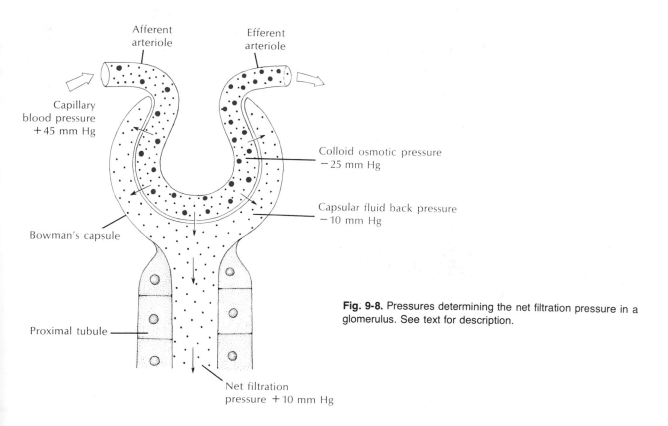

Afferent arteriole

Efferent arteriole

Capillary blood pressure +45 mm Hg

Colloid osmotic pressure −25 mm Hg

Capsular fluid back pressure −10 mm Hg

Bowman's capsule

Proximal tubule

Net filtration pressure +10 mm Hg

Fig. 9-8. Pressures determining the net filtration pressure in a glomerulus. See text for description.

free filtrate resembling plasma is driven by the blood pressure across the capillary walls and into the fluid-filled space of Bowman's capsule. As shown in Fig. 9-8, the net filtration pressure is the difference between the blood pressure in the glomerular capillaries, believed to be approximately 45 mm Hg, and the opposing colloid osmotic and hydrostatic back pressures. Most important of these negative pressures is the colloid (protein) osmotic pressure, which is created because the proteins are too large to pass the glomerular membrane. The unequal distribution of protein causes the water concentration of the plasma to be less than the water concentration of the ultrafiltrate in Bowman's capsule. The osmotic gradient created, approximately 25 mm Hg, opposes filtration. Though small, net filtration pressure of 10 mm Hg is sufficient to force the ultrafiltrate that is formed down the nephric tubule.

The nephric tubule consists of several segments. The first segment, the **proximal convoluted tubule,** leads into a long, thin-walled, hairpin loop called the **loop of Henle** (Fig. 9-7). This loop drops deep into the medulla of the kidney and then returns to the cortex to join the third segment, the **distal convoluted tubule.** The collecting duct empties into the kidney **pelvis,** a cavity that collects the urine before it passes into the **ureter,** on its way to the **urinary bladder** (Fig. 9-6).

Tubular modification of the formative urine

The ultrafiltrate that enters this complex tubular system must undergo extensive modification before it becomes urine. Approximately 200 liters of filtrate are formed each day by the average person's kidneys. Obviously the loss of this volume of body water, not to mention the many other valuable materials present in the filtrate, cannot be tolerated. How does tubular action convert the plasma filtrate into urine?

Two general processes are involved, **tubular reabsorption** and **tubular secretion.** Since the nephric tubules are at all points in close contact with the peritubular capillaries, materials can be transferred from the tubular lumen to the capillary blood plasma (tubular reabsorption) or from the blood plasma to the tubular lumen (tubular secretion).

Tubular reabsorption. The plasma contains a great variety of ions and molecules. With the exception of the plasma proteins, which are too large to pass the glomerular filter, all the plasma components are fil-

tered and most are reabsorbed. Some vital materials, such as glucose and amino acids, are completely reabsorbed. Others, such as sodium, chloride, and most other minerals, undergo variable reabsorption. That is, some are strongly reabsorbed and others weakly reabsorbed, depending on the body's need to conserve each mineral. Much of this reabsorption is by **active transport,** in which cellular energy is used to transport materials from the tubular fluid, across the cell, and into the peritubular blood that returns them to the general circulation.

For most substances there is an upper limit to the amount of substance that can be reabsorbed. This upper limit is termed the **transport maximum** for that substance. For example, glucose is normally completely reabsorbed by the kidney because the transport maximum for the glucose reabsorptive mechanism is poised well above the amount of glucose normally present in the plasma filtrate (approximately 100 mg per 100 ml of filtrate). If the plasma glucose level rises above normal, a condition called hyperglycemia, the concentration in the filtrate rises accordingly and a greater amount of glucose is presented to the proximal tubule for reabsorption. As the level rises, eventually a point is reached (approximately 300 mg per 100 ml of plasma) at which the reabsorptive capacity of the tubular cells is saturated. If the plasma level continues to rise, glucose begins to appear in the urine (glycosuria). This condition happens in the untreated disease diabetes mellitus.

Unlike glucose, which normally never appears in the urine, most of the mineral ions are excreted in the urine in variable amounts. Their excretion is regulated. The reabsorption of sodium, the dominant cation in the plasma, illustrates the flexibility of the reabsorption process. Approximately 600 g of sodium are filtered by the human kidneys every 24 hours. Nearly all of this is reabsorbed, but the exact amount is precisely matched to sodium intake. With a normal sodium intake of 4 g per day, the kidney excretes 4 g and reabsorbs 596 g each day. A person on a low-salt diet of 0.3 g of sodium per day still maintains salt balance because only 0.3 g escapes reabsorption. On a very high salt intake, the kidney can excrete a maximum of approximately 10 g of sodium per day, but not more.

It may seem odd that the maximum sodium excretion is only 10 g per day when approximately 600 g are filtered. We can logically ask, "Why can't more sodium be excreted by simply allowing more to escape tubular reabsorption?'' The answer is that the filtration-reabsorption sequence has a built-in restriction in flexibility. Sodium (and other ions) are reabsorbed in both the proximal and distal portions of the convoluted tubule. Some 85% of the salt and water is reabsorbed in the proximal tubule; this is an **obligatory reabsorption** because it is governed entirely by physical processes (the osmotic pressure of the solutes) and cannot be controlled physiologically. In the distal tubule, however, sodium reabsorption is controlled by **aldosterone,** a steroid hormone of the adrenal cortex. This is called **facultative reabsorption,** meaning that the reabsorption can be adjusted physiologically ac-

cording to need. We can say that proximal reabsorption is involuntary and distal reabsorption is voluntary, although, of course, we are not aware of the adjustments the kidney is performing on our behalf. The flexibility of distal reabsorption varies considerably in different animals: it is restricted in man but very broad in many rodents. These differences have appeared because of selective pressures during evolution that fitted rodents for dry environments where they must conserve water and at the same time excrete considerable sodium.

Tubular secretion. In addition to reabsorbing large amounts of materials from the plasma filtrate, the

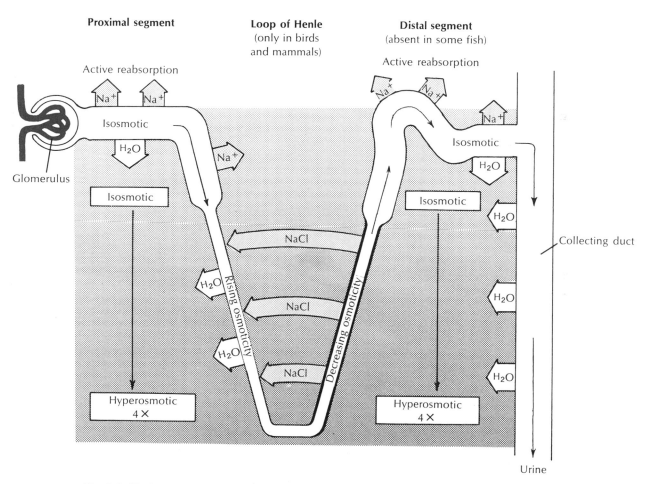

Fig. 9-9. Mechanism of urine concentration in man. Sodium chloride is actively pumped from the ascending limb of loop of Henle and passively reenters the descending limb, building up the osmotic concentration to four times that of blood. This creates an osmotic gradient for the controlled reabsorption of water from the collecting duct. Many rodent and desert mammals have relatively longer loops of Henle and can produce urine much more concentrated than man's.

kidney tubules are able to secrete certain substances into the tubular fluid. This process, which is the reverse of tubular reabsorption, enables the kidney to build up the urine concentrations of materials to be excreted, such as hydrogen and potassium ions, drugs, and various foreign organic materials. The distal tubule is the site of most tubular secretion.

In the kidney of bony marine fishes, reptiles, and birds, tubular secretion is a much more highly developed process than it is in mammalian kidneys. Marine bony fishes actively secrete large amounts of magnesium and sulfate, which are by-products of their mode of osmotic regulation (described in a later chapter, p. 514). Reptiles and birds excrete uric acid instead of urea as their major nitrogenous waste. This material is actively secreted by the tubular epithelium. Since uric acid is nearly insoluble, it forms crystals in the urine and requires little water for excretion. Thus the excretion of uric acid is an important adaptation for water conservation.

Water excretion

The total osmotic pressure of the blood is carefully regulated by the kidney. When fluid intake is high, the kidney excretes a dilute urine, saving salts and excreting water. When fluid intake is low, the kidney conserves water by forming a concentrated urine. A dehydrated man can concentrate his urine to approximately four times his blood concentration.

The capacity of the kidney of mammals and some birds to produce a concentrated urine involves the loop of Henle, the long hairpin loop between the proximal and distal tubules that extends into the renal medulla. Although the loop of Henle was believed to be the locus for urine concentration, the mechanism has only recently been satisfactorily explained. The loops of Henle constitute a **countercurrent multiplier system.** Flow is in opposite directions in the two limbs, hence the name "countercurrent."

The functional characteristics of this system are as follows. Sodium chloride is actively transported out of the ascending limb and into the surrounding tissue fluid (Fig. 9-9). The ascending limb is relatively impermeable to both sodium chloride and water, so there is minimal passive reentry of sodium chloride or water into this limb. However, the descending limb, which does not transport sodium chloride, is permeable to sodium chloride and water. Consequently the sodium chloride that has been actively pumped out the

ascending limb tends to passively enter the descending limb from the tissue fluid. By cycling sodium between the two opposing limbs, the concentration of urine becomes multiplied in the bottom of the loop. A tissue fluid osmotic concentration is established that is greatest at the bottom of the loop deep in the medulla and lowest at the top of the loop in the cortex.

The actual concentrating of the urine, however, does not occur in the loops of Henle, but in the collecting ducts that lie parallel to the loops of Henle. As the urine flows down the collecting duct into regions of increasing sodium concentration, water is osmotically withdrawn from the urine. The amount of water saved and the final concentration of the urine depend on the permeability of the walls of the collecting duct. This is controlled by the **antidiuretic hormone** (ADH, or vasopressin), which is released by the posterior pituitary gland (neurohypophysis). The release of this hormone is governed in turn by special receptors in the brain that constantly sense the osmotic pressure of the blood. When the blood osmotic pressure drops, as during dehydration, the secretion of ADH is increased. ADH increases the permeability of the collecting duct, probably by expanding pore size in the duct membrane. The result is that water diffuses out of the collecting duct into the surrounding interstitial fluid, and the urine becomes more concentrated. Given this sequence of events for dehydration, we can readily imagine the response of this system to overhydration. ADH secretion decreases, the collecting ducts become relatively impermeable, and urine flow is high.

The varying ability of different mammals to form a concentrated urine is closely correlated with the length of the loops of Henle. The beaver, which has no need to conserve water in its aquatic environment, has short loops and can concentrate its urine to only approximately twice that of the blood plasma. Man with relatively longer loops can concentrate his urine 4.2 times that of the blood. As we would anticipate, desert mammals have much greater urine concentrating powers. The camel can produce a urine 8 times the plasma concentration, the gerbil 14 times, and the Australian hopping mouse 22 times. In this creature, the greatest urine concentrator of all, the loops of Henle extend to the tip of a long renal papilla that pushes out into the mouth of the ureter.

SELECTED REFERENCES

(See also general references for Part two, p. 274.)

Edney, E. B. 1957. The water relations of terrestrial arthropods. New York, Cambridge University Press.

Lockwood, A. P. M. 1963. Animal body fluids and their regulation. Cambridge, Mass., Harvard University Press. *This concise volume deals with the physiology of body fluid regulation in both invertebrates and vertebrates.*

Potts, W. T. W., and G. Parry. 1964. Osmotic and ionic regulation in animals. New York, Macmillan, Inc. *An excellent treatise on concepts of homeostatic mechanisms, with special emphasis on osmoregulation and fluid balance.*

Riegel, J. A. 1972. Comparative physiology of renal excretion. New York, Hafner Publishing Co. *Useful review of animal excretion, beginning with vertebrates and ending with the simplest invertebrate systems and a helpful closing chapter on theory of fluid movement.*

SELECTED SCIENTIFIC AMERICAN ARTICLES

Smith, H. W. 1953. The kidney. **188:**40-48 (Jan.). *The structure, physiology, and evolution of the vertebrate kidney.*

Solomon, A. K. 1962. Pumps in the living cell. **207:**100-108 (Aug.). *Active transport processes in kidney tubules are described.*

Too far from water to perform its curious habit of "washing" its food, a raccoon *(Procyon lotor)* turns to a farmer's corn to satisfy its nutritional needs. Intelligent and adaptable, the raccoon is an opportunistic feeder that diets on crayfishes, small birds, and mammals and has no inhibitions about raiding poultry and vegetable gardens. (Photo by L. L. Rue III.)

10 NUTRITION AND CELLULAR METABOLISM

FEEDING MECHANISMS

DIGESTION

    Intracellular versus extracellular digestion
    Vertebrate digestion

NUTRITIONAL REQUIREMENTS

CENTRAL ROLE OF ENZYMES IN THE LIVING PROCESS

CELLULAR METABOLISM

All organisms require energy to maintain their highly ordered and complex structure. This energy is chemical energy that is released by transforming complex compounds acquired from the organism's environment into simpler ones. Obviously, if living organisms must depend on the breakdown of complex foodstuffs to build and maintain their own complexity, these foodstuffs must somehow be synthesized in the first place. Most of the energy for this synthesis is provided by the powerful radiations of the sun, the one great source of energy reaching our planet that is otherwise a virtually isolated system.

At this point, it may be helpful to summarize an earlier discussion (Chapter 1) on energy flow through living communities. Organisms capable of capturing the sun's energy by the process of photosynthesis are, of course, the green plants. Green plants are **autotrophic organisms** capable of synthesizing all the essential organic compounds needed for life. Autotrophic organisms need only inorganic compounds absorbed from their surroundings to provide the raw materials for synthesis and growth. Most autotrophic organisms are the chlorophyll-bearing **phototrophs,** although some, the chemosynthetic bacteria, are **chemotrophs;** they gain energy from inorganic chemical reactions.

Almost all animals are **heterotrophic organisms** that depend on already synthesized organic compounds for their nutritional needs. Animals, with their limited capacities to perform organic synthesis, must feed on plants and other animals to obtain the materials they will use for growth, maintenance, and the reproduction of their kind. The foods of animals, usually the complex tissues of other organisms, can seldom be utilized directly. Food is usually too bulky to be absorbed by the body cells and may contain material of no nutritional value as well. Consequently food must be broken down, or digested, into soluble molecules sufficiently small to be utilized. One important difference then between autotrophs and heterotrophs is that the latter must have digestive systems.

Animals may be divided into a number of categories on the basis of dietary habits. **Herbivorous** animals feed mainly on plant life. **Carnivorous** animals feed mainly on herbivores and other carnivores. **Omnivorous** forms eat both plants and animals. A fourth category is sometimes distinguished, the **insectivorous** animals, which are those birds and mammals that subsist chiefly on insects.

The ingestion of foods and their simplification by digestion are only initial steps in nutrition. Foods reduced by digestion to soluble, molecular form are **absorbed** into the circulatory system and **transported** to the tissues of the body. There they are **assimilated** into the protoplasm of the cells. Oxygen is also transported by the blood to the tissues, where food products are **oxidized,** or burned to yield energy and heat. Much food is not immediately utilized but is **stored** for future use. Then the wastes produced by oxidation must be **excreted.** Food products unsuitable for digestion are rejected by the digestive system and are **egested** in the form of feces.

The sum total of all these nutritional processes is called **metabolism.** Metabolism includes both constructive and tearing-down processes. When molecular fragments are formed into larger molecules or when substances are built into new tissues or stored for later use, such processes are called **synthesis** or **anabolic reactions.** The breaking down of complex materials into simpler ones, usually for the release of energy, is called a **degradative** or **catabolic reaction.** Both processes occur simultaneously in all living cells.

FEEDING MECHANISMS

Only a few animals can absorb nutrients directly from their external environment. Blood parasites and intestinal parasites may derive all their nourishment as primary organic molecules by surface absorption; some aquatic invertebrates may soak up part of their nutritional needs directly from the water. For most animals, however, working for their meals is the main business of living, and the specializations that have evolved for food procurement are almost as numerous as species of animals. In this brief discussion we consider some of the major food-gathering devices.

Feeding on particulate matter

Drifting microscopic particles fill the upper few hundred feet of the ocean. Most of this uncountable multitude is **plankton,** plant and animal microorganisms too small to do anything but drift with the ocean's currents. The rest is organic debris, the disintegrating remains of dead plants and animals. Altogether this oceanic swarm forms the richest life domain on earth. It is preyed on by numerous larger animals, invertebrates and vertebrates, using a variety of feeding mechanisms. Some protozoans, such as the ameboid sarcodines, ingest particulate food by a process called **phagocytosis.** The animal, stimulated by the proximity

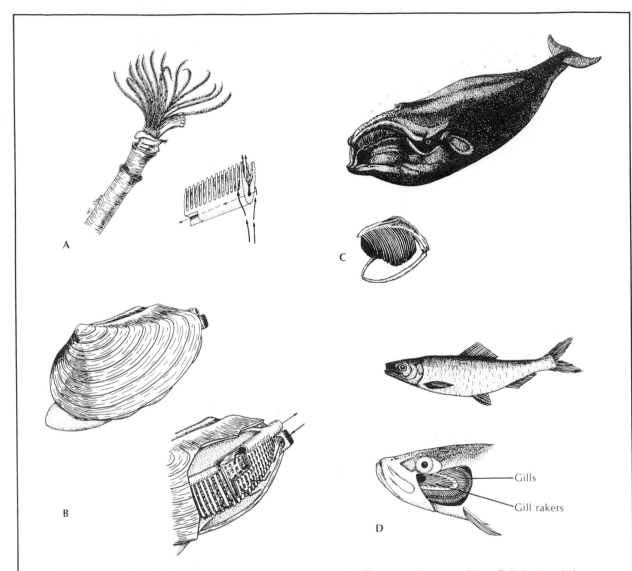

Fig. 10-1. Some filter feeders and their feeding mechanisms. **A,** The marine fan worm (class Polychaeta, phylum Annelida) has a crown of tentacles. Numerous cilia on the edges of the tentacles draw water *(solid arrows)* between pinnules where food particles are entrapped in mucus; particles are then carried down a "gutter" in the center of the tentacle to the mouth *(broken arrows).* **B,** Bivalve molluscs (class Pelecypoda, phylum Mollusca) use their gills as feeding devices, as well as for respiration. Water currents created by cilia on the gills carry food particles into the inhalant siphon and between slits in the gills where they are entangled in a mucous sheet covering the gill surface. Particles are then transported by ciliated food grooves to the mouth (not shown). Arrows indicate direction of water movement. **C,** Whalebone whales (class Mammalia, phylum Chordata) filter out plankton, principally large copepods called "krill," with whalebone, or baleen. Water enters the swimming whale's open mouth by the force of the animal's forward motion and is strained out through the more than 300 horny baleen plates that hang down like a curtain from the roof of the mouth. Krill and other plankters caught in the baleen are periodically wiped off with the huge tongue and swallowed. **D,** Herring and other filter-feeding fishes (class Osteichthyes, phylum Chordata) use gill rakers, which project forward from the gill bars into the pharyngeal cavity to strain off plankters. Herring swim almost constantly, forcing water and suspended food into the mouth; food is strained out by the gill rakers, and the water passes on out the gill openings.

of food, pushes out armlike extensions of the plasma-lemma (cell membrane) and engulfs the particle into a food vacuole, in which it is digested. Other protozoans have specialized openings, called **cytostomes,** through which the food passes to be enclosed in a food vacuole.

By far the most important method to have evolved for particle feeding is **filter feeding** (Fig. 10-1). It is a primitive, but immensely successful and widely employed mechanism. The majority of filter feeders employ ciliated surfaces to produce currents that draw drifting food particles into their mouths. Most filter-feeding invertebrates, such as the tube-dwelling worms and bivalve molluscs, entrap the particulate food in mucus sheets that convey the food into the digestive tract. Filter feeding is characteristic of a sessile way of life; the ciliary currents serve to bring the food to the immobile or slow-moving animal. However, active feeders such as tiny copepod crustaceans and herring are also filter feeders, as are immense baleen whales. The vital importance of one component of the plankton, the diatoms, in supporting a great pyramid of filter-feeding animals is stressed by N. J. Berrill*:

A humpback whale . . . needs a ton of herring in its stomach to feel comfortably full—as many as five thousand individual fish. Each herring, in turn, may well have 6,000 or 7,000 small crustaceans in its own stomach, each of which contains as many as 130,000 diatoms. In other words, some 400 billion yellow-green diatoms sustain a single medium-sized whale for a few hours at most.

Filter feeding utilizes the abundance and extravagance of life in the sea.

Filter feeders are as a rule nonselective and omnivorous. Sessile filter feeders take what they can get, having only the options of continuing or ceasing to filter. Active filter feeders, however, such as fish and baleen whales, are much more selective in their feeding.

Feeding on food masses

Some of the most interesting animal adaptations are those that have evolved for procuring and manipulating solid food. Such adaptations and the animals bearing them are partly shaped by what the animal eats.

Predators must be able to locate prey, capture it, hold it, and swallow it. Most animals use teeth for this purpose (Fig. 10-2). Although teeth are variable in size, shape, and arrangement, vertebrates as different

*Berrill, N. J. 1958. You and the universe, New York, Dodd, Mead & Co.

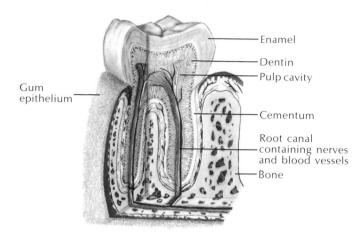

Fig. 10-2. Structure of human molar tooth. Tooth is built of three layers of calcified tissue covering: enamel, which is 98% mineral and hardest material in the body; dentin, which composes the mass of the tooth and is approximately 75% mineral; and cementum, which forms a thin covering over the dentin in the root of the tooth and is very similar to dense bone in composition. Pulp cavity contains loose connective tissue, blood vessels, nerves, and tooth-building cells. Roots of the tooth are anchored to the wall of the socket by a fibrous connective tissue layer called the "periodontal membrane." (After Netter, F. H. 1959. The Ciba collection of medical illustrations, vol. 3. Summit, N.J., Ciba Pharmaceutical Products, Inc.)

as fish and mammals sometimes have remarkably similar tooth arrangements for seizing the prey and cutting it into pieces small enough to swallow.

Mammals characteristically have four different types of teeth, each adapted for specific functions. **Incisors** are for biting and cutting; **canines** are designed for seizing, piercing, and tearing; **premolars** and **molars,** at the back of the jaws, are for grinding and crushing (Fig. 10-3). This basic pattern is often greatly modified (see Fig. 29-6, p. 579). Herbivores have suppressed canines but have well-developed molars with enamel ridges for grinding. Such teeth are usually high crowned, in contrast to the low-crowned teeth of carnivores. The well-developed, self-sharpening incisors of rodents grow throughout life and must be worn away by gnawing to keep pace with growth. Some teeth have become so highly modified that they are no longer useful for biting or chewing food. An elephant's tusk is a modified upper incisor used for defense, attack, and rooting, whereas the male wild boar has modified canines that are used as weapons.

Many carnivores among the fishes, amphibians, and

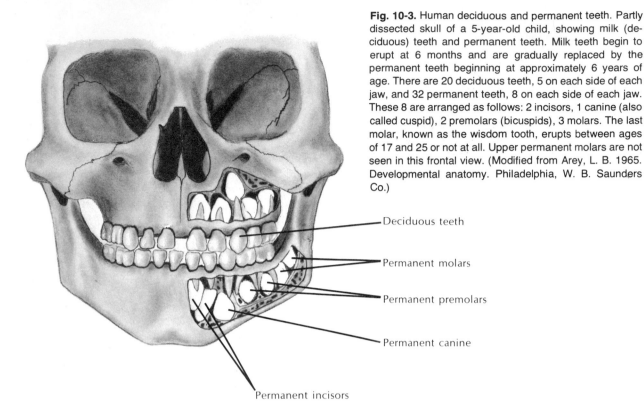

Fig. 10-3. Human deciduous and permanent teeth. Partly dissected skull of a 5-year-old child, showing milk (deciduous) teeth and permanent teeth. Milk teeth begin to erupt at 6 months and are gradually replaced by the permanent teeth beginning at approximately 6 years of age. There are 20 deciduous teeth, 5 on each side of each jaw, and 32 permanent teeth, 8 on each side of each jaw. These 8 are arranged as follows: 2 incisors, 1 canine (also called cuspid), 2 premolars (bicuspids), 3 molars. The last molar, known as the wisdom tooth, erupts between ages of 17 and 25 or not at all. Upper permanent molars are not seen in this frontal view. (Modified from Arey, L. B. 1965. Developmental anatomy. Philadelphia, W. B. Saunders Co.)

Deciduous teeth

Permanent molars

Permanent premolars

Permanent canine

Permanent incisors

reptiles swallow their prey whole. Snakes and some fishes can swallow enormous meals. This, together with the absence of limbs, is associated with some striking feeding adaptations in these groups—recurved teeth for seizing and holding the prey and distensible jaws and stomachs to accommodate their large and infrequent meals.

Teeth are not vertebrate innovations; biting, scraping, and gnawing devices are common in the invertebrates. Insects, for example, have three pairs of appendages on their heads that serve variously as jaws, teeth, chisels, tongues, or sucking tubes. Usually the first pair serves as crushing teeth; the second, grasping jaws; and the third, a probing and tasting tongue.

Herbivorous, or plant-eating, animals, whether vertebrate or invertebrate, have evolved special devices for crushing and cutting plant material. Despite its abundance on earth, the woody cellulose that encloses plant cells is to many animals an indigestible and useless material; herbivores, however, make use of intestinal microorganisms to digest cellulose, once it is ground up. Certain invertebrates such as snails have

rasplike, scraping mouthparts. Insects such as locusts have grinding and cutting mandibles; herbivorous mammals such as horses and cattle use wide, corrugated molars for grinding. All these mechanisms serve to disrupt the tough cellulose cell wall, to accelerate its digestion by intestinal microorganisms, as well as to release the cell contents for direct enzymatic breakdown.

Feeding on fluids

Fluid feeding is especially characteristic of parasites, but is certainly practiced among free-living forms as well. Most internal parasites (endoparasites) simply absorb the nutrient surrounding them, unwittingly provided by the host. External parasites (ectoparasites) such as leeches, lampreys, parasitic crustaceans, and insects use a variety of efficient piercing and sucking mouthparts to feed on blood or other body fluid. Unfortunately for man and other warm-blooded animals, the ubiquitous mosquito excels in its bloodsucking habit. Alighting gently, the mosquito sets about puncturing its prey with an array of six needle-

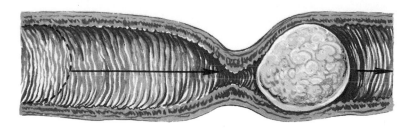

Fig. 10-4. Peristalsis. Food is pushed along before a wave of circular muscle contraction. (From Schottelius, B. A., and D. D. Schottelius. 1973. A textbook of physiology, ed. 17. St. Louis, The C. V. Mosby Co.)

like mouthparts (Fig. 21-23, p. 431). One of these is used to inject an anticoagulant saliva (responsible for the irritating itch that follows the "bite" and serving as vector for microorganisms causing malaria, yellow fever, encephalitis, and other diseases); another mouthpart is a channel through which the blood is sucked. It is of little comfort that only the female of the species dines on blood. Far less troublesome to man are the free-living butterflies, moths, and aphids that suck up plant fluids with long, tubelike mouthparts.

DIGESTION

In the process of digestion, which means literally "carrying asunder," organic foods are mechanically and chemically broken down into small units for absorption. Animal foods vary enormously, having in common only that all are organic in composition. Even though food solids consist principally of carbohydrates, proteins, and fats, the very components that make up the body of the consumer, these components must nevertheless be reduced to their simplest molecular units before they can be utilized. Each animal reassembles some of these digested and absorbed units into organic compounds of the animal's own unique pattern. Cannibals enjoy no special metabolic benefit from eating their own kind; they digest their victims just as thoroughly as they do food of another species!

The digestive tract is actually an extension of the outside environment into or through the animal. Since most animals eat all manner of organic and inorganic materials, the gut's lining must be something like the protective skin; yet it must be permeable so that foodstuffs can be absorbed. Once in the gut the foods are digested and absorbed as they are slowly moved through it.

Movement is either by **cilia** or by **musculature.** In general, the filter feeders that use cilia to feed, such as bivalve molluscs, also use cilia to propel the food through the gut. Animals feeding on bulky foods rely

upon well-developed gut musculature. As a rule the gut is lined with two opposing layers of muscle—a longitudinal layer, in which the smooth muscle fibers run parallel with the length of the gut, and a circular layer, in which the muscle fibers embrace the circumference of the gut. This arrangement is ideal for mixing and propelling foods. The most characteristic gut movement is **peristalsis** (Fig. 10-4). In this movement a wave of circular muscle contraction sweeps down the gut for some distance, pushing the food along before it. The peristaltic waves may start at any point and move for variable distances. Also characteristic of the gut are **segmentation** movements that divide and mix the food.

Intracellular versus extracellular digestion

Man and other vertebrates and the higher invertebrates digest their food **extracellularly** by secreting digestive juices into the intestinal lumen. There, foodstuffs are enzymatically split into molecular units small enough to be selectively absorbed by the intestinal epithelium, transported by the circulation, and utilized by all body cells. Digestion then occurs outside the body's tissues.

Intracellular digestion is a primitive process typical of the lower invertebrates. This type of digestion is best illustrated by the single-celled Protozoa, which capture food particles by phagocytosis, enclose these particles within food vacuoles, and then digest them. Obviously the big limitation to intracellular digestion is that only small particles of food can be handled. Nevertheless many multicellular invertebrates have intracellular digestion. Intracellular digestion is typical of filter-feeding marine animals such as brachiopods, rotifers, bivalves, and cephalochordates, as well as the coelenterates and flatworms. In all of these forms the food particle, phagocytized by the cell, is enclosed within a membrane as a food vacuole (Fig. 10-5). Digestive enzymes are then added. The products of

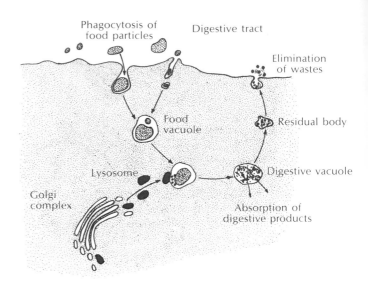

Fig. 10-5. Intracellular digestion. Lysosomes containing digestive enzymes (lysozymes) are produced within the cell, possibly by the Golgi complex. Lysosomes fuse with food vacuoles and release enzymes that digest the enclosed food. Usable products of digestion are absorbed into the cytoplasm, and indigestible wastes are expelled.

digestion, the simple sugars, amino acids, and other molecules, are absorbed into the cell cytoplasm where they may be utilized directly or may be transferred to other cells. The inevitable food wastes are extruded from the cell.

It is believed that cellular enzymes are packaged into membrane-bound vacuoles called **lysosomes** (Fig. 10-5). These somehow join with and discharge their enzymes into the food vacuoles. All cells seem to contain lysosomes, even those of higher animals that do not practice intracellular digestion. Lysosomes have been called "suicide-bags" because they rupture spontaneously in dying or useless cells, digesting the cell contents. Lysosomes also play a role in the lives of healthy cells in cleaning up residues left by growth processes.

It is probable that the obvious limitations of intracellular digestion were responsible for shaping the evolution of extracellular digestion. Extracellular digestion offers several advantages: bulky foods may be ingested; the digestive tract can be smaller, more specialized, and more efficient; and food wastes are more easily discarded. Only with extracellular digestion could the enormous variation in feeding methods of the higher animals have evolved.

Vertebrate digestion

The vertebrate digestive plan is similar to that of the higher invertebrates. Both have a highly differentiated alimentary canal with devices for increasing the surface area, such as increased length, inside folds, and diverticula. The more primitive fishes (lampreys and sharks) have longitudinal or spiral folds in their intestines. Higher vertebrates have developed elaborate folds and small fingerlike projections (**villi**). Also the electron microscope reveals that each cell lining the intestinal cavity is bordered by hundreds of short, delicate processes called **microvilli** (Fig. 10-6). These processes, together with larger villi and intestinal folds, may increase the internal surface of the intestine more than a million times compared to a smooth cylinder of the same diameter. The absorption of food molecules is enormously facilitated as a result.

Digestion in man, an omnivore

Anatomy of digestive system. The structural plan of the digestive system of man is shown in Fig. 10-6. The **mouth** is provided with teeth and a tongue for grasping, masticating, manipulating, and swallowing the food. Three pairs of salivary glands lubricate the food and, in man at least, perform limited digestion. In man and other mammals two sets of teeth are formed during life—the temporary, or "milk," teeth (also called deciduous teeth) and the permanent teeth (Fig. 10-3).

The **pharynx** is the throat cavity that serves for the passage of food. It is actually a complex reception chamber, receiving openings from (1) the nasal cavity,

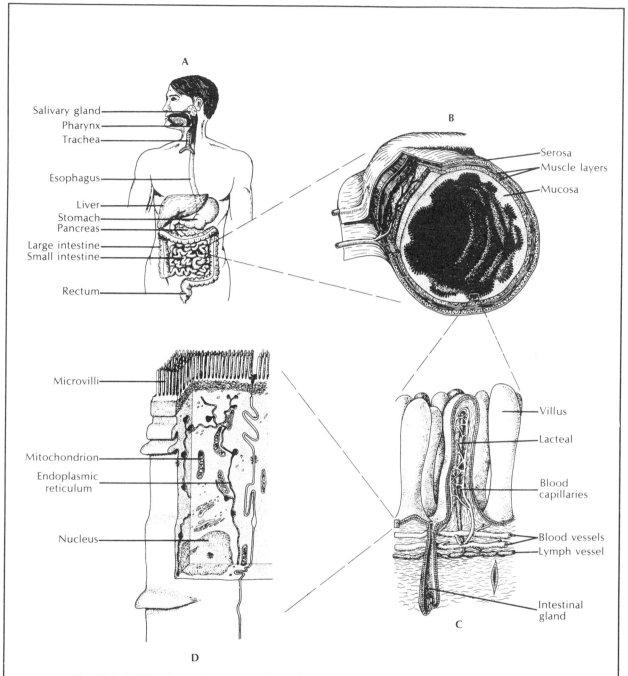

Fig. 10-6. A, Digestive system of man. **B,** Portion of small intestine. **C,** Portion of mucosal lining of intestine, showing fingerlike villi. **D,** Optical section of single lining cell, as shown by electron microscope.

(2) the mouth, (3) the middle ear by way of two eustachian tubes, (4) the esophagus, and (5) the trachea via the glottis.

The **esophagus** is a muscular tube connecting pharynx and stomach. It opens into the stomach at the cardiac opening.

The **stomach** is an enlargement of the gut between esophagus and intestine. In man it is divided into the **cardiac** region (adjacent to esophagus), **fundus** (central region), and **pyloric antrum** region (adjacent to intestine). The stomach is principally a storage organ but aids in digestion.

The **small intestine** is the principal digestive and absorptive area of the gut. It is divided grossly into three regions—**duodenum, jejunum,** and **ilium.** Two large digestive glands, the **liver** and **pancreas,** empty into the duodenum by the **common bile duct.**

The **large intestine (colon)** in man is divided into **ascending, transverse,** and **descending** portions, with the posterior end terminating in the **rectum** and **anus.** At the junction of the large and small intestines is the **colic cecum** and its vestigial **vermiform appendix.** The large intestine lacks villi but contains glands for lubrication.

The stomach, small intestine, and large intestine are all suspended by **mesenteries,** thin sheets of tissue that are modified from the **peritoneum,** or lining, of the coelom and the abdominal organs. Organs such as liver, spleen, and pancreas arc also held in place by mesenteries, which carry blood and lymph vessels as well as nerves to the various abdominal organs.

Action of digestive enzymes. We have already pointed out that digestion involves both mechanical and chemical alterations of food. Mechanical processes of cutting and grinding by teeth and muscular mixing by the intestinal tract are important in digestion. However, the reduction of foods to small absorbable units relies principally on chemical breakdown by **enzymes.** Enzymes are the highly specific organic catalysts essential to the orderly progression of virtually all life processes.

It is well to state that, although digestive enzymes are probably the best known and most studied of all enzymes, they represent but a small fraction of the numerous, perhaps thousands, of enzymes that ultimately regulate all processes in the body. The digestive enzymes are **hydrolytic** enzymes or **hydrolases,** so called because food molecules are split by the process of **hydrolysis,** that is, the breaking of a

chemical bond by adding the components of water across it:

$$R—R + H_2O \xrightarrow[enzyme]{Digestive} R—OH + H—R$$

In this general enzymatic reaction, R—R represents a food molecule that is split into two products, R—OH and R—H. Usually these reaction products must in turn be split repeatedly before the original molecule has been reduced to its numerous subunits. Proteins, for example, are composed of hundreds, or even thousands, of interlinked amino acids, which must be completely separated before the individual amino acids can be absorbed. Similarly carbohydrates must be reduced to simple sugars. Fats (lipids) are reduced to molecules of glycerol and fatty acids, although some fats, unlike proteins and carbohydrates, may be absorbed without being completely hydrolyzed first. There are specific enzymes for each class of organic compounds. These enzymes are located in various regions of the alimentary canal in a sort of "enzyme chain," in which one enzyme may complete what another has started; the product moves along posteriorly for still further hydrolysis.

Digestion in the mouth. In the mouth, food is broken down mechanically by the teeth and is moistened with saliva from the salivary glands. In addition to **mucin,** which helps to lubricate the food for swallowing, saliva contains the enzyme **amylase.** Salivary amylase is a carbohydrate-splitting enzyme that begins the hydrolysis of plant and animal starches (Fig. 10-7). Starches are long polymers of glucose. Salivary amylase does not completely hydrolyze starch, but breaks it down mostly into two glucose fragments called **maltose.** Some free glucose, as well as longer fragments of starch, are also produced. When the food mass (bolus) is swallowed, salivary amylase continues to act for some time, digesting perhaps half of the starch before the enzyme is inactivated by the acid environment of the stomach. Further starch digestion resumes beyond the stomach in the intestine.

Swallowing. Swallowing is a reflex process involving both voluntary and involuntary components. Swallowing begins with the tongue pushing the moistened food bolus toward the pharynx. The nasal cavity is reflexly closed by raising the soft palate. As the food slides into the pharynx, the epiglottis is tipped down over the windpipe, nearly closing it. Some particles of food may enter the opening of the windpipe, but are prevented from going further by contraction of la-

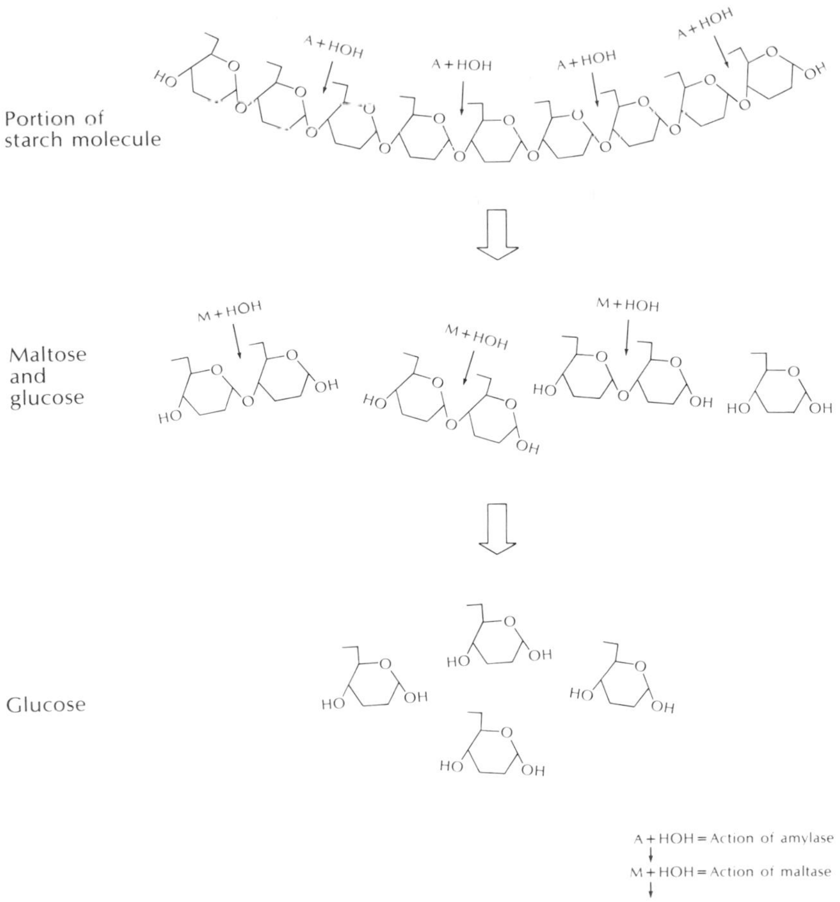

Fig. 10-7. Digestion (hydrolysis) of starch. Long chains of glucose molecules, linked together through oxygen, are first cleaved into disaccharide residues (maltose) by the salivary enzyme amylase. Some glucose may also be split off at the ends of starch chains. The intestinal enzyme maltase then completes the hydrolysis by cleaving the maltose molecules into glucose. A molecule of water is inserted into each enzymatically split bond.

ryngeal muscles. Once in the esophagus, the bolus is forced smoothly toward the stomach by peristaltic contraction of the esophageal muscles.

Digestion in the stomach. When food reaches the stomach, the **cardiac sphincter** opens reflexly to allow entry of the food, then closes to prevent regurgitation back into the esophagus. The stomach is a combination storage, mixing, digestion, and release center. Large peristaltic waves pass over the stomach at the rate of approximately three each minute; churning is most vigorous at the intestinal end where food is steadily released into the duodenum. Approximately 3 liters of **gastric juice** are secreted each day by deep, tubular

glands in the stomach wall. Two types of cells line these glands: (1) **chief cells** secrete an enzyme precursor called **pepsinogen** and (2) **parietal cells** secrete **hydrochloric acid.** Pepsinogen is an inactive form of enzyme that is converted into the active enzyme **pepsin** by hydrochloric acid and by other pepsin already present in the stomach. Pepsin is a **protease** (protein-splitting enzyme) that acts only in an acid medium—pH 1.6 to 2.4. It is a highly specific enzyme that splits large proteins by preferentially breaking down certain peptide bonds scattered along the peptide chain of the protein molecule. Although pepsin, because of its specificity, cannot completely degrade

191

proteins, it effectively breaks them up into a number of small polypeptides. Protein digestion is completed in the intestine by other proteases that can together split all peptide bonds.

Rennin is another enzyme found in the stomachs of the suckling newborn of many mammals, although not of man. Rennin is a milk-curdling enzyme that transforms the proteins of milk into a finely flocculent form that is more readily attacked by pepsin. Rennin extracted from the stomachs of calves is used in cheese-making. Human infants, lacking rennin, digest milk proteins with acidic pepsin, the same as adults do.

The unique ability of the stomach to secrete a strong acid is still an unsolved problem in biology, in part because it has not been possible to collect pure parietal cell secretion or to isolate these cells in culture for study. It is well known that acid solutions can readily destroy organic matter. Since the stomach contains not only a strong acid but also a powerful proteolytic enzyme, it seems remarkable that the stomach mucosa is not digested by its own secretions. That it is not is a result of another protective gastric secretion, **mucin,** a highly viscous organic compound that coats and protects the mucosa from both chemical and mechanical injury. Sometimes, however, the protective mucus coating fails, allowing the gastric juices to begin digesting the stomach. The result is a peptic ulcer. We should note that despite the popular misconception that "acid stomach" is unhealthy, a notion carefully nourished by the makers of patent medicine, stomach acidity is normal and essential.

The secretion of the gastric juices is intermittent. Although a small volume of gastric juice is secreted continuously, even during prolonged periods of starvation, secretion is normally increased by the sight and smell of food, by the presence of food in the stomach, and by emotional states such as anger and hostility.

The most unique and classic investigation in the field of digestion was made by U. S. Army surgeon William Beaumont during the years 1825 to 1833. His subject was a young, hard-living French Canadian voyageur, named Alexis St. Martin, who in 1822 had accidentally shot himself in the abdomen with a musket, the blast "blowing off integuments and muscles of the size of a man's hand, fracturing and carrying away the anterior half of the sixth rib, fracturing the fifth, lacerating the lower portion of the left lobe of the lung and the diaphragm, and perforat-

ing the stomach." Miraculously the wound healed, but a permanent opening, or fistula, was formed that permitted Beaumont to see directly into the stomach. St. Martin became a permanent, although temperamental, patient in Beaumont's care, which included food and housing. Over a period of 8 years, Beaumont was able to observe and record how the lining of the stomach changed under different psychic and physiologic conditions, how foods changed during digestion, the effect of emotional states on stomach motility, and many other facts about the digestive processes of his famous patient.

Digestion in the small intestine. The major part of digestion occurs in the small intestine. Three secretions are poured into this region—**pancreatic juice, intestinal juice,** and **bile.** All of these secretions have a high bicarbonate content, especially the pancreatic juice, which effectively neutralizes the gastric acid, raising the pH of the liquefied food mass, now called **chyme,** from 1.5 to 7 as it enters the duodenum. This change in pH is essential because all the intestinal enzymes are effective only in a neutral or slightly alkaline medium.

Approximately 2 liters of **pancreatic juice** are secreted each day. The pancreatic juice contains several enzymes of major importance in digestion. Two powerful proteases, **trypsin** and **chymotrypsin,** are secreted in inactive form as **trypsinogen** and **chymotrypsinogen.** Trypsinogen is activated in the duodenum by **enterokinase,** an enzyme present in the intestinal juice. Chymotrypsinogen is activated by trypsin. These two proteases continue the enzymatic digestion of proteins begun by pepsin, which is inactivated by the alkalinity of the intestine. Trypsin and chymotrypsin, like pepsin, are highly specific proteases that split apart peptide bonds deep inside the protein molecule. Pancreatic juice also contains **carboxypeptidase,** which splits amino acids off the ends of polypeptides; **pancreatic lipase,** which hydrolyzes fats into fatty acids and glycerol; and **pancreatic amylase,** which is a starch-splitting enzyme identical to salivary amylase in its action.

Intestinal juice from the glands of the mucosal lining furnishes several enzymes. **Aminopeptidase** splits off terminal amino acids from polypeptides; its action is similar to the pancreatic enzyme carboxypeptidase. Three other enzymes are present that complete the hydrolysis of carbohydrates: **maltase** converts maltose to glucose (Fig. 10-7); **sucrase** splits sucrose to glucose and fructose; and **lactase** breaks down lactose (milk sugar) into glucose and galactose.

Bile is secreted by the cells of the **liver** into the **bile duct,** which drains into the upper intestine (duodenum). Between meals the bile is collected into the **gallbladder,** an expansible storage sac that releases the bile when stimulated by the presence of fatty food in the duodenum. Bile contains no enzymes. It is made up of water, bile salts, and pigments. The bile salts (sodium taurocholate and sodium glycocholate) are essential for the complete absorption of fats, which, because of their tendency to remain in large, water-resistant globules, are especially resistant to enzymatic digestion. **Bile salts** reduce the surface tension of fats, so that they are broken up into small droplets by the churning movements of the intestine. This greatly increases the total surface exposure of fat particles, giving the fat-splitting lipases a chance to reduce them. The characteristic golden yellow of bile is produced by the **bile pigments** that are breakdown products of hemoglobin from worn-out red blood cells. The bile pigments also color the feces.

It is well to emphasize the great versatility of the liver. Bile production is only one of the liver's many functions, which include the following: storehouse for glycogen, production center for the plasma proteins, site of protein synthesis and detoxification of protein wastes, destruction of worn-out red blood cells, center for metabolism of fat and carbohydrates, and many others.

Digestion in the large intestine. The liquefied material, called **chyle,** reaching the large intestine, or **colon,** is low in nutrients since most important food materials have already been absorbed into the bloodstream from the small intestine. The main function of the colon is the absorption of water and some minerals from the intestinal chyle that enters. In removing more than half the water from the chyle, the colon forms semisolid feces consisting of undigested food residue, bile pigments, secreted heavy metals, and bacteria. The feces are eliminated from the rectum by the process of **defecation,** a coordinated muscular action that is part voluntary and part involuntary.

The colon contains enormous numbers of bacteria that enter the sterile colon of the newborn infant. In the adult approximately one third of the dry weight of feces is bacteria; these include both harmless bacilli as well as cocci that can cause serious illness if they should escape into the abdomen or bloodstream.

Normally the body's defenses prevent invasion of such bacteria. The bacteria break down organic wastes in the feces and provide some nutritional benefit by synthesizing certain vitamins (biotin and vitamin K), which are absorbed by the body.

Absorption. Most digested foodstuffs are absorbed from the small intestine, where the numerous finger-shaped **villi** provide an enormous surface area through which materials can pass from intestinal lumen into the circulation. Little food is absorbed in the stomach because digestion is still incomplete and because of the limited surface exposure. Some materials, however, such as drugs and alcohol, are absorbed in part there, which explains their rapid action.

The villi (Fig. 10-6) contain a network of blood and lymph capillaries. The absorbable food products (amino acids, simple sugars, fatty acids, glycerol, and triglycerides as well as minerals, vitamins, and water) pass first across the epithelial cells of the intestinal mucosa and then into either the blood capillaries or the lymph system. Small molecules can enter either system, but, since blood flow is several hundred times greater than lymph flow, it is unlikely that the lymph system carries much absorbed material out of the intestine. However, any materials that do enter the lymph system eventually get into the blood by way of the thoracic duct. (See p. 157.)

Carbohydrates are absorbed almost exclusively as simple sugars (for example, glucose, fructose, and galactose) because the intestine is virtually impermeable to polysaccharides. Proteins, too, are absorbed principally as their subunits, amino acids, although it is believed that very small amounts of small proteins or protein fragments may sometimes be absorbed. Simple sugars and amino acids are transferred across the intestinal epithelium by both passive and active processes.

Immediately after a meal these materials are in such high concentration in the gut that they readily diffuse into the blood, where their concentration is initially lower. However, if absorption were passive only, we would expect transfer to cease as soon as the concentrations of a substance became equal on both sides of the intestinal epithelium. This would permit much of the valuable foodstuff to be lost in the feces. In fact, very little is lost because passive transfer is supplemented by an **active transport** mechanism located in the epithelial cells, which pick up the food molecules and transfer them into the blood. Materials are thus moved *against* their concentration gradient, a process that requires the expenditure of energy. Although not all food products are actively transported, those that are, such as glucose, galactose, and most of the amino acids, are handled by transport mechanisms that are specific for each kind of molecule.

NUTRITIONAL REQUIREMENTS

The food of animals must include **carbohydrates, proteins, fats, water, mineral salts,** and **vitamins.** Carbohydrates and fats are required as fuels for energy demands of the body and for the synthesis of various substances and structures. Proteins, or actually the amino acids of which they are composed, are needed for the synthesis of the body's specific proteins and other nitrogen-containing compounds. Water is required as the solvent for the body's chemistry and as the major component of all the body fluids. The inorganic salts are required as the anions and cations of body fluids and tissues and form important structural and physiologic components throughout the body. The vitamins are accessory food factors that are built into the structure of many of the enzymes of the body.

All animals require these broad classes of nutrients, although there are differences in the amounts and kinds of food required. The student should note that of these basic food classes some nutrients are used principally as fuels (carbohydrates and lipids), whereas others are required principally as structural and functional components (proteins, minerals, and vitamins). Any of the basic foods (proteins, carbohydrates, fats) can serve as fuel to supply energy requirements, but no animal can thrive on fuels alone. A **balanced diet** must satisfy all metabolic requirements of the body—requirements for energy, growth, maintenance, reproduction, and physiologic regulation.

The recognition many years ago that many diseases of man and his domesticated animals were caused by or associated with dietary deficiencies led biologists to search for specific nutrients that would prevent such diseases. These studies eventually yielded a list of **"essential" nutrients** for man and other animal species studied. The essential nutrients are those that are needed for normal growth and maintenance and that *must* be supplied in the diet. In other words, it is "essential" that these nutrients be in the diet because the animal cannot synthesize them from other dietary constituents. For man more than 20 organic compounds (amino acids and vitamins) and more than 10

Table 10-1. Nutrients required by man*

Amino acids	Elements	Vitamins
Established as essential		
Isoleucine	Calcium	Ascorbic acid (C)
Leucine	Chlorine	Choline
Lysine	Copper	Folic acid
Methionine	Iodine	Niacin
Phenylalanine	Iron	Pyridoxine (B_6)
Threonine	Magnesium	Riboflavin (B_2)
Tryptophan	Manganese	Thiamine (B_1)
Valine	Phosphorus	Vitamin B_{12}
	Potassium	Vitamins A, D, E,
	Sodium	and K
	Zinc	
Probably essential		
Arginine	Fluorine	Biotin
Histidine	Molybdenum	Pantothenic acid
	Selenium	Polyunsaturated
		fatty acids

*Modified from White, A., P. Handler, and E. L. Smith. 1973. Principles of biochemistry, ed. 5. New York, McGraw-Hill Book Co.

elements have been established as essential (Table 10-1). If we consider that the body contains thousands of different organic compounds, the list in Table 10-1 is remarkably short. Animal cells have marvelous powers of synthesis enabling them to build compounds of enormous variety and complexity from a small, select group of raw materials.

In the average diet of Americans and Canadians, approximately 50% of the total calories (energy content) comes from carbohydrates and 40% comes from lipids. Proteins, essential as they are for structural needs, supply only a little more than 10% of the total calories of the average North American's diet. Carbohydrates are widely consumed because they are more abundant and cheaper than proteins or lipids. Actually man and many other animals can subsist on diets devoid of carbohydrates, provided sufficient total calories and the essential nutrients are present. Eskimos, before the decline of their native culture, lived on a diet that was high in fat and protein and very low in carbohydrate.

Lipids are needed principally to provide energy. Much interest and research has been devoted to lipids in our diets because of the association between fatty diets and the disease **arteriosclerosis** (hardening and narrowing of the arteries). The matter is complex, but evidence suggests that arteriosclerosis may occur when the diet is high in saturated lipids but low in polyunsaturated lipids. For unknown reasons such diets, which are typical of middle-class and affluent North Americans, promote a high blood level of cholesterol, which may deposit in platelike formations in the lining of the major arteries. For this reason the polyunsaturated fatty acids are often considered essential nutrients for man. Generally speaking, animal fat is more saturated, whereas fat from plants is more unsaturated.

Proteins are expensive foods and restricted in the diet. Proteins, of course, are themselves not the essential nutrients, but rather they contain essential amino acids. Of the 20 amino acids commonly found in proteins, 8 and possibly 10 are essential to man (Table 10-1). The rest can be synthesized. We must keep in mind that the terms "essential" and "nonessential" relate only to dietary requirements and to which amino acids can and cannot be synthesized by the body. All 20 amino acids are essential for the various cellular functions of the body. In fact, some of the so-called nonessential amino acids participate in more crucial metabolic activities than the essential amino acids.

In general, animal proteins have more of the essential amino acids than do proteins of plant origin. An adult man would require approximately 67 g of whole wheat bread each day to meet his amino acid requirement, but only 19 g of beefsteak to meet these same requirements. Potato contains only 2% protein; a person would have to eat 3.6 kg (8 pounds) to obtain the minimum daily protein requirement from that source alone. Many plants, however, such as beans and peas, are high-quality protein sources. Even plants deficient in certain amino acids may be used in beneficial combinations. Incaparina, a vegetable mixture distributed in Central and South America, contains 29% whole corn, 29% whole sorghum, 38% cottonseed meal, yeast, $CaCO_3$, and vitamin A and provides a biologic protein value only slightly less than cow's milk.

Because animal proteins are so nutritious, they are in great demand by all countries. North Americans eat far more animal proteins than do Asians and Africans; on the average a North American eats 66 g of animal protein a day, supplemented by milk, eggs, cereals, and legumes. In the Middle East, the individual consumption of protein is 14 g, in Africa 11 g, and in Asia 8 g.

Protein deficiency is a vital factor in the 10,000 deaths estimated by the United Nations as occurring daily from malnutrition. A protein and calorie deficiency disease of children, **kwashiorkor,** is the world's major health problem today and is growing more serious daily. The name literally means "golden boy" and is used because of characteristic changes in pigmentation of skin and hair. It occurs especially in nursing infants displaced from the breast by a newborn sibling. The disease is characterized by retarded growth, anemia, liver and pancreas degeneration, renal lesions, acute diarrhea, and a mortality of 30% to 90%.

Because animal proteins are relatively scarce and expensive, there is presently a great effort by scientists to find cheap, plentiful sources of plant proteins. With the world's population expected to reach 7 billion by the year 2000, the search for protein takes on a desperate urgency. It was altogether fitting that the 1970 Nobel Peace Prize went to the man, Dr. Norman Borlaug, who developed a dwarf wheat variety that has vastly increased the wheat yield in India, West Pakistan, and Mexico, producing what was called the "green revolution." However, Dr. Borlaug himself emphasized that, despite our best efforts to develop more productive grains, mass famine seems inevitable unless the human population is stabilized.

Vitamins are relatively simple organic compounds that are required in small amounts in the diet for specific cellular functions. They are not sources of energy, but are often associated with the activity of important enzymes that have vital metabolic roles. Plants and many microorganisms synthesize all the organic compounds they need; animals, however, have lost certain synthetic abilities during their long evolution and depend ultimately on plants to supply these compounds. Vitamins therefore represent synthetic gaps in the metabolic machinery of animals. We have seen that several amino acids are also dietary essentials. These are not considered vitamins, however, because they usually enter into the actual *structure* of tissues or proteinaceous tissue secretions. Vitamins, on the other hand, are essential *functional* components of enzyme catalytic systems. Nevertheless the distinction between certain vitamins (A, D, E, and K) and other dietary essentials not classified as vitamins is not so clear as it was once believed to be.

Vitamins are usually classified as fat soluble (soluble in fat solvents such as ether) or water soluble. The

195

water-soluble ones include the B complex and vitamin C. The family of B vitamins, so grouped because the original B vitamin was subsequently found to consist of several distinct molecules, tends to be found together in nature. Almost all animals, vertebrate and invertebrate, require the B vitamins; they are "universal" vitamins. The dietary need for vitamin C and the fat-soluble vitamins A, D, E, and K tends to be restricted to the vertebrates, although some are required by certain invertebrates.

CENTRAL ROLE OF ENZYMES IN THE LIVING PROCESS

The whole life process involves numerous chemical reactions occurring within the cells. However, the chemical breakdown of large molecules and the release of energy for cellular activities would not proceed at any meaningful rate without enzymes. Enzymes are **biologic catalysts** that are required for almost every reaction in the body. As every chemist knows, catalysts are chemical substances that accelerate reactions without affecting the end products of the reactions and without being destroyed as a result of the reaction. Enzymes fit this definition, too.

Enzymes are involved in every aspect of life phenomena. They control the reactions by which food is digested, absorbed, and metabolized. They promote the synthesis of structural materials to replace the wear and tear on the body. They determine the release of energy used in respiration, growth, muscle contraction, physical and mental activities, and a host of others. There is little wonder that enzymes are absolutely fundamental to life.

Nature of enzymes. Enzymes are complex molecules varying in size from small, simple proteins with a molecular weight of 10,000 to highly complex molecules with weights up to 1 million. Some enzymes, such as the gastric enzyme pepsin, are pure proteins—delicately folded and interlinked chains of amino acids. Other enzymes contain a special active **prosthetic group** (an attached, nonprotein group) in addition to the protein portion of the molecule. Nearly all the vitamins have been shown to be essential parts of enzyme prosthetic groups. Because vitamins cannot be synthesized by the animal needing them, it is obvious that a dietary vitamin deficiency can be serious.

Naming of enzymes. Enzymes are named for the reactions they catalyze. Usually the suffix *-ase* is added to the root word of the substance, or substrate, on which the enzyme works. Thus sucrase acts on sucrose, lipase acts on lipids, and protease acts on proteins. Enzymes may also be named according to the nature of the reaction. For example, dehydrogenases catalyze dehydrogenations.

Action of enzymes. An enzyme functions by combining in a highly specific way with the substance upon

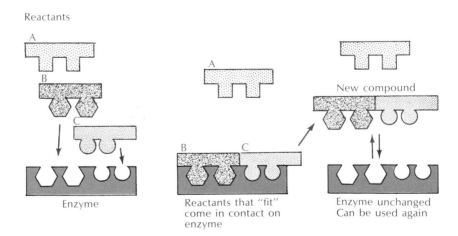

Fig. 10-8. Enzyme action and specificity. Enzymes are thought to have surface configurations that "fit" specific substrates. Here molecules *B* and *C* fit into enzyme surface, but *A* does not. Reactions involving *B* and *C* are speeded up by coming in contact briefly with enzyme. When reaction is complete, the enzyme, still unchanged, can dissociate from the substrate and is free to aid in further reactions. Molecule *A* and others not specific to this enzyme are unaffected by it.

which it acts (the **substrate**). According to the classic **lock-and-key** theory, each enzyme contains an **active site,** which is a unique molecular configuration that is exactly complementary to at least a portion of the specific substrate molecule (Fig. 10-8). By fitting onto the substrate, the enzyme provides a unique chemical environment that somehow makes the substrate molecule more active. Enzyme and substrate combine to form an unstable enzyme-substrate complex.

$$E + S \rightleftharpoons ES \rightleftharpoons ES' \rightleftharpoons E + P$$

With the formation of ES, enough activation energy is introduced to form S', an activated form of the substrate, which splits into its products P. These are liberated from the enzyme, which is restored to its original form.

The classic lock-and-key theory is still largely accepted by biochemists. However, the active site of the enzyme is a flexible surface that infolds and conforms to the substrate, rather than a fixed and nonyielding template as the original theory held. This newer **conformational theory** has not altered a firmly held principle of enzyme action: the enzyme and substrate must combine so that active groups on the enzyme come into precise alignment with reactive sites on the substrate. Only in this way can the substrate be altered chemically. The necessity of correct alignment explains the high specificity of enzymes.

Enzymes that engage in important main-line sequences (for example, the crucial energy-providing reactions of the cell that go on constantly) seem to operate in enzyme sets rather than in isolation. Main-line enzymes are found in relatively high concentrations in the cell, and they may implement quite complex and highly integrated enzymatic sequences. One enzyme carries out one step, then passes the product to another enzyme that catalyzes another step, and so on.

Specificity of enzymes. Most enzymes are highly specific. Such high specificity is a consequence of the exact molecular fit that is required between enzyme and substrate. However, there is some variation in degree of specificity. Some enzymes, such as succinic dehydrogenase, catalyze the oxidation (dehydrogenation) of one substrate only, succinic acid. Others, such as proteases (for example, pepsin and trypsin), act on most any protein. Usually an enzyme takes on one substrate molecule at a time, catalyzes its chemical change, releases the product, and then repeats the

process with another substrate molecule. The enzyme may repeat this process billions of times until it is finally worn out (a few hours to several years) and is broken down by scavenger enzymes in the cell. Some enzymes are able to undergo successive catalytic cycles at dizzying speeds of up to a million cycles per minute; most operate at slower rates. The digestive enzymes, which are secreted into the digestive tract to degrade food materials, are "one-shot" enzymes. After breaking down their substrate, they are themselves digested and lost to the body.

Enzyme-catalyzed reactions. Enzyme-catalyzed reactions are reversible. For example, succinic dehydrogenase catalyzes both the dehydrogenation of succinic acid to fumaric acid and the hydrogenation of fumaric acid to succinic acid.

$$\text{Succinic acid} + H_2O \rightleftharpoons \text{Fumaric acid}$$

Reversibility is signified by the double arrows. Enzymes of this type vastly accelerate reactions in either direction. Many enzymes, however, catalyze reactions almost entirely in one direction. For example, the proteolytic enzyme pepsin can degrade proteins into amino acids, but it cannot accelerate the rebuilding of amino acids into any significant amount of protein. The same is true of most enzymes that catalyze the hydrolysis of large molecules such as nucleic acids, polysaccharides, lipids, and proteins. There is usually one set of reactions and enzymes that break them down, but they must be resynthesized by a different set of reactions that are catalyzed by different enzymes.

This apparent irreversibility exists because the chemical equilibrium usually favors the formation of the smaller degradation products. The net *direction* of any chemical reaction is dependent on the relative energy content of the substances involved, so that other compounds (energy-rich compounds) must participate in the conversion of, for example, amino acids into proteins, thus making the synthetic process different from simply the reverse of the degradation reactions.

Sensitivity of enzymes. Enzyme activity is sensitive to temperature and pH. As a general rule enzymes work faster with increasing temperature, but do so only within certain limits. Moreover this increase in velocity is not proportional to the rise in temperature. Usually the rate is doubled with each 10° C rise, but a change from 20° to 30° C has a greater effect than one from 30° to 40° C. The optimum temperature for animal enzymes is body temperature. Above 40° to 50° C most enzymes become unstable and may be inactivated altogether.

Each enzyme usually works best within a certain range of acidity or alkalinity. Pepsin of the acid gastric juice is most active at approximately pH 1.8; trypsin of the alkaline pancreatic juice is most active at approximately pH 8.2. Many work best when the pH is around neutrality. In strong acid or alkaline solutions, most enzymes irreversibly lose their catalytic power.

CELLULAR METABOLISM

The term "cellular metabolism" refers to the sum total of the chemical processes that are necessary for all the phenomena of life, such as the synthesis of new cell materials, the replacement of that which is destroyed during wear and tear, and whatever is needed by the cell to grow, reproduce, move, etc. For these processes cells require a continuous supply of nutrients obtained from the surrounding extracellular fluid. Living cells, like man-made machines, do work and consequently require fuel. This fuel is in the form of organic molecules, which for animals must be supplied in the diet. Animals, of course, are totally dependent on plants for ready-made fuels. Animal cells tap the stored energy of organic fuels (for example, simple sugars, fatty acids, amino acids) through a series of controlled degradative steps. This process makes use of molecular oxygen from the atmosphere. In return the cells give off carbon dioxide as an end product, which is used by plant cells in making glucose and the more complex molecules. In this way the cellular energy cycle of life involves the harnessing of sunlight energy by green plants directly and by animal cells indirectly.

Metabolic energy

There are certain limited parallels between the combustion of fuel in a fire and metabolic combustion of fuel in a living cell. In both oxygen is consumed, and both are exergonic reactions in that energy is liberated as heat. If the fuel is burned in an internal combustion engine so that work is performed, the parallel is even better. But here the similarity ends.

The burning of gasoline in a cylinder is an explosive event that promotes just one function, the rapid expansion of gas. Many chemical bonds are broken simultaneously, and much energy is lost as heat. In contrast, metabolic energy must be released gradually and coupled to a great variety of energy-consuming reactions. Although metabolic energy exchanges proceed with great efficiency, some heat is inevitably liberated. Heat can be put to use, of course; higher vertebrates use it to elevate and maintain a constant internal body temperature, just as the heat of gasoline combustion is made to warm the occupants of a car. But for the most part, heat is a useless commodity to a cell since it is a nonspecific form of energy that cannot be captured and redistributed to power metabolic processes.

There is actually only one way in which the oxidative release of energy is made available for use by cells: it is coupled to the production of ATP by addition of inorganic phosphate to ADP (adenosine diphosphate). The structure of ATP is shown at the bottom of the page.

The ATP molecule consists of a purine (adenine), a five-carbon sugar (ribose), and three molecules of phosphoric acid linked together by two pyrophosphate bonds to form a triphosphate group. The pyrophosphate bonds are called **high-energy bonds** because they are repositories of a great deal of chemical energy. This energy has been transferred to ATP from other low-energy bonds in the respiratory process. Respiration, by the stepwise oxidation of fuel substrates, redistributes bond energies so that a few high-energy bonds are created and stored in ATP. Obviously this energy is gained at the expense of fuel

ADENINE   RIBOSE   TRIPHOSPHORIC ACID energy; the end products of cellular respiration, CO_2 and H_2O, contain much less bond energy than do the fuel substrates (for example, glucose) that entered the oxidative pathway.

The high-energy pyrophosphate bonds of ADP and ATP are frequently designated by the symbol ~. Thus a low-energy phosphate bond is shown as $-P$, and a high-energy one as $\sim P$. ADP can be represented as $A-P\sim P$ and ATP as $A-P\sim P\sim P$.

The trapping of energy by ATP can be shown as follows:

$$-P \qquad \text{chemical bond energy from fuel substrates}$$
$$\sim P$$
$$A-P\sim P \longrightarrow A-P\sim P\sim P$$

A low-energy phosphate group containing bond energy of 2,000 to 3,000 calories per mole is converted into a high-energy phosphate group containing bond energy of 8,000 to 10,000 calories per mole. Where does the low-energy phosphate group ($-P$) come from? Ultimately it comes from the diet. But in cellular respiration, ATP itself is the immediate source of the phosphate group necessary to start the oxidation process, donating the $-P$ to the fuel substrate molecule (usually glucose).

$$\text{ATP} \qquad \text{ADP}$$
$$\text{Fuel substrate} \longrightarrow \text{Fuel substrate} -P$$

The fuel molecule is phosphorylated with a low-energy phosphate group, and the phosphorylated fuel can then be oxidized to yield energy.

Obviously the fuel molecule must release more energy—actually it provides much more energy—than is loaned to it by ATP at the start. This is a kind of initial deficit financing that is required for an ultimate energy return many times greater than the original energy investment. We will return later to this subject of energy budgeting.

The amount of ATP produced in respiration is totally dependent on its rate of utilization. In other words, ATP is produced by one set of reactions and immediately consumed by another. ATP is a great **energy-coupling** molecule, used to transfer energy from one reaction to another. For example, ATP formed from the oxidation of glucose is used to synthesize proteins or lipids or provide power for some other process such as the contraction of skeletal muscles. The point is, living organisms do not produce and put aside vast amounts of ATP and hoard it for some future energy need. What they do store is the fuel itself in the form of carbohydrates and lipids especially. ATP is formed as it is needed, primarily by oxidative processes in the mitochondria. Oxygen is not consumed unless ADP and phosphate molecules are available, and these do not become available until ATP is hydrolized by some energy-consuming process. *Metabolism is therefore mostly self-regulating.*

There are minor exceptions to the rule that high-energy bonds cannot be stored in cells. Muscle cells contain a type of molecule especially adapted for energy storage. This is phosphocreatine:

$$\begin{array}{c} H \\ | \\ N\sim PO_3^- \\ | \\ H_2N-C \\ | \\ N-CH_3 \\ | \\ CH_2 \\ | \\ COO^- \end{array}$$

Phosphocreatine contains a high-energy bond that can provide instant power to the muscle contractile machinery, which often has sudden energy needs. Phosphocreatine is formed from creatine and ATP and is in direct chemical equilibrium with ATP. A short-term burst of activity rapidly depletes the available ATP, making ADP available. High-energy phosphate is then transferred to ADP from phosphocreatine, providing more ATP for use in muscle contraction.

ATP generated by electron transfer

Having seen that ATP is the one common energy denominator by which all cellular machines are energized, we are in a position to ask, how is this energy captured from fuel substrates? The principal means is by the transfer of electrons from fuel substrates to molecular oxygen through a series of enzymes. This electron transfer system is made of a chain of large molecules localized in the inner mitochondrial membranes. The electron carriers are compounds that can be reduced by accepting electrons from the previous carrier and that can be oxidized again by passing electrons to the next carrier. Each successive carrier is a somewhat stronger oxidizing agent than the one before; that is, *successive carriers are increasingly stronger electron acceptors.* Finally the electrons, as well as the hydrogen protons that accompany them are transferred to molecular oxygen to form water:

$$\text{Fuel-(2H)} \rightarrow A \rightarrow B \rightarrow C \rightarrow D \rightarrow \text{Oxygen} \rightarrow H_2O$$

It is conventional to represent electron transfer through the electron carrier system as the transfer of **hydrogen** atoms, although we must emphasize that it is the energized **electrons** and not the **protons** of the hydrogen that are the important energy packets. The proton of each hydrogen atom simply takes a free ride during this electron shuttle until at the end it bonds with reduced oxygen and forms water.

The whole function of the electron transport chain is the capture of energy from the original fuel substrate and the transformation of it into a form that the cell can use. To do this, the large chemical potential of food molecules is drawn off in small steps (rather than in one explosive burst as in ordinary combustion) by successive electron carriers. At three points along the chain, ATP production takes place by the phosphorylation of ADP. This method of energy capture is called **oxidative phosphorylation** because the formation of high-energy phosphate is coupled to oxygen consumption, and this depends, as we have seen, on the demand for ATP by other metabolic activities within the cell. The actual **mechanism** of ATP formation by oxidative phosphorylation is not yet known; we can only say that the transfer of electrons does something that is translated into the production of high-energy phosphate bonds.

Nature of electron carriers

Oxidative phosphorylation is much too complex to function efficiently, if at all, were the enzymes just floating freely in the cytoplasm of the cell. There is now abundant evidence that the oxidative enzymes and electron carriers are arranged in a highly ordered state on the membranes of the mitochondria.

As previously noted, mitochondria are composed of two membranes. The outside membrane forms a smooth sac enclosing the inner membrane that is turned into numerous ridges called **cristae** (Fig. 3-3, p. 49). The inner membrane is studded with enormous numbers of minute particles, which some investigators think are actually tiny stalked spheres. These particles, or spheres, bear the electron carriers of the respiratory chain responsible for oxidative phosphorylation. A

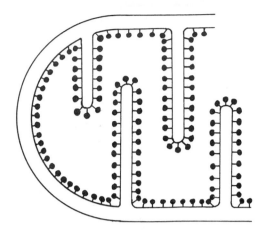

Fig. 10-9. Representation of section of mitochondrion as seen through high-resolution electron microscope, showing the inner membrane spheres that bear enzymes of the respiratory chain. The density of the spheres is actually many times greater than that depicted in this diagram.

section of a mitochondrion, as it might appear under the high-resolution electron microscope, is shown (highly diagrammatically) in Fig. 10-9.

Pairs of electrons, donated initially from food substrates, flow along the electron carriers in succession (Fig. 10-10). For most food substrates the initial electron acceptor is NAD (nicotinamide-adenine dinucleotide, a derivative of the vitamin niacin). The substrate is oxidized in the process (because it loses electrons) and NAD is reduced (because it gains electrons).

Next FAD (flavine-adenine dinucleotide, a riboflavin derivative) oxidizes the reduced NAD by accepting its electrons. FAD becomes reduced (having gained electrons) and NAD is returned to its original oxidized state. In the same way the pair of electrons is passed sequentially to coenzyme Q (chemically related to vitamin K) and then through a series of electron acceptors called **cytochromes.** The cytochromes are large molecules that belong to a class of proteins called chromoproteins because they contain colored prosthetic groups. The prosthetic group of a cytochrome, like the hemoglobin that is closely related to it, is an iron-bearing group that can be reversibly reduced.

$$Fe^{+++} + e^- \rightleftharpoons Fe^{++}$$

As the electrons are passed from cytochrome to cytochrome, each is successively reduced and then oxidized. Finally the electrons are passed to molecular

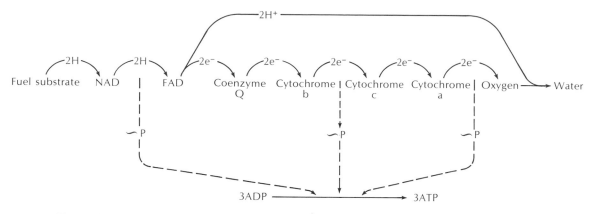

Fig. 10-10. Electron transport system. Electrons are transferred from one carrier to the next, terminating with molecular oxygen to form water. A carrier is reduced by accepting electrons and then is reoxidized by donating its electrons to the next carrier. ATP is generated at three points in the chain. These electron carriers are located on inner membrane spheres of mitochondria.

oxygen. The transfer of electrons and the points of high-energy phosphate bond formation are shown in Fig. 10-10.

Thus for every pair of electrons moved along the carriers to oxygen, a total of three ATP molecules is formed.

Acetyl–coenzyme A: strategic intermediate in aerobic respiration

At this point the student may be forgiven if he thinks that the metabolic energy fixation process consists entirely of an electron transport chain that is capable of passing bond energy to ATP. Actually we have only described the last (but especially crucial) step in **aerobic respiration.** The term "aerobic respiration" describes that kind of respiration requiring atmospheric oxygen, and this, of course, is the familiar sort of respiration practiced by the majority of animals. We have seen that ATP production is coupled to electron transfer, which in turn is completely dependent on oxygen, the final hydrogen and electron acceptor. Without oxygen the process stops because the electrons have nowhere to go. The components of the electron transfer chain would quickly become fully reduced and remain so in the absence of the electron sink, molecular oxygen.

However, aerobic organisms do have a backup system enabling them to respire without oxygen. This is called **anaerobic respiration,** or **fermentation.**

Although anaerobic respiration does support the lives of bacteria, yeasts, and a few other organisms, it is not nearly as efficient as aerobic respiration and consequently is not capable alone of maintaining life of higher animals. But it is useful, indeed essential, as a supplementary source of energy during rapid and intense muscular contraction. We are satisfied now with just noting that animal cells possess the machinery for anaerobic respiration; we discuss this subject in more detail later.

In aerobic respiration most fuel molecules are progressively stripped of their carbon atoms until those carbons are finally converted into molecules of CO_2. During this degradation the hydrogens and their electrons are removed and passed into the important energy-yielding electron transport chain.

In reaching this final sequence, most carbon atoms appear in a two-carbon group, called **acetyl–coenzyme A.** This is a critically important compound. Some two thirds of all the carbon atoms in foods eaten by animals appear as acetyl–coenzyme A at some stage. The strategic metabolic position of acetyl–coenzyme A is illustrated in Fig. 10-11. It is the final oxidation of acetyl–coenzyme A that provides the energized electrons used to generate ATP. Acetyl–coenzyme A is also the source of nearly all the carbon atoms found in the body's fats, as the reverse arrow in Fig. 10-11 indicates. The structure of acetyl–coenzyme A can be shown in abbreviated form

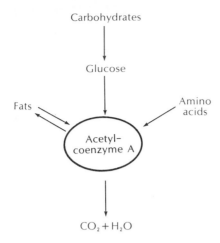

Carbohydrates

Glucose

Fats

Amino
acids

Acetyl–
coenzyme A

$CO_2 + H_2O$

Fig. 10-11. Acetyl–coenzyme A is an important intermediate in oxidation of carbohydrates, proteins (amino acids), and fats.

as:

$$CH_3—C—S—Co\ A$$

with O double-bonded above the C.

ACETYL   COENZYME A
GROUP   GROUP

The right hand side of the molecule is a coenzyme containing the vitamin **pantothenic acid,** another example of how vitamins play important structural roles in critical cellular functions.

Oxidation of acetyl–coenzyme A

The breakdown (oxidation) of the two-carbon acetyl group of acetyl–coenzyme A occurs in a sequence called the **citric acid cycle** (or **Krebs cycle,** after its British discoverer Sir Hans A. Krebs). The cycle is composed of a sequence of nine transformations and oxidations. To simplify an otherwise complex story, we have summarized the cycle in Fig. 10-12.

The citric acid cycle begins with the condensation of acetyl–coenzyme A with oxaloacetate to form a six-carbon compound citrate (citric acid). Citrate enters a series of reactions in which two molecules of CO_2 are produced from the original acetyl group. When the cycle is complete, the four-carbon oxaloacetate is returned to its original form, ready to condense with another molecule of acetyl–coenzyme A. Oxaloacetate therefore acts as a carrier for the two carbons of the acetyl group; it is not itself used up in the cyclic process. As the acetyl group is oxidized carbon atom

by carbon atom, four pairs of electrons and four protons (shown as four pairs of hydrogen atoms in Fig. 10-12) are transferred to the electron transfer chain (shown in the center of the cycle in Fig. 10-12). Three pairs of electrons are passed to NAD; the remaining pair is passed directly to FAD. Each pair of electrons then shuttles down the electron chain to an atom of oxygen.

Three molecules of ATP are generated for *each* molecule of NAD receiving electrons; this yields a total of nine ATP per acetyl group. Two more molecules of ATP are generated from the electrons passed directly to FAD. One more high-energy bond is generated at another point in the cycle; it forms a compound called GTP (guanosine-5′-triphosphate), which has the same energy yield as ATP, and for simplicity's sake, we call it ATP.

Thus the net yield is twelve molecules of ATP for the single acetyl group fed into the cycle. Eleven of these twelve high-energy phosphate bonds are generated by oxidative phosphorylation in the electron transport chain (Fig. 10-10) and not in the citric acid cycle itself. The citric acid cycle simply provides a means for the release of energized electrons during the oxidation of the acetyl group.

All of these reactions occur in mitochondria. But electron release through the citric acid cycle is believed to occur in the *outer* membrane, whereas the electron carriers and the coupling to oxidative phosphorylation occurs in the *inner* mitochondrial membrane. Thus there is a spatial, as well as functional, separation of these processes.

Glucose: major source of acetyl–coenzyme A

All the major fuels (glucose, fats, amino acids) serve as sources of acetyl–coenzyme A. Glucose, however, is a particularly important fuel for most tissues, especially the brain. Glucose is first converted to a three-carbon compound called **pyruvate** (pyruvic acid) through a series of reactions that are called the Embden-Meyerhof pathway. Pyruvic acid is then enzymatically stripped of a carbon atom to form acetyl–coenzyme A. The general outline for this sequence is shown in Fig. 10-13. Again, we shall simplify a rather complex biochemical story by condensing this glucose metabolism pathway, which actually consists of ten consecutive enzymatic reactions, into four major steps.

The metabolism of glucose begins with its phos-

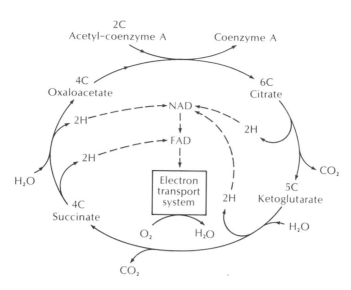

Fig. 10-12. Citric acid cycle. See text for explanation.

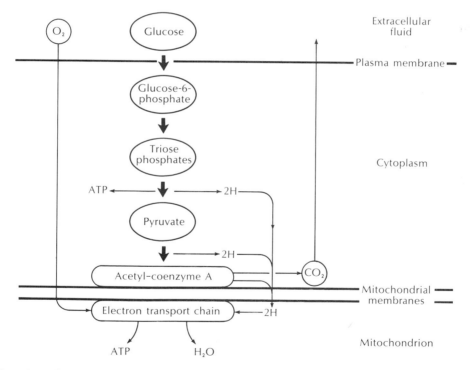

Fig. 10-13. Pathway for oxidation of glucose. Glucose is oxidized to acetyl–coenzyme A by the Embden-Meyerhof pathway *(red).* Acetyl–coenzyme A then enters citric acid cycle (not shown) on outer mitochondrial membrane. Hydrogens removed in cycle are transferred to electron transport chain on inner mitochondrial membrane. See text for details.

$$\text{GLUCOSE} + \text{ATP} \rightarrow \text{GLUCOSE-6-PHOSPHATE} + \text{ADP} + \text{H}^+$$

phorylation by ATP to form **glucose-6-phosphate** (see above).

Glucose-6-phosphate is an important intermediate because it is a "stem" compound that can lead into any of several metabolic pathways. However, the predominant metabolic fate for glucose-6-phosphate is entry into the Embden-Meyerhof sequence. Following still another phosphorylation, the six-carbon glucose molecule is split into two three-carbon sugars, called **triose phosphates.** Each triose phosphate is oxidized and rearranged to form **pyruvate,** resulting in a yield of high-energy phosphate as ATP. A pair of hydrogen atoms and a molecule of CO_2 are then removed from pyruvate, forming acetyl–coenzyme A.

Let us now summarize the entire oxidation of glucose. Glucose first enters the cells of tissues, passing through the plasma membrane by a transport process that requires the presence of the pancreatic hormone **insulin.** Glucose is then phosphorylated and enters the Embden-Meyerhof pathway in the cytoplasm. This is shown in red in Fig. 10-13. Through this sequence it is split in the middle to form two three-carbon sugars (triose phosphates) that are converted to pyruvate. Pyruvate is decarboxylated to form acetyl–coenzyme A. This sets the stage for entry into the citric acid cycle located on the outer mitochondrial membrane. After condensing with oxaloacetic acid to form citric acid, the two-carbon acetyl fragment is oxidized—yielding four pairs of electrons and four protons that are passed along the electron transport chain located on the inner mitochondrial membrane. The electrons finally arrive at oxygen, the ultimate electron acceptor.

What has been accomplished? A molecule of glucose has been completely oxidized to CO_2 and H_2O. ATP has been generated at several points along the way. Let us tabulate the yield. First of all, 2 molecules of ATP are consumed in the initial phosphorylation of glucose. Then, 14 molecules of ATP are generated by the transformations resulting in the formation of acetyl–coenzyme A. (Each glucose molecule is split into

two three-carbon sugars, each of which produces ATP when oxidized to acetyl–coenzyme A. Some ATP is produced directly; the rest results from oxidation followed by oxidative phosphorylation.) Our balance is 12 ATP. To this we add the 12 ATP generated in the complete oxidation of *each* of the acetyl–coenzyme A molecules. This gives a total yield of 36 ATP. The whole sequence can be summarized as follows:

$$\text{Glucose (C}_6\text{H}_{12}\text{O}_6) + 6\text{O}_2 + 36\text{ADP} + 36\text{—P} \longrightarrow$$
$$6\text{CO}_2 + 6\text{H}_2\text{O} + 36\text{ATP}$$

Efficiency of oxidative phosphorylation

It is probably obvious that no energy-transforming process can be 100% efficient, not even the remarkable cellular oxidative machinery produced by organic evolution. If we burn a mole of glucose (180 g) in a bomb calorimeter, it releases approximately 686,000 calories. This is the **potential** energy for forming ATP. It has been determined that 8,000 to 12,000 calories are required to synthesize 1 mole of ATP. Consequently glucose theoretically *could* provide enough energy to generate 50 to 85 moles of ATP from ADP. It actually turns out 38 moles (36 + 2) of ATP. If we assume that each ATP mole represents an average energy equivalent of 10,000 calories, then 38 moles of ATP represents 380,000 calories. Thus the efficiency of glucose oxidation is 380:686, or approximately 55%. Engineers would be delighted if they could build machines that could do as well.

Metabolism of lipids

Animal fats are triglycerides, molecules composed of glycerol and three molecules of fatty acids. These fuels are important sources of energy for many metabolic processes in all animals, not just obese victims of misplaced appetites. Most of fat is fatty acids, carboxylic acids with long hydrocarbon chains. We know that fats enter the mitochondrial metabolic processes through acetyl–coenzyme A (Fig. 10-11). What happens in brief is that the long hydrocarbon chain of a fatty acid is sliced up by oxidation, two carbons at a

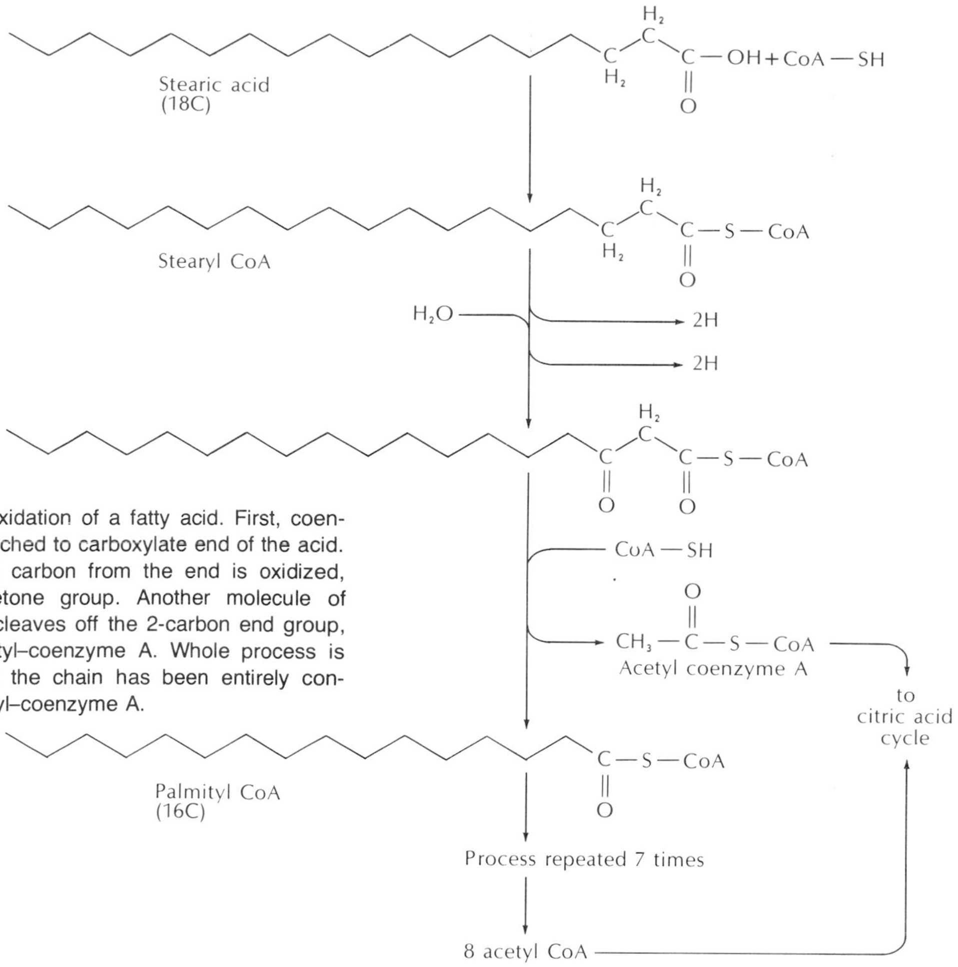

Fig. 10-14. Oxidation of a fatty acid. First, coenzyme A is attached to carboxylate end of the acid. Then, second carbon from the end is oxidized, yielding a ketone group. Another molecule of coenzyme A cleaves off the 2-carbon end group, liberating acetyl–coenzyme A. Whole process is repeated until the chain has been entirely converted to acetyl–coenzyme A.

time; these are released from the end of the molecule as acetyl–coenzyme A. The process is repeated until the entire chain has been reduced to several two-carbon acetyl units.

The oxidation of a fatty acid is diagrammed in Fig. 10-14, using a shorthand representation in which each jog in the chain symbolizes a saturated carbon (−CH$_2$−) of the fatty acid **stearic acid,** one of the abundant naturally occurring fatty acids. First, the fatty acid is combined with coenzyme A. Then, in a three-step process the third carbon from the end is oxidized (stripped of its hydrogens). Next, a molecule of acetyl–coenzyme A is sliced off the end, by *another* molecule of coenzyme A that then adds itself to the chain. Thus the hydrocarbon chain, now two carbons shorter, is left with coenzyme A on its end. The

process is repeated until the whole chain has been chopped up into acetyl–coenzyme A. This material then enters the citric acid cycle to yield ATP in the manner described earlier.

How much ATP is gained from fatty acid oxidation? Five high-energy phosphate bonds are generated for each acetyl–coenzyme A unit split off. Thus oxidation of each two-carbon fragment produces 12 ~P. With allowance for the ATP extended to attach the first coenzyme A, it has been calculated that the complete oxidation of 18-carbon stearic acid yields 147 ATP molecules. By comparison, 3 molecules of glucose (also totaling 18 carbons) yields 108 ATPs. Little wonder that fat is considered the king of animal fuels! Fats are more concentrated fuels than carbohydrates because fats are almost pure hydrocarbons; they con-

205

tain more hydrogen per carbon atom than sugars do, and it is the energized electrons of hydrogen that generate high-energy bonds when they are carried through the mitochondrial electron transport system.

Glycolysis: generating ATP without oxygen

Up to this point, we have been describing **oxidative,** or **aerobic,** metabolism, the kind of respiration that predominates in the majority of animals. It hardly needs emphasizing that the availability of oxygen is an obvious basic necessity of animal life. Nevertheless there are microorganisms, notably the yeasts and certain bacteria, that multiply happily with no oxygen at all. These organisms are called **anaerobes.** They occupy important ecologic niches, some of the niches created by man. For example, oxygen-depleted streams are becoming regrettably common appendages to our industrialized society. Anaerobic organisms use carbon compounds as fuel, breaking them down by a process commonly called **fermentation.** This term, meaning "cause to rise," was originally used to describe the action of yeasts that break down glucose into alcohol (ethanol) and CO_2. It is now applied to any microorganism that metabolizes foodstuffs without oxygen. The end products, which vary with the nature of the fermentive process, include butanol, acetone, lactic acid, and hydrogen gas.

Most higher organisms also have the capacity to ferment glucose, that is, break it down to produce high-energy phosphate in the absence of oxygen. The process is called **glycolysis.** It is used as a backup system for aerobic metabolism, providing a means for short-term generation of ATP during brief periods of heavy energy expenditure when the slow rate of O_2 diffusion would be a limiting factor.

In glycolysis, glucose is split eventually into two molecules of pyruvate (pyruvic acid), yielding two molecules of ATP in the process. The glycolytic pathway is shown in Fig. 10-15. Glycolysis utilizes the same Embden-Meyerhof pathway that in oxidative metabolism directs glucose into the citric acid cycle via acetyl–coenzyme A (compare Figs. 10-13 and 10-15). But, in the absence of oxygen, both pyruvate and hydrogen accumulate in the cytoplasm because neither can proceed into their oxidative channels without oxygen. The problem is neatly solved by forming lactate. Pyruvate is converted into lactate that accepts the hydrogen, as shown below:

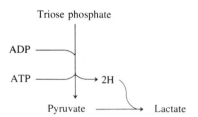

Lactate then diffuses out into the blood, where it is later disposed of in the liver. Thus lactate formation prevents the cytoplasm from being swamped with pyruvate and allows *some* ATP formation. Of course, glycolysis is not an efficient producer of ATP; only 2 moles of ATP per mole glucose are generated by glycolysis compared to 36 moles by oxidative phosphyorylation. Nevertheless the capacity to produce a little extra ATP during an emergency may mean the difference between life and death for an animal. Some animals rely heavily on glycolysis during normal activities. For example, diving birds and mammals fall back on glycolysis almost entirely to give them the needed energy to sustain a long dive. And salmon would never reach their spawning grounds were it not for glycolysis, which provides almost all the ATP used in the powerful muscular bursts needed to carry them over falls and up rapids.

Fig. 10-15. Glycolysis and formation of lactate. See text for details.

SELECTED REFERENCES

(See also general references for Part two, p. 274.)

Giese, A. C. 1973. Cell physiology, ed. 4. Philadelphia, W. B. Saunders Co. *An excellent cell physiology text for students who have had good introductory courses in chemistry and physics.*

Hochachka, P. A., and G. N. Somero. 1973. Strategies of biochemical adaptation. Philadelphia, W. B. Saunders Co. *Important treatment of comparative biochemistry.*

Jennings, J. B. 1965. Feeding, digestion and assimilation in animals. Oxford, Pergamon Press, Inc. *A general, comparative approach. Excellent account of feeding mechanisms in animals.*

Lehninger, A. L. 1975. Biochemistry: the molecular basis of cell structure and function, ed. 2. New York, Worth Publishers. *A very lucidly written and amply illustrated undergraduate biochemistry text, particularly suitable for the student who leans more toward chemistry than biology.*

McDonald, P., R. A. Edwards, and J. F. D. Greenhalgh. 1973. Animal nutrition. New York, Hafner Press. *Restricted to farm animals, but thorough treatment of this area.*

McGilvery, R. W. 1970. Biochemistry: A functional approach. Philadelphia, W. B. Saunders Co. *Well-written mammalian biochemistry.*

Morton, G. 1967. Guts. The form and function of digestive systems. Institute of Biology's Studies in Biology, no. 7, New York, St. Martin's Press.

Yost, H. T. 1972. Cellular physiology. Englewood Cliffs, N. J., Prentice-Hall, Inc. *Advanced treatment of molecular physiology and biochemistry.*

SELECTED SCIENTIFIC AMERICAN ARTICLES

Loewenstein, W. R. 1970. Intercellular communication. **222:**79-86 (May).

Margaria, R. 1972. The sources of muscular energy. **226:**84-91 (March).

Rogers, T. A. 1958. The metabolism of ruminants. **198:**34-38 (Feb.). *How cellulose is digested in the four-chambered stomach of cows and other ruminants.*

The eyes of this red-tailed hawk, with a visual acuity eight times better than man's, gather more detailed information about the environment than all other special senses combined. (Photo by C. P. Hickman, Jr.)

11 NERVOUS COORDINATION
Nervous system and sense organs

The nervous system originated in a fundamental property of protoplasm—irritability. Each cell responds to stimulation in a manner characteristic of that type of cell. But certain cells have become highly specialized for receiving stimuli and for conducting impulses to various parts of the body. Through evolutionary changes, these cells have become integrated into the most complex of all body systems—the nervous system. The endocrine system is also important in coordination, but the nervous system has a wider and more direct control of body functions than does the endocrine system.

The evolution of the nervous system has been correlated with the development of bilateral symmetry and cephalization. Along with this development, animals acquired exteroceptors and associated ganglia. The basic plan of the nervous system is to code sensory information, and transmit it to regions of the central nervous system where it is processed into appropriate action. This action may be of several types, such as simple reflexes, automatic behavior patterns, conscious perception, or learning processes.

NERVOUS SYSTEMS OF INVERTEBRATES

Nervous systems as such are lacking in protozoans, which depend upon primitive membrane irritability and its conduction across the cell surface to respond to stimuli. Nevertheless there are instances of remarkable neural development in certain protozoans. The relatively complex **neuromotor apparatus** of ciliates coordinates the beat of the cilia, and certain species have neurofibrils passing from an anterior motor mass—the beginnings of a central nervous system.

The Metazoa show a progressive increase in nervous system complexity that we believe recapitulates to some extent the evolution of the nervous system. The coelenterates have a **nerve net** containing bipolar and multipolar cells (protoneurons). These may be separated from each other by synaptic junctions, but they form an extensive network that is found in and under the epidermis over all the body. An impulse starting in one part of this net is conducted in all directions, since the synapses do not restrict transmission to one-way movement, as they do in higher animals. There are no differentiated sensory, motor, or connector components in the strict meaning of those terms. Branches of the nerve net connect to receptors in the epidermis and to epitheliomuscular cells. Most responses tend to be

generalized, yet many are astonishingly complex for so simple a nervous system. Such a type of nervous system is retained among higher animals in the form of nerve plexuses in which such generalized movements as intestinal peristalsis are involved.

Flatworms are provided with two anterior **ganglia** of nerve cells from which two main nerve trunks run posteriorly, with lateral branches extending to the various parts of the body (Fig. 17-2, p. 340). This is the true beginning of a differentiation into a **peripheral nervous system,** a communications network extending to all parts of the body, and a **central nervous system,** which coordinates everything. It is also the first appearance of the **linear** type of nervous system, which is more developed in higher invertebrates. These have a more centralized nervous system with the two longitudinal nerve cords fused (although still recognizable) and many ganglia present. The annelids have a well-developed nervous system consisting of distinctive **afferent** (sensory) and **efferent** (motor) neurons. At the anterior end, the ventral nerve cord divides and passes upward around the digestive tract to join the bi-lobed brain. In each segment the double nerve cord bears a double ganglion, each with two pairs of nerves. Arthropods have a system similar to that of earthworms, except that the ganglia are larger and the sense organs better developed.

Molluscs have a system of three pairs of ganglia; one pair is near the mouth, another pair at the base of the foot, and one pair in the viscera. The ganglia are joined by connectives. The molluscs also have a number of sense organs, especially well developed in the cephalopods. Among the echinoderms the nervous system is radially arranged (Fig. 23-8, p. 472).

The nerve cord in all invertebrates is ventral to the alimentary canal and is solid. This arrangement is in pronounced contrast to the nerve cord of vertebrates, which is dorsal to the digestive system, single, and hollow.

NERVOUS SYSTEMS OF VERTEBRATES

Vertebrates have, as a rule, a brain much larger than the spinal cord. In lower vertebrates this difference is not significant, but higher in the vertebrate kingdom the brain increases in size, reaching its maximum in mammals, especially man. Along with this enlargement has come an increase in complexity, bringing better patterns of coordination, integration, and intelligence. The nervous system is commonly divided into central and peripheral parts. The **central nervous system** is housed within the skull and vertebral column and is concerned with integrative activity. The **peripheral nervous system** consists of nerve cells or extensions of nerve cells that lie outside the skull and vertebral column. It is a communications system for the conduction of sensory and motor information to all parts of the body.

Neurons: functional units of the nervous system

The vertebrate nervous system is composed of **neurons** and **neuroglial cells.** Although neurons are the basic functional units of the nervous system, they are outnumbered 10 to 1 by the neuroglial cells, which physically support the neurons and nourish them metabolically (p. 217).

The neuron is a cell body and all of its processes. Although neurons assume many shapes depending upon their function and location, a ''typical'' type is shown diagrammatically in Fig. 11-1. From the nucleated cell body extends an **axon,** which carries impulses *away* from the cell. Typically several branching **dendrites** surround the cell body. These carry impulses *toward* the cell body. Axons are usually covered with a soft, white lipid-containing material called **myelin.** This insulating material is often laid down in concentric rings by specialized **Schwann cells** to form a **myelin sheath.** This is enclosed by an outer membrane called the **neurolemma.**

Neurons are commonly divided into **afferent,** or sensory, neurons; **efferent,** or motor, neurons; and **association** neurons (interneurons). Afferent and efferent neurons lie mostly outside the skull and vertebral column; association neurons, which in man comprise 99% of all the nerve cells in the body, lie entirely within the central nervous system. Afferent neurons are connected to receptors. When these respond to some environmental change, they generate **action potentials** in the afferent neurons, which are carried into the central nervous system. Here the impulses may be perceived as conscious sensation. The impulses also move to efferent neurons, which carry them out by the peripheral system to **effectors** (muscles or glands).

Nerves (not to be confused with neurons) are actually made up of many neuronal processes—axons or dendrites or both—bound together with connective tissue (Fig. 11-2). The cell bodies of these nerve

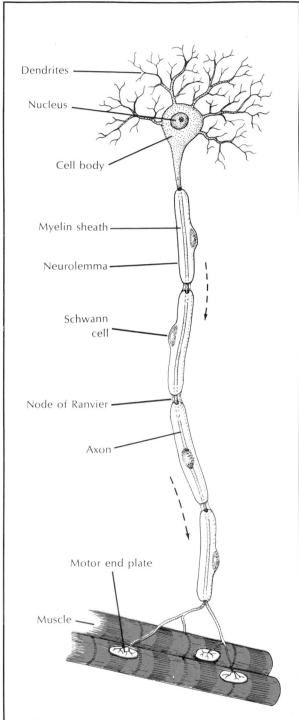

Fig. 11-1. Structure of a motor (efferent) neuron.

Dendrites

Nucleus

Cell body

Myelin sheath

Neurolemma

Schwann cell

Node of Ranvier

Axon

Motor end plate

Muscle

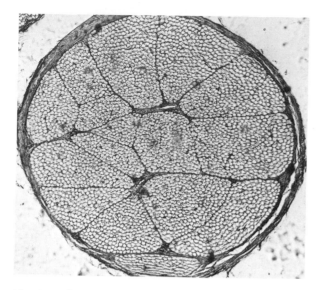

Fig. 11-2. Cross section of nerve showing cut ends of nerve processes *(small white circles)*. Such a trunk may contain thousands of both afferent and efferent fibers.

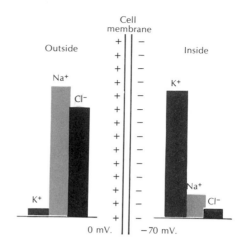

Fig. 11-3. Ionic composition inside and outside a resting nerve cell. An active sodium pump located in the cell membrane drives sodium to the outside, keeping its concentration low inside. Potassium concentration is high inside, and, although the membrane is "leaky" to potassium, this ion is held inside by repelling the positive charge outside the membrane.

bundles are located either in ganglia or somewhere in the central nervous system (brain or spinal cord).

Nature of the nerve impulse

The nerve impulse is the chemical-electrical message of nerves, the common functional denominator of all nervous system activity. Despite the incredible complexity of the nervous system of advanced animals, nerve impulses are basically alike in all nerves and in all animals. It is an **all-or-none** phenomenon; either the axon is conducting an impulse, or it is not. All impulses are alike, and the only way an axon can vary its effect on the tissue it innervates is by changing the **frequency** of impulse conduction. Frequency change is the language of a nerve fiber. A fiber may conduct no impulses at all or a very few per second up to a maximum approaching 1,000 per second. The higher the frequency (or rate) of conduction, the greater is the level of excitation.

Resting potential. To understand what happens when an impulse is conducted down a fiber, we need to know something about the resting, undisturbed fiber. Nerve cell membranes, like all cell membranes, have special permeability properties that create ionic imbalances. Sodium and chloride predominate on the outside, whereas potassium ions are more common inside (Fig. 11-3). These differences are quite dra-

matic; there is approximately 10 times more sodium outside than in, and 25 to 30 times more potassium inside than out. However, the nerve cell membrane is 50 to 70 times more permeable to potassium than to sodium. The result is that potassium ions tend to leak outward, moving down their concentration gradient. This movement of positively charged ions outward creates a **diffusion potential,** with the outside of the membrane positive with respect to the inside.

If the membrane is highly permeable to potassium, why doesn't all of the potassium escape from the cell, allowing the potential to disappear? This does not happen because the potential difference created by outward movement of potassium begins to influence this movement. Since potassium ions are positively charged, they are attracted by the negatively charged inside membrane and repelled by the positively charged outside membrane. As potassium flows outward the electric force across the membrane becomes large enough to prevent any further net outward movement of potassium. This membrane potential is called an **equilibrium potential** because it is a permanent bioelectrical potential produced by a balance between a concentration force favoring the outward flow of potassium and an electrical force that opposes it.

Thus, this is the origin of the **resting transmem-**

211

brane potential, which is positive outside and negative inside. It is created, as we have seen, by two important characteristics of the living nerve cell : (1) the potassium concentration is much greater inside the cell than outside, and (2) the cell membrane is far more permeable to potassium than to sodium. The resting potential is usually −70 mV, with the inside of the membrane negative to the outside.

Action potential. The nerve impulse is a rapidly moving change in electrical potential called the **action potential** (Fig. 11-4). It is a very rapid and brief depolarization of the axon membrane; in fact, not only is the resting potential abolished, but in most nerves the potential actually reverses for an instant so that the outside becomes negative as compared to the inside. Then, as the action potential moves ahead, the membrane returns to its normal resting potential ready to conduct another impulse. The entire event occupies approximately a millisecond. Perhaps the most significant property of the nerve impulse is that it is **self-propagating;** that is, once started the impulse moves ahead automatically, much like the burning of a fuse.

What causes the reversal of polarity in the cell membrane during passage of an action potential? We have seen that the resting potential depends upon the high membrane permeability to potassium, some 50 to 70 times greater than the permeability to sodium. When the action potential arrives at a given point, the permeability of the membrane to potassium and so-

dium is markedly changed. The membrane suddenly becomes approximately 600 times more permeable to sodium, whereas potassium permeability changes very little. Sodium rushes in.

Actually only an extremely small amount of sodium traverses the membrane in that instant—less than 1 millionth of the sodium outside—but this brief shift of positive ions inward causes the membrane potential to disappear, even reverse. An electrical "hole" is created. Potassium, finding its electrical barrier gone, begins to move out. Then, as the action potential passes on, the membrane quickly regains its resting properties. It becomes once again practically impermeable to sodium, and the outward movement of potassium is checked.

The rising phase of the action potential is associated with the rapid influx (inward movement) of sodium (Fig. 11-4). When the action potential reaches its peak, the sodium permeability is restored to normal and potassium permeability briefly increases above the resting level. This causes the action potential to decrease rapidly toward the resting membrane level.

Sodium pump. The resting cell membrane has a very low permeability to sodium. Nevertheless some sodium ions leak across, even in the resting condition. When the axon is active sodium flows inward with each passing impulse, and, although the amount is very small, it is obvious that the ionic gradient would eventually disappear if the sodium ions were not moved back out again. This is done by a **"sodium pump"** located in the axon plasma membrane. Although no one has ever actually seen the "pump," we do know quite a bit about it because it has been the object of intense biochemical and biophysical studies.

The sodium pump is an active transport device capable of combining with sodium on the inside surface of the membrane and then moving to the outside surface where the sodium is released. It is probably composed of phosphate-containing protein molecules. The sodium pump requires energy since it is moving sodium "uphill" against the sodium electrical and concentration gradient. This energy is supplied by ATP through cellular metabolic processes. There is evidence that in some cells sodium transport outward is linked to potassium transport inward; the same carrier molecule may act as a two-way shuttle, carrying ions on both trips across the membrane. This kind of pump is called a sodium-potassium pump. Active transport was discussed earlier on p. 59.

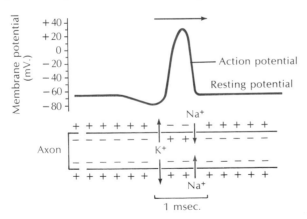

Fig. 11-4. Action potential of nerve impulse. The electrical event, moving from left to right, is associated with rapid changes in membrane permeability to sodium and potassium ions. When the impulse arrives at a point, sodium ions suddenly rush in, making the axon positive inside and negative outside. Then the sodium holes close and potassium holes open up. This restores the normal negative resting potential.

Synapses: junction points between nerves

A synapse is found at the end of a nerve axon where it connects to the dendrites or cell body of a second neuron. Neurons bringing impulses toward synapses are called **presynaptic neurons;** those carrying impulses away are **postsynaptic neurons.** At the synapse, the membranes are separated by a narrow gap, the **synaptic cleft,** having a very uniform width of approximately 20 nm. The synapse is of great functional importance because it acts as a one-way valve that allows nerve impulses to move in one direction only. It is also through the many synapses that information is modulated from one nerve to the next.

The axon of most nerves divides at its end into many branches, each of which bears a synaptic knob that sits on the dendrites or cell body of the next nerve (Fig. 11-5). The axon terminations of several nerves may almost cover a nerve cell body and its dendrites with thousands of synaptic clefts. Because a single impulse coming down a nerve axon sprays out into the many branches and synaptic endings on the next nerve cell, many impulses converge at the cell body at one moment. But these may not excite the cell body enough to fire off an impulse. A neuron requires much prompting to fire. Usually many impulses must arrive at the cell body simultaneously from several presynaptic nerve cells, or within a very brief interval, to raise the cell body to its firing threshold level. This is called **summation;** it is the cumulative excitatory effect of many incoming impulses that pushes a nerve cell up to firing threshold.

The synapse is a kind of chemical bridge. The electron microscope shows that each synaptic knob contains numerous membrane-limited **synaptic vesicles.** These are filled with molecules of **chemical transmitter** (Fig. 11-5, *B*). For most synapses this

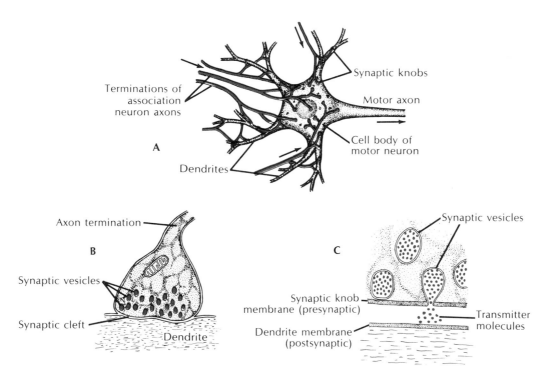

Fig. 11-5. Transmission of impulses across nerve synapses. **A,** Cell body of a motor nerve is shown covered with the terminations of association neurons. Each termination ends in a synaptic knob; hundreds of synaptic knobs may be on a single nerve cell body and its dendrites. **B,** Synaptic knob enlarged 60 times more than in **A.** An impulse traveling down the axon causes some synaptic vesicles to move down to the synaptic cleft and rupture, releasing transmitter molecules into the cleft. **C,** Synaptic cleft as it might appear under a high-resolution electron microscope. Transmitter molecules from a ruptured synaptic vesicle move quickly across the gap to produce an electrical potential change in the postsynaptic membrane.

transmitter substance is either **acetylcholine** or **norepinephrine.** When an impulse arrives at the knob, it induces some of the vesicles to move to the base of the knob and release their contents of transmitter molecules into the synaptic cleft (Fig. 11-5, *C*). These move rapidly across the narrow cleft to combine with the postsynaptic membrane, producing a small potential. If reinforced by the arrival of more impulses and by the release of more packets of transmitter molecules, either at the same or adjacent synapses, the small potential may build into a large one, sufficient to fire off an impulse in the postsynaptic nerve cell.

Synapses, then, are critical determinants in nervous system function. Although a nerve impulse is an all-or-none event, the synapses act like variable gates that may or may not allow impulses to proceed from one neuron to the next.

Components of the reflex arc

Neurons work in groups called **reflex arcs** (Fig. 11-6). There must be at least two neurons in a reflex arc, but usually there are more. The parts of a typical reflex arc consist of (1) a **receptor,** a sense organ in the skin, muscle, or other organ; (2) an **afferent** or sensory neuron, which carries the impulse toward the central nervous system; (3) a **nerve center,** where synaptic junctions are made between the sensory neurons and the association neurons; (4) the **efferent** or motor neuron, which makes synaptic junction with the association neuron and carries impulses out from the central nervous system; and (5) the **effector,** by which the animal responds to its environmental changes. Examples of effectors are muscles, glands, cilia, nematocysts of coelenterates, electric organs of fish, and chromatophores.

A reflex arc at its simplest consists of only two neurons—a sensory (afferent) neuron and a motor (efferent) neuron. Usually, however, association neurons are interposed (Fig. 11-6). Association neurons may connect afferent and efferent neurons on the same side of the spinal cord, connect them on opposite sides of the cord, or connect them on different levels of the spinal cord, either on the same or opposite sides. In almost any reflex act a number of reflex arcs are involved. For instance, a single afferent neuron may make synaptic junctions with many efferent neurons. In a similar way an efferent neuron may receive impulses from many afferent neurons. In this latter case the efferent neuron is referred to as the **final common path.**

A **reflex act** is the response to a stimulus acting over a reflex arc. It is **involuntary** and may involve the cerebrospinal or the autonomic nervous divisions of the nervous system. Many of the vital processes of the body, such as control of breathing, heartbeat, diameter of blood vessels, and sweat gland secretion, are reflex

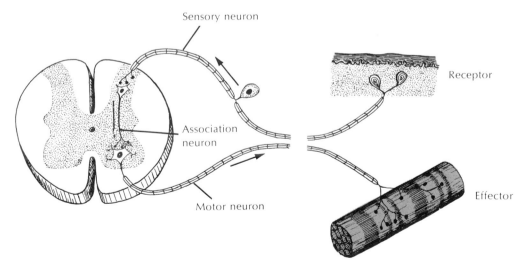

Fig. 11-6. Reflex arc. Impulse generated in the receptor is conducted over an afferent (sensory) nerve to the spinal cord, relayed by an association neuron to an efferent (motor) nerve cell body and by the efferent axon to an effector.

actions. Some reflex acts are inherited and innate; others are acquired through learning processes (conditioning).

ORGANIZATION OF THE VERTEBRATE NERVOUS SYSTEM

The basic plan of the vertebrate nervous system is a dorsal longitudinal hollow nerve cord that runs from head to tail. During early embryonic development, the central nervous system begins as an ectodermal **neural groove,** which by folding and enlarging becomes a long, hollow, **neural tube.** The cephalic end enlarges into the brain vesicles and the rest becomes the spinal cord. The spinal nerves (31 pairs in man) have a dual origin. The **spinal ganglia** (dorsal root ganglia in Fig. 11-7), containing the sensory neurons differentiate from specialized cells, called **neural crest cells,** that pinch off from the edges of the neural groove as it closes to form a tube. The ventral roots contain motor fibers that originate in the spinal cord. Both dorsal (sensory) and ventral (motor) roots meet some distance beyond the cord to form a mixed **spinal nerve** (Fig. 11-7).

Central nervous system

The central nervous system is composed of the brain and spinal cord.

Spinal cord

The spinal cord is enclosed by the vertebral canal and additionally protected by three layers collectively called the **meninges** (men-in'jeez). The three layers are a tough outer **dura mater,** a thin spider weblike **arachnoid,** and a delicate innermost sheath, the **pia mater.** Between the arachnoid and the pia mater is a space containing **cerebrospinal fluid,** a secreted fluid forming a protective cushion and thermal insulation for the cord. The meninges and cerebrospinal fluid blanket are continuous with those covering the brain.

In cross section the cord shows two zones. An inner H-shaped zone of **gray matter** is made up of association neurons and the cell bodies of motor neurons (Fig. 11-6). The outer zone of **white matter** contains nerve bundles of axons and dendrites linking different levels of the cord with each other and with the brain. The fibers are bundled into **ascending tracts** carrying impulses to the brain and **descending tracts** carrying

Fig. 11-7. Spinal cord and meninges with relation to spinal nerves, sympathetic system, and vertebrae. Three coats of meninges have been partly cut away to expose spinal cord. Only two vertebrae are shown in position.

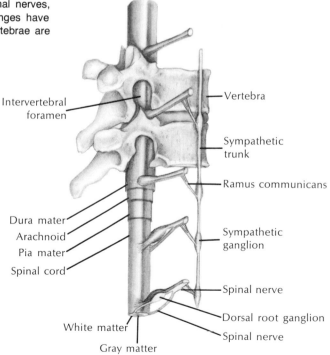

Intervertebral foramen

Vertebra

Sympathetic trunk

Ramus communicans

Dura mater
Arachnoid
Pia mater
Spinal cord

Sympathetic ganglion

Spinal nerve

Dorsal root ganglion

Spinal nerve

White matter

Gray matter

215

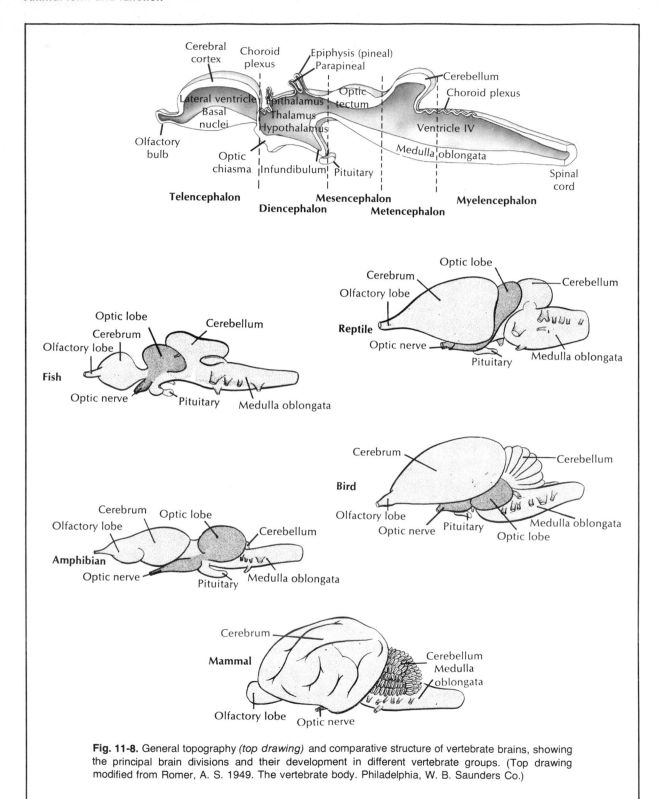

Fig. 11-8. General topography *(top drawing)* and comparative structure of vertebrate brains, showing the principal brain divisions and their development in different vertebrate groups. (Top drawing modified from Romer, A. S. 1949. The vertebrate body. Philadelphia, W. B. Saunders Co.)

impulses away from the brain. The sensory (ascending) tracts are located mainly in the dorsal part of the cord; the motor (descending) tracts are found ventrally and laterally in the cord. Fibers also cross over from one side of the cord to the other, with the sensory fibers crossing at a higher level than the motor fibers. Although the different tracts cannot be distinguished in a sectioned cord even with a microscope, their position is known from painstaking mapping experiments.

Brain

The brain in vertebrates shows an evolution from the linear arrangement in lower forms (fish and amphibians) to the much-folded and enlarged brain found in higher vertebrates (birds and mammals) (Fig. 11-8). The brain is really the enlarged anterior end of the spinal cord. The ratio between the weight of the brain and spinal cord affords a fair criterion of an animal's intelligence. In fish and amphibians this ratio is approximately 1:1, in man the ratio is 55:1—in other words, the brain is 55 times heavier than the spinal cord. Although man's brain is not the largest (the elephant's brain is four times heavier) nor the most convoluted (that of the porpoise is even more wrinkled), it is by all odds the best. Indeed the human brain is the most complex structure known to man. It has no parallel in the living or nonliving world. This "great ravelled knot," as the British physiologist Sir Charles Sherrington called man's brain, in fact, may be so complex that it will never be able to understand its own function.

The primitive three-part brain is made up of prosencephalon, mesencephalon, and rhombencephalon (forebrain, midbrain, and hindbrain) (Fig. 11-8; Table 11-1). The prosencephalon and rhombencephalon each divide again to form the five-part brain characteristic of the adults of all vertebrates. The five-part brain includes the telencephalon, diencephalon, mesencephalon, metencephalon, and myelencephalon. From these divisions the different functional brain structures arise.

The impressive evolutionary improvement of the vertebrate brain has accompanied the increased powers of locomotion and greater environmental awareness of the more advanced vertebrates. In the primitive vertebrate brain each of the three parts was concerned with one or more special senses: the prosencephalon with the sense of smell, the mesencephalon with vision, and the rhombencephalon with hearing and balance. These primitive but very fundamental concerns of the brain have been in some instances amplified and in others reduced or over-shadowed during continued evolution as sensory priorities were shaped by the animal's habitat and way of life.

The brain is made up of both white and gray matter, with the gray matter on the outside (in contrast to the spinal cord in which the gray matter is inside). The gray matter of the brain is mostly in the convoluted **cortex.** In the deeper parts of the brain the white matter of nerve fibers connects the cortex with lower centers of the brain and spinal cord or connects one part of the cortex with another. Also in deeper portions of the brain are collections of nerve cell bodies (gray matter) that provide synaptic junctions between the neurons of higher centers and those of lower centers.

There are also nonnervous elements in the nervous system such as connective, supporting, and capsule cells and the **neuroglia** ("nerve glue"). Neuroglial cells, or simply "glial" cells, greatly outnumber the neurons and play various vital roles in the functioning of the neurons. They bind together the nervous tissue proper, support it metabolically, and serve in the regenerative processes that follow injury or disease. Neuroglial cells are unfortunately the chief source of tumors of the central nervous system.

The main divisions of the brain are given in Table

Table 11-1. Divisions of the vertebrate brain

Embryonic vesicles	Main components in adults
Forebrain	
Telencephalon	Olfactory bulbs
	Cerebrum
	Lateral ventricles
Diencephalon	Thalamus
	Hypothalamus
	Infundibulum
	Third ventricle
Midbrain	
Mesencephalon	Optic lobes (tectum)
	Cerebral peduncles
	Red nucleus
	Aqueduct of Sylvius
Hindbrain	
Metencephalon	Cerebellum
	Pons
	Fourth ventricle
Myelencephalon	Medulla

11-1. The **medulla,** the most posterior division of the brain, is really a conical continuation of the spinal cord. The medulla, together with the more anterior midbrain, constitutes the "brainstem," an area in which numerous vital and largely subconscious activities are controlled, such as heartbeat, respiration, vasomotor tone, and swallowing. The brainstem contains the roots of all the cranial nerves except the first and is traversed by many sensory and motor fiber tracts. Although it is small in size and largely hidden from view by the much enlarged "higher" centers, it is, in fact, the most vital brain area; whereas damage to higher centers may result in severely debilitating loss of sensory or motor function, damage to the brainstem usually results in death.

The **pons,** between the medulla and the midbrain, is made up of a thick bundle of fibers that carries impulses from one side of the cerebellum to the other.

The **cerebellum,** lying above the medulla, is concerned with equilibrium, posture, and movement. Its development is directly correlated with the animal's mode of locomotion, agility of limb movement, and balance. It is usually weakly developed in amphibians and reptiles, which are relatively clumsy forms that stick close to the ground, and well developed in the more agile bony fishes. It reaches its apogee in birds and mammals in which it is greatly expanded and folded. The cerebellum does not initiate movements, but operates as a precision error-control center, or servomechanism, that programs a movement initiated somewhere else, such as in the motor cortex. Primates and especially man, which possess a manual dexterity far surpassing that of other animals, have the most complex cerebellum of all since hand and finger movements may involve the simultaneous contraction and relaxation of hundreds of individual muscles.

Between the medulla and diencephalon is the **midbrain.** This is the anterior portion of the brainstem. The white matter of the midbrain consists of ascending and descending tracts that go to the thalamus and cerebrum. On the upper side of the midbrain are the rounded **optic lobes,** serving as centers for visual and auditory reflexes. The midbrain has undergone little evolutionary change in size among vertebrates but has changed in function. It is responsible for the most complex behavior of fishes and amphibians; the midbrain serves the higher integrative functions in these lower vertebrates that the cerebrum serves in higher vertebrates.

The **thalamus,** above the midbrain, contains masses of gray matter surrounded by the cerebral hemispheres on each side. This is the relay center for the sensory tracts from the spinal cord. Centers for the sensations of pain, temperature, and touch are supposedly located in the thalamus. In the **hypothalamus** are centers that regulate body temperature, water balance, sleep, and a few other body functions. The hypothalamus also has neurosecretory cells that produce pituitary-regulating neurohormones. These pass down fiber tracts to the anterior and posterior pituitary where the hormones are released into the circulation.

The anterior region of the brain, the **cerebrum,** can be divided into two anatomically distinct areas, the **paleocortex** and the **neocortex.** As its name implies, the paleocortex is the ancient telencephalon. Originally concerned with smell, it became well developed in the advanced fishes and early terrestrial vertebrates, which depend on this special sense. In mammals and especially in primates the paleocortex is a deep lying area called the rhinencephalon ("nose brain"), which actually has little to do with the sense of smell. Instead it seems to have acquired a variety of ill-defined functions concerned with consciousness, sleep, memory, emotional control, and sex. Together with a portion of the midbrain it is often called the **limbic-midbrain system.**

Though a late arrival in vertebrate evolution, the neocortex completely overshadows the paleocortex and has become so expanded that it envelopes the diencephalon and midbrain (Fig. 11-9). Almost all the integrative activities primitively assigned to the midbrain were transferred to the neocortex, or cerebral cortex as it is usually called.

The cerebral cortex is incompletely divided into two hemispheres by a deep longitudinal fissure. The right and left hemispheres are bridged through the **corpus callosum,** a neural mass lying between and connecting both hemispheres. Via the corpus callosum the two hemispheres are able to transfer information and coordinate mental activities.

Until recently it was thought that one hemisphere, almost always the left, becomes functionally dominant over the other during childhood. This concept is now recognized as misleading. We know now that the left and right brain hemispheres are specialized for entirely different functions: the left brain (controlling right side

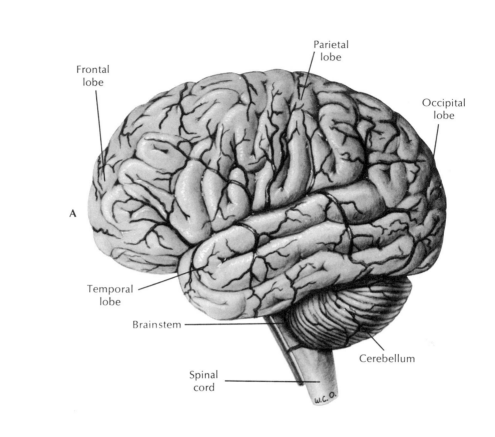

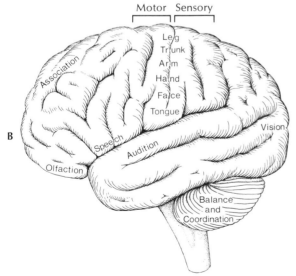

Fig. 11-9. Human brain. **A,** External view of the left side of brain showing lobes of the cerebrum, the cerebellum, and the brainstem. **B,** Localization of function on left cerebral cortex and cerebellum.

of body) for language development, mathematical and learning capabilities, and sequential thought processes; and the right brain (controlling left side of body) for spatial, musical, artistic, intuitive, and perceptual activities. It has long been known that even extensive damage to the right hemisphere may cause varying degrees of left-sided paralysis but has little effect on intellect. Conversely damage to the left hemisphere usually has disastrous effects on intellect. Since these differences in brain symmetry and function exist at birth, they appear to be inborn rather than the result of developmental or environmental effects as previously thought. Most people are right-handed (left hemisphere), but handedness is determined separately since it is not always related to left cerebral speech dominance.

It has been possible to localize function in the cerebrum by direct stimulation of exposed brains of people and experimental animals, by postmortem examination of persons suffering from various lesions, and by surgical removal of specific brain areas in experimental animals. The cortex contains discrete motor and sensory areas (Fig. 11-9) as well as large "silent" regions, called **association areas,** concerned with memory, judgment, reasoning, and other integrative functions. These regions are not directly connected to sense organs or muscles.

Peripheral nervous system

The peripheral nervous system, the nerve processes connecting the central nervous system to receptors and effectors, can be broadly subdivided into afferent and efferent components. As shown in the following outline, the efferent system is considerably more complex, consisting of a somatic nervous system and an autonomic nervous system.

  I. Afferent system (sensory)
 II. Efferent system (motor)
    A. Somatic nervous system
    B. Autonomic nervous system
      1. Sympathetic nervous system
      2. Parasympathetic nervous system

Afferent division

Afferent (sensory) neurons carry signals from receptors in the periphery of the body to the central nervous system. The afferent neuron of a reflex arc (Fig. 11-6) is representative of all afferent pathways in the peripheral nervous system. One long nerve process extends from the cell body in the dorsal root ganglion just outside the spinal cord to innervate receptors; another process passes from the cell body into the central nervous system where it connects with other neurons.

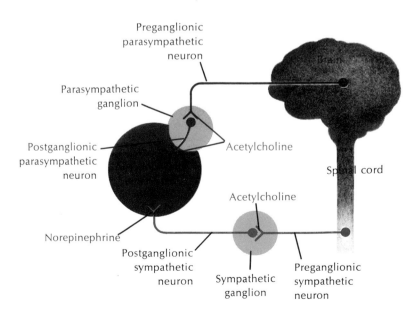

Fig. 11-10. Parasympathetic and sympathetic divisions of the autonomic nervous system, showing location of ganglia. Short preganglionic fibers of sympathetic division *(red)* pass from spinal cord to sympathetic ganglion, which lies close to the spinal cord; long postganglionic fibers continue to effector organ. In contrast, parasympathetic ganglion *(black)* lies in wall of effector organ, and the preganglionic fiber is much longer than the postganglionic. The sympathetic transmitter substance between the postganglionic fiber and effector organ is norepinephrine; at the equivalent parasympathetic ending the transmitter is acetylcholine.

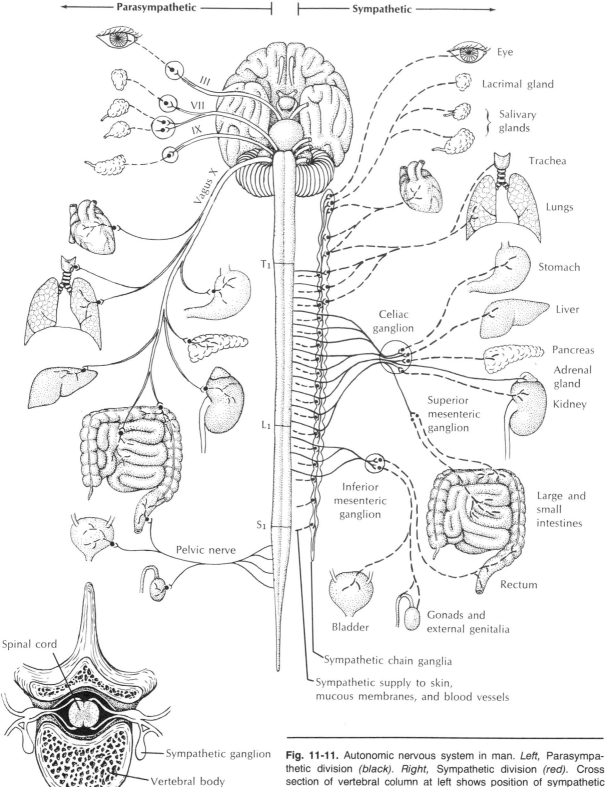

Parasympathetic | Sympathetic

Eye

Lacrimal gland

Salivary glands

Trachea

Lungs

Stomach

Liver

Pancreas

Adrenal gland

Kidney

Celiac ganglion

Superior mesenteric ganglion

Inferior mesenteric ganglion

Large and small intestines

Rectum

III

VII

IX

Vagus X

Pelvic nerve

Bladder

Gonads and external genitalia

Sympathetic chain ganglia

Sympathetic supply to skin, mucous membranes, and blood vessels

T₁

L₁

S₁

Spinal cord

Sympathetic ganglion

Vertebral body

Fig. 11-11. Autonomic nervous system in man. *Left*, Parasympathetic division *(black)*. *Right*, Sympathetic division *(red)*. Cross section of vertebral column at left shows position of sympathetic ganglia.

Efferent division

Somatic nervous system. Nerve fibers of the somatic division of the peripheral nervous system pass from the brain or the spinal cord to skeletal muscle fibers. These are called motor neurons because they control muscle movement. They release acetylcholine as the transmitter substance at the nerve endings. Motor neurons in the reflex arc are representative of the functional position of the somatic nervous system.

Autonomic nervous system. The autonomic nerves govern the involuntary functions of the body that do not ordinarily affect consciousness. The cerebrum has no direct control over these nerves; thus we cannot by volition stimulate or inhibit their action. Autonomic nerves control the movements of the alimentary canal and heart, the contraction of the smooth muscle of the blood vessels, urinary bladder, iris of eye, etc., and the secretions of various glands.

Subdivisions of the autonomic system are the **parasympathetic** and the **sympathetic** (Fig. 11-10). Most organs in the body are innervated by both sympathetic and parasympathetic fibers, and their actions are antagonistic (Fig. 11-11). If one fiber speeds up an activity, the other slows it down. However, neither kind of nerve is exclusively excitatory or inhibitory. For example, parasympathetic fibers inhibit heartbeat but excite peristaltic movements of the intestine; sympathetic fibers increase heartbeat but slow down peristaltic movement.

The **parasympathetic** system consists of motor nerves some of which emerge from the brain by certain cranial nerves and others from the pelvic region of the spinal cord by certain spinal nerves. Parasympathetic fibers *excite* the stomach and intestine, urinary bladder, bronchi, constrictor of iris, salivary glands, and coronary arteries. They *inhibit* the heart, intestinal sphincters, and sphincter of the urinary bladder.

In the **sympathetic** division the nerve cell bodies are located in the thoracic and upper lumbar areas of the spinal cord. Their fibers pass out through the ventral roots of the spinal nerves, separate from these, and go to the sympathetic ganglia, which are paired and form a chain on each side of the spinal column. From these ganglia some of the fibers run through spinal nerves to the limbs and body wall, where they innervate the blood vessels of the skin, the smooth muscles of the hair follicles, the sweat glands, etc.; other fibers run to the abdominal organs as the splanchnic nerves. Sympathetic fibers *excite* the heart, blood vessels, sphincters of the intestines, urinary bladder, dilator muscles of the iris, and others. They *inhibit* the stomach, intestine, and bronchial muscles and coronary arterioles.

All preganglionic fibers, whether sympathetic or parasympathetic, release **acetylcholine** at the synapse for stimulating the ganglion cells. The terminations of the parasympathetic and sympathetic nervous systems release different types of chemical transmitter substances (Fig. 11-10). The parasympathetic fibers release **acetylcholine** at their endings, whereas the sympathetic fibers release **norepinephrine** (also called noradrenaline). These chemical substances produce characteristic physiologic reactions. Since there is some physiologic overlapping of sympathetic and parasympathetic fibers, it is customary to describe nerve fibers as either adrenergic (norepinephrine effect) or cholinergic (acetylcholine effect).

SENSE ORGANS

Animals require a constant inflow of information from the environment to regulate their lives. Sense organs are specialized receptors designed for detecting environmental status and change. An animal's sense organs are its first level of environmental perception; they are data input channels for the brain.

A **stimulus** is some form of energy—electrical, mechanical, chemical, or radiant. The task of the sense organ is to transform the energy form of the stimulus it receives into nerve impulses, the common language of the nervous system. In a very real sense, then, sense organs are **biologic transducers.** A microphone, for example, is a man-made transducer that converts mechanical (sound) energy into electrical energy. Like the microphone that is sensitive only to sound, sense organs are, as a rule, specific for one kind, or **modality,** of stimulus energy. Thus eyes respond only to light, ears to sound, pressure receptors to pressure, and chemoreceptors to chemical molecules. But again, all of these different forms of energy are converted into nerve impulses.

Since all nerve impulses are qualitatively alike, how do animals perceive and distinguish the different **sensations** of varying stimuli? The answer is that the real perception of sensation is done in localized regions of the brain, where each sense organ has its own hookup. Impulses arriving at a particular sensory area of the brain can be interpreted in only one way. This is why pressure on the eye causes us to see "stars" or other visual patterns; the mechanical distortion of the eye initiates impulses in the optic nerve fibers that are perceived as light sensations. Although the operation hopefully could never be done, the deliberate surgical switching of optic and auditory nerves would cause the recipient to literally see thunder and hear lightning!

Classification of receptors

Receptors are classified on the basis of their location. Those near the external surface are called **exteroceptors** and are stimulated by changes in the external environment. Internal parts of the body are provided with **interoceptors,** which pick up stimuli from the internal organs. Muscles, tendons, and joints have **proprioceptors,** which are sensitive to changes in the tension of muscles and provide the organism with a sense of position.

Another way of classifying receptors is on the basis of the energy form used to stimulate them, such as **chemical, mechanical, photo,** or **thermal.**

Chemoreception

Chemoreception is the most primitive and most universal sense in the animal kingdon. It probably guides the behavior of animals more than any other sense. The most primitive animals, protozoans, use **contact chemical receptors** to locate food and adequately oxygenated water and to avoid harmful substances. These receptors elicit a simple trial-and-error behavior, called **chemotaxis.** More advanced animals have specialized **distance chemical receptors.** These are often developed to a truly amazing degree of sensitivity. Distance chemoreception, usually referred to as sense of smell or olfactory sense, guides feeding behavior, location and selection of sexual mates, territorial and trail marking, and alarm reactions of numerous animals. The social insects produce species-specific odors, called **pheromones,** which comprise a highly developed chemical language. Pheromones are a diverse group of organic compounds released by epithelial glands that serve either to initiate specific patterns of behavior, such as attracting mates or marking trails (releaser pheromones), or to trigger some internal physiologic change, such as metamorphosis (primer pheromones). Insects have a variety of chemoreceptors on the body surface for sensing specific pheromones, as well as other nonspecific odors.

In all vertebrates and in insects as well, the senses of taste and smell are clearly distinguishable. Although there are similarities between taste and smell receptors, in general the sense of taste is more restricted in response and is less sensitive than the sense of smell. Taste and smell centers are also located in different parts of the brain. In higher forms, **taste buds** are found on the tongue and in the mouth cavity (Fig. 11-12). A taste bud consists of a few sensitive cells surrounded by supporting cells and is provided with a

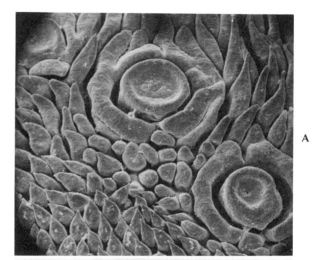

A

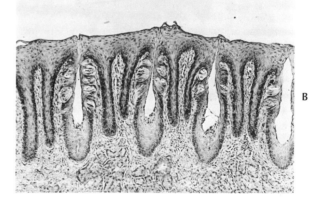

B

Fig. 11-12. Taste buds. **A,** Scanning electron micrograph of circumvallate papillae on surface of tongue of puppy. Taste buds (not visible) are located in walls of circular trench surrounding papilla. The numerous filiform papillae surrounding the two circumvallate papillae lack taste buds. (×55.) **B,** Light micrograph of section of rabbit's tongue. Taste buds are little oval bodies on sides of slitlike recesses. (×400.) (**A** courtesy P. P. A. Graziadei, Florida State University, Tallahassee, Fla.; **B** courtesy J. B. Bamberger, Los Angeles, Calif.)

small external pore through which the slender tips of the sensory cells project. The basal ends of the sensory cells contact nerve endings from cranial nerves. Taste bud cells in vertebrates have a short life of approximately 10 days and are continually being replaced.

The four basic taste sensations of man—sour, salt, bitter, and sweet—are each attributable to a different kind of taste bud. The tastes for salt and sweet are found mainly at the tip of the tongue, bitter at the base of the tongue, and sour along the sides of the tongue. Of these, bitter taste is by far the most sensitive since it protects the body against toxic substances. Taste buds are more numerous in ruminants (mammals that chew the cud) than in man. They tend to degenerate with age; the child has more buds widely distributed over the mouth.

Sense organs of **smell** (olfaction) are found in a specialized epithelium located either in the nasal cavity (terrestrial vertebrates) or in pouches on the snout (aquatic vertebrates). The sense of smell is much more complex than that of taste. There are millions of olfactory receptor cells in the nasal epithelium, and as many as a thousand of these may converge on a single neuron. This allows great summation and vastly improves sensitivity.

Some people can detect many thousands of different odors, and it is obvious that many other vertebrates can easily outdo man. Gases must be dissolved in a fluid to be smelled; therefore the nasal cavity must be moist.

The sensory cells with projecting hairs are scattered singly through the olfactory epithelium. Their basal ends are connected to fibers of the olfactory cranial nerve that runs to one of the olfactory lobes. The sensitivity to certain odors approaches the theoretical maximum for the chemical sense. The human nose can detect 1/25-millionth of 1 mg of mercaptan, the odoriferous principle of the skunk. This averages out to approximately 1 molecule per sensory ending.

Since taste and smell are stimulated by chemicals in solution, their sensations may be confused. The taste of food is dependent to a great extent on odors that reach the olfactory membrane through the throat. All the various ''tastes'' other than the four basic ones (sweet, sour, bitter, salt) are really the result of the flavors' reaching the sense of smell in this manner.

The sense of smell is the least understood sense. Of the numerous theories that have been proposed, the favored ones today postulate some kind of **physical interaction** between the odor molecule and a protein receptor site on a cell membrane. This interaction somehow alters membrane permeability and results in depolarization in the receptor cell, which triggers a nerve impulse. One theory (J. E. Amoore) proposes that odor molecules have specific stereochemical shapes and that the range of detectable odors is attributable to differences in the way the smelled molecule fits the receptor site.

Mechanoreception

Mechanoreceptors are sensitive to quantitative forces such as touch, pressure, stretching, sound, and gravity. Many receptors in and on the body constantly monitor information about conditions within the body (muscle position, body equilibrium, blood pressure, pain, etc.) and conditions in the environment (sound and other vibrations such as water currents).

Touch and pain. Although superficial touch receptors are distributed over all the body, they tend to be concentrated in the few areas especially important in exploring and interpreting the environment. In most animals these areas are on the face and limb extremities. Of the more than half a million separate sensitive spots on man's body surface, most are found on his lips, tongue, and fingertips. Many touch receptors are bare nerve-fiber terminals, but there is an assortment of other kinds of receptors of varying shapes and sizes. Each hair follicle is crowded with receptors that are sensitive to touch.

The sensation of deep touch and pressure is registered by relatively large receptors called **pacinian corpuscles.** They are common in deep layers of skin (Fig. 7-1, *B,* p. 130), in connective tissue surrounding muscles and tendons, and in the abdominal mesenteries. Each corpuscle, easily visible to the naked eye, is built of numerous layers like an onion. Any kind of mechanical deformation of the pacinian corpuscles is converted into nerve impulses that are sent to sensory areas of the brain.

Pain receptors are relatively unspecialized nerve fiber endings that respond to a variety of stimuli that signal possible or real tissue damage. It is still uncertain whether pain fibers respond directly to injury or indirectly to some substance such as histamine, which is released by damaged cells.

Lateral line system of fishes. The lateral line is a distant touch reception system for detecting wave vibrations and currents in water. The receptor cells,

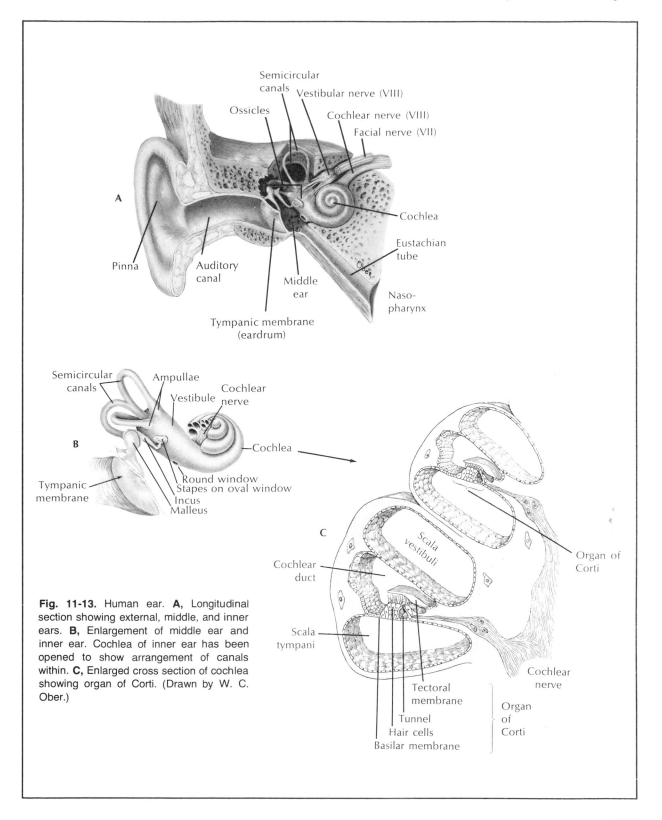

Fig. 11-13. Human ear. **A,** Longitudinal section showing external, middle, and inner ears. **B,** Enlargement of middle ear and inner ear. Cochlea of inner ear has been opened to show arrangement of canals within. **C,** Enlarged cross section of cochlea showing organ of Corti. (Drawn by W. C. Ober.)

called **neuromasts,** are located free on the body surface in primitive fishes and aquatic amphibians, but in the advanced fishes they are located within canals running beneath the epidermis; these open at intervals to the surface. Each neuromast is a collection of hair cells with the sensory hairs embedded in a gelatinous, wedge-shaped mass known as a **cupula.** This projects into the center of the lateral line canal so that it bends in response to any disturbance of water on the body surface. The lateral line system is one of the principal sensory systems that guide fishes in their movements.

Hearing. The ear is a specialized receptor for detecting sound waves in the surrounding environment (Fig. 11-13). Another sense, equilibrium, is also associated with the ears of all vertebrate animals. Among the invertebrates, only certain insects have true sound receptors.

In its evolution the ear was at first associated more with equilibrium than with hearing. Hearing sense is found only in the internal ear, which is the only part of the ear present in many of the lower vertebrates; the middle and the external ears were added in later evolutionary developments. The internal ear is considered to be a development of part of the lateral line system of fishes. Some fishes apparently can transmit sound from their swim bladders by the Weberian ossicles (series of small bones) to some part of the inner ear since they lack a cochlea.

The ear found in higher vertebrates is made up of three parts: (1) the **inner ear,** which contains the essential organs of hearing and equilibrium and is present in all vertebrates; (2) the **middle ear,** an airfilled chamber with one or more ossicles for conducting sound waves to the inner ear, present in amphibians and higher vertebrates only; and (3) the **outer ear,** which collects the sound waves and conducts them to the tympanic membrane lying next to the middle ear and is present only in reptiles, birds, and mammals, but most highly developed in mammals.

Outer ear. The outer, or external, ear of higher vertebrates is made up of two parts: (1) the **pinna,** or skin-covered flap of elastic cartilage and muscles, and (2) the **auditory canal** (Fig. 11-13, *A*). In many mammals, such as the rabbit and cat, the pinna is freely movable and is effective in collecting sound waves. The auditory canal condenses the waves and passes them to the tympanic membrane. The walls of the auditory canal are lined with hair and wax-secreting glands as a protection against the entrance of foreign objects.

Middle ear. The middle ear is separated from the external ear by the tympanic membrane (eardrum), which consists of a tightly stretched connective membrane. Within the air-filled middle ear a remarkable chain of three tiny bones, **malleus** (hammer), **incus** (anvil), and **stapes** (stirrup), conducts the sound waves across the middle ear (Fig. 11-13, *B*). This bridge of bones is so arranged that the force of sound waves pushing against the tympanic membrane is amplified as many as 90 times where the stapes contacts the oval window of the inner ear. Muscles attached to the middle ear bones contract when the ear receives very loud noises, thus protecting the inner ear from damage. However, these muscles cannot contract quickly enough to protect the inner ear from the damaging effects of a sudden blast. The middle ear communicates with the pharynx by means of the eustachian tube, which acts as a safety device to equalize pressure on both sides of the tympanic membrane.

Inner ear. The inner ear consists essentially of two labyrinths, one within the other. The inner one is called the **membranous labyrinth** and is a closed ectodermal sac filled with the fluid, **endolymph.** The part involved with hearing **(cochlea)** is coiled like a snail's shell, making two and a half turns in man (Fig. 11-13, *B*). Surrounding the membranous labyrinth is the **bony labyrinth,** which is a hollowed-out part of the temporal bone and conforms to the shape and contours of the membranous labyrinth. In the space between the two labyrinths, **perilymph,** a fluid similar to endolymph, is found.

The cochlea is divided into three longitudinal canals that are separated from each other by thin membranes (Fig. 11-13, B and C). These canals become progressively smaller from the base of the cochlea to the apex. One of these canals is called the **vestibular canal** (scala vestibuli); its base is closed by the oval window. The **tympanic canal** (scala tympani), which is in communication with the vestibular canal at the tip of the cochlea, has its base closed by the round window. Between these two canals is the **cochlear duct,** which contains the organ of hearing, the **organ of Corti** (Fig. 11-13, *C*). The latter organ is made up of fine rows of hair cells that run lengthwise from the base to the tip of the cochlea. There are at least 24,000 of these hair cells in the human ear, each cell with many hairs projecting into the endolymph of the

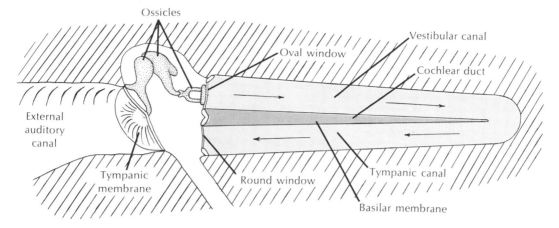

Fig. 11-14. Mammalian ear as it would appear with cochlea stretched out. Sound waves transmitted to oval window produce vibration waves that travel down basilar membrane. High-frequency vibrations cause membrane to resonate at end near oval window before dying out; low-frequency tones travel farther down basilar membrane.

cochlear canal and each connected with neurons of the auditory nerve. The hair cells rest on the **basilar membrane,** which separates the tympanic canal and cochlear duct, and are covered over by the **tectorial membrane** found directly above them. These relationships are shown diagrammatically in Fig. 11-14.

Sound waves picked up by the external ear are transmitted through the auditory canal to the tympanic membrane, which causes it to vibrate. These vibrations are conducted by the chain of ear bones to the oval window, which transmits the vibrations to the fluid in the vestibular and tympanic canals. The vibrations of the endolymph cause the basilar membrane with its hair cells to vibrate so that the hair cells are bent against the tectorial membrane. This stimulation of the hair cells causes them to initiate nerve impulses in the fibers of the auditory nerve, with which they are connected.

According to the **place theory** of pitch discrimination, it is stated that when sound waves strike the inner ear the entire basilar membrane is set in vibration by a traveling wave of displacement, which increases in amplitude from the oval window toward the apex of the cochlea. This displacement wave reaches a maximum at the region of the basilar membrane, where the natural frequency of the membrane corresponds to the sound frequency. Here, the membrane vibrates with such case that the energy of the traveling wave is

completely dissipated. Hair cells in that region are stimulated and the impulses conveyed to the fibers of the auditory nerve. Those impulses that are carried by certain fibers of the auditory nerve are interpreted by the hearing center as particular tones. The **loudness** of a tone depends on the number of hair cells stimulated, whereas the **timbre,** or quality, of a tone is produced by the pattern of the hair cells stimulated by sympathetic vibration. This latter characteristic of tone enables us to distinguish between different human voices and different musical instruments, even though the notes in each case may be of the same pitch and loudness.

Sense of equilibrium. Closely connected to the inner ear and forming a part of it are two small sacs, the **saccule** and **utricle,** and three **semicircular canals.** Like the cochlea, the sacs and canals are filled with endolymph. They are concerned with the sense of balance and rotation. They are well developed in all vertebrates, and in some lower forms they represent about all there is of the internal ear, for the cochlea is absent in fishes. They are innervated by the nonacoustic branch of the auditory nerve.

The utricle and saccule are hollow sacs lined with sensitive hairs on which are deposited a mass of minute calcium carbonate crystals called **otoconia.** In bony fishes, these crystals are formed into compact stonelike structures, the **otoliths.** Similar stony accretions are

227

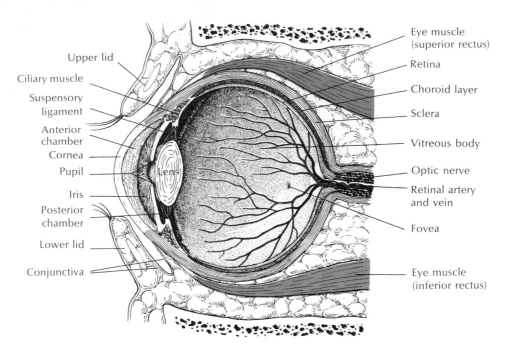

Upper lid

Ciliary muscle

Suspensory
ligament

Anterior
chamber

Cornea

Pupil

Iris

Posterior
chamber

Lower lid

Conjunctiva

Lens

Eye muscle
(superior rectus)

Retina

Choroid layer

Sclera

Vitreous body

Optic nerve

Retinal artery
and vein

Fovea

Eye muscle
(inferior rectus)

Fig. 11-15. Section through human eye.

found within statocysts, the balance organs of many invertebrates. Although the anatomic nature of these static balance organs varies in different groups, they all function in the same basic way: the weight of the stony accretion presses on the hair cells to give information about the position of the head (or entire body) relative to the force of gravity. As the head is tilted in one direction or another, different groups of hair cells are stimulated; conveyed to the brain, this information is interpreted with reference to position.

The semicircular canals of vertebrates are designed to detect changes in movement: acceleration or deceleration. The three semicircular canals are at right angles to each other, one in each plane of space (Fig. 11-13, B). They are filled with fluid, and at the opening of each canal into the utricle there is a bulblike enlargement, the **ampulla,** which contains hair cells but no otoconia. Whenever the fluid moves, these hair cells are stimulated. Rotating the head causes a lag, because of inertia, in certain of these ampullae. This lag produces consciousness of movement. Since the three canals of each internal ear are in different planes, any kind of movement stimulates at least one of the ampullae.

Vision

Light sensitive receptors are called **photoreceptors.** These receptors range all the way from simple light-sensitive cells scattered randomly on the body surface of the lowest invertebrates (dermal light sense) to the exquisitely developed vertebrate eye. Although dermal light receptors contain little photochemical substance and are far less sensitive than optic receptors, they are important in locomotory orientation, pigment distribution in chromatophores, photoperiodic adjustment of reproductive cycles, and other behavioral changes in many lower invertebrates.

The arthropods, however, have **compound** eyes composed of many independent visual units called **ommatidia.** The eye of a bee contains about 15,000 of these units, each of which views a separate narrow sector of the visual field. Such eyes form a mosaic of images from the separate units. The compound eye probably does not produce a very distinct image of the visual field, but it is extremely well suited to picking up motion, as any one knows who has tried to swat a fly.

The vertebrate eye is built like a camera—or rather we should say a camera is modeled somewhat after the vertebrate eye. It contains a light-tight chamber with a lens system in front, which focuses an image of the visual field on a light-sensitive surface (the retina) in back (Fig. 11-15). Because eyes and cameras are based on the same laws of optics, we can wear glasses to correct optical defects in our eyes. But here the similarity between eye and camera ends.

The human eye is actually replete with optic shortcomings; projected on the retina of the normal eye are more colored fringe halos, apparitions, and distortions than would be produced by even the cheapest camera lens. Yet man's brain corrects for this "poor design" so completely that we perceive a perfect image of the visual field. It is in the retina and the optic center of the brain that the marvel of vertebrate vision can be understood.

The spheric eyeball is built of three layers: (1) a tough outer white **sclerotic** coat (sclera) serving for support and protection, (2) the middle **choroid** coat containing blood vessels for nourishment, and (3) the light-sensitive **retinal** coat (Fig. 11-15). The **cornea** is a transparent modification of the sclera. A circular curtain, the **iris,** regulates the size of the light opening, the **pupil.** Just behind the iris is the **lens,** a transparent, elastic ball that bends the rays and focuses them on the retina. In land vertebrates the cornea actually does most of the bending of light rays, whereas the lens adjusts the focus for near and far objects. Between the cornea and the lens is the outer chamber filled with the watery **aqueous humor;** between the lens and the retina is the much larger inner chamber filled with the viscous **vitreous humor.** Surrounding the margin of the lens and holding it in place is the **suspensory ligament.** This, together with the **ciliary muscle,** a ring of radiating muscle fibers attached to the suspensory ligament, makes possible the stretching and relaxing of the lens for close or distant vision (accommodation).

The **retina** is composed of photoreceptors, the **rods** and **cones.** Approximately 125 million rods and 7 million cones are present in each human eye. Cones are primarily concerned with color vision in ample light; rods, with colorless vision in dim light. The retina is actually made up of three sets of neurons in series with each other: (1) photoreceptors (rods and cones), (2) intermediate neurons, and (3) ganglionic neurons whose axons form the optic nerve.

The **fovea centralis,** the region of keenest vision, is located in the center of the retina, in direct line with the center of the lens and cornea. It contains only cones. The acuity of an animal's eyes depends on the density of cones in the fovea. The human fovea and that of a lion contain approximately 150,000 cones per square millimeter. But many water and field birds have up to 1 million cones per sq mm. Their eyes are as good as man's eyes would be if aided by eight-power binoculars.

At the peripheral parts of the retina only rods are found. This is why we can see better at night by looking out of the corners of our eyes because the rods, adapted for high sensitivity with dim light, are brought into use.

Chemistry of vision. Each rod contains a light-sensitive pigment known as **rhodopsin.** Each rhodopsin molecule consists of a large, colorless protein, **opsin,** and a small carotenoid molecule, **retinal** (formerly called retinene), a derivative of vitamin A. When a quantum of light strikes a rod and is absorbed by the rhodopsin molecule, the latter undergoes a chemical bleaching process that causes it to split into separate opsin and retinal molecules. In some way not yet understood this change triggers the discharge of a nerve impulse in the receptor cell. The impulse is relayed to the optic center of the brain. Rhodopsin is then enzymatically resynthesized so that it can respond to a subsequent light signal.

The amount of intact rhodopsin in the retina depends on the intensity of light reaching the eye. The dark-adapted eye contains much rhodopsin and is very sensitive to weak light. Conversely most of the rhodopsin is broken down in the light-adapted eye. It takes approximately half an hour for the light-adapted eye to accommodate to darkness, while the rhodopsin level is gradually built up. The remarkable ability of the eye to adapt to darkness and light vastly increases the versatility of the eye; it enables us to see by starlight as well as by the noonday sun, 10 billion times brighter.

The light-sensitive pigment of cones is called **iodopsin.** It is similar to rhodopsin, containing **retinal** combined with a special protein, **cone opsin.** Cones function to perceive color and require 50 to 100 times more light for stimulation than do rods. Consequently night vision is almost totally rod vision; this is why the landscape illuminated by moonlight appears in shades of black and white only. Unlike man, who has both day and night vision, some vertebrates specialize for one or the other. Strictly nocturnal animals, such as

229

bats and owls, have pure rod retinas. Purely diurnal forms, such as the common gray squirrel and some birds, have only cones. They are, of course, virtually blind at night.

Color vision. How does the eye see colors? According to the trichromatic theory of color vision, there are three different types of cones that react most strongly to red, green, and violet light. Colors are perceived by comparing the levels of excitation of the three different kinds of cones. This comparison is made both in nerve circuits in the retina and in the visual cortex of the brain. Color vision is present in all vertebrate groups with the possible exception of the amphibians. Bony fishes and birds have particularly good color vision. Surprisingly most mammals are color blind; exceptions are primates and a very few other species such as squirrels.

SELECTED REFERENCES

(See also general references for Part two, p. 274.)

Bachelard, H. S. 1974. Brain biochemistry. New York, John Wiley & Sons, Inc. *Brief but concentrated account of brain chemistry (neurotransmitter substances, hormones, enzymes, drug effects) with helpful diagrams. Paperback.*

Burton, M. 1972. The sixth sense of animals. New York, Taplinger Publishing Co., Inc. *Wide ranging comparative account. Undergraduate level.*

Droscher, V. B. 1969. The magic of the senses. New York, E. P. Dutton & Co., Inc. *Readable comparative account of the senses, ending with consideration of animal migration and navigation. Undergraduate level.*

Mellon, D., Jr. 1968. The physiology of sense organs. San Francisco, W. H. Freeman & Co., Publishers. *Advanced treatment of sense organ physiology.*

Milne, L., and M. Milne. 1972. The senses of animals and men. New York, Atheneum Publishers. *Beautifully written account of animal senses by these prolific authors. Undergraduate level.*

Mountcastle, V. B. 1974. Medical physiology, vol. 2, ed. 13. St. Louis, The C. V. Mosby Co. *The section on the nervous system is especially well written in this outstanding medical physiology book. Advanced level.*

Nathan, P. 1969. The nervous system. Philadelphia, J. B. Lippincott Co. *Highly readable account of the nervous system and special senses of higher vertebrates.*

Sherrington, C. S. 1947. The integrative action of the nervous system, rev. ed. New Haven, Conn., Yale University Press. *A classic work on the structural and functional plan of the nervous system.*

SELECTED SCIENTIFIC AMERICAN ARTICLES

Amoore, J. E., J. W. Johnston, Jr., and M. Rubin. 1964. The stereochemical theory of odor. **210:**42-49 (Feb.).

Axelrod, J. 1974. Neurotransmitters. **230:**58-71 (June).

Baker, P. F. 1966. The nerve axon. **214:**75-82 (March). *Describes techniques used to study nerve impulse conduction.*

Dowling, J. E. 1966. Night blindness. **215:**78-84 (Oct.). *The importance of vitamin A in vision.*

Eccles, J. 1965. The synapse. **212:**56-66 (Jan.). *A famous neurophysiologist explains how nerve impulses are transmitted from cell to cell.*

Heimer, L. 1971. Pathways in the brain. **225:**48-60 (July). *A new staining technique has vastly improved studies of neural pathways and connections in the nervous sytem.*

Hendricks, S. B. 1968. How light interacts with living matter. **219:**174-186 (Sept.). *Special pigments mediate the interaction of light with matter in photosynthesis, vision, and photoperiodism.*

Hodgson, E. S. 1961. Taste receptors. **204:**135-144 (May). *The mechanism of taste reception is studied with blowflys.*

Jerison, H. J. 1976. Paleoneurology and the evolution of mind. **234:**90-101 (Jan.) *The relationship between brain size and body size during animal evolution provides insight into intelligence.*

Katz, B. 1961. How cells communicate. **205:**209-220 (Sept.). *Nervous communication and nature of the nerve impulses.*

Kennedy, D. 1967. Small systems of nerve cells. **216:**44-52 (May). *The simple nervous systems of invertebrates facilitate studies of nervous integration.*

Llinás, R. R. 1975. The cortex of the cerebellum. **232:**56-71 (Jan.). *The pattern of neuronal connections in the cerebellum can now be related to its role in motor coordination.*

Luria, A. R. 1970. The functional organization of the brain. **222:**66-78 (March).

Melzack, R. 1961. The perception of pain. **204:**41-49 (Feb.). *Pain is greatly modified by experience and "state of mind."*

Miller, W. H., F. Ratliff, and H. K. Hartline. 1961. How cells receive stimuli. **205:**222-238 (Sept.). *Structure and functional properties of receptor cells.*

Nathanson, J. A., and P. Greengard. 1977. "Second messengers" in the brain. **237:**108-119 (Aug.). *A survey of the various neurotransmitters of the vertebrate nervous system and how their chemical messages are translated into physiologic actions.*

Stent, G. S. 1972. Cellular communication. **227:**43-51 (Sept.). *Information processing and communication is discussed with particular emphasis on vision.*

Werblin, F. S. 1973. The control of sensitivity in the retina. **228:**70-79 (Jan.). *Recent studies of neuron interactions in the retina help to explain its versatility over widely ranging light conditions.*

Young, R. W. 1970. Visual cells. **223:**80-91 (Oct.). *How rods and cones renew themselves.*

Two bull moose fight for dominance during the rutting season. Male aggressiveness during rut is associated with a rising blood level of the male sex hormone testosterone. The entire sex cycle is controlled by changing activity of endocrine glands, all orchestrated by neurosecretory centers in the brain. (Photo by L. L. Rue III.)

12 CHEMICAL COORDINATION
Endocrine system

MECHANISMS OF HORMONE ACTION

INVERTEBRATE HORMONES

VERTEBRATE ENDOCRINE GLANDS AND HORMONES

Hormones of the pituitary gland and
  hypothalamus
Hormones of metabolism
Hormones of digestion

The endocrine system is the second great integrative system controlling the body's activities. Endocrine glands, or specialized tissues, secrete **hormones** (from the Greek root meaning "to excite") that are transported by the blood for variable distances to some part of the body where they produce definite physiologic effects. Hormones are effective in minute quantities; some are active when diluted several billion times in the blood. The endocrine system is a slow-acting integrative system as compared to the nervous system, and in general hormonal effects are long lasting. Some hormones are excitatory, others inhibitory. Many physiologic processes are governed by antagonistic hormones (one that stimulates, the other that inhibits the process). Such combinations are very effective in maintaining homeostatic conditions.

Endocrinology is a young field. Its birthdate is usually given as 1902, the year two English physiologists, W. H. Bayliss and E. H. Starling, demonstrated the action of an internal secretion. They were interested in determining how the pancreas secreted its digestive juice into the small intestine at the proper time of the digestive process. In an anesthetized dog they tied off a section of the small intestine beyond the duodenum (the part of the intestine next to the stomach) and removed all nerves leading to this tied-off loop, but left its blood vessels intact. Bayliss and Starling found that the injection of hydrochloric acid into the blood serving this intestinal loop had no effect upon the secretion of pancreatic juice; but, when they introduced 0.4% hydrochloric acid directly inside the intestinal loop, a pronounced flow of pancreatic juice into the duodenum occurred through the pancreatic duct. When they scraped off some of the mucous membrane lining of the intestine and mixed it with acid, they found that the injection of this extract into the blood caused an abundant flow of pancreatic juice. They concluded that, when the partly digested and slightly acid food from the stomach arrives in the small intestine, the hydrochloric acid reacts with something in the mucous lining to produce an internal secretion, or chemical messenger, which is conveyed by the bloodstream to the pancreas, causing it to secrete pancreatic digestive juices. They called this messenger **secretin.** In a 1905 Croonian lecture at the Royal College of Physicians, Starling first used the word "hormone," a general term to describe all such chemical messengers since he correctly surmised that secretin was only the first of many hormones that remained to be described.

Since hormones are transported in the blood, they reach virtually all body tissues. This makes it possible for certain hormones, such as the growth hormone of the pituitary gland, to have a very wide-spread action affecting most, if not all, cells during the growth of an animal. However, most hormones, despite their ubiquitous distribution, are highly specific in their action. Usually only certain cells respond to the presence of a given hormone. For example, only the pancreatic cells respond to secretin, even though secretin is carried throughout the body by the circulation. All other cells, lacking specific receptors for secretin, are totally unresponsive to its presence. The cells that respond to a particular hormone are called **target-organ cells.**

Recently it has become evident that the classic definition of a hormone as the bloodborne product of a discrete ductless gland no longer perfectly applies to the heterogeneous endocrine system recognized today. Some cells secrete hormones that may diffuse only a short distance to neighboring cells to exert their effect without ever entering the bloodstream. These are the "local hormones"—prostaglandins and kinins. **Prostaglandins,** for example, are a family of 20-carbon fatty acids, synthesized in numerous body tissues from polyunsaturated fatty acids in the diet, that have a diverse list of physiologic effects. These include regulation of smooth-muscle contraction and stimulation of specific metabolic processes. Prostaglandins are unfortunately misnamed since the source of their high concentration in semen, in which they were first discovered, is the seminal vesicles rather than the prostate gland. Furthermore, they are produced by both males and females. They are among the most potent hormones known. They produce their effect extremely rapidly and are just as quickly metabolized and inactivated. Prostaglandins have ascended to recent prominence because they are being used very successfully to facilitate labor contractions and birth; they also can be used to promote abortion and thus show promise as birth control agents.

Even before local hormones were disovered, it had become evident that certain nerve cells are capable of secreting hormones. Such specialized nerve cells are called **neurosecretory cells** and their secreted products are called **neurosecretory hormones.** Subsequent studies demonstrated that these neurosecretory cells are crucial links between the body's two great integrative systems. Such knowledge made it possible to understand how, for example, increasing day length in

233

the spring stimulates the breeding cycle of birds. Increasing amounts of light received by the eyes are relayed via nerve tracts in the brain to neurosecretory cells that release hormones, which set the reproductive cycle into motion.

Neurosecretory cells are known to be the main source—in some instances the only source—of hormones of many invertebrate groups. Neurosecretion is also a widespread phenomenon among the vertebrates. Because neurosecretion is obviously a very ancient physiologic activity and because it serves as a crucial link between the nervous system and the ductless gland system, we believe that hormones first evolved as nerve cell secretions. Later, nonnervous endocrine

glands appeared in other parts of the body. These remote glands remained chemically linked to the nervous system, however, by the neurosecretory hormones. The vertebrate hypothalamic-hypophyseal complex mentioned above in connection with the regulation of breeding in birds is a much-studied example.

MECHANISMS OF HORMONE ACTION

How do hormones exert their effects? This question has been the object of intense research in recent years. We can readily appreciate that it is much easier to observe the physiologic effect of a hormone than to determine what the hormone does to produce the

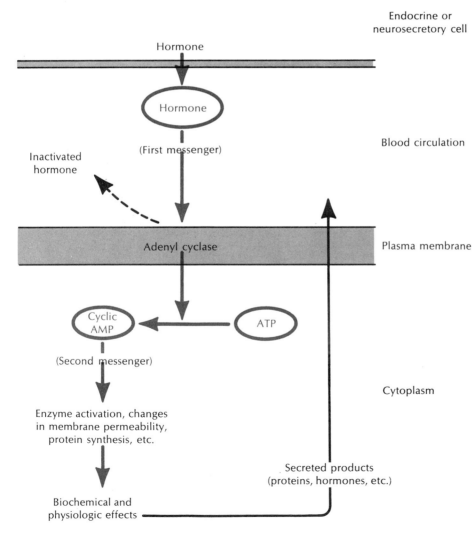

Fig. 12-1. Second-messenger concept of hormone action. Many hormones exert their action through cyclic AMP. Hormone (first messenger) is carried by bloodstream from endocrine gland to target cell where it combines with unidentified receptor in plasma membrane. This interaction stimulates enzyme adenyl cyclase to catalyze formation of cyclic AMP (second messenger). Cyclic AMP acts intracellularly to initiate any of several changes, depending on kind of cell.

effect. Although we have known for years that insulin lowers the blood glucose level, we are still uncertain as to *how* insulin does this.

It seems that there may be no more than two basic mechanisms of hormone action.

1. *Stimulation of protein synthesis.* Several hormones, such as the thyroid hormones and the insect molting hormone, ecdysone, stimulate the synthesis of specific enzymes and other proteins by causing the transcription of particular kinds of messenger RNA. These hormones therefore act directly on specific genes. It is possible that the hormone may activate genes by somehow antagonizing a repressor, according to the Jacob-Monod model for gene action (see p. 35). With the repressor removed by the hormone, RNA polymerase then begins building the enzymes (proteins) that set in motion the observed action of the hormone.

2. *Activation of a "second messenger," cyclic AMP* (adenosine 3′,5′-monophosphate). **Cyclic AMP** is formed from ATP by the action of a special enzyme **adenyl cyclase** (also called adenylyl cyclase) located in the cell membrane (Fig. 12-1). There is rapidly accumulating evidence that the mechanism of action for many, if not most, vertebrate hormones is as follows: When a hormone (the "first messenger") reaches a target cell, it binds to receptor sites on the cell membrane. Such sites are highly specific and recognize only one hormone of the many circulating in the bloodstream. Hormone binding in some way increases adenyl cyclase activity in the membrane, which in turn transforms ATP into cyclic AMP.

Cyclic AMP then diffuses into the cell where it acts as a "second messenger" to alter (usually stimulate) cellular processes. Since cyclic AMP is such a powerful regulator (it is 1,000 times less abundant than ATP), it is rapidly degraded into inert AMP by another enzyme, phosphodiesterase, so that its effect is short lived.

Endocrinologists have long searched for a single, fundamental mechanism through which all hormones act. The second messenger concept has been shown to apply to many but not all hormonal actions. Some hormones seem to act by means of both mechanisms described; others by one or the other. Still other hormonal actions may be exerted through yet unidentified intermediates. We have the feeling that we have read only the first few chapters of a detective story in which most of the central characters have made their appearance but none has yet revealed his full part in the intrigue. Certainly the unveiling of cyclic AMP in the 1960s has had far-reaching consequences on subsequent endocrinologic research. But as E. W. Sutherland, discoverer of cyclic AMP, has said, "Our present understanding of the biological role of cyclic AMP is probably very small compared to what it will be in the future."*

INVERTEBRATE HORMONES

Over the last 40 years physiologists have shown that the invertebrates have endocrine integrative systems that approach the complexity of the vertebrate endocrine system. Not surprisingly, however, there are few, if any, homologies between invertebrate and vertebrate hormones. Invertebrates have different functional systems, different growth patterns, and different reproductive processes from vertebrates and have been separated from them phylogenetically for a vast span of time. Most studies have been concentrated on the huge phylum Arthropoda, especially the insects and crustaceans. However, recent research has revealed

*Sutherland, E. W. 1972. Studies on the mechanism of hormone action. Science 177:401-408.

ATP

CYCLIC AMP

hormonal systems, especially neurosecretory systems, in most of the other invertebrate phyla too.

Neurosecretions are known to influence growth, asexual reproduction, and regeneration of hydra (phylum Coelenterata). Neurosecretory hormones also regulate regeneration, training, reproduction, and other aspects of flatworm physiology. We have known for some time that molluscs have neurosecretory hormones, especially among the gastropods and pelecypods. In the polychaete annelids, amputation of part of the worm body causes neurosecretory cells in the cerebral ganglia to secrete hormones that trigger regeneration. The cerebral ganglia of young worms produce a ''juvenile hormone'' that has a braking effect on metamorphosis; if the brain is removed, the worms become sexually premature.

The chromatophores (pigment cells) of shrimp and crabs are controlled by hormones from the **sinus gland** in the eyestalk or in regions close to the brain. Many crustaceans are capable of remarkably beautiful color patterns that change adaptively in relation to their environment; these changes are governed by an elaborate system of endocrine glands and hormones.

Growth and metamorphosis of arthropods are under endocrine control. In arthropods, growth is a series of steps in which the rigid, nonexpansible exoskeleton is periodically discarded and replaced with a new larger one. This process is especially dramatic in insects. In the type of development called **holometabolous,** seen in many insect orders (for example, butterflies, moths, ants, bees, wasps, and beetles), there is a series of wormlike larval stages, each requiring the formation of a new exoskeleton; each stage ends with a molt. The last larval stage enters a state of quiescence (pupa) during which the internal tissues are dissolved and rearranged into adult structures **(metamorphosis).** Finally the transformed adult emerges.

Insect physiologists have discovered that molting and metamorphosis are controlled by the interaction of two hormones, one favoring growth and the differentiation of adult structures and the other favoring the retention of larval structures. These two hormones

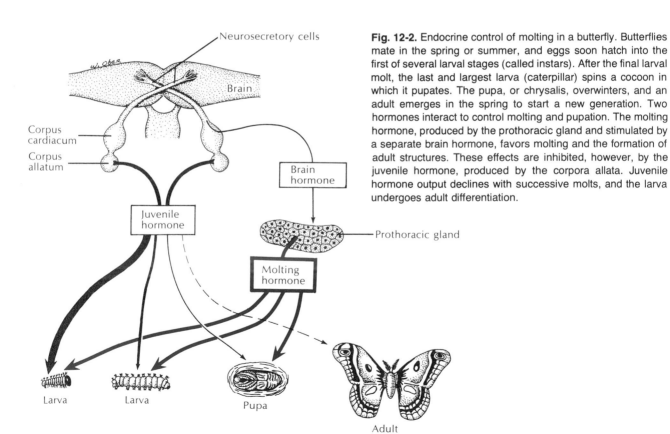

Fig. 12-2. Endocrine control of molting in a butterfly. Butterflies mate in the spring or summer, and eggs soon hatch into the first of several larval stages (called instars). After the final larval molt, the last and largest larva (caterpillar) spins a cocoon in which it pupates. The pupa, or chrysalis, overwinters, and an adult emerges in the spring to start a new generation. Two hormones interact to control molting and pupation. The molting hormone, produced by the prothoracic gland and stimulated by a separate brain hormone, favors molting and the formation of adult structures. These effects are inhibited, however, by the juvenile hormone, produced by the corpora allata. Juvenile hormone output declines with successive molts, and the larva undergoes adult differentiation.

Neurosecretory cells

Brain

Corpus cardiacum

Corpus allatum

Brain hormone

Juvenile hormone

Prothoracic gland

Molting hormone

Larva

Larva

Pupa

Adult

are the **molting hormone** (also referred to as **ecdysone** [ek′duh-sone]), produced by the prothoracic gland, and the **juvenile hormone,** produced by the corpora allata (Fig. 12-2). The structure of both hormones has recently been determined. It required the extraction of 1,000 kg (4 tons) of silkworm pupae to show that the molting hormone is a steroid.

MOLTING HORMONE (α-ECDYSONE) OF SILKWORM

The juvenile hormone has an entirely different structure (see below).

The molting hormone is under the control of a neurosecretory hormone from the brain, called **brain hormone** (or ecdysiotropin). At intervals during larval growth, brain hormone is released into the blood and stimulates the release of molting hormone. Molting hormone appears to act directly on the chromosomes to set in motion the changes resulting in a molt. The molting hormone favors the formation of a pupa and the development of adult structures. It is held in check, however, by the juvenile hormone, which favors the development of larval characteristics. During larval life the juvenile hormone predominates and each molt yields another larger larva. Finally the output of juvenile hormone decreases and the final pupal molt occurs.

Chemists have synthesized several potent analogs of the juvenile hormone, which hold great promise as insecticides. Minute quantities of these synthetic analogs induce abnormal final molts or prolong or block larval development. Unlike the usual chemical insecticides, they are highly specific and do not contaminate the environment.

The pattern of endocrine regulation in crustaceans shows certain parallels to that in insects. Some crustaceans, such as the crayfish, reach a definite adult size after a final molt; others, such as the lobster, molt continuously and keep growing ever larger until death. Molting is controlled by a neurosecretory hormone called **molt-inhibiting hormone,** which is produced by the X-organ in the eyestalk and by a **molting hormone** (ecdysone) produced by the Y-organ located in the thorax. During intermolt, the molt-inhibiting hormone inhibits activity of the Y-organ. But at intervals just before the molt, the output of molt-inhibiting hormone decreases; this permits the Y-organ to secrete molting hormone, which stimulates molting.

VERTEBRATE ENDOCRINE GLANDS AND HORMONES

The vertebrate **endocrine** glands are ductless groups of cells arranged in cords or plates; their hormonal secretions enter the bloodstream and are carried throughout the body. **Exocrine** glands, in contrast, are provided with ducts for discharging their secretions onto a free surface. Examples of exocrine glands are sweat glands and sebaceous glands of skin, salivary glands, and the various enzyme-secreting glands lining the wall of the stomach and intestine. Since the endocrine glands have no ducts, their only connection with the rest of the body is by the bloodstream; they must capture their raw materials from the blood and secrete their finished hormonal products into it. Consequently it is not surprising that the endocrine glands receive enormous blood flows. The thyroid gland is said to have the highest blood flow per unit of tissue weight of any organ in the body.

In the remainder of this section we describe some of the best understood and most important of the vertebrate hormones. The hormones of reproduction are discussed in the next chapter. Space does not permit us to deal with all the hormones and hormonelike substances that have been discovered. The mammalian hormonal mechanisms are the best understood since laboratory mammals and man have always been the objects of the most intensive research. Research with the lower vertebrates has revealed that all vertebrates

JUVENILE HORMONE OF SILKWORM

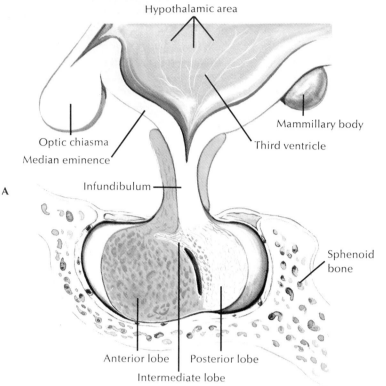

Hypothalamic area

Mammillary body

Third ventricle

Optic chiasma

Median eminence

Infundibulum

A

Sphenoid bone

Anterior lobe

Posterior lobe

Intermediate lobe

Fig. 12-3. Lateral view of the structure of human pituitary gland and its relationship to hypothalamus. **A,** Posterior lobe is connected directly to hypothalamus by neurosecretory fibers. Anterior lobe is indirectly connected to hypothalamus. **B,** Neurosecretory fibers end in the median eminence, in contact with a portal circulation that conveys hormones to the anterior pituitary. (**A** from Schottelius, B. A., and D. D. Schottelius. 1973. Textbook of physiology, ed. 17. St. Louis, The C. V. Mosby Co.)

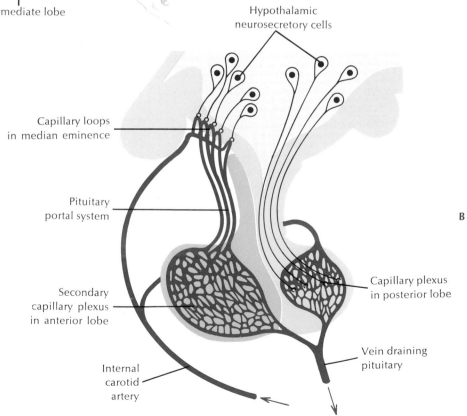

Hypothalamic neurosecretory cells

Capillary loops in median eminence

Pituitary portal system

B

Secondary capillary plexus in anterior lobe

Capillary plexus in posterior lobe

Vein draining pituitary

Internal carotid artery

share similar endocrine organs. All vertebrates have a pituitary gland, for example, and all have thyroid glands, adrenal glands (or the special cells of which they are composed), and gonads. Nevertheless there are some important differences in the functional roles that the hormones of these glands have among the different vertebrates, as we shall seek to point out.

Hormones of the pituitary gland and hypothalamus

The pituitary gland, or **hypophysis,** is a small gland (0.5 g in man) lying in a well-protected position between the roof of the mouth and the floor of the brain (Fig. 12-3). It is a two-part gland having a double embryologic origin. The **anterior pituitary** (adenohypophysis) is derived embryologically from the roof of the mouth. The **posterior pituitary** (neurohypophysis) arises from a ventral portion of the brain, the **hypothalamus,** and is connected to it by a stalk, the **infundibulum** (Fig. 12-3, *A*). Although the anterior pituitary lacks any *anatomic* connection to the brain, it is nonetheless *functionally* connected to it by a special portal circulatory system (Fig. 12-3, *B*).

The anterior pituitary consists of an **anterior lobe** (pars distalis) and an **intermediate lobe** (pars intermedia) as shown in Fig. 12-3. The anterior lobe, despite its minute dimensions, produces at least six protein hormones. All but one of these six are **tropic hormones,** that is, they regulate other endocrine glands (Fig. 12-4 and Table 12-1). Because of the strategic importance of the pituitary in influencing most of the hormonal activities in the body, the pituitary has been called the body's "master gland." This name is misleading, however, since the tropic hormones are themselves regulated by neurosecretory hormones from the hypothalamus, as well as by hormones from the target glands they stimulate.

The **thyrotropic hormone** (TSH) regulates the production of thyroid hormones by the thyroid gland. The **adrenocorticotropic hormone** (ACTH) stimulates the adrenal cortex. Two of the tropic hormones are commonly called **gonadotropins** because they act on the gonads (ovary of the female, testis of the male). These are the **follicle-stimulating hormone** (FSH) and the **luteinizing hormone** (LH). The fifth tropic hormone is **prolactin,** which stimulates milk production by the female mammary glands and has a variety of other effects in the lower vertebrates. The functions of the two gonadotropins and prolactin are discussed in the next chapter in connection with the hormonal control of reproduction.

The sixth hormone of the anterior lobe is the **growth hormone** (also called somatotropic hormone). This hormone performs a vital role in governing body growth through its stimulatory effect on cellular mitosis and protein synthesis, especially in new tissue of young animals. If produced in excess, the growth hormone causes giantism. A deficiency of this hormone in the child or young animal causes dwarfism. Growth hormone and prolactin are so closely related chemically that growth hormone contains considerable prolactin activity; that is, it tends to stimulate milk production (like prolactin) as well as promote growth.

The intermediate lobe of the pituitary (Fig. 12-3) produces **intermedin** (also called melanophore-stimulating hormone [MSH]), which controls the dispersion of melanin within the melanophores of amphibians. In other vertebrates, intermedin appears to perform no important physiologic role, even though it causes darkening of the skin in man if injected into the circulation. Intermedin and ACTH are chemically very similar and ACTH also causes skin darkening when it is secreted in abnormally large amounts.

As pointed out previously, the pituitary gland is not the top director of the body's system of endocrine glands as endocrinologists once believed. The pituitary serves higher masters, the neurosecretory centers of the hypothalamus; and the hypothalamus is itself under the ultimate control of the brain. The hypothalamus contains groups of neurosecretory cells, which are specialized giant nerve cells. Polypeptide hormones are manufactured in the cell bodies and then travel down the nerve fibers to their endings where the hormones are stored until released into the blood. The discharge of neurosecretory hormones may occur when a nerve impulse travels down the same neurosecretory fiber (these specialized nerve cells can in most cases still perform their original impulse-conducting function), or the release may be activated by ordinary fibers traveling alongside them.

Both the anterior and posterior lobes of the pituitary are under hypothalamic control, but in different ways. Neurosecretory fibers serving the posterior lobe travel down the infundibular stalk and into the posterior lobe, ending in proximity to blood capillaries into which the hormones enter when released. In a sense the posterior lobe is not a true endocrine gland, but a storage and

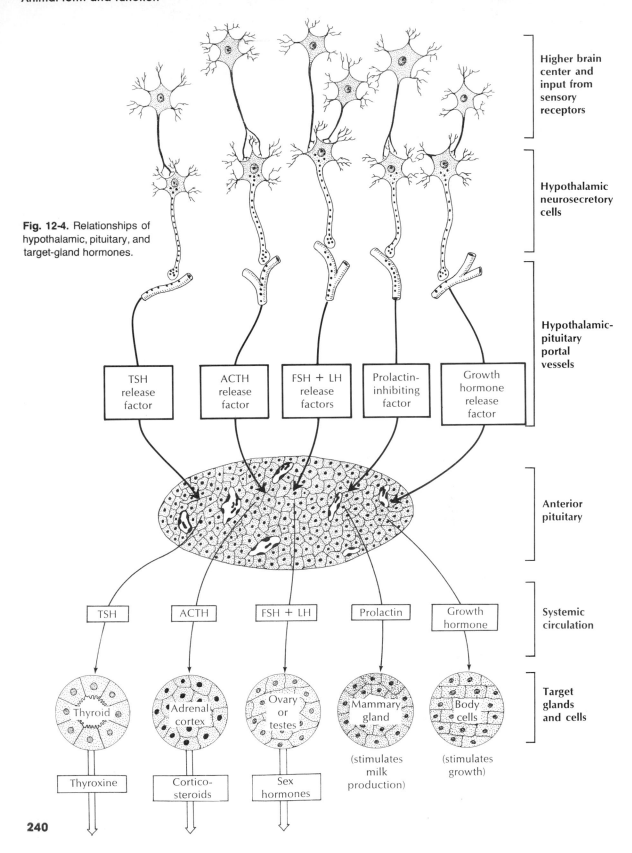

Fig. 12-4. Relationships of hypothalamic, pituitary, and target-gland hormones.

Higher brain center and input from sensory receptors

Hypothalamic neurosecretory cells

Hypothalamic-pituitary portal vessels

TSH release factor

ACTH release factor

FSH + LH release factors

Prolactin-inhibiting factor

Growth hormone release factor

Anterior pituitary

TSH

ACTH

FSH + LH

Prolactin

Growth hormone

Systemic circulation

Thyroid

Adrenal cortex

Ovary or testes

Mammary gland

Body cells

Target glands and cells

Thyroxine

Cortico-steroids

Sex hormones

(stimulates milk production)

(stimulates growth)

Table 12-1. Hormones of the vertebrate pituitary—chemical nature and actions

Hormone	Chemical nature	Action
Adenohypophysis		
Anterior lobe		
Thyrotropin (TSH)	Glycoprotein	Stimulates thyroid to secrete thyroid hormones
Adrenocorticotropin (ACTH)	Polypeptide	Stimulates adrenal cortex to secrete steroid hormones
Gonadotropins		
1. Follicle-stimulating hormone (FSH)	Glycoprotein	Stimulates gamete production and secretion of sex hormones
2. Luteinizing hormone (LH, ICSH)	Glycoprotein	Stimulates sex hormone secretion and ovulation
Prolactin (LTH)	Protein	Stimulates mammary gland growth and secretion in mammals; various reproductive and nonreproductive functions in lower vertebrates
Growth hormone (GH)	Protein	Stimulates growth
Intermediate lobe Intermedin (MSH)	Polypeptide	Pigment dispersion in amphibian melanophores
Neurohypophysis		
Posterior lobe		
Vasopressin (ADH)	Octapeptide	Antidiuretic effect on kidney
Oxytocin	Octapeptide	Stimulates milk ejection and uterine contraction
Vasotocin	Octapeptide	Antidiuretic activity
Isotocin and three others in lower vertebrates	Octapeptides	Functions unknown
Median eminence Thyrotropin-releasing factor (TRF)		
Corticotropin-releasing factor (CRF)		
Follicle-stimulating hormone–releasing factor (FSH-RF)	Probably all polypeptides	Control release of anterior lobe hormones
Luteinizing hormone–releasing factor (LH-RF)		
Prolactin-inhibiting factor (PIF)		
Growth hormone–releasing factor (GH-RF)		

release center for hormones manufactured entirely in the hypothalamus.

The anterior lobe's relationship to the hypothalamus is quite different. Neurosecretory fibers do not travel to the anterior lobe, but end some distance above it in the **median eminence** at the base of the infundibular stalk (Fig. 12-4). Neurosecretory hormones released here enter a capillary network and complete their journey to the anterior lobe via a short but crucial pituitary portal system. These hormones are called **releasing factors** because they govern the release of the anterior lobe hormones (Table 12-1).

There appears to be a specific releasing factor for each pituitary tropic hormone. Two of the releasing factors, TRF (thyrotropin-releasing factor) and LH-RF (luteinizing hormone–releasing factor), have recently been isolated and characterized chemically. Both are polypeptides. The structure of TRF, a tripeptide, is shown at the top of p. 242.

The hormones of the **posterior lobe** are chemically similar polypeptides consisting of eight amino acids (referred to as octapeptides). All vertebrates, except the most primitive fishes, secrete two posterior lobe octapeptides. However, their chemical structure has

PYROGLUTAMIC ACID–HISTIDINE–PROLINE NH₂
(THYROTROPIN-RELEASING FACTOR)

changed slightly in the course of evolution. The two posterior lobe hormones secreted, for example, by fish are not identical to those secreted by mammals. Altogether seven different posterior lobe hormones have been identified from the different vertebrate groups (Table 12-1).

Vasotocin is found in all vertebrate classes except mammals. It is a water balance hormone in amphibians, especially toads, in which it acts to conserve water by (1) increasing permeability of the skin (to promote water absorption from the environment), (2) stimulating water reabsorption from the urinary bladder, and (3) decreasing urine flow. The action of vasotocin is best understood in amphibians, but it appears to play some water-conserving role in birds and reptiles as well.

The two mammalian posterior lobe hormones are **oxytocin** and **vasopressin** (Fig. 12-5). They are formed, as we have seen, in the cell bodies of neurosecretory cells in the hypothalamus and then transported down the nerve cell axons to the posterior lobe. These hormones are among the fastest-acting hormones in the body since they are capable of producing a response within seconds of their release from the posterior lobe.

Oxytocin has two important specialized reproductive functions in adult female mammals. It causes contraction of uterine smooth muscles during parturition (birth of the young). Doctors sometimes use oxytocin clinically to induce labor and facilitate delivery and to prevent uterine hemorrhage after birth. The second and most important action of oxytocin is that of

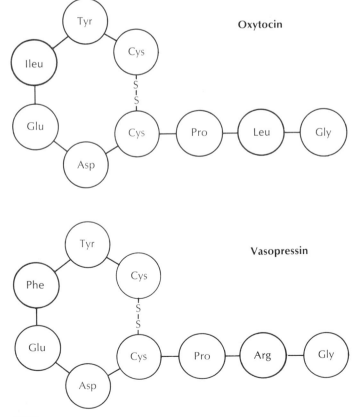

Fig. 12-5. Posterior lobe hormones of man. Both oxytocin and vasopressin consist of eight amino acids (the two sulfur-linked cysteine molecules are considered to be a single amino acid, cystine). Oxytocin and vasopressin are identical except for amino acid substitutions in the red positions.

milk ejection by the mammary glands in response to suckling. Though present, oxytocin has no known function in the male.

Vasopressin, the second posterior lobe hormone, acts on the kidney to restrict urine flow, as already described on p. 180. It is therefore often referred to as the **antidiuretic hormone** (ADH). Vasopressin has a second, weaker effect of increasing the blood pressure through its generalized constrictor effect on the smooth muscles of the arterioles. Although the name "vasopressin" unfortunately suggests that the vasoconstrictor action is the hormone's major effect, it is probably of little physiologic importance, except perhaps to help sustain the blood pressure during a severe hemorrhage.

Hormones of metabolism

Many hormones act to adjust the delicate balance of metabolic activities in the body. Metabolism includes the **anabolic** activities of tissue synthesis, the building up of energy reserves, and maintenance of tissue organization and the **catabolic** activities of energy release and tissue destruction. Such activities are mediated almost entirely by enzymes. The numerous enzymic reactions proceeding within cells are complex, but each step in a sequence is in large part self-regulating as long as the equilibrium between substrate, enzyme, and product remains stable. However, hormones may alter the activity of crucial enzymes in a metabolic process, thus accelerating or inhibiting the entire process. We must emphasize that hormones never initiate enzymic processes. They simply alter their rate, speeding them up or slowing them down. The most important hormones of metabolism are those of the thyroid, parathyroid, and adrenal glands and the pancreas.

Thyroid hormones

The two thyroid hormones **thyroxine** and **triiodothyronine** are secreted by the thyroid gland. This largest of endocrine glands is located in the neck region of all vertebrates; in many animals, including man, it is a bi-lobed structure. The thyroid is made up of thousands of tiny spheres called **follicles;** each follicle is composed of a single layer of epithelial cells enclosing a hollow, fluid-filled center. This fluid contains stored thyroid hormone that is released into the bloodstream as it is needed.

One of the unique characteristics of the thyroid is its high concentration of **iodine;** in most animals this single gland contains well over half the body store of iodine. The epithelial cells actively trap iodine from the blood and combine it with the amino acid tyrosine, creating the two thyroid hormones. Each molecule of thyroxine contains four atoms of iodine as indicated by the following structural formula:

$$HO - \overset{\overset{\displaystyle I}{|}}{\underset{\underset{\displaystyle I}{|}}{\bigcirc}} - O - \overset{\overset{\displaystyle I}{|}}{\underset{\underset{\displaystyle I}{|}}{\bigcirc}} - CH_2 - \underset{\underset{\displaystyle NH_2}{|}}{CH} - COOH$$

THYROXINE

Triiodothyronine is identical to thyroxine except that it has three instead of four iodine atoms. Thyroxine is formed in much greater amounts than triiodothyronine, but both hormones have two important similar effects. One is to promote the normal growth and development of growing animals. The other is to stimulate the metabolic rate.

We do not know exactly how the thyroid hormones promote growth, although there is evidence that they stimulate protein synthesis through their effect on messenger RNA. Certainly the undersecretion of thyroid hormone dramatically impairs growth, especially of the nervous system. The human **cretin,** a mentally retarded dwarf, is the tragic product of thyroid malfunction from a very early age. Conversely the oversecretion of thyroid hormones causes precocious development, particularly in lower vertebrates. In one of the earliest demonstrations of hormone action, J. F. Gudernatsch in 1912 induced precocious metamorphosis of frog tadpoles by feeding them bits of horse thyroid. The tadpoles quickly resorbed their tails, grew limbs, changed from gill to lung respiration, and became froglets approximately one-third normal size. The result of a similar experiment is shown in Fig. 12-6.

The control of oxygen consumption and heat production in birds and mammals is the best known action of the thyroid hormones. The thyroid enables warm-blooded animals to adapt to cold by increasing their heat production. Responding to directives from the hypothalamus, the body's "thermostat," the thyroid gland releases more thyroxine when the blood temperature begins to decrease. Thyroxine in some way causes cells to produce more heat and store less chemical energy (ATP); in other words, thyroxine *reduces* the efficiency of the cellular oxidative phosphorylation system (p. 204). This is why cold-adapted

Fig. 12-6. Precocious metamorphosis of frog tadpoles, *Rana pipiens,* caused by adding thyroid hormone to the water. When frog tadpoles had developed hindlimb buds, small amounts of thyroxine were added to aquarium water. In only 3 weeks tadpoles metamorphosed to normal, but miniature, adults approximately one third of the size of the mother. (From Turner, C. D., and J. T. Bagnara. 1971. General endocrinology, ed. 5. Philadelphia, W. B. Saunders Co.)

animals eat more food than do warm-adapted ones, even though they are doing no more work; the food is being converted directly to heat, thus keeping the body warm.

The synthesis and release of thyroxine and triiodothyronine are governed by **thyrotropic hormone** (TSH) from the anterior pituitary gland (Fig. 12-4). Thyrotropic hormone controls the thyroid through a **negative feedback mechanism.** If the thyroxine level in the blood decreases, more thyrotropic hormone is released. Should the thyroxine level rise too high, less thyrotropic hormone is released. This sensitive feedback mechanism normally keeps the blood thyroxine level very steady, but certain neural stimuli, as might arise from exposure to cold, can directly increase the release of thyrotropic hormone.

Some years ago, a condition called **goiter** was common among people living in the Great Lakes region of the United States and Canada, as well as in other parts of the earth such as the Swiss Alps. Goiter is an enlargement of the thyroid gland caused by a deficiency of iodine in the food and water. In striving to produce thyroid hormone with not enough iodine available, the gland hypertrophies, sometimes so much that the entire neck region becomes swollen. Goiter is seldom seen today because of the widespread use of iodized salt.

Parathyroid hormone, calcitonin, and calcium metabolism

Closely associated with the thyroid gland and often buried within it are the parathyroid glands. These tiny glands occur as two pairs in man but vary in number and position in other vertebrates. They were discovered at the end of the nineteenth century when the fatal effects of "thyroidectomy" were traced to the unknowing removal of the parathyroids as well as the thyroid.

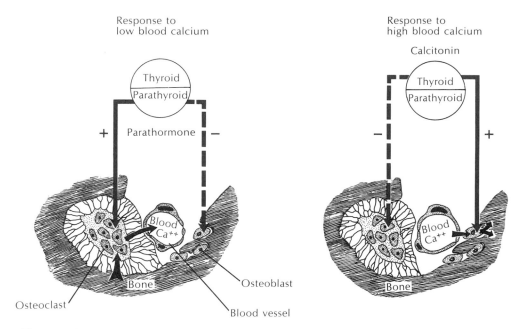

Response to
low blood calcium

Response to
high blood calcium

Fig. 12-7. Action of parathormone and calcitonin on calcium resorption and deposition in bone. When blood calcium is low *(left)*, parathyroid gland secretes parathormone, which stimulates large bone-destroying osteoclasts to resorb calcium. Bone-building osteoblasts are inhibited. Calcium and phosphate (not shown) enter the blood and restore the blood calcium level to normal. When blood calcium rises above normal *(right)*, thyroid gland secretes calcitonin that inhibits osteoclasts. Osteoblasts then remove calcium (and phosphate) from the blood and use it to build new bone. Calcitonin may directly stimulate osteoblastic activity. A third hormone, calciferol (vitamin D), is necessary for calcium deposition in bones.

In many animals, including man, removal of the parathyroids causes the blood calcium to decrease rapidly. This results in a serious increase in nervous system excitability, severe muscular spasms and tetany, and finally death.

The parathyroid glands are vitally concerned with the maintenance of the normal level of calcium in the blood. Actually two hormones are involved: **parathyroid hormone** (parathormone), produced by the parathyroid glands, and **calcitonin,** produced by special cells within the thyroid gland (not the same follicle cells that synthesize thyroxine and triiodothyronine). These two hormones have opposing but cooperative actions. Between them they stabilize both the calcium and phosphorus levels in the blood through their action on bone.

Bone is a densely packed storehouse of these elements, containing approximately 98% of the body calcium and 80% of the phosphorus. Although bone is second only to teeth as the most durable material in the body, as evidenced by the survival of fossil bones for

millions of years, it is in a state of constant turnover in the living body. Bone-building cells **(osteoblasts)** withdraw calcium and phosphorus (as phosphate) from the blood and deposit them in a complex crystalline form around previously formed organic fibers (Fig. 12-7). Bone-resorbing cells **(osteoclasts),** present in the same bone, tear down bone by engulfing it and releasing the calcium and phosphate into the blood.

These conflicting activities are not as pointless as they may seem. First, they allow bone to constantly remodel itself, especially in the growing animal, for structural improvements to counter new mechanical stresses on the body. Second, they provide a vast and accessible reservoir of minerals that can be withdrawn as the body needs them for its general cellular requirements.

If the blood calcium should decrease slightly, the parathyroid gland increases its output of parathormone. This stimulates the osteoclasts to destroy bone adjacent to these cells, thus releasing calcium and phosphate into the bloodstream and returning the blood calcium level to normal. Parthormone also acts on the kidney to decrease the excretion of calcium, and this, of course, also helps to increase the blood calcium level. Should the calcium in the blood rise above normal, the parathyroid gland decreases its output of parathormone. In addition, the thyroid is stimulated to release calcitonin. These relationships are shown in Fig. 12-7.

Although the action of this recently discovered (1962) hormone, calcitonin, is yet imperfectly understood, evidence suggests that it inhibits bone resorption by the osteoclasts. Calcitonin thus protects the body against a dangerous increase in the blood calcium level, just as parathormone protects it from a dangerous decrease in blood calcium. The two act together to smooth out oscillations in blood calcium.

A third hormone, **calciferol** (commonly called **vitamin D**), is also necessary for calcium absorption from the gut. Calciferol is a steroid hormone in skin formed from a steroid precursor by the action of ultraviolet rays of sunlight.

Adrenocortical hormones

The vertebrate adrenal gland is a double gland consisting of two very different kinds of tissue: **interrenal** tissue, called **cortex** in mammals, and **chromaffin** tissue, called **medulla** in mammals (not to be confused with the cortex and medulla of the vertebrate brain). The mammalian terminology of cortex (meaning "bark") and medulla (meaning "core") arose because in this group of vertebrates the interrenal tissue completely surrounds the chromaffin tissue like a cover. Although in the lower vertebrates the interrenal and chromaffin tissue are usually separated, the mammalian terms "cortex" and "medulla" are so firmly fixed in our vocabulary that we commonly use them for all vertebrates instead of the more correct terms "interrenal" and "chromaffin."

Biochemists have found that the adrenal cortex contains at least 30 different compounds, all of them closely related lipoid compounds known as steroids. Only a few of these compounds, however, are true steroid *hormones;* most are various intermediates in the synthesis of steroid hormones from **cholesterol** (Fig. 12-8). The corticosteroid hormones are commonly classified into three groups, according to their function:

1. **Glucocorticoids,** such as **cortisol** (Fig. 12-8) and **corticosterone,** have a number of important

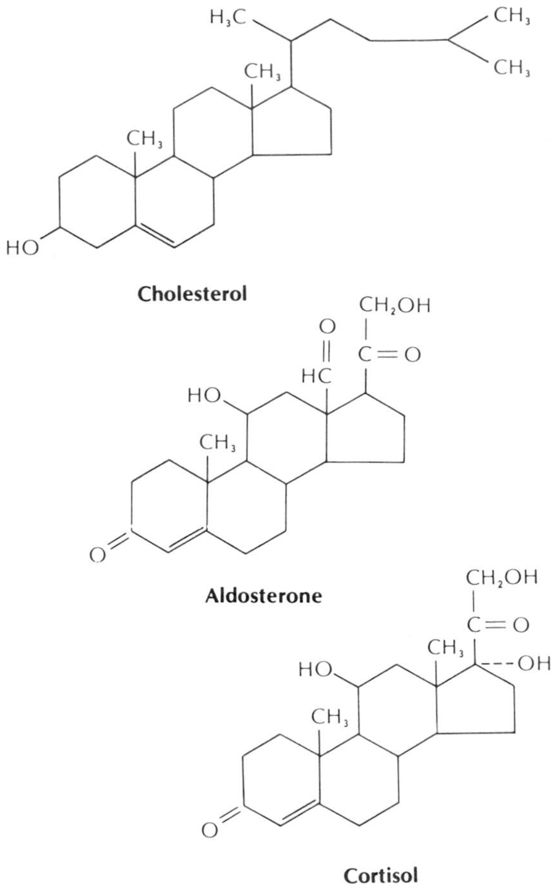

Fig. 12-8. Hormones of the adrenal cortex. Cortisol (a glucocorticoid) and aldosterone (a mineralocorticoid) are two of several steroid hormones synthesized from cholesterol in the adrenal cortex.

effects concerned with food metabolism and inflammation. They cause the conversion of nonglucose compounds, particularly amino acids and fats, into glucose. This process, called **gluconeogenesis,** is extremely important since most of the body's stored energy reserves are in the form of fats and proteins that must be converted to glucose before they can be burned for energy. Cortisol, cortisone, and corticosterone are also **anti-inflammatory.** Because several diseases of man are inflammatory diseases (for example, allergies, hypersensitivity, arthritis), these corticosteroids have important medical applications. They must be used with great care, however, since, if administered in excess, they may suppress the body's normal repair processes and lower resistance to infectious agents.

2. **Mineralocorticoids,** the second group of corticosteroids, are those that regulate salt balance. **Aldosterone** (Fig. 12-8) and **deoxycorticosterone** are the most important steroids of this group. They promote the tubular reabsorption of sodium and chloride and the tubular excretion of potassium by the kidney. Since sodium usually is in short supply in the diet and potassium in excess, it is obvious that the mineralocorticoids play vital roles in preserving the correct balance of blood electrolytes. We may also note that the mineralocorticoids *oppose* the anti-inflammatory effect of cortisol and cortisone. In other words, they promote the *inflammatory* defense of the body to various noxious stimuli. Although these opposing actions of the corticosteroids seem self-defeating, they actually are not. They are necessary to maintain readiness of the body's defenses for any stress or disease threat, yet prevent these defenses from becoming so powerful that they turn against the body's own tissues.

3. **Sex hormones,** such as testosterone, estrogen, progesterone, are produced primarily by the ovaries and testes (see p. 269). The adrenal cortex is also a minor source of certain steroids that mimic the action of testosterone. These sex hormone–like secretions are of little physiologic significance, except in certain disease states of man.

The synthesis and secretion of the corticosteroids are controlled principally by the **adrenocorticotropic hormone** (ACTH) of the anterior pituitary (Fig. 12-4). As with pituitary control of the thyroid, a negative feedback relationship exists between ACTH and the adrenal cortex: an increase in the level of corticosteroids suppresses the output of ACTH; a decrease in the blood steroid level increases ACTH output.

Adrenal medulla hormones

The adrenal medulla secretes two structurally similar hormones, **epinephrine** (adrenaline) and **norepinephrine** (noradrenaline). Their structures are as follows:

EPINEPHRINE

NOREPINEPHRINE

Both are derived from catechol:

and each bears an amine group on two-carbon side chains. Consequently both belong to a class of compounds called **catecholamines.**

Norepinephrine is also released at the endings of sympathetic nerve fibers throughout the body, where it serves as a "transmitter" substance to carry neural signals across the gap that separates the fiber and the organ it innervates. The adrenal medulla has the same embryologic origin as sympathetic nerves; in many respects the adrenal medulla is nothing more than a giant sympathetic nerve ending.

It is not surprising then that the adrenal medulla hormones have the same general effects on the body that the sympathetic nervous system has. These effects center around emergency functions of the body, such as fear, rage, fight, and flight, although they have important integrative functions in more peaceful times as well. We are all familiar with the increased heart rate, tightening of the stomach, dry mouth, trembling muscles, general feeling of anxiety, and the increased awareness that attends sudden fright or other strong emotional states. These effects are attributable both to the rapid release into the blood of epinephrine from the adrenal medulla and to increased activity of the sympathetic nervous system.

Epinephrine and norepinephrine have many other effects that we are not so aware of, including constriction of the arterioles (which, together with the increased heart rate, increases the blood pressure), mobilization of liver glycogen to release glucose for energy, increased oxygen consumption and heat production, hastening of blood coagulation, and inhibition of the gastrointestinal tract. All of these changes in one way or another tune up the body for emergencies.

Epinephrine and norepinephrine are among those hormones that activate the enzyme **adenyl cyclase,** causing increased production of **cyclic AMP.** Cyclic AMP then becomes the second messenger, which produces the many observed effects of the adrenal medulla hormones.

Insulin from the islet cells of the pancreas

The pancreas is both an exocrine and an endocrine organ. The *exocrine* portion produces pancreatic juice,

247

a mixture of digestive enzymes that is conveyed by ducts to the digestive tract. Scattered within the extensive exocrine portion of the pancreas are numerous small islets of tissue, called **islets of Langerhans.** This is the *endocrine* portion of the gland. The islets are without ducts and secrete their hormones directly into blood vessels that extend throughout the pancreas.

Two polypeptide hormones are secreted by different cell types within the islets: **insulin,** produced by the **beta cells,** and **glucagon,** produced by the **alpha cells.** Insulin and glucagon have antagonistic actions of great importance in the metabolism of carbohydrates and fats. Insulin is essential for the utilization of blood glucose by cells, especially skeletal muscle cells. Insulin somehow allows glucose in the blood to be transported into the cells. Without insulin, the blood glucose levels rise (hyperglycemia) and sugar appears in the urine. Insulin also promotes the uptake of amino acids by skeletal muscle and inhibits the mobilization of fats in adipose tissue.

Failure of the pancreas to produce enough insulin causes the disease **diabetes mellitus,** which afflicts 2% of the population. It is attended by serious alterations in carbohydrate, lipid, protein, salt, and water metabolism, which, if left untreated, may cause death.

The first extraction of insulin in 1921 by two Canadians, Frederick Banting and Charles Best, was one of the most dramatic and important events in the history of medicine. Many years earlier two German scientists, J. Von Mering and O. Minkowski, discovered that surgical removal of the pancreas of dogs invariably caused severe symptoms of diabetes, resulting in the animal's death within a few weeks. Many attempts were made to isolate the diabetes preventive factor, but all failed because powerful protein-splitting digestive enzymes in the exocrine portion of the pancreas destroyed the hormone during extraction

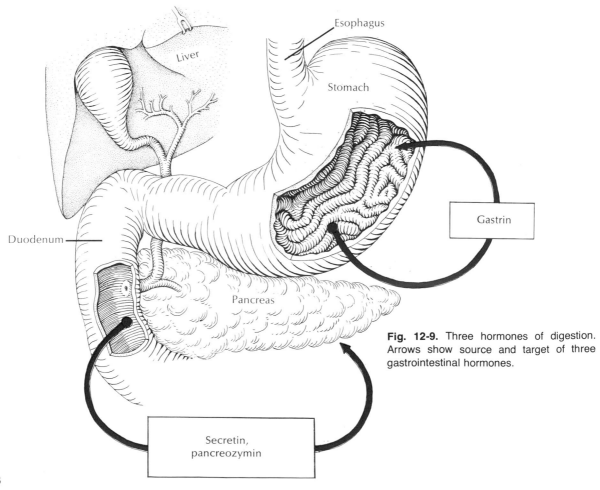

Fig. 12-9. Three hormones of digestion. Arrows show source and target of three gastrointestinal hormones.

procedures. Following a hunch, Banting in collaboration with Best and his physiology professor J. J. R. Macleod tied off the pancreatic ducts of several dogs. This caused the exocrine portion of the gland with its hormone-destroying enzyme to degenerate, but left the islets' tissue healthy since they were independently served by their own blood supply. Banting and Best then successfully extracted insulin from these glands. Injected into another dog, the insulin immediately lowered the blood sugar level. Their experiment paved the way for the commercial extraction of insulin from slaughterhouse animals. It meant that millions of diabetics, previously doomed to invalidism or death, could look forward to nearly normal lives.

Glucagon, the second hormone of the pancreas, has several effects on carbohydrates and fat metabolism that are opposite to the effects of insulin. For example, glucagon raises the blood glucose level, whereas insulin lowers it. Glucagon and insulin do not have the same effects in all vertebrates, and in some glucagon is lacking altogether. Glucagon is another example of a hormone that operates through the cyclic AMP second-messenger system.

Hormones of digestion

Several hormones assist in coordinating the secretion of digestive enzymes. Of these, we will discuss three of the best understood (Fig. 12-9). **Gastrin** is a small polypeptide hormone produced in the mucosa of the pyloric portion of the stomach. When food enters the stomach, gastrin stimulates the secretion of hydrochloric acid by the stomach wall. Gastrin is an unusual hormone in that it exerts its action on the same organ from which it is secreted. Two other hormones of digestion are **secretin** and **pancreozymin.** Both are polypeptide hormones secreted by the intestinal mucosa in response to the entrance of acid and food into the duodenum from the stomach; both stimulate the secretions of pancreatic juice, but their effects differ somewhat. Secretin stimulates a pancreatic secretion rich in bicarbonate that rapidly neutralizes stomach acid. Pancreozymin stimulates the pancreas to release an enzyme-rich secretion. Although secretin was the first hormone discovered (by the British scientists Bayliss and Starling in 1903), the gastrointestinal hormones as a group have received much less attention than the other vertebrate endocrine glands. Only secretin has been obtained in pure form.

SELECTED REFERENCES

(See also general references for Part two, p. 274.)

Bentley, P. J. 1976. Comparative vertebrate endocrinology. Cambridge, Cambridge University Press. *A comprehensive, up-to-date, and clearly written account that relates endocrine function to the animals' ecology and evolutionary background.*

Highnam, K. C., and L. Hill. 1969. The comparative endocrinology of the invertebrates. New York, American Elsevier Publishing Co., Inc. *Clearly written, well-illustrated comparative account emphasizing hormonal action.*

Holmes, R. L., and J. N. Ball. 1974. The pituitary gland: a comparative account. New York, Cambridge University Press. *Advanced comprehensive survey.*

Tombes, A. S. 1970. An introduction to invertebrate endocrinology. New York, Academic Press Inc.

Turner, C. L., and J. T. Bagnara. 1976. General endocrinology, 6th ed. Philadelphia, W. B. Saunders Co. *Popular undergraduate-level text.*

SELECTED SCIENTIFIC AMERICAN ARTICLES

Gillie, R. B. 1971. Endemic goiter. **224:**92-101 (June). *In some parts of the world, people still suffer from this thyroid disorder caused by insufficient iodine in the diet.*

Guillemin, R., and R. Burgus. 1972. The hormones of the hypothalamus. **227:**24-33 (Nov.). *Two of the brain's "releasing factors" have been isolated and synthesized. Their function is described.*

Levine, R., and M. S. Goldstein. 1958. The action of insulin. **198:**99-106 (May). *Describes insulin's role in promoting the passage of sugar across cell membranes.*

Li, C. H. 1963. The ACTH molecule. **209:**46-53. *Its composition and what it does.*

Loomis, W. F. 1970. Rickets. **223:**76-91 (Dec.). *The role of calciferol (vitamin D), a sunlight-dependent hormone, in preventing this oldest of air-pollution diseases.*

McEwen, B. S. 1976. Interactions between hormones and nerve tissue. **235:**48-58 (July). *Steroid hormones secreted by gonads and adrenal cortex can be traced to target cells in brain.*

O'Malley, B. W., and W. T. Schrader. 1976. The receptors of steroid hormones. **234:**32-43 (Feb.). *The action of steroid hormones is mediated by receptor molecules that reside only in the target organs.*

Pastan, I. 1972. Cyclic AMP. **227:**97-105 (Aug.). *The hormonal "second messenger" and some of the roles it plays.*

Pike, J. E. 1971. Prostaglandins. **225:**84-92. *These recently isolated substances have numerous physiologic effects.*

Rasmussen, H., and M. M. Pechet. 1970. Calcitonin. **223:**42-50 (Oct.). *Discovery and action of this calcium-regulating hormone.*

Wilkins, L. 1960. The thyroid gland. **202:**119-129 (March). *How the gland is constructed, what it produces, and how it is controlled.*

Williams, C. M. 1958. The juvenile hormone. **198:**67-74 (Feb.). *This insect hormone prevents metamorphosis of the larva into a pupa until growth is complete.*

Four young opossums, *Didelphis marsupialis,* peer curiously at their new world from the safety of their mother's back. Like other marsupials, the opossum bears tiny, poorly developed young no larger than honeybees, after a gestation period of only 12 to 13 days. Each newborn finds its own way to the pouch by wriggling along a path of fur dampened by the mother's tongue, attaches itself to a nipple, and remains attached for 60 to 70 days. When they have reached the size of these four, they leave the pouch to move about freely on the mother but return at intervals to suckle for another month. Marsupials represent an evolutionary intermediate between the monotreme mammals that lay eggs and the placental mammals that nourish their young with an advanced fetal-maternal structure, the placenta. (Photo by F. M. Blake.)

13 REPRODUCTION AND DEVELOPMENT

NATURE OF THE REPRODUCTIVE PROCESS

> Asexual reproduction
> Sexual reproduction

PLAN OF REPRODUCTIVE SYSTEMS

> Human reproductive system

ORIGIN OF REPRODUCTIVE CELLS

> Meiosis: germ cell division

DEVELOPMENTAL PROCESS

> Oocyte maturation
> Fertilization and activation
> Cleavage and early development
> Gastrulation
> Differentiation

EXTRAEMBRYONIC MEMBRANES OF THE AMNIOTES

HUMAN DEVELOPMENT

> Early development of human embryo
> Placenta
> Pregnancy and birth

CONTROL OF DEVELOPMENT

All living organisms are capable of giving rise to new organisms similar to themselves. If we admit that all living things are mortal, that every organism is endowed with a life-span that must eventually terminate, we must also acknowledge the indispensability of reproduction. Without reproduction, the evolution of a species is impossible. Like Samuel Butler who concluded that a chicken is just an egg's way of making another egg, many biologists consider the ability to reproduce to be the ultimate objective of all life processes.

The word "reproduction" implies replication, and it is true that biologic reproduction almost always yields a reasonable facsimile of the parent unit. However, sexual reproduction, practiced by the majority of animals, produces the *diversity* needed for survival in a world of constant change. At least for multicellular animals sexual reproduction offers enormous advantages over asexual reproduction, as we strive to point out later.

Reproduction must have developed very early in animal evolution. The process, whether sexual or asexual, embodies a basic pattern: (1) the conversion of raw materials from the environment into the offspring or sex cells that develop into offspring of a similar constitution and (2) the transmission of a hereditary pattern or code from the parents. The code, of course, is in the deoxyribonucleic acid (DNA) of the genes. The chemical nature of DNA and the genetic code were described in Chapter 2.

NATURE OF THE REPRODUCTIVE PROCESS

The two fundamental types of reproduction are sexual and asexual. In **asexual** reproduction there is only one parent and no special reproductive organs or cells. Each organism is capable of producing identical copies of itself as soon as it becomes an adult. The production of copies is simple and direct and typically rapid. **Sexual** reproduction involves two parents as a rule, each of which contributes special **sex cells,** or **gametes,** that in union develop into a new individual. The **zygote** formed from this union receives genetic material from **both** parents and accordingly is different from both. The combination of genes produces a genetically unique individual, still bearing the characteristics of the species but also bearing traits that make it different from its parents.

Sexual reproduction, by recombining the parental characters, tends to multiply variations and makes possible a richer and more diversified evolution. Asexual reproduction can only produce carbon copies and must await mutations to introduce variation into the line. This would seem to explain why asexual reproduction is restricted mostly to unicellular forms, which can multiply rapidly enough to offset the disadvantages of relentless replication of identical products.

Of course, in those asexual organisms such as molds and bacteria that are haploid (bear only one set of genes), mutations are immediately expressed and evolution can proceed quickly. In sexual animals, on the other hand, a gene mutation is seldom expressed immediately since it is masked by its normal partner on the homologous chromosome. (Homologous chromosomes are those that pair during mitosis and have genes controlling the same characteristics.) There is only a remote chance that both members of a gene pair will mutate in the same way at the same moment.

Asexual reproduction

As previously pointed out, asexual reproduction is found only among the simpler forms of life, such as protozoans, coelenterates, bryozoans, bacteria, and a few others. It is absent among the higher invertebrates (molluscs and arthropods) and all vertebrates. Even in the phyla in which it occurs, most of the members employ the sexual method as well. In these groups, asexual reproduction ensures rapid increase in numbers when the differentiation of the organism has not advanced to the point of forming highly specialized gametes.

The forms of asexual reproduction are fission, budding (both internal and external), fragmentation, and sporulation. **Fission** is common among protozoans and to a limited extent among metazoans. In this method the body of the parent is divided into two approximately equal parts, each of which grows into an individual similar to the parent. Fission may be either transverse or longitudinal. **Budding** is an unequal division of the organism. The new individual arises as an outgrowth (bud) from the parent. This bud develops organs like those of the parent and then usually detaches itself. If the bud is formed on the surface of the parent, it is an external bud; but in some cases internal buds, or **gemmules,** are produced. Gemmules are collections of many cells surrounded by a dense covering in the body wall. When the body of

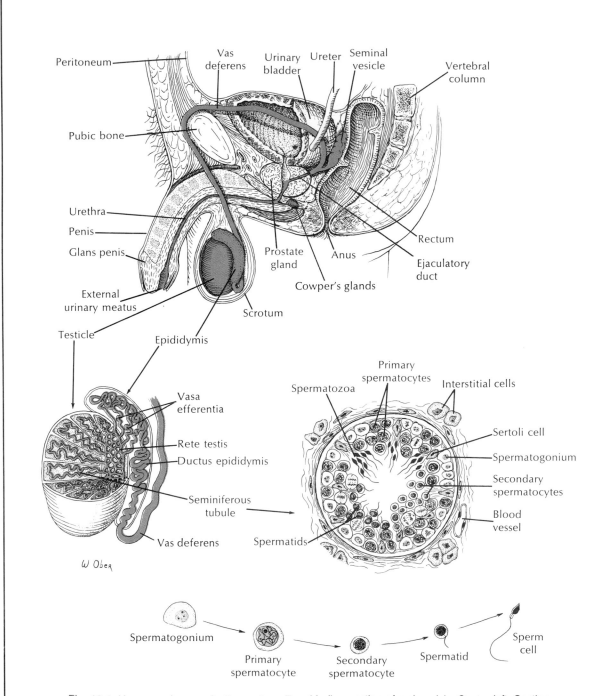

Fig. 13-1. Human male reproductive system. *Top,* Median section of male pelvis. *Center left,* Section of left testis. *Center right,* Cross section of one seminiferous tubule, showing different stages in spermatogenesis. *Bottom,* Sequential stages in spermatogenesis.

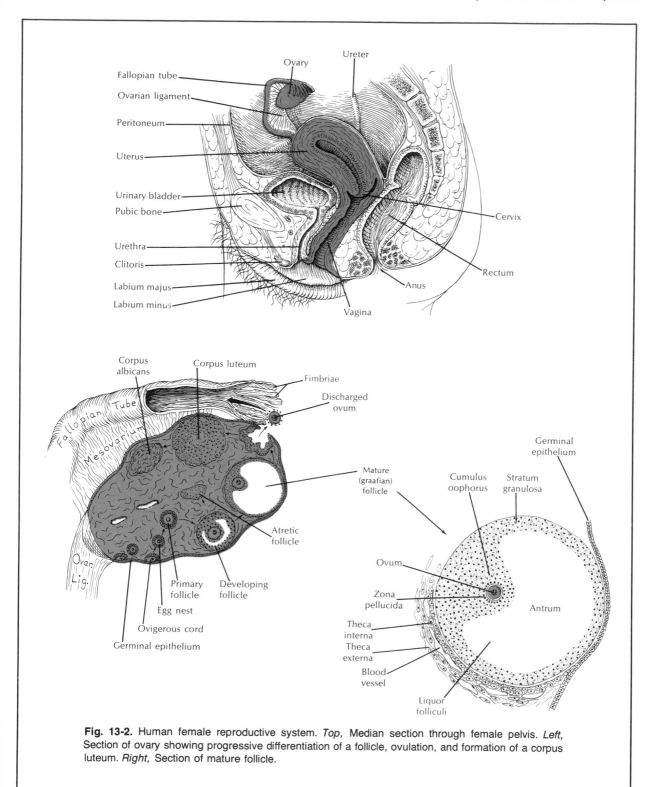

Fig. 13-2. Human female reproductive system. *Top,* Median section through female pelvis. *Left,* Section of ovary showing progressive differentiation of a follicle, ovulation, and formation of a corpus luteum. *Right,* Section of mature follicle.

the parent disintegrates, each gemmule gives rise to a new individual. External budding is common in the hydra and internal budding in the freshwater sponges. Bryozoans also have a form of internal bud called statoblast. **Fragmentation** is a method in which an organism breaks into two or more parts, each capable of becoming a complete animal. This method is found among the Platyhelminthes, Nemertinea, and Echinodermata. **Sporulation** is a method of multiple fission in which many cells are formed and enclosed together in a cystlike structure. Sporulation occurs in a number of protozoan forms.

Sexual reproduction

Sexual reproduction is the general rule in the animal kingdom. It involves two parents, each of which produces special sex cells, called **gametes.** Only one sex, the female, can produce offspring, and this must depend on the intervention of the male. There are two kinds of gametes: the **ovum** (egg), produced by the female, and the **spermatozoon** (sperm), produced by the male. Ova are nonmotile, are produced in relatively small numbers, and may contain a large amount of yolk, the stored food material that sustains early development of the new individual. Sperm are motile, are produced in enormous numbers, and are small. The union of egg and sperm is called **fertilization,** and the resulting cell is known as a **zygote,** which develops into a new individual.

Sexual reproduction appears in certain protozoans, although most protozoans reproduce asexually by cell division. When sexual reproduction is found it may involve the formation of male and female gametes that unite to form a zygote or it may involve two mature sexual individuals that join together to exchange nuclear material or merge cytoplasm. In some cases it is difficult to distinguish sex, for although two parents are involved, they cannot be designated as male and female.

The male-female distinction is clearly evident in the Metazoa. Organs that produce the germ cells are known as **gonads.** The gonad that produces the sperm is called the **testis** (Fig. 13-1) and that which forms the egg, the **ovary** (Fig. 13-2). The gonads represent the **primary sex organs,** the only sex organs found in certain groups of animals. Most metazoans, however, have various **accessory sex organs.** In the primary sex organs the sex cells undergo many complicated changes during their development, the details of which

are described in a later section. In our present discussion we shall point out the various types of sexual reproduction—biparental reproduction, parthenogenesis, paedogenesis, and hermaphroditism.

Biparental reproduction. Biparental reproduction is the common and familiar method of sexual reproduction involving separate and distinct male and female individuals. Each has its own reproductive system and produces only one kind of sex cell, spermatozoon or ovum, but never both. Nearly all vertebrates and many invertebrates have separate sexes, and such a condition is called **dioecious.**

Parthenogenesis. Parthenogenesis is a modification of sexual reproduction in which an unfertilized egg develops into a complete individual. It is found in rotifers, plant lice, certain ants, bees, and crustaceans. Usually parthenogenesis occurs for several generations and is followed by a biparental generation in which the egg is fertilized. In some cases parthenogenesis appears to be the only form of reproduction. The queen bee is fertilized only once by a male (drone) or sometimes by more than one drone. She stores the sperm in her seminal receptacles, and as she lays her eggs she can either fertilize the eggs or allow them to pass unfertilized. The fertilized eggs become females (queens or workers), and the unfertilized eggs become males (drones).

Paedogenesis. The production of eggs by larval forms is called **paedogenesis.** A striking example of this is the tiger salamander *(Ambystoma tigrinum mexicanum),* which in certain parts of its range is found to mate in a larval (axolotl) form. Such larvae can be transformed into adults under the right conditions.

Hermaphroditism. Animals that have both male and female organs in the same individual are called hermaphrodites, and the condition is called **hermaphroditism.** In contrast to the dioecious state of separate sexes, hermaphrodites are **monoecious.** Many lower animals (flatworms and hydra) are hermaphroditic. Most avoid self-fertilization by exchanging germ cells with each other. For example, although the earthworm bears both male and female organs, its eggs are fertilized by the copulating mate and vice versa. Another way of preventing self-fertilization is by developing the eggs and sperm at different times.

PLAN OF REPRODUCTIVE SYSTEMS

The basic plan of the reproductive systems is similar in all animals, although differences in reproductive habits, methods of fertilization, etc. have produced many variations. In vertebrate animals the reproductive and excretory systems are often referred to as the **urinogenital system** because of their close anatomic connection. This association is very striking in their embryonic development.

The male urinogenital system usually has a more intimate connection than has the female. This is the case with those forms (some fish and amphibians) that have an opisthonephric kidney. In male fishes and amphibians the duct that drains the kidney **(wolffian duct)** also serves as the sperm duct. In male reptiles, birds, and mammals in which the kidney develops its own independent duct **(ureter)** to carry away waste, the old wolffian duct becomes exclusively a sperm duct **(vas deferens).** In all these forms, with the exception of mammals, the ducts open into a **cloaca.** In higher mammals there is no cloaca; instead the urinogenital system has its own opening separate from the anal opening. The **oviduct** of the female is an independent duct that, however, does open into the cloaca in forms that have a cloaca.

The plan of the reproductive system in vertebrates includes (1) **gonads** that produce the sperm and eggs, (2) **ducts** to transport the gametes, (3) **special organs** for transferring and receiving gametes, (4) **accessory glands** (exocrine and endocrine) to provide secretions necessary to synchronize the reproductive process, and (5) **organs** for storage before and after fertilization. This plan is modified among the various vertebrates, and some of the items may be lacking altogether.

Human reproductive system
Male reproductive system

The human male reproductive system (Fig. 13-1) includes testes, vasa efferentia, vasa deferentia, penis, and glands.

The **testes** are paired and are responsible for the production and development of the sperm. Each testis is made up of numerous **seminiferous tubules,** which produce the sperm (Fig. 13-1), and the **interstitial** tissue lying among the tubules, which produces the male sex hormone (testosterone). The two testes are housed in the scrotal sac, which in many mammals hangs down as an appendage of the body. This arrangement protects the sperm against high temperature since in at least some forms sperm apparently do not form at body temperatures.

The sperm are conveyed from the seminiferous tubules to the **vasa efferentia,** small tubes passing to a coiled **vas epididymis** (one for each testis). The epididymis is connected by a **vas deferens** to the **urethra.** From this point the urethra serves to carry both sperm and urinary products through the penis, or external intromittent organ.

Three pairs of glands open into the reproductive channels—**seminal vesicles, prostate glands,** and **Cowper's glands.** Fluid secreted by these glands furnishes food to the sperm, lubricates the passageways for the sperm, and counteracts the acidity of the urine so that the sperm are not harmed.

Female reproductive system

The female reproductive system (Fig. 13-2) contains ovaries, oviduct, uterus, vagina, and vulva.

The paired ovaries, each approximately the size of an almond, contain many thousands of developing eggs (ova). Each egg develops within a **graafian follicle** that enlarges and finally ruptures to release the mature egg (Fig. 13-2). During the fertile period of the woman approximately 13 eggs mature each year, and usually the ovaries alternate in releasing an egg. Since the female is fertile for only some 30 years, only approximately 400 eggs have a chance to reach maturity; the others degenerate and are absorbed.

The **oviducts,** or fallopian tubes, are egg-carrying tubes with funnel-shaped openings for receiving the eggs when they emerge from the ovary. The oviduct is lined with cilia for propelling the egg in its course. The two ducts open into the upper corners of the **uterus,** or womb, which is specialized for housing the embryo during the 9 months of its intrauterine existence. It is provided with thick muscular walls, many blood vessels, and a specialized lining—the **endometrium.** The uterus varies with different mammals. It was originally paired but tends to fuse in higher forms.

The uterus is connected to the outside of the body by the **vagina.** This muscular tube is adapted for receiving the male's penis and for serving as the birth canal during expulsion of the fetus from the uterus. Where the vagina and the uterus meet, the uterus projects down into the vagina to form the **cervix.**

The external genitalia of the female, or vulva, include folds of skin, the **labia majora** and **labia minora,** and a small erectile organ, the **clitoris.** The opening into the vagina is normally reduced in size in the virgin state by a membrane, the **hymen.**

255

Homology of sex organs

For every structure in the male system, there is a homologous one in the female. They do not have identical functions in both sexes, however. Many of them are functional in one sex, whereas their homologues in the other sex may be vestigial and nonfunctional.

Although sex is probably determined at the time of fertilization, it is not until many weeks later that the distinct sex characters associated with one or the other sex are recognized.

ORIGIN OF REPRODUCTIVE CELLS

The animal body has two obvious contrasting components, the somatic cells, which die with the individual, and the germinal cells, some of which may contribute to the formation of a zygote and thereby to a new generation. The germinal cells are set aside early in development, usually in the endoderm, and migrate to the gonads. The germinal cells, or primordial germ cells, develop into eggs and sperm—nothing else. The other cells of the gonads are somatic cells. They cannot themselves form eggs or sperm, but they are necessary aids in the gametogenesis of the germinal cells.

Meiosis: germ cell division

The gametes are formed by a special kind of maturation division called **meiosis.** Although it superficially resembles ordinary cell division, or **mitosis,** the two kinds of divisions are different in one crucially important way: mitosis maintains the parental chromosome number, whereas meiosis reduces the parental chromosome number to one-half. Meiosis was described in Chapter 4 (p. 71), but it will be helpful to review meiosis broadly at this time.

In ordinary cell division, or mitosis, each of the two daughter cells receives exactly the same number and kind of chromosomes. All body (somatic) cells have two sets of chromosomes, one of paternal and the other of maternal origin. We earlier defined the members of such a pair as **homologous chromosomes.**

In sexual reproduction, however, the formation of the **gametes,** or **germ cells,** requires a different process than that of somatic cells. The fusion of two gametes (egg and sperm) produces the zygote or fertilized egg from which the new organism arises. If each sperm and egg had the same number of chromosomes as somatic cells, there would be a doubling of chromosomes in each successive generation. To pre-

vent this from happening, germ cells are formed whereby the chromosome number is reduced by one half. The result is that mature gametes have only *one* member of each homologous pair, or a **haploid** (n) number of chromosomes. In man the zygotes and all body cells normally have the **diploid** number (2n) of 46; the gametes (eggs and sperm) have the haploid number (n) of 23.

Meiosis consists of two successive nuclear divisions, called meiosis I and meiosis II (See Fig. 4-4, p. 72). In the first meiotic division the homologous chromosomes, each already duplicated into two halves or sister chromatids, come to lie side by side. The homologous chromosomes then separate to opposite poles of the cell. In the second meiotic division, the sister chromatids of each chromosome separate and move to opposite poles. The final result is four gametes, each with the haploid number of chromosomes.

Gametogenesis

The series of transformations that results in the formation of mature gametes (germ cells) is called gametogenesis.

Although the same essential processes are involved in the maturation of both sperm and eggs, there are some important differences. Gametogenesis in the testis is called **spermatogenesis** and in the ovary it is called **oogenesis.**

Spermatogenesis (Fig. 13-1). The walls of the seminiferous tubules contain the differentiating sex cells arranged in a stratified layer five to eight cells deep. The outermost layers contain **spermatogonia** (Fig. 3-1), which have increased in number by ordinary mitosis. Each spermatogonium increases in size and becomes a **primary spermatocyte.** Each primary spermatocyte then undergoes the first meiotic division, as described above, to become two **secondary spermatocytes.**

Each secondary spermatocyte enters the second meiotic division, without the intervention of a resting period. The resulting cells are called **spermatids,** and each contains the haploid number (23) of chromosomes. A spermatid may have all maternal, all paternal, or both maternal and paternal chromosomes in varying proportions. Without further divisions the spermatids are transformed into mature sperm by losing a great deal of cytoplasm, by condensing the nucleus into a head, and by forming a whiplike tail.

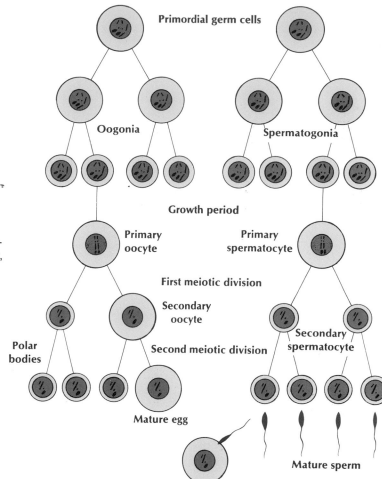

Fig. 13-3. Process of gametogenesis, or formation of germ cells. Oogenesis shown on left, spermatogenesis on right.

From following the divisions of meiosis it can be seen that each primary spermatocyte gives rise to four functional sperm, each with the haploid number of chromosomes (Fig. 13-3).

Oogenesis (Fig. 13-3). The early germ cells in the ovary are called **oogonia,** which increase in number by ordinary mitosis. Each oogonium contains the diploid number of chromosomes. In females after puberty, one of these oogonia typically develops each menstrual month into a functional egg. After the oogonia cease to increase in number, they grow in size and become **primary oocytes.** Before the first meiotic division, the chromosomes in each primary oocyte meet in pairs, paternal and maternal homologues, just as in spermatogenesis. When the first maturation (reduction) division

occurs, the cytoplasm is divided unequally. One of the two daughter cells, the **secondary oocyte,** is large and receives most of the cytoplasm; the other is very small and is called the **first polar body.** Each of these daughter cells, however, has received half the nuclear material or chromosomes.

In the second meiotic division, the secondary oocyte divides into a large **ootid** and a small polar body. If the first polar body also divides in this division, which sometimes happens, there are three polar bodies and one ootid. The ootid grows into a functional **ovum;** the polar bodies are nonfunctional and disintegrate. The formation of the nonfunctional polar bodies is necessary to enable the egg to get rid of excess chromosomes, and the unequal cytoplasmic division makes

257

possible a large cell with sufficient yolk for the development of the young. Thus the mature ovum has the haploid number of chromosomes, the same as the sperm. However, each primary oocyte gives rise to only *one* functional gamete instead of four as in spermatogenesis.

DEVELOPMENTAL PROCESS

The phenomenon of development is a remarkable and in many ways awesome process. How is it possible that a tiny, spheric fertilized human egg, scarcely visible to the naked eye, can unfold into a fully formed, unique person, consisting of thousands of billions of cells, each cell performing a predestined functional or structural role? How is this marvelous unfolding controlled? Obviously all the information needed is contained within the egg, principally in the genes of the egg's nucleus. The fabric of genes is deoxyribonucleic acid (DNA). Thus all development originates from the structure of the nuclear DNA molecules and in the egg cytoplasm surrounding the nucleus. But knowing where the blueprint for development resides is very different from understanding how this control system guides the conversion of a fertilized egg into a fully differentiated animal. This remains a major—many consider *the* major—unsolved problem of biology. It has stimulated a vast amount of research on the processes and phenomena involved; from it have emerged some early and in many cases tentative answers to the questions of how development is controlled.

Oocyte maturation

During oogenesis the egg becomes a highly specialized, very large cell containing condensed food reserves for subsequent growth. The nucleus also grows rapidly in size during egg maturation, although not as much as the cell as a whole. Large amounts of both DNA and RNA accumulate during oogenesis. Early in the oogenesis of large amphibian eggs, the chromosomes become vastly expanded by thin loops thrown out laterally from the chromosomal axis; because of their fuzzy appearance, the chromosomes at this time are called **lampbrush chromosomes.** It is believed that the messenger RNA is being rapidly synthesized on DNA templates as each chromomere puffs out in lampbrush loops, exposing the double helix of DNA. The messenger RNA subsequently controls the synthesis of protein in the oocyte. Ribo-

somal RNA is also undergoing intense synthesis in the nucleolus during oocyte growth. Toward the end of oocyte maturation, this intense nucleic acid and protein synthesis gradually winds down. The lampbrush chromosomes contract and migrate to just beneath the egg surface, in preparation for the reduction divisions. The ribosomal RNA and proteins mix with the egg cytoplasm.

Most of this enormous food accumulation and nucleic acid synthesis occurs before the meiotic, or maturation, divisions begin. When the maturation divisions do occur, they are, as already described, highly unequal: the single mature ovum retains a haploid set of chromosomes and the vast bulk of the cytoplasm, whereas the other three sets of haploid chromosomes are cast off as small, cytoplasm-starved polar bodies. Because all three polar bodies lack stored nutrients, none are capable of further development. This obviously undemocratic hoarding of all accumulated food reserves by just one of four otherwise genetically equal cells is, of course, a device for avoiding a decrease in ovum size once all the nutrients have been packaged inside at the end of the growth phase.

In most vertebrates, the egg does not actually complete all the meiotic divisions before fertilization occurs. The general rule is that the egg completes the first meiotic division and proceeds to the metaphase stage of the second meiotic division, at which point further progress stops. The second meiotic division is completed and the second polar body extruded only if the egg is activated by fertilization.

Fertilization and activation

Fertilization is the union of male and female gametes to form a *zygote*. This process accomplishes two things: it activates the process of development and it provides for the recombination of paternal and maternal genes. Thus it restores the original diploid number of chromosomes characteristic of the species.

For a species to survive, it must ensure that fertilization occurs and that enough progeny survive to maintain a healthy population. Many marine fishes simply set their eggs and sperm adrift in the ocean and rely on the random swimming movements of sperm to make chance encounters with eggs. Even though an egg is a large target for a sperm, the enormous dispersing effect of the ocean, the short life-span of the gametes (usually just a few minutes for fish gametes), and the limited range of the tiny sperm all conspire against an egg and a sperm coming together. Accordingly each male releases countless millions of sperm at spawning. The odds against fertilization are further reduced by coordinating the time and place of spawning of both parents. Ensuring that some eggs are fertilized, however, is not enough. The ocean is a perilous environment for a developing fish, and most never make it to maturity. Thus the females produce huge numbers of eggs. The common gray cod of the North American east coast regularly spawns 4 to 6 million eggs, of which only two or three, on the average, reach maturity.

Fishes and other vertebrates that provide more protection for their young produce fewer eggs than do the oceanic marine fishes. The chances of the eggs and sperm meeting are also increased by courtship and mating procedures and the simultaneous shedding of the gametes in a nest or closely circumscribed area. Internal fertilization, characteristic of the sharks and rays as well as reptiles, birds, and mammals, avoids dispersion of the gametes and protects them. However, even with internal fertilization vast numbers of sperm must be released by the male into the female tract. Furthermore the events of ovulation and insemination must be closely synchronized and the gametes must remain viable for several hours to accomplish fertilization. Sperm may have to travel a considerable distance to reach the egg in the female genital tract, many parts of which may be hostile to sperm. Experiments with rabbits have shown that of the approximately 10 million sperm released into the female vagina, only approximately 100 reach the site of fertilization.

Activation—restoration of metabolic activity in the quiescent egg—is a dramatic event. In sea urchin eggs, the contact of the spermatozoan with the egg surface sets off almost instantaneous changes in the egg cortex. At the point of contact, a **fertilization cone** appears into which the sperm head is later drawn.

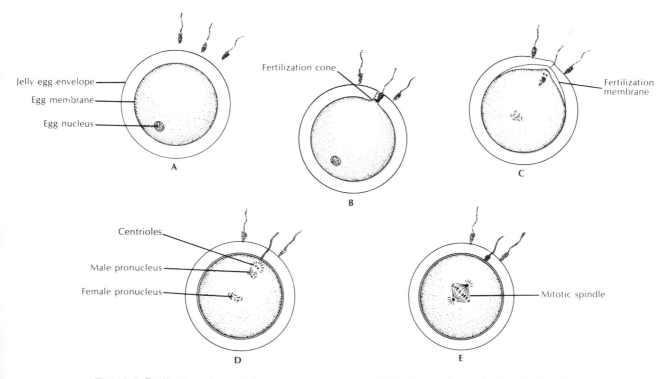

Fig. 13-4. Fertilization of egg. **A,** Many sperm swim to egg. **B,** First sperm to penetrate protective jelly envelope and contact egg membrane causes fertilization cone to rise and engulf sperm head. **C,** Fertilization membrane begins to form at site of penetration and spreads around entire egg, preventing entrance of additional sperm. **D,** Male and female pronuclei approach one another, lose their nuclear membranes, swell, and fuse. **E,** Mitotic spindle forms, signaling creation of a zygote and heralding first cleavage of new embryo.

From this point, a visible change travels wavelike across the egg surface, causing immediate elevation of a **fertilization membrane** (see Fig. 13-4). At the same time **cortical granules,** which form a layer beneath the plasma membrane, explode, releasing materials that fuse together to build up a new egg surface, the **hyaline layer.** This lies between the inner plasma membrane and the outer fertilization membrane. The **cortical reaction,** as these changes are called, is a crucial event in development. Within seconds it seems to produce a complete molecular reorganization of the egg cortex. It also serves to remove one or more inhibitors that have blocked the energy-yielding systems and protein synthesis in the egg and kept the egg in its quiescent, suspended-animation state. Almost immediately following the cortical reaction, stored nucleic acids begin producing protein. Normal metabolic activity is restored. The reduction divisions, if not completed, are brought to completion. The male and female pronuclei then fuse, and the zygote enters into cleavage.

Cleavage and early development

During cleavage, the zygote divides repeatedly to convert the large, unwieldy cytoplasmic mass into a large number of small, maneuverable cells (called **blastomeres**) clustered together like a mass of soap bubbles. There is no growth during this period, only subdivision of mass, which continues until normal cell size and nucleocytoplasmic ratios are attained. At the end of cleavage the zygote has been divided into many hundreds or thousands of cells (approximately 1,000 in polychaete worms, 9,000 in amphioxus, and 700,000 in frogs). There is a rapid increase in DNA content during cleavage as the number of nuclei and the amount of DNA are doubled with each division. Apart from this, there is little change in chemical composition or displacement of constituent parts of the egg cytoplasm during cleavage. **Polarity,** that is, a polar axis, is present in the egg, and this establishes the direction of cleavage and subsequent differentiation of the embryo. Usually cleavage is very regular although enormously affected by the quantity of yolk present and by whether cleavage is radial or spiral.

Patterns of cleavage

The zygote of most animals divides regularly, first into two cells and those two into four. The planes of division are usually vertical, with the second plane at right angles to the first. The third division plane forms horizontally, at right angles to the first two planes of division. Thus, of the eight blastomeres, four lie on top of the other four.

Radial and spiral cleavage. Two different kinds of cleavage symmetry are recognized among different groups of animals. In **radial cleavage,** the cleavage planes are symmetric to the polar axis and produce tiers, or layers, of blastomeres on top of each other. Radial cleavage is typical of the Deuterostomia (echinoderms, protochordates, and chordates; see inside front cover and p. 284). In **spiral cleavage,** the cleavage planes are diagonal to the polar axis and produce alternate clockwise and counterclockwise quartets of unequal cells around the axis of polarity. With few exceptions, radial cleavage is found in the Protostomia (acoelomates, pseudocoelomates, annelids, molluscs, and arthropods).

Determinate and indeterminate cleavage. In many species, especially those showing radial cleavage (Deuterostomia), early blastomeres, if separated from each other, each give rise to a whole larva. This is called **indeterminate cleavage,** meaning that the first blastomeres to form are equipotent; there is no segregation into potentially different histogenetic regions. Thus, if a sea urchin embryo at the four-cell stage is placed in calcium-free seawater and gently shaken, the blastomeres fall apart. Replaced in normal seawater, each subsequently develops into a complete, though small, pluteus larva fully capable of growing into an adult sea urchin (Fig. 13-5). Early blastomeres of frog and rabbit embryos also can be separated and yield complete embryos. Indeterminate cleavage is also called **regulative.** The reason is that each of the first four or eight blastomeres is capable of regulating its developmental fate to produce a portion of a larva if it develops in the company of other blastomeres or to produce a whole larva if it is forced to develop alone.

The eggs of many invertebrates having spiral cleavage lack this early versatility. If the early blastomeres of a mollusc, annelid, flatworm, or ascidian are separated, each gives rise only to a part of an embryo in accord with its original fate. This is called **determinate** (or mosaic) **cleavage.** Such blastomeres appear to have a fixed informational content as soon as cleavage begins. The terms "determinate" and "indeterminate" can be used only in a provisional sense, since there are many eggs that do not fit clearly into either category.

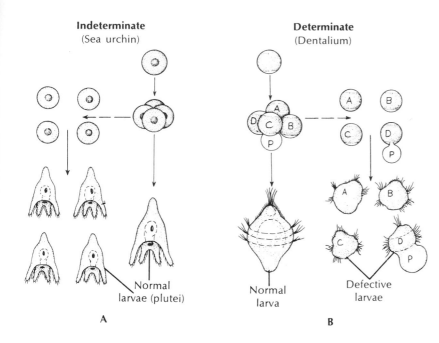

Indeterminate
(Sea urchin)

Determinate
(Dentalium)

Normal
larvae (plutei)

A

Normal
larva

Defective
larvae

B

Fig. 13-5. Indeterminate and determinate (mosaic) cleavage. **A,** Indeterminate cleavage. Each of the early blastomeres (such as that of the sea urchin) when separated from the others develops into a small pluteus larva. **B,** Determinate (mosaic) cleavage. In the mollusc (such as *Dentalium*), when the blastomeres are separated, each gives rise to only a part of an embryo. The larger size of one of the defective larvae is the result of the formation of a polar lobe *(P)* composed of clear cytoplasm of the vegetal pole, which this blastomere alone receives.

The explanation behind these two types of cleavage seems to lie in the extent to which early blastomeres depend exclusively on information segregated by the early divisions (determinate cleavage) or whether they are able to fall back upon a supply of information from the nucleus (indeterminate cleavage). In the determinate cleavage type the information-containing material is localized so early that each blastomere receives a portion that is qualitatively unique. With such mosaic (determinate) eggs it is often possible to map out the fates of specific areas on the egg surface that are known to be presumptive for specific structures, such as germ layers, notochord, and nervous system.

From the evidence of identical twins, man apparently has the indeterminate type of cleavage, for identical twins come from the same zygote. Of course, the totipotency (the capacity to develop into a complete embryo) of the blastomeres in indeterminate cleavage is strictly limited. After the first three or four cleavages, the fate of the blastomeres becomes fixed.

Significance of the cortex

Cleavage proceeds independent of nuclear genetic information, guided instead by information deposited in the egg during its maturation. It was once believed that the visible particulate material in the cytoplasm had determinative properties. However, it was soon discovered that if the egg was strongly centrifuged so

that everything inside—nucleus, mitochondria, lipid droplets, yolk, and other inclusions—was thoroughly displaced, the embryo still developed perfectly. If sea urchin eggs are examined by electron microscope after being centrifuged for 5 minutes at several thousand times the force of gravity, the only thing not affected is the plasma membrane and a gellike layer just beneath the plasma membrane **(plasmagel layer).** Yet development proceeds normally. This and similar experiments show conclusively that the plasmagel (or cortical) layer of the egg contains an invisible but dynamic organization that determines the pattern of cleavage. Cortical organization is at first labile (especially so in indeterminate eggs) but soon becomes regionally fixed and irreversible. Thus, as cleavage progresses, the cortex becomes segregated into territories having specific determinative properties. This explains why different blastomeres bear different cytodifferentiation properties.

Gastrulation

Gastrulation is a regrouping process in which new and important cell associations are formed. Up to this point the embryo has divided itself into a multicellular complex; the cytoplasm of these numerous cells is nearly in the same position it was in the original undivided egg. In other words, there has been no significant movement or displacement of the cells from their place of origin. As gastrulation begins the cells become rearranged in an orderly way by morphogenetic movements. In amphibian embryos, as in most forms, cells on the surface begin to sink inward at one point, the **blastopore.** Through the curved groove of the blastopore, surface cells move as a sheet to the interior to form a two-layered embryo. A rodlike **notochord** forms at this time, growing forward to run lengthwise along the dorsal side of the embryo. Continued rearrangements of cells form a third layer; these three layers, called **germ layers,** are the primary structural layers that play crucial roles in the further differentiation of the embryo. The outer layer, or **ectoderm,** gives rise to the nervous system and outer epithelium of the body. The middle layer, or **mesoderm,** gives rise to the circulatory, skeletal, and muscular structures. The inner layer, or **endoderm,** develops into the digestive tube and its associated structures.

Differentiation

With formation of the three primary germ layers, cells continue to regroup and rearrange themselves into primordial cell masses. These masses continue to differentiate, ultimately resulting in the formation of specific organs and tissues. During this process, cells become increasingly committed to specific directions of differentiation. Cells that previously had the potential to develop into a variety of structures lose this diverse potential and assume commitments to become, for example, kidney cells, intestinal cells, or brain cells.

Derivatives of ectoderm: nervous system and nerve growth

The brain, spinal cord, and nearly all the outer epithelial structures of the body develop from the primitive ectoderm. They are among the earliest organs to appear. Just above the notochord, the **ectoderm** thickens to form a **neural plate** (Fig. 13-6). The edges of this plate rise up, fold, and join together at the top to create an elongated, hollow **neural tube.** The neural tube gives rise to most of the nervous system: anteriorly it enlarges and differentiates into the brain, cranial nerves, and eyes; posteriorly it forms the spinal cord and spinal motor nerves. Sensory nerves arise from special **neural crest** cells pinched off from the neural tube before it closes.

How are the billions of nerve axons in the body

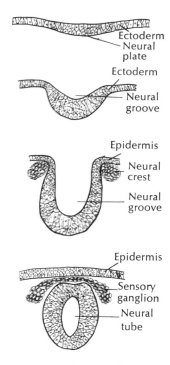

Fig. 13-6. Development of neural tube and neural crest from neural plate ectoderm (cross section).

formed? What directs their growth? Biologists were intrigued with these questions that seemed to have no easy solutions. Since a single nerve axon may be many feet in length (for example, motor nerves running from the spinal cord to the toes), it seemed impossible that a single cell could spin out so far. It was suggested that nerve fibers grew from a series of preformed protoplasmic bridges along its route. The answer had to await the development of one of the most powerful tools available to biologists, the cell culture technique.

In 1907 embryologist Ross G. Harrison discovered that he could culture living neuroblasts (embryonic nerve cells) for weeks outside the body by placing them in a drop of frog lymph hung from the underside of a cover slip. Watching nerves grow for periods of days, he saw that each nerve fiber was the outgrowth of a single cell. As the fibers extended outward, materials for growth flowed down the axon center to the growing tip where they were incorporated into new protoplasm.

The tissue culture technique is now used extensively by scientists in all fields of active biomedical research, not just by embryologists. The great impact of the

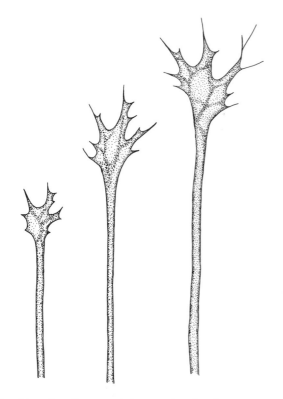

Fig. 13-7. Growth cone at the growing tip of a nerve axon. Materials for growth flow down axon to growth cone from which numerous threadlike pseudopodial processes extend. These appear to serve as a pioneering guidance system for the developing axon.

technique has been felt only in recent years. Harrison was twice considered for the Nobel Prize (1917 and 1933), but he failed ever to receive the award because, ironically, the tissue culture method was then believed to be "of rather limited value."

The second question—what directs nerve growth—has taken longer to unravel. An idea held well into the 1940s was that nerve growth is a random, diffuse process. It was thought that the nervous system developed as an equipotential network, or blank slate, that later would be shaped by usage into a functional system. The nervous system just seemed too incredibly complex for us to imagine that nerve fibers could find their way selectively to predetermined destinations. Yet it appears that this is exactly what they do! Recent work indicates that each of the billions of nerve cell axons acquires a chemical identification tag that in some way directs it along a correct path. Many years ago Ross Harrison observed that a growing nerve axon

terminated in a "growth cone," from which extend numerous tiny threadlike processes (Fig. 13-7). These are constantly reaching out and testing the environment in all directions to guide the nerve chemically to its proper destination. This chemical guidepost system, which must, of course, be genetically directed, is just one example of the amazing precision that characterizes the entire process of differentiation.

Derivatives of endoderm: digestive tube and survival of gill arches

In the frog embryo the primitive gut makes its appearance during gastrulation with the formation of an internal cavity, the **archenteron.** From this simple endodermal cavity develop the lining of the digestive tract, the lining of the pharynx and lungs, most of the liver and pancreas, the thyroid and parathyroid glands, and the thymus.

The **alimentary canal** is early folded off from the yolk sac by the growth and folding of the body wall (Fig. 13-8). The ends of the tube open to the exterior and are lined with ectoderm, whereas the rest of the tube is lined with endoderm. The **lungs, liver,** and **pancreas** arise from the foregut.

Among the most intriguing derivatives of the digestive tract are the pharyngeal (gill) arches and pouches, which make their appearance in the early embryonic stages of all vertebrates (Fig. 5-8, p. 96). In fishes, the gill arches develop into gills and supportive structure and serve as respiratory organs. When the early vertebrates moved onto land, gills were unsuitable for aerial respiration and were replaced by lungs.

Why then do gill arches persist in the embryos of terrestrial vertebrates? Certainly not for the convenience of biologists who use these and other embryonic structures to reconstruct lines of vertebrate descent. Even though the gill arches serve no respiratory function in either the embryos or adults of terrestrial vertebrates, they remain as necessary primordia for a great variety of other structures. For example, the first arch and its endoderm-lined pouch (the space between adjacent arches) form the upper and lower jaws and inner ear of higher vertebrates. The second, third, and fourth gill pouches contribute to the tonsils, parathyroid gland, and thymus. We can understand then why gill arches and other fishlike structures appear in early mammalian embryos. Their original function has been abandoned, but the structures are retained for new

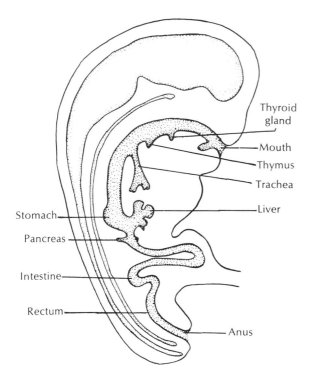

Fig. 13-8. Derivatives of alimentary canal.

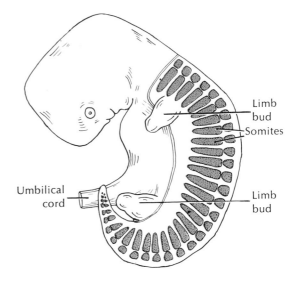

Fig. 13-9. Embryo showing muscle somites.

purposes. It is the great conservatism of early embryonic development that has so conveniently provided us with a telescoped evolutionary history.

Derivatives of mesoderm: support, movement, and beating heart

The intermediate germ layer, the mesoderm, forms the vertebrate skeletal, muscular, and circulatory structures and the kidney. As vertebrates have increased in size and complexity, the mesodermally derived supportive, movement, and transport structures make up an ever greater proportion of the body bulk.

Most **muscles** arise from the mesoderm along each side of the spinal cord (Fig. 13-9). This mesoderm divides into a linear series of somites (38 in man), which by splitting, fusion, and migration become the muscles of the body and axial parts of the skeleton. The **limbs** begin as buds from the side of the body. Projections of the limb buds develop into digits.

Although the primitive mesoderm appears after the ectoderm and endoderm, it gives rise to the first functional organ, the embryonic heart. Guided by the underlying endoderm, clusters of precardiac mesoder-

mal cells move amebalike into a central position between the underlying primitive gut and the overlying neural tube. Here the heart is established, first as a single, thin tube.

Even while the cells group together, the first twitchings are evident. In the chick embryo, a favorite and nearly ideal animal for experimental embryology studies, the primitive heart begins to beat on the second day of the 21-day incubation period; it begins beating before any true blood vessels have formed and before there is any blood to pump. As the ventricle primordium develops, the spontaneous cellular twitchings become coordinated into a feeble but rhythmic beat. Then, as the atrium develops behind the ventricle followed by the development of the sinus venosus behind the atrium, the heart rate quickens. Each new heart chamber has an intrinsic beat that is faster than its predecessor.

Finally a specialized area of heart muscle called the **sinoatrial** node develops in the sinus venosus and takes command of the entire heartbeat. This becomes the heart's **pacemaker.** As the heart builds up a strong and efficient beat, vascular channels open within the embryo and across the yolk. Within the vessels are the first primitive blood cells suspended in plasma.

The early development of the heart and circulation is crucial to continued embryonic development because without a circulation the embryo could not obtain materials for growth. Food is absorbed from the yolk

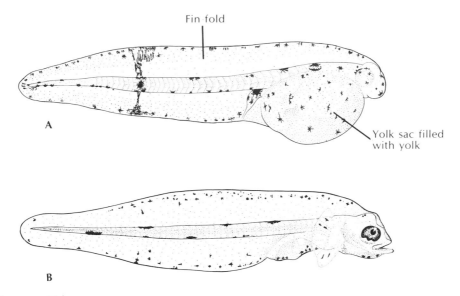

Fin fold

Yolk sac filled with yolk

A

B

Fig. 13-10. Fish embryos showing yolk sac. **A,** The 1-day-old larva of a marine flounder has a large yolk sac. **B,** By the time the 10-day-old larva has developed a mouth and primitive digestive tract, its yolk supply has been exhausted. It must now catch its own food to survive and continue growing. (From Hickman, C. P., Jr.: The larval development of the sand sole, *Psettichthys melanostictus*, Washington State Fisheries Research Papers **2**:38-47, 1959.)

and carried to the embryonic body; oxygen is delivered to all the tissues, and carbon dioxide and other wastes are carried away. The embryo is totally dependent on these and other extraembryonic support systems, and the circulation is the vital link between them.

EXTRAEMBRYONIC MEMBRANES OF THE AMNIOTES

Reptiles, birds, and mammals form a natural grouping of vertebrates called **amniotes,** meaning that they develop an amnion, one of the extraembryonic membranes that make the development of these forms unique among animals.

As rapidly growing living organisms, embryos have the same basic animal requirements as adults—food, oxygen, and disposal of wastes. For the embryos of marine invertebrates, living a contact existence with their environment, gas exchange is a simple matter of direct diffusion. Food can be acquired as soon as the embryo develops a mouth and begins feeding on plankton. All eggs of aquatic animals are provided with just enough stored yolk to allow growth to this critical stage.

Beyond this point, the embryo (called a free-swimming **larva**) is on its own. Yolk enclosed in a

membranous **yolk sac** is a conspicuous feature of all fish embryos (Fig. 13-10). The yolk is gradually used up as the embryo grows; the yolk sac shrinks and finally is enclosed within the body of the embryo. The mass of yolk is an **extraembryonic** structure since it is not really a part of the embryo proper; the yolk sac is an **extraembryonic membrane.** Bird and reptile eggs are also provided with large amounts of yolk to support early development. In birds the yolk reaches relatively massive proportions since it must build a baby bird in a much more advanced stage of growth at hatching than a larval fish.

In abandoning an aquatic life for a land existence the first terrestrial animals had to evolve a sophisticated egg containing a complete set of life-support systems. Thus appeared the egg of amniotes, equipped to protect and support the growth of embryos on dry land. In addition to the yolk sac containing the nourishing yolk are three other membranous sacs—amnion, chorion, and allantois. All are referred to as extraembryonic membranes because they are accessory structures that develop beyond the embryonic body and are discarded when the embryo hatches.

The **amnion** is a fluid-filled bag that encloses the embryo and provides a private aquarium for develop-

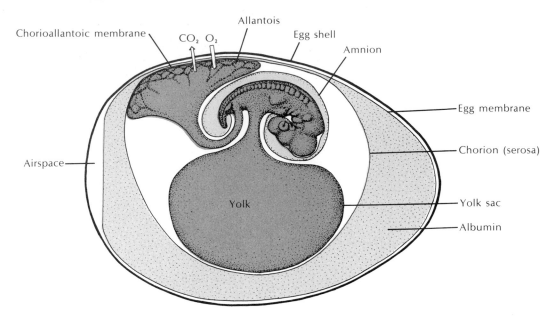

Fig. 13-11. Egg of amniote at early stage of development showing a chick embryo and its extraembryonic membranes. Porous shell allows gaseous exchange of oxygen and carbon dioxide. Circulatory channels from embryo's body to allantois and yolk sac are not shown.

ment (Fig. 13-11). Floating freely in this aquatic environment, the embryo is fully protected from shocks and adhesions. The evolution of this structure was crucial to the successful habitation of land.

The **allantois,** another component in the support system for embryos of land animals, is a bag that grows out of the hindgut of the embryo (Fig. 13-11). It collects the wastes of metabolism (mostly uric acid). At hatching, the young animal breaks its connection with the allantois and leaves it and its refuse behind in the shell.

The **chorion** (also called **serosa**) is an outermost extraembryonic membrane that completely encloses the rest of the embryonic system. It lies just beneath the shell (Fig. 13-11).

As the embryo grows and its need for oxygen increases, the allantois and chorion fuse together to form a **chorioallantoic membrane.** This double membrane is provided with a rich vascular network, connected to the embryonic circulation. Lying just beneath the porous shell, the vascular chorioallantoic membrane serves as a kind of lung across which oxygen and carbon dioxide can freely exchange. Although nature did not plan for it, the chorioallantoic membrane of the chicken egg has been used exten-

sively by generations of experimental embryologists as a place to culture chick and mammalian tissues.

The great importance of the amniotic egg to the establishment of a land existence cannot be overemphasized. Amphibians must return to water to lay their eggs. But the reptiles, even before they took to land, developed the amniotic egg with its self-contained aquatic environment enclosed by a tough outer shell. Protected from drying out and provided with yolk for nourishment, such eggs could be laid on dry land, far from water. Reptiles were thus freed from aquatic life and could become the first true terrestrial tetrapods.

Incidentally the sexual act itself comes from the requirement that the egg be fertilized *before* the egg shell is wrapped around it, if it is to develop. Thus the male must introduce the sperm into the female tract so that the sperm can reach the egg before it passes to that part of the oviduct where the shell is secreted. Hence, as one biologist puts it, it is the egg shell and not the devil that deserves blame for the happy event we know as sex.

HUMAN DEVELOPMENT

The amniotic egg, for all its virtues, has one basic flaw: placed neatly in a nest, it makes fine food for other animals. The mammals evolved the best solution for early development: allow the embryo to grow within the protective confines of the mother's body. This has resulted in important modifications in development of mammals as compared with other vertebrates. The earliest mammals, descended from early reptiles, were egg layers. Even today the most primitive mammals, the monotremes (for example, duckbill platypus, spiny anteater), lay large yolky eggs that closely resemble bird eggs. In the marsupials (pouched mammals such as the opossum and kangaroo), the embryos develop for a time within the mother's uterus. But the embryo does not "take root" in the uterine wall, as do the embryos of the more advanced **placental mammals,** and consequently it receives little nourishment from the mother. The young of marsupials are therefore born immature and are sheltered and nourished in a pouch of the abdominal wall.

All other mammals, the placentalians, nourish their young in the uterus by means of a **placenta.**

Early development of human embryo

The eggs of all placental mammals, though relatively enormous on a cellular scale, are small by egg standards. The human egg is approximately 0.1 mm in diameter and barely visible with the unaided eye. It contains very little yolk. After fertilization in the funnel-shaped opening of the oviduct, the dividing egg begins a 5-day journey down the oviduct toward the uterus, propelled by a combination of ciliary action (especially in the ampullary region) and muscular peristalsis. Cleavage is very slow: 24 hours for the first cleavage and 10 to 12 hours for each subsequent cleavage. By comparison, frog eggs cleave once every hour. Cleavage produces a small ball of 20 to 30 cells (called the morula) within which a fluid-filled cavity appears, the **segmentation cavity** (Fig. 13-12). This is comparable to the blastocoele of a frog's egg. The embryo is now called a **blastocyst.**

At this point, development of the mammalian embryo departs radically from that of lower vertebrates. A mass of cells, called the **inner cell mass,** appears on one side of the peripheral cell, or **trophoblast,** layer (Fig. 13-12). The inner cell mass forms the embryo, whereas the surrounding trophoblast forms the placenta. When the blastocyst is approximately 6 days old and composed of approximately 100 cells, it contacts and implants into the uterine endometrium (Fig. 13-

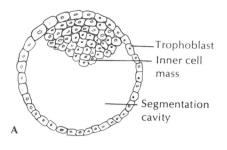

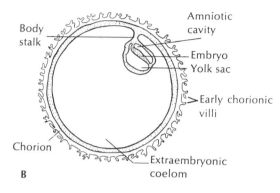

Fig. 13-12. Early development of human embryo. **A,** Blastocyst at approximately 7 days. Trophoblast gives rise to chorion, which later becomes part of placenta; inner cell mass becomes embryo. **B,** Blastocyst at approximately 15 days. Note amniotic cavity and yolk sac; primary villi are developing from chorion to begin establishment of placenta.

13). Very little is known of the forces involved in implantation, or why incidentally the intrauterine birth control devices are so effective in preventing successful implantation. Upon contact, the trophoblast cells proliferate rapidly and produce enzymes that break down the epithelium of the uterine endometrium. This allows the blastocyst to sink into the endometrium. By the eleventh or twelfth day the blastocyst, totally buried, has eroded through the walls of capillaries and small arterioles; this releases a pool of blood that bathes the embryo. At first the minute embryo derives what nourishment it requires by direct diffusion from the surrounding blood. But very soon a remarkable fetal-maternal structure, the **placenta,** develops to assume these exchange tasks.

Placenta

The placenta is a marvel of biologic engineering. Serving as a provisional lung, intestine, and kidney for the embryo, it performs elaborate selective activities without ever allowing the maternal and fetal blood to

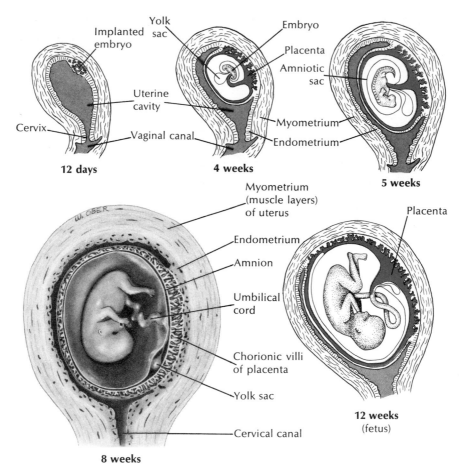

Fig. 13-13. Development of human embryo from 12 days to 12 weeks. The 8-week-old embryo and uterus are drawn to actual size. As the embryo grows, placenta develops into a disclike structure attached to one side of the uterus. Note vestigial yolk sac.

intermix (very small amounts of fetal blood may regularly escape into the maternal system without causing harm). The placenta permits the entry of foodstuffs, hormones, vitamins, and oxygen and the exit of carbon dioxide and metabolic wastes. Its action is highly selective since it allows some materials to enter that are chemically quite similar to others that are rejected.

The two circulations are physically separated at the placenta by an exceedingly thin membrane only 2 μm thick, across which materials are transferred by diffusive interchange. The transfer occurs across thousands of tiny fingerlike projections, called **chorionic villi,** which develop from the original chorion membrane (Figs. 13-12 and 13-13). These projections sink

like roots into the uterine endometrium after the embryo implants. As development proceeds and embryonic demands for food and gas exchange increase, the great proliferation of villi in the placenta vastly increases its total surface area. Although the human placenta at term measures only 18 cm (7 inches) across, its total absorbing surface is approximately 13 square meters—50 times the surface area of the skin of the newborn infant.

Since the mammalian embryo is protected and nourished by the placenta, what becomes of the various embryonic membranes of the amniotic egg whose functions are no longer required? Surprisingly perhaps, all of these special membranes are still present, although they may be serving a new function.

The yolk sac is retained, empty and purposeless, a vestige of our distant past (Fig. 13-13). Perhaps evolution has not had enough time to discard it. The amnion remains unchanged, a protective water jacket in which the embryo weightlessly floats. The remaining two extraembryonic membranes, the allantois and chorion, have been totally redesigned. The allantois is no longer needed as a urinary bladder. Instead it becomes the stalk, or **umbilical cord,** that links the embryo physically and functionally with the placenta. The chorion, the outermost membrane, forms most of the placenta itself.

One of the most intriguing questions the placenta presents is why it is not rejected by the mother's tissues. The placenta is a uniquely successful foreign transplant, or **allograft.** Since the placenta is an embryonic structure, containing both paternal and maternal antigens, we should expect it to be rejected by the uterine tissues, just as a piece of a child's skin is rejected by the child's mother when a surgeon attempts a grafting transplant. The placenta in some way circumvents the normal rejection phenomenon, a matter of the greatest interest to immunologists seeking ways to transplant tissues and organs successfully.

Pregnancy and birth

Pregnancy may be divided into four phases. The first phase of 6 or 7 days is the period of cleavage and blastocyst formation; it ends when the blastocyst implants in the uterus. The second phase of 2 weeks is the period of gastrulation and formation of the neural plate. The third phase, called the **embryonic period,** is a crucial and sensitive period of primary organ system differentiation. This phase ends at approximately the eighth week of pregnancy. The last phase, known as the **fetal period,** is characterized by rapid growth, proportional changes in body parts, and final preparation for birth.

During pregnancy the placenta gradually takes over most of the functions of regulating growth and development of the uterus and the fetus. As an endocrine gland it secretes estradiol and progesterone, hormones that are secreted by the ovaries and corpus luteum in the early periods of pregnancy. The placenta also produces **chorionic gonadotropin,** a hormone that assumes the role of the LH and FSH pituitary hormones, which cease their secretions about the second month of pregnancy. Chorionic gonadotropin maintains the corpus luteum so that it may continue

secreting the progesterone and estradiol that is necessary for the integrity of the placenta. In the later stages of pregnancy the placenta becomes a totally independent endocrine organ requiring support from neither the corpus luteum nor the pituitary.

What stimulates birth? Why does not pregnancy continue indefinitely? What factors produce the onset of labor (the rhythmic contractions of the uterus)? Thus far we have only tentative answers to these questions. It has long been known that estrogens stimulate uterine contractions, whereas progesterone, also secreted by the placenta, blocks uterine activity. Just before birth there is a very sharp increase in the plasma level of estrogens and a decrease in progesterone. This appears to remove the "progesterone block" that keeps the uterus quiescent throughout pregnancy. Thus labor contractions can proceed.

Recently much more information on the control of birth has emerged, involving **prostaglandins,** a group of hormonelike long-chain fatty acids. Prostaglandins have complex actions on virtually all body tissues and organs. They are especially powerful in stimulating contractions and, when used pharmacologically, are very effective in inducing labor or abortion. There is increasing evidence from studies on sheep (which, like human females, have an endocrine placenta during pregnancy) that the fetus itself controls the onset of labor. In some way not yet understood, the fetus, when it reaches normal term weight, releases an endocrine signal that stimulates the placenta to produce prostaglandins and thus initiates labor.

The first major signal that birth is imminent is the so-called labor pains, caused by the rhythmic contractions of the uterine musculature. These are usually slight at first and occur at intervals of 15 to 30 minutes. They gradually become more intense, longer in duration, and more frequent. They may last anywhere from 6 to 24 hours, usually longer with the first child. The object of these contractions is to force the baby from the uterus and birth canal (vagina).

Childbirth occurs in three stages. In the first stage the neck (cervix), or opening of the uterus into the vagina, is enlarged by the pressure of the baby in its bag of amniotic fluid, which may be ruptured at this time. In the second stage the baby is forced out of the uterus and through the vagina to the outside (Fig. 13-14). In the third stage the placenta, or afterbirth, is expelled from the mother's body, usually within 10 minutes after the baby is born.

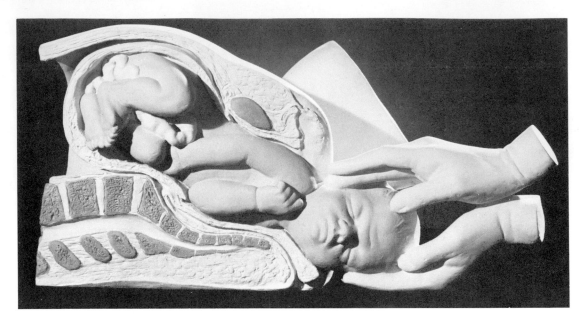

Fig. 13-14. Birth of a baby. (Courtesy American Museum of Natural History.)

Many mammals give birth to more than one off-spring at a time or to a litter, each member of which has come from a separate egg. Most higher mammals, however, have only one offspring at a time, although occasionally they may have plural young. The armadillo *(Dasypus)* is almost unique among mammals in giving birth to four young at one time—all of the same sex, either male or female, and all derived from one zygote.

Human twins may come from one zygote (identical twins) or two zygotes (nonidentical, or fraternal, twins). Triplets, quadruplets, and quintuplets may include a pair of identical twins. The other babies in such multiple births usually come from separate zygotes. Fraternal twins do not resemble each other more than other children born separately in the same family, but identical twins are, of course, strikingly alike and always of the same sex. Embryologically, each member of fraternal twins has its own placenta, chorion, and amnion. Usually (but not always) identical twins share the same chorion and the same placenta, but each has its own amnion. Sometimes identical twins fail to separate completely and form Siamese twins, in which the organs of one may be a mirror image of the organs of the other. The frequency of twin births in comparison to single births is approximately 1 in 86, that of triplets approximately 1 in 86^2, and that of quadruplets approximately 1 in 86^3.

CONTROL OF DEVELOPMENT

Earlier in this chapter we asked what it was that enabled the fertilized egg to unfold into billions of differentiated cells, organized into a functional animal. Although this remains one of the great unanswered riddles of biology, some elements of the mystery have been cleared away.

First of all, we know that a fertilized egg contains a full complement of maternal and paternal genes. The DNA contained within the egg nucleus carries all the information needed to form not just another egg cell, heart cell, or liver cell, but a complete animal. What happens to this genetic information as the egg cell divides again and again? Does each daughter cell receive a full set of genes, or are the original genes, some 90,000 in man, parcelled out among the cleaving cells according to the ultimate fate of these cells? Simple arithmetic tells us that if genes were parcelled out—the first two daughter cells each receiving half the genes, and these being split again in half by the next division—then all the genes would be exhausted long before a billion-cell embryo could form. But there is direct proof as well that many cells

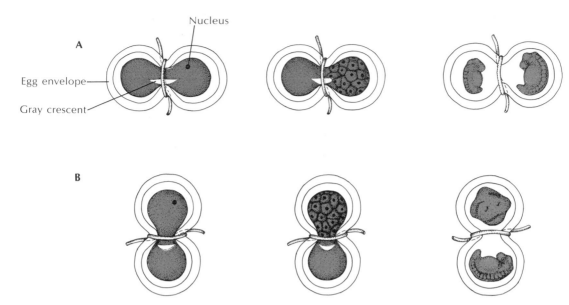

Fig. 13-15. Spemann's delayed nucleation experiments. Two kinds of experiments were performed. **A,** Hair ligature was used to partly divide an uncleaved fertilized newt egg. Both sides contained part of the gray crescent. Nucleated side alone cleaved until a descendent nucleus crossed over the cytoplasmic bridge. Then both sides completed cleavage and formed two complete embryos. **B,** Hair ligature was placed so that the nucleus and gray crescent were completely separated. Side lacking the gray crescent became an unorganized piece of belly tissue; other side developed normally.

contain all the genes necessary to grow a complete organism.

Recently two Oxford scientists, J. B. Gurdon and R. Laskey, were able to grow a normal, reproductive adult frog from an unfertilized egg containing the nucleus of a differentiated intestinal cell of frog tadpoles. The technique they used had been developed by R. Briggs and T. J. King at Indiana University in 1952. Using minute glass tools, they were able to pluck out the nucleus from an egg and replace it with a nucleus from a fully differentiated cell. These experiments are of great significance because they demonstrate that a cell, or actually the nucleus from a cell, can be forced backward from its specialized state and once again make available all of its genetic information.

The basic problem then is **gene expression.** As cells differentiate, they use only a part of the genetic instructions that their nuclei contain. The unneeded genes are in some way switched off but not destroyed. But what switches off genes (or turns them on) at the right moment in development? What determines that a particular cell of, for example, a 100-cell embryo, differentiates into muscle or skin or thyroid gland? Presumably a gene, or set of genes, responsible for the development of a thyroid gland is set in motion by the chemical environment found *only* in the region of the future thyroid gland. But how can such a unique chemical environment be created unless some *previous* genetic action made the thyroid region different from the rest of the body? Even this earlier genic action must have been expressed in a unique chemical environment, or else thyroid glands would grow all over the body.

It is easy to see that this kind of argument quickly takes us back to the fertilized egg itself. If genes are the same in all nuclei of the early embryo, then the only way differences can develop is through some interaction between these nuclei and the surrounding cytoplasm. This is the basis for the great present-day interest in nucleocytoplasmic interactions.

Role of cytoplasm

The importance of the egg cytoplasm in differentiation was demonstrated many years ago by Hans Spemann, a German embryologist. Spemann put ligatures of human hair around newt eggs (amphibian eggs similar to frog eggs) just as they were about to divide, constricting them until they were almost but not quite separated into two halves (Fig. 13-15). The nucleus lay in one half of the partially divided egg; the other side was anucleate, containing only cytoplasm. The egg completed its first cleavage division on the side containing the nucleus; the anucleate side remained undivided. Eventually, when the nucleated side had divided into approximately 16 cells, one of the cleavage nuclei wandered across the narrow cytoplasmic bridge to the anucleate side. Immediately this side began to divide.

With both halves of the embryo containing nuclei, Spemann drew the ligature tight, separating the two halves of the embryo. He then watched their development. Usually two complete embryos resulted. Although the one embryo possessed only $1/16$ the original nuclear material and the other contained $15/16$ they both developed normally. The $1/16$ embryo was initially smaller, but caught up by 140 days. This proves that every nucleus of the 16-cell embryo contains a complete set of genes; all are equivalent.

Sometimes, however, Spemann observed that the nucleated half of the embryo developed only into an abnormal ball of "belly" tissue, although the half that received the delayed nucleus developed normally. This was odd. Why should the more generously endowed $15/16$ embryo fail to develop and the small $1/16$ embryo live? The explanation, Spemann discovered, depended on the position of the **gray crescent,** a crescent-shaped, pigment-free area on the egg surface. In amphibian eggs the gray crescent forms at the moment of fertilization and determines the plane of bilateral symmetry in the future animal. If one half of the constricted embryo lacked a portion of the gray crescent, it would not develop.

Obviously then, there must be cytoplasmic inequalities involved. The gray crescent cytoplasm contains substances that are essential for normal development. Since all the nuclei of the 16-cell embryo are equivalent, each capable of supporting full development, it is clear that the cytoplasmic environment is crucial to nuclear expression. The nuclei are all alike, but the cytoplasm throughout the embryo is not all alike. In some way chemically different regions of the unfertilized egg, created during the early maturation of the unfertilized egg (oogenesis), are segregated out into specific cells during early cleavage. Thus, al-though all nuclei have the same information content, cytoplasmic substances surrounding the nucleus determine what part of the genome is expressed and when.

Differential gene action

Attention is now turned to the nuclear DNA, from which genes are made. It is clear that not all genes are active all of the time. In fact, for any cell at any given time, only a very small part of the genetic information present is being used.

Let us briefly summarize what is presently known of the transmission of genetic information. Genetic information is coded in the sequence of nucleotides in DNA molecules. DNA serves as a template for the synthesis of messenger RNA in the nucleus. Messenger RNA then migrates out through nuclear pores into the cytoplasm, where it attaches to a ribosome. Here the messenger RNA serves as a template for the synthesis of specific proteins. In this way cytoplasmic proteins are formed that may be specific for that cell.

Evidence to date suggests that at the beginning of development most nuclear genes are inactive. Only small amounts of messenger RNA are being produced on the DNA templates. As development proceeds, new cytoplasmic proteins appear, indicating that more genic DNA is producing messenger RNA. Evidently different genes are activated (or "derepressed") in different parts of the young embryo, and this differential gene activity is responsible for embryonic differentiation.

What is the mechanism by which genes are repressed and then derepressed at specific times during development? We presently do not know. Whatever the mechanism is, it seems certain that the kinds of cytoplasm present in different cells determine what genes come into action. Nucleocytoplasmic interactions form the basis of the organized differentiation of tissues that characterizes animal development.

SELECTED REFERENCES

(See also general references for Part two, p. 274.)

Austin, C. R., and R. V. Short. (ed.) 1972-1976. Reproduction in mammals. Cambridge, Cambridge University Press. 6 vol. *Highly readable and well illustrated series dealing with germ cells, fertilization, embryonic and fetal development; reproductive hormones and patterns; the artificial control of reproduction and evolution of reproduction. Paperback.*

Balinsky, B. I. 1975. An introduction to embryology, 4th ed. Philadelphia, W. B. Saunders Co. *One of the best animal embryology texts at the advanced undergraduate level. Stresses mechanisms of development.*

Berrill, N. J. and G. Karp. 1976. Development. New York, McGraw-Hill Book Co. *Balanced developmental biology text, embracing the entire scope of this complex field, yet readable and well ordered.*

Carr, D. E. 1970. The sexes. Garden City, N. Y., Doubleday & Co. *Deals with the evolution of sex from protozoans to man, sex taboos, birth control, and population problems. Written in a popular and frequently amusing style but containing a wealth of information.*

Nelsen, O. E. 1953. Comparative embryology of the vertebrates. New York, The Blakiston Co. *This treatise is a comprehensive study of the comparative morphology of the vertebrates and protochordates. Though dated, it contains a wealth of comparative information difficult to find in other texts.*

Van Tienhoven, A. 1968. Reproductive physiology of vertebrates. Philadelphia, W. B. Saunders Co.

Wickler, W. 1972. The sexual cycle: The social behavior of animals and men. Garden City, N. Y., Doubleday & Co. *A wide-ranging and interesting exploration of sexual behavior in animals, and the origins of human ethics and morals.*

Wood, C. 1969. Human fertility: Threat and promise. World of Science Library, New York, Funk & Wagnalls. *Succinct, beautifully illustrated account of human reproduction, birth control, population problems and human destiny. Highly recommended.*

SELECTED SCIENTIFIC AMERICAN ARTICLES

Allen, R. D. 1959. The moment of fertilization. **201:**124-134 (July). *Studies with sea urchin eggs have revealed the complexity of this critical biologic event.*

Bryant, P. J., S. V. Bryant, and V. French. 1977. Biological regeneration and pattern formation. **237:**66-81 (July). *Studies of regeneration reveal much about basic principles of development.*

Ebert, J. D. 1959. The first heartbeats. **201:**87-96 (March). *The formation and early function of the heart is studied with chick embryos.*

Edwards, R. G., and R. E. Fowler. 1970. Human embryos in the laboratory. **223:**44-54 (Dec). *This article describes recent successes in culturing human and other mammalian eggs and embryos for observation.*

Fischberg, M., and A. W. Blackler. 1961. How cells specialize. **205:**124-140 (Sept.). *Early steps in differentiation of the embryo are programmed into the egg before fertilization.*

Gurdon, J. B. 1968. Transplanted nuclei and cell differentiation. **219:**24-35 (Dec.). *An extension of the work first successfully done by Briggs and King.*

Jacobson, M., and R. K. Hunt. 1973. The origins of nerve-cell specificity. **228:**26-35 (Feb.). *How nerves find direction during growth.*

Singer, M. 1958. The regeneration of body parts. **199:**79-88 (Oct). *Studies on the remarkable ability of amphibians to regrow lost limbs.*

Wessells, N. K., and W. J. Rutter. 1969. Phases in cell differentiation. **220:**36-44 (March). *Cultivation of embryonic pancreas tissues reveals stages of specialization.*

Wessells, N. K. 1971. How living cells change shape. **225:**76-82 (Oct). *How microtubules and microfilaments make possible the cell movements so important in development.*

REFERENCES TO PART TWO

The books listed in the following selection are mainly textbooks covering wide areas of physiology. They vary considerably in depth and in the level of background in biology and chemistry required of the reader for a full understanding, as indicated in the annotations.

Barrington, E. J. W. 1968. The chemical basis of physiological regulation. Glenview, Ill., Scott, Foresman & Co. *A selection of topics in comparative physiology, with emphasis on experimental approach. Clearly written.*

Florey, E. 1966. General and comparative physiology. Philadelphia, W. B. Saunders Co. *A detailed graduate-level text, numerous illustrations, good invertebrate-vertebrate balance.*

Gordon, M. S. (ed.). 1972. Animal function: principles and adaptations, ed. 2. New York, Macmillan, Inc. *Graduate-level vertebrate physiology.*

Guyton, A. C. 1976. Textbook of medical physiology, ed. 5. Philadelphia, W. B. Saunders Co. *A detailed but readable treatment of medical physiology.*

Hill R. W. 1976. Comparative physiology of animals: an environmental approach. New York, Harper & Row, Publishers. *Highly selective treatment, stressing environmental interactions.*

Hoar, W. S. 1975. General and comparative physiology, ed. 2. Englewood Cliffs, N.J., Prentice-Hall, Inc. *The best balanced of college comparative physiology texts.*

Prosser, C. L. (ed.). 1973. Comparative animal physiology, ed. 3. Philadelphia, W. B. Saunders Co. *Advanced treatise.*

Schmidt-Nielsen, K. 1975. Animal physiology: adaptation and environment. New York, Cambridge University Press. *Interestingly written selective treatment of comparative physiology, emphasizing physiologic adaptations to the environment.*

Schottelius, B. A., and D. D. Schottelius. 1973. Textbook of physiology, ed. 17. St. Louis, The C. V. Mosby Co. *Intermediate-level college human physiology text. Clearly written and illustrated.*

Vander, A. J., J. H. Sherman, and D. S. Luciano. 1975. Human physiology: the mechanisms of body function, ed. 2. New York, McGraw-Hill Book Co. *Excellent intermediate level human physiology text.*

Vertebrate structures and functions. Readings from Scientific American with introductions by N. K. Wessels. 1974. San Francisco, W. H. Freeman & Co., Publishers. *Excellent collection of articles on vertebrate structure and physiology with helpful introductions. Highly recommended supplementary reading.*

Wilson, J. A. 1972. Principles of animal physiology. New York, Macmillan, Inc. *Clearly written comparative physiology at the advanced undergraduate level.*

PART THREE THE INVERTEBRATE ANIMALS

Carolus Linnaeus, the great Swedish naturalist who founded our modern system of classification in the mideighteenth century. This statue of a youthful Linnaeus stands before his home in the old university town of Uppsala, Sweden. (Photo by C. P. Hickman, Jr.)

14 CLASSIFICATION AND PHYLOGENY OF ANIMALS

ANIMAL CLASSIFICATION

Linnaeus and the development of classification
Species
Basis for formation of taxa

ANIMAL PHYLOGENY

Kingdoms of life
Divisions of the animal kingdom

ANIMAL CLASSIFICATION

Zoologists have named more than 1.5 million species of animals, and thousands more are added to the list each year. Yet some evolutionists believe that presently named species make up less than 20% of all living animals and less than 1% of all those that have existed in the past.

To communicate with each other about the diversity of life, biologists have found it a practical necessity not only to name living organisms but to classify them. It is not just that the desire to put things into some kind of order is a fundamental activity of man's mind. A system of classification is a storage, retrieval, and communication mechanism for biologic information. The science of **systematics** embraces the studies of speciation, classification, and phylogeny and is concerned with the identification and naming of each kind of organism by a uniformly adopted system that best expresses the degree of similarity between organisms.

Linnaeus and the development of classification

Although the history of man's efforts to distinguish and name plants and animals must have been rooted in the beginnings of his language, the great Greek philosopher Aristotle was the first to attempt seriously the classification of organisms on the basis of structural similarities. Following the Dark Ages in Europe, the English naturalist John Ray (1627-1705) introduced a more comprehensive classification system and a modern concept of species. The flowering of systematics in the eighteenth century culminated in the work of Carolus Linnaeus (1707-1778), who gave us our modern scheme of classification.

Linnaeus was a Swedish botanist at the University of Uppsala. He had a great talent for collecting and classifying objects, especially flowers. Linnaeus worked out a fairly extensive system of classification for both plants and animals. His scheme of classification, published in his great work *Systema Naturae,* emphasized morphologic characters as a basis for arranging specimens in collections. Actually his classification was largely arbitrary and artificial, and he believed strongly in the fixity of species. He divided the animal kingdom down to species, and according to his scheme each species was given a distinctive name. He recognized four classes of vertebrates and two classes of invertebrates. These classes were divided into orders, the orders into genera, and the genera into

species. Since his knowledge of animals was limited, his lower groups, such as the genera, were very broad and included animals that are now placed in several orders or families. As a result, much of his classification has been drastically altered, yet the basic principle of his scheme is followed at the present time.

Linnaeus's scheme of arranging organisms into an ascending series of groups of ever-increasing inclusiveness is the **hierarchic system** of classification. Species were grouped into genera, genera into orders, and orders into classes. This taxonomic hierarchy has been considerably expanded since Linnaeus's time. The major categories, or **taxa** (sing., **taxon**), now used are, in descending series, kingdom, phylum, class, order, family, genus, and species. This hierarchy of seven ranks can be subdivided into finer categories, such as superclass, subclass, infraclass, superorder, suborder, and so on. In all, more than 30 taxa are recognized. For very large and complex groups, such as the fishes and insects, these additional ranks are required to express recognized degrees of evolutionary divergence. Unfortunately they also contribute complexity to the system.

Linnaeus' system for naming species is known as **binomial nomenclature.** Each species has a latinized name composed of two words (hence binomial). The first word is the genus, written with a capital initial letter; the second word is the specific name that is peculiar to the species and is written with a small initial letter (Table 14-1). The genus name is always a noun, and the specific name is usually an adjective that must agree in gender with the genus. For instance, the scientific name of the common robin is *Turdus migratorius* (L. *turdus,* thrush; *migratorius,* of the migratory habit).

There are times when a species is divided into subspecies, in which case a **trinomial nomenclature** is employed (see katydid example, Table 14-1). Thus to distinguish the southern form of the robin from the eastern robin, the scientific term *Turdus migratorius achrustera* (duller color) is employed for the southern type. Taxa lower than subspecies are sometimes employed when four words are used in the scientific name, the last one usually standing for **variety.** In this latter case the nomenclature is **quadrinomial.** The trinomial and quadrinomial nomenclatures are really additions to the Linnaean system, which is basically binomial.

It is important to recognize that *only* the species is

277

Table 14-1. Examples of classification of animals

	Man	Gorilla	Southern leopard frog	Katydid
Phylum	Chordata	Chordata	Chordata	Arthropoda
Subphylum	Vertebrata	Vertebrata	Vertebrata	
Class	Mammalia	Mammalia	Amphibia	Insecta
Subclass	Eutheria	Eutheria		
Order	Primates	Primates	Salientia	Orthoptera
Suborder	Anthropoidea	Anthropoidea		
Family	Hominidae	Simiidae	Ranidae	Tettigoniidae
Subfamily			Raninae	
Genus	*Homo*	*Gorilla*	*Rana*	*Scudderia*
Species	*Homo sapiens*	*Gorilla gorilla*	*Rana pipiens*	*Scudderia furcata*
Subspecies			*Rana pipiens sphenocephala*	*Scudderia furcata furcata*

binomial. All ranks above the species are uninomial nouns, written with a capital initial letter. We must also note that the second word of a species is an epithet that has no meaning by itself. The scientific name of the white-breasted nuthatch is *Sitta carolinensis*. The "carolinensis" alone has no meaning since it may be and is used in combination with other genera to mean "of Carolina," as for instance *Parus carolinensis* (Carolina chickadee) and *Dumetella carolinensis* (catbird). The genus name, on the other hand, may stand alone to designate a taxon that may include several species.

Species

Despite the central importance of the species concept in biology, biologists do not agree on a single rigid definition that applies to all cases. Before Darwin's time, the species was considered a primeval pattern, or archetype, divinely created. With gradual acceptance of the concept of organic evolution scientists realized that species were not fixed, immutable units but have evolved one from another. Sometimes the gaps between species are so subtle that they can be distinguished only by the most careful examination. In other instances, a species is so distinctive in every way that it is clearly unique and only remotely related to other species. Consequently the criteria of taxonomy have undergone gradual changes.

At first each species was supposed to have been represented by a **type** that was used as a fixed standard. The **typologic specimen** was duly labeled and deposited in some prestigious center such as a museum. Anyone classifying a particular group would always take the pains to compare his specimens with the available typologic specimens. Since variations from the type specimen nearly always occurred, these differences were supposed to be attributable to imperfections during embryonic development and were considered of minor significance.

The typologic method of classifying persisted for a long period (and still does to some extent); during this time, though, the idea that species represent lineages in evolutionary descent was becoming more firmly established. Gradually taxonomists began to think of a species in terms of a **group of organisms of a reproducing population or populations that are reproductively isolated from other populations.** This concept of **genetic incompatibility** between different species therefore replaced the earlier idea of **character discontinuity** for species separation. Taxonomists began to think of species as populations in which every individual is unique and in which every individual may change to a greater or lesser extent when placed in a different environment.

This modern viewpoint is, of course, the antithesis of that of the typologist to whom any variations from the type specimen are illusions caused by small mistakes during embryonic development. In population studies, the type is considered an abstract average of *real* variations that occur within the interbreeding population. Thus the species must be regarded as an **interbreeding population** made up of individuals of common descent and sharing intergrading characteristics. A species is set apart from all other species by a distinct evolutionary role.

Basis for formation of taxa

Classification emphasizes the natural relationships of animals. Descent from a common ancestor explains similarity in character; the more recent the descent, the more closely the animals are grouped in taxonomic units. The genera of a particular family show less diversity than do families of an order. This is because the common ancestor of families within an order is more remote than the common ancestor of different genera within a family. The same applies to higher categories. The common ancestor of the various vertebrate classes, for example, must be much older than the common ancestor of orders of mammals within the class Mammalia.

It is apparent that this criterion of evolutionary ancestry is not a very definite one for use in setting up taxa. There is no way to define a class that does not apply equally well to an order or a family. The genera of certain ancient groups, such as the molluscs, may actually be much older than orders and classes of other groups. The taxonomist must arrange his classifications so that all members of a taxon resemble one another more closely than they do the members of any other taxon of the same rank. Obviously the assignment of taxonomic rank depends upon the opinion of the taxonomist making the study. As more is learned about animals and their relationships, changes in classification are required. This brings instability to biologic nomenclature and reduces its efficiency as a reference system.

The **law of priority** also brings about frequent changes. The first name proposed for a taxonomic unit that is published and meets other proper specifications has priority over all subsequent names proposed. The rejected duplicate names are called **synonyms.** It is disturbing sometimes to find that a species that has been well established for years must undergo a change in terminology when some industrious systematist discovers that on the basis of priority or for some other reason the species is, according to this "law," misnamed.

Despite such difficulties the hierarchic system of classification is the only accepted system we have. To reduce confusion in the field of taxonomy and to lay down a uniform code of rules for the classification of animals, the International Commission on Zoological Nomenclature was established in 1898. This Commission meets from time to time to formulate rules and to make decisions in connection with taxonomic work.

ANIMAL PHYLOGENY

Phylogeny ("tribe origin") is the science of genealogic relationships among lineages of life forms. The phylogenist's attempts to reconstruct lines of descent resulting in living organisms are not unlike our efforts to ferret out the ancestral history (genealogy) of our own family. Neither species nor members of our family arose spontaneously, but rather they represent branching of ancestral forms. Although the representation of ancestral relationships of animals in the form of a family tree seems obvious to us today, it was not apparent to biologists before classification became founded on evolutionary theory. Even Darwin never attempted a pictorial diagram of animal relationships. Yet, despite all the shortcomings that any phylogenetic tree must possess, especially the danger of depicting and accepting highly tentative relationships as dogmatic certainty, such a scheme is a valuable tool. A family tree serves to tie the taxa together in an evolutionary blueprint. Constructing a family tree is a creative activity based on judgment and experience and as such is always subject to modification as new information becomes available.

In the following chapters, we begin each with a summary of that group's origins and relationships within the group. The student is encouraged to make frequent reference to the geologic time scale on the inside back cover of this book and to the family tree of multicellular animals on the inside front cover. We have tried to base conclusions about group histories on recent paleontologic opinion. Although we recognize that family trees can give false impressions, we have nonetheless used them in the absence of a better alternative. They may be viewed as educated speculations and as such with a certain measure of skepticism; at the same time they are not science fiction! They are derived from close morphologic reasoning and a thorough understanding of general biologic principles by scientists who have devoted their lives to this form of detective work. Evolution, with its idea of life transforming itself through the ages, is supported by a vast wealth of fossil and living evidence. It is after all the framework of biology.

Kingdoms of life

Since Aristotle's time, it has been traditional to assign every living organism to one of two kingdoms—plant and animal. However, the two-kingdom system has outlived its usefulness. Although it is easy to place rooted, photosynthetic organisms such as trees, flowers, mosses, and ferns among the plants and food-ingesting, motile forms such as worms, fishes, and mammals among the animals, unicellular organisms present difficulties. Some forms are claimed both for the plant kingdom by botanists and for the animal kingdom by zoologists. An example is *Euglena* (p. 295) and its phytoflagellate kin, which are motile, like animals, but have chlorophyll and photosynthesize, like plants. Other groups such as the bacteria were rather arbitrarily assigned to the plant kingdom.

It was inevitable that biologists would try to resolve these problems by separating problem groups into new kingdoms. This was first done in 1866 by Ernst Haeckel who proposed the new kingdom Protista to include all single-celled organisms. At first the bacteria and blue-green algae (forms that lack nuclei) were included with nucleated unicellular organisms. Finally the important differences between the anucleate bacteria and blue-green algae (**prokaryotes**) and all other organisms that have nucleated cells (**eukaryotes**) were recognized. The prokaryote-eukaryote distinction is actually much more profound than the plant-animal distinction of the traditional system. The many differences between prokaryotes and eukaryotes are summarized in Chapter 2 (pp. 42 and 43).

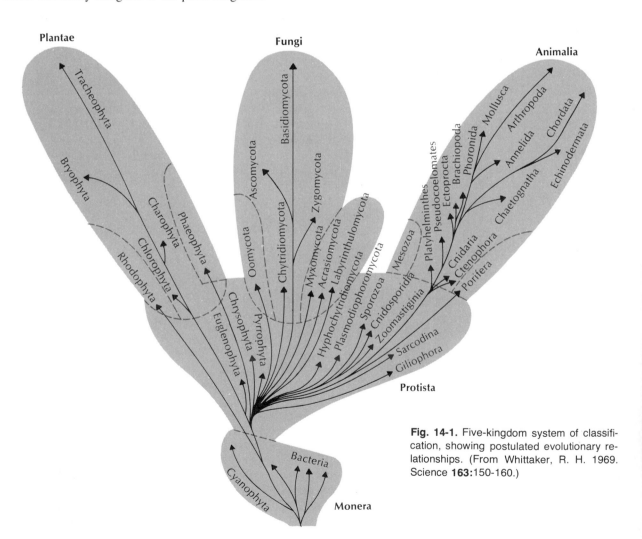

Fig. 14-1. Five-kingdom system of classification, showing postulated evolutionary relationships. (From Whittaker, R. H. 1969. Science **163:**150-160.)

In 1969 R. H. Whittaker proposed a five-kingdom system that incorporated the basic prokaryote-eukaryote distinction (Fig. 14-1). The kingdom Monera contains the prokaryotes; the eukaryotes are divided among the remaining four kingdoms. The kingdom Protista contains the unicellular eukaryotic organisms (protozoans and eukaryotic algae). The multicellular organisms are split into three kingdoms on the basis of mode of nutrition and other fundamental differences in organization. The kingdom Plantae includes multicellular photosynthesizing organisms, higher plants, and multicellular algae. Kingdom Fungi contains the molds, yeasts, and fungi that obtain their food by absorption. The invertebrates (except the Protozoa) and the vertebrates comprise the kingdom Animalia. Most of these forms ingest their food and digest it internally although some parasitic forms are absorptive. The supposed evolutionary relationships of the five kingdoms are shown in Fig. 14-1. The Protista are believed to have given rise to all three multicellular kingdoms, which therefore have evolved independently.

Exclusion of the unicellular protozoans from the animal kingdom, in which they have been traditionally included, presents a didactic problem for the zoologist. In the five-kingdom system, phylum Protozoa has been divided into seven groups that are each elevated to phylum status. This change emphasizes the fluid nature of the hierarchic system of classification pointed out earlier in this chapter. Since many protozoologists had already advocated splitting the protozoans into even more phyla on the basis of fundamental differences between groups, taxonomic inflation of the protozoans, as treated in the five-kingdom system, may be legitimate and defensible. On the other hand, the protozoans share many animal-like characteristics. Most ingest their food; many have specialized organelles and advanced locomotory systems, portending tissue differentiation in multicellular forms; many reproduce sexually; and some flagellate forms are colonial with division of labor among cell types, again suggestive of a metazoan pattern.

Whether or not the protozoans should be classified as a single animal phylum, as their name suggests, or split into several phyla is a matter of opinion. At the present time it is reasonable and workable to retain the protozoans as an animal phylum. This relationship is depicted in the family tree of animals on the inside front cover of this book.

Divisions of the animal kingdom

Traditionally the animal kingdom has been divided into two subkingdoms: (1) the Protozoa, made up of unicellular (acellular) animals, and (2) the Metazoa, containing the multicellular animals (all other phyla). As seen in the following outlines, the metazoan phyla in turn are divided into three more or less natural groupings, the Mesozoa, Parazoa, and Eumetazoa. The first two are small, primitive groups, each comprising a single phylum, phylum Mesozoa (the mesozoans) and phylum Porifera (the sponges). Members of these phyla are little more than cell aggregates and lack true tissues and organs. The Eumetazoa include all of the remaining phyla (Fig. 14-2).

Subkingdom Protozoa—unicellular (acellular) animals (phylum Protozoa)
Subkingdom Metazoa—multicellular animals (all other phyla)
  Branch A: Mesozoa—phylum Mesozoa
  Branch B: Parazoa—phylum Porifera
  Branch C: Eumetazoa—all other phyla
    Grade I: Radiata—phyla Cnidaria (Coelenterata), Ctenophora
    Grade II: Bilateria—all other phyla
      Acoelomates
        Phyla Platyhelminthes, Nemertina (Rhynchocoela), Gnathostomulida
      Pseudocoelomates
        Phyla Rotifera, Gastrotricha, Kinorhyncha, Nematoda, Nematomorpha, Acanthocephala, Entoprocta
      Eucoelomates
        Lophophorates
          Phyla Phoronida, Ectoprocta, Brachiopoda
        Protostomia
          Phyla Mollusca, Annelida, Arthropoda, Priapulida, Echiurida, Sipunculida, Tardigrada, Pentastomida, Onychophora
        Deuterostomia
          Phyla Echinodermata, Chaetognatha, Hemichordata, Pogonophora, Chordata

To understand why the Eumetazoa have been subdivided in this manner we need to discuss some of them in connection with animal symmetry, body cavity formation, body form, and embryologic origins of certain body characteristics.

Symmetry and body plan

Animal symmetry. Symmetry refers to balanced proportions, or the correspondence in size and shape of parts on opposite sides of a median plane (Fig. 14-3). **Spheric symmetry** means that any plane passing through the center divides the body into equivalent, or mirrored, halves. This type of symmetry is found

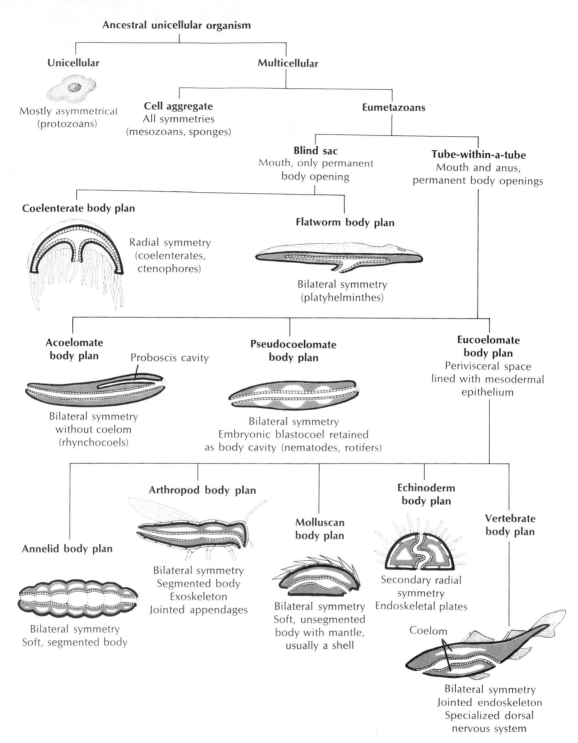

Fig. 14-2. Architectural patterns of animals. The basic body plans are unicellular (acellular), cell aggregate, blind-sac, and tube-within-a-tube. These, especially the last, have been variously modified during evolutionary descent to fit animals to a great variety of life styles. Ectoderm is shown in solid black, endoderm as open blocks, mesoderm in red.

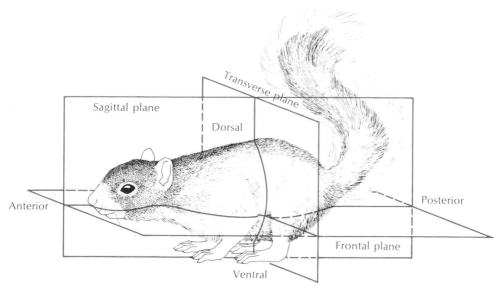

Fig. 14-3. The planes of symmetry as illustrated by a bilateral animal.

chiefly among protozoans and is rare in other animals. Spheric forms are best suited for floating or rolling.

Radial symmetry applies to forms that can be divided into similar halves by any plane passing through the longitudinal axis. These are the tubular, vase, or bowl shapes found in some sponges and in the hydras, jellyfishes, sea urchins, etc., in which one end of the longitudinal axis is usually the mouth. A variant form is **biradial symmetry** in which, because of some part that is single or paired rather than radial, only one or two planes passing through the longitudinal axis produce mirrored halves. Sea walnuts, which are more or less globular in form but have a pair of tentacles, are an example. Radial animals are usually sessile or slow moving. The two phyla that are primarily radial, Cnidaria and Ctenophora, are called the **Radiata.**

In **bilateral symmetry** only a sagittal plane can divide the animal into two mirrored portions—right and left halves (Fig. 14-3). Bilateral animals make up all of the higher phyla and are collectively called the **Bilateria.** They are better fitted for forward movement than radially symmetric animals.

Body cavities. The bilateral animals can be grouped according to their body-cavity type or lack of body cavity (Fig. 14-2). In higher animals the main body cavity is the **coelom,** a fluid-filled space that surrounds the gut. The true coelom develops from the mesoderm and is lined with mesodermal epithelium called the

peritoneum. The coelom is of great significance in animal evolution. It provides increased body flexibility and space for visceral organs and permits greater size and complexity by exposing more cells to surface exchange. The fluid-filled space also serves as a hydrostatic skeleton in some forms, aiding them in movement, burrowing, etc.

Acoelomate Bilateria. The more primitive bilateral animals do not have a true coelom. In fact, the flatworms and a few others have no body cavity surrounding the gut. The region between the ectodermal epidermis and the endodermal digestive tract is completely filled with mesoderm in the form of a primitive type of connective tissue called parenchyma.

Pseudocoelomate Bilateria. Nematodes and related phyla have a cavity surrounding the gut, but it is not lined with mesodermal peritoneum. It is derived from the blastocoel of the embryo; it represents a persistent blastocoel. This type of body cavity is called a **pseudocoel.**

Eucoelomate Bilateria. Animals possessing a true coelom lined with mesodermal peritoneum include the remainder of the bilateral animals. The coelom, or true body cavity that contains the viscera, may be formed by one of two methods—**schizocoelous** or **enterocoelous** (Fig. 14-4). (The two terms are descriptive, for *schizo* comes from the Greek *schizein,* to split; *entero*

283

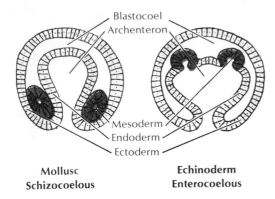

Mollusc
Schizocoelous

Echinoderm
Enterocoelous

Fig. 14-4. Two types of mesoderm and coelom formation—schizocoelous, in which mesoderm originates from wall of archenteron near lips of blastopore, and enterocoelous, in which mesoderm and coelom develop from endodermal pouches.

is a Greek form from *enteron,* gut; coelous comes from the Greek *koilos,* hollow or cavity.) In schizocoelous formation the coelom arises, as the word implies, from the splitting of mesodermal bands that originate from the blastopore region and grow between the ectoderm and endoderm; in enterocoelous formation the coelom comes from pouches of the archenteron, or primitive gut.

Protostomes and deuterostomes

Since coelom formation occurs very early in embryonic development, the appearance of two quite different methods of formation among animals is believed to signal a fundamental division in metazoan evolution. The **deuterostome** division of the Metazoa

(echinoderms, chaetognaths, pogonophorans, hemichordates, and chordates) follows the enterocoelous method of coelom formation. The **protostome** division (almost all other Metazoa) follows the schizocoelous method.

The terms "protostome" (first mouth) and "deuterostome" (second mouth) refer to the embryonic origin of the future mouth of the organism. In the Protostomia the first embryonic opening, the blastopore, becomes subdivided into two openings, one becoming the mouth and the other the anus. In the Deuterostomia the blastopore contributes to the formation of the anus only, whereas the mouth is derived from a second, independent perforation of the ventral body wall. Other characteristics that distinguish these two phylogenetic divisions of the bilateral animals are included in the following summary:

Protostomia
Mouth and anus formation by subdivision of blastopore
Schizocoelous formation of body cavity (coelom)
Mostly spiral cleavage
Determinate or mosaic pattern of egg cleavage
Ciliated larva (when present) a trochophore or trochosphere type
Acoelomates, pseudocoelomates, and, among the eucoelomates, phyla Annelida, Mollusca, Arthropoda, and a number of minor phyla

Deuterostomia
Anus formation from or near the blastopore
Enterocoelous formation of coelom
Mostly radial cleavage
Indeterminate pattern of egg cleavage
Ciliated larva (when present) a pluteus type
Phyla Echinodermata, Hemichordata, Pogonophora, Chaetognatha, and Chordata

SELECTED REFERENCES

Clifford, H. T., and W. Stephenson. 1975. An introduction to numerical classification. New York, Academic Press, Inc. *Methods of numerical approaches to classification are clearly described and compared to classic taxonomy. Less detailed than the Sneath and Sokal book.*

Crowson, R. A. 1970. Classification and biology. New York, Atherton Press. *A thorough treatment of the methods and problems of zoologic, botanic and paleontologic classification.*

Jeffrey, C. 1973. Biological nomenclature. London, Edward Arnold. *A concise, practical guide to the principles and practice of biologic nomenclature and a useful interpretation of the Codes of Nomenclature.*

Leone, C. A. (editor). 1964. Taxonomic biochemistry and serology. New York, The Ronald Press Co. *Collection of 47 papers treating the principles of systematics, molecular taxonomy, and taxonomic biochemistry and serology of animals, plants, and microorganisms.*

Margulis, L. 1970. Origin of eukaryotic cells. New Haven, Yale University Press. *Detailed consideration of theories of origins and evolution of microbial, plant, and animal cells.*

Margulis, L. 1974. Five-kingdom classification and the origin and evolution of cells. Evolutionary Biology **7**:45-78. *Describes a modified version of the Whittaker five-kingdom system.*

Ross, H. H. 1974. Biological systematics. Reading, Mass., Addison-Wesley Publishing Co., Inc. *Theory and practice of systematics is presented and exemplified from animals, plants, and microorganisms. Very comprehensive and useful.*

Sneath, P. H. A., and R. R. Sokal. 1973. Numerical taxonomy. San Francisco, W. H. Freeman & Co. Publishers. *Extensive treatment of taxonomic theory, methods of classification, and the implications and application of numeric taxonomy.*

Whittaker, R. H. 1969. New concepts of the kingdoms of organisms. Science **163**:150-160. *The author's five-kingdom system.*

A lovely little radiolarian, approximately 3 mm in diameter, that floats near the surface of tropical Atlantic seas. Like other radiolarians, *Aulonia hexagonia* has an exquisitely meshed skeleton of siliceous material through which its slender pseudopodia extend. (Photo of glass model, courtesy American Museum of Natural History.)

15 PROTOZOANS AND SPONGES

The protozoans and sponges are the most primitive forms of animal life that we know today. In the protozoans all activities necessary to the life and reproduction of an animal are efficiently carried out within the limits of a single plasma membrane—a large task, we might think, for so small a creature! Specialization (division of labor) in a protozoan involves the organization of its basic protoplasm into functional units, or organelles—a **protoplasmic level** of organization.

Sponges (phylum Porifera), on the other hand, are many-celled animals, or metazoans. However, they, along with a small phylum of minute parasitic organisms called the Mesozoa, have the simplest type of metazoan organization—a **cellular level** of organization. Since the sponge lacks true tissues or organs, the division of labor is confined to a few types of cells that have become specialized for certain functions.

Some protozoans form **colonies** consisting of a few members or thousands of members. Some members of a colony may be reproductive in function and other members purely vegetative (somatic). Some large, highly integrated protozoan colonies closely resemble multicellular organisms in both appearance and behavior. What then is the distinction between a protozoan colony and a multicellular (metazoan) animal? An arbitrary line of distinction has been selected by biologists, namely that, if no more than one type of nonreproductive cell is present, the organism is considered a protozoan colony. If there is more than one kind of nonreproductive cell, it is considered a metazoan. By this definition then, *Volvox,* which is made up of reproductive cells and one kind of nonreproductive member (vegetative), is a protozoan colony. A sponge, which has several types of specialized cells, is a metazoan organism.

PROTOZOANS—PHYLUM PROTOZOA

More than 250 years ago Anthony van Leeuwenhoek, using lenses he had ground himself, discovered tiny "animalcules" in a drop of water. He described them with great enthusiasm and accuracy. Ever since zoologists, equally enthused, have been studying the form, function, and behavior of these tiny protozoans.

As pointed out in the preceding chapter, the protozoans are often included in a larger group of unicellular organisms, kingdom Protista, which includes only those organisms with a diploid nucleus and a nuclear membrane—the eukaryotic cells. This would exclude the prokaryotic cells such as the bacteria and blue-green algae of the kingdom Monera.

Is a protozoan a single cell (unicellular), or is it "acellular"? An ameba, covered by a plasmalemma and containing a single nucleus and the typical cellular components, conforms to our concept of a cell. But it is also a complete animal that carries out all the complex activities that in multicellular animals are divided among numerous cells. Also, not all protozoans have a single nucleus—some contain a large and a small nucleus, some two similar nuclei, and some at certain stages may have dozens of nuclei. Here the resemblance to a typical cell ends. Consequently many zoologists prefer to think of protozoans as acellular rather than unicellular animals.

Whether we call protozoans unicellular or acellular, it is a mistake to consider them "simple" organisms. Although confined within a single membrane, they perform all the life processes of higher animals and are physiologically very complex. In lieu of cells and organs, the cytoplasm is highly organized into specialized components, called **organelles,** each fitted for a specific function. Organelles may serve as skeletons, sensory systems, contractile systems, or means of locomotion, or they may be mechanisms for conduction, feeding and digestion, fluid regulation, defense, etc.

Most of the 30,000 to 50,000 species of protozoans are small or microscopic, usually from 3 to 300 μm long. Some amebae, however, may be as much as 4 to 5 mm in diameter, and some of the foraminiferans have shells 100 to 125 mm in diameter.

Evolution and adaptive radiation

With the exception of the shell- and test-bearing forms, the protozoans have left no fossil record. However, radiolarian shells have been found in Precambrian rocks, and the naked protozoans are presumed to be much older. The flagellates are usually considered to be the oldest of the protozoans. That some flagellates bear chlorophyll and resemble plant algae seems to indicate that plants and animals have had a common origin. Possibly different kinds of flagellate stock have given rise to the different orders of ameboid and ciliated protozoans.

In spite of the small size and apparent simplicity of the protozoans, they have been successful in adapting to many types of habitat and in occupying most of the ecologic niches of the ecosystem. Most of their adaptive diversity has had to do with the ectoplasm, for it is here that the various organelles of locomotion and feeding arise. Flagella, cilia, infraciliature, trichocysts, myonemes, photoreceptors, and the various body coverings are all modifications of the ectoplasm. Other diversities have evolved in the nuclear apparatus. In the various adaptive specializations that have occurred within their cell membranes, the protozoans and their ancestors have furnished the roots for all functional units of the metazoan form.

Ecologic relationships

Protozoans seem to be found wherever life exists. They need moist habitats and are found in seawater, fresh water, soil and decaying organic matter and plants and as parasites in every kind of animal, including other protozoans. They may be free living or symbiotic, solitary or colonial, sessile or free swimming.

All forms of symbiosis occur among protozoans. A green alga harbored by *Paramecium bursaria* photosynthesizes carbohydrates for its host and receives shelter in return (mutualism). Many species of protozoans live as commensals and many others as parasites either inside or on the surface of other animals. One protozoan class, Sporozoa, is entirely parasitic.

Protozoans make up a large part of the plankton of both fresh and marine waters and, along with other planktonic forms, are an important food source for many aquatic animals.

Some protozoans contribute to the contamination of water, affecting its taste and odor, and some disease-producing protozoans are transmitted in water. There are also certain dinoflagellates (*Gonyaulax* and *Gymnodinium*) that, when occurring in unusually large numbers, cause the discoloration of seawater known as "red tides" and produce toxins that cause widespread destruction of fish and other sea life. *Gonyaulax* releases saxitoxin, a potent neurotoxin identified as a purine derivative, which blocks nerve transmission by preventing necessary sodium ions from entering nerves. The toxin apparently accumulates in the gills and digestive tracts of shellfish and becomes poisonous to humans and other mammals that eat the shellfish.

Characteristics

1. **Acellular** (unicellular); some colonial
2. Locomotion by **pseudopodia, flagella, cilia,** and direct cell movements
3. Mostly microscopic; some can be seen by the unaided eye
4. **Specialized organelles,** but no organs or tissues; single or multiple nuclei
5. Shape, variable or constant
6. Mostly free living, some sessile; all symbiotic relationships are represented
7. Mostly naked, but some with protective exoskeletons
8. **All types of nutrition:** autotrophic (manufacturing own nutrients by photosynthesis), heterotrophic (depending on other plants or animals for food), saprozoic (absorbing nutrients dissolved in surrounding medium; saprobic; saprophagous)
9. **Asexual** reproduction by fission or budding, or **sexual** reproduction by conjugation or gametes

Classification

Traditionally four main groups of protozoans—flagellate, ameboid, spore-forming, and ciliate protozoans—have been represented as taxonomic classes. However, in recent years the tendency has been to raise the rank of these groups to superclass, subphylum, or even phylum. The following is one version that has been widely accepted.

Subphylum Sarcomastigophora (sar′ko-mas-ti-gof′o-ra) (Gr. *sarkos,* flesh, + *mastix,* whip, + *phora,* pl. of bearing). Flagella, pseudopodia, or both types of locomotory organelles; usually with one type of nucleus; typically no spore formation; sexuality, when present, involves syngamy.

Superclass Mastigophora (mas-ti-gof′o-ra) (Gr. *mastix,* whip, + *phora,* bearing). One or more flagella typically present in adult stages; autotrophic or heterotrophic or both. Reproduction usually asexual by fission. *Chilomonas, Euglena, Volvox, Ceratium, Trichomonas, Trichonympha, Trypanosoma, Leishmania.*

Superclass Opalinata (o′-pa-lin-a′ta) (NF. *opaline,* like opal in appearance, + *-ata,* group suffix). Body covered with longitudinal oblique rows of cilia-like organelles; parasitic; cytosome (cell mouth) lacking; two to many nuclei of one type. *Opalina, Protoopalina.*

Superclass Sarcodina (sar-ko-di′na) (Gr. *sarkos,* flesh, + *-ina,* belonging to). Pseudopodia typically present; flagella present in developmental stages of some; cortical zone of cytoplasm relatively undifferentiated as compared with other major taxa; body naked or with external or internal skeleton; free-living or parasitic. *Amoeba, Entamoeba, Arcella, Actinosphaerium, Actinophrys, Thalassicola.*

Subphylum Apicomplexa (a′pi-com-plex′a) (L. *apex,* tip or summit, + *complex,* twisted around, + *a,* suffix). Characteristic set of organelles (apical complex) associated with anterior end present in some developmental stages; cilia and flagella absent except for flagellated microgametes in some groups; cysts often present; all species parasitic.

Class Sporozoa (spor-o-zo′a) (Gr. *sporos,* seed, + *zoon,* animal). Spores typically present, without polar filaments, and with one to many sporozoites. *Monocystis, Gregarina, Eimeria, Plasmodium, Toxoplasma.*

Class Piroplasmea (pir-o-plaz′me-a) (L. *pirum,* pear, + Gr.

plasma, a thing molded). Spores absent; no flagella or cilia at any stage; two-host life cycle, known hosts are vertebrates and ticks. *Babesia.*

Subphylum Myxospora (mix-os'por-a) (Gr. *myxa,* slime, mucus, + *sporos,* seed). Spores of multicellular origin, and enclosed in two or three valves; parasites of invertebrates and lower vertebrates. *Myxosoma, Henneguya.*

Subphylum Microspora (mi-cros'por-a) (Gr. *micro,* small, + *sporos,* seed). Spores of unicellular origin; single valve; parasites of invertebrates, especially arthropods, and lower vertebrates. *Nosema, Glugea, Plistophora.*

Subphylum Ciliophora (sil-i-off'or-a) (L. *cilium,* eyelash, + Gr. *phora,* bearing). Cilia or ciliary organelles in at least one stage of life cycle; two types of nucleus; asexual reproduction by binary fission across rows of cilia; sexuality involving conjugation, autogamy, and cytogamy.

Class Ciliata (sil-i-aht'a) (L. *cilium,* eyelash, + *ata,* group suffix). Characteristics of subphylum. *Paramecium, Tetrahymena, Stentor, Blepharisma, Vorticella, Podophrya.*

Form and function
Body coverings

There is infinite variety among the protozoans' outer coverings. Most of them, like the amebae, are "naked," that is, covered only by the thin **cell membrane,** or **plasmalemma.** But many others also have protective coverings, or **exoskeletons,** which range from simple gelatinous envelopes to armor of cellulose or chitin. The armor may be further strengthened by the addition of sand grains, minerals, or diatom shells taken from the environment. Other protozoans secrete elaborate shells, or tests, which may be siliceous (the radiolarians) or calcareous (the foraminiferans) and are often of surprising delicacy and beauty.

Locomotion

Locomotion is achieved either by means of **pseudopodia,** which are temporary extensions of the cell, or by the activity of special organelles called **cilia** and **flagella,** which are also extensions of the cell surface but are more or less permanently fixed in position.

Pseudopodia. Pseudopodia are temporary extensions of a cell that form and retract at different areas of the surface so that the shape of the cell is continuously changing. Pseudopodia may form at any point and extend in any direction, but to be effective in locomotion the cell membrane must be in contact with a substratum.

Beneath the thin outer membrane (plasmalemma) of an ameba is a nongranular layer, the **ectoplasm,** which encloses the granular **endoplasm.** The ectoplasm is firm and gelatinous in nature (sometimes called plasmagel). The endoplasm is more fluid (sometimes called plasmasol).

When an ameba moves, a stream of endoplasm flows in the direction of movement, forming a **pseudopodium.** At the advancing end of the pseudopodium the endoplasmic stream everts, turning inside out, like the cuff of a sleeve. Around the periphery the endoplasm converts to ectoplasm, thus forming a lengthening tube, or sleeve, around the pseudopodium (Fig. 15-1). At some point the tube becomes anchored

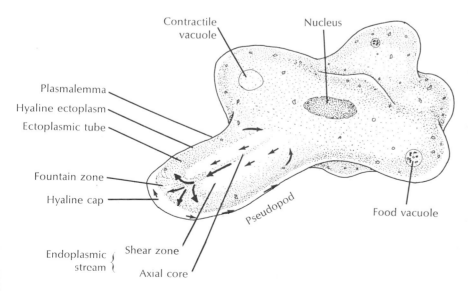

Fig. 15-1. Ameba in active locomotion. Arrows indicate direction of streaming protoplasm. First sign of new pseudopodium is thickening of ectoplasm to form clear hyaline cap, into which the fluid endoplasm flows. As endoplasm reaches forward tip, it fountains out and is converted into ectoplasm, forming a stiff outer tube that lengthens as the forward flow continues. Posteriorly the ectoplasm is converted into fluid endoplasm, replenishing the flow. Substratum is necessary for ameboid movement. See text for further explanation.

Contractile vacuole
Nucleus
Plasmalemma
Hyaline ectoplasm
Ectoplasmic tube
Fountain zone
Hyaline cap
Pseudopod
Food vacuole
Endoplasmic stream
Shear zone
Axial core

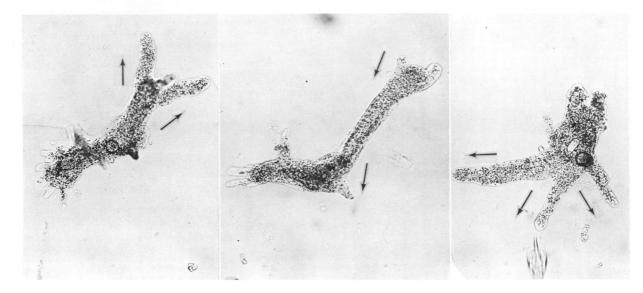

Fig. 15-2. Ameba "changes its mind." This series of a single ameba shows the change of direction in protoplasmic flow as tentative pseudopodia are advanced and retracted. Note at far right the pseudopodium advancing toward a retreating rotifer. (Photos by F. M. Hickman.)

through its plasmalemma to the substratum, and the animal is drawn forward. At the temporary "posterior" end of the pseudopod the ectoplasmic tube is converted to streaming protoplasm, thus replenishing the forward flow. In essence then, the ameba creates a tube, anchors it, and flows through it as it moves forward. At any time the direction of the pseudopodium can be reversed, its length shortened, and new pseudopodia started (Fig. 15-2).

But what is the mechanism by which this process occurs? Several theories have been advanced, all of which depend upon the contraction of cellular proteins.

One theory maintains that this movement results from high internal pressure caused by the contraction, or folding, of chains of protein molecules at the posterior end of the pseudopodium and the consequent squeezing out of fluid, which pushes the endoplasm forward. This is sometimes called the tail-contraction theory.

Another hypothesis, which is probably more widely accepted today, suggests that the protein molecules are in an extended condition in the endoplasm and that they contract, or fold, at the advancing tip of the pseudopodium to form the stiff ectoplasmic tube. At the rear of the pseudopodium as the ectoplasm is converted back to flowing endoplasm, the protein

chains unfold and become extended. This front-contraction theory maintains that the protoplasm is *pulled* forward by the contraction at the advancing end. Recent studies, in which strong suction applied to one pseudopodium failed to change the direction of flow of other pseudopodia in the animal, seem to indicate that the flow is not the result of internal pressure; rather they add support to the front-contraction theory. This type of ameboid movement is best seen in the naked amebae.

In some of the shelled sarcodines, in which the pseudopodia are thin and rigid, a different kind of ameboid movement that involves two opposing streams of cytoplasm occurs. In *Allogromia*, for instance, with the light microscope, particles can be seen to move up one side of the pseudopodium and down the other. This suggests a kind of shearing, or sliding effect, of opposing cytoplasmic units. There is some evidence that ATP is the source of energy for this action.

Cilia and flagella. The terms "cilium," meaning eyelash, and "flagellum," meaning whip, are descriptive of the slender, protoplasmic extensions by which some protozoans swim and by which many fixed cells in higher animals create water currents. Cilia and flagella are of similar construction, differing mainly in degree—the cilia are shorter, usually 5 to 10 μm, and

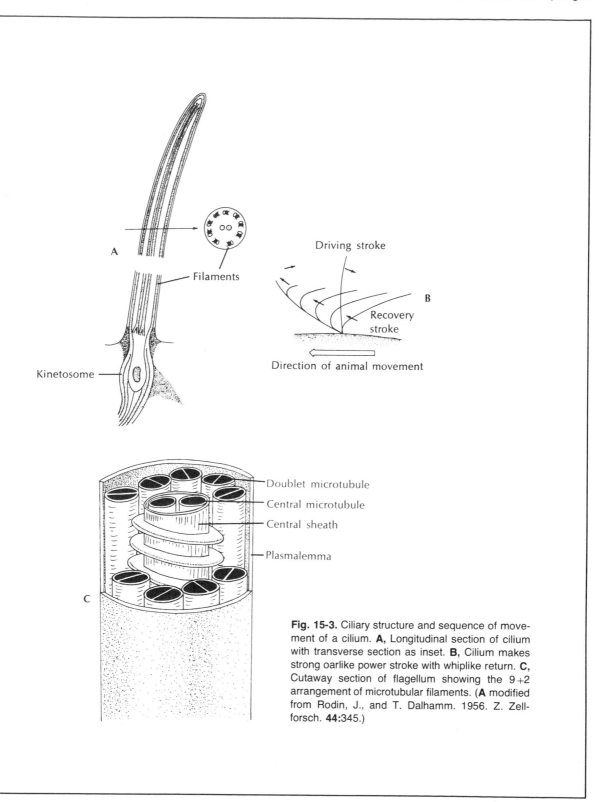

Fig. 15-3. Ciliary structure and sequence of movement of a cilium. **A,** Longitudinal section of cilium with transverse section as inset. **B,** Cilium makes strong oarlike power stroke with whiplike return. **C,** Cutaway section of flagellum showing the 9 +2 arrangement of microtubular filaments. (**A** modified from Rodin, J., and T. Dalhamm. 1956. Z. Zellforsch. **44:**345.)

more numerous, whereas the flagella may be as much as 150 μm long and are fewer in number, sometimes only one or two to an animal. Ciliary or flagellar locomotion is effective only in a fluid environment.

The electron microscope shows both the flagellum and the cilium to be composed of an **axoneme** (a sheaf of microtubules) enclosed in an extension of the cell membrane (Fig. 15-3). In the axoneme the microtubules, made up of small protein molecules, are arranged in a peripheral circle of 9 doublets surrounding 2 singlets in a central sheath. This 9 + 2 pattern is found not only in protozoans but also in the cilia and flagella of higher animals as well. The subunits of the peripheral tubules are unequal in size and bear tangential arms. Flagella also have some circular fibrils that run around the central bundle (Fig. 15-3, *C*). At the root of the cilium or flagellum, just beneath the plasma membrane, is a basal body, which in ciliates is usually called a kinetosome and in flagellates a blepharoplast. The basal body contains extensions of the microtubules.

There is evidence that when the microtubules,

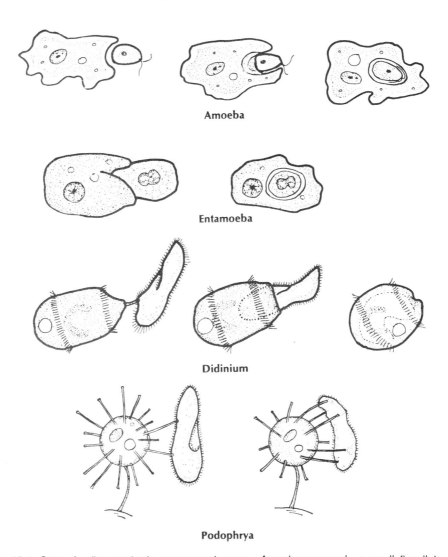

Amoeba

Entamoeba

Didinium

Podophrya

Fig. 15-4. Some feeding methods among protozoans. *Amoeba* surrounds a small flagellate with pseudopodium. *Entamoeba,* a parasite, engulfs a leucocyte. *Didinium* feeds only on paramecium, which it pierces before swallowing it whole. Suctorians have protoplasmic tentacles with funnel ends that suck protoplasm from prey.

powered by ATP, slide past one another, bending occurs in the cilium or flagellum. The structure of the microtubules is reminiscent of the structure of the sliding filaments of muscle fibrils. The rhythmic beating of the cilia often involves an oarlike power stroke followed by a whiplike return stroke in which the cilium swings out to one side as it returns to its original position. Flagellar movement may consist of successive waves propagated toward the tip.

Nutrition and digestion

Protozoans exhibit a wide variety of food-getting habits. The phytoflagellates, such as *Euglena* and *Volvox,* are chiefly manufacturers. Endowed with the necessary chlorophyll, they can produce their own food by photosynthesis in the same manner as green plants **(autotrophic, holophytic nutrition).** Some of them are also "beggars" with the ability, when deprived of light, to absorb nutrients from the water in which they live **(saprozoic nutrition).** Some saprozoic protozoans are parasites, or thieves, that absorb nutrients from their hosts.

The majority of protozoans, however, ingest solid or liquid food through a temporary or permanent oral opening, the cytostome, or "cell mouth," **(holozoic nutrition).** These are the browsers, hunters, and trappers. All have some adaptation for ingestion, whether by mouth or tentacles, as seen in the ciliates, or by the use of pseudopodia that surround and engulf *(phagocytize)* the food, as seen in the amebae and other sarcodines (Fig. 15-4).

The hunters may move by flagella, cilia, or pseudopodia. A slow-moving but surprisingly adept ameba captures with its pseudopodia much more swiftly moving ciliates or rotifers. On the other hand, *Didinium,* a ciliate, conducts an active chase to capture a *Paramecium.* Trapping is done by means of water currents created by cilia or flagella that sweep food

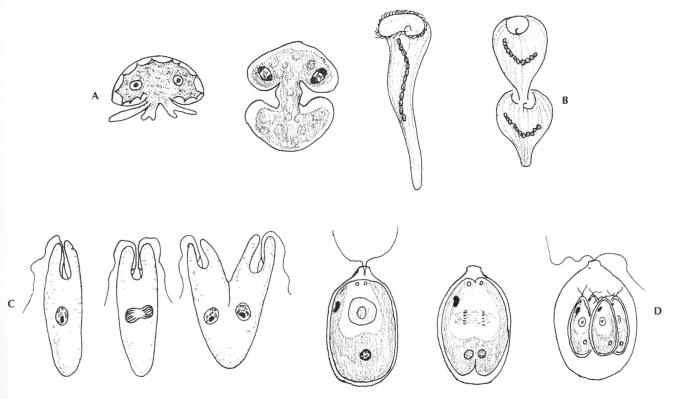

Fig. 15-5. Binary fission in protozoans. **A,** *Arcella* extrudes cytoplasm and one of its two nuclei; cytoplasm secretes new shell for daughter cell. **B,** *Stentor* divides across the rows of cilia. **C,** *Euglena* divides longitudinally. **D,** *Chlamydomonas* has a sexual phase in which two gametes join to form zygote. Zygote encysts, as shown here, and produces four to eight daughter cells by fission.

particles toward the mouth **(filter-feeding).** Filter-feeding is used by both motile and sessile forms, but some of the sessile forms are especially well adapted for it. Other trappers are found among the shelled sarcodines, which have especially adapted pseudopodia that anastomose to form protoplasmic nets for trapping their prey.

Food is digested within **food vacuoles,** which generally also contain some of the environmental water. Lysosomes, which are produced by Golgi bodies and contain digestive enzymes, apparently attach to the food vacuoles and release enzymes therein. The digested products are absorbed into the surrounding cytoplasm, and indigestible wastes are expelled to the outside.

Excretion and osmoregulation

Water balance, or osmoregulation, is a function of the one or more **water-expulsion vesicles** (also called **contractile vacuoles**) (Figs. 15-1 and 15-2) possessed by most protozoans, particularly freshwater forms, which live in a hypotonic environment. These vesicles are often absent in marine or parasitic protozoans, which live in a nearly isotonic medium.

The water-expulsion vesicles are usually located in the ectoplasm and act as pumps to remove excess water from the cytoplasm. They are filled by droplets fed by a system of collecting canals in some species and from smaller vesicles in others. When full they empty though a canal to the outside. The rate of pulsation varies; in *Paramecium* the posterior vacuole may

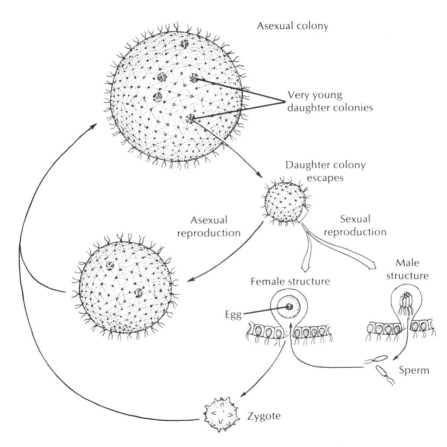

Fig. 15-6. Life cycle of *Volvox*. Asexual reproduction occurs in spring and summer when specialized diploid reproductive cells divide to form young colonies that remain in the mother colony until large enough to escape. Sexual reproduction occurs largely in autumn when haploid sex cells develop. The fertilized ova may encyst and so survive the winter, developing into mature asexual colony in the spring. In some species the colonies have separate sexes; in others both eggs and sperm are produced in the same colony.

pulsate faster than the anterior one because of water delivered there along with ingested food. Vesicles in marine forms pulsate more slowly than in similar freshwater forms.

Nitrogenous wastes from metabolism are apparently eliminated by diffusion through the cell membrane, but it is likely that some may also be emptied by way of the water-expulsion vesicles.

Respiration

There are no respiratory organelles in protozoans. Exchange of oxygen and carbon dioxide between the organism and its environment occurs by diffusion through the cell membrane.

Reproduction and life cycles

Asexual reproduction. The method of reproduction varies among protozoans, but the most common method is **binary fission,** an asexual method involving ordinary cell division and resulting in two daughter organisms similar to the mother (Fig. 15-5). Both the nucleus and the cytoplasm divide; the nucleus divides by mitosis. As in higher animals, the chromosome number appears to be constant for a species.

Budding involves unequal cell division; the parent retains its own identity while forming internally or externally, one or more small cells, which when free assume the parental form.

In **multiple fission (sporulation)** the nucleus divides a number of times, followed by the division of the organism into as many parts as there are nuclei. This rapid multiplication is characteristic of parasitic protozoans, and it is often associated with complicated life cycles.

Sexual reproduction. Sexual reproduction may involve either the formation of male and female gametes, which unite to form a zygote, or the union of two mature sexual individuals, which merge their cytoplasm and nuclei to form a zygote. By division the zygote gives rise to new individuals. Another method, common to ciliates, involves a temporary union of two individuals **(conjugation)** during which the mates exchange nuclear materials. They then separate and undergo a series of divisions.

There are advantages to both asexual and sexual types of reproduction. Asexual reproduction eliminates the need to find a mate and ensures continuation of the species even under adverse circumstances. Sexual reproduction, through gene recombinations, promotes greater variability within a population and thus encourages natural selection and stimulates evolutionary change.

Protozoan colonies

Colonies are formed when, following division, the daughter individuals **(zooids)** remain associated together instead of moving apart and living a separate existence. Colonies may be typically small—perhaps a few individuals embedded in a gelatinous substance—or they may involve hundreds or even thousands of zooids (Fig. 15-6). Colonies may be linear, spheric, discoid, or arboroid in shape. The individuals may be all alike, or, as in *Volvox,* they may involve reproductive members and members adapted for the feeding and motility of the colony.

Protozoan responsiveness

When an ameba is struck by a beam of light, its protoplasmic streaming is immediately slowed or stopped. Yet the ameba has no light-sensitive organelle

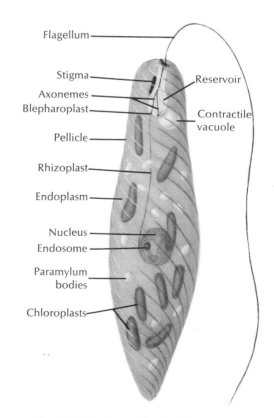

Flagellum

Stigma

Axonemes

Blepharoplast

Pellicle

Rhizoplast

Endoplasm

Nucleus

Endosome

Paramylum bodies

Chloroplasts

Reservoir

Contractile vacuole

Fig. 15-7. Structure of the flagellate *Euglena.*

and no specialized equipment for the conduction of stimuli. It responds because protoplasm itself is sensitive to light.

Protozoans respond to physical, chemical, and biotic factors, adjusting physiologically by changing their growth rates, selection of habitats, rate of reproduction, and so on. Their behavior patterns usually involve an **avoiding reaction,** which is the basis of trial-and-error behavior. Through their general sensitivity to light, temperature, salinity, touch, etc. and thus through an avoiding reaction, they seek optimal conditions.

Cilia and flagella are highly sensitive to touch; and some ciliates have special dorsal bristles that appear to be mechanoreceptors. Green flagellates usually have a special photosensitive organelle, the **stigma,** or **eye spot,** which helps them orient to light (Fig. 15-7). Obviously stimuli are conducted from one point to another in the protozoan, but just how this occurs is still uncertain.

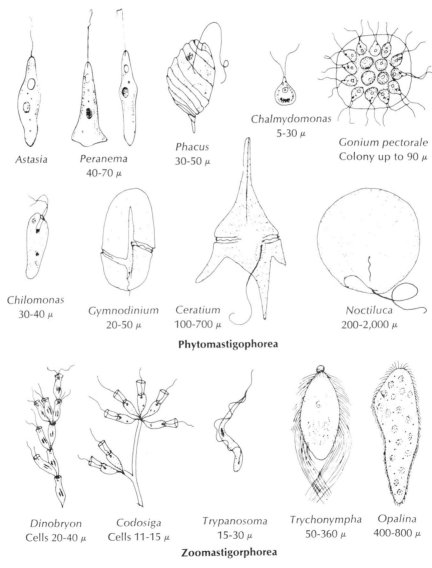

Astasia

Peranema
40-70 μ

Phacus
30-50 μ

Chalmydomonas
5-30 μ

Gonium pectorale
Colony up to 90 μ

Chilomonas
30-40 μ

Gymnodinium
20-50 μ

Ceratium
100-700 μ

Noctiluca
200-2,000 μ

Phytomastigophorea

Dinobryon
Cells 20-40 μ

Codosiga
Cells 11-15 μ

Trypanosoma
15-30 μ

Trychonympha
50-360 μ

Opalina
400-800 μ

Zoomastigorphorea

Fig. 15-8. Some flagellate protozoans.

Flagellates—superclass Mastigophora

The flagellates are protozoans that move by means of flagella. The name of the group, "Mastigophora," means "whip-bearing." These organisms are common in both fresh and marine waters. The group is subdivided into the **phytoflagellates** (Phytomastigophorea), most of which contain chlorophyll and are thus plantlike, and the **zooflagellates** (Zoomastigophorea), which, lacking chlorophyll, are holozoic or saprozoic and are thus animal-like (Fig. 15-8).

The **phytoflagellates** are commonly called the "green flagellates," although they may appear green, yellow, brown, or even colorless, according to the pigments present. They are manufacturers, producing most of what they need themselves as well as most of what is needed in the food chains of pelagic communities. In the process of **photosynthesis,** they use light as the energy source and usually carbon dioxide, and nitrate or ammonium ions are the carbon and nitrogen sources from which they synthesize their carbon compounds. Under the microscope these forms are seen to have one or more characteristic **chromoplasts,** which contain the chlorophyll and sometimes other pigments. Most of the green flagellates have a light-sensitive **stigma,** or eye spot (Fig. 15-7). This is a shallow pigment cup that allows light from only one direction to strike a light-sensitive receptor.

Although phytoflagellates are primarily **autotrophic,** some are also **saprozoic,** some **holozoic,** and some use a combination of methods.

Phytoflagellates are mostly free living and include such familiar forms as *Euglena, Chlamydomonas, Peranema,* and the red-tide dinoflagellates. Some, such as *Noctiluca,* are luminescent; some are colonial, such as *Gonium* (Fig. 15-8) and *Eudorina,* which have 16 and 32 zooids, respectively, or *Volvox,* which has thousands (Fig. 15-6).

The **zooflagellates,** colorless because they lack chromoplasts, are **holozoic** or **saprozoic.** Many are **parasitic,** such as the various species of *Trypanosoma,* a blood parasite that causes African sleeping sickness. Some have pseudopodia as well as flagella.

Euglena (Fig. 15-7) and many other flagellates have a protective pellicle that covers the plasmalemma. In *Euglena* and *Paranema* it is thin and flexible, but it may be thicker in other species so that the body form remains constant. Some are even encased in an armor of cellulose plates, as is *Ceratium* (Fig. 15-8).

Flagellar locomotion

The movement of the flagellum is varied, making it a versatile tool. Its movement may be fast or slow, and it may direct its owner forward, backward, or laterally. In some species—*Euglena,* for example—spiral, propellar-like undulations of the flagellum force the water backward and pull the animal forward in a spiral course. In others the effective stroke may be a more lateral rowing type of stroke. There may be more than one flagellum; *Trichomonas,* for example, has three of them extending forward and one trailing backward, which apparently helps anchor and steer the animal. Flagella may also form rows somewhat like ciliary rows.

Reproduction

Most flagellates reproduce asexually by longitudinal binary fission (Fig. 15-5, *C*). However, multiple fission is seen in trypanosomes and others. The sexual process is rare except in some of the colonial forms such as *Volvox* (Fig. 15-6). In *Volvox* certain reproductive zooids develop into eggs or bundles of sperm. Each fertilized egg (zygote) undergoes repeated divisions to produce a small colony, which is released in the spring. A number of asexual generations may follow before sexual reproduction occurs again.

Ameboid protozoans—superclass Sarcodina

The sarcodines are the ameboid protozoans that characteristically move and feed by means of pseudopodia (Fig. 15-9). Classification of sarcodines is based largely on the kind of pseudopodia they have and whether they are naked (covered only by a cell membrane) or possess protective tests or internal skeletons. Sarcodines are found in both fresh and salt water and in moist soils. Some are planktonic; some prefer a substratum. A few are parasitic.

Nutrition

Nutrition in amebae and other sarcodines is **holozoic;** that is, they ingest and digest liquid or solid foods. Most amebae are omnivorous, living on algae, bacteria, protozoans, rotifers, and other microscopic organisms. An ameba may take in food at any part of its body surface merely by putting out a pseudopodium to enclose the food **(phagocytosis).** The enclosed food particle, along with some of the environmental water, becomes a food vacuole, which is carried about by the

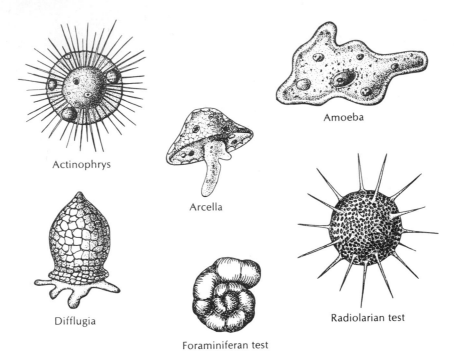

Fig. 15-9. A group of sarcodines.

streaming movements of endoplasm. As digestion occurs within the vacuole by enzymic action **(intracellular digestion),** water and digested materials pass into the cytoplasm. Fecal particles are eliminated through the cell membrane.

Reproduction

Most sarcodines reproduce by binary fission. Mitosis in an ameba typically takes approximately 30 minutes. Normally newly divided amebae reach the proper size for division within 3 days or so. Sporulation and budding occur in some of the sarcodines.

Shelled sarcodines

Not all sarcodines are naked. Some are covered with a protective **test,** or shell, with openings for the pseudopodia. *Arcella* and *Difflugia* have tests of secreted siliceous material reinforced with sand grains (Fig. 15-9). The **foraminiferans** are mostly marine forms that secrete complex many-chambered tests of calcium carbonate (Fig. 15-10, *A*). They usually add sand grains to the secreted material, using great selectivity in choosing colors. Slender pseudopodia extend through openings in the test and then run

together to form a protoplasmic net, in which they ensnare their prey. *Actinosphaerium* and *Globigerina* are beautiful little creatures with many slender stiffened pseudopodia radiating out from a central test. The **radiolarians** are marine forms, mostly living in plankton, that have intricate and beautiful siliceous skeletons (Fig. 15-10, *B,* and p. 286). A central capsule that separates the inner and outer cytoplasm is perforated to allow cytoplasmic continuity.

The radiolarians are the oldest known group of animals, and they and the foraminiferans have left excellent fossil records. For millions of years the tests of dead protozoans have been dropping to the sea floor, forming deep-sea sediments that are estimated to be from 600 to 3,600 m deep (approximately 2,000 to 12,000 feet). Many limestone and chalk deposits on land were laid down by these small creatures when the land was covered by the sea.

Parasitic sarcodines

Many of the amebae are endoparasitic, mostly in the intestine of man or other animals. *Entamoeba histolytica,* which is the only amebic parasite harmful to man, causes amebic dysentery. *Escherichia coli* causes

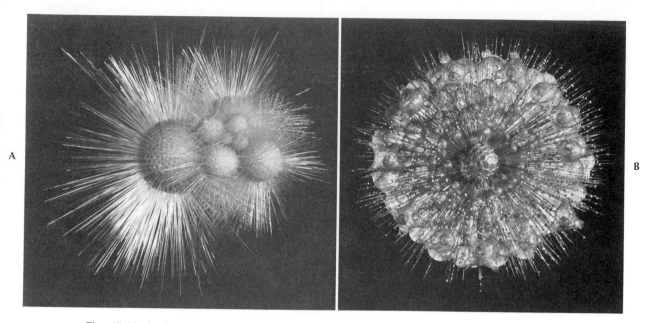

Fig. 15-10. A, *Globigerina bulloides,* a pelagic foraminiferan that builds a many-chambered test, largely of calcium carbonate. **B,** *Trypanosphaera regina,* a pelagic radiolarian, has an internal siliceous skeleton. (Photos of glass models, courtesy American Museum of Natural History.)

intestinal disturbances, and *Escherichia gingivalis* causes pyorrhea in the mouth.

Sporozoans—class Sporozoa

All sporozoans are endoparasites, and the hosts in which they are parasitized belong to all of the animal phyla. They are a heterogeneous group, often with little in common except that (1) they bear **spores** during some stage of the life cycle and (2) the adult stages have no organelles of locomotion, such as pseudopodia, cilia, or flagella.

Hosts are infected by means of spores or sometimes by the transmission of naked young. There are often intermediate hosts such as mosquitoes, leeches, flies, or other vectors. Nutrition is mostly saprozoic. Reproduction is by spore formation and may include both a sexual and an asexual cycle.

Plasmodium is the sporozoan parasite that causes **malaria.** The vectors (carriers) of the parasites are female *Anopheles* mosquitoes. The infected mosquito introduces the parasites from its salivary glands into the human blood in the form of **sporozoites** (Fig. 15-11). The sporozoites enter the cells of the liver, where they undergo asexual multiple fission (schizogeny), the

products of which enter the red blood cells. During this incubation period in the liver antimalarial drugs have little effect. In the red blood cells they again undergo multiple fission. In 10 to 14 days after inoculation the number of parasites **(merozoites)** is so great that the toxins they release cause the characteristic chills and fever in the host.

Some of the merozoites develop into sexual forms **(gametes).** When these are sucked up by the feeding mosquito, the gametes unite to form **zygotes.** From these develop thousands of sporozoites that migrate to the salivary gland to await transference to a human by a bite of the mosquito. The cycle in the mosquito takes 7 to 18 days.

Other common sporozoan parasites are the **coccidians,** which infect epithelial tissues in both vertebrates and invertebrates, and the **gregarines,** which live mainly in the digestive tract and body cavity of certain invertebrates.

Ciliates—subphylum Ciliophora

The ciliates are the most complex and diversely specialized of all the protozoans. They live in both fresh and salt water, and most of them are free living,

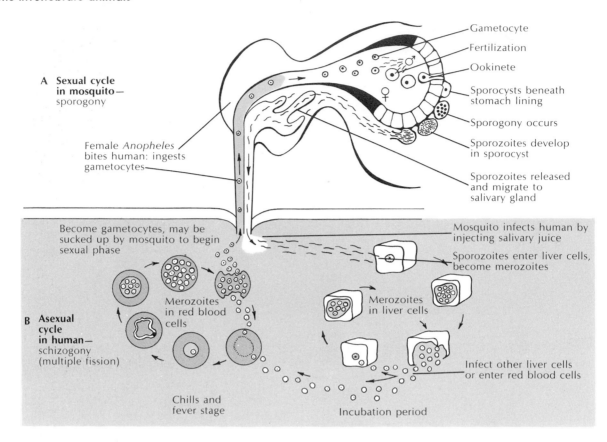

A Sexual cycle in mosquito— sporogony

Gametocyte

Fertilization

Ookinete

Sporocysts beneath stomach lining

Sporogony occurs

Sporozoites develop in sporocyst

Sporozoites released and migrate to salivary gland

Female *Anopheles* bites human: ingests gametocytes

Become gametocytes, may be sucked up by mosquito to begin sexual phase

Mosquito infects human by injecting salivary juice

Sporozoites enter liver cells, become merozoites

Merozoites in red blood cells

Merozoites in liver cells

B Asexual cycle in human— schizogony (multiple fission)

Infect other liver cells or enter red blood cells

Chills and fever stage

Incubation period

Fig. 15-11. Life cycle of *Plasmodium vivax*, protozoan (class Sporozoa) that causes malaria in man. **A,** Sexual cycle produces sporozoites in body of mosquito. **B,** Sporozoites infect man and reproduce asexually, first in the cells of liver sinusoids and finally in red blood cells. Malaria is spread by the mosquito, which sucks up gametocytes along with human blood and later, when biting another victim, leaves sporozoites in the new wound.

solitary, and motile. They are generally larger than most other protozoans, but range from 10μm to 3 mm long.

Ciliates differ from other protozoans by (1) the presence at some stage of **cilia** for locomotion and (2) the fact that they are all **multinucleate,** having at least one large, primarily vegetative **macronucleus** and one small, primarily reproductive **micronucleus** (Fig. 15-12). Ciliates are covered by a **pellicle,** which in the various species ranges from very thin to a thickened armor through which the cilia extend.

Cilia and motor coordination

Cilia may cover the animal surface or may be restricted to the region around the cytostome (cell mouth) or to certain bands. Each cilium terminates in a **basal body** (kinetosome) just beneath the pellicle (Fig. 15-13). Root fibrils from each of the basal bodies of a row of cilia join to form a basal fiber that passes beneath the row. Waves of effective beat pass along adjacent rows of cilia in unison, creating the typical swimming or feeding movements of the organism. How these waves of movement are coordinated— whether by mechanical or nervous means or both—has been the subject of much study and experimentation. Whether the basal fibrils are involved in the conduction of impulses or whether they merely serve to anchor the cilia in the cytoplasm is not really known.

The body cilia are usually shorter than the oral cilia and seem to be under separate control. In some forms the rows of oral cilia are fused to form **membranelles** that sweep food toward the cytostome. In creeping

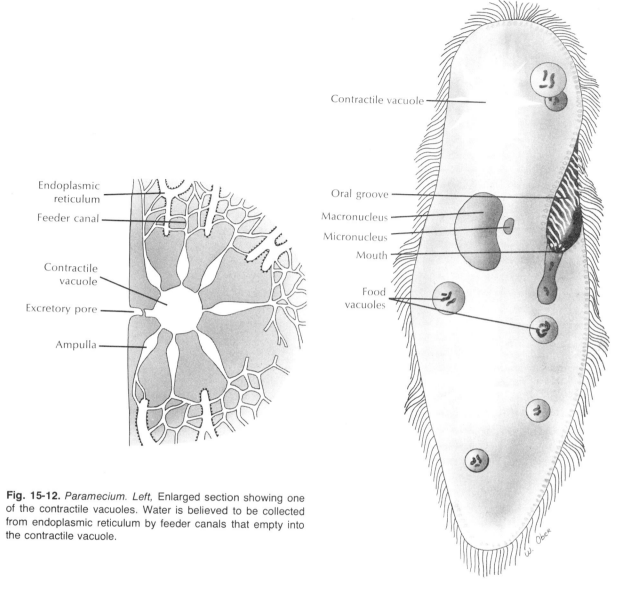

Endoplasmic
reticulum

Feeder canal

Contractile
vacuole

Excretory pore

Ampulla

Contractile vacuole

Oral groove

Macronucleus

Micronucleus

Mouth

Food
vacuoles

W. Ober

Fig. 15-12. *Paramecium. Left,* Enlarged section showing one of the contractile vacuoles. Water is believed to be collected from endoplasmic reticulum by feeder canals that empty into the contractile vacuole.

ciliates, tufts of cilia may be fused into stiff **cirri** that are used in locomotion.

Trichocysts and myonemes

Trichocysts are explosive organelles embedded in the ectoplasm just under the pellicle of many ciliates (Fig. 15-13). Those of *Paramecium,* when discharged to the exterior, form a long threadlike shaft with a barb at the tip. They may be used in anchoring the animal while it is feeding. In *Dileptus,* which is a carnivore,

the trichocysts are apparently inverted as they are discharged, and their contents can paralyze the rotifers and protozoans upon which it preys.

Many ciliates have contractile fibrils, called **myonemes,** located in the ectoplasm. They run parallel to the longitudinal axis and permit shortening or bending in the animal. *Stentor* (Fig. 15-14), whose trumpet-shaped body may be as much as 3,000 μm long, when fully extended, can contract into a 200 μm ball. The bell-shaped bodies of *Vorticella* and *Carchesium* are

301

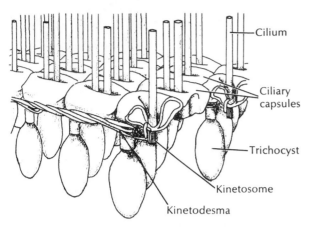

Fig. 15-13. The infraciliature of a ciliate, showing the structure of the pellicle and its relation to the basal fibrils. (Redrawn with some modification from Ehret, C. F., and E. L. Powers. 1959. The cell surface of *Paramecium*. Int. Rev. Cytol. **8:**97-133.)

attached to long, contractile stalks that can contract like coiled springs (Fig. 15-14).

Nuclei

The **macronucleus** is usually large and single and is apparently responsible for metabolism and development. The **micronucleus,** containing the chromosomes, is minute and primarily reproductive. The number of micronuclei varies in different species from 1 to as many as 80. The macronucleus is found in a variety of shapes. It is somewhat kidney shaped in *Paramecium* (Fig. 15-12), C shaped in *Euplotes*, horseshoe shaped in *Vorticella* (Fig. 15-14), and resembling beads on a string in *Spirostomum* and *Stentor* (Fig. 15-15).

Nutrition

Most ciliates are holozoic and possess a **mouth** (cytostome), which may lie at the base of a ciliated groove and **gullet** (cytopharynx). Their cilia create currents that bring in food particles (Fig. 15-15). The oral cilia sweep the food particles along the groove to the mouth and gullet, where a food vacuole is formed and dropped off into the endoplasm. The food vacuole is kept moving by the streaming protoplasm as digestion and absorption occur. Fecal material is eliminated through a temporary anal pore.

Osmoregulation

Water-expulsion vesicles (contractile vacuoles) are found in all freshwater and some marine ciliates; the number varies from one to many among the different species. In most ciliates the vesicle is fed by one or more collecting canals (Fig. 15-12). The vesicles occupy a fixed position, and each discharges through a more or less permanent pore.

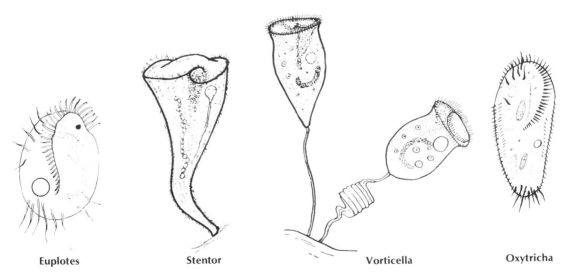

Euplotes   Stentor   Vorticella   Oxytricha

Fig. 15-14. Some representative ciliates. *Euplotes* and *Oxytricha* can use stiff cirri for crawling about. Contractile myonemes in ectoplasm of *Stentor* and in stalks of *Vorticella* allow them to expand and contract.

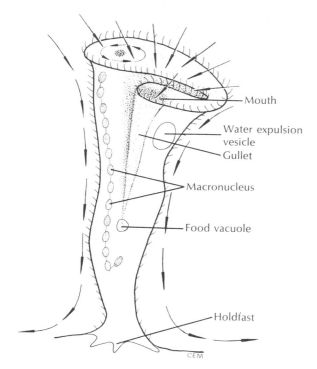

Fig. 15-15. *Stentor.* Arrows indicate ciliary feeding currents that bring food toward mouth. Rejected materials are rolled up and dropped off the edge.

Reproduction and life cycles

The life cycles of ciliates usually involve both **asexual binary fission** and a type of **sexual reproduction.** In binary fission the micronucleus divides by mitosis and the macronucleus divides amitotically. Under ideal conditions paramecia can divide from one to four times each day with a possibility of as many as 600 generations in a year. If all the descendants lived to reproduce, paramecia would soon cover the earth!

At intervals **conjugation** occurs. Conjugation is the temporary union of two individuals for the purpose of exchanging chromosomal material. During union the micronucleus of each individual undergoes meiosis, giving rise to four haploid micronuclei, three of which degenerate. The remaining micronucleus then divides into two haploid pronuclei, one of which is exchanged for a pronucleus of the conjugant partner. When the exchanged pronucleus unites with the pronucleus of the partner, the diploid number of chromosomes is restored. The two partners, each with fused pronuclei (now comparable to a zygote), separate, and each

divides twice by mitosis, each one thereby giving rise to four daughter paramecia.

Conjugation always involves two individuals of different **mating types.** This prevents inbreeding. Most ciliate species are divided into several varieties, each variety made up of two mating types. Mating occurs only between two mating types of each variety.

The advantage of sexual reproduction is that it permits gene recombinations, thus stimulating evolutionary change. Seasonal changes or a deteriorating environment usually stimulates sexual reproduction.

Color in ciliates

Although most ciliates are colorless, some are green, pink, blue or brown. The color may be caused by the presence of symbiotic green or brown algae, as, for example, in the green *Paramecium bursaria*. In others the color results from pigments in the ectoplasm. *Stentor coeruleus* has a blue-green pigment called stentorin; *Blepharisma* is pink because of the pigment purpurin.

Review of contributions of the protozoans

1. **Protoplasmic specialization** (division of labor within the cell) involves the organization of protoplasm into functional organelles.
2. The earliest indication of **division of labor between cells** is seen in certain colonial protozoans that have both somatic and reproductive zooids.
3. **Asexual reproduction** by mitotic division is first developed in the protists.
4. The behavior of conjugant mates in certain protozoans may indicate the earliest differentiation of sex. **True sexual reproduction** with zygote formation is found in some protozoans.
5. The responses of protozoans to stimuli represent the **beginning of reflexes and instincts** as we know them in metazoans.
6. The first appearance of an **exoskeleton** is indicated in certain shelled protozoans.
7. The beginning of future **contractile systems** is indicated by the presence of myonemes (contractile fibrils).
8. **All types of nutrition** are developed in the protozoans—autotrophic, saprozoic, and holozoic.

SPONGES—PHYLUM PORIFERA

The phylum name "Porifera" means "pore-bearing," a fitting name for these strange creatures, which have no organs, no digestive tract, not even well-defined tissue. Their bodies, indeed, seem little more than systems of canals and pores held together by cells embedded in jelly and stiffened by a skeleton of minute spicules of calcium or silica or by fibers of spongin. They bear little resemblance to the usual concept of animal life.

Although sponges are metazoans, they are so different from other metazoans that they are often called the Parazoa. "Para" means beside; so these are the "beside animals," an aberrant group whose evolutionary line apparently leads nowhere but to themselves. The sponges have given rise to no other known groups.

Most sponges are colonial; the colonies range from 1 to 2 mm to as much as 1 to 2 meters in diameter. They vary in color from dull gray, brown, or green to brilliant scarlet and orange. Many living sponges appear as slimy gelatinous masses. Most of the dried sponges that we see consist only of the skeletal framework from which the protoplasmic mass has been removed.

Evolution of sponges

Although sponges, along with other metazoans, are believed to have arisen from some sort of flagellate, perhaps by way of a colonial form, the flagellate was probably of a different group than that which led to the Eumetazoa, or "true metazoans." The fossil record is incomplete; however, there is evidence that sponges, in particular glass sponges, flourished throughout the Cambrian period.

The fact that sponges evolved along very different lines from other metazoans in no way indicates their lack of "success" as a group. More than 5,000 species have been described so far. They have been thriving since Cambrian times and may well occupy our oceans and streams long after many other "successful" groups become extinct.

Ecologic relationships

Sponges are abundant in shallow coastal waters, attached to the bottom or to submerged rocks or other objects (Fig. 15-16). Some also occur in deep water, and a few have adapted to fresh water. Although

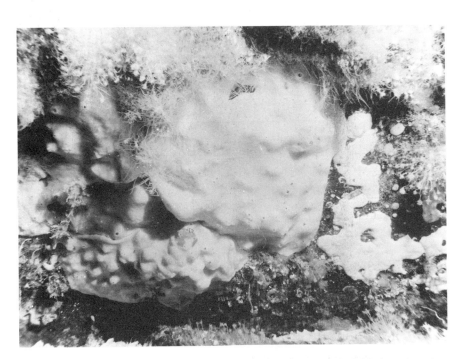

Fig. 15-16. Large marine sponge (class Demospongiae) growing on underside of overhanging rock, surrounded by clusters of flowerlike hydrozoans. Note the many small oscular openings. (Courtesy T. Lundalv, Kristinebergs Zoological Station, Sweden.)

sponge embryos are free swimming, the adults are always attached, are practically motionless, and show little or no reaction to handling.

Many animals have associations with sponges either as parasites or for protective purposes. Some crabs attach pieces of sponge to their carapaces for camouflage. Although most fishes tend to avoid sponges, some reef fishes graze upon shallow-water sponges.

Characteristics

1. **Multicellular;** body a loose aggregation of cells of mesenchymal origin
2. Body with **pores** (ostia), **canals,** and **chambers** that serve for the passage of water
3. Mostly marine; **all aquatic;** adults **sessile** and **attached**
4. Radial symmetry or none
5. Epidermis of flat **pinacocytes;** most interior surfaces lined with flagellate collar cells **(choanocytes)** that create water currents; gelatinous mesenchyme **(mesoglea)** contains amebocytes, collencytes, and skeletal elements
6. Skeleton of calcareous or siliceous crystalline **spicules** or protein **spongin** or a combination
7. **No organs or true tissues;** intracellular digestion; excretion and respiration by diffusion

8. Apparently local and independent reactions to stimuli; nervous system probably absent
9. Asexual reproduction by buds or gemmules; sexual reproduction by eggs and sperm; free-swimming, ciliated larvae

Classification

There are three classes of sponges, divided mainly by the kinds of skeletons that they possess.

Class Calcarea (kal-ka′re-a) (L. *calcis*, lime). **(Calcispongiae).** Spicules of carbonate of lime often form a fringe around the osculum and are needle shaped or three or four rayed; all three types of canal systems represented; all marine. *Scypha, Leucosolenia.*

Class Hexactinellida (hek-sak-ti-nel′-i-da) (Gr. *hex*, six + dim. of *aktis, aktinos*, a ray). **(Hyalospongiae).** Three-dimensional, six-rayed, siliceous spicules, often united to form network; body often cylindric or funnel shaped; flagellated chambers in simple syconoid or leuconoid arrangement; all marine. Venus's flower basket (*Euplectella*), *Hyalonema.*

Class Demospongiae (de-mo-spun′je-e) (Gr. *dēmos*, population + *spongos*, sponge). Siliceous spicules that are not six rayed or spongin or a combination of the two; leuconoid-type canal systems; one family found in fresh water; all other families marine. *Thenea, Cliona, Spongilla, Meyenia,* and all bath sponges.

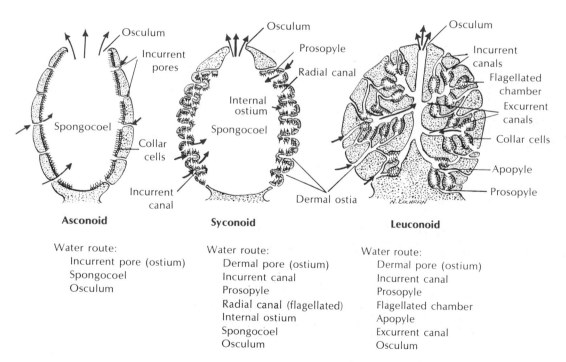

Asconoid

Water route:
    Incurrent pore (ostium)
    Spongocoel
    Osculum

Syconoid

Water route:
    Dermal pore (ostium)
    Incurrent canal
    Prosopyle
    Radial canal (flagellated)
    Internal ostium
    Spongocoel
    Osculum

Leuconoid

Water route:
    Dermal pore (ostium)
    Incurrent canal
    Prosopyle
    Flagellated chamber
    Apopyle
    Excurrent canal
    Osculum

Fig. 15-17. Three types of sponge structure. Degree of complexity from simple asconoid type to complex leuconoid type has involved mainly the water and skeletal systems, accompanied by outfolding and branching of collar cell layer. Leuconoid type considered major plan for sponges, for it permits greater size and more efficient water circulation.

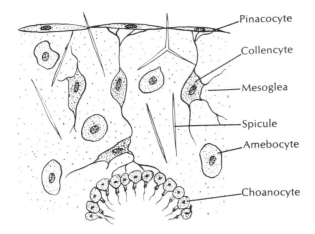

Fig. 15-18. Small section through sponge wall showing four types of sponge cells.

Form and function

The only body openings of these unusual animals are **pores,** usually many tiny ones for incoming water and a few large ones, called **oscula** (sing., **osculum**), for outgoing water. These are connected by a system of **canals** (Fig. 15-17), some of which are lined with peculiar flagellated collar cells called **choanocytes,** which have flagella that maintain a current of seawater in the canals. The choanocytes not only keep the water moving, but trap and phagocytize food particles that are carried in it. Collapse of the canals is prevented by the **skeleton,** which depending upon the species may be made up of needlelike **spicules** (Fig. 15-18) or a meshwork of **spongin** fibers or a combination of both.

The sponge's system of flagellated canals serves adequately as a **filter-feeding system,** and, because most of its body cells are in contact with the environmental water either externally or as it passes through the canals, individual cells can feed by ameboid (phagocytic) methods, making a digestive tract unnecessary.

Because sessile animals make few movements, they need few nervous, sensory, or locomotor parts. Sponges apparently have lived as sessile animals from their earliest appearance and have never acquired specialized nervous or sensory structures; they have only the simplest of contractile systems. They are all either asymmetric or radial. Radial symmetry is fitting for a sessile animal since it allows it to face its environment in any direction.

Types of canal systems

Most sponges have one of three types of canal systems—asconoid, syconoid, or leuconoid (Fig. 15-17).

The **asconoid** type is characterized by a flagellated spongocoel and is the simplest type of organization. These sponges are small and tube shaped. Water enters through microscopic dermal pores into a large cavity, or **spongocoel,** which is lined with choanocytes whose flagella pull the water in through the pores and expel it through a single large **osculum** (Fig. 15-17). *Leucosolenia* is probably the best known example of an asconoid sponge. Its slender, tubular individuals grow in groups attached by a common stolen, or stem, to objects in shallow seawater (Fig. 15-20, *C*).

The **syconoid** type is characterized by flagellated canals. Syconoid sponges look somewhat like asconoid sponges. They have the tubular body and the single osculum, but the body wall, which is thicker and more complex, contains choanocyte-lined **radial canals** that empty into the spongocoel. The spongocoel in syconoid sponges is lined with epidermis rather than choanocytes, as in the asconoid type. Water enters through dermal pores into incurrent canals and filters through tiny openings into the choanocyte-lined radial canals. Food is ingested by the choanocytes, whose flagella force the water on through internal pores into the spongocoel and through the osculum (Fig. 15-17). Syconoid sponges do not form highly branched colonies as do the asconoid sponges. *Scypha* is a commonly studied example of the syconoid type of sponge.

The **leuconoid** type contains flagellated chambers. Its organization is the most complex of the sponge types and is the best adapted for increasing the size of the sponge. Most leuconoid sponges form large colonial masses in which each member of the mass has its own osculum, but individual members are poorly defined and often impossible to distinguish. Clusters of flagellated chambers are filled from incurrent canals. The water is discharged into excurrent canals, which may join to form larger excurrent canals that eventually lead to the osculum (Fig. 15-17). Most of the sponges are of the leuconoid type; they occur in both class Calcarea and class Demospongiae.

These three types of canal systems, asconoid, syconoid, and leuconoid, are correlated with the evolution of sponges from simple to complex forms. Throughout their evolution the main trend has been an increasing of flagellated surfaces in proportion to the volume so that there would be enough collar cells to

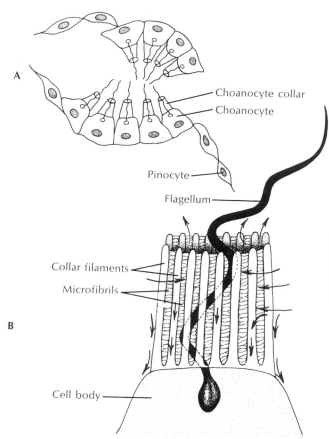

Choanocyte collar

Choanocyte

Pinocyte

Flagellum

Collar filaments

Microfibrils

Cell body

A

B

Fig. 15-19. A, Small portion of choanocyte-lined canal. **B,** Collar of a choanocyte as revealed by electron microscopy. Fine protoplasmic processes connected by extremely fine microfibrils form a filter through which water is drawn by flagellar action.

meet the food demands. The problem was solved by the outpushing of the spongocoel of a simple sponge, such as the asconoid type, to form the radial canals (lined with choanocytes) of the syconoid type. Further increase in the body wall foldings produced the complex canals and chambers (with collar cells) of the leuconoid type.

Types of cells

Sponge cells are loosely arranged in a gelatinous matrix called **mesoglea** (Fig. 15-18). There are several types of cells.

Pinacocytes. The closest resemblance to a true tissue in sponges is found in the arrangement of the pinacocyte cells of the epidermis (Fig. 15-18). These are flat, epithelial-type cells that cover the exterior surface and some interior surfaces. Pinacocytes are somewhat contractile and help regulate the surface area of the sponge.

Choanocytes. The choanocytes, which line the

flagellated canals and chambers, are ovoid cells with one end embedded in the mesoglea and the other exposed. The exposed end bears a flagellum surrounded by a collar (Fig. 15-19). The electron microscope shows the collar to be made up of adjacent protoplasmic processes, or fibrils, connected to each other by delicate microvilli, so that the collar forms a fine filtering device for straining food particles from the water.

The beat of the flagellum pulls water through the sievelike collar and forces it out through the open top of the collar. Particles too large to enter the collar become trapped in secreted mucus and slide down the collar to the base where they are phagocytized by the cell body. Larger particles have already been screened out by the small size of the dermal pores and prosopyles. The food engulfed by the cells may be digested in food vacuoles within the choanocyte or passed on to neighboring amebocytes for digestion.

Amebocytes. Various ameboid cells, called ame-

307

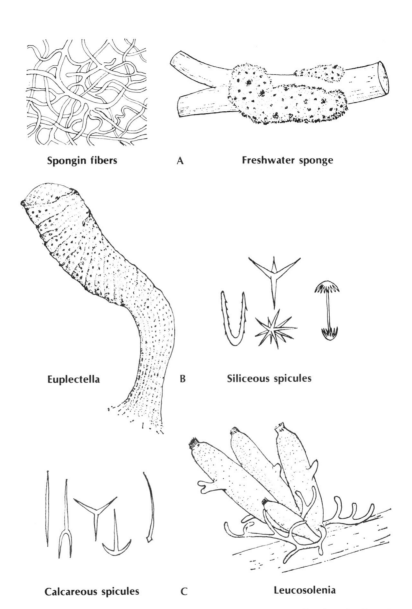

Spongin fibers   A   **Freshwater sponge**

Euplectella   B   **Siliceous spicules**

Calcareous spicules   C   **Leucosolenia**

Fig. 15-20. Types of skeletal structure found in sponges, with example of each. Amazing diversity, complexity, and beauty of form exist among the many types of spicules.

bocytes, lie embedded in the mesoglea (Fig. 15-18) and carry out a number of functions. Some are filled with food reserves; others aid in digestion, carry pigments, form ova or sperm, or secrete the spicules or the spongin fibers of the skeleton. Some, called collencytes, appear to be contractile, although some authorities have suggested a nervous function for them.

Skeletons

The skeleton, which is the basis for classifying sponges, gives support to the sponge, preventing the collapse of the canals and chambers. In many small marine sponges of class Calcispongiae the skeleton consists of **spicules** of calcium carbonate (Figs. 15-18 and 15-20, *C*). Glass sponges of class Hyalospongiae and sponges of class Demospongiae have spicules formed of siliceous material (Fig. 15-20, *B*). The spicules are of many shapes. Some are entirely embedded in the mesoglea; others protrude to the exterior, giving the sponge a prickly or rough surface.

Many sponges contain **spongin,** a proteinlike substance, which forms a branching, fibrous network that supports the soft living cells (Fig. 15-20, *A*). Both spicules and the spongin are secreted by special amebocytes in the mesoglea.

Sponge physiology

Sponges probably feed on fine detritus particles and planktonic organisms, which are screened by the dermal pores and prosopyles and again by the choanocyte collars. Choanocytes ingest most of the food, but pinacocytes can also phagocytize food. Food is often passed from cell to cell. Digestion may be started in a collar cell and completed in an amebocyte or transferred to another cell for the final stage. Thus the wandering amebocytes perform a carrier service, act as digesters, and serve as storage warehouses. Although the choanocyte-lined chambers are the chief areas for feeding activity, they are not in any way comparable to the digestive tracts of other metazoans.

There are no respiratory or excretory organs; both functions are apparently carried out by diffusion in individual cells. Some wastes may accumulate in certain amebocytes, which release them into the canals.

All life activities of the sponge depend upon the current of water flowing through the body. A sponge pumps a remarkable amount of water. *Leuconia,* for example, is a small leuconoid sponge approximately 10 cm tall and 1 cm in diameter. It is estimated that water enters through some 81,000 incurrent canals at a velocity of 0.1 cm/sec. However, *Leuconia* has more than 2 million flagellated chambers whose combined diameter is much greater than that of the canals, so that in the chambers the water slows down to 0.001 cm/sec, allowing ample opportunity for food capture by the collar cells. All of the water is expelled through a single osculum at a velocity of 8.5 cm/sec—a jet force capable of carrying waste products some distance away from the sponge. Some large sponges have been found to filter 1,500 liters of water per day.

Reproduction and development

Sponges may reproduce asexually by budding or by gemmules, or sexually. Asexual buds that form externally may become detached, or they may remain to form colonies. Internal buds, or gemmules, are produced by freshwater and some marine sponges. The gemmules survive during periods of drought or freezing, and later the cells in the gemmules escape and develop into new sponges.

In sexual reproduction ova are fertilized in the mesoglea; there the zygotes develop into flagellated larvae (called amphiblastulae in calcareous sponges), which break loose and are carried away by water currents. The larvae swim about for some time, then settle, become attached, and grow into adults.

Some sponges are **monoecious** (having both male and female sexes in one individual), and some are **dioecious** (having separate sexes). Sponges have high regenerative powers: pieces of sponge can regenerate into whole sponges.

Class Calcarea (Calcispongiae)

Calcarea are the calcareous sponges, so-called because their spicules are composed of calcium carbonate. The spicules are straight monaxons or three or four rayed (Fig. 15-20, *C*). The sponges tend to be small—10 cm or less in height—and tubular or vase shaped. They may be asconoid, syconoid, or leuconoid in structure. Though many are drab in color, some are bright yellow, red, green, or lavender. *Leucosolenia* (Fig. 15-20, *C*) and *Scypha* are common examples.

Class Hexactinellida (Hyalospongiae)

The glass sponges are nearly all deep-sea forms. Most of them are radially symmetric, range from 7 to

Fig. 15-21. **A,** Large marine sponge surrounded by algae, scallops *(lower right),* and a large contracted sea anemone *(upper right).* **B,** Football-shaped sponges *Geodia baretti* occur in great numbers in Scandanavian waters below 60 m. (Courtesy T. Lundälv, Kristinebergs Zoological Station, Sweden.)

10 cm to more than 1 meter in length, and are attached to a substratum. Their distinguishing feature, reflected in the class name, is the skeleton of six-rayed siliceous spicules that are bound together in an exquisite glass-like latticework (Fig. 15-20, *B*). The living tissue is a network produced by the fusion of the pseudopodia of many types of amebocytes, forming chambers lined with choanocytes. The chambers open into the spongocoel, and the osculum is unusually large. Glass sponges include both syconoid and leuconoid types. Their structure is adapted to the slow, constant current of the sea bottom, for the channels and pores are relatively large and permit an easy flow of water. *Euplectella,* the "Venus's flower basket," is an example of this class.

Class Demospongiae

Demospongiae contain approximately 80% of all sponge species, including most of the larger sponges. The spicules are siliceous but are not six rayed; they may be bound together by spongin, or they may be absent altogether. All members of the class are leuconoid, and all are marine except one family, the Spongillidae, or freshwater sponges. Freshwater sponges are widely distributed in well-oxygenated ponds and streams, where they are found encrusting plant stems and old pieces of submerged wood (Fig. 15-20, *A*). They resemble a bit of wrinkled scum, pitted with pores, and are brownish or greenish in color. Common genera are *Spongilla* and *Meyenia*. Freshwater sponges die and disintegrate in late autumn, leaving gemmules to survive the winter.

The marine Demospongiae are varied in both color and shape. Some are encrusting; some are tall and fingerlike; and some are shaped like fans, vases, cushions, or balls (Fig. 15-21). Some are boring sponges that bore into molluscan shells and encrust. Loggerhead sponges may grow several meters in diameter. The so-called bath sponges belong to the group called horny sponges, which have only spongin skeletons. They can be cultured by cutting out pieces of the individual sponges, fastening them to a weight, and dropping them into the proper water conditions. It takes many years for them to grow to market size.

Review of contributions of the sponges

1. The poriferans have a somewhat higher level of organization than the protozoans in both form and physiology. They have several types of cells differentiated for various functions, some of which are

organized into incipient tissues of a primitive nature. They might be said to have a cellular level of organization.

2. Poriferans have developed a unique system of water currents for filter feeding and oxygen supply.

SELECTED REFERENCES

Allen, R. D., D. Francis, and R. Zek. 1971. Direct test of the positive pressure gradient theory of pseudopod extension and retraction in amoebae. Science **174**:1237-1240. (For additional information see also Science **177**:636-638. [1972].)

Barrington, E. J. W. 1967. Invertebrate structure and function. Boston, Houghton Mifflin Co.

Brien, P. 1968. The sponges, or Porifera. In Florkin, M., and B. J. Sheer (eds.). Chemical Zoology, vol 2. New York, Academic Press, Inc.

Cheng, T. C. 1964. The biology of animal parasites. Philadelphia, W. B. Saunders Co. *A parasitology text containing a good account of protozoan parasites.*

Corliss, J. O. 1961. The ciliated Protozoa: characterization, classification and guide to the literature. New York, Pergamon Press, Inc.

Edmondson, W. T. (ed.). 1959. Ward and Whipple's fresh-water biology, ed. 2. New York, John Wiley & Sons, Inc. *In this handbook there are useful keys to the major groups of Protozoa and freshwater sponges.*

Florkin, M., and B. T. Scheer, eds. 1967, 1968. Chemical zoology. Vol. 1, Protozoa. Vol. 2, Porifera. New York, Academic Press, Inc.

Fry, W. G. (ed.). 1970. The biology of Porifera. New York, Academic Press, Inc. *A collection of papers presented at a symposium of the Zoological Society of London.*

Gardiner, M. S. 1972. The biology of invertebrates. New York, McGraw-Hill Book Co.

Gosner, K. L. 1971. Guide to identification of marine and estuarine invertebrates: Cape Hatteras to the Bay of Fundy. New York, Interscience Div., John Wiley & Sons, Inc.

Harrison, F. W., and R. R. Cowden (ed.). 1976. Aspects of sponge biology. New York, Academic Press, Inc.

Hickman, C. P. 1973. Biology of the invertebrates, ed. 2. St. Louis, The C. V. Mosby Co.

Hyman, L. H. 1940. The invertebrates: Protozoa through Ctenophora, vol. 1. New York, McGraw-Hill Book Co. *An extensive and exhaustive section is devoted to the morphology and physiology of protozoans.*

Jahn, T. L., and F. F. Jahn. 1949. How to know the Protozoa. Dubuque, Iowa, William C. Brown Co., Publishers. *A manual on the identification and description of protozoan forms.*

Jeon, K. W. (ed.). 1973. The biology of amoeba. New York, Academic Press, Inc.

Jones, A. R. 1974. The ciliates. New York, St. Martins Press. Inc.

Jørgensen, C. B. 1966. The biology of suspension feeding. New York, Pergamon Press, Inc.

Kaestner, A. 1967. Invertebrate zoology, vol. 1. New York, Interscience Div., John Wiley & Sons, Inc.

Kudo, R. R. 1966. Protozoology, ed. 5. Springfield, Ill., Charles C Thomas, Publisher.

Leedale, G. F. 1967. Euglenoid flagellates. Englewood Cliff, N. J., Prentice-Hall, Inc.

Levine, N. D. 1973. Protozoan parasites of domestic animals and man. Minneapolis, Burgess Publishing Co.

Light, S. F., R. I. Smith, F. A. Pitelka, D. P. Abbott, and F. M. Weesner. 1967. Intertidal invertebrates of the central California coast. Berkeley, University of California Press. *Contains keys to common west coast sponges.*

Mackinnon, D. L., and R. S. J. Hawes. 1961. An introduction to the study of Protozoa. Oxford, The Clarendon Press.

Manwell, R. D. 1961. Introduction to protozoology. New York, St. Martin's Press, Inc.

Minchin, E. A. 1900. Porifera. In Lankester's treatise on zoology, part II. London, A. & C. Black Ltd. *A classic account of the general morphology of sponges.*

Murray, J. W. 1973. Distribution and ecology of living benthic foraminiferids. New York, Crane, Russak, & Co., Inc.

Noble, E. R., and G. A. Noble. 1972. Parasitology, ed. 3. Philadelphia, Lea & Febiger.

Pennak, R. W. 1953. Fresh-water invertebrates of the United States. New York, The Ronald Press Co. *A reference work with considerable attention devoted to Protozoa and freshwater sponges.*

Pettersson, H. 1954. The ocean floor. New Haven, Conn., Yale University Press. *The role that the Foraminifera and Radiolaria have played in building up the sediment carpet of the ocean floor is described in this little book.*

Pitelka, D. R. 1963. Electron-microscopic structure of Protozoa. New York, The Macmillan Co. *A fine account of the ultrastructure of Protozoa by one of the great authorities in the field.*

Pitelka, D. R., 1970. Ciliate ultrastructure: some problems in cell biology. J. Protozool. **17**(1):1-10.

Sleigh, M. A. 1962. The biology of cilia and flagella. New York, The Macmillan Co.

Sleigh, M. 1973. The biology of protozoa. New York, American Elsevier Publishing Co., Inc. *A well-written text with many micrographs and diagrams of life cycles.*

Schmidt, G. D., and L. S. Roberts. 1977. Foundations of parasitology. St. Louis, The C. V. Mosby Co. *A parasitology text containing a good account of protozoan parasites.*

Tartar, V. 1961. The biology of Stentor. New York, Pergamon Press, Inc.

SELECTED SCIENTIFIC AMERICAN ARTICLES

Allen, R. D. 1962. Amoeboid movement. **206**:112-122 (Feb.).

Sater, P. 1974. How cilia move. **231**:44-81 (Oct.).

Jellyfishes are weak swimmers that are carried about by the ocean currents. They swim by rhythmic pulsations of their umbrella-shaped bodies. This Mediterranean form, *Cotylorhiza,* is similar to a related genus, *Cassiopeia,* common to the Florida waters.

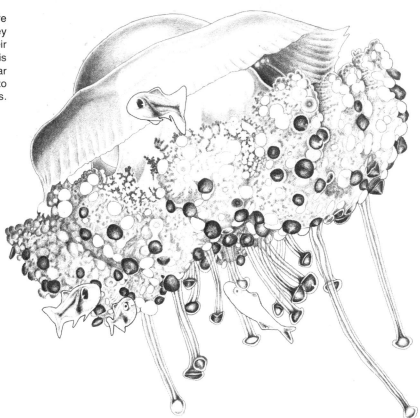

16 RADIATE ANIMALS

The "radiate phyla" contain some of nature's strangest, loveliest, and most fragile creatures. All of them are aquatic. Phylum Cnidaria (Coelenterata), which includes the hydroids, jellyfishes, sea anemones, and corals, and phylum Ctenophora, commonly known as the comb jellies, are called the **radiate phyla**—or **Radiata**—because their members have radial or biradial symmetry.

In **radial symmetry** the body parts are arranged concentrically around an oral-aboral axis. Radially symmetric animals are best adapted for a sessile or sedentary life. Most cnidarians are either attached or very slow moving. Even the free-swimming medusae and comb jellies are weak swimmers that are carried about by currents. For sessile or slow-moving animals there is a natural advantage in having a radiate form since it is equally receptive, responsive, and protected on all sides.

The disadvantage to true radial symmetry is the necessity for repetition of parts around the axis, and this apparently has restricted the evolutionary development of specialized structures in these phyla. Instead of an infinite repetition of parts, many of the cnidarians have evolved a **tetramerous** or **hexamerous** plan, that is, a plan of four or six or multiples of either. The ctenophorans have developed a form of **biradial symmetry,** evidenced by the arrangement of their tentacles and internal canals. Both of these adaptations would seem to be logical evolutionary steps toward the bilateral symmetry found in all other eumetozoans.

In contrast to the sponges, which have developed to the cellular level of organization, the Radiata have developed to the **tissue level of organization;** they are formed primarily of tissues and incipient tissues.

ADVANCEMENTS OF THE RADIATE PHYLA

In contrast to the loosely organized cells of the sponges, the cnidarians and ctenophores have developed a saclike body plan that involves a three-layered body wall enclosing an internal space for digestion, called the **coelenteron,** or **gastrovascular cavity.** This is the only body cavity, and it has a single opening to the outside, which serves as both mouth and anus. Digestion occurs in the cavity **(extracellular digestion),** as well as in the cells lining the cavity **(intracellular digestion,** also called endocytosis).

Both phyla have two well-defined **germ layers.** The body wall is composed of an outer epidermis (from ectoderm) and an inner gastrodermis (from endoderm). Between these layers lies a third layer, called the mesoglea, which may range from a thin noncellular matrix to a thick, fibrous jellylike layer with or without wandering cells (largely from ectoderm).

Although there are a number of types of cells, these, with the exception of the nerve cells, are usually not grouped into the true tissues typical of higher forms. The first true **nerve cells (protoneurons)** occur in the radiate phyla; they are arranged in a **nerve net** that is the forerunner of the centralized nervous system of higher forms.

Most of the radiates possess extensible **tentacles** around the oral end, which aid in food capture—a distinct advantage for a group of carnivores that are ill fitted for rapid movement. Further aids for capturing prey are specialized cells on the tentacles. Cnidarians are equipped with **cnidocytes** that contain stinging organoids called **nematocysts;** ctenophores have **adhesive** cells. Both are able to capture prey on contact. Such capture, although perhaps more accidental than purposeful, satisfies the low nutritional demands of these attached or leisurely moving animals.

EVOLUTION OF THE RADIATE PHYLA

The radiate phyla are probably closer to the primitive ancestral stock of eumetazoans than any other phyla. The transitional form between the protozoans and the higher metazoans may have been a ciliated, free-swimming, gastrula-like animal, somewhat similar to the planula larva that is found in the life history of most cnidarians. The planula is a ciliated larva, cylindric or ovoid in shape, which has an outer layer of ectodermal cells, and an inner core of endodermal cells.

According to some authorities, the most primitive organisms of the modern cnidarians are the trachyline medusae—an order of class Hydrozoa in which the polyp stage is much reduced. We cannot know with certainty, of course, whether the polyp or the medusa evolved first. One theory is that the medusoid form may have been the ancestral form, whereas the polyp may have represented the retention and diversification of the larval stage. The colonial hydroid may have evolved from the polyplike larva of the ancestral medusa, which, by budding off more larvae that remained attached, could in time have given rise to a colony of polyps. The stages of development of living hydrozoans illustrate this sequence.

Because of the similarities between the cnidarians and the ctenophores, we might expect that they came from the same ancestral stock, although no fossil evidence is available. The biradial symmetry of the ctenophores may have been derived from the tetramerous radial symmetry of the hydroid medusa. The ancestral metazoan may have lacked either the stinging cells or the adhesive cells that have evolved in modern radiates.

PHYLUM CNIDARIA (COELENTERATA)

Phylum Cnidaria (ny-dar'e-a) derives its name from the Greek *knide* (nettle) and refers to the specialized stinging cells (cnidocytes) that are characteristic of the phylum. The fact that no other animals are capable of producing cnidocytes and that no other animals bear them except one species of Ctenophora and a few flatworms and molluscs that acquire them by feeding on

cnidarians is considered reason enough for calling the phylum Cnidaria. The name was first used by Aristotle and later revived (1888) by the zoologist Hatchek. The phylum is also known as Coelenterata (se-len'te-ra'ta) because of the large cavity, or coelenteron, that serves as a digestive tract. Coelenteron means "hollow intestine" and comes from the Greek *koilos* (hollow) and *enteron* (gut). Since both cnidarians and ctenophores have a coelenteron, the term Coelenterata is sometimes used to refer to the Radiata.

There are more than 9,000 species of cnidarians, all of them aquatic and most of them marine. They include the branching plantlike hydroids, the flowerlike sea anemones, the medusoid jellyfishes, and those architects of the ocean floor, sea whips, sea fans, sea pansies, and hard corals. The latter, through eons of constructing tiny, calcareous homes, have produced great reefs and coral islands.

Fig. 16-1. A hermit crab with its cnidarian commensals. The crab has moved into an empty snail shell and *Hydractinia,* a tiny colonial hydroid, has formed a velvety cover over the outside of the shell. *Adamsia,* a sea anemone, perches atop the shell. Hermits often seek out suitable anemones and hold them with their pincers till they attach themselves to their shells. (Photo by R. C. Hermes.)

Ecologic relationships

Marine cnidarians occur most widely in shallow tropical or subtropical waters, although they are also found in deeper and colder waters in all oceans. A few live in freshwater streams, ponds, and lakes. Colonial hydroids attach to shells, rocks, wharves, and even to other animals (Fig. 16-1), usually in shallow coastal waters. Floating and free-swimming medusae (jellyfishes) are found in the open seas. Floating colonies, such as *Physalia,* the Portuguese man-of-war (Fig. 16-10), and *Velella,* have floats by which they are blown about in the breeze or carried in currents. Reef-building corals prefer warm water near continental and island shores. They prefer depths of 20 m or less and rarely live deeper than 100 m.

Economically the cnidarians as a group are relatively unimportant. However, the precious corals are used for jewelry and ornaments, and coral rock is used for road building and other building purposes. The coral reefs supply habitation and shelter for many other kinds of animal life. Cnidarian larvae and medusae become a part of the ocean plankton and serve as food for fish and other animals. They themselves are carnivores and feed upon fish and planktonic forms.

Polymorphism in cnidarians

One of the most interesting—and sometimes puzzling—aspects of this phylum is the almost universal dimorphism and often polymorphism displayed among its members. In general, all cnidarian forms fit into one of two morphologic types—the **polyp,** or hydroid form, which is adapted to a sedentary or sessile life, or the **medusa,** or jellyfish form, which is adapted for a floating or free-swimming existence (Fig. 16-2).

Most polyps have tubular bodies with a mouth at one end surrounded by tentacles. The aboral end is usually attached to a substratum by a pedal disc or other device. Polyps may live singly, or they may live in colonies of more than one kind of individual, each specialized for a certain function, such as feeding, reproduction, or defense.

Medusae are free swimming and have bell-shaped or umbrella-shaped bodies and tetramerous symmetry (body parts arranged in fours). The mouth is usually centered on the concave side, and tentacles extend from the rim of the umbrella.

The sea anemones and corals (class Anthozoa) are all polyps, and the true jellyfishes (class Scyphozoa) are all medusae but may have a polypoid larval stage. The colonial hydroids of class Hydrozoa, however, have unique life histories that feature both the polyp, or hydroid stage, and the free-swimming medusa stage—similar to a Jeckyll-and-Hyde existence. A species that has both the attached polyp and the floating medusa within its life history can take advantage of the feeding and distribution possibilities of both the pelagic (open-water) and the benthonic (bottom) types of environment.

Superficially the polyp and medusa seem very different. But actually each has retained the saclike body plan that is basic to the phylum (Fig. 16-2). The medusa is essentially an unattached, upside-down polyp with the tubular portion widened and flattened into the bell shape; the mouth, in this inverted

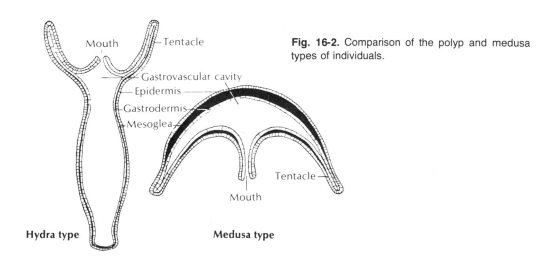

Fig. 16-2. Comparison of the polyp and medusa types of individuals.

Mouth — Tentacle

Gastrovascular cavity
Epidermis
Gastrodermis
Mesoglea

Tentacle

Mouth

Hydra type   **Medusa type**

condition, faces downward, and the circle of tentacles fringing the rim also hangs downward.

Both the polyp and the medusa possess the three body wall layers typical of the cnidarians, but the jellylike layer of mesoglea is much thicker in the medusa, constituting the bulk of the animal and making it more buoyant. It is because of this mass of mesogleal "jelly" that the medusae are commonly called jellyfishes.

Characteristics

1. **Radial** or **biradial symmetry** around a longitudinal axis with oral and aboral ends
2. Polymorphism common, featuring two main types of individuals—attached **polyps** and free **medusae;** colonies, often with a variety of specialized individuals (zooids)
3. Entirely aquatic, mostly marine
4. Body of three layers: outer **epidermis** (derived from ectoderm), inner **gastrodermis** (derived from endoderm), and between them a **mesoglea** with or without cells (largely from ectoderm)
5. **Gastrovascular cavity** (coelenteron) with single opening that serves as both mouth and anus; extensible oral **tentacles**
6. **Exoskeleton** of chitin or lime in some
7. Special stinging cell organoids called **nematocysts** in epidermis and/or gastrodermis
8. **Nerve net** of synaptic and nonsynaptic patterns; diffuse conduction; some sensory organs; no excretory or respiratory organs
9. Muscular system (epitheliomuscular type) of longitudinal and circular fibers; higher cnidarians have separate bundles of independent muscle fibers in the mesoglea
10. **Asexual** reproduction by budding (in polyps) or **sexual** reproduction by gametes (in all medusae and some polyps); monoecious or dioecious sexual forms; **planula larva;** holoblastic cleavage; mouth from blastopore

Classification

Class Hydrozoa (hy-dro-zo′a) (Gr. *hydra,* mythological nine-headed serpent, + *zoon,* animal). Solitary or colonial; asexual type with polyps and sexual type with medusae, although one type may be suppressed; hydranths with no mesenteries; medusae (when present) with a velum; both freshwater and marine. *Hydra, Obelia, Physalia.*

Class Scyphozoa (si-fo-zo′a) (Gr. *skyphos,* cup, + *zoon,* animal). Solitary; polyp stage reduced or absent; medusae without velum; gelatinous mesoglea much enlarged; margin of umbrella typically scalloped, with notches that are provided with sense organs; all marine. *Aurelia, Cassiopeia.*

Class Anthozoa (an-tho-zo′a) (Gr. *anthos,* flower, + *zoon,* animal). All polyps; no medusae; solitary or colonial; some provided with skeleton; coelenteron (gastrovascular cavity) subdivided by mesenteries (septa) with nematocysts; endodermal gonads; all marine. Sea anemones (*Metridium, Adamsia, Condylactus*), corals, sea pens.

Form and function
Body wall

The cnidarian body wall is made up of three layers. The **epidermis** and **gastrodermis** are epithelial layers with muscle fibers and nervous tissue occurring in the base of each layer, and between them is a layer of **mesoglea** (Fig. 16-3).

Epidermis. The outer, or epidermal, layer is made up of epitheliomuscular cells, gland cells, sensory and nerve cells, cnidocytes, and interstitial cells (Fig. 16-3).

Epitheliomuscular cells make up the greater portion of the layer. These are tall cells in hydras and sea anemones but are cuboid or flat in some other forms. Usually the bases of the cells are drawn out into long strands containing **contractile fibers** (myonemes) that run longitudinally, close to the mesoglea. These fibers correspond in function to a layer of longitudinal muscle but are not true muscle cells. Their contraction causes shortening of the body wall or tentacles.

Gland cells secrete a sticky mucus, which is useful for protection, for attachment, or for entanglement of prey; they also aid in swallowing.

Interstitial cells are undifferentiated cells that can transform into cnidocytes, sex cells, etc. as needed.

Cnidocytes are modified interstitial cells that contain the stinging organelles, or **nematocysts,** that are unique to this phylum (Fig. 16-3). The nematocyst is a tiny capsule containing a coiled tubular filament, or "thread," which is a continuation of the narrowed end of the capsule (Fig. 16-4). The cell has a projecting hairlike "trigger" (cnidocil), which when stimulated causes the coiled filament to turn inside out with explosive force. Having discharged its nematocyst, the rest of the cnidocyte is digested and replaced from the interstitial cells. Cnidocytes are most abundant on the tentacles.

Nervous connections play no direct part in the discharge of nematocysts but may play an indirect part in affecting the threshold of discharge. It is believed that their discharge is stimulated by a combination of chemical and mechanical stimuli. Thus chemical action, such as the presence of food, lowers the threshold of response to a mechanical stimulus, such as direct contact. The presence of food does not trigger the discharge of nematocysts in a sated hydra, and many cnidarians harbor commensals that can move about freely over the cnidocytes, even bending down their cnidocils without causing discharge.

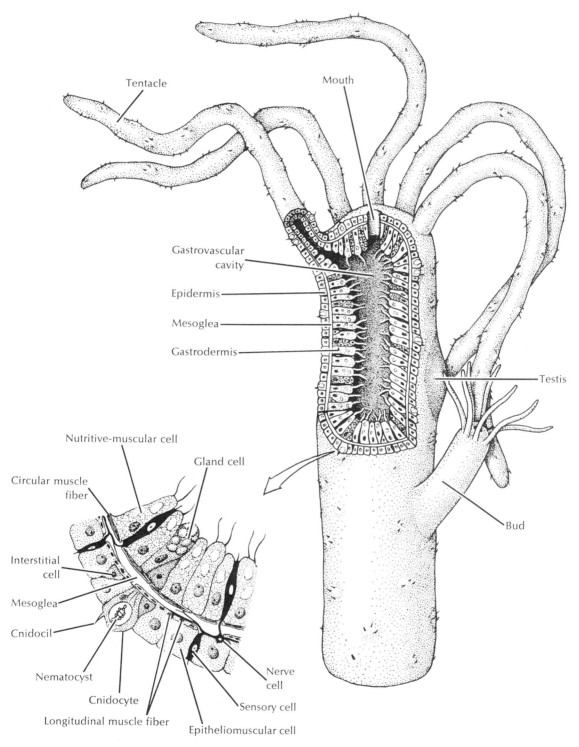

Tentacle

Mouth

Gastrovascular cavity

Epidermis

Mesoglea

Gastrodermis

Testis

Nutritive-muscular cell

Gland cell

Circular muscle fiber

Bud

Interstitial cell

Mesoglea

Cnidocil

Nematocyst

Nerve cell

Cnidocyte

Sensory cell

Longitudinal muscle fiber

Epitheliomuscular cell

Fig. 16-3. Structure of a hydra. Both bud and developing gonad are shown, but in life they rarely develop simultaneously. Portion of the body wall is enlarged at lower left to show cellular detail.

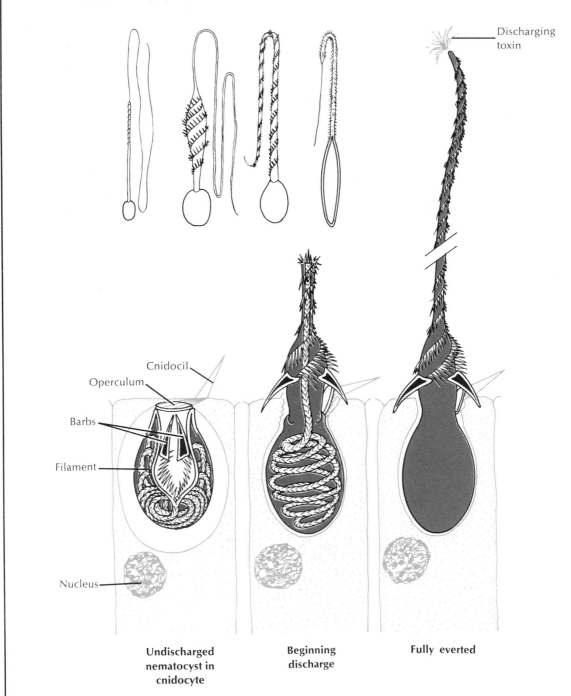

Cnidocil

Operculum

Barbs

Filament

Nucleus

Discharging toxin

Undischarged nematocyst in cnidocyte

Beginning discharge

Fully everted

Fig. 16-4. How a cnidocyte discharges its nematocyst. A penetrant nematocyst, the largest and most familiar type, is shown undischarged and during discharge. The nematocyst can discharge only once. The inset above shows four other types of nematocyst. (Redrawn from Wells, M. 1968. Lower animals. World University Library.)

Three common types of nematocysts are volvent, penetrant, and glutinant. Discharged volvents wrap around an object that is touched. Small crustaceans, such as *Daphnia,* are quickly immobilized by a volley of discharged volvents wound around their appendages. Penetrants can actually penetrate tissues of the prey and inject from their tips secretions that may be irritating, toxic, or narcotic (Fig. 16-4). Glutinants discharge adhesive substances, which may serve to anchor the animal to some object or to entangle prey organisms. The nematocysts and the substances released by them are apparently synthesized by the cnidocytes.

Sensory cells for detecting touch, temperature, chemicals, etc. are scattered among the epitheliomuscular cells, especially on the tentacles. The free ends are flagellated and the other ends branch into fibrils attached to the nerve net.

Epidermal **nerve cells** (protoneurons) each have two or more processes, or neurites, that lie at the base of the epidermis. These connect with the sensory cells, muscle fibers, and other cells and form a **nerve plexus.** They can conduct impulses in either direction.

Gastrodermis. The inner layer, or gastrodermis, lines the gastrovascular cavity and has a plan similar to that of the epidermis (Fig. 16-3).

Nutritive-muscular cells are tall, flagellated cells with irregular flat bases, and they make up approximately 70% of the gastrodermal layer. Their bases are drawn out into contractile fibers that run circularly around the body or tentacles. When these fibers contract, they lengthen the body by decreasing its diameter. The cells are often filled with food vacuoles. The flagella ensure continual movement of foods and fluids in the digestive cavity.

Interstitial cells and **gland cells** are also found in this layer. The gland cells secrete digestive enzymes and mucus. Cnidocytes are absent.

Mesoglea. The mesoglea lying between the two epidermal layers supports and gives rigidity to the body, acting as a type of hydrostatic skeleton. It is made up largely of jelly—mostly water—which may enclose certain mesenchymal cells and connective tissue fibers. In hydras and hydrozoan polyps it is usually thin and without cells or fibers. The scyphozoans, or true medusae, have a thicker mesoglea that is more complex, containing cells and fibers. In the sea anemones this layer becomes a connective tissue layer containing large numbers of mesenchyme cells and muscle fibers suspended in a small amount of jelly.

Nutrition and digestion

Cnidarians are all carnivorous. Most of them are tentacular feeders that capture and hold prey organisms with their tentacles (Fig. 16-5). Some cnidarians, however, are ciliary feeders that trap minute organisms in mucus on their bodies or tentacles and convey them to their mouths by ciliary action.

Digestion is both extracellular and intracellular. Foods are broken down into small particles in the gastrovascular cavity; then the particles are engulfed by the nutritive-muscular cells of the gastrodermis where digestion is completed in food vacuoles. There is no anal opening; indigestible parts are regurgitated through the mouth. Digested food nutrients are diffused from cell to cell to all parts of the body, eliminating the need for a vascular transportation system.

Nerve net

The **nerve net** of cnidarians is the forerunner of the central nervous system in higher forms. It appears in the body wall not only of the cnidarians, but also of echinoderms and balanoglossids and in the digestive tracts of many higher invertebrates and vertebrates, including man.

The nerve net is a diffuse network of neurons lying under the epithelial layers. The neurons may be bipolar (with two processes) or multipolar (with many processes). Sensory cells, which are especially numerous in the tentacles and oral disc, feed into the nerve plexus, which also connects with the muscle fibers.

In a nerve net, conduction in any direction is possible. The excitation spreads among the neurons in a nondirectional, nonpolarized manner. This was demonstrated a century or so ago by G. J. Romanes, an English biologist, who found that deep cuts made in the disc of a medusa *(Aurelia)* did not interfere with the wave of contraction that spread around the bell following a local stimulation; the impulse was conducted around the cuts. The impulse is not transmitted along a linear chain of neurons but radiates out from its point of origin. This contrasts with a centralized system in which cutting a nerve trunk or cord would prevent the passage of an impulse to the tissues. A diffuse, nonpolarized system that provides alternative pathways

319

is obviously advantageous to a radially symmetric animal that is equally sensitive all around.

There seem to be two nerve net conduction systems in most cnidarians, one for potentials of small amplitude and one for those of large amplitude. The latter provides a less diffuse and more rapid through pathway. The scyphozoan medusa *Cyanea* has two through pathways—one controlling symmetric contractions of the bell and the other capable of increasing the frequency of the pulsations—and there is evidence of a third, slower one also.

Cells designated as **neurosecretory cells** have been identified in certain cnidarians, and there is some evidence that they are involved in the control of growth and regeneration.

The nerve net system and the contractile fibers of the epithelial layers are often referred to as a neuromuscular system, the first important landmark in the evolution of the nervous system.

Reproduction

Both sexual and asexual reproduction occur in cnidarians. Most cnidarians are **dioecious** (have separate sexes), and the sex cells originate from interstitial cells in the epithelium. Fertilization is usually external in the surrounding water. In **asexual reproduction** polyps may bud from the stems of colonies or from other polyps, and medusae may bud from polyps, stems, or other medusae.

Polyps have great powers of **regeneration,** and most parts cut from a parent stock have the potential of developing into new polyps.

Respiration and excretion

There are no respiratory or excretory organs in cnidarians. Since most of the body cells are in contact with or close to either the external environment or the digestive cavity, which takes in quantities of water along with food, gas exchange and elimination of liquid wastes can be accomplished by diffusion across the membranes of individual cells.

Class Hydrozoa

The majority of hydrozoans are marine and colonial in form; the typical life cycle is composed of both the asexual polyp and the sexual medusa stages. The sessile polyps bud off medusae, which produce gametes; the zygotes from the medusae develop into polyps, and the cycle is repeated. Common examples of hydroids are *Obelia* (Fig. 16-6) and *Campanularia,* which form bushy colonies of tiny, flowerlike polyps but have microscopic medusae; *Sertularia,* which have colonies that may form fernlike or plumelike branches, and *Gonionemus* (Fig. 16-9) and *Craspedacusta,* which have minute polyp stages but larger medusae that range from 10 to 25 mm in diameter.

There are well-known exceptions to this dimorphic polyp-medusa plan. The hydras, for instance, have no medusoid stage at all. Neither does *Hydractinia,* a colonial hydroid that forms a velvety mat attached to the snail shells inhabited by certain hermit crabs (Fig. 16-1). *Liriope* is a little medusa that has no hydroid stage but develops directly from a free-swimming larva.

The medusae of the class Hydrozoa are usually called hydromedusae to distinguish them from those of class Scyphozoa, which are called scyphomedusae.

Hydra—a freshwater hydrozoan

Probably the best known members of class Hydrozoa—and certainly the most often studied by beginning students—are the freshwater hydras (Figs. 16-3 and 16-5). They are found throughout the world in fresh clean pools and streams. The brown hydra *Pelmatohydra* and the green hydra *Chlorohydra* are very common. In their tiny tubular bodies, which can extend up to 25 to 30 mm in length, the typical layers of the body wall are easily seen. The tentacles surrounding the mouth are covered with batteries of nematocysts. Their transparent bodies are ideal for watching the engulfment of food or the development of asexual buds or of ovaries or testes.

Unlike colonial polyps, which are permanently attached, the hydra can move about freely by gliding on its basal disc, aided by mucus secretions. Or using a "measuring worm" movement it can loop along by bending over and attaching its tentacles to the substratum. It may even turn handsprings or detach itself and, by forming a gas bubble on its basal disc, float to the surface.

Hydras feed upon a variety of small crustacea, insect larvae, and annelid worms. The hydra awaits its prey with tentacles extended (Fig. 16-5). The food organism that brushes against its tentacles may find itself harpooned by scores of nematocysts that render it helpless, even though it may be larger than the hydra. The tentacles move toward the mouth, which slowly widens. Well moistened with mucus secretions, the

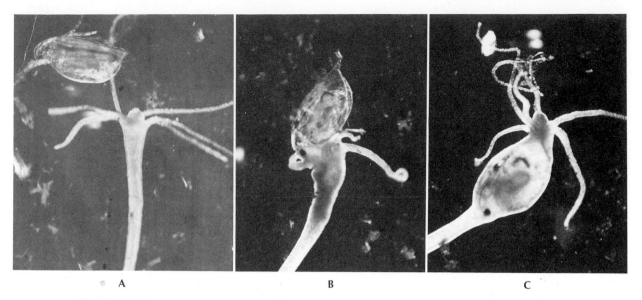

Fig. 16-5. A, Hungry hydra catches an unwary water flea with the stinging cells of its tentacle and, **B,** swallows it whole. **C,** Hydra is full, but not too full to capture a protozoan for dessert. (Photos by F. M. Hickman.)

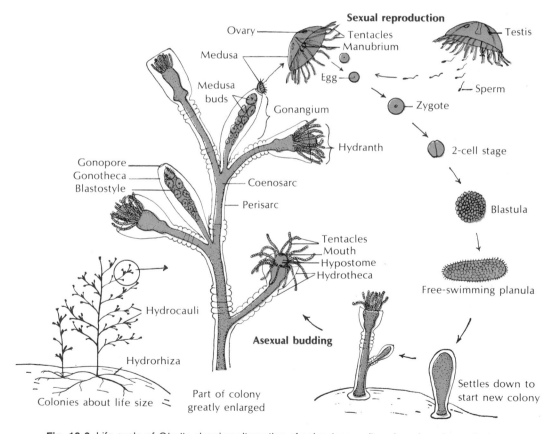

Fig. 16-6. Life cycle of *Obelia* showing alternation of polyp (asexual) and medusa (sexual) stages.

mouth glides over and around the prey, totally engulfing it.

The stimulus that causes the mouth to open was found to be the reduced glutathione released by the prey when it is pierced by nematocysts. Glutathione is a tripeptide found to some degree in all living cells but more abundantly in some than in others. Glutathione released in the water causes the hydras to go through the feeding motions, even when no prey are present.

The hydrozoan digestive system is a simple sac into which the mouth opens. Digestion (extracellular) is started in the digestive cavity by enzymes discharged by gland cells, which make up approximately one fourth of the gastrodermal cells. Food particles are then drawn into the nutritive muscular cells, where digestion is completed (intracellular digestion). The

surface of these cells is equipped with many fine processes, or microvilli, that facilitate phagocytosis. Protein droplets in various stages of digestion, as well as lipid droplets, accumulate in the bases of the cells, thus furnishing a food reserve available to the animal when under stress. Excess or unusable materials are regurgitated several hours after the food is ingested.

Hydras, like other polypoids, have many sensory cells (Fig. 16-3) but no specialized sense organs. These seem to be limited to the medusae, which, as free-swimming organisms, have more need of them.

The hydra reproduces sexually and asexually. Asexually, buds appear as outpocketings of the body wall (Fig. 16-3) and develop into young hydras that eventually detach from the parent. Most species are dioecious. Temporary gonads (Fig. 16-3) usually ap-

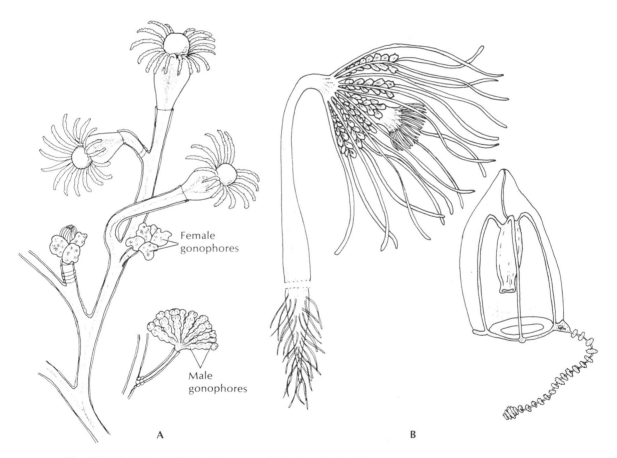

Female gonophores

Male gonophores

A   B

Fig. 16-7. Hydroids. **A,** *Eudendrium* grows in bushy colonies that range, in different species, from 1 to 300 cm tall. A large colony may bear thousands of delicate pink hydranths; they have gonophores but no medusa stage. **B,** *Corymorpha* is a solitary hydroid that grows more than a centimeter tall. Its gonophores produce free-swimming medusae, each with a single trailing tentacle.

pear in the autumn, stimulated by the lower temperatures and perhaps also by the reduced aeration of stagnant waters. Eggs in the ovary usually mature one at a time and are fertilized by sperm shed into the water.

The zygotes undergo holoblastic cleavage to form a hollow blastula. The inner part of the blastula delaminates to form the endoderm (gastrodermis), and the mesoglea is laid down between the ectoderm and endoderm. A cyst forms around the embryo before it breaks loose from the parent, enabling it to survive the winter. Young hydras hatch out in spring when the weather is favorable.

Hydroid colonies

Far more representative of class Hydrozoa than the hydras are those hydroids that have a medusa stage in their life cycle. *Obelia* is often used in laboratory exercises for beginning students to illustrate the hydroid type (Fig. 16-6).

A typical hydroid has a base, a stalk, and one or more terminal zooids. The base by which the colonial

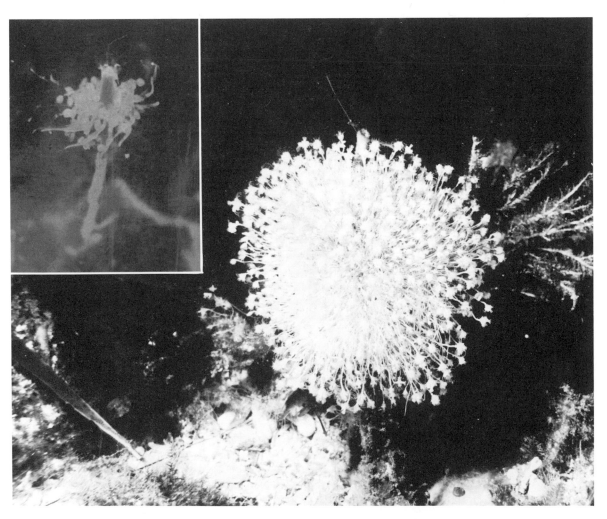

Fig. 16-8. Colony of *Tubularia larynx*. Each polyp *(inset)* has a circlet of tiny tentacles around the mouth and a lower circlet of larger tentacles, with gonophores between them. Medusas do not detach but shed gametes while in place on parent. (Photo by T. Lundälv, Kristinebergs Zoological Station, Sweden; inset photo by B. Tallmark, Uppsala, Sweden.)

323

hydroids are attached to the substratum is a rootlike stolon, or hydrorhiza, which gives rise to one or more stalks called hydrocauli. The living cellular part of the hydrocaulus is the tubular coenosarc, composed of the three typical cnidarian layers surrounding the coelenteron (gastrovascular cavity). The protective covering of the hydrocaulus is a nonliving chitinous sheath, or perisarc. Attached to the hydrocaulus are the individual polyp animals, or zooids. Most of the zooids are feeding polyps called **hydranths,** or gastrozooids. They may be tubular, bottle shaped, or vaselike, but all have a terminal mouth and a circlet of tentacles. In some forms, such as *Obelia,* the perisarc continues as a protective cup around the polyp into which it can withdraw for protection (Fig. 16-6). In others, such as

Tubularia, Eudendrium, and *Corymorpha* (Fig. 16-7), the polyp is naked.

The hydranths, much like miniature hydras, capture and ingest prey, such as tiny crustaceans, worms, and larvae, thus providing nutrition for the entire colony. After partial digestion in the hydranth, the digestive broth passes into the common coelenteron where intracellular digestion occurs.

Circulation within the coelenteron is a function of the flagellated gastrodermis but is also aided by rhythmic contractions and pulsations of the body, which occur in many hydroids.

Just as hydras reproduce asexually by budding, the colonial hydroids bud off new individuals, thus increasing the size of the colony. New feeding polyps

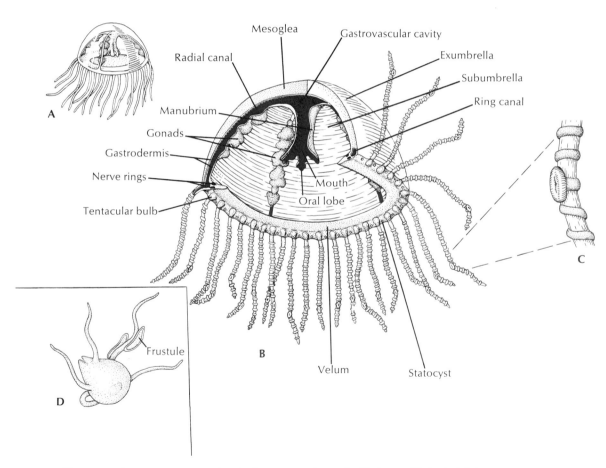

Fig. 16-9. *Gonionemus,* a marine hydrozoan medusa. **A,** Medusa with typical tetramerous arrangement. **B,** Cutaway view showing morphology. **C,** Portion of a tentacle with its adhesive pad and ridges of nematocysts. **D,** Tiny polyp, or hydroid stage, that develops from the planula larva. It can bud off more polyps (frustules) or produce medusa buds.

arise by budding, and medusoid buds also arise on the colony. In *Obelia* these medusae are budded from a reproductive polyp called a gonangium. The young medusae leave the colony as free-swimming individuals that mature and produce gametes (eggs and sperm) (Fig. 16-6). *Corymorpha* produces young medusae from gonophores attached to the feeding polyps (Fig. 16-7, *B*). In *Tubularia* (Fig. 16-8) medusae also arise from the feeding polyps but remain attached to the colony and shed their gametes there. In *Eudendrium* no medusae develop; male and female gonophores produce gametes that, when fertilized, develop into planula larvae.

Hydroid medusae are usually smaller than scyphozoan medusae, ranging from 2 or 3 mm to several centimeters in diameter. The margin of the bell projects inward as a shelflike **velum,** which partly closes the open side of the bell and is used in swimming (Fig. 16-9). Muscular pulsations that alternately fill and empty the bell propel the animal forward, aboral side first, with a sort of "jet propulsion." The tentacles attached to the bell margin are richly supplied with nematocysts.

The mouth opening at the end of a suspended **manubrium** leads to a stomach and four radial canals that connect with a ring canal around the margin. This in turn connects with the hollow tentacles. Thus the coelenteron is continuous from mouth to tentacles, and the entire system is lined with gastrodermis. Nutrition is similar to that of the hydranths.

The gelatinous mesoglea is thick in the hydromedusae; it lacks cells but contains fibers.

The muscular system is restricted to the contractile fibers of the epidermal layer and is absent in the gastrodermal layer. It is best developed around the bell margin and velum and in the subumbrellar surface where contractions produce the rhythmic pulsation of the bell. Swimming movement is mostly vertical in direction, with the water currents providing the power for horizontal movement.

The nerve net is usually concentrated into two nerve rings at the base of the velum. The bell margin is liberally supplied with sensory cells. It usually also bears two kinds of specialized sense organs—**statocysts,** which are small organs of equilibrium (Fig. 16-9, *B*), and **ocelli,** which are light-sensitive organs.

The medusae reproduce sexually. Gonads are usually suspended beneath each radial canal. Fertilization may be external, as in *Obelia,* with both the eggs and sperm shed into the water, or fertilization and development of the egg may occur in the medusa. The zygote develops into a ciliated planula larva that swims about for a time, then settles down to a substratum to develop into a minute polyp that gives rise, by asexual budding, to the hydroid colony, thus completing the life cycle.

Floating colonies

Floating colonies, such as *Physalia* (the Portuguese man-of-war) (Fig. 16-10) and *Velella,* are made up of several types of modified medusae and polyps. *Physalia* has a rainbow-hued float, said to be a modified polyp, which carries it along at the mercy of the winds and tides. It contains an air sac filled with secreted gas and acts as a carrier for the generations of individuals

Fig. 16-10. Portuguese man-of-war, *Physalia physalis* (order Siphonophora, class Hydrozoa), eating fish. This colony of medusa and polyp types is integrated to act as one individual. As many as a thousand zooids may be found in one colony. Although a drifter, the colony has restricted directional movement. Their stinging organoids secrete a powerful neurotoxin. Colonies often drift onto southern ocean beaches, where they are a hazard to bathers. (Courtesy New York Zoological Society.)

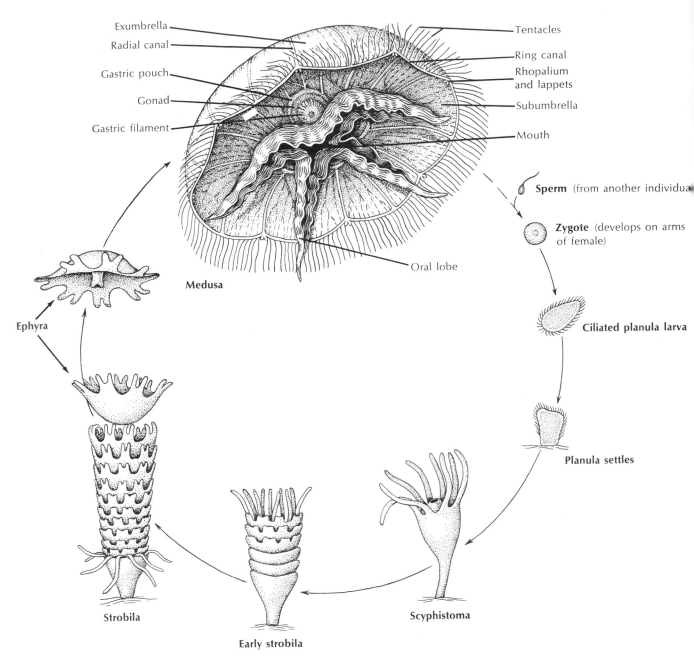

Exumbrella

Radial canal

Gastric pouch

Gonad

Gastric filament

Tentacles

Ring canal

Rhopalium
and lappets

Subumbrella

Mouth

Sperm (from another individual)

Zygote (develops on arms
of female)

Oral lobe

Medusa

Ciliated planula larva

Ephyra

Planula settles

Strobila

Early strobila

Scyphistoma

Fig. 16-11. Life cycle of *Aurelia,* a marine medusa of class Scyphozoa.

that bud from it and hang suspended in the water. There are several types of individuals, including the feeding polyps, reproductive polyps, long stinging tentacles, and the so-called jelly polyps. Many swimmers have experienced the uncomfortable sting that these colonial floaters can inflict.

Class Scyphozoa

Class Scyphozoa (si-fo-zo′a) includes most of the larger jellyfishes, or "cup animals" (from *skyphos,* cup). A few, such as *Cyanea,* may attain a bell diameter of several meters and tentacles 60 to 70 m long. Most scyphozoans, however, range from 2 to 40 cm in diameter. Most are found floating in the open sea, some even at depths of 3,000 m, but one unusual order is sessile and attaches by a stalk to seaweeds and other objects on the sea bottom. Their coloring may range from colorless to striking orange and pink hues.

Scyphomedusae, unlike the hydromedusae, have no velum at all. The bell ranges from shallow to a deep helmet shape, and in many the margin is usually scalloped, each notch bearing a sense organ called a **rhopalium** and a pair of lobelike projections called lappets. *Aurelia* has eight such notches (Fig. 16-11); others may have four or sixteen. The mesoglea is thick and contains cells as well as fibers. The stomach is usually divided into pouches containing small tentacles with nematocysts.

The large size and abundant nematocysts make these medusae unpopular with and often dangerous to swimmers who call them "stinging nettles." The most venomous is the sea wasp *Chironex,* which is far more deadly than the dreaded Portuguese man-of-war. Its sting not only produces excruciating pain, but the venom has a neurotoxic effect and death can occur within 3 to 20 minutes after stinging. Several deaths have been reported from Australian coasts. Because they are only approximately 7.5 cm in diameter and because their translucent bodies are the same color as the waters they frequent, the sea wasps are not easily seen by swimmers and divers.

Locomotion of scyphomedusae is effected by a band of powerful circular muscles around the margin of the bell, which can cause its contraction. The thick mesoglea acts as an antagonist to the muscles to restore the bell shape between pulsations. Some of the muscle fibers in medusae are striated, although as a rule cnidarian muscle is smooth.

The mouth is centered on the subumbrellar side. The manubrium is usually drawn out into four frilly oral arms that are used in food capture and ingestion. The marginal tentacles may be many or few and may be short, as in *Aurelia,* or long, as in *Cyanea.* Scyphozoans feed on all sorts of small organisms, from protozoans to fishes. Capture of prey involves stinging and manipulation with tentacles and oral arms, but the methods vary. In *Chrysaora* the tentacles contract so that the food they have caught can be swept off by the lips. In *Aurelia* food particles collect on the bell, are licked off by the ciliated oral arms, and then are carried by cilia to the mouth. *Cassiopeia* lies lazily on the bottom, oral side up; its pulsations create currents to draw food organisms toward its ample arms, there to be trapped in mucus and conveyed along ciliary tracts to the mouth.

Internally four **gastric pouches** containing nematocysts connect with the stomach, and a complex system of **radial canals** that branch from the pouches to the **ring canal** (Fig. 16-11) complete the gastrovascular cavity, through which nutrients circulate.

The nerve net system in scyphozoans seems to be of synaptic type. Sense organs are concentrated in rhopalia located in the marginal notches. Each rhopalium bears a statocyst for balance, two sensory pits containing concentrations of sensory cells, and sometimes an ocellus for photoreception.

The sexes are separate, and fertilization and early development occur in the gastric pouches or on the frilled oral arms. In most scyphozoans, a ciliated planula larva develops into a little polypoid larva called a **scyphistoma,** which looks somewhat like a hydra. During the summer the scyphistoma produces more scyphistomas by budding. In winter and spring it buds off by a process called **strobilation,** which involves a series of minute, saucer-shaped buds, called **ephyrae,** that break loose, swim away, and grow into mature, sexual medusae (Fig. 16-11).

Class Anthozoa

Anthozoans, or "flower animals," were so named because of their flowerlike appearance (*anthos,* flower). They are all polypoid with no medusa stages. They include more than 6,000 species of sea anemones, stony corals, horny, black, and soft corals, sea pens, sea pansies, sea feathers, and others. Some are solitary, and some are colonial; all are marine. Most anthozoans live attached to rocks, shells, or timbers;

327

Fig. 16-12. A, *Condylactus,* a large exotic sea anemone common to the Florida Keys area. Such anemones may be seen in delicate pinks, greens, and lavenders. **B,** Burrowing anemones *(Cerianthus),* that live in protective tubes in the mud. (**A,** Photo by F. M. Hickman; **B,** photo by C. P. Hickman, Jr.)

some sea anemones burrow into mud or sand (Fig. 16-12, *B*). Several live as commensals on the shells of hermit crabs (Fig. 16-1). Some hermit crabs actively seek out certain species of sea anemones and hold them on their shells until they attach. The sea anemones provide camouflage for the hermit crab and in return get free transportation and share in the food caught by the crabs.

Although polypoid, anthozoans are different from the hydrozoan polyps. They have a large coelenteron that is partitioned by longitudinal **mesenteries,** or **septa,** into radiating compartments (Fig. 16-13). The walls and mesenteries contain both circular and lon-gitudinal muscles. The fibers are often completely separated from the epithelial cells and may lie in bundles in the mesoglea. The mesoglea is cellular and mesodermal. Some anthozoans are provided with skeletons.

There are two subclasses. **Zoantharians,** which include the sea anemones and stony corals, are hexamerous—their tentacles and mesenteries are arranged on a plan of six or multiples of six, and their tentacles are unbranched. **Alcyonarians** include the sea fans, sea plumes, sea pansies, and other soft corals. They are octamerous and always have eight pinnately branched tentacles.

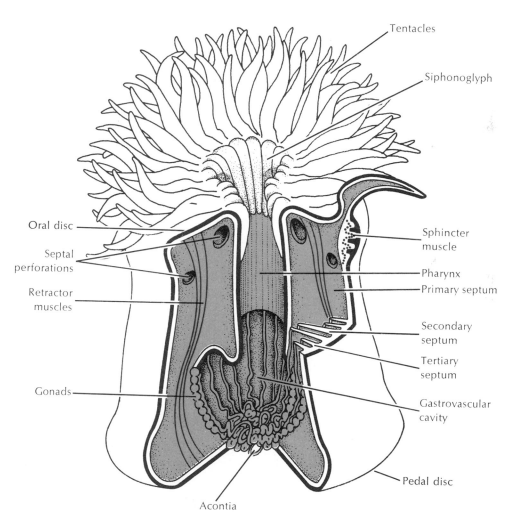

Fig. 16-13. Structure of a sea anemone. The free edges of the septa and the acontia threads are equipped with nematocysts to complete the paralyzation of prey begun by the tentacles.

Sea anemones

Most sea anemones range from 10 to 50 mm in diameter, but some are much larger. They are cylindric, with a crown of hollow tentacles arranged in circlets around the slit-shaped mouth (Fig. 16-13). At one or both ends of the mouth is a ciliated groove (**siphonoglyph**) leading into the pharynx. The grooves create currents to bring in oxygen-rich water, whereas cilia on the rest of the pharynx carry water and waste products out. The coelenteron is divided and subdivided by **septa,** or **mesenteries,** arranged by a hexamerous plan (Fig. 16-13). The edges of the septa bear nematocysts. At the base of the septa are threadlike **acontia** (also laden with nematocysts), which can be protruded through the mouth or body pores. These gastrodermal nematocysts subdue any prey that are swallowed still struggling.

Sea anemones are muscular and can glide on their pedal discs. They can expand in length and can stretch out their tentacles to find and overpower the prey and convey it to the mouth. They feed on various invertebrates and small fishes. When they are disturbed, the tentacles retract, retractor muscles in the mesenteries shorten the body column, and a sphincter muscle (in many species) causes the oral disc to be covered over (Fig. 16-13). Some anemones, to escape from predatory sea stars or nudibranchs, detach and move away with rhythmic creeping or swimming movements. Irritating substances given off by the enemy stimulate this escape reaction.

Anemones reproduce asexually either by budding or by moving away and leaving bits of their pedal discs, which regenerate into complete anemones. They also reproduce sexually and produce a ciliated **planula larva.**

Corals

The zoantharians, or hard corals, belong to the same subclass (Zoantharia) as the sea anemones, have a very similar hexamerous structure, but are smaller and are usually colonial. The stony coral animal looks much like a miniature sea anemone. However, it lives permanently in a stony cup with radial ridges into which it can withdraw for protection during the day and from which it can stretch forth at night to feed (Fig. 16-14, *A*). The cup is an exoskeleton of calcium carbonate that has been secreted by the epidermis around the polyp and between its paired septa. Thus the radial ridges of the cup bear the pattern of the polyp's septa (Fig. 16-14, *B*).

Fig. 16-14. A, Group of coral polyps, *Astrangia danae,* protruding from their shallow cups. This is a common coral of Atlantic coast. **B,** Stony coral *Oculina* showing cups in which polyps once lived. (**A** courtesy General Biological Supply House, Inc., Chicago.)

Most of the corals live in colonies, which assume a great variety of shapes. Colonies grow asexually by the budding of new polyps, especially along the edges of the colony. Each new individual secretes its own limy cup. They also reproduce sexually, producing ciliated larvae that may start new colonies elsewhere. New generations build over the skeletons of previous generations so that the living polyps are found only on the surface layers of coral masses. Over long periods of time great reefs are built by these tiny architects. Reefs generally grow at a rate of from 10 to 200 mm a year. Most existing reefs could have been formed in 15,000 to 30,000 years.

Fig. 16-15. *Alcyonium digitatum,* a colony-forming alcyonarian coral, abundant in Scandinavian waters. (Photo by T. Lundälv, Kristinebergs Zoological Station, Sweden.)

Alcyonarian corals are also called octocorallians because, unlike the stony corals, they always have **eight** tentacles and eight mesenteries (octamerous body plan). Each tentacle is branched in a featherlike fashion called pinnate. These are the horny and soft corals and include a wide variety, such as sea pens, sea whips, sea fans, sea pansies, and pipe corals. The polyps are usually smaller than those of the stony corals and are always colonial (Fig. 16-15). The polyps are embedded in a mass of mesoglea and connected by a system of canals. The mesoglea and polyps cover a secreted core of horny material or limy spicules that provide support for the colony.

Coral reefs are of great ecologic importance for they serve as habitats for a great variety of organisms, such as sponges, worms, echinoderms, molluscs, crustaceans, and many kinds of fishes. They also serve as barriers to surf and storm action, creating lagoons and sheltered coastlines, and thus are of economic importance.

PHYLUM CTENOPHORA

The ctenophores, a group of only a hundred or so species, all marine, are widely distributed but are especially abundant in warm waters. The name "Ctenophora" (te-nof'o-ra) comes from the Greek *ktenos,* meaning comb, and refers to the radially arranged ciliated comb plates (ctenes) that are used for swimming. Ctenophores are commonly called the comb jellies, sea walnuts, or sea gooseberries.

The ctenophores are nearly all free swimmers, using their ciliated comb plates to propel themselves oral-end forward. However, they are feeble swimmers and are often at the mercy of the tides and strong currents. They avoid storms by swimming downward in the water. Modified forms such as *Cestum* use sinuous body movements as well as comb plates to achieve a creeping type of locomotion. Ctenophores are fragile and beautiful (Fig. 16-16). Their transparent bodies glisten like glass, are brilliantly iridescent by day, and give off vivid flashes of luminescence by night.

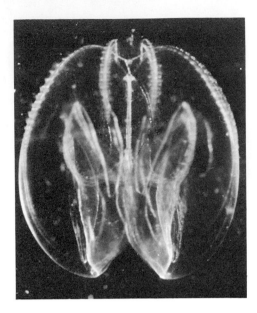

Fig. 16-16. The sea walnut, *Mnemiopsis*, a ctenophore. Its fragile beauty is particularly apparent at night when large numbers can be seen pulsating in the water. Common in Atlantic and Gulf waters. (Photo by Roman Vishniac; courtesy Encyclopaedia Britannica Films, Inc.)

Like the cnidarians, they have a form of radial symmetry (biradial) and the same three layers with the mesoglea layer well developed and cellular; also they have a diffuse nerve net. Unlike the cnidarians, they lack nematocysts (except in a single species, *Euchlora rubra*) but have comb plates and a type of adhesive cell called colloblasts. There is no polymorphism in this phylum as in the cnidarians.

In spite of their general resemblances there is no convincing evidence that ctenophores have evolved from cnidarians, although some kinship seems obvious. They appear to be a blind offshoot from the ancestral medusoid cnidarian and not in direct evolutionary line with the higher forms.

Characteristics

1. **Biradial symmetry** because of the arrangement of canals and tentacles
2. Usually ovoid or spheric in shape (a few are belt shaped) with radially arranged **rows of comb plates** for swimming
3. Ectoderm, endoderm, and ectomesoderm (mesoglea with cells and muscle fibers)
4. Nematocysts absent (except in one species) but with **adhesive cells (colloblasts)** present for food capture
5. **Gastrovascular system** consisting of a mouth, pharynx, stomach, and a series of canals

6. Nervous system consisting of a **subepidermal plexus** concentrated around the mouth and under the comb rows; an **aboral sense organ** (statocyst)
7. Monoecious reproduction; gonads (endodermal) on the walls of the digestive canals, which are under the rows of comb plates; determinate cleavage; cydippid larva
8. Luminescence common
9. No polymorphism or attached stages; no coelomic cavity

Classification

Class Tentaculata (ten-tak-yu-la'ta) (L. *tentaculum,* feeler; + *-ata,* group suffix). With tentacles, which may or may not have sheaths into which they retract; mostly ovoid or spheric; some flattened for creeping; others compressed into a bandlike form; in some species comb rows often missing in adults. *Pleurobrachia* (Fig. 16-17), *Mnemiopsis* (Fig. 16-16), *Cestum.*

Class Nuda (nu'da) (L. *nudus,* naked). Without tentacles; conic form; wide mouth and pharynx; branched gastrovascular canals. *Beroe.*

Form and function

Pleurobrachia, a pretty little sea walnut and a member of class Tentaculata, is often used as a representative example of the ctenophores (Fig. 16-17). Its surface bears eight longitudinal rows of transverse plates bearing long fused cilia and called **comb plates.** The beating of the cilia in each row starts at the aboral end and proceeds along the rows to the oral end, thus propelling the animal forward. All rows beat in unison. A reversal of the wave direction drives the animal backward. Ctenophores may be the largest animals that swim exclusively by cilia.

Two long tentacles are carried in a pair of tentacle sheaths (Fig. 16-17, *B*) from which they can stretch to a length of perhaps 15 cm. The surface of the tentacles bears specialized glue cells called **colloblasts,** which secrete a sticky substance that facilitates the catching of small prey organisms. When covered with food, the tentacles contract and the food is wiped off on the mouth. The gastrovascular cavity consists of a pharynx, stomach, and a system of gastrovascular canals. Rapid digestion occurs in the pharynx, then partly digested food is circulated through the rest of the system where digestion is completed intracellularly. Residues are regurgitated or expelled through small pores at the aboral end.

A nerve net system similar to that of the cnidarians includes a subepidermal plexus that is concentrated under each comb plate.

The sense organ at the aboral pole is a **statocyst,** or organ of equilibrium, and is also concerned with the beating of the comb rows but does not trigger their

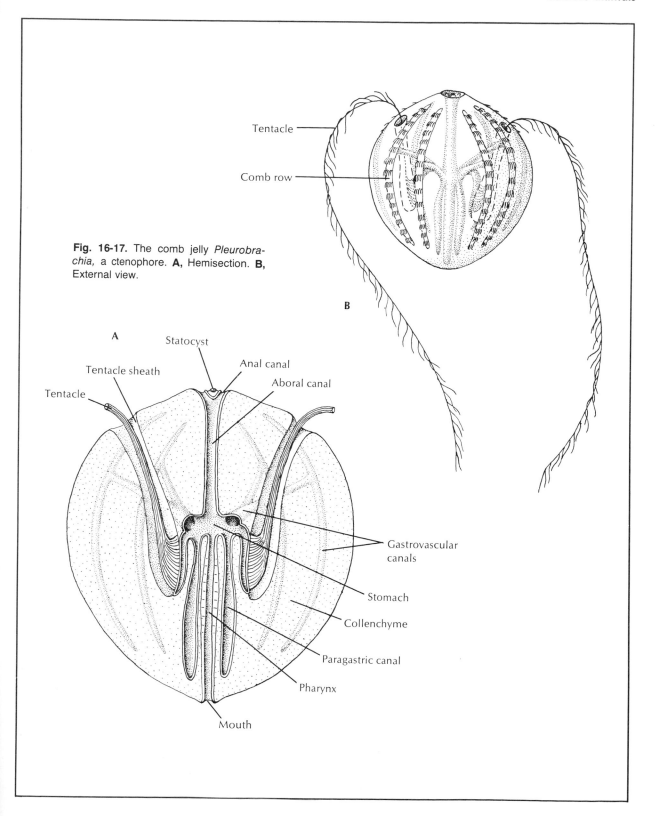

Tentacle

Comb row

Fig. 16-17. The comb jelly *Pleurobra-chia,* a ctenophore. **A,** Hemisection. **B,** External view.

B

A

Statocyst

Tentacle sheath

Anal canal

Aboral canal

Tentacle

Gastrovascular canals

Stomach

Collenchyme

Paragastric canal

Pharynx

Mouth

beat. Other sensory cells are abundant in the epidermis.

All ctenophores are monoecious, bearing both an ovary and a testis. Gametes are shed into the water, except in a few species that brood their eggs. Cleavage is total and determinate. The free-swimming larva, called a cydippid larva, is ovoid or spheric, resembling the ovoid or spheric adults of such ctenophores as *Pleurobrachia*. This shape further supports the belief that primitive ctenophores were ovoid or spheric and that the evolutionary tendency has been toward expanding or lengthening along the tentacular meridians. As a result there are such forms as *Mnemiopsis,* which is moderately flattened, and *Velamen* and *Cestum* (Venus's girdle), which are long and belt shaped.

Light production (bioluminescence) occurs in the walls of the canals under the comb rows, so that the light appears to come from the comb rows themselves. Ctenophores are noted for their luminescence.

REVIEW OF CONTRIBUTIONS OF THE RADIATE PHYLA

1. The body wall is of three layers, derived from ectoderm and endoderm.
2. A gastrovascular cavity (coelenteron) involves a mouth (no anus). There is both extracellular and intracellular digestion.
3. Tentacles aid in food capture.
4. The first nerve cells (protoneurons) have a nerve net arrangement.
5. The first sense organs are apparent (statocysts and ocelli).
6. Locomotion is by muscular contractions (cnidarians) or ciliary comb plates (ctenophores).
7. Biradial symmetry is present and is a forerunner of bilateral symmetry.
8. Polymorphism, which widens the ecologic possibilities, allows a species to occupy both a benthic (bottom) and a pelagic (open-water) habitat.

SELECTED REFERENCES

Barnes, R. D. 1974. Invertebrate zoology, 3rd ed. Philadelphia, W. B. Saunders Co.

Bayer, F. M. 1971. Coral. Natural History **80**(3):42-47 (March). *Excellent color photographs of reef corals.*

Bloom, A. L. 1974. Geomorphology of reef complexes. In L. F. LaPorte (ed.). Reefs in time and space. Tulsa, Society of Economic Paleontologists and Minerologists, Special publication No. 18.

Bourne, G. C. 1900. Ctenophora, In E. R. Lankester: A treatise on zoology. London, A. & C. Black, Ltd. *Classic detailed descriptions of the morphology of ctenophorans.*

Buchsbaum, R., and L. J. Milne. The lower animals. New York, Doubleday & Co., Inc. *Beautiful color illustrations.*

Bullough, W. S. 1950. Practical invertebrate anatomy. London, Macmillan, Inc. *A practical manual for certain selected types.*

Burnett, A. L. (ed.). 1973. Biology of hydra. New York, Academic Press, Inc.

Fraser, C. M. 1937. Hydroids of the Pacific coast of Canada and the United States. Toronto, University of Toronto Press. *This monograph and a similar one on the hydroids of the Atlantic coast are the most comprehensive taxonomic studies yet made on the American group.*

Gardiner, M. S. 1972. The biology of invertebrates. New York, McGraw-Hill Book Co.

Gosner, K. L. 1971. Guide to identification of marine and estuarine invertebrates, Cape Hatteras to the Bay of Fundy. New York, Interscience Div., John Wiley & Sons, Inc.

Hardy, A. C. 1956. The open sea. Boston, Houghton Mifflin Co. *Many beautiful plates of medusae and other coelenterates in this outstanding treatise on sea life.*

Hickman, C. P. 1973. Biology of the invertebrates, ed. 2. St. Louis, The C. V. Mosby Co.

Hyman, L. H. 1940. The invertebrates: Protozoa through Ctenophora. New York, McGraw-Hill Book Co. *Extensive accounts are given of the coelenterates (Cnidaria) and the ctenophorans in the last two chapters of this authoritative work.*

Jones, O. A., and R. Endean (eds.). 1976. Biology and geology of coral reefs. New York, Academic Press, Inc.

Lenhoff, H. M., and W. F. Loomis. 1961. The biology of hydra and some other coelenterates. Coral Gables, Fla., University of Miami Press.

Lenhoff, H. M., L. Muscatine, and L. V. Davis. 1971. Experimental coelenterate biology. Honolulu, University of Hawaii Press.

Meglitsch, P. A. 1972. Invertebrate zoology, 2nd ed. New York, Oxford University Press.

Miner, R. W. 1950. Field book of seashore life. New York, G. P. Putnam's Sons. *Good for identification of Atlantic coast forms.*

Muscatine, I., and H. M. Lenhoff. 1974. Coelenterate biology. New York, Academic Press, Inc.

Pennak, R. W. 1953. Freshwater invertebrates of the United States. New York, The Ronald Press Co. *Chapter 4 is devoted to the freshwater cnidarians, including a good description of the rare freshwater jellyfish.*

Rees, W. J. (ed.). 1966. The Cnidaria and their evolution. New York, Academic Press, Inc. *A symposium devoted entirely to the cnidarians of Great Britain.*

Russell, F. S. 1953. The medusae of the British Isles. New York, Cambridge University Press. *In this fine monograph the many species of British medusae are described in text and by beautiful plates, many in color.*

Smith, F. G. W. 1948. Atlantic reef corals. Miami, University of Miami Press.

Windsor, M. P. 1976. Starfish, jellyfish and the order of life. New Haven, Yale University Press.

SELECTED SCIENTIFIC AMERICAN ARTICLES

Berrill, N. J. 1957. The indestructible hydra. **197:**118-125 (Dec.). *Describes the remarkable regenerative powers of this animal.*

Feder, H. 1972. Escape responses in marine invertebrates. **227:**92-100 (July). *Sea anemones, molluscs, and other slow-moving sea creatures become remarkably active when approached by predatory starfish.*

Gierer, A. 1974. Hydra as a model for the development of biological form. **231:**44-54 (Dec.). *Dissection of a polyp provides clues to morphogenesis.*

Lane, C. E. 1960. The Portuguese man-of-war. **202:**158-168 (March). *This beautiful but dangerous jellyfish represents a remarkable colonial development.*

Loomis, W. F. 1959. The sex gas of hydra. **200:**145-156 (Apr.). *The carbon dioxide tension of the water appears to play an important role in controlling sexual reproduction in the hydra.*

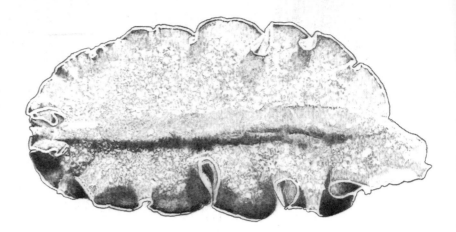

A marine turbellarian flatworm. Free-living flatworms of this type are flat and very thin, are usually 2 to 5 cm long, have elegantly ruffled edges, and swim or creep with graceful, undulating movements.

17 ACOELOMATE ANIMALS— FLATWORMS AND RIBBON WORMS

EVOLUTION AND ADAPTIVE RADIATION

PHYLUM PLATYHELMINTHES

    Characteristics
    Classification
    Class Turbellaria
    Class Trematoda—flukes
    Class Cestoda—tapeworms

PHYLUM NEMERTINA (RHYNCHOCOELA)— RIBBON WORMS

PHYLUM GNATHOSTOMULIDA

REVIEW OF CONTRIBUTIONS OF THE ACOELOMATES

All organisms in the animal phyla, with the exception of those we have already discussed, have **bilateral symmetry** and are collectively called the **Bilateria.** Even though in some forms, such as the echinoderms, bilateral symmetry is not present in the adults, it is always present in the embryo and therefore is considered the primary symmetry of the group.

Bilateral symmetry is much more suitable for active movement than is radial symmetry. The evolutionary trend toward active movement and bilateral symmetry is also associated with certain other selective pressures. Increased activity resulted in the need for better locomotor means, which usually meant better muscle development and coordination. This in turn created the need for sense organs and for a more centralized nervous system. If movement was to be in a forward direction, it was advantageous to have the sensory equipment concentrated in the forward end—thus evolution toward cephalization. As more complex neuromuscular systems evolved, there was an increase in metabolic requirements, so that more specialized systems arose for the functions of nutrition, respiration, and excretion. Thus bilateral symmetry was an important factor in initiating the evolutionary development of the organ systems.

The most primitive bilateral animals are those that have a single body cavity—that of the digestive tract. In phylum Platyhelminthes (flatworms), phylum Nemertina (nemertines, or ribbon worms), and the much smaller phylum Gnathostomulida, the space between the digestive tract and the body wall is not a cavity as in higher forms but is filled with a spongy mass of vacuolated mesenchymal cells called **parenchyma.** Thus these phyla are called **acoelomate,** or "without a coelom."

The acoelomate animals are worms—but "worms" is a very general term that might also apply to several other phyla. At one time zoologists considered all worms (all elongate invertebrate animals without appendages and with bilateral symmetry) to be a single group, which they called Vermes, but this has long since been broken up and the "worms" reclassified. By tradition, however, the various groups are still referred to as "flatworms" (Platyhelminthes), "ribbon worms" (Nemertina), "roundworms" (Nematoda), "segmented worms" (Annelida), etc.

Although primitive, the acoelomate phyla illustrate well some of the earliest effects of bilateral symmetry. These groups are elongated, and, though some still move largely by ciliary action, their wriggling movements are evidence of the much stronger **mesenchymal system of muscles,** which are arranged in circular and longitudinal layers, as well as dorsoventral fibers through the parenchyma. Acoelomates have a **central nervous system** of the "ladder type," cerebral ganglia (brain), and sensory eye spots. A primitive excretory system of **protonephridia** is present. The flatworms and gnathostomulids still possess a gastrovascular system with no anus, but the nemertines have gone a step further and have a complete digestive system—that is, a tubular gut with both mouth and anus. The nemertines are also the most primitive phylum to possess a **circulatory system** for more efficient internal transportation. Internal transport in flatworms and gnathostomulids, however, is still achieved entirely by diffusion.

The space occupied by jellylike mesoglea in the radiates is in the acoelomates filled with mesenchymal parenchyma. Whereas the mesoglea, often lacking cells, is derived largely from ectoderm, the parenchyma arises largely from endoderm and thus is a "true" mesoderm. It is a network of cells with long processes that tend to anastomose, and its fluid-filled spaces contain wandering ameboid cells.

EVOLUTION AND ADAPTIVE RADIATION

The three phyla—Platyhelminthes (flatworms), Nemertina (ribbon worms), and Gnathostomulida—appear to be closely related, with the flatworms being the most primitive. The earliest flatworms are thought to have been a primitive form of free-living turbellarian flatworm from which the more advanced turbellarians, the parasitic flatworms (flukes and tapeworms), and the nemertines diverged. The turbellarian ancestor may have come from a planula-like cnidarian form.

The flatworm body has become adapted in many ways for a parasitic, as well as a free-living, existence. In the nemertines, or ribbon worms, adaptive radiation is best seen in their peculiar proboscis apparatus. Originally a highly sensitive sensory organ used for exploring the environment, the proboscis has evolved into an efficient tool for capturing prey.

PHYLUM PLATYHELMINTHES

The name "Platyhelminthes" (plat-y-hel-min′theez) comes from the Greek *platys,* meaning flat, and *helmins,* meaning worm. Some 10,000 to 15,000 species of flatworms comprise this phylum. They range in length from less than 1 mm to 13 or 14 m.

The flatworms include the **turbellarians,** which are mostly free living, and the **flukes** and **tapeworms,** which are entirely parasitic. Of the turbellarians (class Turbellaria), some are symbiotic, some are parasitic, a few are terrestrial, and a number live in fresh water, but by far the majority of them are marine and free living. Only the larvae are pelagic as a rule; the adults hide under stones or in crevices in streams or in the littoral zones of the ocean.

The flukes and tapeworms (classes Trematoda and Cestoda) are largely endoparasitic in a variety of animals and usually have two or three hosts in the life cycle.

Characteristics

1. Three germ layers (triploblastic)
2. Bilateral symmetry
3. Body flattened dorsoventrally; mouth and genital pore usually ventral

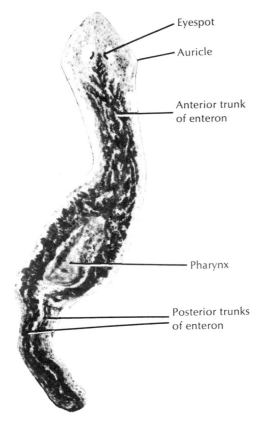

Fig. 17-1. A living planarian, *Dugesia.* (Photo by F. M. Hickman.)

4. Unsegmented body except in tapeworms
5. No coelom (acoelomate); spaces between organs filled with parenchyma, a form of connective tissue
6. Fluid-filled spaces in parenchyma serving as hydrostatic skeleton and aiding in internal transport
7. Ciliated epidermis in turbellarians; epidermis absent in flukes and tapeworms, which are covered instead by a protective tegument
8. Muscular system of mesenchymal origin; circular, longitudinal, and oblique fibers
9. Gastrovascular-type digestive system; digestive organs absent in tapeworms
10. Centralized nervous system consisting of cerebral ganglia and longitudinal nerve cords, connected by transverse nerves ("ladder type"); nerve plexus in primitive turbellarians (Acoela); eye spots in some
11. Excretory system consisting of protonephridia bearing flame bulbs
12. Complex reproductive systems; most forms monoecious; complicated life cycles in internal parasites

Classification

Traditionally phylum Platyhelminthes has been divided into the following three classes. However, because of a number of basic differences within these groups, the more recent trend has been to divide these classes by raising certain of their orders to the rank of class, resulting, according to some authorities, in eight classes. These new divisions are given in the footnotes.

Class Turbellaria* (tur-bel-lar'e-a) (L. *turbellae* pl., stir, bustle, + *-aria,* like or connected with). The turbellarians; usually free living, with soft, flattened bodies, ciliated epidermis containing secreting cells and rhabdites; mouth usually ventral. *Dugesia, Microstomum, Planocera.*

Class Trematoda† (trem-a-to'da) (Gr. *trematodes,* with holes, *eidos,* form). The flukes; body covered with thick nonciliated, living cuticle (tegument); leaflike or cylindric shape; suckers and sometimes hooks for attachment; all parasitic, often with complex life cycles. *Fasciola, Opisthorchis (Clonorchis), Schistosoma.*

Class Cestoda‡ (ses-to'da) (Gr. *kestos,* girdle, + *eidos,* form). The tapeworms; body covered with thick, nonciliated tegument; scolex with suckers or hooks or both for attachment; body divided into series of proglottids; no digestive organs; general tapelike form of body; all parasitic, usually with alternate hosts. *Diphyllobothrium, Taenia, Echinococcus.*

Class Turbellaria

Most of the turbellarians are free living, but a few are parasitic or commensal in aquatic hosts. Although

*Some authorities separate the Temnocephalida, which are commensal on the gills of crustaceans, into a separate class.
†Divided by some authorities into classes Monogenea, Aspidogastrea, Digenea, and Didymozoonidea.
‡Divided by some authorities into classes Cestodaria and Cestoda

most of the free-living turbellarians are marine, there are a number of freshwater forms and a few that are terrestrial. All are dorsoventrally flattened, some are leaf shaped, and some are long and almost cylindric. Most of them range from 0.5 mm to 5 cm, but some are longer. Covered with ciliated epidermis, they often combine muscular movements with ciliary action to achieve their creeping style of locomotion. The mouth is located on the ventral side.

Of the five orders of turbellarians, the Acoela are the most primitive, probably having changed little from the ancestral form. They have a mouth but no gastrovascular cavity or excretory system. The other orders are characterized by the shape of their gastrovascular cavities. In the two more primitive groups the digestive tract is either sac-shaped or straight and unbranched. In the planarians it is three pronged, and in the free-swimming marine forms it is greatly branched.

The most familiar of the turbellarians are the freshwater planarians such as *Dugesia* (Fig. 17-1) and *Dendrocelopsis;* the marine planarian *Bdelloura,* which lives on the gills of horseshoe crabs; and *Bipalium* and *Terricola,* which are terrestrial planarians found in tropical and subtropical climates—and sometimes in greenhouses. The freshwater planarians are used extensively in introductory laboratories. Marine polyclads are usually larger than freshwater planarians (2 to 5 cm), are more leaflike in shape, and often have ruffled edges (p. 336). Some are colorfully striped.

Form and function

Locomotion. Free-living flatworms have a **ciliated epidermis,** and smaller species can swim by ciliary action. **Mucous glands** secrete a slime track in which the beating of cilia provides the propulsive force for gliding or creeping. These slime glands also secrete a protective mucous layer over the animal's body. In larger forms ciliary action alone is inadequate, and waves of muscular action in conjunction with a **hydrostatic skeleton** create a much greater locomotory force for digging and burrowing activities. This fluid skeleton depends on the action of antagonistic muscles in the body wall contracting against an enclosed volume of fluid in a fixed space. In flatworms the parenchyma with its fluid-filled spaces serves as the hydrostatic skeleton; muscle fibers develop freely in the parenchyma instead of being restricted to

the epidermis and gastrodermis, as in cnidarians.

Adhesive glands aid the animal in anchoring itself to a substrate. Adhesive glands are numerous around the head, margin, and tail, and their sticky secretions aid in food capture as well as locomotion. In the epidermis of most turbellarians are numerous rod-shaped bodies known as rhabdoids, which are produced by gland cells and, when secreted, dissolve rapidly. They are thought to supplement the mucous glands in producing a slime covering that protects the animal against chemical irritants.

Nutrition, digestion, and circulation. Most turbellarians are carnivorous. They prey on various invertebrates that are small enough to handle, ranging from copepods, nematodes, and earthworms to small molluscs and crustaceans. Prey detection is by chemoreception or by feeling water disturbances caused by movements of prey. Their diet also includes dead animals. A few take in some plant material, and some contain symbiotic zoochlorella, which they digest if necessary.

With the exception of the acoels, turbellarians have a **muscular pharynx,** adapted for sucking, and an extensive gastrovascular cavity. The pharynx in the planarians is tubular and can be extended out through the ventral mouth (Fig. 17-2, *B* and *C*). The hungry animal grasps the prey with its pharynx and adhesive organs and clings to the substrate with whatever adhesive and muscular power it has, while it secretes a mucous and adhesive substance to entangle and subdue the prey. The animal may wrap itself around the prey, if the prey is small enough. Some species swallow the prey whole; others ingest small fragments at a time. The planarians partially predigest large prey by secreting proteolytic enzymes through the tubular pharynx thrust into the prey's tissue. The softened tissue is then sucked into the gastrovascular cavity by muscular action. *Phagocata* differs from most planarians in having not one pharynx, but many (Fig. 17-3, *A*). The genus *Stylochus* moves about inside an oyster shell and nips off bits of living oyster tissue. Land planarians feed upon nematodes, slugs, and earthworms.

Pharyngeal enzymes initiate disintegration of the food, which is continued by proteolytic enzymes in the digestive cavity (**extracellular digestion**). Fragmentation may be aided by the use of parenchymal muscles. Phagocytic cells in the lining of the gastrovascular cavity then engulf food particles and complete the digestion (**intracellular digestion**). In some turbellar-

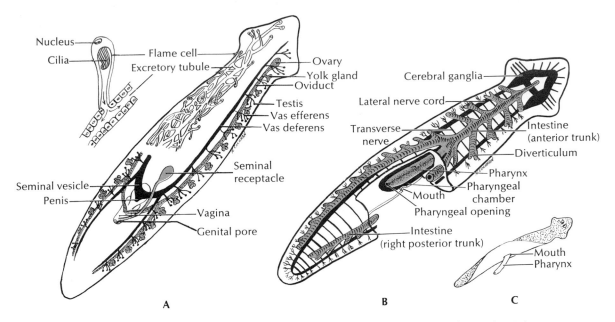

Fig. 17-2. Structure of a planarian. **A,** Reproductive and excretory systems, shown in part. Inset shows enlargement of a flame cell. **B,** Digestive tract and ladder-type nervous system. Pharynx shown in resting position. **C,** Pharynx extended through ventral mouth.

ians most of the digestion occurs in the gastrovascular cavity; in others digestion seems to be wholly intracellular. Protein and fat-digesting enzymes have been recovered from turbellarians, and starch digestion has been reported in the triclads.

In the primitive Acoela, which have no gastrovascular cavity at all, food passes directly into the parenchyma from the mouth and is digested intracellularly.

There is no circulatory system, but movement of materials through the fluid spaces of the parenchyma is aided by muscular body movements. Since the body is thin and the distance from one part of the body to another is not great, diffusion of materials is an adequate substitute for a transport system.

Excretion, osmoregulation, and respiration. A **tubular excretory system** that is also found in one form or another in a number of more advanced phyla first occurs in the flatworms. A protonephridial system consisting of protonephridial tubules and terminal flame bulbs is characteristic of the entire phylum, with the exception of the Acoela (Fig. 17-2, *A*).

A **protonephridium** is a tubule that branches into fine capillaries, each of which terminates in a **flame bulb.** The flame bulb bears the "flame," which is simply a tuft of cilia that flickers flamelike within the capillary. Usually the capillary and flame bulb together make up a tubular cell known as a "flame cell." The flame bulb cilia circulate the fluid in the tubules and perhaps regulate the rate of entry of substances through the tubular wall. The protonephridium opens to the outside by a **nephridiopore.** Among the various flatworms, there may be a single protonephridium or from one to four pairs. In planarians they anastomose into a network along each side of the animal (Fig. 17-2, *A*) and empty through a hundred or more nephridiopores.

The function of the protonephridia is still in question, and it may well be that they are more osmoregulatory than excretory in function, although there is evidence that the tubules function also in reabsorption of salts and possibly in secretion of waste materials. They are best developed in freshwater forms, in which ridding the body of excess water is always a problem; they may be poorly developed or absent in marine forms.

No special respiratory organs are present in flatworms. The small size and flattened shape promote **exchange of gases through the body surface.** Gas exchange is also encouraged by the epidermal cilia that keep the surface ventilated with moving water. A few

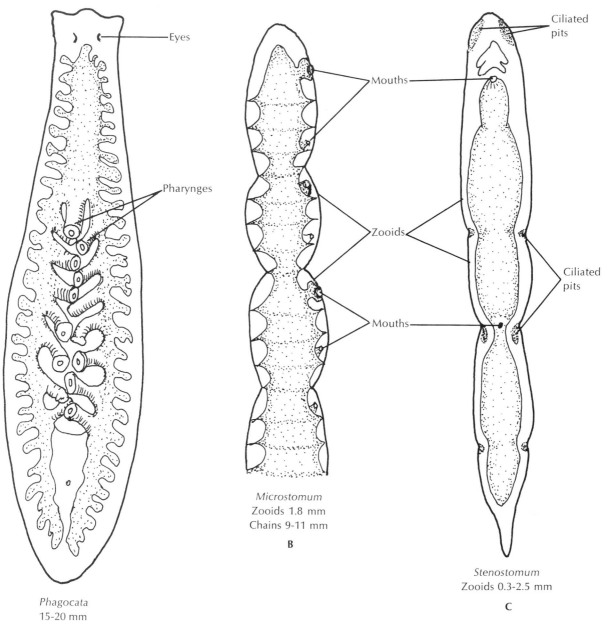

Microstomum
Zooids 1.8 mm
Chains 9-11 mm

B

Stenostomum
Zooids 0.3-2.5 mm

C

Phagocata
15-20 mm

A

Fig. 17-3. Some small freshwater turbellarians. **A,** *Phagocata* has numerous pharynges. **B** and **C,** Incomplete fission results for a time in strings of zooids.

turbellarians have been reported to have hemoglobin.

Nervous system. The most primitive flatworm nervous system, found in some of the acoels, is a **subepidermal nerve plexus** resembling the nerve net of the cnidarians. Other flatworms have, in addition to a nerve plexus, one to five pairs of **longitudinal nerve cords** lying under the muscle layer. The more advanced flatworms tend to have the lesser number of nerve cords. Freshwater planarians have one ventral pair (Fig. 17-2, *B*). Connecting nerves form a "ladder-type" pattern. The brain is a bilobed mass of ganglion cells arising anteriorly from the ventral nerve cords.

Except in the acoels, which have a diffuse system, the neurons are organized into sensory, motor, and association types—an important advance in the evolution of the nervous system.

Flatworms demonstrate some ability to learn, though the process is a slow one. Experiments have shown that *Dugesia* can be taught to avoid conditions it would not normally avoid but cannot be taught to turn toward conditions it would normally avoid.

Sense organs. Active locomotion in flatworms has favored not only cephalization in the nervous system but also advancements in the development of sense organs. **Ocelli,** or light-sensitive organs, are common in the turbellarians. Some have a simple pigment-spot ocellus. In *Dugesia,* the brown planarian (Fig. 17-1), the pigment cups are turned away from the light. Retinal cells extend from the brain to dip into the pigment cups; the photosensitive cells are inside the cup. They are sensitive to light direction and intensity but cannot form images. Planarians are negatively phototactic and feed mostly at night.

Tactile cells and chemoreceptive cells are abundant over the body, and in planarians they form definite organs on the auricles (the earlike lobes on the sides of the head). Tactile receptors usually have one or more sensory bristles exposed, and chemoreceptors form ciliated pits or grooves. Some also have statocysts for equilibrium and rheoreceptors for sensing water current direction.

Reproduction. Many turbellarians reproduce both asexually (by fission) and sexually. Asexually, the freshwater planarian merely constricts behind the pharynx and separates into two animals, each of which regenerates the missing parts—a quick means of population increase. There is evidence that a reduced population density results in an increase in the rate of fissioning. In some forms, such as *Stenostomum* and *Microstomum,* in which fissioning occurs, the individuals do not separate at once but remain attached, forming chains of zooids (Fig. 17-3, *B* and *C,*).

All turbellarians are monoecious (hermaphroditic). During the breeding season each individual develops both male and female organs, which usually open through a common genital pore (Fig. 17-2, *A*).

In the male system of the planarian a row of testes on each side is connected by ducts to a seminal vesicle, where mature sperm are stored until discharged through a muscular penis during copulation (the mating act). The female system consists of ovaries and yolk glands, which discharge eggs and yolk into a pair of oviducts that join to form a vagina. Connected to the vagina is a seminal receptacle, which during copulation receives and stores sperm from the mating partner.

Turbellarians practice cross-fertilization, mutually exchanging sperm during copulation. After copulation one or more fertilized eggs and some yolk cells become enclosed in a small cocoon. The cocoons are attached by little stalks to the underside of stones or plants. Embryos emerge as juveniles that resemble mature adults. In some marine forms the egg develops into a ciliated free-swimming larva (called Müller's larva).

Turbellarians display remarkable powers of **regeneration.** When cut, free cells (neoblasts) migrate to the cut surface to form a blastema (a mass of undifferentiated cells), which develops into the new part. Apparently chemical secretions from the wound attract the neoblasts.

Class Trematoda—flukes

The flukes, or trematodes, are all **parasitic,** and they parasitize every class of vertebrates. Nearly 6,000 species have been described. Some are ectoparasites, living on the body surface or occupying the mouth, gills, or cloaca; some are endoparasites, preferring certain internal organs. They are much more common in aquatic animals than in terrestrial ones. Most flukes are from 1 to 25 mm long, but some are larger; one from the ocean sunfish *Mola mola* is said to reach 7 m in length.

Trematodes are similar in structure to the free-living turbellarians, and their organ systems are also similar. Their sense organs are poorly developed; only a few have simple eyes. Flukes are unsegmented and usually somewhat leaf shaped, but, unlike the turbellarians, the adults are not ciliated. Movement is entirely muscular.

Two of the ways in which the trematodes have adapted to a parasitic life are shown in their lack of a true epidermis but presence of a protective **tegument** and the development of a variety of special **attachment organs** such as hooks, muscular or glandular discs, and true suckers. The tegument of both tapeworms and flukes was once thought to be a nonliving cuticle, but now it is known to be cytoplasmic and to contain mitochondria. It is made up of the protoplasmic extensions of deeper-lying cells, and is resistant to digestion by the enzymes in its environment. The outer surface of the tegument bears many microscopic fingerlike extensions (microtriches) that tend to mesh with the microscopic extensions (microvilli) of the lining of the host intestine. This provides a greater surface for absorption of nutritive materials and a convenient insurance against expulsion from the intestine by peristalsis.

Most flukes ingest tissue cells or fluid, blood cells, or mucus; some food is probably absorbed through the body surface. Digestion is primarily extracellular. Most flukes are hermaphroditic.

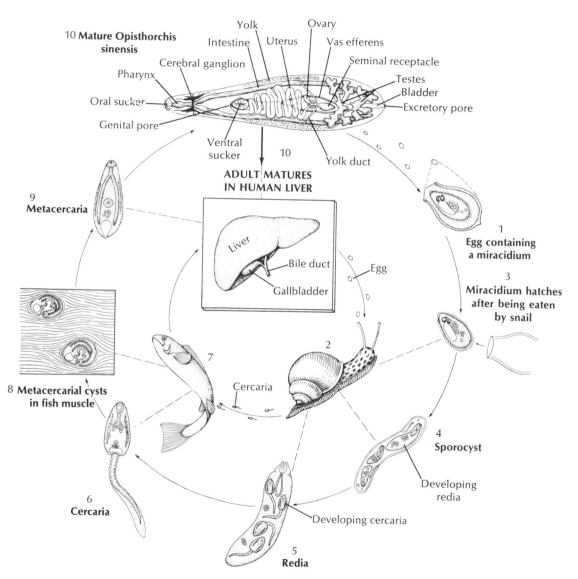

Fig. 17-4. Life cycle of human liver fluke, *Clonorchis sinensis*. Egg, *1,* shed from adult trematode, *10,* in bile ducts of man, is carried out in feces and ingested by snail, *2,* in which miricidium, *3,* hatches and becomes mother sporocyst. *4,* Young rediae are produced in sporocyst, grow, *5,* and produce young cercariae. Cercariae leave snail, *6,* find a fish host, *7,* and burrow under scales to encyst in muscle, *8.* When fish containing live cysts is eaten by man, metacercaria is released, *9,* and enters bile duct, where it matures, *10,* to shed eggs in feces, *1,* thus starting another cycle.

The trematodes are usually divided into three subclasses, or orders. The Monogenea are mostly ectoparasitic on vertebrates and occasionally on crustaceans and cephalopods, to which they cling tenaciously by means of powerful glandulomuscular organs that may include both hooks and suckers. They have only one host. Examples are *Polystoma,* found in the frog urinary bladder, and *Gyrodactylus,* which lives on the skin and gills of freshwater fish.

The Aspidogastrea are endoparasites of molluscs and cold-blooded vertebrates. Their ventral side is modified into an enormous compartmented sucker or a series of suckers. They have direct development and only one host.

The Digenea include most of the endoparasites. They have complicated life histories with two to four hosts in the life cycle. The sexual adults usually have vertebrates as their definitive hosts. The larval stages multiply asexually and are passed in intermediate hosts, often molluscs. Adults usually possess an oral sucker around the mouth at the anterior end and a large ventral sucker (Fig. 17-4). Each type of fluke is specific for a certain organ, such as the intestine, gall bladder, liver, lungs, bladder, kidneys, or blood vessels.

Larval stages of flukes

Trematodes usually have four larval stages, **miracidium, sporocyst, redia,** and **cercaria** (Fig. 17-4). Another stage, the **metacercaria,** is considered a juvenile fluke. Eggs released from the adult develop into ciliated larvae called miracidia, which may be ingested by an intermediate host. There the miracidium develops into a sporocyst, which is a sac of germinal cells that may multiply to form more sporocysts or may produce the next larval stage, the rediae (absent in many forms). Either the sporocysts or the rediae may, by internal budding, produce free-swimming cercariae, which leave the intermediate host and encyst on vegetation or in a third host. When the cyst and the enclosed metacercaria (juvenile) is eaten by a vertebrate it develops into an adult fluke.

Common fluke parasites of man

Man, as a vertebrate, is not exempt from trematode infestation, and, although some of these human parasites are less common in North America than in the Far East, the tropics, and much of the southern hemisphere, today's world travel dictates the need to be watchful. Hygienic conditions and thorough cooking of foods are the best preventives.

Clonorchis (Opisthorchis) sinensis is the most common of the liver flukes in man (Fig. 17-4). It is most prevalent in the Far East where the custom of eating raw fish is common. The adult lives in the human bile passages, and eggs containing the miracidia are shed in the feces. If ingested by certain freshwater snails, the sporocyst and redia stages are passed, and free-swimming cercariae emerge. Cercariae that manage to find a suitable fish encyst in the skin or muscles. If eaten raw by man, the juveniles migrate up the bile duct to mature and may survive there for 15 to 30 years. Heavy infestations cause cirrhosis of the liver and other ailments.

Schistosoma is a blood fluke for which the alternate host is a snail. It is spread when eggs shed in human feces and urine get into water containing host snails. Cercariae that contact human skin penetrate through the skin to enter blood vessels, which they follow to certain favorite regions depending upon the type of fluke. The inch-long adults may live for years in the human host, causing such disturbances as severe dysentery, anemia, liver enlargement, bladder inflammation, and brain damage. More than 200 million persons in Asia, Africa, South America, and the Caribbean are said to be infected by one or more of the three species of schistosomes that infect humans. Millions of new infections are being caused by the spread of infected snails through irrigation ditches. An avian schistosome common in our northern lakes often enters the skin of human bathers in its search for a suitable bird host, causing a skin irritation known as "swimmer's itch." In this case the human is a dead end in the fluke's life cycle.

Paragonimus is a lung fluke that can cause lung tissue damage and hemorrhaging. Though common in the Orient, it is also found in the United States. The three hosts are snails, crustaceans (freshwater crabs or crayfishes), and humans (or other carnivorous mammals). By eating poorly cooked or pickled crabs or crayfishes, humans can be infected by the parasite, whose eggs are coughed up in the sputum.

Class Cestoda—tapeworms

The tapeworms differ from turbellarians and trematodes in a number of ways. Their long, flat bodies are usually divided into segments, or **proglottids,** and they have **no digestive system** at all. Like the flukes, they

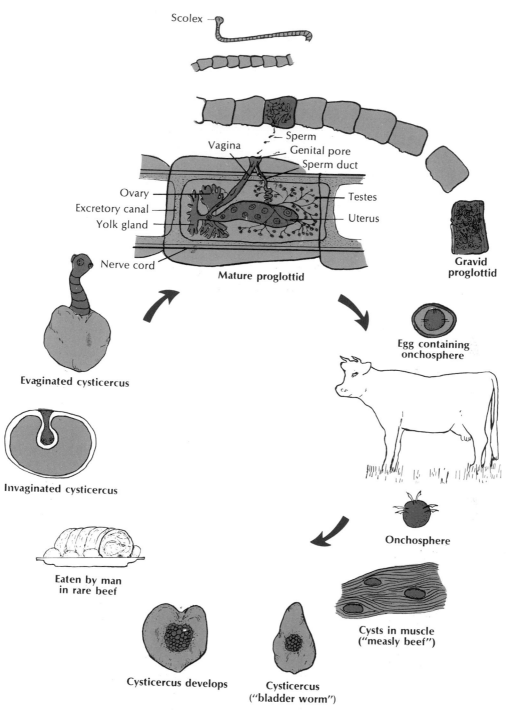

Scolex

Vagina
Sperm
Genital pore
Sperm duct

Ovary
Excretory canal
Yolk gland

Testes
Uterus

Nerve cord
Mature proglottid

Gravid proglottid

Evaginated cysticercus

Egg containing onchosphere

Invaginated cysticercus

Onchosphere

Eaten by man in rare beef

Cysts in muscle ("measly beef")

Cysticercus develops

Cysticercus ("bladder worm")

Fig. 17-5. Life cycle of beef tapeworm, *Taenia saginata.* Ripe proglottids break off in man's intestine, pass out in feces, and are ingested by cows. Eggs hatch in cow's intestine, freeing onchospheres, which penetrate into muscles and encyst, developing into "bladder worms." Man eats infected rare beef and cysticercus is freed in intestine where it develops, forms a scolex, attaches to intestine wall, and matures.

have no cilia or epidermis but are covered with a living **tegument.** One of their most characteristic and specialized structures is the **scolex,** or holdfast, an organ of attachment provided with suckers and sometimes hooks.

There are over a thousand species of tapeworms. They are all endoparasites, and all, with a few exceptions, involve at least two hosts. Life cycles include the definitive host, always a vertebrate, and one or two (sometimes more) intermediate hosts. Intermediate hosts may be invertebrates (often arthropods) or vertebrates.

One subclass, Cestodaria, has **unsegmented bodies that lack scolices;** but they do have some organ of attachment such as a rosette or proboscis. Most of them parasitize lower fishes. In the other subclass, Eucestoda, the body is usually divided into proglottids and is provided with a scolex, or holdfast, that may or may not bear hooks. This group contains the typical tapeworms, such as the beef tapeworm, *Taenia saginata* (Fig. 17-5), and the pork tapeworm, *T. solium*.

The anterior scolex is purely an organ of attachment for clinging to the host intestine. There is no mouth. Behind the scolex is a narrow neck, which produces by transverse constrictions the series of proglottids that make up the body of the worm (Fig. 17-6). The youngest and smallest proglottids are always at the neck end, and they increase in size and maturity toward the posterior end. Each proglottid is a sexually complete unit, the chief function of which is the production of gametes (Fig. 17-7).

The scolex is usually small and provided with suckers, hooks, or both for attachment. In *Taenia* there are four suckers arranged around the scolex, but the type and arrangement of suckers vary greatly in other tapeworms. The pork tapeworm, *T. solium,* and the dog tapeworm, *T. pisiformis,* have a circlet of hooks that is lacking in the beef tapeworm, *T. saginata.*

Tapeworms absorb food through the body surface since there is a complete absence of digestive organs. The tegument with its microtriches is considered an important factor in food absorption. Living as it does in the host's intestine, the tapeworm finds predigested products readily available.

A pair of nerve cords extends from a nerve ring in the scolex and runs through all the proglottids. There are no sense organs, but free sensory nerve endings are present. The protonephridial system consists of one or two pairs of excretory ducts running parallel to the

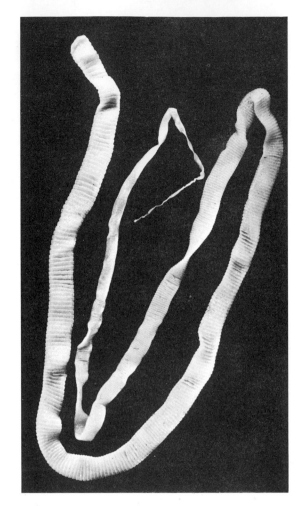

Fig. 17-6. Sheep tapeworm, *Moniezia expansa.* Note progressive increase in size. Young proglottids are budded from scolex and neck *(center);* oldest (gravid) proglottids shown at upper left. (Photo by F. M. Hickman.)

nerve cords (Fig. 17-7). The long ducts are connected by a transverse duct in each proglottid. Movement in tapeworms is entirely muscular.

Cestodes are all monoecious. Each proglottid is an effective little gamete factory, containing a complete set of both male and female sex organs (Fig. 17-7). Cross-fertilization is the rule in most species, but self-fertilization between two different proglottids of the same tapeworm or even within the same proglottid is known to occur.

When the terminal proglottids become greatly enlarged (gravid) with embryos, they break off and pass out of the host with the feces (Fig. 17-5). In some

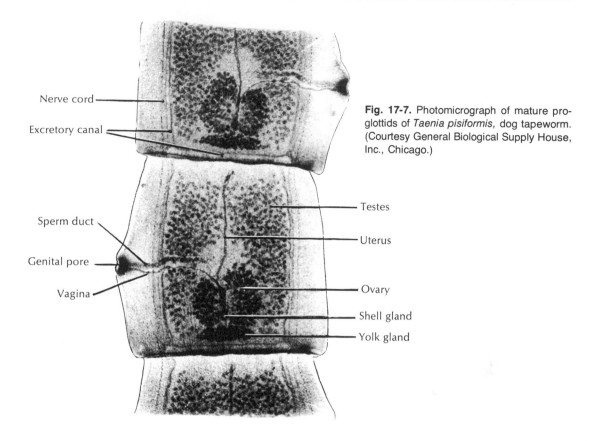

Nerve cord

Excretory canal

Sperm duct

Genital pore

Vagina

Testes

Uterus

Ovary

Shell gland

Yolk gland

Fig. 17-7. Photomicrograph of mature proglottids of *Taenia pisiformis*, dog tapeworm. (Courtesy General Biological Supply House, Inc., Chicago.)

species the young embryos are merely shed continuously from the gonopores of the proglottids. The embryos are thus scattered on grass or soil or in water, to be picked up by an intermediate host.

Examples of human parasites

In the beef tapeworm, *Taenia saginata,* eggs shed from the human host are ingested by cattle (Fig. 17-5). The six-hooked larvae (onchospheres) hatch, burrow into blood or lymph vessels, and migrate to skeletal muscle where they encyst to become "bladder worms" (cysticerci). Each larva develops an invaginated scolex and remains quiescent until the uncooked muscle is eaten by man or another suitable host. In the new host the scolex evaginates, attaches to the intestine, and matures in 2 or 3 weeks; then ripe proglottids may be expelled daily for many years. Man becomes infected by eating raw or rare infested ("measly") beef.

The pork tapeworm, *Taenia solium,* uses man as the definitive host and the pig as the intermediate host. Man can also serve as intermediate host by ingesting the eggs from contaminated hands or food or, in persons harboring an adult worm, by regurgitating segments into the stomach. The larvae may encyst in the central nervous system where great damage may result.

The fish tapeworm, *Diphyllobothrium latum (Dibothriocephalus latus),* often called the broad tapeworm of man, reaches a length of 20 to 40 m with 3,000 to 4,000 proglottids. The larvae develop in copepod crustaceans, which are eaten by a variety of fishes, including yellow perch, trout, northern pike, and salmon. The larvae encyst in the fish muscle, which, if eaten insufficiently cooked, may result in human infection. Heavy infections may cause intestinal obstruction and possibly toxemia. The fish tapeworm is found all over the world and is increasing in northern United States and Canada. Several deaths have been reported.

The dwarf tapeworm, *Hymenolepis nana,* needs no intermediate host but is spread by poor sanitary habits.

347

Dipylidium caninum, found in dogs, cats, and sometimes children, is carried by the louse or flea. Another dog tapeworm, *Taenia pisiformis,* uses the liver of rabbits for its larval encystment.

PHYLUM NEMERTINA (RHYNCHOCOELA) — RIBBON WORMS

The nemertine worms are often called ribbon worms or proboscis worms. They are soft, slender, flattened, free-living worms with a long, eversible proboscis. The proboscis is not usually connected with the digestive system. It is used as a tactile and defensive organ. It is in reference to the swift, unerring aim of this organ that the phylum is named, for Nemertina (nem-er-ti′na) comes from the Greek *nemertes* meaning unerring one. The alternate name, "Rhynchocoela," (ring-ko-se′la) means hollow beak and comes from the Greek *rhynchos,* beak, and *koilos,* hollow.

Of the nearly 600 species of nemertines, most range between a few millimeters and several inches in length, but some are longer; *Lineus longissimus,* a long, threadlike worm, is said to reach the incredible length of 30 m! They are very elastic and extensile.

Most nemertines are pale, but some have bright hues and patterns of red, yellow, orange, or green.

Most nemertines are marine, living in mud, under rocks, and among seaweed. Some, such as *Cerebratulus,* seek out empty mollusc shells. One genus lives in fresh water (*Prostoma rubrum* is a well-known example), and one lives in soil. A few are commensals.

Comparison with flatworms

The nemertines resemble the turbellarians in (1) having a ciliated epidermis, with rhabdoids in some, (2) being flattened in shape, (3) having the space between the body wall and enteron filled with parenchyma, partially gelatinous, (4) having a protonephridial system with flame bulbs, and (5) that the marine forms have a ciliated helmet-shaped larva called a pilidium (Gr. *pilidion,* little cap) similar to the larva of certain marine turbellarians.

There are, however, important differences between the two phyla. Nemertines have a **circulatory system,** a **complete digestive system** with mouth and anus, a unique **proboscis** carried in a special chamber, called a **rhynchocoel,** and **nervous and muscular systems** that are more highly developed than those of the flatworms.

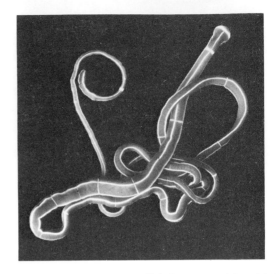

Fig. 17-8. The ribbon worm *Tubulanus annulatus* (phylum Nemertina) may be several feet long. The anterior end has the enlarged, rounded cephalic lobe. (Courtesy Encyclopedia Britannica Films, Inc.)

Evolution

That nemertines are an evolutionary offshoot from the free-living flatworms is strongly suggested by their acoelomate construction and their notable similarities to the turbellarians. Although not in the direct line leading to the coelomates, they are more highly organized than the flatworms and show a number of advancements that point to conditions found in higher groups.

Form and function

Many nemertines are difficult to examine because they are so long and fragile (Fig. 17-8). *Amphiporus,* one of the smaller genera that ranges from 2 to 10 cm in length, is fairly typical of the nemertine structure (Fig. 17-9). Its body wall consists of ciliated epidermis and layers of circular and longitudinal muscles. Locomotion consists largely of gliding over a slime track, though larger species move more by muscular contractions.

The mouth is anterior and ventral, and the digestive tract extends the length of the body, ending at the anus. The development of an anus marks a significant advancement over the gastrovascular systems of the lower phyla. Regurgitation of wastes is no longer necessary; ingestion and egestion can occur simultaneously. Food is moved through the intestine by cilia. Digestion is largely extracellular.

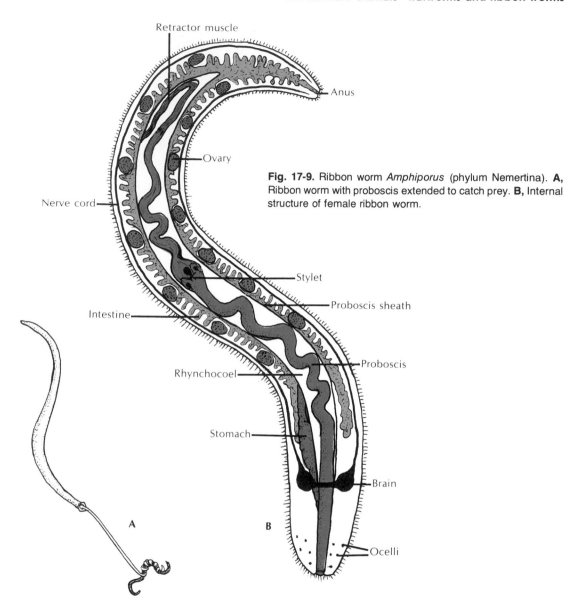

Retractor muscle

Anus

Ovary

Nerve cord

Fig. 17-9. Ribbon worm *Amphiporus* (phylum Nemertina). **A,** Ribbon worm with proboscis extended to catch prey. **B,** Internal structure of female ribbon worm.

Stylet

Proboscis sheath

Intestine

Proboscis

Rhynchocoel

Stomach

Brain

A   B

Ocelli

Nemertines are carnivorous, feeding primarily on annelids and other small invertebrates. They seize their prey with a proboscis that lies in an interior cavity of its own, the rhynchocoel, above the digestive tract (but not connected with it). The proboscis itself is a long, blind muscular tube that opens at the anterior end at a proboscis pore above the mouth. (In a few nemertines the esophagus opens through the proboscis pore rather than through a separate mouth.) Muscular pressure on the fluid in the rhynchocoel causes the long tubular proboscis to be everted through the proboscis pore. Eversion of the proboscis exposes a sharp barb, called a stylet (absent in some nemertines). The sticky, slime-covered proboscis coils around the prey and stabs it repeatedly with the stylet, which pours a toxic secretion on the prey. As the proboscis is withdrawn by a retractor muscle, the prey is brought close to the mouth so that it can be engulfed and swallowed. The proboscis withdraws to its own cavity, ready to be shot out again in defense or in search of food.

349

Unlike other acoelomates, the nemertines have a true circulatory system, usually consisting of one middorsal and two lateral blood vessels connected by transverse vessels. Since it lacks a heart, movement of the blood is maintained by the contractile walls of the vessels, but the flow is irregular and may be backward or forward.

The excretory system consists of a pair of protonephridia with flame bulbs. The nervous system is composed of a pair of ganglia, and one or more pairs of longitudinal nerve cords are connected by lateral nerves. Sense organs include ocelli, cerebral organs that are chemoreceptive, and ciliated grooves and pits that are both tactile and chemoreceptive.

Some species reproduce asexually by fragmentation and regeneration. *Amphiporus* is dioecious, as are the majority of nemertines, and has simple gonads and ducts for the discharge of gametes. In some forms a helmet-shaped, ciliated pilidium larva develops from the zygote. Other nemertines lack a free-swimming larval stage.

PHYLUM GNATHOSTOMULIDA

The phylum Gnathostomulida (nath-o-sto-myu′lid-a, Gr. *gnathos,* jaw, + *stoma,* mouth) makes up a small group of 12 or so genera of minute wormlike animals that live in the fine sand and silt of coastal sediments. Their phylogeny is still obscure. Like the Platyhelminthes, they are acoelomate and lack a circulatory system and an anus. In fact, they were at first included in the turbellarians. However, in contrast to the turbellarians, the pharynx is armed with lateral jaws used in scraping up fungi and bacteria, and the epidermal cells have only one cilium per cell. Gnathostomulids can glide, swim, and bend the head from side to side.

REVIEW OF CONTRIBUTIONS OF THE ACOELOMATES

1. The acoelomates have developed the basic **bilateral plan of organization** that has been widely exploited in the animal kingdom.
2. The **mesoderm** is now developed into a well-defined embryonic germ layer (triploblastic), thus making available a great source of tissues, organs, and systems and pointing the way to greater complexity of animal structure.
3. Along with bilateral symmetry, **cephalization** has been established. There is some centralization of the nervous system evident in the ladder type of system found in flatworms.
4. Along with the subepidermal musculature, there is also a mesenchymal system of muscle fibers.
5. An **excretory system** appears for the first time.
6. In the nemertines the first **circulatory system** with blood and the first **one-way alimentary canal** are developed.

SELECTED REFERENCES

Baer, J. G. 1971. Animal parasites. New York, McGraw-Hill Book Co.

Brown, F. A. (ed.). 1950. Selected invertebrate types. New York, John Wiley & Sons, Inc. *Directions for laboratory study of representative tubellarians.*

Burt, D. R. R. 1970. Platyhelminthes and parasitism: an introduction to parasitism. New York, American Elsevier Publishing Co., Inc.

Cheng, T. C. 1964. The biology of animal parasites. Philadelphia, W. B. Saunders Co.

Edmondson, W. T., H. B. Ward, and G. C. Whipple, (eds.). 1959. Freshwater biology, 2nd ed. New York, John Wiley & Sons, Inc. *Contains key and figures of the common freshwater turbellarians.*

Erasmus, D. A. 1972. The biology of trematodes. London, Edward Arnold (Publishers) Ltd.

Fallis, A. M. (ed.). 1971. Ecology and physiology of parasites. Toronto, University of Toronto Press.

Hyman, L. H. 1951. The invertebrates: Platyhelminthes and Rhynchocoela, vol. 2. New York, McGraw-Hill Book Co. *A comprehensive account of this group, with excellent figures and bibliography.*

Light, S. F., R. I. Smith, F. A. Pitelka, D. P. Abbott, and F. M. Weenser. 1967. Intertidal invertebrates of the central California coast. Berkley, University of California Press.

Lumsden, R. D. 1975. Surface ultrastructure and cytochemistry of parasitic helminths. Exp. Parasitol. **37**:267-339.

Pennak, R. W. 1953. Fresh-water invertebrates of the United States. New York, The Ronald Press Co. *Includes an excellent description of the structure and life history of the freshwater nemertine Prostoma rubrum.*

Riser, N. W., and M. P. Morse (eds.). 1974. Biology of the Turbellarea. New York, McGraw-Hill Book Co.

Schell, S. C. 1970. How to know the trematodes. Dubuque, William C. Brown Co., Publishers.

Schmidt, G. D. 1970. How to know the tapeworms. Dubuque, William C. Brown Co., Publishers.

Schmidt, G. D., and L. S. Roberts. 1977. Foundations of parasitology. St. Louis, The C. V. Mosby Co.

Smyth, J. D. 1962. Introduction to parasitology. London, English Universities Press.

Smyth, J. D. 1966. The physiology of trematodes. San Francisco, W. H. Freeman & Co. Publishers.

Smyth, J. D., 1969. The physiology of cestodes. San Francisco, W. H. Freeman & Co. Publishers.

Swellengrebel, N. H., and M. N. Sterman. 1961. Animal parasites in man. New York, D. Van Nostrand Co. *An English edition of a well-established Dutch text.*

Wardle, R. A., and J. A. McLeod. 1952. The zoology of tapeworms. Minneapolis, University of Minnesota Press. *A comprehensive account.*

Wardle, R. A., J. A. McLeod, and S. Radinovsky. 1974. Advances in the zoology of tapeworms, 1950-1970. Minneapolis, University of Minnesota Press.

Wells, M. 1968. Lower animals. New York, McGraw-Hill Book Co.

Wilson, R. A., and L. A. Webster. 1974. Protonephridia. Biol. Rev. **49**:127-160.

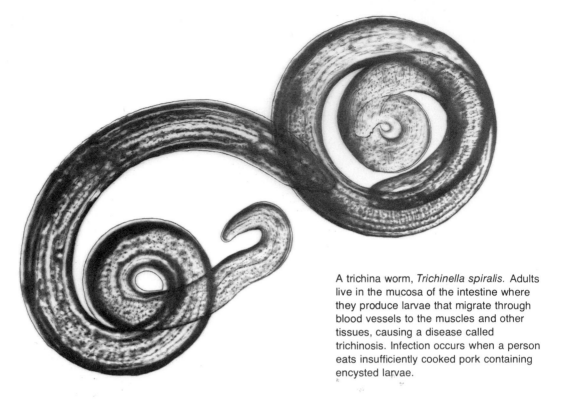

A trichina worm, *Trichinella spiralis.* Adults live in the mucosa of the intestine where they produce larvae that migrate through blood vessels to the muscles and other tissues, causing a disease called trichinosis. Infection occurs when a person eats insufficiently cooked pork containing encysted larvae.

18 PSEUDOCOELOMATE ANIMALS

PHYLUM ROTIFERA

PHYLUM GASTROTRICHA

PHYLUM KINORHYNCHA

PHYLUM NEMATODA—ROUNDWORMS

PHYLUM NEMATOMORPHA

PHYLUM ACANTHOCEPHALA—SPINY-HEADED WORMS

PHYLUM ENTOPROCTA

In contrast to the radiate and acoelomate animals in which the space between the body wall and the digestive tract is filled with mesoglea or with a solid mesenchymal parenchyma, the remaining bilateral animals have a **body cavity** of some sort in which the organs are located. In higher invertebrates and vertebrates the body cavity is a **coelom,** which forms in the mesoderm of the embryo and is lined with a membrane, the **peritoneum.** In the pseudocoelomate phyla the embryonic blastocoel persists as a body cavity that is *not* lined with peritoneum. Therefore the cavity is called a "false cavity," or **pseudocoel.**

The pseudocoel may be filled with fluid or with a gelatinous substance, but it has the advantage over the parenchyma-filled space of (1) permitting greater freedom of movement, (2) providing ample space for the development and differentiation of digestive, excretory, and reproductive systems, (3) providing a simple means of circulation or distribution of materials throughout the body, (4) acting as a storage space for waste products that are later discharged to the outside by excretory ducts, and (5) often playing an important role as a hydrostatic organ.

The pseudocoelomates are a heterogeneous assemblage of animals that seem to have little in common except the pseudocoel. Five of the groups, Rotifera, Gastrotricha, Kinorhyncha, Nematoda, and Nematomorpha, are often grouped together as a phylum or superphylum called Aschelminthes (as-kel-min'theez), or "cavity worms," but many authorities believe that they are different enough to be considered as separate phyla. The Entoprocta used to be grouped together with the Ectoprocta and called collectively the Bryozoa (moss animals), but the Ectoprocta have a true coelom and the Entoprocta a pseudocoel. The Acanthocephala, or "spiny-headed worms," are entirely parasitic.

As diversified as they are, the pseudocoelomates do share a few other characteristics. They all have a body wall of epidermis, dermis, and muscles, and they all have a complete mouth-to-anus digestive tract. In a number of them there is an emphasis upon the longitudinal muscle layer.

PHYLUM ROTIFERA

The rotifers, or "wheel animals" (L. *rota,* wheel, + *fera,* those that bear), derive their name from the beating of an anterior crown of cilia that gives the appearance of a revolving wheel. Rotifers are mostly microscopic (0.5 to 1.5 mm long), though a few are

longer, and they are usually transparent. They have a world-wide distribution, living predominantly in fresh water.

The body, which is covered by a thin cuticle, is usually made up of a head, trunk, and foot. The head region bears a ciliated organ called the **corona.** The corona, or wheel organ, may form a ciliated funnel with its upper edges folded into lobes bearing bristles, or, as in the familiar *Philodina* (Fig. 18-1), the corona may be made up of a pair of ciliated discs. The cilia create currents of water toward the mouth that draw in small planktonic forms for food. The corona may be retractile. The **trunk** contains the visceral organs, and the terminal **foot,** when present, is segmented and in some, such as *Philodina,* is ringed into joints that can telescope to shorten. The one to four toes secrete a sticky substance for attachment.

The mouth, surrounded by some part of the corona, opens into a modified muscular pharynx called a **mastax,** which is a unique characteristic of the rotifers. The mastax is equipped with a set of intricate jaws composed of seven hard pieces called trophi and used for grasping and chewing. Some rotifers are

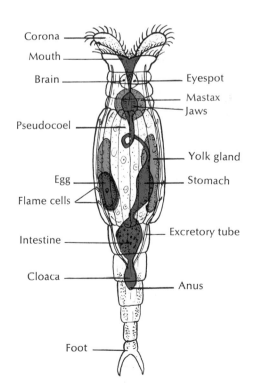

Fig. 18-1. Structure of *Philodina,* a common rotifer.

trappers, using a funnel-shaped area around the mouth to collect food, and lobes that fold in to hold the food until it is swallowed. Others are hunters whose trophi can be projected and used as forceps to seize the prey, bring it into the pharynx, and break it up. Salivary and gastric glands are believed to secrete enzymes for extracellular digestion.

Rotifers have a pair of protonephridial tubules with flame bulbs; a bilobed brain; and sense organs that include eyespots, sensory pits, and papillae.

Rotifers are dioecious. The males, however, are few in number and with few exceptions are degenerate, having neither mouth nor digestive organs.

Philodina and others of the same family reproduce parthenogenetically; that is, females produce only diploid eggs that have not undergone reduction division, cannot be fertilized, and develop only into females. Such eggs are called **amictic eggs.** Most other rotifers can produce two kinds of eggs—amictic eggs, which develop parthenogenetically into females, and **mictic eggs,** which have undergone meiosis and are haploid. Mictic eggs, if unfertilized, develop quickly and parthenogenetically into males; if fertilized, they secrete a thick shell and become dormant for several months before developing into females. Such dormant eggs can withstand desiccation and other adverse conditions and permit rotifers to live in temporary ponds that dry up during certain seasons.

PHYLUM GASTROTRICHA

Phylum Gastrotricha (gas-trot're-ka) is a small phylum made up of microscopic animals, approximately 65 to 500 μm long, that are flattened on the ventral side and are usually bristly or scaly in appearance. The outer cuticle is modified into patterns of bristles, scales, or plates. The head is ciliated, and the tail is often forked.

Gastrotrichs are common in lakes and ponds and in seashore sands. There are two orders, one marine and one largely freshwater. *Chaetonotus* is a common freshwater genus (Fig. 18-2). Gastrotrichs are probably closely related to the rotifers.

Some gastrotrichs move by gliding on ventral cilia, as their name implies (*gastr-*, from Gr. *gastēr*, belly; *trich-*, from *thrix*, hair). Others move in a leechlike fashion by briefly attaching the posterior end by means of adhesive glands. They feed on bacteria, diatoms, and small protozoans, which are drawn in by cilia or sucked in by the pharynx. They have a pair of

Fig. 18-2. *Chaetonotus*, a common gastrotrich.

protonephridia, a ganglionic brain mass, and sensory hairs and pits. Most marine forms are hermaphroditic; freshwater forms are parthenogenetic females.

PHYLUM KINORHYNCHA

The kinorhynchs comprise approximately 100 species of minute marine worms, usually less than 1 mm long, that prefer shallow mud bottoms. Unlike the rotifers and gastrotrichs, they have no external cilia. The cuticle is characteristically divided into 13 or 14 segments (zonites). The head, or first zonite, has circlets of recurved spines and can be retracted into the neck segments (Fig. 18-3). The mouth is on the end of a protrusible cone, the tip of which is surrounded by spines. The name "Kinorhyncha" (kin-o-ring'ka), which comes from the Greek *kineo,* meaning to move, and *rhynchos,* meaning beak or snout, refers to the retractile head cone, or proboscis, which is characteristic of the group.

Kinorhynchs burrow into the mud by extending the head, anchoring it by its recurved spines, and drawing the body forward until the head is retracted; they then repeat the process. They feed on organic sediment in the mud in which they live. A few feed on diatoms.

The kinorhynchs share anatomic features with several groups without being closely related to any. With the rotifers and gastrotrichs they share protonephridia, spines, and retractile head ends; with nematodes they share longitudinal cords, the pattern of the nervous system, and copulatory spicules. They are segmented, but their segmentation is superficial, rather than complete as it is in annelids.

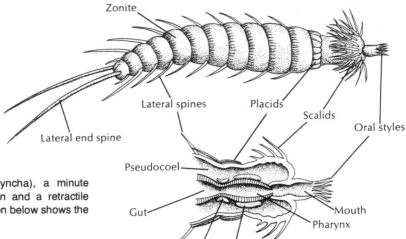

Fig. 18-3. *Echinoderella* (phylum Kinorhyncha), a minute marine worm with superficial segmentation and a retractile head with a circlet of spines. The hemisection below shows the internal structure of the anterior end.

PHYLUM NEMATODA — ROUNDWORMS

It has been said that, if the earth were to disappear, leaving only the nematode worms, the general contour of the earth's surface would be outlined by the worms, for they are present in nearly every conceivable kind of ecologic niche. Approximately 10,000 species have been named, but it has been estimated that, if all species were known, the number would be nearer 500,000. They live in the sea, in fresh water, and in soil, from polar regions to the tropics, and from mountain tops to the depths of the sea. Good topsoil may contain billions of nematodes per acre. Nematodes also parasitize virtually every type of plant and animal. The effects of nematode infestation on crops, domestic animals, and man make this phylum one of the most important of all parasitic animal groups. Nematode eggs are especially resistant and can be carried far by animals and winds.

Form and function

Most nematodes are less than 5 cm long; many of them are microscopic, but some parasitic nematodes are more than 1 m long. Their phylum name, which comes from the Greek *nema,* meaning thread, and *eidos,* meaning form, is descriptive of their long, slender shape and of the fact that they are perfectly cylindric in cross-section. Cilia are entirely lacking in nematodes, except in certain sense organs.

Nematodes are all basically very much alike, although there are some differences in structure and in life histories. Because they are large and readily available, some species of the genus *Ascaris* (Fig. 18-4) is usually selected as a representative type for study of the phylum.

Body wall. The body wall of the typical roundworm is made up of a thick, tough cuticle, a syncytial epidermis (which secretes the cuticle), and a layer of longitudinal muscles arranged in four bands (circular muscles are lacking entirely). Between the muscle bands the epidermis projects inward as four longitudinal epidermal cords, which enclose the dorsal and ventral nerve cords and the lateral excretory canals. The muscle cells have innervation processes that extend to the nerve trunks in the epidermal cords.

Movement involves undulatory waves of contraction along the longitudinal muscle fibers, producing a snakelike gliding movement. The pliable quality of the cuticle and the hydrostatic skeleton provided by the fluid-filled pseudocoel are important factors in locomotion.

Nutrition. Many free-living nematodes are carnivores and feed on small metazoans; others feed on diatoms, algae, or the cells of plant roots; still others feed on dead organic matter or on the bacteria and fungi found in dead matter. The cuticle lining the mouth cavity is often thickened to form ridges, plates,

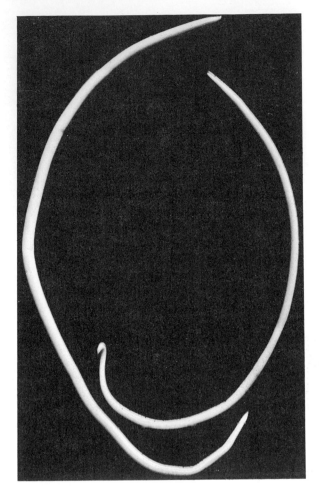

Fig. 18-4. Intestinal roundworm *Ascaris lumbricordes*, male and female (phylum Nematoda). Male *(right)* is smaller and distinguished by the kink in its tail.

that inhabit deep mud bottoms of lakes and seas, the respiration is anaerobic; that is, glucose is broken down to produce ATP in the abscence of oxygen. Many nematodes have been found to have hemoglobin in the fluid of the pseudocoel, which may provide a small reserve of oxygen although its function is not fully understood.

Excretion and osmoregulation. Protonephridia are absent, but some forms (primitive marine forms) have one or two large gland cells called **renettes,** which are located in the pseudocoel and empty through an excretory pore. Others have an H-shaped **canal system** evolved from the renettes. Either type is probably chiefly osmoregulatory in function.

Nervous and sensory system. A nerve ring connects with dorsal and ventral nerve cords in the epidermal cords. The chief sense organs are papillae of the lips and cephalic bristles. A few aquatic nematodes have ocelli. Most free-living nematodes possess a pair of pits or slits called amphids in the head region. Most parasitic nematodes have a pair of glandular phasmids in the tail region. Both amphids and phasmids are thought to be chemoreceptors.

Reproduction. Most nematodes are dioecious. The male is smaller than the female, and its posterior end, which is sharply curved, bears a pair of copulatory spicules. Fertilization is internal, and eggs are usually stored in the uterus until deposition. Embryonic development may begin before the eggs are discharged.

Nematode parasites of man

Nearly all vertebrates and many invertebrates are parasitized by nematodes. One authority estimated that more than 3 billion pets and farm animals in the United States have nearly 8 billion worm infestations a year, most of these being nematodes. Another estimate listed 44 million humans as harboring some sort of parasitic worm—again mostly nematodes. With United States government inspection of meats and with proper cooking methods, the United States citizen should have little fear of parasites. However, world travel today brings us into contact with the most remote parts of the earth so that the possibility of exposure to parasites is still a very real problem. Only some of the more common nematode parasites of man are described in this discussion.

Ascaris—intestinal roundworm. *Ascaris lumbricoides* (Fig. 18-4) is usually acquired by swallowing embryonated ova. Unsanitary habits and consumption

or teeth, or in plant parasites it is modified to form a stylet to puncture plant cells. Enzymic secretions may be ejected from the mouth to partially predigest food, so that the food, when ingested, is often largely fluid. The gut is a simple tube in which digestion is completed extracellularly. Glycogen and fat reserves can be stored in the intestinal cells for use during periods of starvation or molting.

Gaseous exchange. Respiratory organs are absent, yet the nematodes have adapted well to low-oxygen environments such as mud, wet soil, and parasitic conditions. Nematode parasites living in blood and tissues have an aerobic type of metabolism. However, in those that live in the intestinal lumen, such as *Ascaris* and *Oxyuris,* and in those free-living forms

of contaminated food and vegetables are frequent causes of infection.

A large female ascaris, which can easily reach 30 cm in length, may lay 200,000 eggs per day. These are eliminated in the host feces. On the ground under suitable conditions, larvae may develop in the shells within 2 or 3 weeks. If ingested at this stage, the embryonated eggs hatch and the larvae burrow through the intestinal wall into the veins or lymph vessels. In the blood they are carried through the heart to the lungs, move up the trachea, cross over into the esophagus, and then go down the alimentary canal to the intestine where they grow to maturity in approximately 2 months. They copulate, and the female begins her egg laying. No intermediate host is required in the life cycle. The greatest damage to the host occurs while the juvenile worms are migrating.

Hookworm. Hookworms are so named because the male has a hook-shaped body; actually they have no hooks. The adult females of *Necator americanus* are 9 to 15 mm long; males are shorter. Cutting plates in their mouths (Fig. 18-5) cut into the intestinal mucosa of the host where they suck up blood and body fluids. They secrete an anticoagulant to prevent blood clotting while they feed; they often leave a bleeding wound where they have fed.

Eggs are passed with the host's feces and hatch in a day or so on the warm soil. The larvae can live several weeks in the soil, feeding mainly on bacteria. When human skin comes in contact with infested soil, the larvae burrow through the skin into the blood. The bare foot is a common point of entry where a mild irritation known as "ground itch" may occur. Hookworms can live for years in a host and can cause anemia in the patient. Heavy infections may result in retarded mental and physical growth and general loss of energy. Sanitary disposal of feces and the wearing of shoes are preventives.

Trichina worm. *Trichinella spiralis* is the tiny nematode responsible for the serious disease trichinosis. Adults burrow into the mucosa of the small intestine where the female may produce as many as 1,500 live larvae. The larvae penetrate into blood vessels and are carried to the skeletal muscle where they coil up and form cysts that become calcified (Fig. 18-6). The worms may live in the cysts for years, if not disturbed. When meat containing live cysts is eaten, the liberated larvae mature and produce living larvae. Besides man, trichina worms infest hogs, rats, cats,

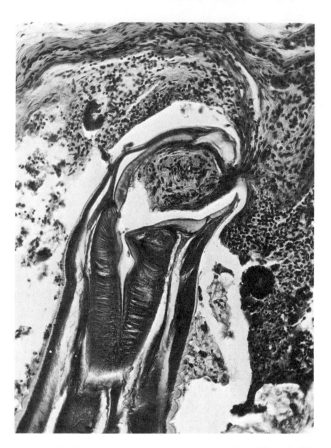

Fig. 18-5. Section through hookworm (phylum Nematoda) attached to human intestine. Note cutting plates of mouth pinching off bit of mucosa from which muscular pharynx sucks blood. Mouth secretes anticoagulant. (AFIP No. 33810.)

dogs, and many wild animals. Man usually acquires them by eating improperly cooked pork. Hogs acquire them by eating garbage containing pork scraps with cysts or by eating infested rats. The best preventive is thorough cooking of all pork.

Pinworms. The pinworm *Enterobius* is very common throughout the world (Fig. 18-7). Reported surveys indicate that 35% to 41% of the population and 37% to 57% of the children in the United States were infected. The adult pinworms, measuring as much as 13 mm long, live in the large intestine. Females with eggs migrate to the anal region at night to lay their eggs. Since this causes irritation, scratching at night is common, and fingers, clothing, bedding, and even the air become contaminated. When ova are swallowed, they hatch in the duodenum and mature in the large intestine. No intermediate host is necessary. Each generation lasts 3 to 4 weeks. If reinfection does not occur, they die.

Filarial worms. The filarial worms *Wuchereria* are tropical nematodes, 3 to 10 cm long, that live in the lymphatic glands where they often obstruct the flow of lymph. Larvae are passed from person to person by mosquitoes. The World Health Organization in 1963 estimated that there were 200 million persons with filariasis in the world. Repeated infections cause enormous swelling of tissues, a condition known as elephantiasis (Fig. 18-8).

PHYLUM NEMATOMORPHA

The old popular name for Nematomorpha (nem-a-to-mor'fa) was "horsehair worms" because of the superstition that they arose spontaneously in water from hairs fallen from horses' tails. They do not, but hairworms, whose cylindric bodies are only approxi-

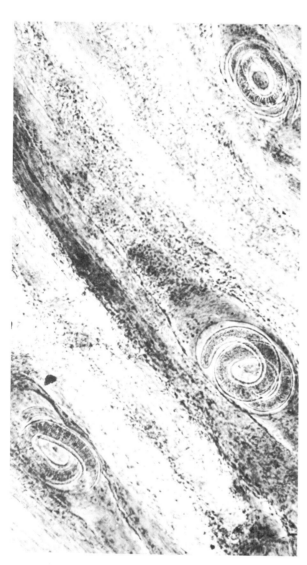

Fig. 18-6. Muscle infected with trichina worm, *Trichinella spiralis* (phylum Nematoda). Larvae may live up to 20 years in these cysts. If eaten in insufficiently cooked meat, larvae are liberated in human intestine, where they quickly mature and release many larvae into blood of host.

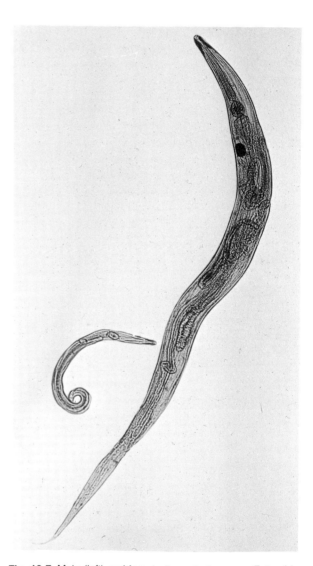

Fig. 18-7. Male *(left)* and female (larger) pinworms, *Enterobius vermicularis* (phylum Nematoda). This worm may be the most common and most widely distributed of human helminth parasites. (Courtesy Indiana University School of Medicine, Indianapolis.)

mately 1 mm thick and as much as 30 cm or more in length, do resemble long hairs or threads. Their name comes from the Greek *nēma, nēmatos*, meaning thread, and *morphē*, meaning form.

Nematomorpha often coil themselves up into tight knots around an aquatic plant. Most hairworms, as juveniles, parasitize insects but, as adults, emerge into fresh water. The members of genus *Nectonema*, as juveniles, parasitize crabs and emerge, as adults, into salt water. The adults do not feed, since their digestive tract is vestigial.

The adult females deposit their eggs in water. After hatching, the young enter suitable hosts—cockroaches, beetles, crickets, grasshoppers, crabs, etc.—where development is completed in the hemocoel. The young

parasites absorb food materials through the body wall. After several molts the worms leave the host near water and soon gain sexual maturity as free-living aquatic animals. *Gordius* and *Paragordius* are common freshwater forms; *Nectonema* is a marine form.

PHYLUM ACANTHOCEPHALA— SPINY-HEADED WORMS

The Acanthocephala (a-kan-tho-sef'a-la) include approximately 800 species of spiny-headed worms, all of which are endoparasites. They range in size from 1.5 mm to more than 0.5 m in length, but most species are less than 25 mm. The adults inhabit fish, birds, and mammals; their larval stages are found in arthropods. There is no free-living stage in their life history.

Their technical name comes from the Greek *akantha,* meaning spine, and *kephale,* meaning head, and refers, as does also their common name, to the distinctive eversible **proboscis** that bears rows of recurved spines (Fig. 18-9). When the proboscis is retracted into the proboscis receptacle, the spines point anteriorly, but, when everted, they point backward, making the proboscis a formidable and dangerous weapon for attachment to a host's intestinal wall where it may cause severe damage. The body, too, is often covered with spines. Muscles attach the proboscis to the receptacle, and the receptacle attaches to the body wall, which prevents overextension of the proboscis.

Internal organs are so degenerate that it is difficult to determine acanthocephalan affinities with other phyla, although there are some links with various other pseudocoelomates.

Probably the best known of the acanthocephalans is *Macracanthorhynchus,* an intestinal worm of pigs common all over the world; occasionally but rarely they are found in man. Eggs can survive in soil for 2 or 3 years until swallowed by grubs or June beetles. Pigs become infested by eating the grubs or beetles. Great damage can be done to the pig's intestine by the perforations caused by the spiny head. *Moniliformis* is a common parasite of house rats and spends its adult life in cockroaches. Humans can acquire them by unwittingly eating roaches or beetles, but such instances are fortunately rare.

Form and function

Body wall. The body wall consists of a thin cuticle, a thick syncytial epidermis, a thin dermis, and a thin layer of circular and longitudinal muscles. Two flaps

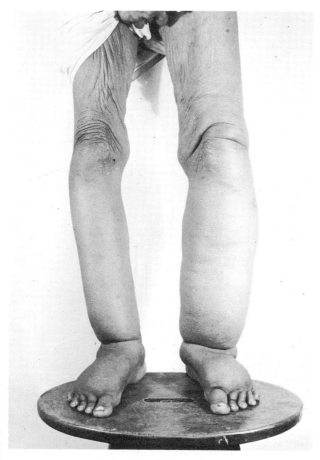

Fig. 18-8. Elephantiasis of leg caused by adult filarial worms, *Wuchereria bancrofti* (phylum Nematoda), which live in lymph passages and block the flow of lymph. Larvae are transmitted by mosquitoes. (AFIP No. 44430-1.)

359

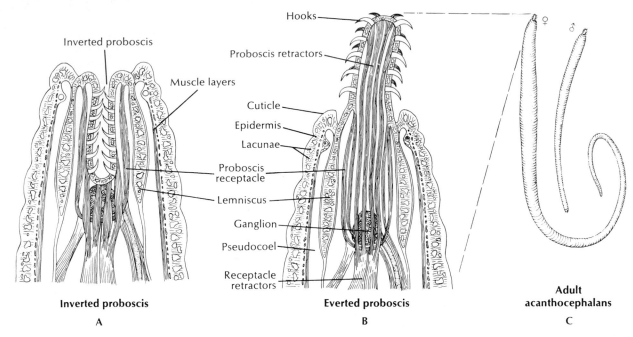

Fig. 18-9. Structure of a spiny-headed worm (phylum Acanthocephala). **A** and **B,** Eversible spiny proboscis by which the parasite attaches to the intestine of the host, often doing great damage. Since there is no digestive system, lacunae in the epidermis absorb and distribute nutrients from the host. **C,** Male is typically smaller than female.

of the body wall project inward into the pseudocoel forming two bags called **lemnisci,** which connect with a system of spaces **(lacunae)** in the epidermis. The lacunae are thought to serve as storage areas for food reserves. The lemnisci are probably reservoirs for fluid from the lacunar system when the proboscis is retracted (Fig. 18-9).

Nutrition, respiration, and excretion. The acanthocephalan has **no digestive tract** at any stage of its life history. All food materials must be absorbed through the body surface. The proboscis has no food-getting function but is purely an organ of attachment. Energy metabolism is probably anaerobic. Some forms have a pair of protonephridia with flame bulbs.

A simple ganglion in the proboscis receptacle, connected to lateral trunks along the body walls, serves as a brain. Sense organs are reduced to a sensory pit on the proboscis and nerve endings in the reproductive organs.

Reproduction and life cycle. Sexes are separate. The male has a protrusible penis, and at copulation the sperm travel up the genital duct and escape into the pseudocoel of the female. The zygotes develop into shell-covered, six-hooked acanthor larvae. The female possesses a peculiar apparatus that sorts the shelled larvae; this apparatus allows the ripe ones to pass through the uterus and genital pore, and returns the unripe ones to the pseudocoel for further development. Shelled larvae escape from the vertebrate host in the feces, and, if eaten by a suitable arthropod, they hatch and work their way into the blood spaces where they grow to juvenile acanthocephalans. Either development ceases until the arthropod is eaten by a suitable host, or they may pass through several transport hosts in which they encyst until eaten.

PHYLUM ENTOPROCTA

The entoprocts are a small group of less than 100 species of tiny, sessile animals. They have a round, flowerlike body crowned by an oval circlet of tentacles and supported on a stalk (Fig. 18-10). Only 5 mm or less in height, they look far more like cnidarian hydroids than the wormlike creatures that make up the rest of the pseudocoelomate groups.

There has been some confusion about their classification since they were once grouped together with the

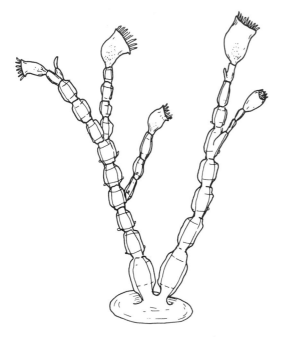

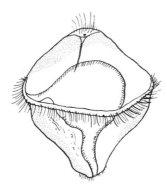

Fig. 18-11. A trochophore larva. Trochophores of some type are found in several phyla and are thought to indicate a phylogenetic relationship.

Fig. 18-10. A freshwater entoproct, *Urnatella,* forms small colonies of two or three stalks arising from a basal plate.

Ectoprocta in a phylum called Bryozoa (the moss animals). However, because the ectoprocts have a true coelom and because the entoprocts are generally considered to be pseudocoelomates, it seems best to place them in separate phyla. The entoprocts are probably closest in affinity to the ectoprocts, but little evidence of that relationship is actually available. They may have been an offshoot of the evolutionary line leading to the ectoprocts.

Most of these tiny creatures are colonial, and, except for one genus, *Urnatella,* the entoprocts are all marine, found from polar regions to the tropics, usually in coastal areas attached to shells or algae. Some are commensal on annelid worms.

Form and function

Like the hydroids, entoprocts possess a circlet of tentacles about the mouth. Unlike the hydroids, the tentacles, 8 to 30 in number, are ciliated and not retractable; instead they tend to roll inward, protecting the mouth and anus, which lie within the circlet.

The crown of tentacles is located on a cup-shaped body called the calyx, which is attached to a substratum by means of an attachment disc with adhesive glands (Fig. 18-10). Colonial forms may have several

stalks springing from one attachment disc, or many stalks may arise from a creeping stolon or from upright branching stems. *Loxosoma,* which lives in the tubes of marine annelids, seems able to move about freely within the tube, but most entoprocts are attached and restricted in movement.

The U-shaped gut lies in the calyx, and both the mouth and the anus open *within* the circlet of tentacles—thus no doubt leading to the phylum name "Entoprocta" (Gr. *entos,* within or inner, + *proctos,* anus) meaning "inner anus." This contrasts with the phylum Ectoprocta, meaning "outside anus," in which the anus lies *outside* the circlet of tentacles. The pseudocoel is filled with a gelatinous mesenchyme containing a few cells.

Entoprocts are **filter feeders.** Long cilia on the sides of the tentacles create currents that bring protozoans, diatoms, and detritus particles toward the calyx to be trapped by the shorter cilia on the inner surface of the tentacles. These cilia direct the food downward toward ciliated food grooves in the base of the crown that carry the food to the mouth. A disturbed entoproct ceases feeding and folds its tentacles over the food grooves.

Certain areas of the stomach produce enzymes for extracellular digestion. The anus is often mounted on a projection called an anal cone, an adaptation that tends to direct the feces away from the mouth. There are two nephridial ducts with flame bulbs.

A large ganglion between the stomach and vestibule gives rise to pairs of nerves that extend to the tentacles, to the stalk, and to the calyx. Sensory cells with bristles are common on the body surface, particularly on the tentacles.

Large colonies are formed by asexual budding. There are also both dioecious and monoecious species, which produce free-swimming **trochophore larvae** that superficially resemble the trochophores of annelids and molluscs. The term "trochophore," which means "wheel-bearing," refers to a generalized type of minute, translucent larva (Fig. 18-11). It is more or less pear shaped and has a prominent circlet of cilia (the "wheel") and sometimes one or two accessory circlets. Some form of trochophore larva is found in several of the invertebrate phyla, including some marine turbellarians, nemertines, brachiopods, phoronids, sipunculids, marine annelids, and molluscs. This similarity among protostome larvae seems to reflect some sort of phylogenetic relationships, not yet fully analyzed, between a number of protostome phyla.

SELECTED REFERENCES

Baer, J. G. 1971. Animal parasites, New York, McGraw-Hill Book Co.

Barnes, R. D. 1974. Invertebrate zoology, 3rd ed. Philadelphia, W. B. Saunders Co.

Bird, A. F. 1971. The structure of nematodes. New York, Academic Press. Inc.

Cheng, T. S. 1964. The biology of animal parasites. Philadelphia, W. B. Saunders Co.

Chitwood, B., and M. B. Chitwood. 1974. Introduction to nematology. Baltimore, University Park Press. *Revision and consolidation of earlier editions of an authoritative work.*

Crofton, H. D. 1966. Nematodes. London, Hutchinson University Library.

Croll, N. A. 1970. The behaviour of nematodes. New York, St. Martin's Press, Inc.

Donner, J. 1966. Rotifers. New York, Frederick Warne & Co., Inc.

Dougherty, E. C. (ed.). 1963. The lower Metazoa. Berkeley, University of California Press. *This is a text comparing phylogeny, morphology, and physiology of the invertebrates lower than the annelids.*

Edmondson, W. T., H. B. Ward, and G. C. Whipple (eds.). 1959. Freshwater biology, 2nd ed. New York, John Wiley & Sons, Inc. *An excellent handbook for identification of the freshwater fauna and flora of North America.*

Gardiner, M. S. 1972. The biology of invertebrates. New York, McGraw-Hill Book Co.

Gosner, K. L. 1971. Guide to identification of marine and estuarine invertebrates: Cape Hatteras to the Bay of Fundy. New York, Interscience Div., John Wiley & Sons, Inc.

Hickman, C. P. 1973. Biology of the invertebrates, 2nd ed. St. Louis, The C. V. Mosby Co.

Hyman, L. H. 1951. The invertebrates: Acanthocephala, Aschelminthes, and Entoprocta, vol. 3. New York, McGraw-Hill Book Co. *Miss Hyman has described these phyla with great accuracy.*

Lees, D. L. 1965. The physiology of nematodes. San Francisco, W. H. Freeman and Co. Publishers.

Light, S. F., R. I. Smith, F. A. Pitelka, D. P. Abbott, and F. M. Weesner. 1967. Intertidal invertebrates of the central California coast. Berkeley, University of California Press. *An identification guide.*

Meglitsch, P. A. 1972. Invertebrate zoology, 2nd ed. New York, Oxford University Press.

Nicholas, W. L. 1975. The biology of free nematodes. New York, Oxford University Press.

Noble, E. R., and G. A. Noble. 1972. Parasitology, 3rd ed. Philadelphia, Lea and Febiger.

Pennak, R. W. 1953. Freshwater invertebrates of the United States. New York, The Ronald Press Co. *An identification guide dealing with the free-living, freshwater invertebrates and omitting the parasitic forms.*

Schmidt, G. D., and L. S. Roberts, 1977. Foundations of parasitology. St. Louis, The C. V. Mosby Co.

SELECTED SCIENTIFIC AMERICAN ARTICLES

Crowe, J. H., and A. F. Cooper, Jr. 1971. Cryptobiosis. **225:**30-36 (Dec.).

Edwards, C. A. 1969. Soil pollutants and soil animals. **220:**88-99 (April).

Mating octopuses. After a courtship ritual, the male *(above)* uses a specialized arm to insert sperm into the female's mantle cavity. (Drawn from photo by R. Sisson.)

19 MOLLUSCS

PHYLUM MOLLUSCA

Next to the arthropods the Mollusca (mol-lus′ka) have the most named species in the animal kingdom—probably approximately 100,000 living species, as well as at least 35,000 fossil species. It is a diverse group of animals, some familiar, some beautiful, some bizarre. It includes the chitons, tooth shells, clams, mussels, oysters, snails, slugs, nudibranchs, sea butterflies, sea hares, squids, octopuses, and nautiluses.

The molluscs have a true coelom, although the coelomic space is usually reduced to areas around the heart, gonads, and kidneys. This phylum, along with Annelida, Arthropoda, and several smaller phyla, make up the Protostomia, which, as previously noted, have spiral and determinate cleavage, schizocoelous formation of the coelom, and formation of the mouth from the blastopore or near the site of the closed blastopore. (See Chapter 14.)

The name "Mollusca," from the Latin *molluscus,* meaning soft, describes one of the distinctive features of the molluscs, a soft body. Also characteristic of the phylum are the folds of dorsal body wall, called the **mantle,** which enclose a space between the mantle and body, called the **mantle cavity.** The mantle cavity houses the **gills** or **lungs.** The mantle serves a number of functions, one of which is to secrete a protective outer **shell** in most molluscs. Unique to this phylum are a muscular **foot,** usually used in locomotion, and a tonguelike rasping organ called a **radula,** used in feeding. All body systems are present and well developed in molluscs.

Molluscs range from fairly simple organisms to some of the most complex invertebrates, and in size they range from almost microscopic organisms to the giant clams that may be 4 feet long and weigh more than 500 pounds; giant squids have been found up to 55 feet long and are the largest of all the invertebrates. Molluscs include some of the most sluggish animals and some of the swiftest and most active invertebrates. Nutritionally they range from primitive filter feeders to herbivorous grazers to predaceous carnivores.

Ecologic relationships

Molluscs are found in a great range of habitats, from the tropics to polar seas, at altitudes exceeding 20,000 feet, in ponds, lakes, and streams, on mud flats, in pounding surf, and in open ocean from the surface to the abyssal depths. Most of them live in the sea, and they represent a variety of life-styles, including bottom feeders, burrowers, borers, and pelagic forms.

According to the fossil evidence, the molluscs originated in the sea, and most of them have remained there. Much of their evolution occurred along the shores where food was abundant and habitats were varied. Only the bivalves and gastropods moved on to brackish and freshwater habitats. As filter feeders, the bivalves were unable to leave aquatic surroundings. Only the snails (gastropods) actually invaded the land. Terrestrial snails are limited in their range by their need for humidity, shelter, and the presence of calcium in the soil.

Evolution of molluscs

The molluscs, which date from the Precambrian era, have left a continuous fossil record since Cambrian times. Some of the marine molluscs have a ciliated **trochophore** type of larva (Figs. 18-11 and 19-15, pp. 361 and 380) similar to that of marine annelids, and the type of egg cleavage is also similar in molluscs and annelids, which seems to indicate a relationship between the two phyla. The ladderlike nervous system of some molluscs resembles that of the turbellarians. It is conceivable that a flatworm type of ancestor gave rise to the two main protostome groups—the nonsegmented molluscs and the segmentally arranged annelid-arthropod stem. The discovery in 1952 of the monoplacophoran *Neopilina* (Fig. 19-1) with its somewhat superficial segmentation (the only segmented mollusc) strengthened the belief that molluscs and annelids were related. Although many zoologists believe that the segmentation in *Neopilina* is different from that of the annelids and does not necessarily represent a link, there does seem to be a real, though remote, relationship.

Zoologists have tried to draw up a hypothetical plan of what the primitive ancestral mollusc might have been like (Fig. 19-1). It was probably a creature with a dorsal calcareous shell, a ventral muscular foot that served as a platform for a soft body containing the viscera, and a mantle formed from the soft folds of dorsal skin that hung down around the body and formed a mantle cavity to contain the gills. It would have used a rasping radula in feeding. Such a basic plan could have been modified by adaptive radiation to form all the various classes of molluscs (Fig. 19-1). The mantle, shell, and radula have all undergone seemingly endless functional adaptations.

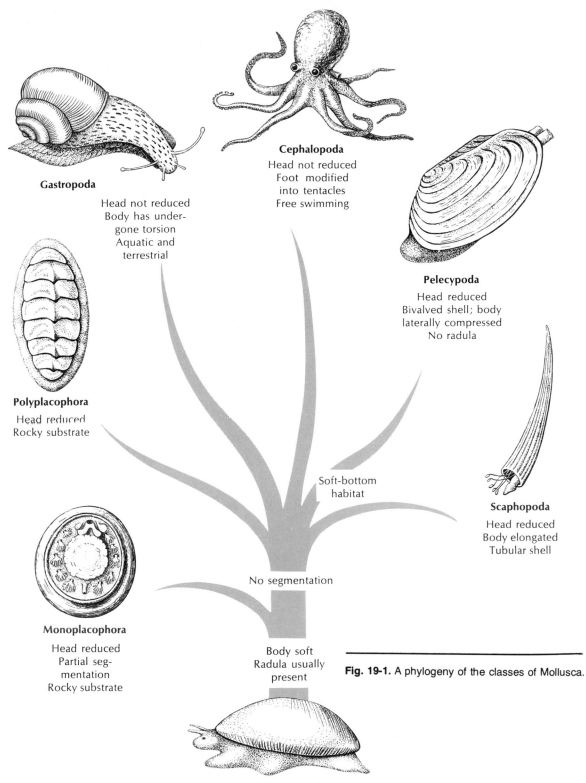

Gastropoda

Head not reduced
Body has under-
gone torsion
Aquatic and
terrestrial

Cephalopoda

Head not reduced
Foot modified
into tentacles
Free swimming

Pelecypoda

Head reduced
Bivalved shell; body
laterally compressed
No radula

Polyplacophora

Head reduced
Rocky substrate

Soft-bottom
habitat

Scaphopoda

Head reduced
Body elongated
Tubular shell

No segmentation

Monoplacophora

Head reduced
Partial seg-
mentation
Rocky substrate

Body soft
Radula usually
present

Fig. 19-1. A phylogeny of the classes of Mollusca.

Hypothetical ancestral mollusc

Significance of the coelom

Heretofore we have been studying animals that lacked a true coelom. These followed several patterns. In the radiates, with gelatinous mesoglea between the body surface and the enteron, diffusion of substances was a simple matter. In flatworms body spaces were filled with cellular parenchyma of endomesoderm; here the need for a better transport method was met by the extensive branching of the gastrovascular cavity and of the protonephridial system throughout the body. In the nemertines this was aided by a system of blood vessels.

In the pseudocoelomates the parenchyma leads into spongy or open spaces—the pseudocoel. Fluid in the pseudocoel bathes the organs, thus providing a means of internal transport serviceable enough for small animals. The gut is a simple straight tube, and the protonephridial system, connected with the pseudo-coel, is more diffuse. But, since there are no mes-enteries, the organs lie loose in the body cavity. The pseudocoelomates are all small—obviously such an arrangement would be unsuitable for more sizeable forms.

In the coelomates the coelom develops as a second-ary cavity within the mesoderm. The coelomic cavity is completely surrounded by mesodermal epithelium, called **parietal peritoneum.** There is not only ample room in the coelom for organs, but the organs are held in place by **mesenteries,** which are continuations of the peritoneum, and the organs are themselves covered with visceral peritoneum. This ensures a more stable arrangement of organs with less crowding. The ali-mentary canal can become more muscular, more highly specialized, and diversified without getting in the way of other organs, such as the heart, liver, or lungs. Thus, with organs securely and advantageously placed, greater growth and complexity are possible in the organism.

The coelom performs other important functions. It is filled with **coelomic fluid,** and its lining is often ciliated to keep the fluid moving. Thus **it aids in the movement of materials,** such as absorbed foods and metabolic wastes, from one place to another. In many smaller coelomates no other transport system is neces-sary. In animals with a vascular system the mesenteries provide an ideal location for the network of blood vessels necessary to reach every body organ.

Another function is to serve as a **repository for sex cells.** The gonads are usually associated with the peritoneum, and their products are discharged into the coelomic cavity. The coelom is connected to the outside by ducts for the discharge of both sex cells and nitrogenous wastes of metabolism.

The coelom can also serve as a hydrostatic skele-ton—an especially important function in animals lack-ing rigid and jointed skeletons. Circular and longitudi-nal body wall muscles, acting as antagonists, can contract or relax to vary the force exerted on the coelomic fluid and thus produce a variety of body movements. In earthworms, for example, the coelom is divided into a series of compartments, in each of which the fluid volume is constant. By varying and controlling the muscular pressure on the various com-partments, the animal can swim, crawl, burrow, thrust out a proboscis or draw one in, etc.

Altogether the development of the coelom must be considered an important stepping stone in the evolution of larger and more complex forms. There were three main adaptive lines of protostomes that resulted in the three major phyla of coelomate protostomes—the molluscs, in which segmentation was lacking except in one branch, and the annelids and arthropods, which have exploited the segmented body. All three phyla are highly developed and enormously successful. The coelomates also include a number of smaller inverte-brate phyla and the vertebrates.

Characteristics

1. Body bilaterally symmetric (bilateral asymmetry in some); unsegmented (except in Monoplacophora); usually with definite head
2. Ventral body wall specialized as a muscular **foot,** variously modified but used chiefly for locomotion
3. Dorsal body wall forms pair of folds called the **mantle,** which encloses the **mantle cavity;** modified into **gills** or **lungs;** secretes the **shell** (shell absent in some)
4. Surface epithelium usually ciliated and bearing mucous glands and sensory nerve endings
5. Coelom mainly reduced to areas around heart, gonads, and kidney
6. Complex digestive system; rasping organ **(radula)** usually present; anus usually emptying into mantle cavity
7. **Open circulatory system** of heart (usually three-chambered), blood vessels, and sinuses; respiratory pigments in blood
8. Gaseous exchange by **gills, lungs, mantle,** or **body surface**
9. One or two kidneys **(metanephridia)** opening into the peri-cardial cavity and usually emptying into the mantle cavity
10. Nervous system of paired cerebral, pleural, pedal, and visceral ganglia, with nerve cords and subepidermal plexus; ganglia centralized in nerve ring in gastropods and cephalopods
11. Sensory organs of touch, smell, taste, equilibrium, and vision (in some); eyes highly developed in cephalopods

12. **Spiral and determinate cleavage;** characteristic larva the **trochophore**

Classification

Class Monoplacophora—segmented molluscs. Single symmetric shell; replication of certain organs (apparent segmentation); separate sexes. *Neopilina.*

Class Polyplacophora (Amphineura)—chitons. Large flat foot; reduced head; shell of eight dorsal plates; sexes usually separate. *Chiton, Chaetopleura, Cryptochiton.*

Class Aplacophora (Solenogastres). Wormlike body with mantle completely investing the body; no shell. *Chaetoderma, Proneomenia, Neomenia.*

Class Scaphopoda—tusk shells, or tooth shells. Body enclosed in a one-piece tubular shell open at both ends; sexes separate. *Dentalium, Cadulus.*

Class Gastropoda—snails, nudibranchs, and others. Body usually asymmetric in a coiled shell (shell absent in some); head well developed; foot large and flat; dioecious or monoecious. *Littorina, Physa, Helix, Dendrodoris.*

Class Pelecypoda (Bivalvia, Lamellibranchia)—bivalves. Body enclosed in a two-lobed mantle; shell of two lateral valves of variable size and form, with dorsal hinge; foot usually wedge shaped; sexes usually separate. *Anodonta, Teredo, Venus.*

Class Cephalopoda—nautiluses, squids, and octopuses. Body with shell often reduced or absent; head well developed with complex eyes; foot modified into arms or tentacles; sexes separate. *Nautilus, Loligo, Octopus, Sepia.*

Molluscan body plan

The modern mollusc, like its hypothetical ancestor, typically has a **head,** a **visceral mass** enclosed in a **soft skin,** a protective **mantle,** a calcareous **shell,** and a ventral, muscular **foot** that is the organ of locomotion. These structures are modified in numerous ways among the different groups, and in some the head, the shell, or the mantle may be lacking entirely. The epidermis is soft and glandular, and much of it is ciliated.

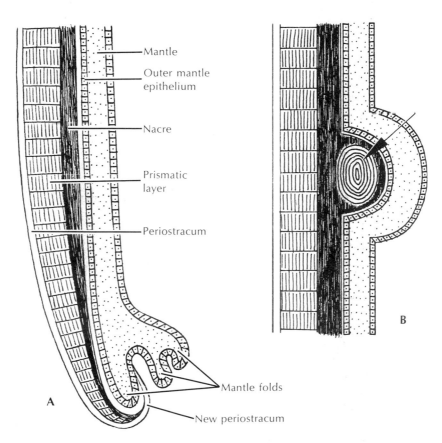

Fig. 19-2. A, Diagram of vertical section of the shell and mantle of a bivalve. The epithelium on the outer side of the mantle secretes the shell; the inner epithelium is usually ciliated. **B,** Formation of a pearl between the mantle and shell. A parasite or irritating bit of sand under the mantle becomes covered with secreted layers of nacre (mother-of-pearl).

Mantle. The mantle is a sheath of skin that hangs down in two folds around the soft body, enclosing the mantle cavity space. The outer side of the mantle secretes the shell, the inner side is usually ciliated (Fig. 19-2). The mantle, along with the gills or lungs that develop from it, is usually active in gaseous exchange. In cephalopods it is muscular and used for locomotion.

Shell. The shell of the mollusc, when present, is secreted by the mantle and is lined by it. It has three typical layers (Fig. 19-2): (1) the **periostracum,** an outer horny layer, (2) a middle **prismatic layer** of crystalline calcium carbonate, and (3) an inner layer of irridescent mother-of-pearl, called **nacre.** The nacre is secreted by cells scattered over the outer surface of the mantle; the prismatic layer is secreted by the mantle edge, and the periostracum is secreted by a groove along the edge. Calcium for the shell comes from the surrounding water, soil, or food. The first shell appears during the larval period and grows continuously throughout life.

Foot. The molluscan foot may be adapted for locomotion, for attachment to a substratum, for food capture (in cephalopods), or for a combination of functions. It is usually a ventral, solelike structure in which waves of muscular contraction effect a creeping locomotion. However, there are many modifications,

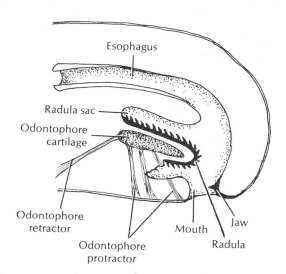

Fig. 19-3. Diagram of section through head of snail showing the mouth cavity with radula in radula sac. As the animal grazes, the mouth opens, the odontophore cartilage is thrust forward, and the radula scrapes backward to bring food into the pharynx. The movement of the radula is rhythmic.

such as the attachment disc of the limpets, the laterally compressed "hatchet foot" of the bivalves, and the division of the foot into the suckered arms and tentacles of the squids and octopuses. Secreted mucus is often used as an adhesive aid or is used as a slime track by small molluscs that glide on cilia.

The foot is manipulated by a combination of muscles and hydrostatic skeleton. Some forms glide by rhythmic waves of muscular action. Burrowing forms can extend the foot into the mud by muscular action and then anchor it by enlargement with blood pressure while the body is drawn forward. In swimming forms the foot may be modified into winglike parapodia.

Radula. The radula is a rasping organ used in scraping, tearing, and rasping food. Found only in molluscs, it is common to every class except the bivalves. The radula is a chitinous ribbon that bears many rows of fine teeth and is moved rhythmically back and forth over a cartilaginous tongue (odontophore) by tiny muscles (Fig. 19-3). The radula rasps off fine particles of food and then serves as a conveyer belt to carry the particles toward the digestive tract. As the radula wears away anteriorly, it is continuously replaced at its posterior end. The pattern of the teeth varies and is used as a diagnostic feature in classification of molluscs.

Internal structure and function. In the molluscs oxygen–carbon dioxide exchange occurs not only through the body surface, particularly that of the **mantle,** but in specialized respiratory organs such as gills or lungs, which are derivatives of the mantle. There is an **open circulatory system** with a pumping **heart,** blood vessels, and blood sinuses. The digestive tract is complex and highly specialized, according to the feeding habits of the various molluscs. Most molluscs have a pair of **kidneys** (nephridia), which connect with the coelom; the ducts of the kidneys in many forms serve also for the discharge of eggs and sperm. The **nervous system,** consisting of ganglionic masses, nerve cords, and sense organs, is in general simpler than that of the annelids and arthropods, but there are a number of types of highly specialized sense organs. Most molluscs are dioecious, although some of the gastropods are hermaphroditic. Many aquatic molluscs pass through free-swimming trochophore and veliger larval stages. The **veliger** is the free-swimming larva of most marine snails, tusk shells, and bivalves. It develops from the trochophore and has the beginning of a foot, shell, and mantle (Fig. 19-15, p. 380).

SEGMENTED MOLLUSCS—CLASS MONOPLACOPHORA

Until 1952 it was thought that the Monoplacophora (mon-o-pla-kof'-o-ra) consisted only of Paleozoic shells. However, in that year living specimens of *Neopilina* were dredged up from the ocean bottom near the west coast of Mexico. These molluscs are small and have a low rounded shell and a creeping foot (Fig. 19-1). They have a superficial resemblance to the chitons and limpets, but, unlike other molluscs, they are internally segmented. Five or six pairs each of gills, nephridia, and gill hearts are arranged segmentally. The mouth bears the characteristic radula. Many zoologists believe that the monoplacophorans are evidence of a relationship between annelids and molluscs.

CHITONS—CLASS POLYPLACOPHORA (AMPHINEURA)

The chitons are somewhat flattened and have a convex dorsal surface that bears eight articulating limy **plates,** or **valves,** which give them their name (Fig. 19-4). The term "Polyplacophora" comes from the Greek *poly,* meaning many, *plax,* meaning anything flat (plate), and *phora,* meaning bearing—in other words, "bearing many plates" in contrast to the Monoplacophora, which bear one shell (*mono,* single), and the Aplacophora, which bear none (*a,* without). The plates overlap posteriorly and are usually dull colored to match the rocks to which the chitons cling.

Most chitons are small (2 to 5 cm); the largest, *Cryptochiton,* rarely exceeds 30 cm. They prefer rocky

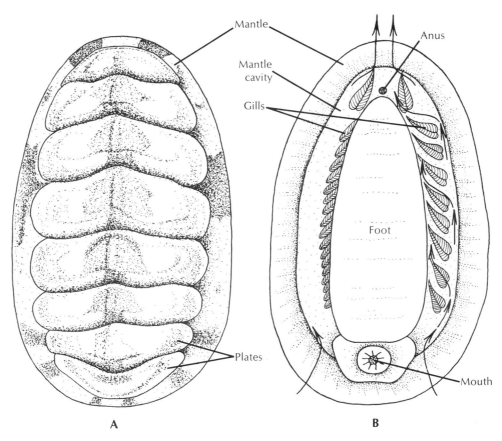

Fig. 19-4. A, Dorsal view of chiton, showing its eight overlapping plates and surrounding mantle girdle. **B,** Ventral view showing direction of respiratory currents in the gill grooves between the foot and the mantle. Gills are shown diagrammatically on the right.

surfaces in intertidal regions, though some live at great depths. Chitons are stay-at-home organisms, straying only very short distances for feeding. In feeding, a sensory subradular organ protrudes from the mouth to explore for algae. When some are found, the radula is then projected to scrape them off. They cling tenaciously to their rock with the broad flat foot. If detached, they can roll up like an armadillo for protection.

The mantle forms a girdle around the margin of the plates, and in some species mantle folds cover part or all of the plates. On each side of the broad ventral foot and lying between the foot and the mantle is a row of gills suspended from the roof of the mantle cavity. With the foot and the mantle margin adhering tightly to the substrate, these grooves become closed chambers, open only at the ends. Water enters the grooves anteriorly, flows across the gills, and leaves posteriorly, thus bringing to the gills a continuous supply of oxygen (Fig. 19-4, *B*). Blood pumped by the heart reaches the gills by way of an aorta and a pair of sinuses.

Sexes are separate in chitons. Sperm shed by males in the exhalent currents enter the gill grooves of the females by inhalent currents. Eggs are shed into the sea singly or in strings.

CLASS APLACOPHORA

The class Aplacophora includes less than 100 species of strange aberrant molluscs. They are small wormlike creatures with a terminal mouth and anus. Most are approximately 2.5 cm long, but some range up to 30 cm. There is no shell, but the skin is embedded with spicules. The mantle encloses the body except for a ventral groove; the foot is vestigial; and the gills are located posteriorly. Some have a radula. They live in fairly deep water where they feed on hydroids and corals.

TUSK SHELLS—CLASS SCAPHOPODA

The Scaphopoda (ska-fop'o-da), commonly called the tusk shells or tooth shells, have a slender body covered with a mantle and a **tubular shell** open at both ends. Most are 2.5 to 5 cm long, though they range from 4 mm to 25 cm. *Dentalium* is a common Atlantic coast form (Fig. 19-5). A **burrowing foot** protrudes through the larger end of the shell. The mouth near the foot has a radula and **contractile tentacles** that are sensory and prehensile. Scaphopods are sedentary

marine animals that burrow through mud or sand feeding on protozoans and other small forms. Adhesive knobs on the tentacles aid in food capture. The ciliated mantle cavity keeps water circulating for gaseous exchange through the mantle surface. The sexes are separate, and the larva is a trochophore.

SNAILS AND THEIR RELATIVES—CLASS GASTROPODA

The Gastropoda are the "belly-footed" molluscs. The largest and most diverse class of molluscs, they include the snails, limpets, slugs, whelks, conchs, periwinkles, sea slugs, sea hares, sea butterflies, etc. They are basically bilateral animals, which during a larval stage undergo torsion and become asymmetric. Their one-piece (univalve) shell may be coiled or uncoiled or may be lacking entirely.

Since Cambrian times the gastropods have spread to nearly all parts of the earth. They are common in both littoral and abyssal zones of the sea, and some are

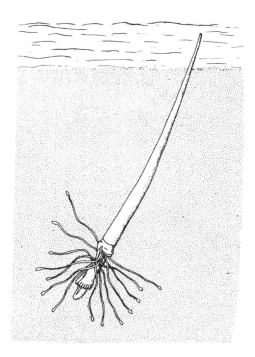

Fig. 19-5. The tusk shell, *Dentalium,* a scaphopod. It burrows into soft mud or sand and feeds by means of its prehensile tentacles. Respiratory currents of water are drawn in by ciliary action through the small open end of the shell, then expelled through the same opening by muscular action.

pelagic. They live in brackish and fresh water and in many terrestrial habitats. They range from microscopic forms to giant sea snails more than 2 feet long. Approximately 15,000 fossil and nearly 40,000 modern species have been named.

Most gastropods are slow moving and sedentary, and the foot is adapted to creeping or gliding movements; but some can also climb, some use the foot to burrow, a few can leap, and some have the foot or mantle modified for swimming.

Shell. Gastropods have a one-piece (univalve) shell composed of the three typical layers (Fig. 19-2). The opening (aperture) of the shell is proportionate to the size of the animal. Some snails have a horny **operculum** for closing the aperture when the snail is withdrawn (Fig. 19-10, *A*). The apex of the shell is the oldest part; new shell is laid down at the aperture, and, since the snail is a living growing animal, the growth of the shell reflects the growth of the body.

Torsion. The gastropod body, like that of most molluscs, is composed of the head, foot, mantle, and visceral mass, and, in the larval stage at least, it is bilaterally symmetric with the anus and mantle cavity opening posteriorly. However, during the larval stage the symmetry is modified by a **torsion,** or twisting, that brings the anus and mantle cavity downward, forward, and then around a half-circle to a position above the head.

The process begins in the veliger larva (Fig. 19-6, *A*) and is in two stages. First there is the ventral flexure in the sagittal plane that brings the openings of the anus and mantle cavity downward and forward. They lie ventrally and face forward (Fig. 19-6, *B*). Then the second stage, at right angles to the first, rotates the entire mantle and visceral mass 180° so that the ventral structures shift upward and become dorsal, and the dorsal structures rotate downward to a ventral position. Thus the left gill, kidney, etc. shift to the right, and the

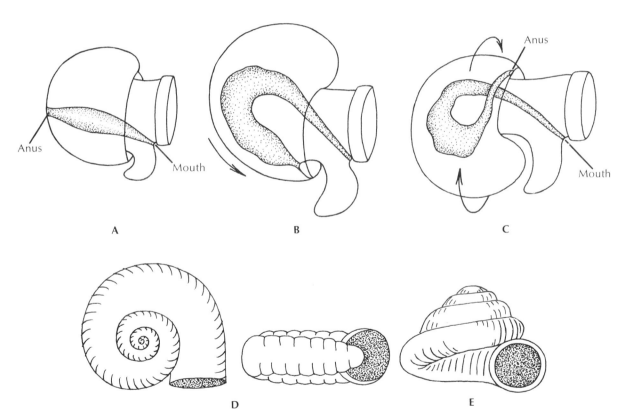

Fig. 19-6. A to **C,** Torsion in gastropod larva. **B,** Anus and mantle cavity move downward and forward. **C,** Rotation of visceral mass counterclockwise through 180° brings mantle cavity above the head. **D** and **E,** Coiling of the shell, which is independent of torsion. **D,** Two views of planospiral coiling. **E,** Conispiral coiling, producing a cone-shaped shell.

right structures shift to the left. The mantle cavity lies above the head, which can be withdrawn within it for protection (Fig. 19-6, *C*). The nerve cords are twisted into a figure eight. The crowding caused by torsion often results in the disappearance of one set of gills, kidneys, etc.

Some gastropod forms later undergo some degree of **detorsion,** usually in the juvenile stage, which partially restores the bilateral symmetry. In many gastropods the detorsion carries the anus, mantle cavity, and gills back to the right side or even back to the posterior position. In nudibranchs the mantle cavity and shell, although present in the veliger, disappear in the adult.

There are various theories about the possible advantage of torsion for the gastropod, such as the ability of the larva to withdraw the head into the mantle cavity, the advantage of facing inhalent currents that flush out the mantle cavity, or the advantage of bringing the sensory parts of the mantle forward. Torsion also introduced a sanitation problem by placing the anus and excretory pores forward where wastes might be washed back over the gills. Some forms have adapted by having only one gill, which is able to produce a suitable exhalent current to carry away the wastes. Others have lost both gills and developed a lung from the mantle.

Coiling. The coiling, or spiral winding, of the shell and visceral hump is an entirely separate process from torsion and does not seem to result from it. It may occur in the larval stage at the same time as torsion, but historically the fossil record shows that coiling occurred before torsion. Coiling is achieved by a more rapid growth of one side of the visceral mass than the other.

Early gastropods had a bilaterally symmetric **planospiral** shell; that is, all the whorls lay in a single plane (Fig. 19-6, *D*). Such a shell is unwieldly, and the space within the inner whorls is very small for the visceral mass. Only a few modern forms have this type of shell. Growth of the whorls in both outward and downward directions produced the more compact cone-shaped, or **conispiral,** type of shell that is easier to balance and provides more space for the visceral hump (Fig. 19-6, *E*). Differing rates of growth of the various whorls results in an enormous number of variations of this type of coiled shell.

Feeding and nutrition. Most gastropods are herbivorous, living on seaweed and algae scraped off with the radula. The abalone *Haliotis* holds seaweed with

Fig. 19-7. Pulmonate snail. Note the large second pair of tentacles bearing the eyes on the tips. (Photo by C. P. Hickman, Jr.)

the foot and breaks off pieces with the radula. Some snails are scavengers, living on dead and decayed flesh; others are carnivorous, tearing their prey apart with their radular teeth. Some, such as *Urosalpinx,* the oyster borer, and *Busycon,* the whelk, have an extensible proboscis for drilling holes in the shells of the bivalves whose soft parts they find delectable. Some even have a spine for opening the shells. Most of the pulmonates (air-breathing snails) (Fig. 19-7) are herbivorous, but some live on earthworms and other snails.

Some of the sessile gastropods, such as the limpets, are ciliary feeders that use the gill cilia to draw in particulate matter, which is rolled into a mucous ball

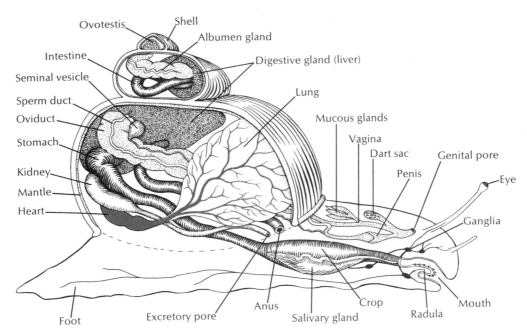

Fig. 19-8. Anatomy of a pulmonate snail.

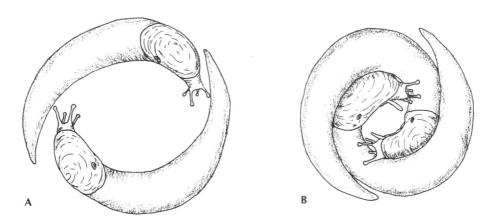

Fig. 19-9. Courtship and copulation of pulmonate slugs. **A,** Courtship, in which they trail each other in circles, each licking the caudal gland of the other. **B,** Copulation, in which an exchange of spermatophores occurs.

and carried to the mouth. Some of the sea butterflies secrete a mucous net to catch small planktonic forms and then draw the web into the mouth.

After maceration by the radula or by some grinding device, such as the so-called gizzard in the sea hare *Aplysia* and in others, digestion is usually extracellular in the lumen of the stomach or digestive glands. In ciliary feeders the stomachs are sorting regions and most of the digestion is intracellular in the digestive glands.

Internal form and function. Respiration in most gastropods is carried out by **gills,** although some aquatic forms, lacking gills, depend upon the **skin.** The pulmonates have a highly vascular area in the mantle that serves as a **lung** (Fig. 19-8). Freshwater pulmonates must surface in order to expel a bubble of gas from the lung and curl the edge of the mantle around the pneumostome to form a siphon to take in air.

Gastropods have a single **nephridium.** The cir-

A

B

Fig. 19-10. The whelk, *Busycon.* **A,** Ventral view showing the operculum closed to protect the soft body. **B,** Egg laying. Eggs are laid in a string of double-edged, parchmentlike discs, one end of which will be fastened to the substrate when completed. The string of discs may be a meter long and have as many as 100 capsules in which the eggs develop into minute snails. (Photos by C. P. Hickman, Jr.)

culatory and nervous systems are well developed. The latter includes three pairs of ganglia connected by nerves. Sense organs include eyes, statocysts, tactile organs, and chemoreceptors.

There are both dioecious and hermaphroditic gastropods. During copulation in hermaphroditic species there is an exchange of spermatophores (bundles of sperm). Many forms perform courtship ceremonies (Fig. 19-9). Most land snails lay their eggs in holes in the ground or under logs. Some aquatic gastropods lay their eggs in gelatinous masses; others enclose them in gelatinous capsules or in parchment egg cases (Fig. 19-10, *B*). Marine gastropods go through a free-swimming veliger larval stage during which torsion and coiling occur.

Major groups of gastropods

There are three major groups of gastropods.

Prosobranchia. In the prosobranchs (Gr. *prosō*, in front, +*branchia*, gills) the gills are located anteriorly in front of the heart. They have one pair of tentacles, and the sexes are separate. There is usually an **operculum,** or horny plate (Fig. 19-10, *A*), on the foot sealing the shell opening when the head and foot are withdrawn into the shell. This group includes most of the marine snails and a few of the freshwater ones. Familiar examples are the periwinkles, limpets, whelks, conchs, abalones, slipper shells, oyster borers, rock shells, and certain freshwater forms.

Opisthobranchia. In the opisthobranchs (Gr. *opistho,* at the back, +*branchia,* gills) the gill is displaced

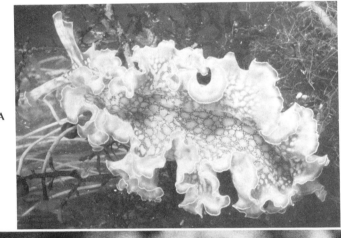

Fig. 19-11. Nudibranchs. **A,** The frilled sea slug, *Tridachia,* is a dainty green and white nudibranch from Florida waters. **B,** *Dendrodoris,* a Pacific coast nudibranch, is a rich yellow and has a dorsal cluster of gills. (**A** courtesy R. C. Hermes, Homestead, Fla.; **B,** photo by F. M. Hickman.)

by detorsion to the right side or rear of the body. There are usually two pairs of tentacles, and the shell is reduced or absent. All are marine, and all are hermaphroditic. They include the large sea hares *(Aplysia)*, which may be a foot or more long; the sea butterflies, or pteropods, in which the foot is modified into fins for swimming; and the nudibranchs, or sea slugs, which rank among the most beautiful and colorful of the molluscs (Fig. 19-11). In undergoing detorsion nudibranchs have lost both the mantle cavity and the gill, but the body surface is often increased for gaseous exchange by small projections, called cerata, as in *Eolis,* by secondary gills around the anus as in *Dendrodoris* (Fig. 19-11, *B*), or by a fluting of the mantle edge as in *Tridachia* (Fig. 19-11, *A*). Nudibranchs are carnivorous, many of them feeding mainly on hydroids and sea anemones.

Pulmonata. The pulmonates (L. *pulmo,* lung) are the terrestrial and freshwater snails and slugs (and a few marine forms) in which the anterior mantle cavity has developed into an air-breathing lung instead of a gill (Fig. 19-8). Aquatic species have one pair of non-retractile tentacles with a pair of eyes at the base of the tentacles. Land forms have two pairs of retractile tentacles, with the eyes located on the ends of the posterior pair (Fig. 19-7).

Bivalved molluscs—class Pelecypoda

The Pelecypoda (pel-e-sip'o-da), or "hatchet-footed" animals, as their name implies (Gr. *pelekus,* hatchet, + *pous, podos,* foot) are the bivalved molluscs. They include the mussels, clams, scallops, oysters, and shipworms and range in size from tiny seed shells 1 to 2 mm in length to the giant South Pacific clam *Tridacna,* which may reach more than 1 m in length and as much as 225 kg (500 pounds) in weight. Most bivalves are sedentary **filter feeders** that depend upon ciliary currents produced by the gills to bring in food materials. Unlike the gastropods, they have no head, no radula, and very little cephalization.

Most pelecypods are marine, but many live in brackish water and in streams, ponds, and lakes.

Shell. Bivalves are laterally compressed, and their two shells **(valves)** are held together dorsally by a hinge ligament that causes the valves to gape ventrally.

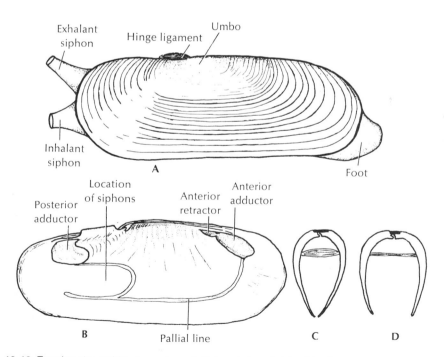

Fig. 19-12. *Tagelus,* the stubby razor clam. **A,** External view of right side. **B,** Inside of left shell showing where muscles were attached. **C** and **D,** Sections showing function of adductors and hinge ligament. In **C** adductor is contracted, pulling valves together. In **D** adductor is relaxed, allowing the hinge ligament to pull valves apart.

The valves are drawn together by adductor muscles that work in opposition to the hinge ligament (Fig. 19-12). The valves function largely for protection, but those of the shipworms *(Teredo)* have microscopic teeth for rasping wood, and the rock borers *(Pholas)* use spiny valves for boring into rock. A few bivalves, such as scallops, swim about jerkily by clapping their shells together, creating a sort of jet propulsion.

The umbo is the oldest part of the shell, and growth occurs in concentric lines around it (Fig. 19-12, *A*).

Pearl production is the by-product of a protective device used by the animal when a foreign object (grain of sand, parasite, etc.) becomes lodged between the shell and mantle. The mantle secretes layers of nacre around the irritating object (Fig. 19-2). Pearls are cultured by inserting particles of nacre between the shell and mantle of a certain species of oyster and by keeping the oysters in enclosures for several years.

Mantle. The mantle hangs down on each side of the visceral mass, shielding a pair of gills on each side (some species have a single gill). The posterior edges of the mantle folds are modified to form dorsal **excurrent** and ventral **incurrent apertures** for regulating water flow (Figs. 19-12 and 19-13, *A*). In some marine bivalves the mantle is drawn out into long muscular siphons that allow the clam to burrow into the mud or sand and extend the siphons to the water above (Fig. 19-13, *B* to *D*).

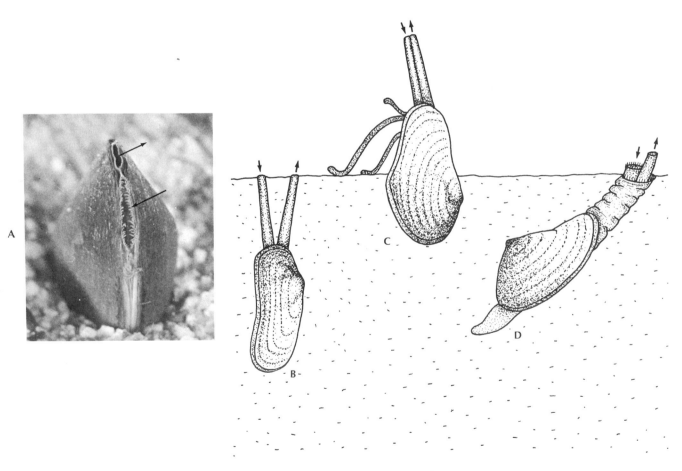

Fig. 19-13. Adaptations for inhalant and exhalant water currents. **A,** In freshwater clam the mantle edges form apertures. **B to D,** In many marine forms the mantle is drawn out into long siphons. In **A, B,** and **D** the inhalant current brings in both food and oxygen. **C,** In *Yoldia* the siphons are respiratory; long ciliated palps feel about over the mud surface and convey food to the mouth. (**A** from *Adaptive Radiation—the Mollusks,* an Encyclopaedia Britannica film.)

Foot. The ventral foot is an organ of locomotion. The end of the foot can be extended and swollen with blood to anchor it in the mud and then shortened by muscular action to pull the animal forward or in burrowing forms to pull it downward.

Some pelecypods are sessile; oysters attach their shells to a surface by secreting cement, and mussels attach themselves by secreting a number of slender byssus threads that harden upon exposure to air.

Feeding and digestion. Bivalves are filter feeders that secrete mucus to trap food particles brought in by the gill currents. In the stomach the mucus and food particles are kept whirling by a rotating gelatinous rod, called a crystalline style. As layers of the rotating style

dissolve, certain digestive enzymes are freed for extracellular digestion. Food particles detached from the spinning mass are sorted in the ciliated ridges of the stomach from which suitable particles are directed to the digestive gland for intracellular digestion.

Internal features and reproduction. Bivalves have a three-chambered heart that pumps blood through the gills and mantle for oxygenation and to the kidneys for waste elimination (Fig. 19-14). The three pairs of ganglia are widely separated, and sense organs are poorly developed. A few pelecypods have ocelli. The steely blue eyes of the scallops, located around the mantle edge, are equipped with cornea, lens, and retina.

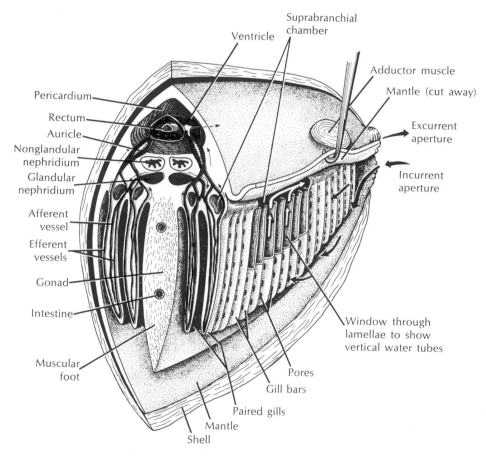

Fig. 19-14. Section through heart region of clam showing relation of circulatory and respiratory systems. *Blood circulation:* Heart ventricle pumps blood to sinuses of foot, viscera, and mantle; on its return to the heart, blood passes through kidney or gills. *Water currents:* Water drawn in by cilia enters gill pores, passes up gill tubes, and out excurrent aperture. Blood in gills exchanges carbon dioxide for oxygen.

Sexes are separate, and fertilization is usually external. Marine embryos go through three free-swimming larval stages—**trochophore, veliger larva,** and young **spat**—before reaching adulthood (Fig. 19-15). In freshwater clams some of the gill tubes become temporary brood chambers where the zygotes develop into tiny bivalved **glochidium larvae,** which are discharged with the exhalant current (Fig. 19-16). If the larvae come in contact with passing fishes, they hitchhike a ride as parasites for the next 20 to 70 days before sinking to the bottom to become sedentary adults.

SQUIDS AND OCTOPUSES—CLASS CEPHALOPODA

The Cephalopoda (sef-a-lop'o-da) are the most advanced of the molluscs—in fact, in some respects

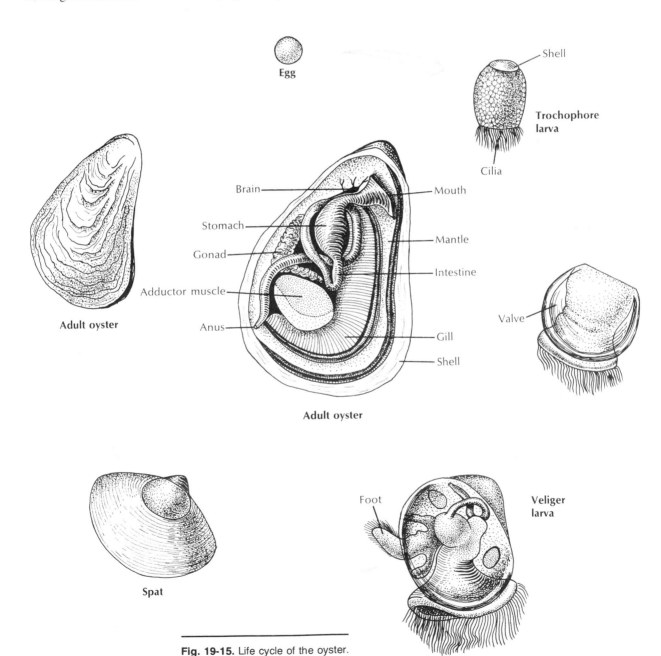

Fig. 19-15. Life cycle of the oyster.

they are the most advanced of all the invertebrates. They include the squids, octopuses, nautiluses, devilfishes, and cuttlefishes. All are marine, and all are active predators.

Cephalopods are the "head-footed" molluscs (Gr. *kephalē*, head, + *pous, podos*, foot) in which the modified foot is concentrated in the head region. The edges of the foot are drawn out into arms and tentacles that bear sucking discs for seizing prey; also part of the foot is modified to form a funnel for expelling water from the mantle cavity.

Cephalopods range upward in size from 2 or 3 cm. The giant squid *Architeuthis* has been found measuring as much as 17 m (55 feet), including the 11 m (35-foot) tentacles. They are the largest invertebrates known.

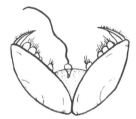

Fig. 19-16. Glochidium, or larval form, of freshwater clam. When the larva is released from brood pouch of mother, it may become attached to a fish by clamping its valves closed. It remains as parasite on the fish for several weeks. Its size is approximately 0.3 mm.

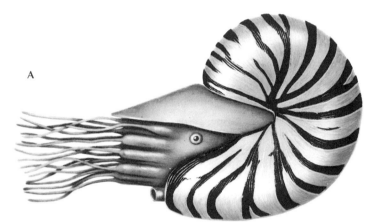

Fig. 19-17. *Nautilus,* a cephalopod. **A,** External appearance. **B,** Longitudinal section, showing gas-filled chambers of shell and diagram of body structure.

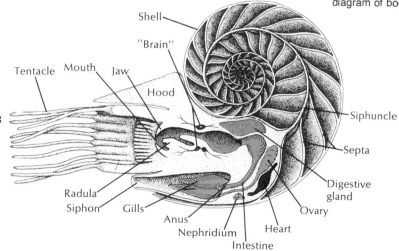

Fossil records of cephalopods go back to Cambrian times. The earliest shells were straight cones; others were curved or coiled, culminating in the coiled shell similar to that of the modern *Nautilus*—the only remaining member of the once flourishing nautiloids (Fig. 19-17). Cephalopods without shells or with internal shells (such as octopuses and squids) are believed to have evolved from some early straight-shelled nautiloid.

Shell. Although early nautiloid shells were heavy, they were made buoyant by a series of **gas chambers,** as is that of *Nautilus* (Fig. 19-17, *B*). The shell of *Nautilus,* although coiled, is quite different from that of a gastropod. Its shell is divided by transverse septa into internal chambers (Fig. 19-17, *B*). The living animal inhabits only the last chamber. As it grows, it moves forward, secreting behind it a new septum. The chambers are connected by a tube called the **siphuncle,** which extends from the visceral mass and secretes gas into the empty chambers. The resulting buoyancy allows the animal to swim. Cuttlefishes also have a small coiled or curved shell, but it is entirely enclosed by the mantle. In the squids most of the shell has disappeared, leaving only a thin, horny strip called a pen, which is enclosed by the mantle. In *Octopus* the shell has disappeared entirely.

Locomotion. Most cephalopods swim by forcefully expelling water from the mantle cavity through a ventral **funnel**—a sort of jet propulsion method. The funnel is mobile and can be pointed forward or backward to control direction; speed is controlled by the force with which water is expelled.

Squids and cuttlefishes are excellent swimmers. The squid body is streamlined and built for speed (Fig. 19-18). The cuttlefish is slower but can regulate its buoyancy by regulating the relative amounts of fluid and gas in the narrow spaces of its shell. Both squids and cuttlefishes have lateral fins that can serve as stabilizers, but they are held close to the body for rapid swimming.

Nautilus is active at night; its gas-filled chambers keep the shell upright. Though not as fast as the squid, it moves surprisingly well.

Octopus has a globular body and no fins (Fig. 19-19, *A*). The octopus can swim backwards by spurting jets of water from its funnel, but it is better adapted to crawling about over the rocks and coral, using the suction discs on its arms to pull or to anchor itself. Some deep-water octopods have the arms webbed like an umbrella and swim in a sort of medusa fashion.

External features. During the larval development of the cephalopod, the head and foot become indistinguishable. The ring around the mouth, which bears the arms, or tentacles, is considered to be derived from the foot.

In *Nautilus* the head with its 60 to 90 or more tentacles can be extruded from the opening of the

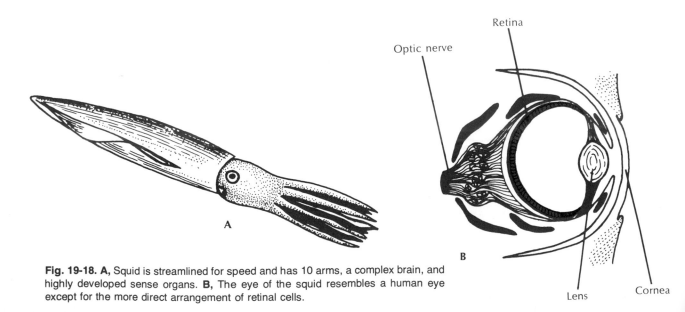

Fig. 19-18. A, Squid is streamlined for speed and has 10 arms, a complex brain, and highly developed sense organs. **B,** The eye of the squid resembles a human eye except for the more direct arrangement of retinal cells.

living compartment of the shell (Fig. 19-17). Its tentacles have no suckers but are made adhesive by secretions. They are used in searching for, sensing, and grasping food. Beneath the head is the funnel. The mantle, mantle cavity, and visceral mass are sheltered by the shell. Two pairs of gills are located in the mantle cavity.

Cephalopods other than nautiloids have either eight or ten appendages around the mouth, no external shell, and only one pair of gills. Octopods have eight suckered arms; squids and cuttlefishes (decapods) have ten arms: eight suckered arms and a pair of long retractile tentacles. The thick mantle covering the trunk fits loosely at the neck region allowing intake of water into the mantle cavity. When the mantle edges contract closely about the neck, water is expelled through the funnel. The water current thus created provides oxygenation for the gills in the mantle cavity, jet power for locomotion, and a means of carrying wastes and sexual products away from the body.

The head of the cephalopod bears a pair of large and complex eyes (Fig. 19-18).

Color changes. There are special pigment cells called **chromatophores** in the skin of most cephalopods, which by expanding and contracting produce color changes. They are controlled by the nervous system and perhaps by hormones. Some color changes are protective to agree with background hues; most are behavioral and are associated with alarm or with courtship. Many deep-sea squids are also bioluminescent.

Ink production. All cephalopods, except *Nautilus*, have an ink sac that empties into the rectum. The sac contains an ink gland that secretes into the sac **sepia,** a dark fluid containing the pigment melanin. When the animal is alarmed, it releases a cloud of ink through the anus to form a ''smokescreen'' to confuse the enemy or perhaps to dull its senses.

Feeding and nutrition. Cephalopods are predaceous, feeding chiefly upon small fishes, molluscs, crustaceans, and worms. Their arms, which are used in food capture and handling, have a complex musculature and are capable of delicately controlled movements. Except in *Nautilus* whose tentacles secrete a sticky substance, the inner surfaces of cephalopod arms bear powerful suction cups. The longer tentacles

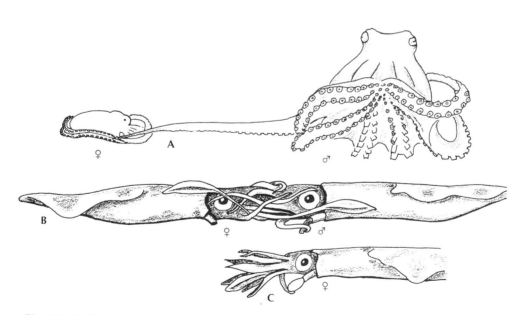

Fig. 19-19. Copulation in cephalopods. **A,** Large male octopus uses modified arm to deposit spermatophores in female mantle cavity to fertilize her eggs (see drawing on p. 364). Octopuses often tend their eggs during development. **B,** Male squid grasps spermatophore as it emerges from his funnel and will thrust it between the ventral arms of the female near sperm reservoir. **C,** After copulation, female reaches for string of fertilized eggs emerging from her funnel and will attach them to rock or other base.

of squids and cuttlefishes are suckered only at the ends. They are highly mobile and used for swiftly seizing the prey and bringing it to the mouth. Strong, beaklike **jaws** can bite or tear off pieces of flesh, which are then pulled into the mouth by the tonguelike action of the **radula.** Octopods and cuttlefishes have salivary glands that secrete a poison for immobilizing prey. Digestion is extracellular and occurs in the stomach and cecum.

Internal features and reproduction. The circulatory and nervous systems of cephalopods are more advanced than in other molluscs. They have the most complex brain among the invertebrates. Their best developed sense organs are the eyes, which, except in *Nautilus,* which has relatively simple eyes, are able to form images. Cephalopod eyes are remarkably like vertebrate eyes, with cornea, lens, chambers, and retina (Fig. 19-18).

Sexes are separate in cephalopods. In the male seminal vesicle the spermatozoa are encased in spermatophores and stored in a sac that opens into the mantle cavity. One arm of the adult male is modified as an intromittant organ, which, during copulation, plucks a spermatophore from his own mantle cavity and inserts it into the mantle cavity of the female near the oviduct opening (Fig. 19-19). Before copulation males often undergo color displays, apparently directed against rival males. Eggs are fertilized as they leave the oviduct and are usually attached to stones or other objects to develop. Some octopods tend their eggs, and *Argonauta*, the paper nautilus, secretes a fluted "shell," or capsule in which she broods her eggs.

REVIEW OF CONTRIBUTIONS OF THE MOLLUSCS

1. In molluscs gaseous exchange occurs not only through the body surface, as in lower invertebrates, but also in specialized **respiratory organs** in the form of gills or lungs.
2. They have an open **circulatory system** with pumping **heart,** vessels, and blood sinuses.
3. The efficiency of their respiratory and circulatory systems has made greater body size possible. Invertebrates reach their largest size in some of the molluscs.
4. They introduce for the first time a fleshy **mantle** that, in most cases, secretes a shell.
5. Several features unique to the phylum are found, for example, the **radula** and the muscular **foot.**
6. The highly developed direct (camera-type) **eye** of higher molluscs is similar to the indirect eye of the vertebrates but arises as a skin derivative in contrast to the eye of vertebrates, which is an outgrowth of the brain (an example of convergent evolution).

SELECTED REFERENCES

Abbott, R. T. 1974. American sea shells, 2nd ed. Princeton, D. Van Nostrand Co. *A semitechnical guide to the marine molluscs of the Atlantic and Pacific coasts of North America.*

Barnes, R. D. 1974. Invertebrate zoology, 3rd ed. Philadelphia, W. B. Saunders Co.

Buchsbaum, R. M., and L. J. Milne. 1960. The lower animals: living invertebrates of the world. Garden City, N.Y., Doubleday & Co., Inc. *Good photographs (many in color) and concise accounts of the invertebrates.*

Burch, J. B. 1962. How to know the eastern land snails. Dubuque, Iowa, William C. Brown Co., Publishers.

Corning, W. C., J. A. Dyal, and A. O. D. Willows (eds.). 1973, 1975. Invertebrate learning: vol 2, Arthropods and gastropod molluscs; and vol. 3, Cephalopods and echinoderms. New York, Plenum Publishing Corp.

Edmondson, W. T. (ed.). 1959. Ward and Whipple's fresh-water biology, ed. 2. New York, John Wiley & sons, Inc. *A taxonomic key to the freshwater families of molluscs is included.*

Fretter, V. (ed.). 1968. Studies in the structure, physiology, and ecology of molluscs. New York, Academic Press, Inc.

Fretter, V., and J. Peake (eds.). 1975. Pulmonates: functional anatomy and physiology. New York, Academic Press, Inc.

Gardiner, M. S. 1972. The biology of the invertebrates. New York, McGraw-Hill Book Co.

Hickman, C. P. 1973. Biology of the invertebrates, 2nd ed. St. Louis, The C. V. Mosby Co.

Hyman, L. H. 1967. The invertebrates: Mollusca (vol. 6). New York, McGraw-Hill Book Co. *This volume covers four groups of the molluscs—Aplacophora, Polyplacophora, Monoplacophora, and Gastropoda—and upholds the fine traditions of the other volumes in this series.*

Jorgensen, C. B. 1966. Biology of suspension feeding. New York, Pergamon Press, Inc.

Kaestner, A. 1964. Invertebrate zoology, vol. 1. New York, Interscience Div., John Wiley & Sons, Inc.

Keen, A. M. 1971. Marine molluscan genera of western North America, 2nd ed. Stanford, Calif., Stanford University Press.

Keen, A. M., and J. H. McLean. 1971. Sea shells of tropical west America. Stanford, Calif., Stanford University Press.

Light, S. F., R. I. Smith, F. A. Pitelka, D. P. Abbott, and F. M. Weesner. 1967. Intertidal invertebrates of the central California coast. Berkeley, University of California Press.

Meglitsch, P. A. 1972. Invertebrate zoology, 2nd ed. New York, Oxford University Press.

Morton, J. E. 1967. Molluscs, ed. 4. London, Hutchinson & Co.

Pennak, R. W. 1953. Fresh-water invertebrates of the United States. New York, The Ronald Press Co.

Potts, W. T. W. 1967. Excretion in the mollusks, Biol. Rev. **42:**1-41.

Purchon, R. D. 1968. The biology of the Mollusca. New York, Pergamon Press, Inc.

Runham, W. W., and P. J. Hunter. 1971. Terrestrial slugs. London, Hutchinson University Library. *A general biology text of pulmonate slugs.*

Solem, A. 1974. The shell makers: introducing mollusks. New York, John Wiley & Sons, Inc.

Wells, M. J. 1962. Brain and behavior in cephalopods. Stanford, Calif., Stanford University Press.

Wilbur, K. M., and C. M. Yonge (eds.). 1964, 1966. Physiology of Mollusca, vols. 1 and 2. New York, Academic Press Inc. *A monograph that summarizes much of the research work on molluscs.*

SELECTED SCIENTIFIC AMERICAN ARTICLES

Boycott, B. B. 1965. Learning in the octopus. **212:**42-50 (March).

Feder, H. M. 1972. Escape responses in marine invertebrates. **227:**92-100 (July).

Korringa, P. 1953. Oysters. **189:**86-91 (Nov.). *Their life history is described.*

Lane, C. E. 1961. The teredo. **204:**132-142 (Feb.). *Biology of the shipworm.*

Willows, H. O. D. 1971. Giant brain cells in mollusks. **224:**68-75 (Feb.).

Yonge, C. M. 1975. Giant clams. **232:**96-105 (April).

The "feather duster" worm, *Spirographis,* secretes a sturdy tube from which it can thrust its feathery crown for feeding. Tiny marine organisms are caught on the radioles and carried by cilia to the mouth at the base of the crown. The worm can withdraw into its tube if disturbed. (Photo by C. P. Hickman, Jr.)

20 SEGMENTED WORMS

PHYLUM ANNELIDA

The annelids are the segmented worms. They are a large phylum, numbering approximately 9,000 species, the most familiar of which are the earthworms and freshwater worms (oligochaetes) and the leeches (hirudineans). However, approximately two thirds of the phylum comprise the less familiar marine worms (polychaetes). Among them are many curious members; some are strange, even grotesque, whereas others are graceful and beautiful (see photograph on p. 386). They include the clamworms, plumed worms, parchment worms, scaleworms, lugworms, and many others. The annelids are true coelomates and belong to the protostome branch, with spiral and determinate cleavage. They are a highly developed group in which the nervous system is more centralized and the circulatory system more complex than those of any of the phyla we have studied thus far.

The Annelida (an-nel'i-da) are worms whose bodies are divided into similar rings, or **segments,** externally marked by circular grooves called **annuli.** The name of the phylum, which comes from the Latin *annellus,* meaning a ring, is descriptive of this characteristic. Body segmentation, or **metamerism,** in the annelids is not merely an external feature but is also seen in the repetitive arrangement of organs and systems and in the partitioning off of segments (also called metameres or somites) by septa. Segmentation, however, is not limited to annelids; it is shared by the arthropods (insects, crustaceans, etc.), which are closely related to the annelids, and by the vertebrates, in which it evolved independently.

Annelids are sometimes called "bristle worms" because, with the exception of the leeches, most annelids bear tiny chitinous bristles called **setae** (Gr. *chaite,* hair or bristle). Short needlelike setae help anchor the somites during locomotion to prevent backward slipping; long, hair-like setae aid aquatic forms in swimming. Since many annelids are either burrowers or live in secreted tubes, the stiff setae also aid in preventing the worm from being pulled out or washed out of its home. Robins know from experience how effective the earthworms' setae are.

Ecologic relationships

Annelids are worldwide in distribution, occurring in the sea, fresh water, and terrestrial soil. Some marine annelids live quietly in tubes or burrow into bottom mud or sand. Some of these are mud eaters; others are filter feeders with elaborate ciliary or mucous devices for trapping food. Many are predators, either pelagic or hiding in crevices of coral or rock except when hunting. Freshwater annelids burrow in mud or sand, live among vegetation, or swim about freely. The most familiar annelids are the terrestrial earthworms, which move about through the soil. Some leeches are blood suckers, and others are carnivores; most of them live in fresh water.

Evolution of annelids

There are so many similarities in the early development of the molluscs, annelids, and arthropods that there seems little doubt about their close relationship. It is thought that the common ancestor of the three phyla was some type of flatworm. Many marine annelids and molluscs have a trochophore type of larva similar to that of the marine flatworms, suggesting a real, if remote, relationship. Annelids share with the arthropods an outer secreted cuticle and a similar nervous system, and there is a similarity between the lateral appendages (parapodia) of many marine annelids and the appendages of certain primitive arthropods. The most important resemblance, however, probably lies in the segmented plan of the annelid and the arthropod body structure.

How and why did metamerism originate? We can only guess, of course. Some flatworms, reproducing asexually, form temporary chains of zooids. Perhaps some such ancestral flatworms in time developed structural and functional unity instead of separating. Or segmentation may have begun with mutations that brought about repetition of certain body parts, with partitions later interposed to form segments.

Whatever the origin, the development of a segmented, coelomate animal was an important step in animal evolution, for from this stem came two of the largest and most important of the invertebrate phyla—the annelids and arthropods. Metamerism brought with it more effective means of body movement because the segmented body provides a series of consecutive motor units and is further aided by the coelomic fluid pressure within each segmental compartment. It also affords great possibilities for the specialization of certain segments for particular functions. Such specialization, minimal in the annelids, is much further developed in the arthropods.

Characteristics

1. **Metameric body,** bilateral symmetry
2. Body wall with outer circular and inner longitudinal muscle layers; outer transparent moist cuticle secreted by epithelium
3. **Chitinous setae** present except in leeches
4. **Coelom** (schizocoel) **well developed and divided by septa** except in leeches; coelomic fluid for turgidity
5. **Blood system closed** and segmentally arranged; amebocytes and respiratory pigments in blood plasma
6. Digestive system complete and not metamerically arranged
7. Oxygen–carbon dioxide exchange through skin, gills, or parapodia
8. Excretory system typically a pair of **nephridia** for each metamere
9. Nervous system with a double ventral nerve cord and a pair of ganglia with lateral nerves in each metamere; brain a pair of dorsal cerebral ganglia with connectives to nerve cord
10. Simple sensory receptors usually; eyes with lenses in some
11. Monoecious or dioecious; larvae, if present, of trochophore type; asexual reproduction by budding in some species; spiral and determinate cleavage

Classification of annelids

There are three main groups of annelids: polychaetes, oligochaetes, and leeches (hirudineans). Because the polychaetes and oligochaetes are both provided with setae, some authorities place them under a larger taxon called Chaetopoda (ke-top'o-da), which means bristle footed.

On the other hand, because both the oligochaetes and the hirudineans (L. *hirudo,* leech) bear a saddle-like enlargement involved in reproduction, called a **clitellum** (L. *clitellae,* packsaddle) (Fig. 20-3, *B*), these two groups are often placed under the heading Clitellata (cli-tel-la'ta) and members are called clitellates.

Class Polychaeta. Mostly marine; head distinct, with eyes and tentacles; most segments with parapodia (lateral appendages) bearing tufts of many setae; clitellum absent; separate sexes usually; asexual budding in some; trochophore larvae usually. *Neanthes* (Fig. 20-7), *Arenicola, Sabella* (Fig. 20-3).

Class Oligochaeta —earthworms and freshwater annelids.* Head and parapodia absent; setae usually few in number; spacious coelom; adults with clitellum; hermaphroditic; development direct; chiefly terrestrial and freshwater. *Lumbricus* (Fig. 20-10), *Aelosoma* (Fig. 20-12, *B*), *Tubifex* (Fig. 20-12, *C*).

Class Hirudinea —leeches.† Body with 33 or 34 segments with

*A group of small annelids that are parasitic or commensal on crayfishes and show similarities to both oligochaetes and leeches are usually placed with the oligochaetes, but they are considered by some authorities to be a separate class, Branchiobdellida. They have 14 or 15 segments and bear a head sucker.

†One genus of leech, *Acanthobdella,* is a primitive type, with some characteristics of leeches and some of oligochaetes; it is sometimes separated from the other leeches into a special class, Acanthobdellida, that characteristically has 27 somites, setae on the first five segments, and the anterior sucker absent.

many annuli; parapodia absent; anterior and posterior suckers usually; setae usually absent; coelom reduced; clitellum present; hermaphroditic; direct development. *Hirudo* (Fig. 20-13), *Placobdella.*

General form and function

The annelid body typically has a head (polychaetes) and a segmented trunk. The body wall of strong circular and longitudinal muscles adapted for swimming and crawling is covered with epidermis and a thin outer layer of nonchitinous cuticle (Figs. 20-1 and 20-10, *C* and *D*).

Coelom. In most annelids the coelom develops as a pair of coelomic compartments in each segment. Each compartment is surrounded with **peritoneum** (a layer of mesodermal epithelium), which lines the body wall, forms dorsal and ventral **mesenteries,** and covers all the organs (Fig. 20-1). Where the peritonea of adjacent segments meet, the **septa** are formed. These are perforated by the gut and longitudinal blood vessels. Except in the leeches, the coelom is filled with fluid, which serves as a hydrostatic skeleton and provides the rigidity and resistance necessary for muscular movement. Not only is the coelom metamerically arranged, but practically every body system is affected in some way by this segmental arrangement.

Digestive tract. The long, tubular digestive tract is unsegmented, and its specialized regions vary somewhat with the feeding habits of the different groups. Most annelid digestion is extracellular.

Circulation. Annelids have a double transport system—the coelomic fluid and the circulatory system. Food, wastes, and respiratory gases are carried by both coelomic fluid and blood in varying degrees. The blood is usually red because of the oxygen-combining respiratory pigment hemoglobin, but some marine forms have bright green blood because of another respiratory pigment called chlorocruorin. In still others the blood contains a third type of pigment called hemerythrin, which gives the blood a pinkish color.

The blood is carried in a closed system of vessels, flowing forward in a dorsal vessel and backward in a ventral vessel. Connecting the larger vessels are transverse vessels and systems of capillaries in the tissue.

Gas exchange. In most annelids the exchange of respiratory gases (oxygen and carbon dioxide) occurs through the body wall, but many active aquatic annelids are equipped with gills or thin-skinned, vascular parapodia.

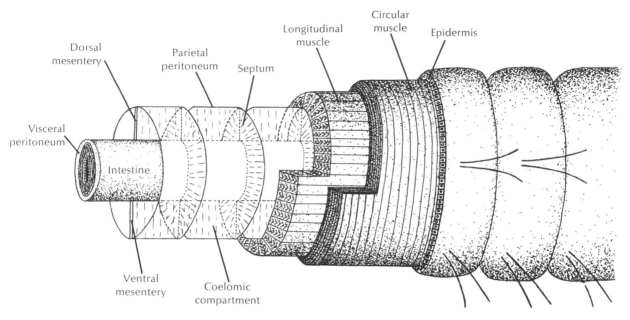

Fig. 20-1. Annelid body plan. The body wall and intestine, a "tube-within-a-tube," are separated by the coelomic cavity, which forms a pair of coelomic spaces in each somite. The peritoneum of each coelomic space lines the body wall (parietal peritoneum), covers the intestine and viscera (visceral peritoneum), and lies against the peritoneum of adjoining coelomic spaces to form the mesenteries and septa.

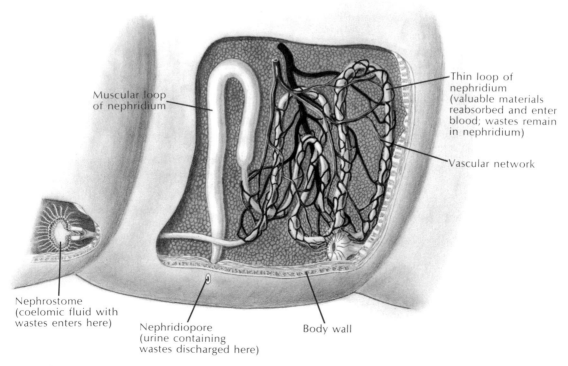

Fig. 20-2. Nephridium of earthworm. Wastes are drawn into ciliated nephrostome in one segment, then passed through loop of nephridium and expelled through nephridiopore of next segment.

Excretion and osmoregulation. Excretion typically occurs by means of **nephridia,** segmentally arranged with a pair to a segment (Fig. 20-2). Each nephridium collects fluid from the coelom through a ciliated funnel (nephrostome), processes it in a nephridial tubule, and empties the resultant urine to the outside through a pore (nephridiopore). What occurs in the tubule depends upon whether the animal needs to conserve or eliminate water, salt, nitrogenous wastes, sugar, etc. Some primitive forms have protonephridia with ciliated flame cells (solenocytes) much like those of the flatworms.

Nervous and sensory systems. Most annelids have a pair of **cerebral ganglia** (the brain) above the pharynx joined to the double **ventral nerve cord** by a pair of connectives around the pharynx. **Neurosecretory cells,** endocrine in function, have been found in the brain and ganglia of annelids and are concerned with growth, reproduction, and regeneration.

Sense organs are well developed in active polychaetes but greatly reduced in sedentary forms. Eyes, such as those in the clam worm *Neanthes* and others are complex, with cornea, lens, and retina. Other annelids may have simple pigment cup ocelli or merely photosensitive cells in the epidermis. Chemoreceptors and touch receptors are numerous. Most setae serve also as touch receptors.

Reproduction and development. Reproduction is sexual in most annelids, although in some forms asexual budding or body division occurs. Most polychaetes are dioecious; oligochaetes and hirudineans are monoecious. Cleavage is **spiral** and **determinate,** and in the polychaetes the early larva is a free-swimming trochophore. In oligochaetes and leeches juveniles similar to the parents hatch from the eggs.

POLYCHAETES

The Polychaeta (pol-e-ke′ta) are the oldest and the largest of the annelid classes. They differ from the other annelids in having better developed head and sense organs; paired, paddlelike appendages on most segments, called **parapodia;** and no clitellum. They also have more setae than other annelids, as indicated by their class name, which comes from the Gr. *polys,* meaning many, and *chaite,* meaning bristle. They are predominantly marine forms, most of which are from 5 to 10 cm long, although they range from less than 1 mm to as much as 3 m long. Some are brightly colored in reds or greens; others are dull or irridescent.

Polychaetes range from ordinary mudworms to the spectacular tubed ''feather duster'' worms and include a great diversity of forms.

Polychaetes live under rocks, in coral crevices, or in abandoned shells, or they burrow into mud or sand; some build their own tubes on submerged objects or in bottom material; some adopt the tubes or homes of other animals; some are pelagic, making up a part of the planktonic population. They are extremely abundant in some areas. For example, a square meter of mud flat may contain thousands of polychaetes. They play a significant part in marine food chains, as they are eaten by fishes, crustaceans, hydroids, and many others.

Ecologically polychaetes are divided into two groups—Errantia and Sedentaria. The **Errantia,** or

Fig. 20-3. Polychaete tubeworm, *Sabella pavonia.* The crown of tentacles is used in ciliary-mucus feeding. Its cylindric, leathery tube is built of secreted material to which sand and bits of debris adhere. This tubeworm was photographed at a depth of 35 m off the west coast of Norway. (Photo by T. Lundälv, Kristinebergs Zoological Station, Sweden.)

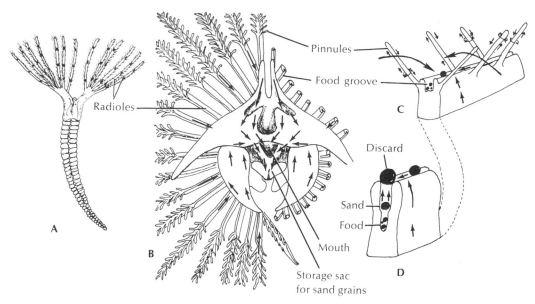

Fig. 20-4. Ciliary feeding in fanworms. **A,** *Sabella,* with its crown of feeding radioles. **B,** Anterior view of base of crown. Cilia direct small food particles along grooved radioles toward mouth. Sand grains are directed to storage sacs and used later for tube building. **C** and **D,** Enlarged portions of radiole. **C,** Distal portion showing ciliary tracts of pinnules and food grooves. **D,** Proximal portion where particles are sorted for size. (**B** to **D** adapted from several sources.)

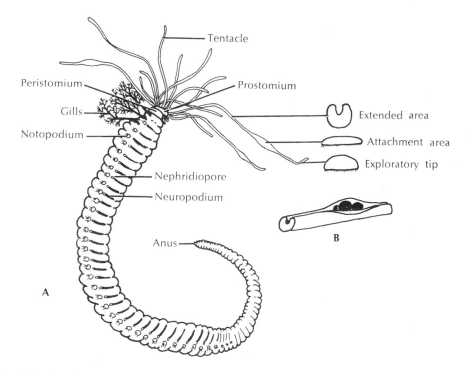

Fig. 20-5. *Amphitrite,* which builds its tubes in mud or sand, extends long grooved tentacles out over the mud to pick up bits of organic matter. Smallest particles are moved along food grooves by cilia, larger particles by peristaltic movement. Its plumelike gills are bloodred.

errant worms (L. *errare,* to wander) include the free-moving pelagic forms, active burrowers, crawlers, and the tube worms that leave their tubes for feeding or breeding. Most of these, like the clam worms (Fig. 20-7), are predatory forms equipped with jaws or teeth. Some have a muscular eversible pharynx armed with teeth that can be thrust out with surprising speed and dexterity for capturing prey. The **Sedentaria** (L. *sedere,* to sit) are the sedentary worms that rarely expose more than the head end from the tubes or burrows in which they live (p. 386 and Fig. 20-3).

Most sedentary tube and burrow dwellers are trappers, using ciliary or mucoid methods of obtaining food. The principal food source is plankton and detritus. Cilia on tentacles or on stiff tentacular crowns create currents of water, filter out food particles, and move them toward the mouth (Fig. 20-4). Some worms, such as *Chaetopterus,* secrete mucous filters through which they pump water to collect edible particles (Fig. 20-8). *Amphitrite,* with its head peeping out of the mud, sends out long extensible tentacles over the mud surface, drawing in bits of detritus along their ciliated grooves (Fig. 20-5). The lugworm *Arenicola* lives in an L-shaped burrow in which, by peristaltic movements, it keeps water filtering down through the sand and out the open end of the burrow. It ingests the food-laden sand brought by the water current.

Tube dwellers secrete many types of tubes. Some are parchmentlike (Fig. 20-8); some are firm, calcareous tubes attached to rocks or other surfaces; some are simply grains of sand or bits of shell or seaweed cemented together by secreted mucus or cement (p. 386) and Fig. 20-3). Many burrowers in sand and mud flats line their burrows with mucus.

Form and function

The polychaete typically has a head, or **prostomium,** which may or may not be retractile and which often bears eyes, antennae, and sensory palps (Fig. 20-7, *A*). The first segment **(peristomium)** surrounds the mouth and may bear setae, palps, or in predatory forms chitinous jaws. Ciliary feeders may bear a tentacular crown that may be opened up like a fan or withdrawn into the tube.

Most polychaete segments bear fleshy appendages called **parapodia,** which may have lobes, cirri, setae, etc. on them (Fig. 20-7). The parapodia are used in crawling, swimming, or anchoring in tubes. They usually serve as the chief respiratory organs, although some polychaetes may also have gills (Fig. 20-5).

Reproductive systems are simple; gonads appear as temporary swellings of the peritoneum and shed their gametes into the coelom. They are carried outside through gonoducts, through nephridia, or by rupture of the body wall. Fertilization is external, and the early larva is a trochophore.

Some polychaetes live most of the year as sexually

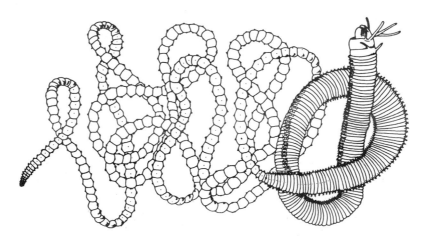

Fig. 20-6. *Eunice viridis,* the Samoan palola worm. The posterior epitokal segments are packed with gametes and provided with eyespots on the ventral side. Once a year, when the worms swarm, the epitokes detach, rise to the surface, and discharge their gametes, leaving the water milky. By next breeding season the epitokes are regenerated.

unripe animals called **atokes,** but during the breeding season a portion of the body develops into a sexually ripe form called an **epitoke,** which is swollen with gametes (Fig. 20-6). An example is the Palolo worm *Eunice viridis,* which lives in burrows among the coral reefs of the South Seas. During the reproductive cycle the posterior somites become swollen with gametes. During the swarming period, which occurs at the beginning of the last quarter of the October-November moon, these epitokes break off and float to the surface. Just before sunrise the sea is literally covered with them, and at sunrise they burst, freeing the eggs and sperm for fertilization. The anterior portions of the worms regenerate new posterior sections. A related form, *Leodice,* swarms in the Atlantic in the third quarter of the June-July moon.

Some interesting polychaetes

Clam worms—Nereis (Neanthes). The clam worms are errant polychaetes that live in mucus-lined burrows in or near low tide. The body grows up to 30 to 40 cm long. The head is made up of a prostomium with sensory palps, tentacles, two pairs of light-sensitive eyes, and a peristomium bearing a ventral mouth and four pairs of sensory tentacles (Fig. 20-7, *A*). Each somite, except the head, bears a pair of

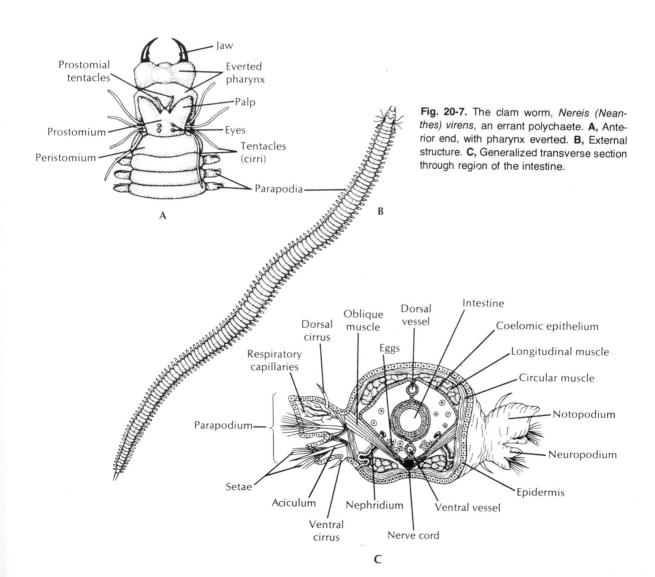

Fig. 20-7. The clam worm, *Nereis (Neanthes) virens,* an errant polychaete. **A,** Anterior end, with pharynx everted. **B,** External structure. **C,** Generalized transverse section through region of the intestine.

Fig. 20-8. *Chaetopterus,* a sedentary polychaete, lives in a parchment U-shaped tube through which it pumps water with three pistonlike fans. The fans beat 60 times per minute to keep water currents moving. The winglike notopodia of the twelfth segment secrete a mucous net that strains out food particles. As the net fills with food, the food cup rolls it into a ball and, when the ball is large enough, it is rolled along a ciliated groove to the mouth and swallowed. (Courtesy American Museum of Natural History.)

fleshy two-lobed paddles, or parapodia, each supported by spines and bearing tufts of setae (Fig. 20-7, *C*). They are abundantly supplied with blood vessels. The animal swims rapidly by undulatory movements.

The clam worm is active at night, feeding upon small invertebrates, larval forms, etc. Its pharynx, equipped with pincerlike jaws, can be everted through the mouth to seize the prey.

Parchment worms—Chaetopterus. *Chaetopterus* lives in a U-shaped parchment tube that is buried,

except for the tapered ends, in sand or mud (Fig. 20-8). Fans (modified parapodia) pump water through the tube by rhythmic movements. A pair of enlarged parapodia secretes a long mucous bag that reaches back to a small food cup in front of the fans. Water is filtered through the mucous bag, which, when full of food particles, is rolled into a ball by cilia in the food cup, directed forward by cilia to the mouth, and swallowed.

Fanworms. The fanworms, or "feather duster" worms, are beautiful tubeworms, fascinating to watch

Fig. 20-9. *Protula,* a marine tube-building polychaete. (Photo by C. P. Hickman, Jr.)

as they emerge from their secreted tubes and unfurl their lovely tentacular crowns to feed (p. 386 and Figs. 20-3 and 20-9). A slight disturbance, sometimes even a passing shadow, causes them to duck quickly into the safety of the homes they have built. Food attracted to the feathery arms, or radioles, by ciliary action, is trapped in mucus and carried down ciliated food grooves to the mouth (Fig. 20-4). Particles too large for the food grooves are carried along the margins and dropped off. Further sorting may occur near the mouth where only the small particles of food enter the mouth, and sand grains are stored in a sac to be used later in enlarging the tube.

OLIGOCHAETES

The Oligochaeta (ol'i-go-ke'ta) vary in size, structure, and habitats but are less diverse than the polychaetes. They differ from polychaetes in having a reduced head, no parapodia, a clitellum, and, as their class name implies, fewer and less conspicuous setae (Gr. *oligos,* few, + *chaite,* bristle). Most of them live in the soil, but some live in freshwater; a few are marine, and a few are parasitic on aquatic animals. They are almost worldwide in their distribution.

Earthworms

The most familiar of the oligochaetes are the earthworms, which burrow in moist, rich soil, emerging at night to explore their surroundings ("night crawlers"). In damp, rainy weather they stay near the surface, often with mouth or anus protruding from the burrow. In very dry weather they may burrow several feet underground, coil up in a slime chamber, and pass into a state of dormancy. *Lumbricus terrestris,* the form commonly studied in school laboratories, is approximately 12 to 30 cm long (Fig. 20-10). *Eisenia,* a small genus, is usually abundant in manure piles. Some giant tropical earthworms grow to as much as 4 m long.

Aristotle called earthworms the "intestines of the soil." Twenty-two centuries later Charles Darwin published his observations of many years in his classic *The Formation of Vegetable Mould Through the Action of Worms.* He showed how worms enrich the soil by bringing subsoil to the surface and mixing it with the topsoil. An earthworm can ingest its own weight in soil every 24 hours, and Darwin estimated that from 10 to 18 tons of dry earth per acre pass through their intestines annually, thus bringing up potassium and phosphorus from the subsoil and also adding to the soil nitrogenous products from their own metabolism. They expose the mold to the air and sift it into small particles. They also drag leaves, twigs, and organic substances into their burrows closer to the roots of plants. Their activities are important in aerating the soil.

Form and function

The prostomium is reduced to a small lobe that overhangs the mouth and bears no sensory appendages

395

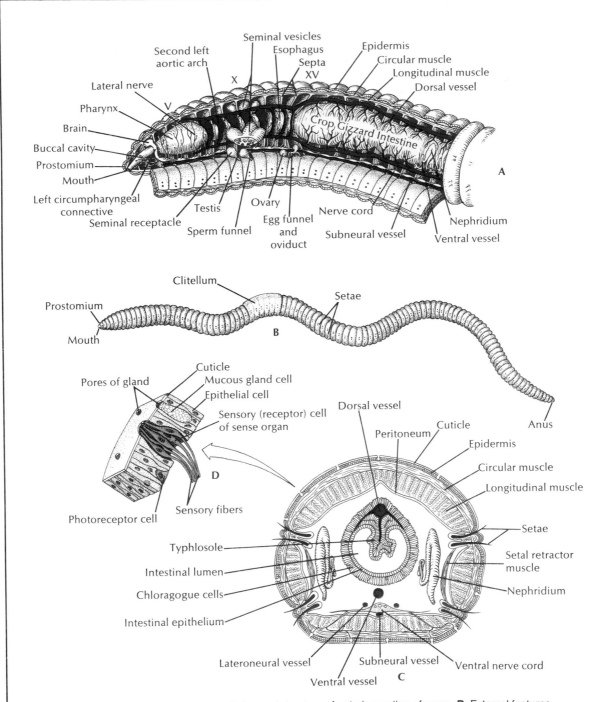

Fig. 20-10. Earthworm anatomy. **A,** Internal structure of anterior portion of worm. **B,** External features, lateral view. **C,** Generalized transverse section through region posterior to clitellum. **D,** Portion of epidermis showing sensory, glandular, and epithelial cells.

such as those found on the prostomium of polychaetes. There are usually 4 pairs of short stout setae to a segment, each one moved by tiny muscles.

Earthworms move by peristaltic contractions, aided by the hydraulic turgor in the coelomic compartments. Burrowers push their way through soft soil and literally eat their way through firmer soil, leaving their castings behind.

Nutrition. The food of the earthworm is mainly decayed organic matter and bits of vegetation drawn in by the muscular pharynx. The calcium from the soil swallowed with the food tends to produce a high blood calcium level. **Calciferous glands** along the esophagus secrete calcium ions into the gut and so reduce the calcium ion concentration of the blood. Food is stored temporarily in the thin-walled crop before being passed on to a thick muscular gizzard for grinding (Fig. 20-10, A).

Circulation. Earthworms have the typical **closed system** of blood vessels and capillaries (Fig. 20-10, A) with a pumping dorsal vessel aided by 5 pairs of aortic arches ("hearts") with muscular walls and a ventral vessel. The blood contains ameboid corpuscles and dissolved hemoglobin, which gives the blood its red color.

Excretion. Paired **nephridia** are found in most segments, each tubule lying in one segment with its ciliated collecting funnel (nephrostome) in the coelomic cavity of the segment just anterior to it. Each nephridium empties through a small pore (nephridiopore) in the body wall (Fig. 20-2).

Specialized **chlorogogue cells** around the intestine

Fig. 20-11. Two earthworms in copulation. Their anterior ends point in opposite direction as their ventral surfaces are held together by mucous bands secreted by the clitella. Mutual insemination occurs during copulation. After separation each worm secretes a cocoon to receive its eggs and sperm. (Courtesy Dr. Guy Carter.)

397

are thought to receive assimilated materials from the gut, store them, and eventually release them to the system. Since they can detach and migrate to the coelom, some authorities believe they convey waste to the nephridia for elimination.

Nervous system. As in other annelids there is a pair of cerebral ganglia above the pharynx joined to the ventral nerve cord by a pair of connectives around the pharynx. The nerve cord has a pair of ganglia in each somite, giving off segmental nerves containing both sensory and motor fibers.

For rapid escape movements the nerve cord usually has one to several large axons known as **giant fibers** that run the length of the cord. Protected by a myelin

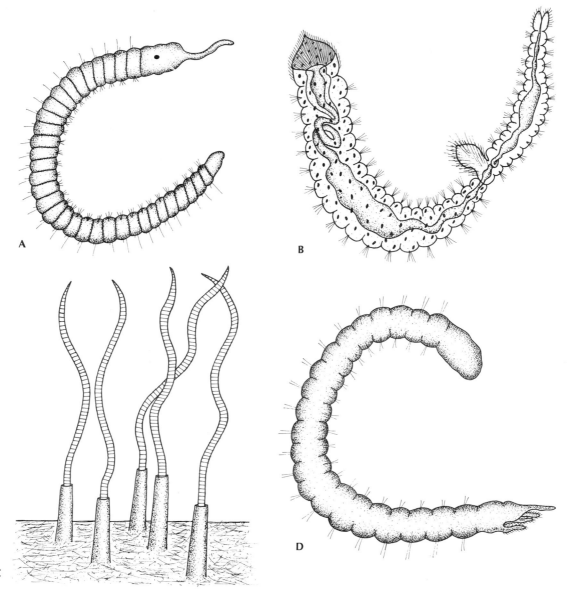

Fig. 20-12. Some freshwater oligochaetes. **A,** *Stylaria* has the prostomium drawn out into a long snout. **B,** *Aeolosoma* uses cilia around the mouth to sweep in food particles, and it buds off new individuals asexually. **C,** *Tubifex* lives head down in long tubes. **D,** *Aulophorus* is provided with ciliated anal gills.

sheath and lacking synaptic barriers, the speed of conduction in these giant nerve fibers is tremendously increased over that of small neurons.

Reproduction. Earthworms are hermaphroditic and exchange sperm during copulation, which usually occurs at night. When mating, the worms extend their anterior ends from their burrows and bring their ventral surfaces together (Fig. 20-11), secreting mucous bands around them. In this position sperm are exchanged. Later each worm secretes a barrel-shaped **cocoon** about its clitellum, which it fills with eggs, albumin, and sperm as the cocoon moves forward. The cocoons slip off over the head and are usually deposited in the earth. In *Lumbricus* only one of the fertilized eggs in a cocoon develops; the others presumably provide nutrition for the developing embryo. Between copulations the earthworm may continue to form cocoons as long as there are sperm from the mate in its seminal receptacles. Juvenile worms escape from the cocoons in 2 to 3 weeks. A clitellum does not develop until the worm is mature.

Freshwater oligochaetes

Freshwater oligochaetes usually are smaller and have more conspicuous setae than do the earthworms. They are more mobile than earthworms and tend to have better developed sense organs. They are generally benthic forms that creep about on the bottom or burrow into the soft mud. Aquatic oligochaetes provide an important food source for fishes. A few are ectoparasitic.

Some aquatic forms have **gills.** In *Branchiura* the gills are long, slender projections from the body surface. Others such as *Dero* and *Aulophorus,* which have ciliated posterior gills (Fig. 20-12), extend them from their tubes and use the cilia to keep the water moving. Most forms breathe through the skin as do the earthworms.

The chief foods are algae and detritus, which they may pick up by extending a mucus-coated pharynx. Burrowers swallow mud and digest the organic material. Some, such as *Aelosoma,* are ciliary feeders that use currents produced by cilia at the anterior end of the body to sweep food particles into the mouth (Fig. 20-12, *B*).

Some common freshwater genera are *Aelosoma,* only approximately 1 mm long; *Stylaria,* 10 to 25 mm long, which has the prostomium drawn out into a long process; and *Tubifex,* 30 to 40 mm long, a red worm that lives head down in a tube in the bottom mud, waving its tail in the water. Living close together, large numbers of *Tubifex* often form reddish patches on the mud. Many of the smaller forms, such as the Aelosomatidae, reproduce asexually by budding, often forming long chains of individuals.

LEECHES—HIRUDINEA

Leeches are found predominantly in freshwater habitats, but a few are marine, and some have even adapted to terrestrial life in moist, warm areas. Most leeches are between 2 and 6 cm in length, but some are smaller, and some reach 20 cm or more. They are found in a variety of patterns and colors—black, brown, red, or olive green. They are usually flattened dorsoventrally.

Like the oligochaetes, leeches have a **clitellum** during the breeding season, but they have **no setae.** They also differ from other annelids in having the number of somites fixed at 33 (34 by one method of counting). However, they appear to have more metameres than they really have, because each somite is marked by transverse grooves to form from 2 to 16 superficial rings called **annuli** (Fig. 20-13).

Coelom. The coelom represents another difference between leeches and oligochaetes; leeches lack distinct coelomic compartments. In all but one species the septa have disappeared and the coelomic cavity is filled with connective tissue and a system of spaces called **sinuses.** The coelomic sinuses form a regular system of channels filled with coelomic fluid, which in some leeches serves as an auxiliary circulatory system.

Nutrition. Leeches are popularly considered to be parasitic, but it would be more accurate to call them predaceous. Even the true bloodsuckers are rarely host specific. Most freshwater leeches are active predators or scavengers equipped with a proboscis that can be extended to draw in small invertebrates or to take blood from cold-blooded vertebrates. Some freshwater leeches are true bloodsuckers, preying on cattle, horses, humans, etc. Some terrestrial leeches feed on insect larvae, earthworms, and slugs, which they hold on to by an anterior sucker while using a strong sucking pharynx to draw in the food. Other terrestrial forms climb bushes or trees to reach warm-blooded vertebrates such as birds or mammals.

Most leeches are fluid feeders. Many prefer to feed on tissue fluids and blood pumped from wounds already open. The true bloodsuckers, which include the so-called medicinal leech *Hirudo* once popularly used in the practice of bloodletting, have cutting plates, or "jaws," for actually cutting through tissues. In feeding, the leech attaches itself to the prey by a posterior sucker, then searches for a vulnerable place on the skin to attach its anterior sucker, which surrounds the mouth. The jaws then cut three fine slits—the characteristic leech bite. Coagulation of flowing blood is prevented by an anticoagulant, called hirudin, secreted into the incisions by salivary glands. The leech may suck enough blood at a single feeding to

quintuple its weight. It may not need to feed again for months.

Locomotion. Most leeches creep with looping movements of the body, by attaching first one sucker and then the other and pulling up the body. Aquatic leeches can also swim with a graceful undulatory movement.

Respiration and excretion. Gas exchange occurs through the skin except in the fish leeches, which have gills. There are 10 to 17 pairs of nephridia, in addition to which coelomocytes and certain other specialized cells may also be involved in excretory functions.

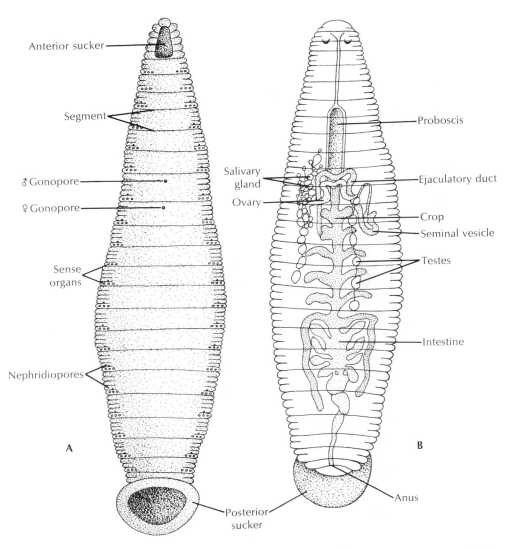

Fig. 20-13. Structure of the leech. **A,** Ventral view of *Hirudo.* **B,** Internal structure of *Helobdella. Hirudo* is an aquatic bloodsucker with five annuli on most of its segments. *Helobdella* is an aquatic leech with three annuli to a segment.

Nervous and sensory system. Leeches have two "brains," one in the head and one in the tail, each made up of 6 to 7 fused ganglia. The additional ganglia are segmentally arranged along the double nerve cord. In addition to free sensory nerve endings and photoreceptor cells in the epidermis, there is a row of sense organs, called sensillae, in the central annulus of each segment and a number of pigment cup ocelli.

Reproduction. Leeches are hermaphroditic but practice cross-fertilization during copulation. Sperm is transferred by a penis or by hypodermic impregnation. After copulation the clitellum secretes a cocoon that receives the eggs and sperm. Cocoons are buried in bottom mud, attached to submerged objects, or in terrestrial species placed in damp soil. Development is similar to that of oligochaetes.

REVIEW OF CONTRIBUTIONS OF THE ANNELIDS

1. The introduction of **metamerism** by the group represents the greatest advancement of this phylum and lays the groundwork for the more highly specialized metamerism of the arthropods.
2. A true coelomic cavity reaches a high stage of development in this group.
3. Specialization of the head region into differentiated organs, such as the tentacles, palps, and eyespots of the polychaetes, is carried further in some annelids than in other invertebrates so far considered.
4. The tendency toward **centralization of the nervous system** is more developed, with cerebral ganglia (brain), two closely fused ventral nerve cords with unique giant fibers running the length of the body, and various ganglia with their lateral branches.
5. The circulatory system is much more complex than any we have so far considered. It is a closed system with muscular blood vessels and aortic arches ("hearts") for propelling the blood.
6. The appearance of the fleshy **parapodia,** with their respiratory function, introduces a suggestion of the paired appendages and specialized gills found in the more highly organized arthropods.
7. The well-developed **nephridia** in most of the somites have reached a differentiation that involves a removal of waste from the blood as well as from the coelom.
8. Annelids are the most highly organized animals capable of complete regeneration. However this ability varies greatly within the group.

SELECTED REFERENCES

Barnes, R. D. 1965. Tube-building and feeding in chaetopterid polychaetes. Biol. Bull. **129:**217-233

Barnes, R. D. 1974. Invertebrate zoology, ed. 3. Philadelphia. W. B. Saunders Co., pp. 233-316.

Clark, L. B., and W. N. Hess. 1940. Swarming of the Atlantic palolo worm, *Leodice fucata.* Tortugas Lab. Papers **332:**21-70.

Dales, R. P. 1967. Annelids. New York, Hutchinson & Co. *A concise up-to-date account of the annelids.*

Darwin, C. R. 1911. The formation of vegetable mould through the action of worms. *A classic account of the way in which earthworms improve and transform the surface of the soil.*

Edmondson, W. T. (ed.). 1959. Fresh-water biology, ed. 2. New York, John Wiley & Sons, Inc. *Contains a guide to identification of freshwater annelids, with keys.*

Gardiner, M. S. 1972. The biology of the invertebrates. New York, McGraw-Hill Book Co.

Giese, A. C. 1974. Reproduction of marine invertebrates: vol. 3, annelids and echiurans. New York, Academic Press, Inc.

Gosner, K. L. 1971. Guide to identification of marine and estuarine invertebrates: Cape Hatteras to the Bay of Fundy. New York, Interscience Div., John Wiley & Sons, Inc., pp. 326-387.

Hickman, C. P. 1973. Biology of the invertebrates, ed. 2. St. Louis, The C. V. Mosby Co.

Kaestner, A. 1967. Invertebrate zoology, vol. 1. New York, Interscience Div., John Wiley & Sons, Inc., pp. 454-566.

Laverack, M. S. 1963. The physiology of earthworms. New York, Macmillan, Inc.

Light, S. F., R. I. Smith, F. A. Pitelka, D. P. Abbott, and F. M. Weesner. 1967. Intertidal invertebrates of the central California coast. Berkeley, University of California Press, pp. 63-108.

Mann, K. H. 1962. Leeches *(Hirudinea),* their structure, physiology, ecology, and embryology. New York, Pergamon Press, Inc.

Meglitsch, P. A. 1972. Invertebrate zoology, ed. 2. New York, Oxford University Press, Inc.

Pennak, R. W. 1953. Freshwater invertebrates of the United States. New York, The Ronald Press Co., pp. 278-320. *A brief account with keys to families and genera of freshwater annelids.*

Russell-Hunter, W. D. 1969. A biology of the higher invertebrates. New York, Macmillan, Inc. *A concise discussion of this group.*

Sawyer, R. T. 1972. North American freshwater leeches, exclusive of the Pisciocolodae, with a key to all species. Illinois Biol. Monogr. No. 46. 154.

Stephenson, J. 1930. The oligochaetes. New York, Oxford University Press, Inc.

Wells, M. 1968. Lower animals. New York, McGraw-Hill Book Co.

SELECTED SCIENTIFIC AMERICAN ARTICLES

Nicholls, J. C., and D. Van Essen. 1974. The nervous system of the leech. **230:**38-48 (Jan.).

Wells, P. 1959. Worm autobiographies. **200:**132-141 (June).

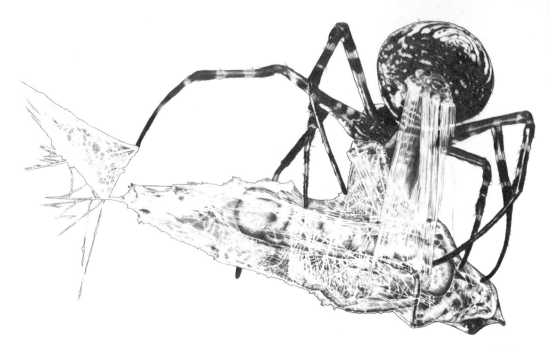

As the spider extrudes silk from her spinnerets, she handles it deftly with her legs to wrap up a fresh meal for future use. She produces several kinds of silk that can be used for different purposes. (Drawn from a photo by J. G. Healey.)

21 ARTHROPODS

PHYLUM ARTHROPODA

Phylum Arthropoda is the most extensive phylum in the animal kingdom, making up more than three fourths of all known species. Approximately 900,000 species have been recorded, and probably as many more remain to be classified. Arthropods include the spiders, scorpions, ticks, mites, crustaceans, millipedes, centipedes, insects, and some others. In addition there is a rich fossil record extending to the very late Precambrian (Fig. 21-1).

Arthropods are eucoelomate protostomes with well-developed organ systems, and they share with the annelids the property of conspicuous segmentation.

Arthropoda (ar-throp′o-da) means joint-footed. (Gr. *arthron,* joint, + *podos,* foot). Arthropods have a chitinous exoskeleton, and their primitive pattern is that of a linear series of similar somites, each with a pair of jointed appendages. However, the pattern of somites and appendages varies greatly in the phylum. There is a tendency for the somites to be combined or fused into functional groups, called **tagmata,** for specialized purposes; the appendages are frequently differentiated and specialized for pronounced division of labor.

Few arthropods exceed 60 cm in length, and most are far below this size. The largest is the Japanese crab *Macrocheira,* which has approximately a 3.7 m span; the smallest is the parasitic mite *Demodex,* which is less than 0.1 mm long.

Arthropods are usually active, energetic animals.

Fig. 21-1. Middle Cambrian diorama. Note the number and variety of arthropods among the sponges and algae. Trilobites (large one, *lower left,* and small ones, *center*) had probably existed millions of years before the Cambrian. (Courtesy American Museum of Natural History.)

Judged by their great diversity and their wide ecologic distribution, as well as by the vast numbers of species, their success is surpassed by no other group of animals.

Although arthropods compete with man for food supplies and spread serious diseases, they also serve as food, cross-pollinate plants, yield useful drugs and dyes, and produce useful products, such as silk, honey, and beeswax.

In diversity of **ecologic distribution,** arthropods are unrivaled. They live in every type of environment—from ocean depths to mountain peaks and from tropics to polar regions. They are adapted for life in the air, on the land, in the sea, in fresh or brackish water, or in or on the bodies of plants or other animals.

Evolution of arthropods

The similarities between the annelids and the arthropods give strong support to the theory that both phyla originated from a line of coelomate segmented protostomes, which in time diverged to form a protoannelid line with laterally located parapodia and a protoarthropod line with more ventrally located parapodia. The protoannelid line gave rise eventually to the polychaetes. The protoarthropod line apparently diverged further into three branches: (1) the trilobites and the chelicerates (spiders and horseshoe crabs), (2) the crustaceans, and (3) the insects, millipedes, and centipedes. The phylum Onychophora may also have come from some part of this protoarthropod line.

Some biologists consider the arthropods to be three separate phyla, each organized at the arthropod level. This is because of the three-stem origin of the arthropods, whereas a phylum is supposed to be monophyletic.

Annelids show little specialization or fusion of somites and little differentiation of appendages. However, in arthropods the adaptive trend has been toward tagmatization of the body by differentiation or fusion of somites, giving rise in more advanced groups to such tagma as head and trunk; head, thorax, and abdomen; or cephalothorax (fused head and thorax) and abdomen. Primitive arthropods tend to have similar appendages, whereas the more advanced forms have appendages specialized for specific functions, or some appendages may be lost entirely.

Much of the amazing diversity in arthropods seems to have developed because of modification and specialization of their chitinous exoskeleton and their

jointed appendages, thus resulting in a wide variety of locomotor and feeding adaptations.

Why have arthropods been so successful?

The success of the arthropods is attested to by their diversity, number of species, wide distribution, variety of habitats and feeding habits, and power of adaptation to changing conditions. Some of the structural and physiologic patterns that have been helpful to them are briefly summarized in the following discussion.

1. A versatile exoskeleton. The arthropods possess an exoskeleton that is highly protective without sacrificing mobility. This skeleton is the **cuticle,** an outer covering secreted by the underlying epidermis. The cuticle is made up of an inner and usually thicker **endocuticle** and an outer, relatively thin **epicuticle.** The endocuticle contains **chitin** bound with protein. Chitin is a tough, resistant, nitrogenous polysaccharide that is insoluble in water, alkalis, and weak acids. Thus the endocuticle is not only flexible and lightweight but also affords protection, particularly against dehydration. In some crustaceans the chitin may make up as much as 60% to 80% of the endocuticle, but in insects it is probably not more than 40%. In most crustaceans, the endocuticle in some areas is also impregnated with **calcium salts,** which reduce its flexibility. In the hard shells of lobsters and crabs, for instance, this calcification is extreme. The outer epicuticle is composed of protein, which is hardened by tanning, adding further protection. Both the endocuticle and epicuticle are laminated, that is, composed of several layers each.

The cuticle may be soft and permeable or may form a veritable coat of armor. Between the segments of appendages it is thin and flexible, creating movable joints and permitting free movements. In crustaceans and insects the cuticle forms ingrowths (apodemes) that serve for muscle attachment. It may also line the fore- and hindgut, form tracheal supports, and be adapted for biting mouthparts, sensory organs, copulatory organs, and ornamental purposes. It is indeed a versatile material.

A chitinous exoskeleton does, however, limit the size of an animal. To grow, an arthropod must shed its outer covering at intervals and grow a larger one—a process called **ecdysis,** or **molting.** Arthropods molt from four to seven times before reaching adulthood, and some continue to molt after that. An exoskeleton is also relatively heavy and becomes proportionately heavier with increasing size. This also limits the ultimate body size.

2. Segmentation and appendages for more efficiency and better locomotion. Typically each somite is provided with a pair of jointed appendages, but this arrangement is often modified, with both segments and appendages specialized for adaptive functions. The limb segments are essentially hollow levers that are moved by internal muscles, most of which are striated for rapid action. The jointed appendages are equipped with sensory hairs and have been modified and adapted for sensory functions, food handling, swift and efficient walking legs, and swimming appendages. This affords greater efficiency and a wider capacity for adjustment to varied habitats.

3. Air piped direct to cells. Most land arthropods have the highly efficient tracheal system of air tubes, which delivers oxygen directly to the tissues and cells and makes high metabolism possible. Aquatic arthropods breathe mainly by some form of gill that is quite efficient.

4. Highly developed sensory organs. Sensory organs are found in great variety, from the compound (mosaic) eye to those simpler senses that have to do with touch, smell, hearing, balancing, chemical reception, etc. Arthropods are keenly alert to what goes on in their environment.

5. Complex behavior patterns. Arthropods exceed most other invertebrates in the complexity and organization of their activities. Innate (unlearned) behavior unquestionably controls much of what they do; learning also plays an important part in the lives of many of them.

6. Reduced competition through metamorphosis. Many arthropods pass through metamorphic changes— larva, pupa, and adult stages—and some go through a series of nymphal stages preceding adulthood. The larval form is often adapted for eating a different kind of food from that of the adult, resulting in less competition within a species.

Characteristics

1. Bilateral symmetry; **metameric body,** often divided into head and trunk; head, thorax, and abdomen; or cephalothorax and abdomen
2. **Jointed appendages;** primitively, one pair to each somite, but number often reduced; often modified for specialized functions
3. **Exoskeleton of chitinous cuticle** secreted by underlying epidermis and shed (molted) at intervals

4. **Complex muscular system,** with exoskeleton for attachment; **striated muscle** for rapid action; smooth muscle for visceral organs; no cilia
5. **Reduced coelom** in adult; most of body cavity consisting of hemocoel filled with blood
6. Complete digestive system; mouthparts modified from appendages and adapted for different methods of feeding
7. Open circulatory system, with dorsal **contractile heart,** arteries, and hemocoel (blood sinuses)
8. Respiration by body surface, **gills, tracheae** (air tubes), or **book lungs**
9. Metameric nephridial system of annelids absent; have excretory organs consisting of **modified nephridia,** called **malpighian tubules**
10. Nervous system of annelid plan, with dorsal brain connected by a ring around the gullet to a double nerve chain of ventral ganglia; fusion of ganglia in some species; well-developed sensory organs
11. Sexes usually separate, with paired reproductive organs and ducts; usually internal fertilization; oviparous or ovoviviparous; direct or indirect **metamorphosis;** parthenogenesis in a few forms

Classification

Subphylum Trilobita (tri′lo-bi′ta) (Gr. *tri-,* three, + *lobos,* lobe)— **trilobites.** All extinct forms; Cambrian to Carboniferous; body divided by two longitudinal furrows into three lobes; distinct head, thorax, and abdomen; biramous (two-branched) appendages.

Subphylum Chelicerata (ke-lis′e-ra′ta) (Gr. *chēlē,* claw, + *keras,* horn, + *ata,* group suffix)—**eurypterids, horseshoe crabs, spiders, ticks.** First pair of appendages modified to form chelicerae with claws; pair of pedipalps and four pairs of legs; no antennae; no mandibles; cephalothorax and abdomen usually unsegmented.

Class Merostomata (mer′o-sto′-ma-ta) (Gr. *mēros,* thigh, + *stoma,* mouth, + *ata,* group suffix)—**aquatic chelicerates.** Cephalothorax and abdomen; compound lateral eyes; appendages with gills; sharp telson. **Subclasses Eurypterida** (all extinct) and **Xiphosura,** the horseshoe crabs.

Class Pycnogonida (pik′no-gon′i-da) (Gr. *pyknos,* compact, + *gony,* knee, angle)—**sea spiders.** Small (3 to 4 mm), but some 500 mm; body chiefly cephalothorax; tiny abdomen; usually eight pairs of long walking legs.

Class Arachnida (ar-ack′ni-da) (Gr. *arachnē,* spider)—**scorpions, spiders, mites, ticks, harvestmen.** Four pairs of legs; segmented or unsegmented abdomen with or without appendages and generally distinct from cephalothorax; respiration by gills, tracheae, or book lungs.

Subphylum Mandibulata (man-dib′u-la′ta) (L. *mandibula,* mandible, + *ata,* group suffix). Head appendages consisting of one or two pairs of antennae, one pair of mandibles, and one or two pairs of maxillae.

Class Crustacea (crus-ta′she-a) (L. *crusta,* shell, + *acea,* group suffix). Mostly aquatic, with gills; hard exoskeleton; cephalothorax with dorsal carapace; biramous appendages, modified for various functions; two pairs of antennae.

Class Diplopoda (di-plop′o-da) (Gr. *diploos,* double, + *pous, podos,* foot)—**millipedes.** Subcylindric body; head with short antennae and simple eyes; body with variable number of somites; short legs, usually two pairs of legs to a somite.

Class Chilopoda (ki-lop'-o-da) (Gr. *cheilos,* lip, + *pous, podos,* foot)—**centipedes.** Dorsoventrally flattened body; variable number of somites, each with one pair of legs; pair of long antennae, jaws, and maxillae.

Class Pauropoda (pau-rop'o-da) (Gr. *pauros,* small, + *pous, podos,* foot)—**pauropods.** Minute (1 to 1.5 mm); cylindric body consisting of double segments and bearing nine or ten pairs of legs; no eyes.

Class Symphyla (sym'fy-la) (Gr. *syn,* together, + *phylon,* tribe)—**garden centipedes.** Slender (1 to 8 mm) with long, filiform antennae; body consisting of 15 to 22 segments with 10 to 12 pairs of legs; no eyes.

Class Insecta (in-sek'ta) (L. *insectus,* cut into)—**insects.** Body with distinct head, thorax, and abdomen and usually marked constriction between thorax and abdomen; pair of antennae; mouthparts modified for different food habits; head with six somites; thorax with three somites; abdomen with variable number, usually eleven somites; thorax with two pairs of wings (sometimes one pair or none) and three pairs of jointed legs; gradual or abrupt metamorphosis.

TRILOBITES

The trilobites probably originated millions of years before the Cambrian period during which they flourished. They have been extinct for some 200 million years. Their name refers to the trilobed shape of the body, caused by a pair of longitudinal grooves (Fig. 21-1). They were bottom dwellers, probably scavengers, equipped with antenna and compound eyes. Most of them could roll up like pill bugs, and they ranged from 2 to 67 cm in length.

CHELICERATE ARTHROPODS

The Chelicerata include the eurypterids (all extinct), horseshoe crabs, spiders, ticks and mites, scorpions, and sea spiders. They have as their most anterior appendages a pair of **chelicerae** and a pair of **pedipalps,** variously adapted for seizing and crushing and for sensory functions. They have **four pairs of walk-**

Fig. 21-2. Silurian diorama. A eurypterid (sea scorpion) is attacking a swimming crustacean. (Courtesy American Museum of Natural History.)

ing legs. There are **no antennae** and **no mandibles.** Most chelicerates suck body fluids from their prey.

Class Merostomata

Eurypterids. The Eurypterida (Fig. 21-2), which lived 200 to 500 million years ago, were the largest of all the fossil arthropods. They resembled the marine horseshoe crabs (Fig. 21-3) and also the scorpions, their land counterparts.

Horseshoe crabs (Xiphosurids). The horseshoe crabs are an ancient marine group that date from the Cambrian period. There are only three genera (five species) living today. *Limulus (Xiphosura),* our common horseshoe crab (Fig. 21-3) goes back practically unchanged to the Triassic period. Horseshoe crabs have an unsegmented, horseshoe-shaped **carapace** (hard dorsal shield) and a broad **abdomen,** which has a long spinelike **telson,** or tailpiece. On some of the abdominal appendages **book gills** (flat, leaflike gills) are exposed. Horseshoe crabs can swim awkwardly by means of the abdominal plates and can walk on their walking legs. They feed at night on worms and small molluscs. They are harmless to man, although they are considered pests by clam and oyster fishermen.

Class Arachnida

The arachnids get their name from the Greek *arachnē.* meaning spider; but, in addition to spiders, the group includes scorpions, ticks and mites, harvestmen, pseudoscorpions, and others. Most of them are free living, and they are most numerous in warm, dry regions. Most arachnids are predaceous and may be provided with claws, fangs, poison glands, or stingers and usually with sucking mouthparts or a strong sucking pharynx.

Spiders—order Araneae

The spiders are a large group of 35,000 species, distributed all over the world. The spider body is compact—a **cephalothorax** and **abdomen,** both unsegmented and joined by a slender pedicel.

The anterior appendages are a pair of **chelicerae** (Figs. 21-4 and 21-5), which have terminal **fangs** provided with ducts from poison glands, and a pair of **pedipalps** with basal parts for chewing (Fig. 21-4). Four pairs of **walking legs** terminate in claws.

All spiders are predaceous. Some chase their prey; others ambush them; but most of them spin a net for trapping them. The spider seizes its prey with chelic-

Fig. 21-3. Mating of horseshoe crabs, *Limulus.* At high tide the female digs a depression in the sand for her eggs, which the male fertilizes externally before the hole fills up with sand. The male *(right)* is following the female *(left)* as she selects the spot for her eggs. Sometimes she is followed by several males. The carapace of the older and larger female is studded with barnacles. (Photo by F. M. Hickman.)

407

Fig. 21-4. A wolf spider strikes a defensive pose that shows its appendages, chelicerae, and eyes to good advantage. (Photo by F. M. Hickman.)

erae and pedipalps, injecting into it venom from the poison glands. It then liquifies the tissues with a digestive fluid and sucks up the resulting broth into the stomach where it may be stored in a digestive gland. Tarantulas, wolf spiders, and others may crush the prey with the chelicerae to facilitate the feeding process.

Spiders breathe by means of **book lungs** or **tracheae** or both. Book lungs, which are unique in spiders, consist of parallel air pockets extending into a blood-filled chamber (Fig. 21-5). Air enters the chamber by a slit in the body wall. The tracheae make up a system of air tubes that carry air directly to the tissues from openings called **spiracles.** Similar tubes are found in the insects.

Spiders and insects have a unique **excretory system of malpighian tubules** (Fig. 21-5), which work in conjunction with specialized rectal glands. Potassium and other solutes and waste materials are secreted into the tubules, which drain the fluid, or "urine," into the intestine. The rectal glands reabsorb most of the potassium and water, leaving behind such wastes as uric acid. By this cycling of water and potassium, forms living in dry environments may conserve body fluids, producing a nearly dry mixture of urine and feces. Many spiders also have **coxal glands,** which are modified nephridia, that open at the coxa, or base, of the first and third walking legs.

Spiders usually have eight **simple eyes,** each provided with a lens, optic rods, and a retina (Fig. 21-5, *B*). They are used chiefly for perception of moving objects, but some, such as those of the hunting and jumping spiders, may form images. Since vision is usually poor, a spider's awareness of its environment

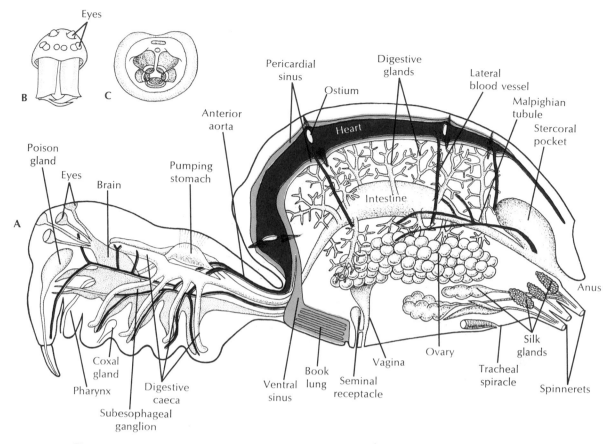

Fig. 21-5. A, Internal structure of the spider. **B,** Anterior view of head, showing eyes and chelicerae with fangs. **C,** Ventral view of spinnerets. The type of chelicerae and the number, arrangement, and/or locations of eyes, lung slits, tracheal spiracles, and spinnerets are all identifying characteristics used in classifying spiders.

depends a great deal upon its **sensory hairs.** Every hair or bristle on its surface, whether or not it is actually connected to receptor cells, is useful in communicating some information about the surroundings, the air currents, or the changing tensions in the spider's web. By sensing the vibrations of its web, the spider judges the size and activity of its entangled prey or receives the message tapped out by a prospective mate.

Web-spinning habits. The ability to spin silk is an important factor in the lives of spiders. Two or three pairs of spinnerets containing hundreds of microscopic tubes run to special abdominal **silk glands** (Fig. 21-5, *A* and *C*). A scleroprotein secretion emitted as a liquid hardens upon contact with air to form the silk thread.

Spiders use the silk threads for many purposes besides web making—to line their nests, form egg cocoons (Fig. 21-7, *A*), spin "balloons," or threads by which the young are carried on the wind, form draglines, and entangle and wrap up their prey (p. 402). Of the several kinds of silk, two in particular are used in web making. An inelastic thread is used in the framework, whereas an elastic thread forms the spirals that run into concentric rows out from the center (Fig. 21-6). The spiral threads bear masses of sticky material for holding the prey. Spiders' silk threads are stronger than steel threads of the same diameter.

Different species make different kinds of nets. Some are simple—a few threads radiating out from the spider's burrow; some make the familiar irregularly

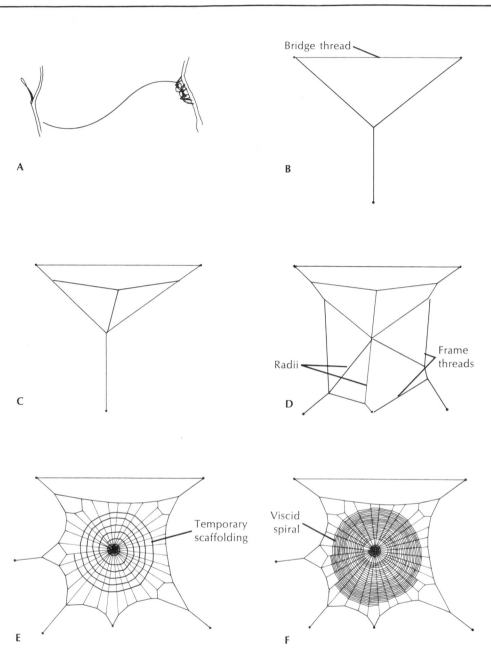

Fig. 21-6. Construction of typical orb web. **A,** Spider establishes a bridge thread, often assisted by a breeze. **B,** A second horizontal line is loosely fixed, then drawn tight by a vertical line fastened below. The juncture of these lines is the central hub, staging point for further progress. **C** and **D,** Frame threads and radii are added. **E,** These are followed by a temporary spiral of nonsticky thread. **F,** Finally the spider lays down a viscid spiral, starting at the outside and working inward, removing and eating the temporary scaffolding as it does. (Modified from Savory, T. H., 1960, Scientific American. **202:**115.)

A

B

Fig. 21-7. **A,** Wolf spider *Lycosa,* carrying egg case. **B,** Tarantula spider, *Dugesiella lentzi,* just captured. Tarantulas are not dangerous; they make good pets and live for years. (**A,** Photo by C. P. Hickman, Jr.; **B,** photo by J. H. Gerard, Alton, Ill.)

Fig. 21-8. **A,** This black widow spider, *Latrodectus,* suspended on her web, has just eaten a large cockroach. Note "hourglass" marking (orange colored) on ventral side of abdomen. **B,** The brown recluse, *Loxosceles reclusa,* is a small brown venomous spider. Note the small violin-shaped marking on its cephalothorax. The venom is hemolytic and dangerous. (**A,** Photo by F. M. Hickman; **B,** photo by J. H. Gerard, Alton, Ill.)

shaped cobwebs; the orb-weavers create beautiful geometric patterns. Not all spiders spin webs, for some, such as the wolf spiders, simply chase and catch their prey. Some spin an anchor thread to warn them of the prey's approach. Most, however, spin some type of net to snare and wrap up their food. The prey are usually small invertebrates, but cases are known of mice being caught, and the fishing spider *Dolomedes* is known to catch small fish.

Reproduction. Before mating, the male spins a small web, deposits a drop of sperm upon it, and then picks the sperm up and stores it in the special cavities of his pedipalps. When he mates, he inserts the pedipalps into the female genital opening to store the sperm in his mate's seminal receptacles. Before mating, there is usually a courtship ritual. The female lays her eggs in a silken net, which she may carry about

(Fig. 21-7, *A*) or may attach to a web or plant. A cocoon may contain hundreds of eggs, which hatch in approximately 2 weeks. The young usually remain in the egg sac for a few weeks and molt once before leaving it. Several molts occur before adulthood.

Are spiders really dangerous? It is truly amazing that such small and helpless creatures as the spiders have generated so much unreasoning fear in the human heart. Spiders are timid creatures, which, rather than being dangerous enemies to man, are actually his allies in his continuing battle with insects. The venom produced to kill the prey is usually harmless to man. Even the most poisonous spiders bite only when tormented or when defending their eggs or young. The American tarantulas (Fig. 21-7, *B*), in spite of their fearsome size, are *not* dangerous. They rarely bite, and their bite is not considered serious.

Fig. 21-9. A, Tropical blue scorpion *Centrurus* (order Scorpionida), which is common in Cuba. Note terminal poison claw. **B,** The wood tick *Dermacentor* (order Acarina), one species of which transmits Rocky Mountain spotted fever and tularemia, as well as producing tick paralysis. (Photos by F. M. Hickman.)

There are, however, two species in the United States that can give severe or even fatal bites—the **black widow** and the **brown recluse.** The black widow, *Latrodectus mactans,* is small and shiny black, with a bright orange or red "hourglass" on the under side of the abdomen (Fig. 21-8, *A*). Their venom is neurotoxic; that is, it acts upon the nervous system. Approximately 4 or 5 out of each 1,000 bites reported have proved fatal.

The brown recluse, *Loxosceles reclusa,* is smaller than the black widow, is brown, and bears a violin-shaped dorsal stripe on its back (Fig. 21-8, *B*). Its venom is hemolytic rather than neurotoxic, producing death of the tissues surrounding the bite. Its bite is serious and occasionally fatal.

Scorpions—order Scorpionida

Although scorpions are more common in tropical and subtropical regions, some also occur in temperate zones. Scorpions are generally secretive, hiding in burrows or under objects by day and feeding at night. They feed largely on insects and spiders, which they seize with the pedipalps and tear up with the chelicerae.

The scorpion body consists of a rather short cephalothorax, which bears the appendages and from 1 to 6 pairs of eyes, a preabdomen, and a long slender postabdomen, or tail, which ends in a stinging apparatus (Fig. 21-9, *A*). The venom of most species is not harmful to man, although that of certain species of *Androctonus* in Africa and *Centruroides* in Mexico can be fatal unless antivenom is available.

Scorpions bring forth living young, which are carried on the back of the mother until after the first molt.

Ticks and mites—order Acarina

Ticks and mites are arachnids in which the cephalothorax and abdomen are fused into an unsegmented ovoid body with eight legs. They are found almost everywhere—in both fresh and salt water, on vegetation, on the ground, and, being parasitic, in animals.

Ticks are usually larger than mites. They pierce the skin of vertebrates and suck up the blood until enormously distended; then they drop off and digest the meal. After molting, they are ready for another meal. Some ticks are important disease vectors. Texas cattle fever is caused by a protozoan parasite transmit-

413

ted by the tick *Boophilus annulatus.* The wood tick *Dermacentor* (Fig. 21-9, *B*) is the vector for Rocky Mountain spotted fever, caused by a rickettsial organism.

Most mites are less than 2 mm long, and they are varied in their habits and habitats. The spider mites, or red spiders, are destructive to plants. The mange mite, *Demodex,* is responsible for mange. Chiggers, *Eutrombicula,* lay their eggs on the ground. When the larvae find a suitable host, they attach to the skin and, with the aid of digestive secretions, form a tiny crater where they feed on partially digested tissues. The adults are not parasitic.

MANDIBULATE ARTHROPODS

The mandibulates are the arthropods that possess **mandibles** (jaws). They include the crustaceans, millipedes, centipedes, and insects. Antennae, mandibles, and maxillae make up the head appendages in the mandibulates. These are sensory, mastigatory, and food-handling organs. The body may consist of a head and trunk or a head, thorax, and abdomen or a cephalothorax (head and thorax fused) and abdomen. Respiration is by gills or tracheae.

Class Crustacea

Crustaceans get their name from the hard shell they bear (L. *crusta,* shell). They include more than 30,000 species of lobsters, crayfishes, shrimps, crabs, water fleas, copepods, sow bugs, wood lice, barnacles, and a few others. There are some terrestrial forms, but the majority by far live in marine or fresh water. Most are free living, but some are sessile, commensal, or parasitic.

The small crustaceans, called the microcrustaceans, have been referred to as the "insects of the sea" although they are by no means restricted to a marine habitat (Fig. 21-10). But in the sea the countless tiny, free-swimming crustaceans are crucial energy converters of the plankton. The position of the microcrustaceans in the food chain is between that of the minute plant life and the filter-feeding fishes and other large oceanic animals. Without the marine microcrustaceans, the existence of advanced forms of life in the ocean in any significant amount would be impossible.

Lobsters, crayfishes, shrimps, and crabs are, of course, highly esteemed as food for man, and the "shellfish" industry is economically important. Un-

fortunately it is especially vulnerable to man's polluting activities in coastal areas since the sensitive larval stages live in coastal estuaries.

Of course, some crustaceans conflict with man's interests and are labeled pests. Barnacles foul the bottom of ships. Certain boring crustaceans are damaging to wharves and pilings. Some carry diseases in tropical countries, some are parasitic on man's food animals, and some destroy crops, especially rice.

External characteristics

The crustaceans are such a highly diversified group that it is very difficult to describe them in terms that apply to the whole group. They are in general the **gill-breathing aquatic mandibulates,** although some are terrestrial or semiterrestrial. They are distinguished from other arthropods by having **five pairs of head appendages**—two pairs of antennae (the first pair are called antennules) for sensing the environment, one pair of mandibles, and two pairs of maxillae for food handling (Fig. 21-11).

The crustacean **cuticle** is usually somewhat harder than that of other arthropods since it is impregnated with calcium salts. The head and thorax are often covered by a hard shieldlike **carapace,** as in the crabs and crayfishes (Fig. 21-11). The number of body segments varies greatly among the different groups. Some have 60 or more somites, whereas some are unsegmented; the majority have from 16 to 20 segments. In most of the larger crustaceans, which belong to the subclass Malacostraca (Fig. 21-13), we find a body plan of 19 segments and 19 pairs of appendages—5 on the head, 8 on the thorax, and 6 on the abdomen. The lobsters, crayfishes, crabs, shrimps, sow bugs, beach fleas, and many others of the malacostracan group follow this general plan, which is often considered to be the primitive, or ancestral (archetypic), plan of the malacostracans (Fig. 21-11).

Resume of crustacean subclasses

Subclass Cephalocarida (sef'a-lo-car'i-da) (Gr. *kephalē,* head, + *karis,* shrimp, + *ida,* pl. suffix) is a small primitive group of one genus *(Hutchinsoniella),* strictly marine and approximately 3 to 4 mm long. They are shrimplike in form and have a horseshoe-shaped head with antennae and no eyes, and 19 or 20 somites of which 9 bear similar biramous (two-branched) appendages.

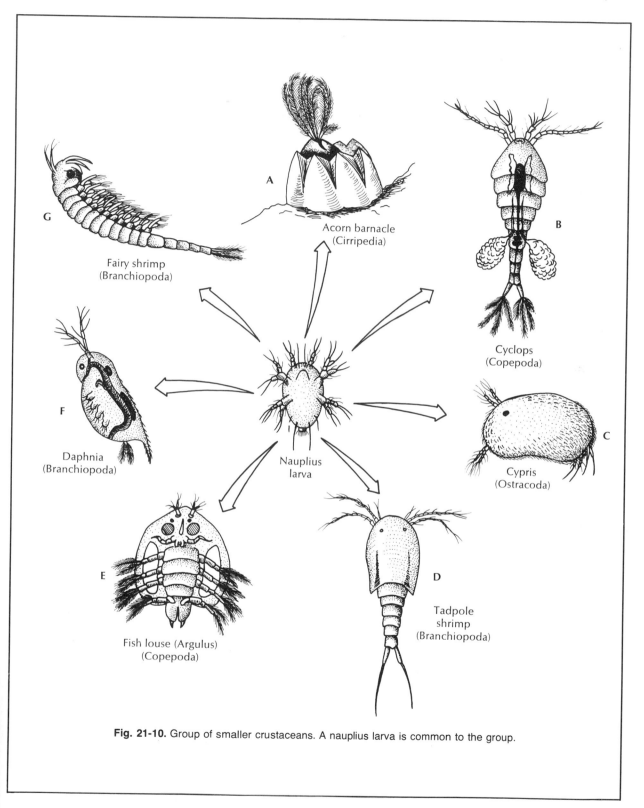

Fig. 21-10. Group of smaller crustaceans. A nauplius larva is common to the group.

Fairy shrimp
(Branchiopoda)

Acorn barnacle
(Cirripedia)

Cyclops
(Copepoda)

Daphnia
(Branchiopoda)

Nauplius
larva

Cypris
(Ostracoda)

Fish louse (Argulus)
(Copepoda)

Tadpole
shrimp
(Branchiopoda)

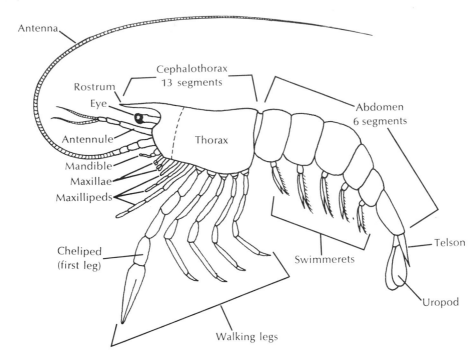

Fig. 21-11. Archetypic plan of the Malacostraca. Note that the maxillae and maxillipeds have been separated diagrammatically to illustrate general plan. Typically in the living animal only the third maxilliped is visible externally.

Subclass Branchiopoda (bran-chi-op′o-da) (NL. *branchio,* gill, + *podos,* foot) is a group of small primitive crustaceans, largely freshwater, ranging from 1 mm or less to 18 cm. They may be long and segmented with many appendages or have a short compact body lacking apparent segmentation and having few appendages. The trunk appendages are usually flat and leaflike and adapted for respiration, filter feeding, locomotion, and sometimes for egg-bearing.

There are four orders: **Anostraca,** the fairy shrimps (Fig. 21-10, *G*), which live in freshwater ponds, and the brine shrimps, which live in salt lakes; **Notostraca,** the tadpole shrimps (Fig. 21-10, *D*); **Conchostraca,** the clam shrimps; and **Cladocera,** the water fleas, such as *Daphnia* (Fig. 21-10, *F*).

Subclass Ostracoda (os-trak′o-da) (Gr. *ostrakōdēs,* testaceous, from *ostrakon,* shell) is the mussel shrimps, or seed shrimps (Fig. 21-10, *C*). They are enclosed in a bivalved carapace, resembling miniature clams. In addition to the head appendages, there are usually two pairs of trunk appendages. When they

move, they thrust out the appendages through the open shell.

Subclass Copepoda (ko-pep′o-da) (Gr. *kōpē,* oar, + *podos,* foot) are composed of small crustaceans with elongated bodies and forked tails. A common form is *Cyclops* (Fig. 21-10, *B*) with a single median eye. *Argulus,* a fish louse (Fig. 21-10, *E*), parasitizes freshwater fish.

Subclass Cirripedia (sir-ri-pe′di-a) (L. *cirrus,* curl, + *pedis,* foot) is made up of barnacles, all marine. Adults are sessile and attached to rocks, pilings, etc.; the larvae are free swimming. The animal, bearing as many as six pairs of thoracic appendages, is surrounded by calcareous shell plates through which it thrusts its feathery legs to scoop in plankton when feeding (Fig. 21-10, *A,* and 21-12). Some barnacles are attached directly; some are stalked; some are parasitic.

Subclass Malacostraca (mal-a-kos′tra-ka) (Gr. *malakos,* soft, + *ostrakon,* shell) includes the lobsters, crayfishes, crabs, shrimps, sow bugs, beach hoppers, wood lice, and others (Fig. 21-13). The trunk usually

416

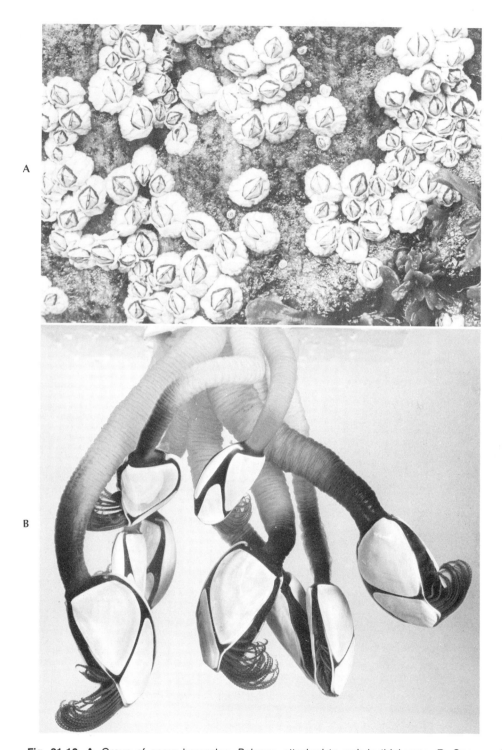

Fig. 21-12. A, Group of acorn barnacles, *Balanus,* attached to rock in tidal zone. **B,** Gooseneck barnacles, *Lepas fascicularis,* attach to submerged rocks and floating objects by long stalks. The barnacle (either kind) is enclosed in a soft mantle and protected by calcareous plates. It sweeps in food particles with its feathery jointed appendages. Adductor muscles pull plates together when animal is not feeding. (**A,** Photo by C. P. Hickman, Jr.; **B,** photo by R. C. Hermes, Homestead, Fla.)

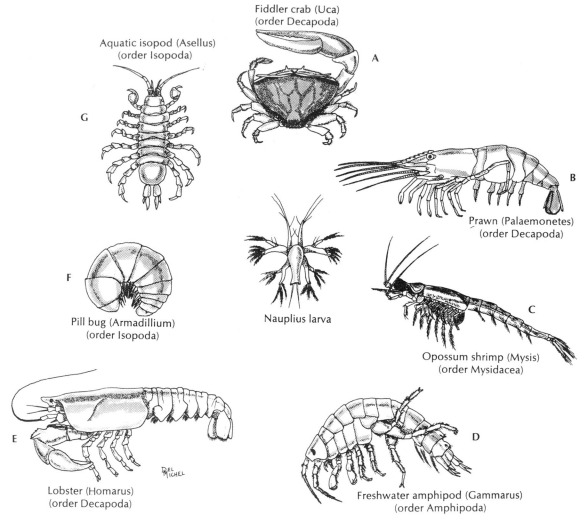

Aquatic isopod (Asellus)
(order Isopoda)

Fiddler crab (Uca)
(order Decapoda)

Prawn (Palaemonetes)
(order Decapoda)

Pill bug (Armadillium)
(order Isopoda)

Nauplius larva

Opossum shrimp (Mysis)
(order Mysidacea)

Lobster (Homarus)
(order Decapoda)

Freshwater amphipod (Gammarus)
(order Amphipoda)

Fig. 21-13. Larger crustaceans (subclass Malacostraca). All members of this subclass have abdominal appendages, gastric mill, eight-segmented thorax, and usually a body of 19 segments. (Not drawn to scale.)

has eight thoracic and six abdominal segments, each with a pair of appendages. There are both marine and freshwater forms.

Of the several orders of malacostracans, the three largest and most familiar are Isopoda, Amphipoda, and Decapoda.

The **Isopoda** (i-sop'o-da) (Gr. *isos*, equal, + *pous*, foot) are the pill bugs, sow bugs, wood lice, and water slaters—both aquatic and terrestrial forms. The body is depressed dorsoventrally and lacks a carapace; the legs are similar except for the first and last pairs (Fig. 21-13, *F* and *G*). *Asellus* is common in freshwater,

and *Ligyda (Lygia)* is abundant on sea beaches and rocky shores.

The **Amphipoda** (am-fip'o-da) (Gr. *amphis*, on both sides, + *pous, podos,* foot) are laterally compressed and have no carapace. They are the sand hoppers, side swimmers, beach fleas, etc. Fourteen pairs of appendages are variously specialized for feeding, crawling, swimming, and jumping (Fig. 21-13, *D*).

The **Decapoda** (de-cap'o-da) (Gr. *deka*, ten, + *pous, podos,* foot) have five pairs of walking legs of which the first is modified to form pincers (chelae)

Fig. 21-14. A, The bright orange "Sally lightfoot" crab, common in the Galápagos Islands. **B,** Hermit crab, which has a very soft abdominal exoskeleton, lives in a snail shell that it carries about and into which it can withdraw for protection. This is the Galapagos hermit crab *Coenlita compressus,* a land hermit that enters the water only to breed. Most hermits are marine. (Photos by C. P. Hickman, Jr.)

(Fig. 21-13, *A, B,* and *E*). These are the lobsters, crayfishes, shrimps, and crabs, the largest of the crustaceans. Crabs differ from the others in having a broader carapace and a much-reduced abdomen (Fig. 21-14, *A*). Familiar examples are the fiddler crabs *Uca,* which burrow in sand just below the high-tide level, the hermit crabs (Fig. 21-14, *B*), which live in snail shells because their abdomens lack a hard

exoskeleton, the decorator crabs, which cover their carapaces with sponges and sea anemones for camouflage, and the spider crabs, such as *Libinia*.

Appendages

Crustacean appendages, which are usually **biramous** (two-branched), have evolved from a common biramous plan best illustrated by the **swimmerets** on

419

the abdomen of the crayfish or lobster. Such an appendage consists of inner and outer branches, called the endopodite and exopodite, which are attached to one or more basal segments collectively called the protopodite (Fig. 21-15).

There are many modifications of this plan. In the more primitive crustaceans, such as the branchiopods (Fig. 21-10, *G*), all of the trunk appendages tend to be similar in structure and adapted for swimming. The evolutionary trend has been toward reduction in number of appendages and toward a variety of modifications that fit them for many functions. Some are **foliaceous** (flat and leaflike), as are the maxillae; some are **biramous,** as are the swimmerets, maxillipeds, uropods, and antennae; some have lost one branch and are **uniramous,** as are the walking legs.

In the crayfish we find the first three thoracic appendages, called maxillipeds, serving along with the two pairs of maxillae as food handlers; the other five pairs of appendages are lengthened and strengthened for walking and defense (Fig. 21-15). The first pair of walking legs, called chelipeds, are enlarged with a strong claw, or chela, for defense. The abdominal swimmerets serve not only for locomotion, but in the male the first pair is modified for copulation and in the female they all serve as a nursery for attached eggs and young. The last pair of appendages, called uropods, are wide and are used as rudders for swift backward movements, and, with the telson, they form a protective device for eggs or young on the swimmerets.

Ecdysis

The problem of growth in spite of a restrictive exoskeleton is solved in crustaceans, as in other

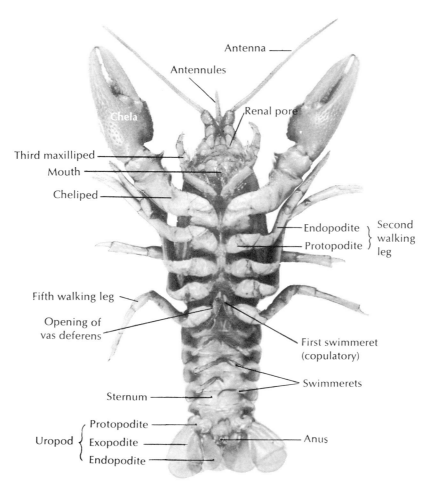

Fig. 21-15. Ventral view of male crayfish.

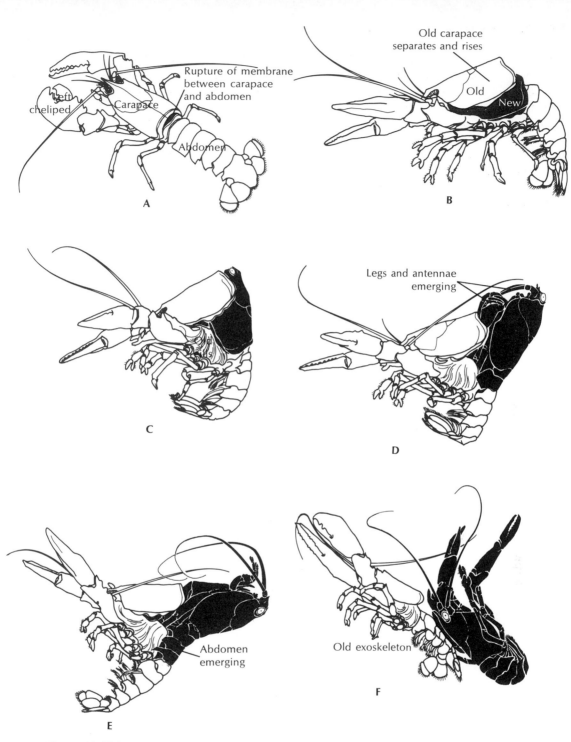

Fig. 21-16. Molting sequence in the lobster, *Homarus americanus*. **A,** Membrane between carapace and abdomen ruptures, and carapace begins slow elevation. This step may take up to 2 hours. **B** to **E,** Head, thorax, and finally abdomen are withdrawn. This process usually takes no more than 15 minutes. **F,** Immediately after ecdysis, chelipeds are dessicated and body is very soft. Lobster now begins rapid absorption of water so that within 12 hours body increases about 20% in length and 50% in weight. Tissue water will be replaced by protein in succeeding weeks. (Drawn from photos by D. E. Aiken, St. Andrews, New Brunswick.)

arthropods, by ecdysis (Gr. *ekdyein,* to strip off), the periodic shedding of the old cuticle and the formation of a larger new one. Molting occurs most frequently during larval stages and less often as the animal reaches adulthood. Although the actual shedding of the cuticle is periodic, the molting process and the preparations for it, involving the storage of reserves and changes in the integument, are a continuous process going on during most of the animal's life.

During each **premolt** period the old cuticle becomes thinner as inorganic salts are withdrawn from it and stored in the tissues. Other reserves, both organic and inorganic, are also accumulated and stored. The underlying epidermis begins to grow by cell division; it secretes first a new inner layer of epicuticle and then enzymes that digest away the inner layers of old endocuticle. Gradually a new cuticle is formed inside the degenerating old one. Finally the actual ecdysis occurs as the old cuticle ruptures, usually along the middorsal line, and the animal backs out of it (Fig. 21-16). By taking in air or water the animal swells to stretch the new larger cuticle to its full size. During the **postmolt** period the cuticle is thickened, the outer layer is hardened by tanning, and the inner layer is strengthened as salvaged inorganic salts and other constituents are redeposited.

That ecdysis is under hormonal control has been demonstrated in both crustaceans and insects. In the decapods **neurosecretory cells** in an **x-organ** located in each eyestalk secrete a **molt-inhibiting hormone** that is stored in a **sinus gland.** This hormone controls the **molting glands,** or **y-organs,** located near the mandibles, which produce a **molt-accelerating hormone.** A reduction in the blood level of the molt-inhibiting hormone permits the y-organs to release their molt-accelerating hormone. The interaction of the two hormones is cyclic. The release of the molt-inhibiting hormone is governed by the central nervous system.

Color changes

Color changes brought about by the redistribution of pigment in pigment cells **(chromatophores)** are also under hormonal control. The chromatophores are irregularly shaped cells with many processes. When pigment is concentrated near the center of the cells, very little of the color is displayed, giving a lightening effect; when it is dispersed throughout the cells, more of the color shows, so that the animal appears darker.

The movement of the pigment is regulated by **chromatophorotropins,** which are produced by neurosecretory cells in the central nervous system or in the x-organs and stored in the sinus gland.

The marine isopod *Ligyda,* which lives on the seashore under seaweed or in crevices, matches its background more or less because of the contraction or dispersal of a type of brown chromatophore controlled by two hormones. A **darkening hormone** acts in response to incident light (light from above), and a **lightening hormone** responds to reflected light. On sand, which reflects light, the animal is pale because more of the lightening hormone is produced. On rotting seaweed or on a dark rock, where little light is reflected, the animal darkens because more of the darkening hormone is being produced. Many different pigment colors are found in crustaceans, and where more than one type is found in an animal the hormonal interplay becomes much more complex.

Feeding habits

Feeding habits and adaptations for feeding vary so much among crustaceans that generalizations have little meaning. Many forms can shift from one type of feeding to another depending upon environment and food availability, but fundamentally the same set of mouthparts are used by all. The mandibles and maxillae are involved in the actual ingestion; maxillipeds hold and crush food. In predators the walking legs, particularly the chelipeds, serve in food capture.

Many crustaceans, both large and small, are predatory, and some have interesting adaptations for killing their prey. One shrimplike form, *Lygiosquilla,* has on one of its walking legs a specialized digit that can be drawn into a groove and released suddenly to pierce a prey passing by its burrow. The pistol shrimp *Alpheus* has one enormously enlarged chela that can be cocked like the hammer of a gun and released with a force that stuns its prey and sounds like the report of a pistol.

The food of crustaceans ranges from plankton, detritus, and bacteria, used by the **filter feeders,** to larvae, worms, crustaceans, snails, and fishes, used by **predators,** and dead animal and plant matter, used by **scavengers.** Filter feeders, such as the fairy shrimps, water fleas, and barnacles, use their legs, which bear a thick fringe of setae, to create water currents that sweep food particles through the setae. The mud shrimp *Upogebia* uses long setae on its first two pairs of thoracic appendages to strain food material from

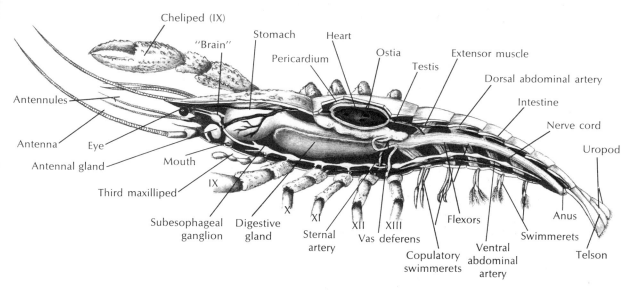

Fig. 21-17. Structure of male crayfish.

water circulated through its burrow by movements of its swimmerets.

Crayfishes and other malacostracans have a two-part stomach. The first part contains a **gastric mill** in which food, already torn up by the mandibles, can be further ground up by three calcareous teeth into particles fine enough to pass through a setose filter in the second part of the stomach; the food particles then pass into the intestine for chemical digestion.

Respiration, excretion, and circulation

The **gills** of crustaceans vary in shape—treelike, leaflike, or filamentous—all provided with blood vessels or sinuses. They are usually attached to the appendages and kept ventilated by the movement of the appendages in the water. The gill chambers are usually protected by the overlapping carapace. Some smaller crustaceans breathe through the general body surface.

Excretory and osmoregulatory organs in crustaceans are paired glands located in the head, with excretory pores opening at the base of either the antennae or the maxillae, and are usually called **antennal glands** or **green glands** (Fig. 21-17). They resemble the coxal glands of the chelicerates. The waste product is mostly ammonia with some urea and uric acid. Some wastes diffuse through the gills as well as through the excretory glands.

Circulation, as in other arthropods, is an **open system** consisting of a heart, either compact or tubular, arteries, and sinuses (hemocoel). Some smaller crustaceans lack a heart. An open circulatory system is less dependent on heartbeats because the movement of organs and limbs circulates the blood more effectively in open sinuses than in capillaries. The blood may contain as respiratory pigments either hemocyanin or hemoglobin, and it has the property of clotting to prevent loss of blood in minor injuries.

Nervous and sensory systems

A cerebral ganglion above the esophagus sends nerves to the anterior sense organs and is connected to a subesophageal ganglion by a pair of connectives around the esophagus. A double ventral nerve cord has a ganglion in each segment that sends nerves to the viscera, appendages, and muscles (Fig. 21-17). Giant fiber systems are common among the crustaceans.

The sensory organs are well developed. There are two types of eyes—the median, or nauplius, eye and the compound eye. The **median eye** is found in the nauplius larvae and in some adult forms and may be the only adult eye, as in the copepods. It is usually a group of three pigment cups containing retinal cells; they may or may not have a lens.

Most crustaceans have **compound eyes** similar to insect eyes. In crabs and crayfishes they are located on the ends of movable eyestalks (Fig. 21-14, *A*). Compound eyes are precise instruments, different from vertebrate eyes, yet especially adept at detecting motion; they are able to analyze polarized light. The convex corneal surface gives a wide visual field, particularly in the stalked eyes where the surface may cover an arc of 200° or more.

The compound eye is composed of many tapering units called **ommatidia** set close together. The facets, or corneal surfaces, of the ommatidia give the surface of the eye the appearance of a fine mosaic. Most crustacean eyes are adapted either to bright or to dim light, depending upon their diurnal or nocturnal habits, but some are able, by means of screening pigments, to adapt, to some extent at least, to both bright and dim light. The number of ommatidia varies from a dozen or two in some small crustaceans to 15,000 or more in a large lobster. Some insects have approximately 30,000.

Other sensory organs include statocysts, tactile hairs on the cuticle of most of the body, and chemosensitive hairs, especially on the antennae, antennules, and mouthparts.

Reproduction and life cycles

Most crustaceans have separate sexes, and there are a variety of specializations for copulation among the different groups. The barnacles are monoecious but generally practice cross-fertilization. In some of the ostracods males are scarce, and reproduction is usually parthenogenetic. Most crustaceans brood their eggs in some manner—branchiopods and barnacles have special brood chambers, the copepods have egg sacs attached to the sides of the abdomen (Fig. 21-10, *B*), and the malacostracans usually carry eggs and young attached to their appendages.

The larvae of crayfishes are **juveniles**—that is, they resemble the adult. But in most marine crustaceans the early larva is called a **nauplius.** It has an unsegmented body, a frontal eye, and three pairs of biramous appendages (Fig. 21-13). In many crustaceans the nauplius is transformed in successive molts to a protozoea, with 7 pairs of appendages, a zoea, with 8 pairs, and then a mysis, with 13 pairs. The adult with 19 pairs of appendages develops from the mysis.

Regeneration and autotomy

Many crustaceans can **regenerate** lost appendages or eyes, particularly during the growing stages. The lost part is usually partially renewed at each successive molt.

Autotomy refers to the power of self-amputation, or the breaking off of an injured leg at a specific breaking point near the base of the leg. A special muscle in the injured leg contracts excessively to cause a rupture at the breaking point. Wounds heal more quickly at this point, preventing loss of blood.

Myriapods

The term **myriapod** means many-footed (Gr. *myrias,* a myriad, + *podos,* foot) and refers to several classes of mandibulates that have a head and trunk, and paired appendages on all trunk segments except the last. They include the Chilopoda (centipedes), Diplopoda (millipedes), Pauropoda (pauropods), and Symphyla (symphylans).

Myriapods and insects are both primarily terrestrial, and they have had a remarkable success story in an environment that has, as a rule, proved most inhospitable to invertebrates. They have evolved along two main lines. One is characterized by a body arrangement of **two tagmata,** the head and trunk, with paired appendages on all but the last trunk segments—the **myriapods.** The other evolutionary branch features a body of **three tagmata,** head, thorax, and abdomen, with the abdominal appendages missing or greatly reduced—the **insects.**

The head of the myriapods and insects resembles the crustacean head but has only **one pair of antennae,** instead of two. It also has the **mandibles** and two pairs of **maxillae** (one pair of maxillae in millipedes). The legs are all **uniramous.**

Respiratory exchange is by body surface and tracheal systems, although juveniles, if aquatic, may have gills.

Centipedes—class Chilopoda

The centipedes are active predators with a preference for moist places such as under logs or stones, where they feed upon earthworms, insects, etc. Their bodies are somewhat flattened dorsoventrally, and they may contain from a few to more than 180 somites. Each somite, except the one behind the head and the last two, bears one pair of appendages (Fig. 21-18, *A*). Those of the first body segment are modified to form poison claws, which they use to kill their prey. Most species are harmless to humans.

The head bears a pair of eyes, each consisting of a

Fig. 21-18. Myriapods. **A,** Centipede *Scolopendra* (class Chilopoda). Most segments have one pair of appendages each. First segment bears pair of poison claws, which in some species can inflict serious wounds. Centipedes are carnivorous. **B,** Millipede *Narceus americanus* (class Diplopoda). Note the typical doubling of appendages on most segments. (**A,** Photo by F. M. Hickman; **B,** photo by C. P. Hickman, Jr.)

group of ocelli. Respiration is by tracheal tubes with a pair of spiracles in each somite. Sexes are separate. Some are oviparous (females release eggs from which young later hatch); some are viviparous (living young, instead of eggs, are released). The young are similar to the adult. The common house centipede *Scutigera,* with 15 pairs of legs, and *Scolopendra* (Fig. 21-18, *A*), with 21 pairs of legs, are familiar genera.

Millipedes—class Diplopoda

Diplopods, or "double-footed" arthropods, are commonly called millipedes, which literally means "thousand-feet" (Fig. 21-18, *B*). Although they do not have that many legs, they do have a great many. Their cylindric bodies are made up of 25 to 100 segments. The four thoracic segments bear only one pair of legs each, but the abdominal segments each have two pairs, a condition that may have evolved from fusion of somites. There are two pairs of spiracles

on each abdominal somite, each opening into an air chamber that gives off tracheal tubes.

Millipedes are less active than centipedes and are generally herbivorous, living on decayed plant and animal matter and sometimes living plants. They prefer dark moist places under stones and logs. Their eggs are laid in a nest and carefully guarded by the female. The larval forms have only one pair of legs to a somite. Common genera are *Spirobolus* and *Julus.*

Class Insecta

The insects are the most successful biologically of all the groups of arthropods. It is estimated that the recorded number of insect species is approximately 750,000, by far the largest class of animals, with thousands of other species yet to be discovered and classified.

The science of entomology occupies the time and resources of skilled men all over the world. The struggle between man and his insect competitors seems to be an endless one, for there are probably more pests among the insects than in all the other invertebrate phyla combined. Yet paradoxically insects are interwoven into the economy of nature in so many useful roles that man would have a difficult time without them.

Insects differ from other arthropods in having **three pairs of legs** and usually **two pairs of wings** on the thoracic region of the body, although some have one pair of wings or none. In size insects range from less than 1 mm to 20 cm in length—the majority being less than 2.5 cm long.

Distribution

Insects have spread into practically every habitat that supports life except the deeper waters of the sea. Marine water striders live on the surface of the sea, and elsewhere insects are abundant in fresh and brackish water, salt marshes, soils and sandy beaches, forests, deserts, and mountain tops; they also live as parasites in and on the bodies of innumerable plants and animals.

Their wide distribution has been greatly aided by their power of flight, a great advantage in escape from predators, in access to food, and in dispersal. Their small size allows them to be carried afar by currents of wind and water. Their well-protected eggs can be carried by birds and other animals. Their agility and aggressiveness enable them to fight for every possible niche in a location.

Insects have a marvelous capacity to adapt to new habitats. This is enhanced by a wide diversity of structural modifications, particularly in the wings, legs, antennae, and mouth parts. Some insects suck the sap of plants; some chew plant foliage; some are predaceous; and some live upon the blood and tissues of various animals.

Insects are well adapted to dry and desert regions because the chitinous exoskeleton prevents evaporation and because they can extract fluid from food and fecal material, as well as utilize moisture from the water byproducts of metabolism.

External morphology

The insect body is made up of the **head, thorax,** and **abdomen.** The cuticle of each segment is typically composed of four plates—a dorsal **notum** (tergum), a ventral **sternum,** and two lateral **pleura** (Fig. 21-19).

The **head** usually bears a pair of compound eyes, a pair of antennae, and usually three ocelli. The mouthparts of chewing insects include three pairs of ap-

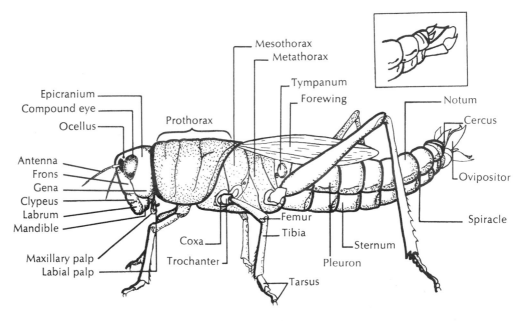

Fig. 21-19. External features of female grasshopper. Terminal segment of male with external genitalia is shown in inset.

pendages—a pair of mandibles covered by an upper lip, or labrum, a pair of maxillae, and a labium (representing the fused second maxillae). A median tonguelike hypopharynx projects from behind the maxillae. Various modifications of this plan for different types of feeding will be discussed later.

The **thorax** is composed of the prothorax, mesothorax, and metathorax, each bearing a pair of legs articulating with the pleura (Fig. 21-19). The legs are usually adapted for walking or running but some may be specialized for swimming (aquatic beetles, aquatic nymphs, and larvae), jumping (grasshoppers, crickets, fleas), digging (mole crickets) (Fig. 21-20, *D*), or grasping prey (praying mantids, dytiscid beetles) (Fig. 21-20, *C*). In the honeybee the legs are highly specialized. The foreleg bears a brush of stiff hairs for cleaning the eyes and a bristle-lined notch and a hinged spine for cleaning the antennae. Each middle leg has a spur to pick up the wax that is secreted between the abdominal segments. Each hindleg has a concave, hair-lined pollen basket. All of the legs bear hairs with which to brush up the pollen that adheres to the body.

Fig. 21-20. A, Female ichneumon wasp (order Hymenoptera) uses long ovipositor to bore into tree and lay an egg near a wood-boring beetle larva, which will become food for the ichneumon larva. This specimen had an overall length of over 6 inches. **B,** Ichneumon head showing compound eyes, antennae, and some mouthparts. **C,** Giant diving beetle, *Dytiscus* (order Coleoptera) about 3 cm long, is streamlined for life as an active aquatic predator. **D,** Mole cricket *Gryllotalpa* (order Orthoptera) has forelegs adapted for digging. (Photos by F. M. Hickman.)

Most insects also bear two pairs of wings, located on the mesothorax and metathorax.

The **abdomen** of insects is composed of 9 to 11 segments—the eleventh, when present, reduced to a pair of cerci (appendages at the posterior end). Larval or nymphal forms have a variety of abdominal appendages, but these are lacking in the adults. The end of the abdomen bears the external genitalia (Fig. 21-19).

There are innumerable variations in body form among the insects. Beetles are usually thick and plump (Fig. 21-20, *C*); damselflies and walking sticks (Fig. 21-36, *A*) are long and slender; aquatic beetles are streamlined; and cockroaches are flat, adapted to living in crevices. The female ovipositor of the ichneumon wasp is extremely long (Fig. 21-20, *A*). The cerci form horny forceps in the earwigs and are long and many jointed in stoneflies and mayflies (Fig. 21-35). Antennae are long in cockroaches and katydids, short in dragonflies and most beetles, knobbed in butterflies, and plumed in most moths.

Wings and the flight mechanism

Insects share the power of flight with birds and flying mammals. However, their wings have evolved in a different manner from that of the limb buds of birds and mammals and are not homologous to them. Insect wings are outgrowths from the body wall of the mesothoracic and metathoracic segments.

Most insects have two pairs of wings, but the Diptera (true flies) have only one pair, the missing pair being represented by a pair of tiny halteres (balancers) that vibrate and are responsible for equilibrium during flight. The males of the scale insects also have one pair of wings but no halteres. Some insects are wingless. Ants and termites, for example, have wings only on males and females during certain periods; workers are always wingless. Lice and fleas are always wingless.

Wings may be thin and membranous, as in flies and many others (Fig. 21-20, *A*), thick and horny, as in the front wings of beetles (Fig. 21-20, *C*), parchment-like, as in the front wings of grasshoppers, or covered with fine scales, as in butterflies and moths, or with hairs, as in caddis flies.

Most membranous wings bear a framework of thickened ridges known as the **veins.** The number and arrangement of the veins are a useful taxonomic tool.

Wing veins are tubular structures, thickened by heavy cuticle to form a supporting skeleton for the wings. Opening into the body wall, the veins contain circulating blood, and the main veins contain also tracheoles (fine tubules from the respiratory system) and sensory nerve branches. The veins are strongest and most numerous near the anterior edge of the wing, and it is the anterior edge that leads in both the up and down strokes during flight.

Wing movements are controlled by a complex of muscles in the thorax. These include some of the largest and strongest of all insect muscles. The muscles of one group, called **direct flight muscles,** run from the lateral and ventral regions of the thorax, insert directly on the body wall at the base of the wings and control usually the tilting, or "feathering," of the wing. However, in the horizontal wings of the dragonflies they also provide the up and down movements (Fig. 21-21, *A*). Another group of muscles, the powerful **indirect flight muscles,** includes both longitudinal muscles and dorsoventral muscles attached to the walls of the thorax. These do not insert directly on the wings, but, by changing the shape of the thoracic walls, they raise or lower the wings (Fig. 21-21, *B*). Contraction of the longitudinal muscles causes the notum to bulge upward and thus lowers the wings. Contraction of the dorsoventral muscles depresses the notum and forces the wings upward. The wing behaves as a lever, pivoting on the pleural process, and the base of the wing, which projects inward from the pivot, moves as the notum moves but in the opposite direction.

Obviously flying entails more than a simple flapping of the wings; a forward thrust is necessary. As the indirect flight muscles alternate rhythmically to raise and lower the wings, the direct flight muscles alter the angle of the wings so that they act as lifting airfoils during both the upstroke and the downstroke, twisting the leading edge of the wings downward during the downstroke and upward during the upstroke. This produces a figure-eight movement that aids in spilling air from the trailing edges of the wings. The quality of the forward thrust depends, of course, upon several factors, such as variations in wing venation, how much the wings are tilted, how they are feathered, and so on.

Flight speeds vary. The fastest flyers usually have narrow, fast-moving wings with a strong tilt and a strong figure-eight component. Sphinx moths and horseflies are said to achieve approximately 30 miles per hour and dragonflies approximately 25 miles per hour. Some insects are capable of long continuous

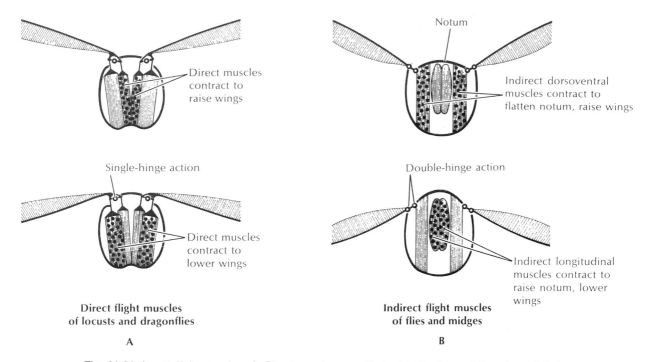

Fig. 21-21. Insect flight muscles. **A,** Direct muscles are attached to the base of the wings at their articulation and raise, lower, or tilt the wings by direct action. **B,** Indirect muscles are attached to the thoracic walls and exert a lever action upon the wings by changing the shape of the thorax.

flights. The migrating monarch butterfly *Danaus plexippus* travels south for hundreds of miles in the fall, flying at a speed of approximately 6 miles per hour.

The frequency of wing beat also varies. Light-bodied insects with large wings, such as butterflies, may beat as few as 4 times per second. The small wings of heavy insects, such as flies and bees, may vibrate at 100 beats per second or more. The fruit fly *Drosophila* can fly at 300 beats per second, and midges have been clocked at more than 1,000 beats per second.

Internal form and function

Nutrition. The digestive system (Fig. 21-24) consists of a foregut (mouth with salivary glands, esophagus, crop for storage, and gizzard for grinding); a midgut (stomach and gastric ceca); and a hindgut (intestine, rectum, and anus). The fore- and hindguts are lined with cuticle, so absorption of food is confined largely to the midgut.

The majority of insects are **phytophagus;** that is, they feed upon the juices and tissues of plants. Some insects, such as grasshoppers, eat almost any plant, but others are specific, restricting their feeding to certain plants. The caterpillars of many moths and butterflies eat the foliage of a single type of plant. Monarch caterpillars feed upon a certain species of milkweed from which they assimilate cardiac glycosides. After metamorphosis these butterflies are toxic to and induce vomiting in the birds that prey upon them. Certain species of ants cultivate fungus gardens as a source of food. Many beetles and insect larvae live upon dead animals **(saprophagus),** and a large number are **predaceous.**

Many insects are **parasitic.** Fleas, for instance, live on the blood of mammals, and the larvae of many varieties of wasps live upon spiders and caterpillars (Fig. 21-22). In turn, many are parasitized by other insects.

The feeding habits of insects are determined to some extent by their mouthparts, which are highly specialized for each type of feeding.

Biting and chewing mouthparts, such as those of the grasshopper and many herbivorous insects, are

429

Fig. 21-22. A, Hornworm, larval stage of a sphinx moth (order Lepidoptera). **B,** Hornworm parasitized by a tiny wasp, *Apanteles,* which laid its eggs inside of it. The wasp larvae have emerged and spun their cocoons on the caterpillar's skin. Young wasps emerge in 5 to 10 days, but the caterpillar usually dies. The more than 100 species of North American sphinx moths are strong fliers and mostly nocturnal feeders. Their larvae, called hornworms because of the large, fleshy posterior spine, are often pests of tomatoes, tobacco, and other plants. (**A,** Photo by C. P. Hickman, Jr.; **B,** photo by O. W. Olsen, Turtox News, August 1967.)

adapted for seizing and crushing food (Fig. 21-23, *A*). The mandibles of chewing insects are strong, toothed plates whose edges can bite or tear while the maxillae hold the food and pass it toward the mouth. Enzymes secreted by the salivary glands add chemical action to the chewing process.

Sucking mouthparts are greatly varied. Houseflies and fruitflies have no mandibles; the labium is modified into two soft lobes containing many small tubules that sponge up liquids with a capillary action much as the holes of a commercial sponge do (Fig. 21-23, *D*). Horseflies, however, are fitted not only to sponge up surface liquids but to bite into the skin with slender, tapering mandibles and then sponge up blood. Mosquitoes combine **piercing** by means of needlelike stylets and sucking through a food channel (Fig. 21-23, *B*). In honeybees the labium forms a flexible and contractile "tongue" covered with many hairs. When the bee plunges its proboscis into nectar, the tip

of the tongue bends upward and moves back and forth rapidly. Liquid enters the tube by capillarity and is drawn up continuously by a pumping pharynx. In butterflies and moths mandibles are usually absent, and the maxillae are modified into a long sucking proboscis (Fig. 21-23, *C*) for drawing nectar from flowers. At rest the proboscis is coiled up into a flat spiral. In feeding it is extended, and fluid is pumped up by pharyngeal muscles.

Circulation. The **open circulatory system** includes a dorsal tubular heart with ostia for receiving the blood (Fig. 21-24), an anterior aorta, and a hemocoel made up of spaces between the body organs. Veins and capillaries are lacking, and with some exceptions there are no respiratory pigments in the blood.

Gas exchange. The **tracheal system** comprises an extensive network of thin-walled tubes that branch into every part of the body. The tracheal trunks, which open to the outside by spiracles on the thorax and

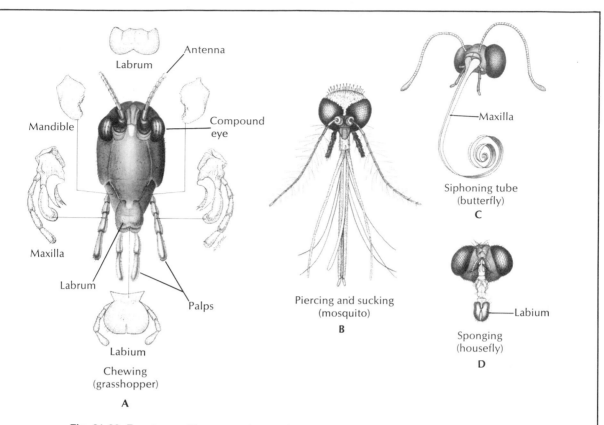

Fig. 21-23. Four types of insect mouthparts. See text for description of types and examples.

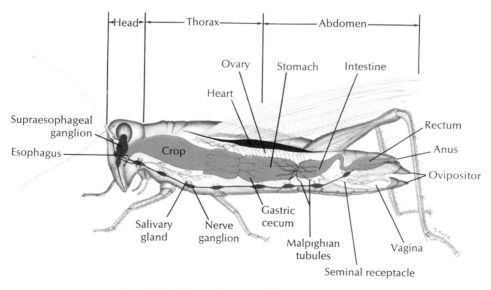

Fig. 21-24. Internal structure of female grasshopper.

abdomen (Fig. 21-19), are lined with a spiral ribbon of chitin that keeps them from collapsing. When the insect molts, these tracheal linings are also shed and replaced. The tracheae, which are composed of a single layer of cells, branch out into smaller tubes, ending in very fine, fluid-filled tubules called tracheoles. Scarcely any living cell is located more than two cell widths away from a tracheole, so oxygen is directly available to all living tissue. Thus insects have an efficient system of oxygen transport without the use of oxygen-carrying pigments in the blood.

In some insects the air moves through the tubes by diffusion only, but most large and active insects employ some ventilation device for moving the air in and out. In the larvae of flies the rhythmic expansion and collapse of the tracheae themselves result in inspiration and expiration. In grasshoppers muscle contraction flattens the abdomen and expels air from the tracheae; relaxation allows air to rush in. In flies and bees it is a telescoping movement, or shortening of the abdomen, that brings about expiration.

The **spiracles** are slits in the body surface designed not only to let air in, but to prevent excess water loss. In some insects the spiracles open and close during the respiratory movements.

The tracheal system, however, is adapted for air breathing. Stoneflies, dragonflies, and mayflies have aquatic larvae that are equipped with tracheal gills.

These are filled with blood and contain fine tracheoles into which the oxygen is diffused and then distributed to the rest of the tracheal system. Although the diving beetle *Dytiscus* (Fig. 21-20, *C*) can fly, it spends most of its life in the water as an excellent swimmer. It uses an ''artificial gill'' in the form of a bubble of air held under its wing covers. The bubble is kept stable by a layer of hairs on top of the abdomen and is in contact with the spiracles on the abdomen. Oxygen from the bubble diffuses into the tracheae and is replaced by diffusion of oxygen from the water. Thus the bubble can last for several hours before the beetle must surface to replace it. The gills of dragonfly nymphs are ridges in the rectum (rectal gills) where gas exchange occurs as water moves in and out. Mosquito larvae are not good swimmers but live just below the surface, putting out short breathing tubes to the surface for air (Fig. 21-25). Spreading oil on the water, a favorite method of mosquito control, clogs the tracheae with oil and so suffocates the larvae. ''Rat-tailed maggots'' of the syrphid flies have an extensible tail that can stretch as much as 15 cm to the water surface.

Excretion and osmoregulation. Malpighian tubules (Fig. 21-24) are typical of most insects. As in the spiders, malpighian tubules are very efficient, both as excretory organs and as a means of conserving body fluids—an important factor in the success of terrestrial animals.

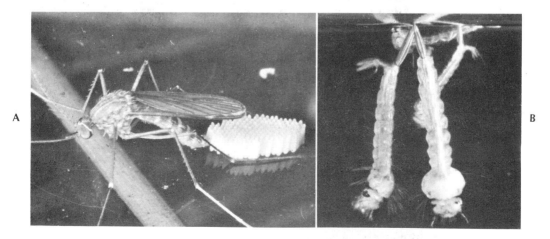

Fig. 21-25. A, Mosquito *Culex* (order Diptera) lays her eggs in small packets or rafts on the surface of standing or slowly moving water. **B,** Mosquito larvae are the familiar wrigglers of ponds and ditches. To breathe they hang head down, with respiratory tubes projecting through the surface film of water. Motion of vibratile tufts of fine hairs on the head brings a constant supply of food. (From *Mosquito,* an Encyclopaedia Britannica film.)

Nervous system. The nervous system in general resembles that of crustaceans, with a similar tendency toward fusion of ganglia (Fig. 21-24). A giant fiber system has been demonstrated in a number of insects. There is also a stomodeal nervous system that corresponds in function with the autonomic nervous system of vertebrates. Neurosecretory cells located in various parts of the brain have an endocrine function, but, except for their role in molting and metamorphosis, little is known of their activity.

Neuromuscular coordination. Insects are active creatures with excellent neuromuscular coordination. Arthropod muscles are typically cross-striated, just as vertebrate skeletal muscles are. A flea can leap a distance of 100 times its own length, and an ant can carry in its jaws a load greater than its own weight. This sounds as though insect muscle were stronger than that of other animals. Actually, however, the force a particular muscle can exert is related directly to its cross-sectional area, not its length. Based upon maximum load moved per square centimeter of cross-section, their strength is relatively the same as that of vertebrate muscle.

Sense organs. The sensory perceptions of insects are unusually keen. Organs receptive to mechanical, auditory, chemical, visual, and other stimuli are well developed. They are scattered over the body but are especially numerous on the appendages.

Photoreceptors include both ocelli and compound eyes. The compound eyes are large and constructed like those of crustaceans. Apparently visual acuity in insect eyes is much lower than that of human eyes, but most flying insects rate much higher than humans in flicker-fusion tests. Flickers of light become fused in the human eye at a frequency of 45 to 55 per second, but in bees and blowflies they do not fuse until 200 to 300 per second. This should be an advantage in analyzing a fast-changing landscape.

Most insects have three ocelli on the head, and they also have dermal light receptors on the body surface, but not much is known about them.

Sounds may be detected by sensitive hair sensilla or by tympanic organs sensitive to sonic or ultrasonic sound. Tympanic organs, found in grasshoppers (Fig. 21-19), crickets, cicadas, butterflies, moths, etc. involve a number of sensory cells extending to a thin tympanic membrane that encloses an air space in which vibrations can be detected.

Chemoreceptors, which are mounted on peglike structures or on hairs, are especially abundant on the antennae, mouthparts, or legs. Mechanical stimuli, such as contact pressure, vibrations, and tension changes in the cuticle, are picked up by hairs or setae or by sensory cells in the epidermis. Insects also sense temperature, humidity, proprioception, gravity, etc.

Reproduction. Sexes are separate in insects, and fertilization is usually internal. Insects have various means of attracting mates. The female moth gives off a powerful scent that can be detected for a great distance by the male. Fireflies use flashes of light; some insects find each other by means of sounds or color signals and by various kinds of courtship behavior.

Sperm are usually deposited in the vagina of the female at the time of copulation (Fig. 21-26). In some orders the sperm are encased in spermatophores that may be transferred at copulation or deposited on the substratum to be picked up by the female. The silverfish deposits a spermatophore on the ground, then spins signal threads to guide the female to it. During evolutionary transition from aquatic to terrestrial life, spermatophores were widely used, with copulation evolving much later. To be able to deposit the sperm directly into the vagina is an adaptive advantage for the male.

Usually the sperm are stored in the spermatheca of the female in numbers sufficient to fertilize more than one batch of eggs. Many insects mate only once during their lifetime, and none mates more than a few times.

Insects usually lay a great many eggs. The queen honeybee, for example, may lay more than 1 million eggs during her lifetime. On the other hand, some flies are viviparous and bring forth only a single offspring at a time. Forms that make no provision for the care of the young may lay many more eggs than those that provide for the young or those that have a very short life cycle.

Most species normally lay their eggs in a particular type of place to which they are guided by visual, chemical, or other clues. Butterflies and moths lay their eggs on the specific kind of plant upon which the caterpillar must feed (Fig. 21-27). The tiger moth may look for a pigweed and the sphinx moth for a tomato or tobacco plant. Insects whose immature stages are aquatic lay their eggs in water (Fig. 21-25, *A*). A tiny braconid wasp lays her eggs on the caterpillar of the sphinx moth where they will pupate in tiny white cocoons (Fig. 21-22). The ichneumon wasp (Fig.

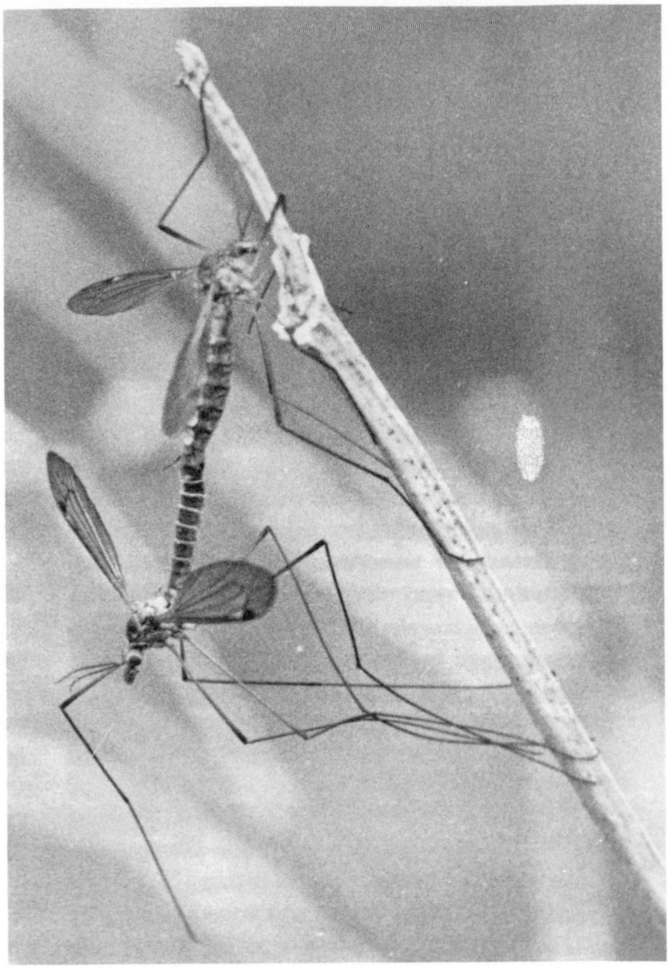

Fig. 21-26. Crane flies (order Diptera) mating. They live in damp places with abundant vegetation. Their larvae are aquatic or semiaquatic. (Photo by C. P. Hickman, Jr.)

21-20, *A*) with unerring accuracy seeks out a certain kind of larva in which her young will live as internal parasites. Her long ovipositors may have to penetrate 1 to 2 cm of wood to find and deposit her eggs in the larva of a wood wasp or a wood-boring beetle.

Metamorphosis and growth

Although metamorphosis is not limited to insects, they illustrate it more dramatically than any other group. The transformation of a caterpillar into a beautiful moth or butterfly is indeed an astonishing change.

Early development occurs within the egg, and the hatching young escape from the egg in various ways. During the postembryonic development most insects change in form—that is, they undergo **metamorpho-sis.** During this period in order to grow they must undergo a number of molts, and each stage of the insect between molts is called an **instar.**

Approximately 88% of insects undergo a **complete metamorphosis,** which separates the physiologic processes of growth (**larva**) from those of differentiation (**pupa**) and reproduction (**adult**). Each stage functions efficiently without competition with the other stages, for the larvae often live in entirely different surroundings and eat different foods from the adults. The wormlike larvae, which usually have chewing mouthparts, are known as caterpillars, maggots, bagworms, fuzzy worms, grubs, etc. After a series of instar stages during which the wings are developing internally, the larva forms a case or cocoon about itself and becomes a pupa, or chrysalis, a nonfeeding stage in which many

Fig. 21-27. Complete metamorphosis. **A,** Monarch butterfly, *Danaus plexippus* (order Lepidoptera), lays eggs on milkweed plant and hatched larvae feed on milkweed leaves. **B,** Larva hangs on milkweed as it prepares to pupate. At this state wings develop internally but are not everted until last larval instar. Larvae have chewing mouthparts but no compound eyes. **C,** Larva has transformed into chrysalis, or pupa, an inactive stage that does not feed and is covered by a cocoon or protective covering. **D,** Adult has emerged, with short, wrinkled wings. Wings soon expand and harden, pigmentation develops, and the butterfly goes on its way. (Photos by J. H. Gerard, Alton, Ill.)

insects pass the winter. When the final molt occurs the full-grown adult emerges (Fig. 21-27), pale and with wings wrinkled. In a short time the wings expand and harden, and the insect is on its way. The stages, then, are **egg, larva, pupa,** and **adult.** The adult undergoes no further molting. Insects that undergo complete metamorphosis are said to **holometabolic.**

Some insects, such as mayflies, dragonflies, and stoneflies, undergo a type of **gradual,** or **incomplete, metamorphosis.** The eggs are laid in water and develop into aquatic **nymphs** (sometimes called **naiads**), which have tracheal gills or other modifications for an aquatic life. The wings develop externally as budlike outgrowths in the early instars and increase in size as the animal grows by successive molts, crawls out of the water, and after the last molt becomes a winged adult (Fig. 21-28). In most other insects with gradual metamorphosis, as, for example,

Fig. 21-28. Gradual metamorphosis. A young dragonfly (order Odonata) has just emerged from its larval case. The aquatic naiad, after several molts, leaves the water for its final molt to become an adult. (Photo by L. L. Rue III.)

the grasshoppers and cicadas, the nymphs and adults are both terrestrial (Fig. 21-29). The stages are **egg, nymph,** and **adult.** Insects that undergo gradual, or incomplete, metamorphosis are called **hemimetabolic** (*hemi,* half, + *metabolē,* change).

A few insects, such as silverfish and springtails, are said to undergo **no metamorphosis.** The young, or juveniles, are similar to the adults except in size. The stages are **egg, juvenile,** and **adult.** Such insects are called **ametabolic** (without change).

Physiology of metamorphosis. Growth and metamorphosis in insects are regulated by three main groups of **hormones.** Neurosecretory mechanisms in the insect bear a remarkable resemblance to those of crustaceans. The hormones that control these processes

are secreted by the brain, the prothoracic glands, and the corpora allata.

Neurosecretory cells in the brain and ganglia of the nerve cord produce an **activation hormone** that stimulates the **prothoracic gland** (an endocrine gland located in the prothorax) to produce the molting hormone, or **ecdysone** (Fig. 12-2, p. 236). Ecdysone sets in motion the processes that lead to the casting off of the old cuticle **(ecdysis)** and to the metamorphic changes that result in a pupa or adult (Fig. 21-30).

The type of cuticle secreted at each molt is influenced by a hormone, the **juvenile hormone,** secreted by the **corpora allata,** located behind the brain. Simple molting, in which the larva molts into a larger larva, is repeated, and final metamorphosis is delayed

Fig. 21-29. A, Young praying mantids (nymphs) emerging from the egg capsule. Egg capsules (oothecae) are glued to shrubbery and other objects in late summer and fall. When eggs hatch in spring, enormous swarms of wingless nymphs emerge from single capsule. **B,** Praying mantis (order Orthoptera), about life size. It gets its name from the way it often holds its forelimbs. (**A,** Photo by F. M. Hickman; **B,** photo by J. W. Bamberger, Los Angeles, Calif.)

as long as the juvenile hormone is secreted. When production of the juvenile hormone ceases and the ecdysone alone is present, the larva metamorphoses into an adult or into a pupa and then an adult. A recent method of insect control involves the use of compounds that mimic the juvenile hormone, thus preventing treated larvae from becoming sexually mature.

What factors initiate the sequential action of these three different hormones? A great deal of experimentation has been done in this area. For instance, in the nymph of the bug *Rhodnius* experiments have shown that molting is stimulated by the stretching of the wall of the gut by the big meal that the nymph ingests during each instar. These impulses to the brain from the gut wall stimulate the release of the activating

hormones, which in turn stimulate the production of the ecdysone by the prothoracic gland, thus initiating the process of ecdysis.

In the American silkworm *Hyalophora* the pupa goes through a period of arrested development, called **diapause,** during which the brain is entirely inactive and no hormonal secretion occurs. Activity is resumed *only* if the pupa is chilled for approximately 10 weeks at 3° to 5° C, after which final metamorphosis occurs. In this instance chilling seems to be the necessary stimulus for the resumption of neurosecretory action, and ecdysone production. No juvenile hormone is produced at the end of diapause.

Adults have no prothoracic gland, so they can no longer grow and molt.

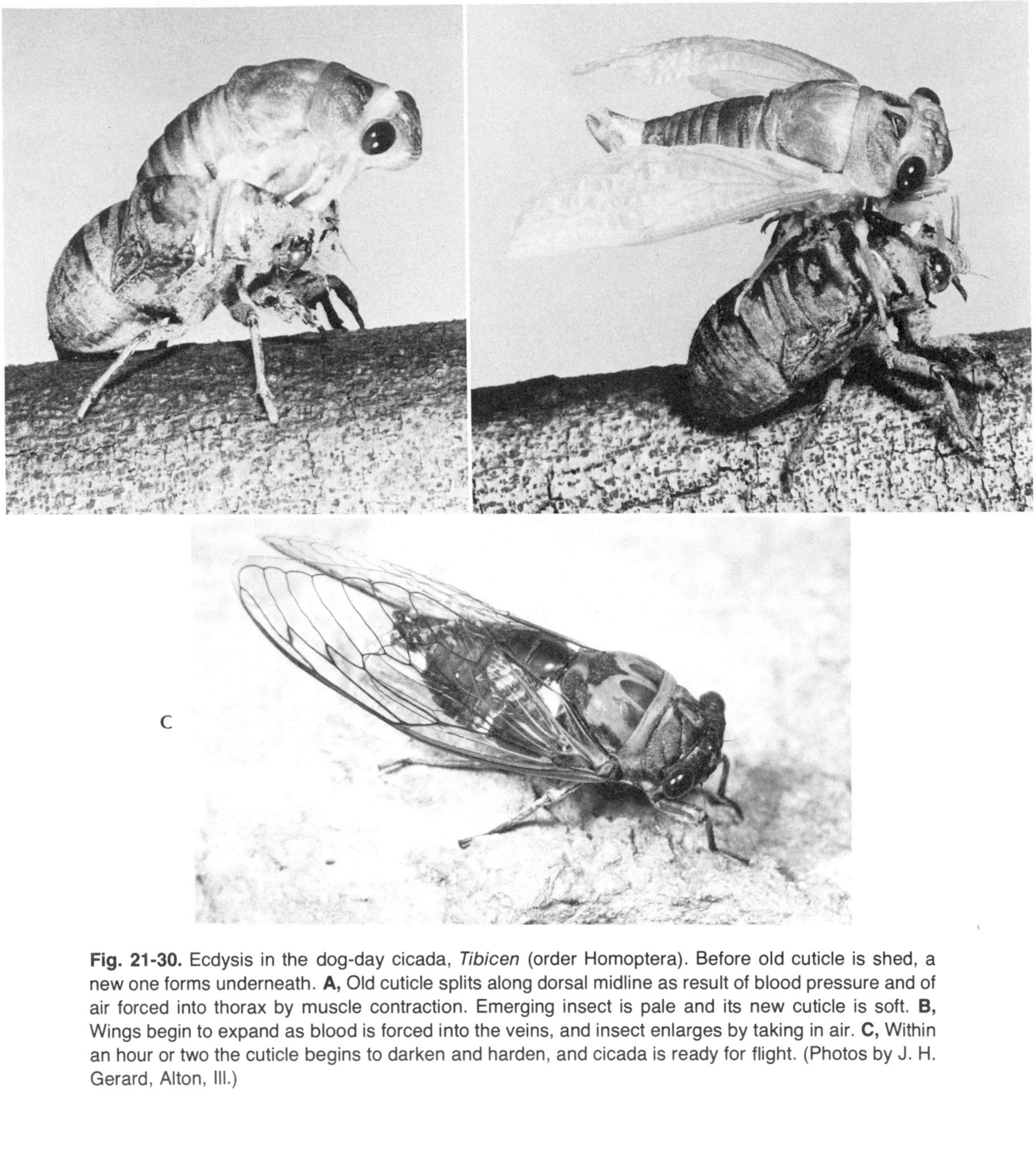

Fig. 21-30. Ecdysis in the dog-day cicada, *Tibicen* (order Homoptera). Before old cuticle is shed, a new one forms underneath. **A,** Old cuticle splits along dorsal midline as result of blood pressure and of air forced into thorax by muscle contraction. Emerging insect is pale and its new cuticle is soft. **B,** Wings begin to expand as blood is forced into the veins, and insect enlarges by taking in air. **C,** Within an hour or two the cuticle begins to darken and harden, and cicada is ready for flight. (Photos by J. H. Gerard, Alton, Ill.)

Insect behavior

Insect behavior consists of responses to stimuli. The nerve pathways involved are largely hereditary, so the responses are largely automatic rather than learned. The response may be a reflex movement toward or away from the stimulus, as, for example, the attraction of the moth to light, the avoidance of light by a cockroach, the attraction of carrion flies to the odor of dead flesh, or the orientation of the caddis fly larva toward a water current or of dragonflies toward an air current. The insect may react positively to certain chemicals in the location of food or negatively to being touched.

A response to a specific stimulus may be modified

Fig. 21-31. Tumble bugs, or dung beetles, *Canthon pilularis* (order Coleoptera), chew off a bit of dung, roll it into a ball, and then roll it to where they will bury it in soil. One beetle pushes while other pulls. Eggs are laid in the ball and the larvae feed on the dung. Tumble bugs are black, an inch or less in length, and common in pasture fields. (Photo by J. H. Gerard, Alton, Ill.)

by other stimuli. For instance, honeybees respond to bright light by leaving the hive—*if* the temperature is high but not if the temperature is low.

Much of the behavior of insects, however, is not a simple matter of orientation but involves a whole series of responses. A pair of tumble bugs, or dung beetles, chew off a bit of dung, roll it into a ball, and roll the ball laboriously to where they intend to bury it, after laying their eggs in it (Fig. 21-31). The cicada slits the bark of a twig and then lays an egg in each of the slits. The female potter wasp *Eumenes* scoops up clay into pellets and carries them one by one to her building site to fashion them into dainty little narrow-necked clay pots into each of which she lays an egg. Then she hunts and paralyzes a number of caterpillars, pokes them into the opening of a pot, and closes up the opening with clay. Each egg, in its own protective pot, hatches to find a well-stocked larder of food awaiting it.

Much of such behavior is innate, an entire sequence of actions being performed instinctively. However, more learning is involved than was once believed. The potter wasp, for example, must learn where she has left her pots if she is to return to fill them with caterpillars one at a time. Experiments have shown that insects have memory sufficient for the establishment of conditioned reflexes. Bees have been taught to make associations between food and color, and ants have learned to associate certain odors and food supplies.

Communication. Insects communicate with other members of their species by means of chemical, visual, auditory, and tactile signals. **Pheromones,** which are substances secreted onto the body surface, serve as **chemical signals.** They usually affect the behavior of other individuals of the same species, although some alarm pheromones are interspecific. Pheromones include sex attractants, releasers of certain behavior patterns, trail markers, alarm signals, territorial markers, etc. Like hormones, pheromones are effective in minute quantities.

439

When a female pine sawfly *Diprion* was caged and placed in a field, its sex attractant attracted more than 11,000 males. Queen honeybees produce a sex attractant in the mandibular glands that attracts males from a considerable distance. Pheromones may attract individuals of the same sex, as well as those of opposite sexes. A queen honeybee removed from her swarm can attract workers to the new location in a short time. In fact, the crushed heads of several queens on a bit of filter paper can attract swarms that have lost their queens.

Worker bees, when they return to the hive after an expedition, discharge an identification pheromone by which they are recognized by other members of the hive who would repel intruders not emitting the scent. Ants lay trails by means of pheromones. An ant that has found a new source of food marks its homeward trail. Caste determination in termites, and to some extent in ants and bees, is determined by pheromones. In fact, pheromones are probably a primary integrating force in populations of social insects. Many insect pheromones have been extracted and chemically identified.

Sound production and reception (phonoproduction and phonoreception) in insects have been studied extensively, and it is evident that this means of communication is meaningful to insects that use it. Sounds serve as warning devices, advertisement of territorial claims, or as courtship songs.

The sounds of crickets and grasshoppers seem to be concerned with courtship. Grasshoppers rub the femur of the third pair of legs over the ridges of the forewings. Male crickets scrape the rough edges of the forewings together to produce their characteristic chirping. The hum of mosquitoes is caused by the rapid vibration of their wings. The long, drawn-out sound of the male cicada is produced by vibrating membranes in a pair of organs located on the ventral side of the basal abdominal segment. Water striders "create" vibrations on the water that are used as signals. Whirligig beetles appear to be the only insects that use echos to detect obstacles, a type of sonar behavior found in bats, porpoises, oil birds, etc.

The production of light, or **bioluminescence,** occurs in a few kinds of flies, springtails, and beetles. The best known of the luminescent beetles are the fireflies, or lightning bugs, in which the flash of light is a means of locating a prospective mate. Each species has its own characteristic flashing rhythm produced on the ventral side of the last abdominal segments. The male flashes his species-specific pattern while flying. If a female flashes the proper answer after the appropriate interval, he flies toward her, giving his signal again. The dialogue usually culminates in copulation.

An interesting instance of mimicry of light signals has been observed in females of several species of *Photuris,* which prey on male fireflies of other species by mimicking the female mating signals of the prey species and then by capturing and devouring the luckless males that court them (Fig. 21-32).

Social behavior. Insects are said to rank first in the entire animal kingdom in their organization of social groups. Social communities are not all as complex as those of the honeybees, however. Some community groups are temporary and uncoordinated, as the hibernating associations of carpenter bees or the feeding gatherings of aphids. Some are coordinated for only

Fig. 21-32. A fatal embrace. The female firefly *Photuris* (order Coleoptera) *(left)* has seized as prey a male of another species, which she had lured by mimicking his species' lighting response. (Photo by J. E. Lloyd, From Science, Feb. 7, 1975, cover. Copyright 1975 by the American Association for the Advancement of Science.)

brief periods, such as the mating swarms of mosquitoes or mayflies. Others cooperate more fully, such as the tent caterpillars *Malacosoma,* which not only gather in sleeping and feeding communities but join in building a home web and a feeding net. However, even these are still open communities, and their social behavior is limited to the larval stage of the life cycle.

In the true societies of the higher orders, such as honeybees, ants, and termites, a complex social life is necessary for the perpetuation of the species. Such societies are closed. In them all stages of the life cycle are involved, the communities are usually permanent, all activities are collective, and there is reciprocal communication. There is a high degree of efficiency in the division of labor. Such a society is essentially a family group in which the mother or perhaps both parents remain with the young, sharing the duties of the group in a cooperative manner. The society is usually characterized by polymorphism, or **caste** differentiation, along with differences in behavior that are associated with the division of labor.

Among the bumblebees the groups are small and last only a season, but in honeybees as many as 60,000 to 70,000 bees may be found in a single hive. Of these there are three castes—a single sexually mature female, or **queen,** a few hundred **drones,** which are sexually mature males, and the **workers,** which are sexually inhibited females. Castes are determined partly by fertilization (males, or drones, come from unfertilized eggs) and partly by what is fed to the larvae ("royal jelly" is fed to female larvae destined to become queens). Female workers are prevented from maturing sexually by a "queen substance," containing inhibiting pheromones, which they lick from the queen's body. If the queen dies or is removed so that the "queen substance" is no longer available, the workers' ovaries develop, and they start enlarging a larval cell and feeding the larva the type of food that produces a new queen.

The workers take care of the young, secrete wax with which they build the six-sided cells of the honeycomb, gather the nectar from flowers, manufacture honey, collect pollen, and ventilate and guard the hive. Each worker appears to be responsible for a specific task, depending upon its age, but during its lifetime of a few weeks it performs all of the various tasks.

One drone, sometimes more, fertilizes the queen, which stores sperm enough in her spermatheca to last her a lifetime. Drones have no stings and are usually driven out or killed by the workers at the end of the summer. A queen may live as long as five seasons, laying thousands of eggs in that time. She is responsible for keeping the hive going through the winter, and only one reigning queen is tolerated in a hive at one time.

Honeybees have evolved an efficient system of communication by which, through certain bodily movements, their scouts inform the workers of the location and quantity of food sources.

Termite colonies contain two main castes, the fertile individuals, both males and females, and the sterile individuals (Fig. 21-33). Some of the fertile in-

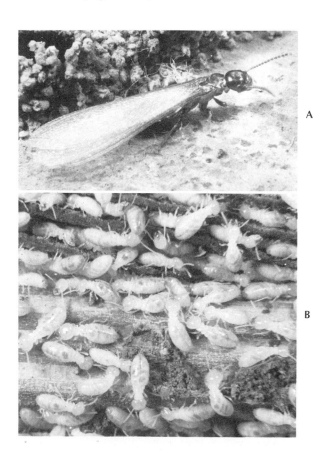

Fig. 21-33. Termites (order Isoptera). **A,** Reproductive adult. After the mating flight, adults shed their wings and then go in pairs to start a new colony. **B,** Workers are wingless sterile adults, which tend the nest, care for the young, etc. Termites are pale, soft bodied, and broad waisted in contrast to ants, which are dark, hard bodied, and narrow waisted. (Photos by J. H. Gerard, Alton, Ill.)

dividuals may have wings and may leave the colony, mate, lose their wings, and as **king** and **queen** start a new colony. Wingless fertile individuals may under certain conditions substitute for the king or queen. Sterile members are wingless and become **workers** and **soldiers.** As in bees, caste differentiation is caused by extrinsic factors. Reproductive individuals and soldiers secrete inhibiting pheromones that are passed to the nymphs through a mutual feeding process, called **trophallaxis,** so that they become sterile workers.

The phenomenon of trophallaxis, or exchange of nutrients, appears to be common among all social insects because it integrates the colony. The process involves feeding of the young by the queen and workers, which in turn may receive a drop of saliva from the young. It may also involve mutual licking, shampooing, etc.

Ants have highly organized societies, some of which involve the capture and use of slaves from other species. Recently an example of slavery within a species has been observed in the honeypot ants *(Myrmecocystus mimicus)*. The honeypot ants have a special caste, called "repletes," that engorge honey, store it in their greatly expandable abdomens, and during the off-season regurgitate it to feed the colony.

The unusually thin cuticle of this species, which allows the abdomen to expand, would be a liability to the ants in the deadly territorial warfare carried on by other species; however, this species has developed a nonlethal type of territorial display in the form of elaborate tournaments between neighboring colonies. When invaded, hundreds of workers rush out. Opponents approach each other on stilt legs, turn sideways, and drum on each other's abdomens with their antennae until the weaker partners yield. If the invaded colony is too small to defend itself, the invaders may rush in and carry off larvae, pupae, workers, and repletes as slaves for their own colony.

Insects and human welfare

Beneficial insects. Some insects are highly beneficial to man's interests. Some of them produce useful materials, such as honey and beeswax from bees, silk from silkworms, and shellac from a wax secreted by the lac insects. Insects are necessary for the cross-fertilization of fruits and other crops. Bees, for example, are indispensable in raising fruits, clover, and other crops, and the Smyrna fig does not grow in California without the help of a small fig wasp,

Blastophaga, which carries pollen from the nonedible caprifig.

Insects and higher plants very early in their evolution formed a relationship of mutual adaptations that have been to each other's advantage. Insects exploit flowers for food, and flowers exploit insects for pollination. Each floral development of petal and sepal arrangement is correlated with the sensory adjustment of certain pollinating insects. Among these mutual adaptations are amazing devices of allurements, traps, specialized structures, precise timing, etc.

Many predaceous insects, such as tiger beetles, aphid lions, ant lions, praying mantids, and ladybird beetles, destroy harmful insects. Some insects control harmful ones by parasitising them or by laying their eggs where their young, when hatched, may devour the host. Dead animals are quickly taken care of by maggots hatched from eggs laid in carcasses.

Insects and their larvae serve as an important source of food for birds, fish, and other animals.

Harmful insects. Harmful insects include those that eat and destroy plants and fruits, such as grasshoppers, chinch bugs, corn borers, boll weevils, grain weevils, San Jose scale, and scores of others. Practically every cultivated crop has some insect pest. Lice, blood-sucking flies, warble flies, botflies, and many others attack our domestic animals. Malaria, carried by the *Anopheles* mosquito, is still one of the world's killers; yellow fever and elephantiasis are also transmitted by mosquitoes. Fleas carry the plague, which at many times in history has almost wiped out whole populations. The housefly is the vector for typhoid and the louse for typhus fever; ticks transmit tularemia, Rocky Mountain spotted fever, and Texas cattle fever; the tsetse fly carries African sleeping sickness; and a blood sucking bug, *Rhodnius,* is a carrier of Chagas fever. In addition there is a tremendous destruction of food, clothing, and property by weevils, cockroaches, ants, clothes moths, termites, and carpet beetles. Not the least of insect pests is the bedbug, *Cimex,* a blood-sucking hemipterous insect that humans acquired, probably early in their evolution, from bats that lived in the same caves.

Control of insects. Because all insects are an integral part of the ecologic communities to which they belong, their total destruction would probably do more harm than good. Food chains would be disturbed, some of our most loved birds would disappear, and the biologic cycles by which dead animal and plant matter disintegrate and return to enrich the soil would be seriously impeded. The beneficial role of insects in our ecology has often been overlooked, and in our zeal to control the pests we have indiscriminately sprayed the landscape with extremely effective insecticides that have tended to eradicate the good, as well as the harmful, insects. We have also found, to our chagrin, that many of the chemicals we have used persist in the environment and accumulate as residues in the bodies of animals higher up in the food chains. Also many strains of insects have developed a resistance to the insecticides in common use.

In recent years an effort has been made to be more selective in the choice and use of pesticides that are specific in their targets. In addition to chemical control, other methods of control have been under intense investigation and experimentation.

The development of insect-resistant crops is one area of investigation. So many factors, such as yield and quality, are involved in developing resistant crops that the teamwork of specialists from many related fields is required. Some progress, however, has been made:

Three general types of **biologic controls** are being developed by the United States Department of Agriculture and others. All of these areas present problems but also show great possibilities. One is the use of **pathogens** such as *Bacillus thurinigensis,* which is used to control the leaf-cutting lepidopteran that takes a heavy toll in California's lettuce crops. This spore-forming bacterium forms a protein crystal that is poisonous to lepidopteran larvae. However, it attacks all lepidopterans (butterflies and moths), not just the specific pest.

A second type of control is the use of various **viruses** that are natural enemies of insects but could be cultivated in large numbers and applied at the most opportune time. Many viruses have been isolated that seem to have potential as insecticides. However, specific viruses are difficult to rear and could be expensive to put into commercial production.

A third method is to interfere with the metabolism or reproduction of the insect pests, for example by introducing **natural predators** or by using the **sterile male approach.** There have already been some successes with natural predators such as the vedalia beetle brought from Australia to counteract the work of the cottony-cushion scale on citrus plants and the parasites introduced from Europe for control of the alfalfa weevil. In Australia, dung beetle larvae, which de-

velop from eggs laid in dung, are effective in controlling the buffalo fly whose larvae develop in the same place. The sterile male approach has been effective in eradicating screwworm flies, a livestock pest. Large numbers of male insects, sterilized by irradiation, are introduced into the natural population; females that mate with the sterile flies lay infertile eggs. Insect **sex attractants** have also been utilized to trap pests. The sex attractant of the gypsy moth, a serious pest of forests and shade trees, has been identified and synthesized.

Brief review of insect orders

Insects are divided into orders on the basis of wing structure, mouthparts, metamorphosis, and so on. Entomologists do not all agree on the names of the orders nor on the limits of each order. Some tend to combine and others to divide the groups. However, the following synopsis of the orders is one that is rather widely accepted.

Subclass Apterygota (ap-ter-y-go'ta) (Gr. *a,* not, + *pterygōtos,* winged) (Ametabola). Primitive **wingless** insects **without metamorphosis.**

1. **Protura** (pro-tura) (Gr. *prōtos,* first + *oura,* tail) Minute (1 to 1.5 mm); no eyes or antennae; appendages on abdomen as well as thorax; live in soil and dark, humid places.
2. **Diplura** (dip-lu'ra) (Gr. *dis,* double, + *oura,* tail)—**japygids.** Usually under 10 mm; pale, eyeless; a pair of long terminal filaments or pair of caudal forceps; live in damp humus or rotting logs.
3. **Collembola** (col-lem'bo-la) (Gr. *kolla,* glue, + *embolon,* peg, wedge)—**springtails and snow fleas.** Small (5 mm or less); no eyes; respiration by trachea or body surface; a springing organ folded under the abdomen for leaping (Fig. 21-34, *A* to *C*); abundant in soil; sometimes swarm on pond surface film or on snow banks in spring.
4. **Thysanura** (thy-sa-nu'ra) (Gr. *thysanos,* tassel, + *oura,* tail)—**silverfish and bristletails** (Fig. 21-34, *D*). Small to medium size; large eyes, long antennae; three long terminal cerci; live under stones and leaves and around human habitations.

Subclass Pterygota (ter-y-go'ta) (Gr. *pterygōtos,* winged) (Metabola). **Winged insects** (some secondarily wingless) **with metamorphosis;** includes 97% of all insects.

Superorder Exopterygota (ek-sop-ter-i-go'ta) (Gr. *exō,* outside, + *pterygōtos,* winged) (Hemimetabola). **Metamorphosis gradual; wings develop externally** on larvae; compound eyes present on larvae; larvae called **nymphs** (or **naiads,** if aquatic).

5. **Ephemeroptera** (e-fem-er-op'ter-a) (Gr. *ephēmeros,* lasting but a day, + *pteron,* wing)—**mayflies** (Fig. 21-35, *F* and *G*). Wings membranous; forewings larger than hindwings; adult mouthparts vestigial; nymphs aquatic, with lateral tracheal gills.
6. **Odonata** (o-do-na'ta) (Gr. *odontos,* tooth, + *ata,* character-

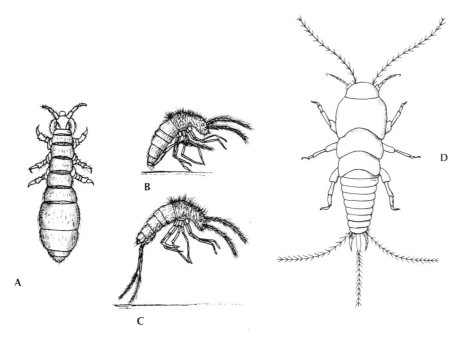

Fig. 21-34. A to **C,** Springtails (order Collembola). **A,** *Anurida.* **B** and **C,** *Orchesetta* in resting and leaping positions. **D,** Silverfish *Lepisma* (order Thysanura), often found in homes.

ized by)—**dragonflies, damselflies** (Figs. 21-28 and 21-35, *C* to *E*). Large; membranous wings are long, narrow, and net veined and similar in size; long and slender body; aquatic nymphs with aquatic gills and prehensile labium for capture of prey.

7. **Orthoptera** (or-thop′ter-a) Gr. *orthos,* straight, + *pteron,* wing)—**grasshoppers, locusts, crickets, cockroaches, walkingsticks** (Fig. 21-36, *A*), **praying mantids** (Fig. 21-29). Forewings thickened; hindwings folded like a fan under forewings; chewing mouth-parts.

8. **Dermaptera** (der-map′ter-a) (Gr. *derma,* skin, + *pteron,* wing)—**earwigs.** Very short forewings; large and membranous hind wings folded under forewings when at rest; biting mouthparts; forcepslike cerci.

9. **Plecoptera** (ple-kop′ter-a) (Gr. *plekein,* to twist, + *pteron,* wing)—**stoneflies** (Fig. 21-35, *A* and *B*). Membranous wings; larger and fanlike hind wings; aquatic nymph with tufts of tracheal gills.

10. **Isoptera** (i-sop′ter-a) (Gr. *isos,* equal, + *pteron,* wing)—**termites** (Fig. 21-33). Small; membranous, narrow wings similar in size with few veins; wings shed at maturity; erroneously called "white ants"; distinguishable from true ants by broad union of thorax and abdomen; complex social organization; exclusively wood diet, which is digested by enzymes secreted by flagellate protozoans or by symbiotic bacteria or fungi living in the intestine.

11. **Embioptera** (em-bi-op′ter-a) (Gr. *embios,* lively, + *pteron,* wing)—**webspinners.** Small; male wings membranous, narrow, and similar in size; wingless females; chewing mouthparts; colonial; make silk-lined channels in tropical soil.

12. **Psocoptera** (so-cop′ter-a) (Gr. *psoco,* rub small, + *pteron,* wing) **(Corrodentia)—psocids, "book lice," "bark lice."** Body as large as 10 mm; membranous, narrow wings with few veins, usually held rooflike over abdomen when at rest; some wingless species; found in books, bark, birdnests, on foliage.

13. **Zoraptera** (zo-rap′ter-a) (Gr. *zōros,* pure, + *apterygos,* wingless)—**zorapterans.** As large as 2.5 mm; membranous, narrow wings usually shed at maturity; colonial and termitelike.

14. **Mallophaga** (mal-lof′a-ga) (Gr. *mallos,* wool, + *phagein,* to eat)—**biting lice.** As large as 6 mm; wingless; chewing mouthparts; legs adapted for clinging to host; live on birds and mammals.

15. **Anoplura** (an-o-plu′ra) (Gr. *anoplos,* unarmed, + *oura,* tail)—**sucking lice.** Depressed body; as large as 6 mm; wingless; mouthparts for piercing and sucking; adapted for clinging to warm-blooded host; includes the head louse, body louse, crab louse, etc.

16. **Thysanoptera** (thy-sa-nop′ter-a) (Gr. *thysanos,* tassel, + *pteron,* wing)—**thrips.** Length, 0.5 to 5.0 mm (a few longer); wings, if present, long, very narrow, with few veins, and fringed with long hairs; sucking mouthparts; destructive plant-eaters, but some feed on insects.

17. **Hemiptera** (he-mip′ter-a) Gr. *hemi,* half, + *pteron,* wing) **(Heteroptera)—true bugs.** Size 2 to 100 mm; wings present or absent; forewings with basal portion leathery, apical portion membranous (Fig. 21-36, *E*); hindwings membranous; at rest, wings held flat over abdomen; piercing-sucking mouthparts; many with odorous scent glands; include water

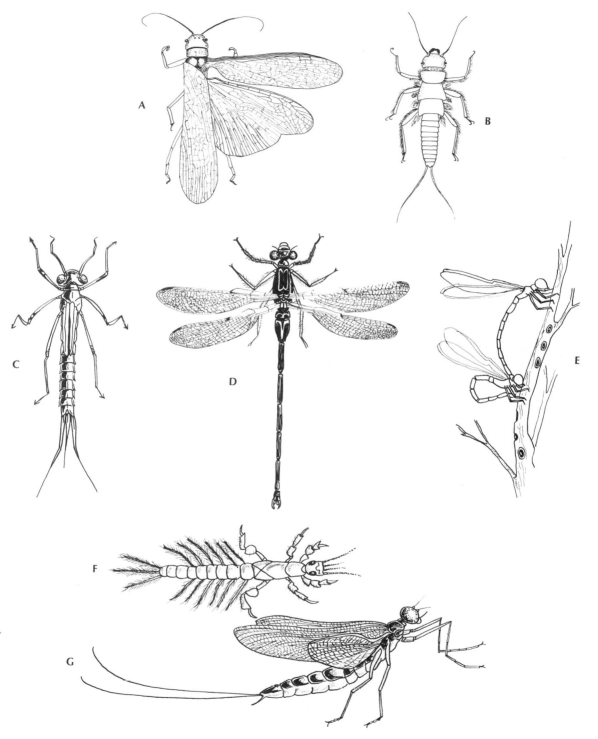

Fig. 21-35. A and **B,** Stonefly adult and naiad (order Plecoptera.) **C** to **E,** Damselfly (order Odonata) naiad, adult, and mating pair. **F** and **G,** Mayfly (order Ephemeroptera) naiad and adult. All have gradual metamorphosis and aquatic larvae.

Fig. 21-36. For legend see opposite page.

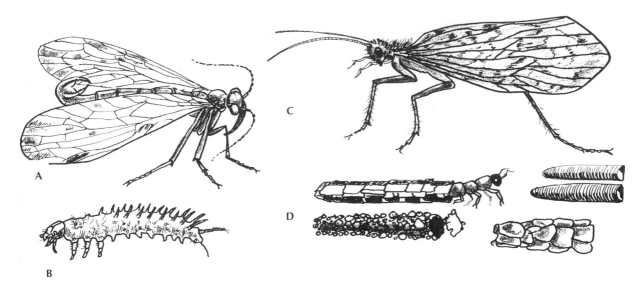

Fig. 21-37. A, Male scorpionfly *Panorpa* (order Mecoptera) has recurved abdomen with scorpion-like claspers. **B,** Scorpionfly nymph. **C,** Caddis fly (order Trichoptera). **D,** Several types of larval cases built by aquatic caddis fly larvae, often on the underside of stones.

scorpions, water striders, bedbugs, squash bugs, assassin bugs, chinch bugs, stinkbugs, plant bugs, lace bugs, etc.

18. Homoptera (ho-mop'ter-a) (Gr. *homos,* same, + *pteron,* wing)—**cicadas, aphids, scale insects, leafhoppers.** (Often included as suborder under Hemiptera.) If winged, membranous or thickened front wings and membranous hindwings; wings held rooflike over body; piercing-sucking mouthparts; all plant-eaters; some destructive; a few serving as source of shellac, dyes, etc; some with complex life histories.

Superorder Endopterygota (en-dop-ter-y-go'ta) Gr. *endon,* inside, + *pterygōtos,* winged). (Holometabola). **Metamorphosis complete; wings develop internally;** larvae without compound eyes.

19. Neuroptera (neu-rop'ter-a) (Gr. *neuron,* nerve, + *pteron,* wing)—**dobsonflies, ant lions, lacewings.** Medium to large size; similar, membranous wings with many cross veins; chewing mouthparts; dobsonflies (Fig. 21-36, *B*) with aquatic larvae; ant lion larvae (doodlebugs) make craters in sand to trap ants.

20. Coleoptera (ko-le-op'ter-a) (Gr. *koleos,* sheath, + *pteron,* wing)—**beetles, fireflies** (Fig. 21-32), **weevils.** The largest order of animals in the world; front wings (elytra) thick, hard, opaque (Fig. 21-36, *C*); membranous hindwings folded under front wings at rest; mouthparts for biting and chewing;

include ground beetles, carrion beetles, whirligig beetles, darkling beetles, stag beetles, dung beetles (Fig. 21-31), diving beetles, boll weevil, etc.

21. Strepsiptera (strep-sip'ter-a) (Gr. *strepsis,* a turning, + *pteron,* wing)—**stylops.** Minute; females with no wings, eyes, or antennae; males with vestigial forewings and fan-shaped hindwings; females and larvae parasitic in bees, wasps, and other insects.

22. Mecoptera (me-kop'ter-a) (Gr. *mēkos,* length, + *pteron,* wing)—**scorpionflies** (Fig. 21-37, *A* and *B*). Small to medium size; wings long, slender, with many veins; at rest, wings held rooflike over back; scorpionlike male clasping organ at end of abdomen; carnivorous; live in moist woodlands.

23. Lepidoptera (lep-i-dop'ter-a) (Gr. *lepidos,* scale, + *pteron,* wing)—**butterflies and moths** (Figs. 21-27 and 21-36, *D*). Membranous wings covered with overlapping scales, coupled at base; mouthparts a sucking tube, coiled when not in use; larvae (caterpillars) with chewing mandibles for plant eating, stubby prolegs on the abdomen, and silk glands for spinning cocoons; antennae knobbed in butterflies and usually plumed in moths.

24. Diptera (dip'ter-a) (Gr. *dis,* two, + *pteron,* wing)—**true flies.** Single pair of wings, membranous, and narrow; hindwings reduced to inconspicuous balancers (halteres); sucking

Fig. 21-36. A, Walking stick (order Orthoptera). Its resemblance to the twigs it lives among is a protective device. **B,** Dobsonfly (order Neuroptera) has aquatic naiad, the hellgrammite, often used by fishermen as bait. **C,** Hercules beetle *Dynastes tityus* (order Coleoptera), 5 to 7 cm long, is common in southeastern states. **D,** Palamedes swallowtail butterflies, *Papilio palamedes* (order Lepidoptera). **E,** Boxelder bug *Leptocoris* (order Hemiptera) is black with red markings; it feeds on box elder and other leaves. (**A** to **D,** Photos by C. P. Hickman, Jr.; **E,** photo by F. M. Hickman.)

mouth parts or adapted for sponging or lapping or piercing; legless larvae called maggots or, when aquatic, called wigglers (Fig. 21-25); includes crane flies (Fig. 21-26), mosquitoes, moth flies, midges, fruit flies, flesh flies, houseflies, horseflies, botflies, blowflies, and many others.

25. **Trichoptera** (tri-kop'ter-a) (Gr. *trichos,* hair, + *pteron,* wing)—**caddis flies** (Fig. 21-37, *C* and *D*). Small, soft-bodied; wings, well veined and hairy, folded rooflike over hairy body; chewing mouthparts; aquatic larvae construct movable cases of leaves, sand, gravel, bits of shell, or plant matter, bound together with secreted silk or cement; some make silk feeding nets attached to rocks in stream.

26. **Siphonaptera** (si-fon-ap'ter-a) (Gr. *siphōn,* a siphon, + *apteros,* wingless)—**fleas.** Small; wingless; bodies laterally compressed; legs adapted for leaping; no eyes; ectoparasitic on birds and mammals; larvae legless and scavengers.

27. **Hymenoptera** (hi-me-nop'ter-a) (Gr. *hymen,* membrane, + *pteron,* wing)—**ants, bees, wasps.** Very small to large; membranous, narrow wings coupled distally; subordinate hindwings; mouthparts for biting and lapping up liquids; ovipositor sometimes modified into stinger, piercer, or saw (Fig. 21-20). Both social and solitary species. Most larvae legless, blind, and maggotlike.

REVIEW OF CONTRIBUTIONS OF THE ARTHROPODS

1. **Cephalization** makes additional advancements, with centralization of fused ganglia and sensory organs in the head.
2. The **somites** have gone beyond the sameness of the annelid type and are **specialized** for a variety of purposes, forming functional groups of somites **(tagmosis).**
3. The presence of paired **jointed appendages** diversified for numerous uses results in greater adaptability
4. Locomotion is by extrinsic limb muscles, in contrast to the body musculature of annelids and other lower forms. **Striated muscles** are emphasized, thus ensuring rapidity of movement.
5. Although **chitin** is found in a few other forms below arthropods, its use is better developed in the arthropods, affording protection without sacrifice of mobility and serving in other capacities as well.
6. The **gills,** and especially the **tracheae,** represent a breathing mechanism more efficient than that of most invertebrates.
7. The alimentary canal shows greater specialization by having chitinous teeth and compartments.
8. Behavior patterns have advanced far beyond those of most invertebrates, with a higher development

of primitive intelligence, methods of communication, and **social** organization.

9. **Metamorphosis** is common in arthropod development, thus enlarging the feeding and habitat possibilities of a species and lessening competition among its members.
10. Many arthropods have well-developed protective coloration and protective resemblances.

SELECTED REFERENCES

Borror, D. J., D. M. Delong, and C. A. Triplehorn. 1976. An introduction to the study of insects, ed. 4. New York, Holt, Rinehart and Winston, Inc.

Bristowe, W. S. 1958. The world of spiders. London, New Naturalist Series.

Butler, C. G. 1975. The world of the honeybee. Rev. ed. Collins (Taplinger). *Updated to add queen substances, pheromones, communication, etc.*

Carthy, J. D. 1965. The behavior of arthropods. Edinburgh, Oliver & Boyd. *A slim paperback; not too difficult for undergraduate students.*

Chu, H. F. 1949. How to know the immature insects. Dubuque, Iowa, William C. Brown Co., Publishers.

Clarke, K. U. 1973. The biology of the Arthropoda. New York, American Elsevier Publishing Co., Inc.

Cloudsley-Thompson, J. E. 1958. Spiders, scorpions, centipedes, and mites. New York, Pergamon Press, Inc. *Emphasizes the behavior and ecology of the group.*

Comstock, J. H., and W. J. Gertsch. 1948. The spider book, Rev. ed. Ithaca, N.Y., Comstock Publishing Co., Inc. *Old but still useful and well-written account.*

Debach, P. 1974. Biological control by natural enemies. New York, Cambridge University Press.

Dethier, V. G. 1976. The hungry fly. Cambridge, Harvard University Press.

Edmondson, W. T. (ed.). 1959. Ward and Whipple's fresh-water biology, ed. 2. New York, John Wiley & Sons, Inc. *Includes keys to freshwater crustaceans.*

Fox, R. M., and J. W. Fox. 1964. Introduction to comparative entomology. New York, Reinhold Publishing Co.

Frisch, K. von. 1971. Bees: their vision, chemical senses, and language. Rev. ed. Ithaca, N. Y., Cornell University Press. *An outstanding work on the way bees communicate with each other and reveal the sources of food supplies.*

Goetsch, W. 1957. The ants. Ann Arbor, The University of Michigan Press. *A concise account of ants and their ways.*

Gosner, K. L. 1971. Guide to identification of marine and estuarine invertebrates: Cape Hatteras to the Bay of Fundy. New York, Interscience Div., John Wiley & Sons, Inc.

Herms, W. B. 1950. Medical entomology, ed. 4. New York, Macmillan, Inc. *A section of this book is devoted to Arachnida. The role arachnids play in disease transmission is emphasized.*

Hickman, C. P. 1973. Biology of the invertebrates, ed. 2. St. Louis, The C. V. Mosby Co.

Jaques, H. E. 1947. How to know the insects, ed. 2. Dubuque, Iowa, William C. Brown Co., Publishers.

Jaques, H. E. 1951. How to know the beetles. Dubuque, Iowa, William C. Brown Co., Publishers. *A useful and compact manual for the coleopterist.*

Kaston, B. J., and E. Kaston. 1972. How to know the spiders, ed. 2. Dubuque, Iowa, William C. Brown Co., Publishers.

Levi, H. W., and L. R. Levi. 1969. A guide to spiders and their kin. New York, Golden Press, Inc.

Light, S. F., et al. 1967. Intertidal invertebrates of the central California coast. Berkeley, University of California Press.

Little, V. A. 1972. General and applied entomology, ed. 3. New York, Harper & Brothers.

Melitsch, P. A. 1972. Invertebrate zoology, ed. 2. New York, Oxford University Press.

Michener, C. D. 1974. The social behavior of the bees. Cambridge, Mass., Harvard University Press.

Moore, R. C., C. G. Lalicker, and A. G. Fisher. 1952. Invertebrate fossils. McGraw-Hill Book Co., Inc. *Four chapters deal with the arthropods.*

Pennak, R. W. 1953. Fresh-water invertebrates of the United States. New York, The Ronald Press Co. *Taxonomic keys and illustrative drawings.*

Price, P. W. 1975. Insect ecology. New York, John Wiley & Sons, Inc.

Ross, H. H. 1965. A textbook of entomology, ed. 3. New York, John Wiley & Sons, Inc.

Russell-Hunter, W. D. 1969. A biology of the higher invertebrates. New York, Macmillan, Inc. *(Paperback.)*

Savory, T. H. 1952. The spider's web. New York, Frederick Warne & Co., Inc.

Savory, T. H. 1964. Arachnida. New York, Academic Press, Inc.

Schmitt, W. L. 1965. Crustaceans. Ann Arbor, The University of Michigan Press.

Skaife, S. H. 1961. The study of ants. New York, Longman, Inc.

Sudd, J. H. 1967. An introduction to the behavior of ants. New York, St. Martin's Press, Inc.

Tombes, A. S. 1970. An introduction to invertebrate endocrinology. New York, Academic Press, Inc.

Waterman, T. H., and F. A. Chase. 1960. General crustacean biology. *In* Physiology of Crustacea, vol. I., New York, Academic Press, Inc. pp. 1-33.

Wigglesworth, V. B. 1974. Insect physiology, ed. 7. New York, John Wiley & Sons, Inc. *A brief and easily understandable summary.*

Wigglesworth, V. B. 1972. Principles of insect physiology, ed. 7. New York, John Wiley & Sons, Inc. *A comprehensive account.*

Wilson, E. O. 1971. The insect societies. Cambridge, Mass., Harvard University Press.

SELECTED SCIENTIFIC AMERICAN ARTICLES

Batra, S. W. T., and L. R. Batra. 1967. The fungus gardens of insects. **217:**112-120 (Nov.).

Bartholomew, G. A. 1972. Temperature control in flying moths. **226:**70-77 (June).

Bennet-Clark, H. C., and A. W. Ewing. 1970. The love song of the fruit fly. **223:**85-92 (July).

Bishop, J. A., and L. M. Cook. 1975. Moths, melanism and clean air. **232:**90-99 (Jan.).

Buck, J. E. 1976. Synchronous fireflies. **234:**74-85 (May).

Burgess, J. W. 1976. Social spiders. **234:**100-107 (March).

Caldwell, R. L., and H. Dingle. 1976. Stomatopods. **234:**80-89 (Jan.). *An account of predatory mantis shrimps.*

Cambi, J. M. 1971. Flight orientation in locusts. **225:**74-81 (Aug.).

Evans, H. E. 1963. Predatory wasps. **208:**144-154 (April).

Evans, H. E., and R. W. Matthews. 1975. The sand wasps of Australia. **233:**108-115 (Dec.).

Frisch, K. von. 1962. Dialects in the language of the bees. **207:**78-86 (Aug.).

Heinrich, B. 1973. The energetics of the bumblebee. **228:**96-102 (April).

Hinton, H. E. 1970. Insect eggshells. **223:**84-91 (Aug.).

Holldobler, B. 1971. Communication between ants and their guests. **224:**86-93 (March).

Hoy, R. R. 1974. The neurobiology of cricket song. **231:**34-52 (Aug.).

Jacobson, M., and M. Beroza. 1964. Insect attractants. **211:**20-27 (Aug.).

Johnson, C. G. 1963. The aerial migration of insects. **209:**132-138 (Dec.).

Jones, J. C. 1968. The sexual life of a mosquito. **218:**108-116 (April).

Milne, L. J., and M. Milne. 1976. The social behavior of burying beetles. **235:**84-89 (Aug.).

Morse, R. A. 1972. Environmental control in the beehive. **226:**93-98 (April).

Palmer, J. D. 1975. Biological clocks of the tidal zone. **232:**70-79 (Feb.).

Roeder, K. D. 1965. Moths and ultrasound. **212:**94-102 (April).

Rothschild, M., et al. 1973. The flying leap of the flea. **222:**92-101 (Nov.).

Saunders, D. S. 1976. The biological clock of insects. **234:**114-121 (Feb.).

Savory, T. H. 1962. Daddy longlegs. **207:**119-128 (Oct.).

Savory, T. H. 1968. Hidden lives. **219:**108-114 (July).

Schneider, D. 1974. The sex-attractant receptor of moths. **231:**28-35 (July).

Topoff, H. R. 1972. The social behavior of army ants. **227:**71-79 (Nov.).

Waterhouse, D. F. 1974. The biological control of dung. **230:**100-109 (April).

Wehner, R. 1976. Polarized-light navigation by insects. **235:**106-115 (July).

Weis-Fogh, T. 1975. Unusual mechanisms for the generation of lift in flying. **233:**80-87 (Nov.).

Wenner, A. M. 1964. Sound communication in honeybees. **210:**116-124 (April).

Williams, C. M. 1967. Third-generation pesticides. **217:**13-17 (July).

Wilson, D. M. 1968. The flight-control system of the locust. **218:**83-90 (May).

Wilson, E. O. 1972. Animal communication. **227:**52-71 (Sept.).

Wilson, E. O. 1975. Slavery in ants. **232:**32-49 (June).

Wright, R. H. 1975. Why mosquito repellents repel. **233:**104-111 (July).

This little fellow, which resembles some prehistoric monster but is only 300 to 500 μm long, is *Echiniscus maucci,* one of the "water bears" of phylum Tardigrada. Unable to swim, it clings to moss or water plants with its claws, and if the environment dries up, it goes into a state of suspended animation and "sleeps away" the drought. (Photo by Diane R. Nelson and Robert O. Schuster.)

22 LESSER PROTOSTOMES AND LOPHOPHORATES

LESSER PROTOSTOMES

> Phylum Sipunculida
> Phylum Echiurida
> Phylum Priapulida
> Phylum Pentastomida
> Phylum Tardigrada
> Phylum Onychophora

LOPHOPHORATES

> Phylum Phoronida
> Phylum Ectoprocta
> Phylum Brachiopoda

This chapter includes a brief discussion of nine coelomate phyla whose position in the phylogenetic lines of the animal kingdom is somewhat problematic, as are their relationships to each other. The great evolutionary flow that began with the appearance of the coelom and led to the three huge phyla of molluscs, annelids, and arthropods also produced some other lines. Some no longer exist, whereas others, though small and lacking in great economic and ecologic importance, have survived. Six of these small phyla, usually grouped together as "lesser protostomes," have probably digressed at different times from the annelid-arthropod stem line, all following different adaptive currents. The three lophophorate phyla, Phoronida, Ectoprocta, and Brachiopoda, are obviously related to each other by the common possession of a crown of ciliated tentacles, called a lophophore, used in food capture and respiration. However, their background is mysterious, for none of the known protostome stocks would seem to be likely ancestors for the lophophorates. Whatever the relationship (or lack of it) may be in these nine phyla, they are grouped together here mainly for convenience.

LESSER PROTOSTOMES

The term "lesser protostomes" refers to a group of small phyla that all belong to the coelomate protostomes. Their relationship to each other and to the other protostomes is puzzling. Some of them have some deuterostome characteristics in their development.

Three of the phyla, Sipunculida, Echiurida, and

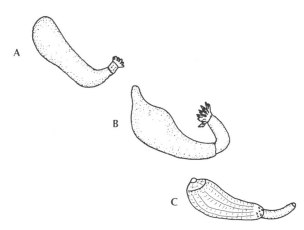

Fig. 22-1. Some types of sipunculids. **A,** *Phascolosoma.* **B,** *Dendrostomum.* **C,** *Aspidosiphon.*

Priapulida, are benthic (bottom-dwelling) marine worms with a variety of proboscis devices used in burrowing and food getting. Until recently (1961) when a true peritoneum was demonstrated, the priapulids were thought to be pseudocoelomates. The echiurids and sipunculids have sometimes been classed with the annelids.

The other three phyla—Pentastomida, Onychophora, and Tardigrada—have often been grouped together and called Pararthropoda because they have unjointed limbs with claws (at some stage) and a cuticle that undergoes molting. One of the phyla—Pentastomida—is parasitic, and another—Onychophora—is terrestrial (but limited to damp areas). Tardigrades are found in marine, freshwater, and terrestrial habitats.

Phylum Sipunculida

The Sipunculida (si′-pun-kyu′li-da) (L. *sipunculus.* little siphon) comprise a phylum of fewer than 350 marine species, but they have a wide geographic distribution.

The sipunculids are often called peanut worms because some of them, when disturbed, can contract into the shape of a peanut (Fig. 22-1). Extended, they are long and cylindric, the majority ranging from 15 to 30 cm in length. They burrow into sand or mud or live in discarded shells, in holes in rocks, or among vegetation.

The narrowed anterior end, called an **introvert,** is retractile and bears the mouth, surrounded by ciliated tentacles. Sipunculids are mostly deposit feeders. They live on organic matter collected by the ciliated tentacles from the mud or sand.

They have a cerebral ganglion, nerve cord, and pair of nephridia; the coelomic fluid contains red blood cells bearing a respiratory pigment (hemerythrin). The larval form is usually a trochophore. Common genera are *Sipunculus, Dendrostomum,* and *Phascolosoma.*

Phylum Echiurida

The Echiurida (ek′ee-yur′i-da) (Gr. *echis,* adder, + *oura,* tail) are marine worms that burrow into mud or sand or live in empty shells and sand dollar tests or in rocky crevices. The number of named species is less than that of sipunculids, but they are usually found in greater densities. They vary in length from a few millimeters to 40 or 50 cm.

The body is cylindric. Anterior to the mouth is a flattened, extensible **proboscis (introvert),** which,

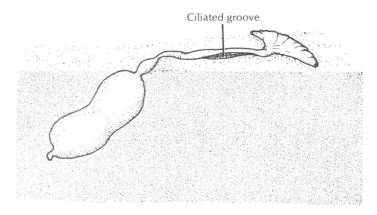

Ciliated groove

Fig. 22-2. *Tatjanellia* (phylum Echiurida) is a detritus feeder. Lying buried in the sand, it explores the surface with its long proboscis, which picks up organic particles and carries them along a ciliated groove to the mouth. (After Zenkevitch; modified from Dawydoff, C. 1959, Classe des Echiuriens. *In* Grassé, P. (ed.). Traité de Zoologie, Vol. 5, Paris, Masson et Cie.)

unlike that of the sipunculids, cannot be retracted into the trunk. Echiurids are often called "spoonworms" because of the shape of the contracted proboscis in some worms. The proboscis has a ciliated groove leading to the mouth. While the animal lies buried, the proboscis can extend out over the mud for exploration and deposit feeding (Fig. 22-2). *Urechis,* however, secretes a mucous net in a U-shaped burrow through which it pumps water and strains out food particles.

Echiurids, with the exception of *Urechis,* have a **closed circulatory system** with a contractile vessel; most have one to three pairs of nephridia (some have many pairs), and all have a nerve ring and ventral nerve cord. A pair of anal sacs arise from the rectum and open into the coelom; they are thought to be respiratory in function and possibly accessory nephridial organs. Sexual dimorphism is pronounced in some species, especially *Bonellia,* in which the males are minute (1 mm) and live as parasites in or on the larger females.

Common genera are *Urechis,* called the "fat innkeeper" because its tube also gives shelter to certain characteristic "guests"; *Thalassema,* which often lives in dead sand dollar tests; and *Echiurus.*

Phylum Priapulida

The Priapulida (prī-a-pyu'li-da) (Gr. *priapos,* phallus) consist of a few species belonging to two genera of marine worms, *Priapulus* (Fig. 22-3) and *Halicryptus,* which live in the bottom muck, chiefly in colder

seawaters. Their cylindric bodies are rarely more than 15 cm long. They are burrowing, predaceous animals that usually orient themselves upright in the mud with the mouth at the surface.

An eversible **proboscis** with rows of curved spines that surround the mouth is used to capture small, soft-bodied prey, which are then swallowed whole. When the proboscis is everted the spines point forward; when invaginated they are directed posteriorly. Most priapulids (pri-ap'u-lids) have one or two hollow **caudal appendages** that communicate with the coelom. These are thought to be respiratory in function and perhaps also chemoreceptive. The epidermis is covered with a cuticle that is molted periodically.

There is no circulatory system, but coelomocytes in the body fluids contain a respiratory pigment (hemerythrin). There is a nerve ring and ventral cord with nerves and a protonephridial tubule that serves also as a gonoduct. The most common species is *Priapulus caudatus.*

Phylum Pentastomida

The wormlike Pentastomida (pen-ta-stom'i-da) (Gr. *pente,* five, + *stoma,* mouth) are bloodsucking **parasites,** 2.5 to 12 cm long, that are found in the lungs and nasal passages of carnivorous vertebrates—most commonly in reptiles. Some human infections have been found in Africa and Europe. The intermediate host is usually a vertebrate that is eaten by the final host.

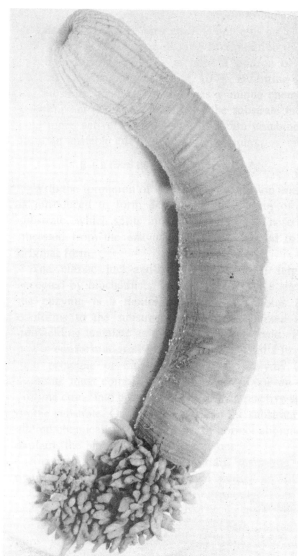

Fig. 22-3. *Priapulus caudatus* (phylum Priapulida) with proboscis *(top)* partially invaginated. The warty trunk is annulated but not truly segmented. The caudal appendages are probably respiratory and chemoreceptive organs. (Courtesy W. L. Shapeero.)

Linguatula tainoides, which live in the nasal passages of carnivorous mammals, discharges its eggs in the mucous secretions of its host. The mitelike larvae may be eaten by rabbits, which in turn may be eaten by a carnivore, in which the parasites mature.

Phylum Tardigrada

The Tardigrada (tar-di-gray'da) (L. *tardus,* slow, + *gradus,* step) are the "slow-steppers." They are often called "water bears" because of their plump little bodies and stubby legs. They are found in both fresh and salt water and in damp soil, moss, lichens, and other terrestrial microhabitats.

The body, usually less than 1 mm long, bears eight short, **unjointed legs,** each with claws (Fig. 22-4). Unable to swim, they creep about awkwardly, clinging to the substrate with their claws. A pair of sharp stylets and a sucking pharynx adapt them for piercing and sucking plant cells or small prey such as nematodes and rotifers.

There is a body covering of nonchitinous **cuticle** that

Fig. 22-4. *Echiniscus gladiator* (phylum Tardigrada) as seen under a scanning electron microscope. (×1250.) Its claws and slow, lumbering movement have earned it the name "water bear." Members of this species are 100 to 150 μm long. (Photo by Diane R. Nelson and Robert O. Schuster.)

is molted several times during the life cycle. As in the arthropods, muscle fibers are attached to the cuticular exoskeleton.

The annelid-type nervous system is surprisingly complex, and in some species there is a pair of eyespots. Circulatory and respiratory organs are lacking.

Females may deposit their eggs in the old cuticle as they molt or attach them to a substrate. Little is known about their reproduction and development, but some genera reproduce parthenogenetically, others sexually.

One of the most intriguing features of terrestrial tardigrades is their capacity to enter a state of suspended animation, called **cryptobiosis** (formerly called anabiosis), during which metabolism is virtually imperceptible; the organism can withstand harsh environmental conditions. Under gradual drying conditions the water content of the body is reduced from 85% to only 3%, movement ceases, and the body becomes barrel shaped. In a cryptobiotic state tardigrades can resist temperature extremes, ionizing radiations, oxygen deficiency, etc., and may survive for years. Activity resumes when moisture is again available.

Phylum Onychophora

Members of Phylum Onychophora (on-i-kof'o-ra) (Gr. *onyx,* claw, + *pherein,* to bear) are called "velvet worms" or "walking worms." They are caterpillar-like animals, 1.4 to 15 cm long, that live in rain forests and other tropical and semitropical leafy habitats. *Peripatus* (Fig. 22-5) is the best known genus.

The fossil record of the onychophorans shows that they have changed little in their 400-million–year history. They have been of unusual interest to zoologists because they share so many characteristics with both the annelids and the arthropods. They have even been called, a bit too hopefully to be sure, the "missing link" between the two phyla.

These curious animals have been dubbed velvet worms because they are covered with a **velvety skin,** and walking worms because they have from 14 to 43 pairs of stumpy, **unjointed legs,** each ending with a flexible pad and two claws. They creep about slowly, swinging their legs forward a few at a time. The head bears a pair of flexible **antennae** with annelid-like eyes at the base.

Onychophorans feed at night, capturing their small

Fig. 22-5. *Peripatus* (phylum Onychophora), one of the velvety-skinned "walking worms." They are air breathers with a tracheal system similar to that of the myriapods and insects. (Courtesy Ward's Natural Science Establishment, Inc.)

prey by spitting out a stream of sticky material from a pair of oral papillae.

They are air breathers, using a **tracheal system** that connects with pores scattered over the body. The tracheal system, though similar to that of the arthropods, has probably evolved independently. Other arthropod characteristics are the open circulatory system with a tubular heart, a hemocoel for a body cavity, and a large brain. Annelid-like characteristics are segmentally arranged nephridia, a muscular body wall, and pigment-cup ocelli.

Onychophorans are dioecious. Most species produce living young, some species being viviparous (having a placental attachment between mother and young) and some ovoviparous (the young develop in the uterus without placental attachment). Two Australian genera are oviparous and use an ovipositor to deposit the eggs in moist places.

LOPHOPHORATES

The three lophophorate phyla might appear upon superficial examination to have nothing in common except that they are all aquatic invertebrates, mostly marine. The **phoronids (Phylum Phoronida)** are wormlike marine forms that live in secreted tubes in sand or mud or attached to rocks or shells. The **ectoprocts (Phylum Ectoprocta)** are minute forms whose protective cases often form encrusting masses on rocks, shells, or plants. The **brachiopods (Phylum Brachiopoda)** are bottom-dwelling marine forms that superficially resemble molluscs because of their bivalved shells.

These three apparently different types of animals are lumped together in a group called Lophophorata. Why? Actually they have more in common than first appears. They are all eucoelomates; all have some protostome characteristics; all are sessile; and none has a distinct head. But these characteristics are also shared by other phyla. What really sets them apart from other phyla is the common possession of a **ciliary feeding device** called a **lophophore,** a term that means crest bearer (Gr. *lophos,* crest or tuft, + *phorein,* to bear).

A lophophore is a unique arrangement of ciliated tentacles borne on a ridge (a fold of the body wall), which surrounds the mouth but not the anus. The lophophore with its crown of tentacles forms an extension of the coelom, and the thin, ciliated walls of the tentacles comprise not only an efficient feeding device but serve also as a respiratory surface for exchange of gases between the environmental water and the coelomic fluid. The lophophore can usually be extended for feeding or withdrawn for protection.

In addition, all three phyla have a **U-shaped alimentary canal,** with the anus placed near the mouth but **outside the lophophore.** All have a **free-swimming larval stage** but are **sessile as adults.**

Phylogeny and adaptive radiation

The ectoprocts and brachiopods have left extensive fossil records, but, although there are still many diverse living species of ectoprocts, there are less than 300 living species of brachiopods and only approximately 15 species of phoronids. No fossil record of phoronids is known.

The possession of a lophophore by all three phyla is evidence of their relationship, but each phylum has specialized along its own lines and developed its own life-style. As a group they seem to occupy a phylogenetic position somewhere between the protostomes and the deuterostomes. The coelom is divided into three regions as in deuterostomes, and in their embryology they display both protostome and deuterostome characteristics. Some authorities consider them the most primitive of the deuterostomes or at least ancestral to the other deuterostomes.

All lophophorates are **filter feeders** and most of their evolutionary diversification has been guided by this function. The tubes of phoronids vary according to their habitats. Various ectoprocts tend to build their protective exoskeletons of chitin or gelatin, which may or may not be impregnated with calcium and sand. Brachiopod variations occur largely in their shells and lophophores.

Phylum Phoronida

The phylum Phoronida (fo-ron′i-da) comprises approximately 15 species of small wormlike animals that live on the bottom of shallow seas. They range from a few millimeters to 20 cm in length. Each worm secretes a leathery or chitinous tube in which it lies free, but which it never leaves (Fig. 22-6). The tubes may be anchored singly or in a tangled mass on rocks, shells, or pilings or buried in the sand. The tentacles on the lophophore are thrust out for feeding, but if the animal is disturbed it can withdraw completely into its tube.

The name "Phoronida" comes from the Greek *phoros,* meaning bearing, and the Latin *nidus,* meaning nest. The "nest," of course, refers to the tentacled lophophore.

The **lophophore** is made up of two parallel ridges curved in a horseshoe shape, the bend located ventrally and the mouth lying between the two ridges. The horns of the ridges are often coiled into twin spirals. Each ridge carries hollow ciliated tentacles, which, like the ridges themselves, are extensions of the body wall.

The cilia on the tentacles direct a water current toward a groove between the two ridges, which leads toward the mouth. Plankton and detritus caught in this current become entangled in mucus and carried by the cilia to the mouth. The anus lies dorsal to the mouth, outside the lophophore, flanked on each side by a nephridiopore. Water leaving the lophophore passes over the anus and nephridiopores, carrying away the wastes. Cilia in the stomach area of the U-shaped gut aid in food movement.

The phoronids have a closed system of contractile blood vessels but no heart; the red blood contains hemoglobin.

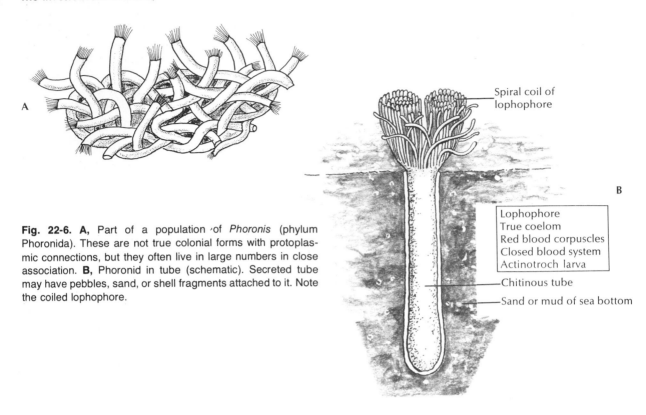

Fig. 22-6. A, Part of a population of *Phoronis* (phylum Phoronida). These are not true colonial forms with protoplasmic connections, but they often live in large numbers in close association. B, Phoronid in tube (schematic). Secreted tube may have pebbles, sand, or shell fragments attached to it. Note the coiled lophophore.

Spiral coil of lophophore

B

Lophophore
True coelom
Red blood corpuscles
Closed blood system
Actinotroch larva

Chitinous tube

Sand or mud of sea bottom

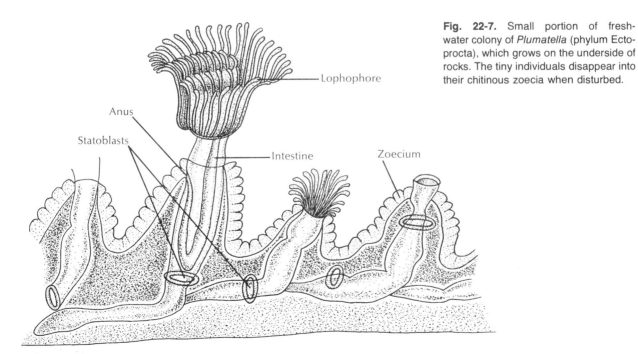

Fig. 22-7. Small portion of freshwater colony of *Plumatella* (phylum Ectoprocta), which grows on the underside of rocks. The tiny individuals disappear into their chitinous zoecia when disturbed.

Lophophore

Anus

Statoblasts

Intestine

Zoecium

There are both monoecious and dioecious species of Phoronida. Cleavage seems to be related to both the spiral and radial types. The free-swimming larva, which is called an actinotroch, metamorphoses into the sessile adult.

Phoronopsis californica is a large, orange-colored West Coast form approximately 30 cm long, and *Phoronis architecta* is a smaller Atlantic coast species (approximately 12 cm long) that has a very wide distribution.

Phylum Ectoprocta

Of the 4,000 species of ectoprocts, few are more than ½ mm long; all are aquatic; and all are colony builders. Each member of the colony lives in a tiny secreted chamber called a **zoecium** (Fig. 22-7). Each individual, or **zooid,** consists of a feeding polypide and a case-forming cystid. The **polypide** includes the lophophore, digestive tract, muscles, and nerve centers. The **cystid** is the trunk of the animal, highly modified, together with its secreted exoskeleton. The exoskeleton, or zoecium, may, according to the species be gelatinous, chitinous, or stiffened with calcium and possibly also impregnated with sand. Some colonies form limy encrustations on seaweed, shells, and rocks; others form fuzzy or shrubby growths. Some ectoprocts might easily be mistaken for hydroids but can be distinguished under a microscope by the fact that their tentacles are ciliated (Fig. 22-8). In some freshwater forms the individuals are borne on finely branching stolons that form delicate tracings on the underside of rocks or plants. Other freshwater ec-

Fig. 22-8. Ciliated lophophore of *Flustrella,* a marine ectoproct. (Photo by J. A. Cooke, American Museum of Natural History.)

toprocts are embedded in large masses of gelatinous material.

The polypide lives a type of jack-in-the-box existence, popping up to feed and then quickly withdrawing into its little chamber, which often has a tiny trapdoor that shuts to conceal its inhabitant. To extend the tentacular crown, certain muscles contract, which increases the hydraulic pressure within the body cavity and pushes the lophophore out. Other muscles can contract with great speed to withdraw the crown to safety.

The name "Ectoprocta" (ek'to-prok'ta), which comes from the Greek *ektos,* meaning outside, and *prōktos,* meaning anus, refers to the location of the anus *outside* of the lophophore. The ectoprocts have traditionally been called Bryozoa, or moss animals (Gr. *bryon,* moss, + *zoon,* animal), a term that originally included the Entoprocta. However, because the entoprocts are pseudocoelomates and have the anus located *within* the tentacular crown, they are now considered a separate phylum.

The lophophore ridge tends to be circular in marine ectoprocts (Fig. 22-8) and U-shaped in freshwater species (Fig. 22-7). When feeding, the animal extends the lophophore and spreads the tentacles out into a funnel. Cilia on the tentacles draw water into the funnel and out between the tentacles. Food particles trapped in the funnel are drawn into the mouth, both by the pumping action of the muscular pharynx and by the action of cilia in the pharynx. Undesirable particles can be rejected by reversing the ciliary action, by drawing the tentacles close together, or by retracting the whole lophophore into the zoecium. Digestion is extracellular in the ciliated, U-shaped digestive tract.

Respiratory, vascular, and excretory organs are absent. Gaseous exchange occurs through the body surface, and, since the ectoprocts are small, the coelomic fluid is adequate for internal transport. There is a ganglionic mass and a nerve ring around the pharynx, but no sense organs are present.

Both sexual and asexual reproduction occur. A colony starts with a single individual, which was sexually produced. It produces new zooids by asexual budding, and they in turn produce more buds. Thus the edges of an encrusting colony or the tips of a branching colony contain the youngest members. Polypides are short lived. Older polypides often contract and degenerate into minute **brown bodies** believed to be accumulations of excretory products. Later new buds

may occupy the old chambers. The brown bodies may remain passive or may be taken up and eliminated by the new digestive tract—an unusual kind of storage excretion.

Freshwater ectoprocts reproduce asexually by producing internal buds known as **statoblasts** (Fig. 22-7), which are released when the zooids die in late autumn and which in the spring start new colonies.

Most colonies are made up of feeding individuals, but polymorphism also occurs. One type of modified zooid resembles a bird beak that snaps at small invading organisms that might foul a colony. Another type has a long bristle that sweeps away foreign particles.

Phylum Brachiopoda

The Brachiopoda (bra-ki-op'o-da), or lamp shells, are an ancient group. Compared with the approximately 300 species now living, some 30,000 fossil species, which once flourished in the Paleozoic and Mesozoic seas, have been described. Modern forms have changed little from the early ones. *Lingula* (Fig. 22-9, *A*) is probably the most ancient of these "living fossils," having existed virtually unchanged since Ordovician times.

Brachiopods are all attached, bottom-dwelling, marine forms that mostly prefer shallow water. Their name means "arm-footed" (Gr. *brachiōn,* arm, + *pous, podos,* foot); the arms, of course, refer to the

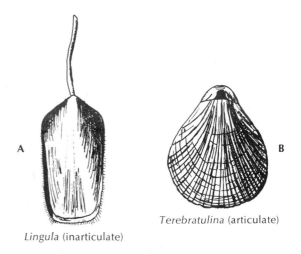

A

Lingula (inarticulate)

Terebratulina (articulate)

B

Fig. 22-9. Phylum Brachiopoda. **A,** *Lingula* has valves held together by muscles only (inarticulate). **B,** Articulate valves of *Terebratulina* have tooth-and-groove articulation.

arms of the **lophophore.** Externally brachiopods resemble the bivalved molluscs in having two shell valves and were, in fact, classed with the molluscs until the middle of the nineteenth century. Brachiopods, however, have **dorsal** and **ventral valves** instead of right and left lateral valves as do the molluscs and, unlike the molluscs, most of them are attached to a substrate by means of a fleshy stalk called a **pedicel** (or pedicle). Muscles open and close the valves and provide movement for the stalk and tentacles.

In most brachiopods the ventral (pedicel) valve is slightly larger than the dorsal (brachial) valve, and one

end projects in the form of a short pointed beak that is perforated where the fleshy stalk passes through (Fig. 22-9, *B*). In many the shape of the pedicel valve is that of the classic oil lamp of Greek and Roman times, so that the brachiopods came to be known as the ''lamp shells.''

There are two classes of brachiopods based upon shell structure. The shell valves of Articulata are connected by a hinge; those of Inarticulata lack the hinge and are held together by muscles only (Fig. 22-9).

The body is enclosed between the two valves but

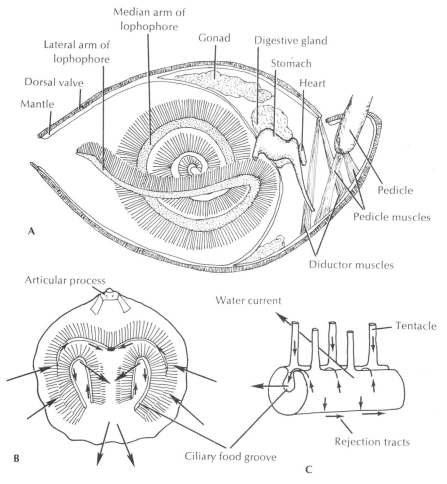

Fig. 22-10. Phylum Brachiopoda. **A,** An articulate brachiopod (longitudinal section). **B** and **C,** Feeding and respiratory currents. **B,** Large arrows show water flow over lophophore; small arrows indicate food movement toward mouth in ciliated food groove. **C,** Portion of lophophore arm showing current direction in feeding tracts and rejection tracts. (**B** and **C** modified from Russell-Hunter, W. D. 1969. A biology of higher invertebrates. New York, Macmillan, Inc.)

occupies only the posterior part of the space, the lophophore taking up the rest (Fig. 22-10, *A*). The valves are lined with a **mantle,** which is an outgrowth of the body wall and which secretes the valves. The horseshoe-shaped lophophore bears long ciliated tentacles used in ciliary feeding and for respiration (Fig. 22-10, *B* and *C*). Most brachiopods lack an anus.

Nephridia, a nerve ring, and a contractile heart and blood vessels are present. Brachiopods are dioecious, and development includes the metamorphosis of a free-swimming, trochophorelike larva.

SELECTED REFERENCES

American Society of Zoology. 1977. Biology of lophophorates. Amer. Zool. **17**(1):3-150. *A collection of 13 papers.*

Barnes, R. D. 1974. Invertebrate zoology, ed. 3. Philadelphia, W. B. Saunders Co.

Barrington, E. J. W. 1967. Invertebrate srructure and function. Boston, Houghton Mifflin Co.

Chapman, G. 1958. The hydraulic skeleton in the invertebrates. Biol. Rev. **33**:338-352.

Edmondson, W. T. (ed.). 1959. Ward and Whipple's fresh-water biology, ed. 2. New York, John Wiley & Sons, Inc.

Fisher, W. K. 1946. Echiuroid worms of the north Pacific Ocean. Proc. U.S. Nat. Mus. **96**:215-292.

Fisher, W. K. 1952. The sipunculid worms of California and Baja California. Proc. U.S. Nat. Mus. **102**:371-450.

Gardiner, M. S. 1972. The biology of invertebrates. New York, McGraw-Hill Book Co.

Giese, A. C. 1974. Reproduction of marine invertebrates. Vol. 2, Entoprocts and lesser coelomates. Vol. 3, Annelids and echiurans. New York, Academic Press, Inc.

Gosner, K. L. 1971. Guide to identification of marine and estuarine invertebrates: Cape Hatteras to the Bay of Fundy. New York, Interscience Div., John Wiley & Sons, Inc.

Hickman, C. P. 1973. Biology of the invertebrates, ed. 2. St. Louis, The C. V. Mosby Co.

Hill, H. R. 1960. Pentastomida. In McGraw-Hill encyclopedia of science and technology, vol. 9. New York, McGraw-Hill Book Co., pp. 623-624.

Hyman, J. H. 1951. The invertebrates, vol. 3. Acanthocephala, Aschelminthes, and Entoprocta. New York, McGraw-Hill Book Co. *Priapulida is treated under Aschelminthes.*

Hyman, L. H. 1959. The invertebrates: smaller coelomate groups, vol. 5. New York, McGraw-Hill Book Co. *Phoronida, Ectoprocta, Brachiopoda, and Sipunculida are covered in this volume.*

Kaestner, A. 1967-1968. Invertebrate zoology, vols. 1 and 2. New York, Interscience Div., John Wiley & Sons, Inc.

Light, S. F., et al. 1967. Intertidal invertebrates of the central California coast. Berkeley, University of California Press.

MacGinitie, G. E., and N. MacGinitie. 1967. Natural history of marine animals, ed. 2. New York, McGraw-Hill Book Co.

Marcus, E. 1959. Tardigrada. *In* Edmondson, W. T., H. B. Ward, and G. C. Whipple (eds.). Freshwater biology, 2nd ed. New York, John Wiley & Sons, Inc. pp. 508-521. *Keys to freshwater species.*

Meglitsch, P. A. 1972. Invertebrate zoology, 2nd ed. New York, Oxford University Press.

Nelson, D. R. 1975. The hundred-year hibernation of the water bear. Nat. Hist. **84**(7):62-65 (Aug.-Sept.).

Pennak, R. W. 1953. Fresh-water invertebrates of the United States. New York, The Ronald Press Co.

Rudwick, M. J. S. 1970. Living and fossil brachiopods. London, Hutchinson University Library, Hutchinson and Co.

Russell-Hunter, W. D. 1969. A biology of higher invertebrates. New York, Macmillan, Inc.

Ryland, J. S. 1970. Bryozoans. London, Hutchinson University Library, Hutchinson & Co.

Stephen, A. C., and S. J. Edmonds. 1972. The phyla Sipuncula and Echiura. London, British Museum.

SELECTED SCIENTIFIC AMERICAN ARTICLE

Crowe, J. H., and A. F. Cooper, Jr. 1971. Cryptobiosis. **225**(6):30-
36 (Dec.).

A group of sea stars, *Asterias rubens,* has congregated on a mussel bed to feast on mussels. (Photo by T. Lundälv, Kristinebergs Zoological Station, Sweden.)

23 ECHINODERMS AND LESSER DEUTEROSTOMES

DEUTEROSTOMES

The echinoderms, along with the chordates and three smaller phyla—Hemichordata (acorn worms and pterobranchs), Chaetognatha (arrowworms), and Pogonophora (beardworms)—belong to the **Deuterostomia division** of bilateral animals. As previously noted, in the protostomes (annelids, molluscs, arthropods, and some minor phyla) the mouth arises from the blastopore. In the deuterostomes the mouth is formed from a new opening some distance from the blastopore, whereas the anus arises from or near the blastopore. This difference may seem a peculiar basis for separating large groups of animals, but it actually underlies a very fundamental difference in embryonic development in these two great evolutionary branches of animals. As we have seen, embryonic development is one of the best clues to invertebrate evolution, apparently representing a telescoped evolutionary history.

Other features that the Deuterostomia have in common are radial and indeterminate cleavage and an enterocoelous method of coelom formation in which the coelomic sacs are pinched off from the enteron. The protostomes, on the other hand, generally have spiral and determinate cleavage and schizocoelous formation of the coelom (see Fig. 14-4, p. 284). Such embryologic similarities have led most biologists to believe that the deuterostome phyla are more closely related to each other than to other animal phyla.

ECHINODERMS—PHYLUM ECHINODERMATA

The Echinodermata (e-ki′no-der′ma-ta) (Gr. *echinos,* sea urchin, hedgehog, + *derma,* skin) are marine animals that include the sea stars, brittle stars, sea urchins, sea cucumbers, sea lilies, and feather stars. They are a bizarre group, sharply distinguished from all other members of the animal kingdom. Their name is derived from the projecting spines and tubercles that give them a rough or spiny appearance.

The echinoderms have at least three uniquely distinctive characteristics. One is their **pentamerous radial symmetry**—a tendency for the body to be arranged around a central axis in five parts. This symmetry, however, is derived secondarily, for, unlike the true radial animals, they begin life as bilaterally symmetric larvae that metamorphose into the adult arrangement.

Another unique feature is their **water-vascular,** or **ambulacral, system** (Fig. 23-1), a system of coelomic canals that send out hollow protrusions called **tube feet,** which function variously in locomotion, feeding, respiratory exchange, sensory perception, or a combination of these functions.

A third feature is the presence of an **internal skeleton of mesodermal origin,** composed of calcareous plates that may be rigidly fused or articulated for flexibility or may be reduced to scattered, minute spicules.

Ecology and distribution

Echinoderms are truly marine animals; only a few species live in brackish water and none in fresh water. Largely bottom dwellers, they live in all oceans, in all climates, and at all depths. They are the most abundant group of animals on the deep-sea floor, making up approximately 90% of the abyssal biomass. Abyssal sea cucumbers range up to 0.5 m long. There are no parasitic echinoderms, and very few are commensal on other animals.

Echinoderms tend to gather in aggregations of huge numbers. Some favorable spots may be literally covered with sea urchins, others with brittle stars or sea stars. Many echinoderms are adapted to rocky bottoms, others to living in sand or mud.

Echinoderms are slow-moving animals, most of which use tube feet (podia) for attachment, for creeping or walking, or for digging or burrowing. The brittle stars are the most active, using their arms (rays) rather than tube feet for locomotion. Some of the crinoids are stalked and sessile; others can use their arms for swimming.

Characteristics

1. Typically **pentamerous,** body usually with five radiating areas **(ambulacra)** containing podia **(tube feet)**
2. **Endoskeleton** of calcareous ossicles with spines; covered by epidermis, usually ciliated; **pedicellariae** (in some)
3. A unique coelomic hydraulic system, the **water-vascular system,** equipped with podia that function in locomotion, respiration, feeding, and sensory perception
4. Extensive coelom, giving rise during development to the hemal and water-vascular systems
5. Respiration by **dermal branchiae** (skin gills), by tube feet, by **respiratory tree** (holothuroids), or by **bursae** (ophiuroids)
6. Reduced circulatory system—a **hemal** or lacunar system enclosed in coelomic channels; **excretory organs absent**
7. Digestive system usually complete, intestine and anus absent in ophiuroids
8. Nervous system diffuse and uncentralized; usually two or three nerve rings around the digestive tract, each with radial nerves and nerve nets; sensory system poorly developed

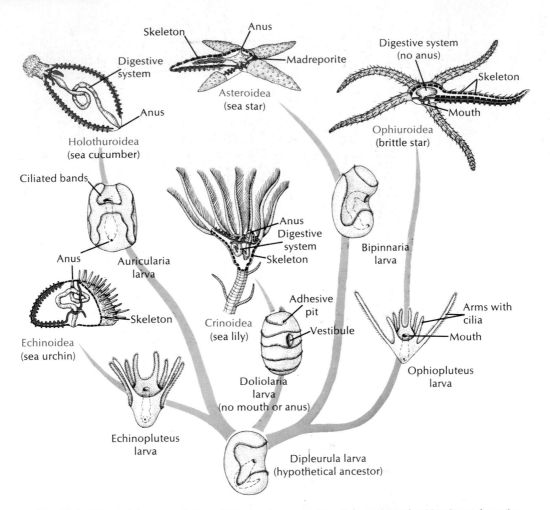

Fig. 23-1. Schematic representation of the development of the five classes of echinoderms from the hypothetic dipleurula larva, one of the theories of their origin. The water-vascular system is shown in red.

9. Usually dioecious, with simple gonads and ducts; external fertilization
10. Development with holoblastic radial cleavage; several types of bilateral free-swimming larvae, which usually metamorphose to radially or biradially symmetric adults

Classification

There are approximately 6,000 living and 20,000 extinct or fossil species. Five classes of existing echinoderms are usually recognized. Some authorities combine Asteroidea and Ophiuroidea as subclasses under one class, called **Stelleroidea,** since both groups are unattached and have a body composed of a flattened central disc and radially arranged arms.

Class Asteroidea (as′ter-oi′de-a)—**sea stars.** Star-shaped echinoderms, with the arms not sharply marked off from the central disc; tube feet with suckers.

Class Ophiuroidea (off′-i-u-roi′de-a)—**brittle stars, basket stars.** Star shaped, with the arms sharply marked off from the central disc; tube feet without suckers.

Class Echinoidea (ek′i-noi′de-a)—**sea urchins, sea biscuits, sand dollars.** More or less globular echinoderms with no arms; compact skeleton or test; movable spines; tube feet with suckers.

Fig. 23-2. Some sea stars (class Asteroidea). **A,** *Dermasterias,* the leather star, has a smooth purplish skin with red markings. **B,** *Hippasteria* has a red, spiny surface with longer spines around the margin; it is a North Atlantic form. **C,** *Pycnopodia,* the sunflower star, has a soft skin of pink or purple, and as many as 24 arms, depending upon age. **D,** *Pisaster,* a large purple star, ranges up to 35 cm in diameter. **A, C,** and **D** are Pacific Coast genera. (**A, C,** and **D,** Photos by C. P. Hickman, Jr.; **B** courtesy Vancouver Public Aquarium, Vancouver, British Columbia.)

Class Holothuroidea (hol′o-thu-roi′de-a)—**sea cucumbers.** Cucumber-shaped echinoderms with no arms; spines absent; tube feet with suckers.

Class Crinoidea (kri-noi′de-a)—**sea lilies, feather stars.** Body attached during part or all of life by an aboral stalk of dermal ossicles; five arms branching at base and bearing pinnules; tentacle-like tube feet for food collecting.

Echinoderm development

The early development of the larval form is similar in all echinoderms. The egg is usually fertilized in seawater or in brood pouches. Cleavage is total and indeterminate. The early larva, called a **dipleurula,** is bilateral, lacks a calcareous skeleton, bears bands of locomotor cilia, and is free swimming. The coelom consists of three pairs of pouches budded off from the archenteron. Later larval stages in some echinoderms involve the development of slender projections (arms), and the larvae of different classes are distinguished by the nature and location of these arms, if present (Fig. 23-1). The metamorphosis of the bilateral larva into the adult is a complicated process involving development of the endoskeleton, torsion, development of one of the coelomic pouches into the water-vascular system, and the assumption of radial symmetry.

Sea stars—class Asteroidea

Although the crinoids (sea lilies and feather stars) are the most primitive of the echinoderms, we take up the asteroids (starfishes, sea stars) first because they are more familiar and because they demonstrate so well the basic features of echinoderm structure. The name "Asteroidea" (as-ter-oi′de-a) means star shaped (Gr. *aster,* star, + *eidos,* form).

Sea stars are common along the shorelines where sometimes large numbers of them congregate on the rocks (p. 462). Sometimes they cling so tenaciously that they are difficult to dislodge without tearing off their tube feet. They also live on muddy or sandy bottoms and among coral reefs. They are found in all oceans, even at great depths and at great distances from the shores. Many are brightly hued in a wide range of colors.

Form and function

The body of asteroids is flattened and flexible, with the oral side down. Most stars have five **arms,** or **rays,** arranged around a **central disc,** but some have many more rays (Fig. 23-2). In some species the rays are very long and slender, and in others they are so short that the body appears pentagonal.

The aboral surface is usually rough and spiny (Fig. 23-3, *A*), although in some species, particularly burrowing forms, it is relatively smooth (Fig. 23-2, *A*) because the specialized spines are short and flattened. The anus and the **madreporite** (opening of the water-vascular system to the outside) are on the aboral side. At the tip of each ray is a tiny sensory tentacle and a pigmented eye spot.

An **ambulacral groove** extends out from the mouth

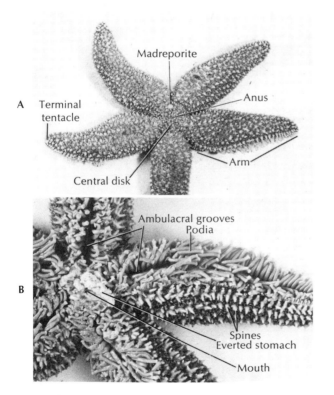

A

B

Fig. 23-3. *Asterias.* **A,** Aboral view. **B,** Oral view. (Photos by F. M. Hickman.)

Madreporite

Terminal tentacle

Anus

Central disk

Arm

Ambulacral grooves
Podia

Spines
Everted stomach

Mouth

along the oral side of each arm. Each groove contains two or four rows of small tubular projections called **tube feet,** or **podia** (Fig. 23-3, *B*). These are a part of the **water-vascular system** and are locomotor organs. Movable **spines** along the margins of the groove guard the tube feet and can close over and cover them.

In two orders of sea stars the body wall bears minute pincerlike appendages, called **pedicellariae,** that keep the body surface free from debris and from small animals or larvae that settle on the body surface (Fig. 23-4). The jaws of the pedicellariae are moved by tiny muscles.

Body wall and endoskeleton. The body wall is covered with **ciliated epidermis,** beneath which is a subepidermal plexus of nerve cells. Muscles in the body wall move the arms.

Beneath the skin is the **endoskeleton** of small calcareous plates, called ossicles, arranged in a lattice network and held together by connective tissue. The spines that protrude through the epidermis are either immovable projections of these ossicles or, in some forms, are separate pieces moved by muscles but resting on the ossicles. The body wall is lined with peritoneum.

Between the ossicles slender extensions of the body wall and coelom form the thin-walled **skin gills (dermal branchiae)** where gaseous exchange occurs (Fig. 23-4).

Water-vascular system. The water-vascular system, which is found only in echinoderms, is a hydraulic system of water-filled canals and specialized tube feet. Water entering a stony sieve (madreporite plate) on the aboral side (Fig. 23-3, *A*) passes through a ring canal around the mouth to a **radial canal** in each ray (Fig. 23-4). Fine lateral canals connect with a series of **tube feet.** Each tube foot is a hollow muscular cylinder. The inner bulblike end **(ampulla)** lies within the coelom; the muscular tube foot extends to the exterior between ossicles of the ambulacral groove. The podia of many species, such as *Asterias,* are tipped with suckers.

Sea stars use the hydraulic system for locomotion and food capture. When an ampulla contracts, a valve prevents back flow of fluid into the radial canal, thus forcing it into the tube foot. The elastic tube foot lengthens and can twist about by the muscles in the wall. When a suckered podium touches a substratum, it adheres by mucous secretions and by the vacuum-cup action of its suckers (Fig. 23-5). The podium shortens when its longitudinal muscles contract, forcing the water back into the ampulla.

The coordinated effort of many tube feet permits the animal to move forward, to climb vertical surfaces, or open the shell of a mollusc. A single podium is estimated to exert a pull of 25 to 30 g. When inverted, the sea star can right itself by twisting its rays until some of its tube feet attach to the substratum, and then it slowly twists and rolls over.

In many species, such as *Luidia* and *Astropecten,* the tube feet are not suckered but have rounded tips that are used for walking on mud or sand, for pushing sand aside in burrowing, or for propelling water currents through the mud or sand in which they have burrowed.

In addition to locomotion, tube feet are used for capturing and holding prey (Fig. 23-5); their thin walls are thought to aid in oxygen–carbon dioxide exchange, and their epidermis is generously supplied with sensory cells.

Nutrition. Asteroids eat bivalves, snails, crustaceans, polychaetes, other echinoderms, and even fishes or dead animals (Fig. 23-5). Some have restricted diets. For example, our Pacific coast sunstar *Solaster stimpsoni* lives on sea cucumbers, whereas *Solaster dawsoni* eats *Solaster stimpsoni*. Some sea stars feed on coral polyps. Recent depredations by the crown of thorns sea star *Acanthaster planci* upon Pacific coral reefs has caused concern. The prevalence of this sea star may have been caused, in part, by shell collectors who have taken so many of the Pacific giant tritons, which are natural predators of the crown of thorns.

Bivalves are a favorite food for many asteroids, which open the shells by attaching their tube feet to both sides of the shell and pulling until the bivalve's adductor muscles tire and allow the valves to gape slightly. Then the sea star everts its stomach into the narrow opening and secretes digestive juices to partly digest the soft parts, which are then sucked up into the stomach. Small molluscs are swallowed whole and the shells regurgitated later.

Some sea stars, such as *Henricia,* are suspension feeders that trap minute animals or detritus in mucus on the ciliated skin and move it by ciliary action to the mouth. *Ctenodiscus* and *Thorocaster* are deep-sea stars that swallow mud, drawn in by ciliary action, to obtain the protozoans, diatoms, and other nutrients in it.

A two-part stomach is located in the central disc. In some species the large, lower part can be everted. The smaller upper stomach connects with digestive ceca

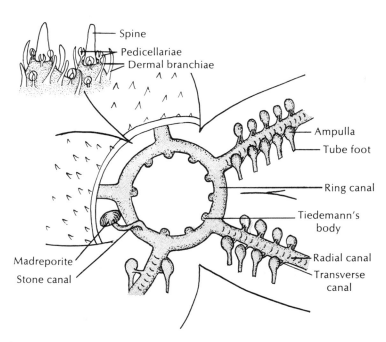

Fig. 23-4. Schematic view of water-vascular system. Upper left, detail of exterior surface showing spines, skin gills (dermal branchiae), and pedicellariae.

Fig. 23-5. Sea star devours a fish. Note use of tube feet to hold the prey. (Courtesy American Museum of Natural History.)

located in the arms. Digestion is largely extracellular, occurring in the digestive ceca.

Internal transport, gas exchange, and excretion. Coelomic fluid circulated by cilia on the peritoneum provides the principal means of internal transport. The blood-vascular, or **hemal, system** in echinoderms consists of a hemal ring around the mouth that connects with a radial sinus in each arm and an ascending axial sinus that passes through a spongy axial gland surrounding the stone canal. The functions of the hemal system and axial gland are poorly known.

The skin gills and tube feet are the main surfaces for gas exchange and for the excretion of nitrogenous wastes by diffusion. Coelomocytes in the coelom and hemal fluid engulf wastes, migrate to the tips of the skin gills or tube feet, and are expelled. Body fluids

are of the same salinity as the surrounding seawater. The inability of echinoderms to osmoregulate prevents them from living in fresh or brackish water.

Nervous and sensory system. The nervous system in echinoderms comprises three subsystems, each made up of a nerve ring and radial nerves placed at different levels in the disc and arms. An epidermal nerve plexus, or nerve net, connects the systems. Neurosecretory cells have been identified in some sea stars, the products of which are believed to affect color changes, water regulation, locomotor activity, and the shedding of gametes. Sense organs include ocelli at the arm tips and sensory cells scattered all over the epidermis.

Reproduction and regeneration. Sexes are separate in most sea stars. A pair of gonads lies in each interradial space, and fertilization is external. Egg

Fig. 23-6. Ophiuroids have slender arms sharply marked off from the central disc. They are quite agile. **A,** Brittle star, *Ophiura*. **B,** Basket star, *Gorgonocephalus*. (**A,** Photo by B. Tallmark, Uppsala University, Sweden; **B** courtesy Vancouver Public Aquarium, Vancouver, British Columbia.)

maturation and spawning are regulated by two neurohormones produced by neurosecretory cells in the radial nerves. One is a shedding substance produced throughout the year; the other is an inhibitor termed "shedhibin," produced in highest concentration just before the period of normal shedding of gametes.

Sea stars can regenerate lost parts. Many can also voluntarily discard an injured arm and regenerate a new one.

Brittle stars and basket stars—class Ophiuroidea

Brittle stars, like sea stars, have a central disc and five or more arms, but the arms are typically long and slender and sharply marked off from the disc (Fig. 23-6, *A*). In some species the rays are very delicate and easily broken. Basket stars are characterized by complex branching of their rays (Fig. 23-6, *B*). Unlike other echinoderms, ophiuroids are agile creatures; their arms are muscular and move with a writhing, serpentlike motion that probably suggested the class name "Ophiuroidea" (off'i-u-roi'de-a) (Gr. *ophis,* snake, + *oura,* tail, + *eidos,* form). Some brittle stars have variegated color patterns.

Brittle stars tend to be smaller than sea stars. Few reach more than 10 to 20 mm in disc diameter or around 100 mm armspread. Small forms may measure only 3 to 5 mm across. The largest ophiuroids are the basket stars; *Gorgonocephalus* (Fig. 23-6, *B*) may have an armspread of 350 mm.

There are more species of brittle stars than of any other class of echinoderms. They are widely distributed in both deep and shallow sea bottoms. Although they are often found in huge aggregations, it is doubtful that they are truly social animals. Their gregariousness is probably linked with ecologic factors.

Form and function

Brittle stars have no pedicellariae, no ambulacral grooves, no skin gills, and no anus. The arms are many jointed and are too slender to contain any visceral organs. Each arm bears two series of tube feet, without suckers, which extend through pores on either side of the vertebrae-like joints of the internal skeleton. The arms and disc are covered by plates. The skin is leathery, and cilia are usually lacking.

The tube feet, although small and without suckers, help the arms in locomotion to grip the substrate by friction or by secretions. They may also be used to walk on or to burrow and are important in passing food particles along to the mouth.

Nutrition. Brittle stars feed on pelagic copepods, tunicates, worms, crustaceans, small clams, and other organisms. Some species attach themselves to soft corals and wave their free arms to catch their prey. Some lie on the sea floor, oral side up, and sweep their arms through the water to collect small organisms; they then draw the arms across the mouth. Some are detritus feeders or scavengers, pushing up mounds of mud with one arm while searching for organic material with the others.

The mouth is equipped with five movable plates bearing teeth (modified spines) that serve as jaws. The saclike stomach lies in the central disc; indigestible material is regurgitated.

Gas exchange. Most ophiuroids have five pairs of invaginated pouches, called **bursae,** opening by slits on the oral disc, through which respiratory water is pumped for gaseous exchange. The tube feet also function in a respiratory capacity.

Reproduction. Most ophiuroids have separate sexes, but some are hermaphroditic, and some hermaphrodites are protandric—that is, the gonads are first male and later female. The larva is called an **ophiopluteus** (Fig. 23-1).

Sea urchins and sand dollars—class Echinoidea

The Echinoidea (ek-i-noi'de-a) are the sea urchins, sea biscuits, heart urchins, and sand dollars. They have a compact body enclosed in an **endoskeletal test** made up of **closely fitted plates.** They lack arms, but their tests show the typical pentamerous plan in the arrangement of the tube feet.

Sea urchins are almost globular and have medium to long **movable spines** (Fig. 23-7). A few have poison glands at the base of the spines. The ovoid heart urchins, the hemispheric sea biscuits, and the disc-shaped sand dollars have short, movable spines. The spines and tube feet are used in locomotion.

Echinoids have a wide distribution. They live on rocky or coral substrate, burrow into mud or sand, or live in beds of algae and grasses.

The **test** is composed of 10 double rows of skeletal plates, or ossicles, bearing tubercles upon which the spines are attached by muscles (Fig. 23-8). Five ambulacral rows have pores through which long tube

Fig. 23-7. The sea urchin, *Lytechinus.* Note the slender suckered tube feet. They often attach to bits of shell, seaweed, etc., for camouflage. Stalked pedicellariae can be seen between spines. (Photo by R. O. Hermes, Homestead, Fla.)

feet extend. Some tube feet bear suckers and are locomotor (Fig. 23-7); others are tapered and sensory in function. Heart urchins have several kinds of podia—some for burrowing, some branchial, and some for gathering food. There are several kinds of pedicellariae, the most common type being stalked and three jawed (Fig. 23-7).

The mouth is surrounded by five converging **teeth,** which are part of a complex chewing mechanism called **Aristotle's lantern.** The teeth are used for scraping up food and for chewing. Most urchins are omnivorous grazers and scavengers, whose main food is algae. Some burrowing urchins leave their burrows at night to feed. Sand dollars and heart urchins are entirely deposit feeders, living on fine particulate matter carried to the mouth by mucus, cilia, and specialized tube feet. Sand dollars have very small Aristotle's lanterns and heart urchins have none at all.

Echinoids differ internally from other echinoderms in having a **siphon** that shuttles incoming water directly to the intestine, bypassing the stomach, and thus concentrating food and enzymes in the stomach.

Echinoids have few special sense organs, but podia, spines, and pedicellariae all have a sensory function.

Sexes are separate and the larval form is a **pluteus** (Fig. 23-1).

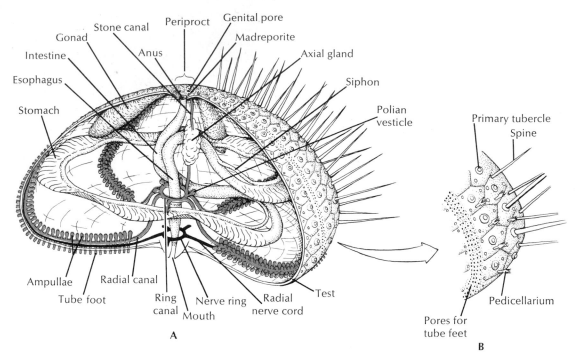

Fig. 23-8. A, Internal structure of the sea urchin; water-vascular system in red. **B,** Detail of portion of test.

Sea cucumbers—class Holothuroidea

The name "Holothuroidea" (hol-o-thu-roi′de-a) comes from the Greek *holothourion,* which means cucumber. Sea cucumbers are soft bodied, bottom-dwelling forms that bear a remarkable resemblance to the vegetable after which they are named. They appear to be more closely related to the echinoids than to the other echinoderms. Ranging in length from 3 cm to 1.5 m, they are found at all depths and in a variety of habitats, such as rock, mud, sand, and seaweeds. They are often found buried in mud or sand of low-tide pools, with their tentacles extending up into clearer water. Common genera are *Stichopus* (Fig. 23-9) and *Cucumaria* on the west coast and *Thyone* (Fig. 23-10) and *Cucumaria* along the east coast.

In sea cucumbers the oral-aboral axis is horizontal, resulting in bilateral symmetry. The bodies are elongated, with mouth and **retractile tentacles** at one end, anus at the other (Fig. 23-10). The muscular body wall is covered with cuticle and a tough, leathery skin in which are embedded scattered **microscopic ossicles.** In some genera, as in *Cucumaria,* the tube feet are arranged in five longitudinal bands, two dorsal and three ventral. However, in some, as in *Thyone,* they are scattered randomly, and in others, as in *Synapta,* they are reduced or even absent.

The coelomic cavity is large and filled with fluid containing coelomocytes. Since holothuroids lack a significant calcareous skeleton, the pressure of coelomic fluid plays an important role as a **hydrostatic skeleton.** Sea cucumbers are sluggish animals, moving partly by tube feet and partly by means of waves of contraction of their powerful muscles.

Sea cucumbers feed upon small organisms, which they usually entangle in the sticky mucus of their tentacles and suck into the mouth. Those with branched tentacles, such as *Cucumaria* and *Thyone,* are mainly suspension feeders that feed in open water or sweep the substrate with their tentacles. Those lacking branched tentacles are detritus feeders that either shovel sediment into the mouth with tentacles or burrow into mud or ooze with open mouths. The digestive tract ends in a widened portion, called the **cloaca,** which empties through the anus (Fig. 23-10).

Fig. 23-9. *Stichopus*, a West Coast sea cucumber.

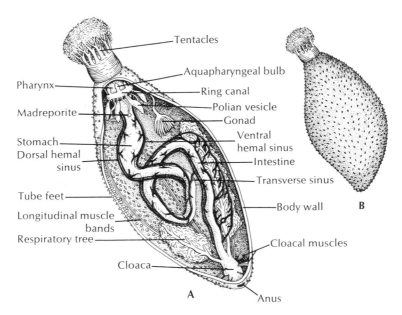

Fig. 23-10. Anatomy of the sea cucumber *Thyone*. **A,** Internal view; hemal system in red. **B,** External view.

Opening into the cloaca are a pair of extensively branched **respiratory trees,** which are evaginations of the cloaca (Fig. 23-10) and into which the muscular cloaca pumps water. Gaseous exchange occurs in the finer branches where the walls are very thin.

The water-vascular system, similar to others of the phylum, also extends into the oral tentacles and has one or more saclike **polian vesicles** hanging in the coelom from the ring canal and serving as expansion chambers for the system. Hemal and nervous systems are similar to others of the phylum. Some species have **statocysts** in addition to tactile, photosensitive, and chemosensitive cells.

When irritated, sea cucumbers may, by muscular contractions, cast out some of the viscera through the anus or through the body wall (autoevisceration) and then regencrate the lost parts. Some species can also divide the body by constriction and regenerate the missing portions—a method of asexual reproduction.

473

Some species brood their young in or on the body. The larval form is the **auricularia** (Fig. 23-1).

Sea lilies and feather stars—class Crinoidea

The crinoids are the most primitive of the echinoderms. This name comes from the Greek *krinon,* meaning lily. The fossil record shows that crinoids were amazingly abundant during the Mississippian period 350 million years ago. There are approximately 600 living and 5,000 extinct species. Mostly deepwater forms, crinoids differ from other echinoderms in being attached for all or part of their lives.

The beautifully colored feather stars (Fig. 23-11) have long many-branched arms and can swim slowly but gracefully by arm movements. Sea lilies are pale and are attached by stalks. Most stalked crinoids are from 15 to 30 cm long; feather stars average approximately 30 cm across.

The pentamerous body (the **crown**) of the sea lily is attached aborally to the **stalk** so that the oral side is up (Fig. 23-12). The base of the crown is the **calyx,** and the upper surface bears the mouth and anus. Attached to the calyx are five flexible **arms** that branch to form 10 arms, and in some species they branch many times. On each side of the arms are pinnules, arranged like barbs on a feather. The stalk appears jointed because its skeleton is a series of ossicles called columnals. Madreporites, spines, and pedicellariae are absent.

Crinoids are ciliary feeders that live on planktonic

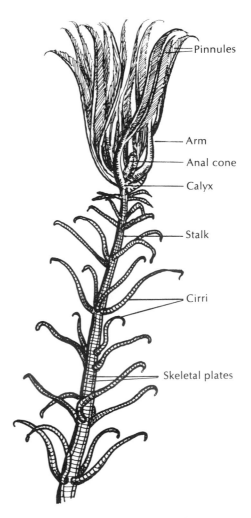

Fig. 23-12. A stalked crinoid with portion of stalk. Modern crinoid stalks rarely exceed 60 cm, but fossil forms were as much as 20 m long.

Fig. 23-11. Feather star (class Crinoidea). Feather stars can crawl by holding onto objects with the adhesive ends of the pinnules and pulling themselves along by contracting their arms. They can also swim by raising and lowering alternate sets of arms. (Courtesy Vancouver Public Aquarium, Vancouver, British Columbia.)

forms. Each arm bears a double row of pinnules whose ciliated grooves lead to the ciliated ambulacral grooves of the arms. In the ambulacral grooves are ciliated tube feet that set up a current from which food particles are trapped in mucus and moved by cilia to the mouth.

The **doliolaria larvae** (Fig. 23-1) are at first free swimming and then become attached and metamorphose to the adult form.

Echinoderm phylogeny

The echinoderms are a very ancient group of animals with a rich fossil record going back to Cambrian times. They probably also had a long Precambrain history. The crinoids are the most primitive of the five living classes. Echinoderms are considered to be the closest to the chordates of all the invertebrates because of their similar embryonic development and the mesodermic origin of their skeletal plates.

The classic concept of echinoderm origin has been that they came from a larval form called the **dipleurula,** a hypothetic bilateral ancestor without a skeleton and with a coelom of three paired sacs. From it came the various larval forms of existing echinoderms (Fig. 23-1). This theory was based on the dipleurula larva, the early larval form that is common to most of the living classes.

Another theory is the **pentactula** concept, which theorizes a bilateral ancestor with five hollow tentacles around the mouth that were extensions of the coelom. These tentacles became the radial canals of the water-vascular system.

A more recent suggestion (Nichols, 1967) is that echinoderms are derived from a **sipunculid-like stem** of the annelid line. This theory compares the hydraulic water-vascular system of the echinoderms to the coelomic compartment of the sipunculids, which hydraulically operates their tentacles. According to this theory, the sipunculids would be ancestral to both the echinoderms and the lophophorates.

In spite of the extensive fossil record of echinoderms, the evidence is not yet sufficient in many areas, and the origin of the echinoderms is still subject to much speculation.

Review of contributions of the Echinoderms

1. They have a **mesodermal endoskeleton** of plates, which may be considered the first indication of the endoskeleton so well developed among vertebrates.
2. They have contributed a pattern of embryonic development similar to that of the highest group, the chordates. This pattern includes (a) an anus derived from the embryonic blastopore, (b) a mouth formed from a stomodeum, which connects to the endodermal esophagus, (c) a mesoderm from evaginations of the archenteron (enterocoel), and (d) a nervous system in close contact with the ectoderm. (Most of the echinoderm characters are so out of line that few of them are copied by other phyla. Some of their unique features are the water-vascular system, tube feet, pedicellariae, dermal branchiae, and calcareous endoskeleton.)

LESSER DEUTEROSTOMES

The Chaetognatha, Pogonophora, and Hemichordata are usually referred to collectively as the "lesser (or minor) deuterostomes," but *only* in the sense that each of these groups has a relatively small number of species. These phyla are all widely distributed and contain some very common and well-known forms. The smaller phyla really deserve more attention than is usually given to them, for they contribute much to our understanding of evolutionary diversity and relationships.

Members of all three phyla apparently have bodies made up of three regions, a **protosome** containing a coelomic space, a **mesosome,** and a **metasome,** each containing a pair of coelomic spaces. This apparent unity of body organization suggests a common ancestry. Recently, however, there have been some doubts among certain specialists as to whether the Pogonophora belong to the deuterostome or the protostome division. Some authorities think that the pogonophorans are more closely related to the annelids than to the hemichordates.

Arrowworms—phylum Chaetognatha

Phylum Chaetognatha (ke-tog′na-tha) is a small phylum of some 65 species of small, transparent, torpedo-shaped animals. They are abundant in ocean plankton. Most of them average approximately 3 cm long, but some range to 10 cm in length. With the exception of one benthic genus, *Spadella,* the chaetognaths are all planktonic forms whose quick, darting movements have earned them the common name of arrowworms.

475

Form and function

The body is divided into a **head** (protosome), **trunk** (mesosome), and **postanal tail** (metasome), with one or two pairs of lateral fins and a tail fin (Fig. 23-13). Arrowworms use the fins as an aid in floating rather than in swimming. When the body begins to sink, a quick contraction of the trunk muscles sends it darting forward, after which it glides or floats until another muscular forward thrust is necessary. The body wall is suggestive of the aschelminths (with which the chaetognaths have sometimes been placed), for circular muscles are absent and the longitudinal muscles are arranged in four bands.

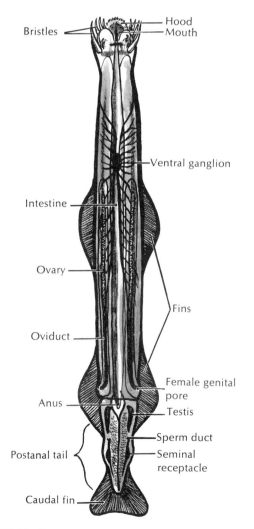

Fig. 23-13. General structure of the arrowworm, *Sagitta* (phylum Chaetognatha).

Labels: Bristles · Hood · Mouth · Ventral ganglion · Intestine · Ovary · Fins · Oviduct · Female genital pore · Anus · Testis · Postanal tail · Sperm duct · Seminal receptacle · Caudal fin

A large vestibule on the ventral side of the head leads to the **mouth.** A number of large curved **spines** hanging down from the sides of the head serve as jaws and give these ''bristle-jawed'' animals their phylum name (Gr. *chaite,* hair, + *gnathos,* jaw). The spines together with rows of smaller spines (teeth) aid in the capture of prey. A peculiar fold of the neck forms a **hood** that can be drawn down over the head and spines. Chaetognaths are carnivores that feed on other planktonic forms, especially copepods. To capture the prey, the animal darts forward with hood retracted and spines spread apart and then pounces with a quick snap of the spines. *Spadella,* the benthic form, is less active. It attaches to the substrate and awaits its prey. Arrowworms have a complete, though simple, digestive tract and extracellular digestion.

There are no respiratory or excretory organs, and the coelomic fluid acts as the circulatory medium. A nerve ring of fused ganglia around the pharynx is connected to a large ventral ganglion in the trunk. Sensory organs include a pair of dorsal eyes, sensory bristles, and a ciliary loop over the neck that is thought to be rheoreceptive and possible chemoreceptive.

Arrowworms are hermaphroditic, with either cross- or self-fertilization. The eggs of *Sagitta* are jelly coated and planktonic, but eggs of other arrowworms may be attached to the parent and carried about for a time. Larvae develop directly.

Although the embryology seems in general to be deuterostome in nature, there are some peculiarities. The coelom resembles a pseudocoel because a peritoneum is lacking, but the coelom is compartmented as in other deuterostomes. Although there are resemblances to both deuterostomes and pseudocoelomates, no specific relationship to any particular phyla is yet apparent.

Beardworms—phylum Pogonophora

The Pogonophora (po-go-nof'e-ra) were unknown before the twentieth century; the first specimens were collected from deep-sea dredgings in 1900 off the coast of Indonesia. Since then more than 80 species have been described, and they seem to be fairly widespread in most seas, especially along the continental slopes. The beardworms live almost exclusively in the bottom ooze of the ocean floor at depths of 100 m or more. They are long, extremely slender, wormlike animals ranging from 5.5 to 85 cm in length and only 0.1 to 2 mm in diameter. They live in secreted chitinous tubes,

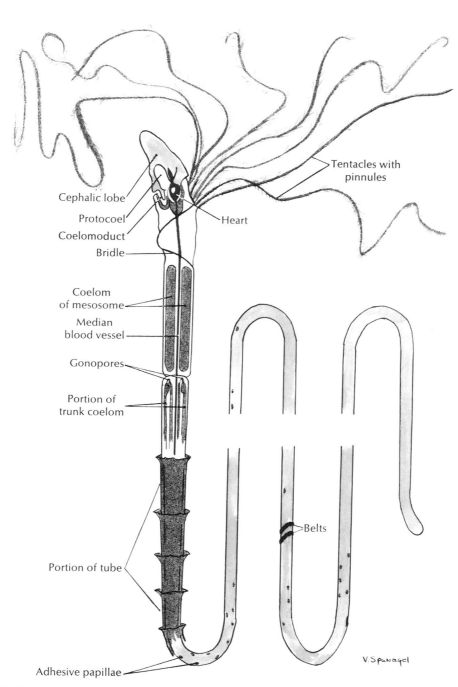

Tentacles with pinnules

Cephalic lobe

Protocoel

Coelomoduct

Heart

Bridle

Coelom of mesosome

Median blood vessel

Gonopores

Portion of trunk coelom

Belts

Portion of tube

V. Spanagel

Adhesive papillae

Fig. 23-14. General structure of a beardworm (phylum Pogonophora), showing principal external and internal features. Portions of trunk are omitted, also the segmented tail end. (Adapted from Ivanov, 1955, and others.)

much longer than the worms themselves. Their name means beard-bearing (Gr. *pogon*, beard, + *phora*, pl. of bearing), and the "beard" refers to their long ciliated tentacles.

Form and function

The body of a beardworm is made up of a short anterior part, a long trunk, and a short segmented tail section (opisthosoma) (Fig. 23-14). The forepart consists of a **cephalic lobe** and a glandular section. At the base of the cephalic lobe grow the long **tentacles.** Depending upon the species and age of individuals, there may be anywhere from one spiral tentacle to more than 250. The tentacles are hollow extensions of the coelom and are "fringed" with minute pinnules. They may be fused together at the base to form a cylinder. The long trunk bears a number of adhesive papillae and, approximately midway down, a pair of girdles (belts) with setae. The tail end, which is fragile and easily broken off, is clearly segmented and supplied with setae. The arrangement of the papillae on the trunk is also suggestive of segmentation. The papillae and setae probably aid in anchoring the animal in its tube.

Beardworms are unique in being the only non-parasitic animals entirely lacking a digestive tract. There is *no sign of a mouth, intestine, or anus* at any time in their life history. It has been suggested that the animal traps food (organic detritus) on the pinnules of the tentacles, then withdraws into the tube, secretes enzymes onto the food for external digestion, and finally absorbs the digested nutrients through the thin walls of the pinnules or tentacles. Another theory is that the beardworms are saprobic, merely absorbing from the environment nutrients already broken down by bacterial action.

There is a closed blood-vascular system, and each tentacle is supplied with two vessels. An epidermal nervous system with a brain is present, with nerves innervating the tentacles. Some forms have groups of photoreceptor cells in the anterior end that resemble those in leeches and oligochaetes. Sexes are separate. Some species brood their eggs in their tubes.

Since 1970 there has been a considerable difference of opinion among authorities as to whether beardworms belong with the deuterostomes or with the protostomes. Some maintain that the enterocoelic coelom formation of the larva places them with the deuterostomes. Others point out that the segmented

terminal part of the body with its setae, as well as the structure of the photoreceptor cells of some pogonophorans, suggest a close relationship with the annelids. It may be some time before the true phylogenetic position of this phylum is known.

Acorn worms and pterobranchs—phylum Hemichordata

The hemichordates are wormlike, marine animals that live in bottom mud or among rocks or plant material. Some live in secreted tubes, and most are sedentary or sessile. Acorn worms are solitary, but some pterobranchs are social and some live in colonies. Their distribution is fairly worldwide.

Until recently hemichordates were considered a subphylum of the chordates because they have pharyngeal gill slits and a diverticulum (stomochord) from the roof of the mouth that was assumed to be a short notochord. Their phylum name, meaning half-cord (Gr. *hemi,* half, + *chorda,* cord), was based upon this assumption. However, it is now known that this structure is neither analogous nor homologous to the chordate notochord, so that the hemichordates are now classed as a separate phylum.

Hemichordate phylogeny

There is strong embryologic evidence indicating a relationship between the hemichordates and both the echinoderms and the chordates. The early embryonic stages and the coelom formation are similar to those of echinoderms, and the free-swimming larva, called a **tornaria,** is very much like the bipinnaria larva of the echinoderms. An affinity with the chordates is indicated by the fact that they are the only two groups that possess pharyngeal gill slits. In addition, in some hemichordates the dorsal nerve is hollow, reminiscent of the dorsal hollow nerve cord of the chordates.

The pterobranchs are considered more primitive than the acorn worms, and some authorities believe that the common ancestor of the echinoderms and hemichordates might have been somewhat similar to the pterobranchs.

Acorn worms—class Enteropneusta

Acorn worms are long, slimy, pinkish worms, very fragile, and very sluggish. They range in length from a few millimeters to 2.5 m and in breadth from 0.3 to 20 cm. The body is composed of a **proboscis,** a short

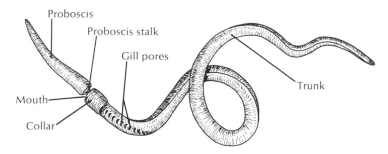

Fig. 23-15. External lateral view of the acorn worm *Saccoglossus* (phylum Hemichordata).

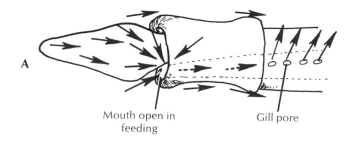

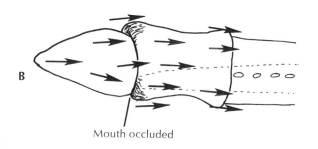

Fig. 23-16. Food currents of acorn worm (phylum Hemichordata), lateral view. **A,** Cilia on proboscis and collar direct food particles toward mouth *(red arrows)* and rejected particles to outside of collar *(black arrows).* Water leaves pharynx by gill pores. **B,** With mouth closed all particles are rejected and passed over collar. (Adapted from Russell-Hunter, W. D. 1969. A biology of the higher invertebrates. Macmillan, Inc.)

collar, and a long **trunk** (protosome, mesosome, and metasome) (Fig. 23-15).

Proboscis. The common name of the group comes from the cone-shaped proboscis capped posteriorly by the edge of the collar, giving it somewhat the appearance of an acorn. The proboscis is the active part of the animal. Burrowers use the proboscis to excavate by thrusting it into the mud or sand and then by lengthening and anchoring it by peristaltic movements. Cilia on the mucus-covered body then move the sand backward. Burrows of such forms as *Balanoglossus* and *Saccoglossus* are U-shaped, with front and back

openings. Movement within the burrow is also by peristalsis of the proboscis.

Nutrition. Most acorn worms are suspension feeders. Plankton and detritus become trapped in mucus on the ciliated proboscis and are directed by ciliary currents to the large mouth on the underside (Fig. 23-16, *A*). The passage of food is aided by water currents flowing into the mouth and out the gill pores. Large particles can be rejected by covering the mouth with the edge of the collar (Fig. 23-16, *B*). Many burrowing species merely swallow the sand or mud, digesting its organic contents and piling up spiral fecal

479

Fig. 23-17. Fecal casting made by acorn worm at back opening of its U-shaped burrow. These are familiar sights on tidal mud flats at low tide. (Photo by C. P. Hickman, Jr.)

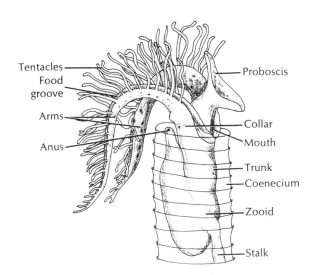

Tentacles
Food groove
Arms
Anus
Proboscis
Collar
Mouth
Trunk
Coenecium
Zooid
Stalk

Fig. 23-18. *Rhabdopleura,* a pterobranch hemichordate in its tube. Individuals live in branching tubes connected to each other by stolons. They protrude the tentacled arms for ciliary-mucus feeding.

mounds at the back door of their burrows. Such telltale castings are familiar sights on exposed mud flats at low tide (Fig. 23-17).

The mouth leads to a short buccal tube in the collar. The diverticulum from the dorsal part of the tube, extending into the proboscis and now called the **stomochord,** was once thought to be a "notochord." Posterior to the buccal tube is the large **pharynx** containing in its dorsal part the U-shaped **gill slits** that open to the outside by **gill pores** (Fig. 23-15). This is followed by an esophagus and long intestine where digestion and absorption occur.

Gas exchange. Gas exchange is assumed to occur through the body surface and in the branchial region of the pharynx. Although there are no gills attached to the gill slits, this region is highly vascular. It was for these apparently respiratory slits in the digestive tract that the class was given a name meaning gut-breathers, (Gr. *entero,* gut, + *pneusta,* from *pnein,* to breathe).

Circulation and excretion. Acorn worms have an open blood-vascular system with two contractile vessels and a system of sinus channels. A sinus network above the stomochord, called the **glomerulus,** is thought to be an excretory organ.

Nervous and sensory systems. There is a subepidermal plexus, thickenings of which form dorsal and ventral nerve cords. The dorsal cord, which continues into the collar, contains giant nerve cells. Neurosensory cells in the epidermis and a ciliary organ near the mouth (probably chemoreceptive) comprise the chief sensory receptors.

Reproduction and development. Sexes are separate, fertilization is external, and some species have a ciliated **tornaria** larval stage similar to the echinoderm bipinnaria larva.

Pterobranchs—class Pterobranchia

The pterobranchs are tiny ectoproct-like colonial or social animals. Individuals are usually 1 to 7 mm long and are either attached to a stalk or connected to each other by a stolon. They are deep-water bottom dwellers, rarely seen since they are collected only by dredging. Two of the three known genera are tube dwellers. Zooids of *Cephalodiscus* live independently in gelatinous tubes; those of *Rhabdopleura* are connected by a common stolon and enclosed in branching tubes (Fig. 23-18). *Atubaria,* a little-known genus, lacks tubes, but individuals are attached by stalks to colonial hydroids.

The zooid is composed of a shield-shaped proboscis, a collar, and a trunk. The arms and tentacles arising from the collar give the pterobranchs their name, which means feather-gilled (Gr. *pteron,* feather, wing, + *branchia,* gills). One pair of arms in *Rhabdopleura* and five to nine pairs in *Cephalodiscus* bear many small ciliated tentacles. They function much as lophophores do in capturing minute organisms and in moving them to the mouth by cilia. Arms and tentacles are hollow and contain extensions of the mesocoel.

Organ systems are similar to those of the acorn worms. Pterobranchs reproduce sexually to start a new colony and then add new members asexually by budding.

SELECTED REFERENCES

Barnes, R. D. 1974. Invertebrate zoology, ed. 3. Philadelphia, W. B. Saunders Co.

Barrington, E. 1965. The biology of Hemichordata and Protochordata. San Francisco, W. H. Freeman & Co., Publishers.

Binyon, J. 1972. Physiology of echinoderms. New York, Pergamon Press, Inc.

Boolootian, R. A. (ed.). 1966. Physiology of Echinodermata. New York, Interscience Div., John Wiley & Sons, Inc.

Buchsbaum, R., and L. Milne. 1960. The lower animals, living invertebrates of the world. Garden City, N.Y., Doubleday & Co., Inc.

Clark, A. M. 1962. Starfishes and their relations. London, British Museum.

Coe, W. R. 1972. Starfishes, serpent stars, sea urchins and sea cucumbers of the Northeast. New York, Dover Publications, Inc. *(Paperback).*

Corning, W. C., J. A. Dyal, and A. O. D. Willows. 1975. Invertebrate learning, vol. 3, cephalopods and echinoderms. New York, Plenum Publishing Corp.

Edmondson, W. T. (ed.). 1966. Marine biology III. New York, The New York Academy of Sciences.

Fell, H. B. 1965. The early evolution of Echinozoa. Breviora, No. 219.

Gardiner, M. S. 1972. The biology of invertebrates. New York, McGraw-Hill Book Co.

Gosner, K. L. 1971. Guide to identification of marine and estuarine invertebrates: Cape Hatteras to the Bay of Fundy. New York, Interscience Div., John Wiley & Sons, Inc.

Harvey, E. B. 1956. The American *Arbacia* and other sea urchins. Princeton, N. J., Princeton University Press.

Hickman, C. P. 1973. Biology of the invertebrates. St. Louis, The C. V. Mosby Co.

Hyman, L. H. 1955. The invertebrates, vol. 4, Echinodermata. New York, McGraw-Hill Book Co.

Hyman, L. H. 1959. The invertebrates, vol. 5, Smaller coelomate groups. New York, McGraw-Hill Book Co.

Ivanov, A. V. 1963. Pogonophora (translated from Russian by D. B. Carlisle). New York, Consultants Bureau Enterprises, Inc.

Light, S. F., R. I. Smith, F. A. Pitelka, D. P. Abbott, and F. M. Weenser. 1967. Intertidal invertebrates of the central California coast. Berkeley, University of California Press.

MacGinitie, G. E., and N. MacGinitie. 1968. Natural history of marine animals, ed. 2. New York, McGraw-Hill Book Co.

Meglitsch, P. A. 1972. Invertebrate zoology. New York, Oxford University Press.

Millott, N. (ed.). 1967. Echinoderm biology. New York, Academic Press, Inc.

Nicol, J. A. C. 1969. The biology of marine animals, ed. 2. New York, Pitman Publishing Corp.

Nichols, D. 1966. Echinoderms. London, Hutchinson & Co., Ltd.

Nichols, D. 1967. The origin of echinoderms. *In* Millot, N. (ed.). Echinoderm biology. New York, Academic Press, Inc., pp. 209-229.

Russell-Hunter, W. D. 1969. A biology of higher invertebrates. New York, Macmillan, Inc.

PART FOUR
THE VERTEBRATE ANIMALS

Anchored throughout its adult life to one spot on the sea floor, there is little in the appearance of this solitary ascidian to suggest that it is a chordate. Yet the free-swimming ascidian "tadpole" larva bears all the right chordate hallmarks—notochord, gill slits, dorsal nerve cord, and postanal tail—and occupies an important position in theories of chordate ancestry. (Photo by C. P. Hickman, Jr.)

24 VERTEBRATE BEGINNINGS
The earliest chordates

INTRODUCTION TO THE PHYLUM CHORDATA

Animals most familiar to the student belong to the great phylum Chordata (kor-da'ta) (L. *chorda,* cord). Man himself is a member and shares one of the common characteristics from which the phylum derives its name—the **notochord** (Gr. *nōton,* back, + L. *chorda,* cord) (Fig. 24-1). This structure is possessed by all members of the phylum, either in the larval or embryonic stages or throughout life. The notochord is a rodlike, semirigid body of vacuolated cells, which extends, in most cases, the length of the body between the enteric canal and the central nervous system. Its primary purpose is to support and to stiffen the body, that is, to act as a skeletal axis.

The structural plan of chordates retains many of the features of invertebrate animals, such as bilateral symmetry, anteroposterior axis, coelom tube-within-a-tube arrangement, metamerism, and cephalization. Yet, whereas the kinship of chordates and invertebrates is obvious, it has not been possible to establish the exact relationship with certainty.

Two possible lines of descent have been proposed. Earlier speculations that focused on the arthropod-annelid-mollusc group (Protostomia branch) of the invertebrates have fallen from favor. It is now believed that only the echinoderm group (Deuterostomia branch) deserves serious consideration as a chordate ancestor. The echinoderms and chordates share several important characteristics: **indeterminate cleavage,** (p. 260) **same type of mesoderm and coelom formation,** and **anus derivation from the first embryonic opening (blastopore) with mouth derivation from an opening of a secondary origin.** These common characteristics indicate a natural, even if remote, kinship between the two groups.

From gilled filter-feeding ancestors to the highest vertebrates, the evolution of chordates has been guided by the specialized basic adaptations of the living endoskeleton, paired limbs, and nervous system.

The **endoskeleton,** which, as we have seen, is first found in the echinoderms, is an internal structure that provides support and serves as a framework for the body. This structure has the advantage of allowing continuous growth without the necessity of shedding a restricting outer shell. For this reason vertebrate animals can attain great size; some of them are the most massive in the animal kingdom. Endoskeletons provide much surface for muscle attachment, and size differences between animals result mainly from the

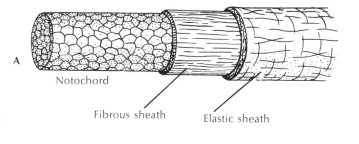

A

Notochord

Fibrous sheath   Elastic sheath

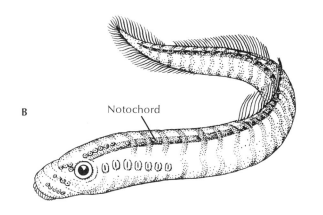

B   Notochord

Fig. 24-1. A, Structure of notochord and its surrounding sheaths. Cells of notochord proper are thick walled, pressed together closely, and filled with semifluid. Stiffness is caused mainly by turgidity of fluid-filled cells and surrounding connective tissue sheaths. This primitive type of endoskeleton is characteristic of all chordates at some stage of life cycle. Notochord provides longitudinal stiffening of main body axis, base for myomeric muscles, and axis around which vertebral column develops. In hagfishes, **B,** and lampreys it persists throughout life, but in higher vertebrates it is largely replaced by the vertebrae. In man slight remnants are found in nuclei pulposi of intervertebral discs. Its method of formation is different in the various groups of animals. In amphioxus it originates from the endoderm; in birds and mammals it arises as an anterior outgrowth of the primitive streak.

amount of muscle tissue they possess. More muscle tissue necessitates greater development of body systems, such as circulatory, digestive, respiratory, excretory, and nervous systems. Thus the endoskeleton is a crucial factor in the development and specialization of the higher animals. Its function has shifted more from a protective to a supportive one. However, endoskeletons still retain some protective functions as, for example, the cranium for the brain and the thorax for important visceral organs.

Paired limbs, pectoral and pelvic, are present in most vertebrates in the form of fins or legs. They originated as swimming stabilizers and later became prominently developed into legs for travel on land.

No single system in the body is more correlated with functional and structural advancement than is the **nervous system.** Throughout the invertebrate phyla we have seen that there has been more or less centralization of nervous systems. This tendency reaches its climax in the higher chordates in which we find the highly efficient tubular nervous system. Such a system allows the greatest possible utilization of space for the nervous units so necessary for well-integrated nervous patterns. Along with the advanced nervous system goes a better sensory system, which partly explains the power of chordates to adapt to a varied environment.

As a whole, there is more fundamental unity of plan throughout all the organs and systems of this phylum than there is in any of the invertebrate phyla. Ecologically the chordates are among the most adaptable of organic forms and are able to occupy most kinds of habitat. From a purely biologic viewpoint, chordates are of primary interest because they illustrate so well the broad biologic principles of evolution, development, and relationship. They represent as a group the background of man himself.

Characteristics

1. Bilateral symmetry; segmented body; three germ layers; well-developed coelom
2. **Notochord** (a skeletal rod) present at some stage in life cycle
3. **Single, dorsal, tubular nerve cord;** anterior end of cord usually enlarged to form brain
4. **Pharyngeal gill slits** present at some stage in life cycle and may or may not be functional
5. **Postanal tail** usually projecting beyond the anus at some stage and may or may not persist
6. **Ventral heart,** with dorsal and ventral blood vessels; closed blood system
7. Complete digestive system
8. Exoskeleton often present; well developed in some vertebrates
9. A cartilage or bony **endoskeleton** present in the majority of members (vertebrates)

The four distinctive characteristics that set chordates apart from all other phyla are the **notochord; single, dorsal, tubular nerve cord; pharyngeal gill slits;** and **postanal tail.** These characteristics are always found in the early embryo, although they may be altered or may disappear altogether in later stages of the life cycle.

Notochord. The notochord is a rigid, yet flexible, rodlike structure, extending the length of the body; it is the first part of the endoskeleton to appear in the embryo. As a rigid axis for muscle attachment, it permits undulatory movements of the body. In most of the protochordates and in primitive vertebrates, the notochord persists throughout life (Fig. 24-1). In all vertebrates a series of cartilaginous or bony vertebrae is formed from the connective tissue sheath around the notochord, which it replaces as the chief mechanical axis of the body.

Dorsal, tubular nerve cord. In the invertebrate phyla the nerve cord (often paired) is ventral to the alimentary canal and is solid, but in the chordates the single cord is dorsal to the alimentary canal and is a tube (although the hollow center may be nearly obliterated during growth). The anterior end becomes enlarged to form the brain. The hollow cord is produced in the embryo by the infolding of ectodermal cells on the dorsal side of the body above the notochord. Among the vertebrates, the nerve cord lies in the neural arches of the vertebrae and the anterior brain is surrounded by a bony or cartilaginous cranium.

Pharyngeal gill slits. Pharyngeal gill slits are perforated slitlike openings that lead from the pharyngeal cavity to the outside. They are formed by the invagination of the outside ectoderm and the evagination of the endodermal lining of the pharynx. The two pockets break through when they meet to form the slit. In higher vertebrates these pockets may not break through, and only grooves are formed instead of slits; most traces of them usually disappear. The slits have in their walls supporting frameworks of gill bars, which, in aquatic vertebrates, develop into gills.

Postanal tail. The postanal tail, together with somatic musculature and the stiffening notochord, provides the motility that larval tunicates and *Amphioxus* need for their free-swimming existence. As a structure added to the body behind the end of the digestive tract, it clearly has evolved specifically for propulsion in water. Its efficiency is later increased in fishes with the addition of fins.

Classification

There are three subphyla under phylum Chordata. Two of these subphyla are small, lack a vertebral column, and are of interest primarily as borderline or first chordates (protochordates). Since these subphyla lack a cranium, they are also referred to as Acrania. The third subphylum is provided with a vertebral column and is called Vertebrata. Since this phylum has a cranium, it is also called Craniata.

Phylum Chordata
  Group Protochordata (Acrania)
    Subphylum Urochordata (u'ro-kor-da'ta) (Gr. *oura*, tail, + L. *chorda*, cord, + *-ata*, characterized by) **(Tunicata).** Notochord and nerve cord in free-swimming larva only; sessile adults encased in tunic. Tunicates.
    Subphylum Cephalochordata (sef'a-lo-kor-da'ta) (Gr. *kephalē* head, + L. *chorda*, cord.) Notochord and nerve cord found along entire length of body and persist throughout life; fishlike in form. Lancelets *(Amphioxus)*.
  Group Craniata
    Subphylum Vertebrata (ver'te-bra'ta) (L. *vertebratus*, back-boned). Bony or cartilaginous vertebrae surrounding spinal cord; notochord in all embryonic stages, persisting in some of the fish; may also be divided into two great groups (superclasses) according to presence of jaws.
      Superclass Agnatha (ag'na-tha) (Gr. *a*, without, + *gnathos*, jaw) **(Cyclostomata)** Without true jaws or appendages. Hagfishes, lampreys.
      Superclass Gnathostomata (na'tho-sto'ma-ta) (Gr. *gnathos*, jaw + *stoma*, mouth). With jaws and (usually) paired appendages. Jawed fishes, all tetrapods.

ANCESTRY AND EVOLUTION

Since the early nineteenth century when the theory of organic evolution became the focal point for ferreting out relationships between groups of living organisms, zoologists have debated the question of vertebrate origins. It has been very difficult to reconstruct lines of descent because the earliest protochordates were in all probability soft-bodied creatures that stood little chance of being preserved as fossils even under the most ideal conditions. Consequently such reconstructions come from the study of living organisms, especially from an analysis of early developmental stages that tend to be more insulated from evolutionary change than the differentiated adult forms that they become.

The earliest speculations understandably focused on the most successful and in many respects most advanced of invertebrate groups, the Arthropoda. It was quickly recognized that, if an arthropod with its segmented body, ventral nerve cord, and dorsal heart were turned over, the basic plan of a vertebrate would result. Later, the ancestral award was transferred to the Annelida, because this group shares a basic body plan with the arthropods and in addition has an excretory system that strikingly resembles that of primitive vertebrates.

Although the annelid-vertebrate theory continued to receive support as late as 1922, it contained unresolvable difficulties. An inverted annelid has its brain and mouth in the wrong relative positions. The annelid's nerve cord is ventral but connects to a dorsal brain via circumpharyngeal connectives through which the digestive tube passes (Fig. 20-10, *A*). When an annelid is inverted, as was required for the annelid-vertebrate theory, the mouth ends up on top of the head and the brain below. Despite efforts to explain this discrepancy, the annelid theory was eventually discarded like the arthropod theory before it.

Early in this century when further theorizing became rooted in developmental patterns of animals, it immediately became apparent that only the echinoderms deserved serious consideration as the vertebrate's ancestor. Echinoderms and chordates belong to the deuterostome branch of the animal kingdom, in which the blastopore of the gastrula becomes the anus and the mouth is formed as a secondary opening, usually on the opposite end. The coelom of both phyla is primitively enterocoelous: it is budded off from the archenteron of the embryo. Furthermore there is a great resemblance between the bipinnaria larvae of certain echinoderms and tornaria larvae of the hemichordates, a phylum bearing some chordate characteristics. Both echinoderm and chordate embryos show indeterminate cleavage; that is, each of the early blastomeres has equivalent potentiality for supporting full development of a complete embryo. These characteristics are shared by brachiopods and pterobranchs (a hemichordate group), as well as by echinoderms, protochordates, amphioxus, and vertebrates. This is probably a natural grouping and almost certainly indicates interrelationships, although remote (Fig. 24-2).

Unable to further narrow the search for a prechordate-chordate connecting link, zoologists are pres-

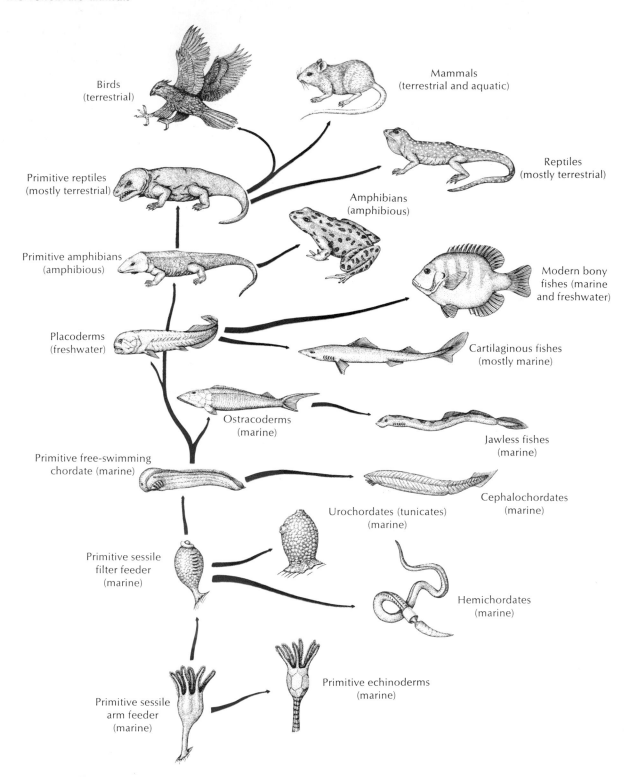

Birds
(terrestrial)

Mammals
(terrestrial and aquatic)

Primitive reptiles
(mostly terrestrial)

Reptiles
(mostly terrestrial)

Amphibians
(amphibious)

Primitive amphibians
(amphibious)

Modern bony
fishes (marine
and freshwater)

Placoderms
(freshwater)

Cartilaginous fishes
(mostly marine)

Ostracoderms
(marine)

Jawless fishes
(marine)

Primitive free-swimming
chordate (marine)

Cephalochordates
(marine)

Urochordates (tunicates)
(marine)

Primitive sessile
filter feeder
(marine)

Hemichordates
(marine)

Primitive echinoderms
(marine)

Primitive sessile
arm feeder
(marine)

Fig. 24-2. Hypothetical family tree of the chordates, suggesting probable origin and relationships. *Red,*
Extinct stem groups. *Black,* Living groups. (Drawn by W. C. Ober.)

ently focusing on groups within the chordate phylum itself, which share enough anatomic and developmental features to make unraveling the evolutionary past a more profitable effort.

Candidates for vertebrate ancestral stock

All members of the phylum Chordata share four anatomic features at some time in their life histories: a notochord, forming a stiff mechanical axis for the body; a dorsal, tubular nerve cord; pharyngeal gill slits, used for breathing and feeding; and a postanal tail, used for larval motility. The three subphyla that bear these characteristics are the Cephalochordata (of which the lancelet, amphioxus, is its famous representative), the Urochordata (tunicates or sea squirts), and the Vertebrata. The vertebrates comprise by far the greatest number of the chordates, so that the terms ''vertebrate'' and ''chordate'' are frequently used (somewhat incorrectly) as synonyms. The backbone-less members of the phylum—cephalochordates and urochordates—are usually referred to as protochordates or prevertebrates and have long been considered good candidates for the vertebrate ancestral stock.

The Hemichordata, described in the preceding chapter, previously were also assigned to the phylum Chordata since they possess a perforated pharynx (gill slits) and a middorsal thickening of nervous tissue, which may be a forerunner of the vertebrate tubular nerve cord. However, the stomochord of the hemichordates cannot be homologized to the notochord as was once believed. Furthermore the gill slits and dorsal nerve center may represent a convergent resemblance to similar chordate structures rather than a true homology. Consequently the hemichordates have been excluded from chordate membership, although most biologists agree that a real, though remote, relationship exists.

Position of amphioxus

The problems of sorting out lines of descent within the phylum Chordata are enormous because most of the prechordates and protochordates are highly specialized remnants of once successful and well-represented groups. The first approach was to place the chordate and prechordate groups in an order of increasing morphologic complexity leading to the vertebrates.

By this analysis, amphioxus becomes the logical structural ancestor of the vertebrates because it possesses as an adult all four chordate characteristics previously mentioned plus several vertebrate hallmarks: segmented musculature, the beginning of optic and olfactory sense organs, a liver diverticulum, beginnings of a ventral heart, and separation of dorsal and ventral spinal roots in the vertebrate style (Fig. 24-8). Little wonder that amphioxus once attained a pinnacle position among zoologists searching for their vertebrate ancestor.

Upon closer scrutiny though, amphioxus fails to meet the qualifications for generalized ancestral type. Its notochord is overdeveloped into a forward extension for its specialized burrowing mode of life, and this effectively prevents the development of a proper brain. Its kidney is a solenocyte type that bears little resemblance to the vertebrate glomerular-tubular nephron. Its unique atrium has no vertebrate counterpart, and there is a non-vertebrate-like proliferation of gill slits. Amphioxus today is usually regarded as a highly specialized and degenerative member of the chordate family: it lies as an offshoot, rather than in the main line, of chordate descent.

Urochordata and recapitulation

After amphioxus, attention then became focused on the alternative protochordate group, the Urochordata (tunicates). The urochordates are divided into three groups of which the ascidians (sea squirts) are the most common and simplest.

At first glance, more unlikely candidates for vertebrate ancestor could hardly be imagined. As adults, ascidians are virtually immobile forms surrounded by a tough, cellulose-containing tunic of variable color. Their adult life is spent in one spot attached to some submarine surface, filtering vast amounts of seawater from which they extract their planktonic food. As adults they lack notochord, tubular nerve cord, postanal tail, sense organs, and segmented musculature. Superficially they resemble sponges far more than they resemble any known vertebrate (Fig. 24-6). Yet the chordate nature of ascidians is abundantly evident in their **tadpole larvae.** These tiny, active, site-seeking forms have all the right qualifications for membership in the prevertebrate club: notochord, hollow dorsal nerve cord, gill slits, postanal tail, brain, and sense organs (otolith balance organ and an eye complete with lens).

The discovery of this form in 1869 not only placed the urochordates squarely in the vertebrate camp but greatly influenced the great German zoologist Ernst Haeckel in formulating his theory of recapitulation. According to this theory, adult stages of ancestors are repeated during the development of their descendants; in other words, the development of a living organism is an accurate record of past evolutionary history.

We recognize now that this record is very slurred and telescoped and must be interpreted with caution. But at the time that the true nature of the ascidian tadpole larva was first understood, it was considered to be a relic of an ancient free-swimming chordate ancestor of the ascidians. Adult ascidians then came to be regarded as degenerate, sessile descendants of the ancient chordate form.

Garstang's theory of chordate larval evolution

W. Garstang in England introduced totally fresh thinking to the vertebrate ancestor debate. In effect, Garstang in 1928 turned the sequence around: rather than the ancestral tadpole larva giving rise to a degenerative sessile ascidian adult, he suggested that the ancestral chordate stock was primarily a filter-feeding, sessile marine group not unlike modern ascidians. The free-swimming tadpole larva was evolved from these attached, bottom-dwelling forms to meet the need for seeking out new habitats. Thus the tadpole larva was visualized as an ascidian creation, evolved within the group to enhance site-seeking capabilities. Garstang next suggested that at some point the tadpole larva became neotenous, that is, became capable of maturing gonads and reproducing in the larval stage. With continued larval evolution, a new group of free-swimming animals would appear.

The best evidences for this theory are found in the living tunicates today, especially among the two planktonic groups, the thaliaceans and the larvaceans. In the latter group, the basic larval form is retained throughout life; they are in effect neotenous tunicates, although extremely specialized.

Garstang departed from previous thinking by suggesting that evolution may occur in the larval stages of animals. Zoologists accepted this idea slowly because they were accustomed to thinking of developmental stages as being largely insulated from change, as embodied in the ''biogenetic law.'' Yet in all likelihood an evolutionary sequence similar to that proposed by Garstang occurred.

The ascidian tadpole larva with its propulsive tail, stiffening notochord, and dorsal nerve cord, which

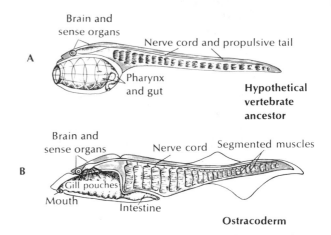

Fig. 24-3. Hypothetical vertebrate ancestor, **A,** compared with an ostracoderm, earliest fossil vertebrate, **B.** According to the Garstang hypothesis, the common vertebrate ancestor was an ascidian tadpole larva that became neotenous and evolved into a new free-swimming species. Continued evolution led to the jawless ostracoderms and jawed placoderms. (Redrawn from Roe, A., and Simpson, G. G. 1958. Behavior and evolution. New Haven, Yale University Press.)

integrates sensory information and motor activity, clearly suggests and foreshadows the vertebrate line. The resemblance of an early vertebrate ostracoderm to this hypothetical ancestor is suggested in Fig. 24-3.

Jawless ostracoderms—earliest vertebrates

The earliest vertebrate fossils are fragments of bony armor discovered in Ordovician rock in Russia and in the United States. They were small, jawless creatures collectively called **ostracoderms** (os-trak′o-derm) (Gr. *ostrakon*, shell + *derma*, skin), which belong to the Agnatha division of the vertebrates.

These earliest ostracoderms lacked paired lateral fins that subsequent fishes found so important for stability. Their swimming movements must have been inefficient and clumsy, although sufficient to propel them from one mud bed to another where they practiced their mud-grubbing search for nutrients in the form of organic debris on the sea bottom. Later ostracoderms improved the efficiency of a benthic life by evolving paired fins. These fins, located just behind the head shield, provided control over pitch and yaw that ensured well-directed forward movement.

One group of ostracoderms, known as the osteostracans (Fig. 24-4), was the subject of a brilliant and

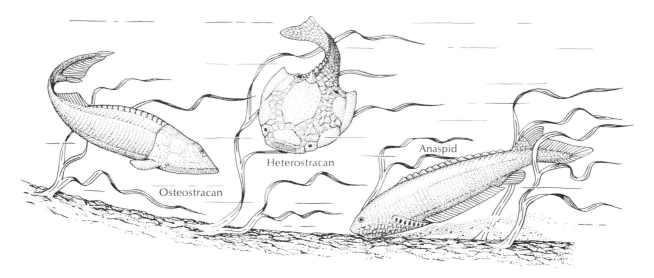

Fig. 24-4. Three ostracoderms, jawless fishes of Silurian and Devonian times. Representatives of three of the best known ostracoderm groups are illustrated as they might have appeared while searching for food on the floor of a Devonian sea. All were filter feeders, drawing water and organic debris in the mouth, straining out the organic matter, and expelling the water through the gill openings and between the ventral head plates.

classic series of studies by the Swedish paleozoologist E. A. Stensiö. A typical osteostracan was a small animal, seldom exceeding 30 cm in length; it was covered by a well-developed armor—the head by a solid shield (rounded anteriorly) and the body by bony plates. It had no axial skeleton or vertebrae. The mouth was ventral and anterior, and it was jawless and toothless. Its paired eyes were located close to the middorsal line. At the lateroposterior corners of the head shield was a pair of flaplike fins. The trunk and tail appeared to be adapted for active swimming. Between the margin of the head shield and the ventral plates, there were ten gill openings on each side. These fishes also had a lateral line system. They were adapted for filter feeding, which may explain the expanded size of the head, made up as it is of a large pharyngeal gill-slit filtering apparatus.

As a group, the ostracoderms were basically fitted for a simple, bottom-feeding life. Yet, despite their anatomic limitations, they enjoyed a respectable radiation in the Silurian and Devonian periods. Their overall contribution was enormous, for they provided a blueprint for subsequent vertebrate evolution. But they could not survive the competition of the more advanced jawed fishes that began to dominate the Devonian, and in the end they disappeared.

Early jawed vertebrates

All jawed vertebrates, whether extinct or living, are collectively called gnathostomes ("jaw mouth") in contrast to jawless vertebrates, the agnathans ("without jaw"). The latter are also often referred to as cyclostomes ("circle mouth").

The first jawed vertebrates to appear in the fossil record were the **placoderms** (plak'o-derm) (Gr. *plax*, plate, + *derma*, skin). The advantages of jaws are obvious; they allow predation on large and active forms of food. Possessors of jaws would enjoy a great advantage over jawless vertebrates, which were restricted to a wormlike existence of sifting out organic debris and small organisms in the bottom mud.

Jaws arose through modifications of the first two of the serially repeated cartilaginous gill arches. The beginnings of this trend can, in fact, be seen in some of the jawless ostracoderms where the mouth became bordered by strong dermal plates that could be manipulated somewhat like jaws with the gill arch musculature. The more anterior arches were gradually modified to permit more efficient seizing, and the skin surrounding the mouth was modified into teeth. Eventually the anterior gill arches became bent into the characteristic position of vertebrate jaws, as seen in the placoderms.

491

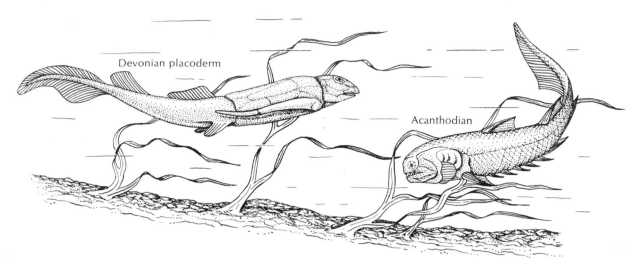

Fig. 24-5. Early jawed fishes of the Devonian period, 400 million years ago. The placoderm *(left)* and a related acanthodian *(right)* were highly mobile and voracious forms. Though more successful than the less maneuverable ostracoderms, they eventually failed in competition with their successors, the bony and cartilagenous fishes.

Placoderms and their relatives (Fig. 24-5) evolved into a great variety of forms, some large and grotesque in appearance. They were armored fish covered with diamond-shaped scales or with large plates of bone. All became extinct by the end of the Paleozoic era.

Evolution of modern fishes and tetrapods

Reconstruction of the origins of the vast and varied assemblage of modern living vertebrates is, as we have seen, based largely on fossil evidence. Unfortunately the fossil evidence for the earliest vertebrates is often incomplete and tells us much less than we would like to know about subsequent trends in evolution. Affinities become much easier to establish as the fossil record improves. For instance, the descent of birds and mammals from reptilian ancestors has been worked out in a highly convincing manner from the relatively abundant fossil record available. By contrast, the ancestry of modern fishes is shrouded in uncertainty.

The Swedish paleontologist E. Jarvik has emphasized that the main vertebrate stem groups (such as cyclostomes, lungfishes, sharks, bony fishes, and stem tetrapods) became anatomically specialized some 400 to 500 million years ago and have changed relatively little since then. Thus main evolutionary lines, as seen in the fossil record, run back almost in parallel; if extended backwards to their illogical extreme, they

would hardly ever meet. Obviously they must meet at some point in the distant past, but this exercise reveals that the crucial separations in vertebrate evolution occurred in the Cambrian, perhaps even the Precambrian, long before the fossil record became established for the convenience of paleozoologists.*

Despite the difficulty of establishing early lines of descent for the vertebrates, they are clearly a natural, monophyletic group, distinguished by a great number of common characters. They have almost certainly descended from a common ancestor, the nature of which we have already discussed. Very early in their evolution, the vertebrates divided into two great stems, the agnathans and the gnathostomes. These two groups differ from each other in many fundamental ways, in addition to the obvious lack of jaws in the former group and their presence in the latter. Thus both groups are very old and of approximately the same age. On this basis we cannot say that agnathans are more "primitive" than gnathostomes, even though the latter have continued on a marvelous evolutionary advance that produced most of the modern fishes, all of the tetrapods, and the reader of this book. Although the agnathans are represented today only by the hagfishes

---

*Jarvik, E. 1967. Aspects of vertebrate phylogeny. *In* T. Ørvig (ed.). Nobel Symposium 4, Stockholm, Almqvist & Wiksell.

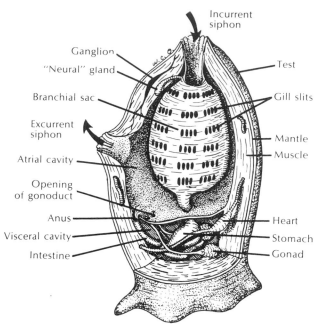

Fig. 24-6. Adult ascidians, or "sea squirts." Two living sea squirts are shown in the photograph. Natural coloration of this species is bright orange. At right is shown the structure of a solitary adult simple ascidian. Arrows indicate direction of water currents.

and the lampreys, these creatures too are successful in their own way.

SUBPHYLUM UROCHORDATA (TUNICATA)

The tunicates are found in all seas from near the shoreline to great depths. Most of them are sessile as adults, although some are free living. The name "tunicate" is suggested by the nonliving tunic that surrounds them and contains cellulose. They vary in size from microscopic forms to several centimeters in length. They may be considered as degenerative or specialized members of the chordates, for they lack many of the common characteristics of chordates.

Urochordata is divided into three classes—**Ascidiacea, Larvacea,** and **Thaliacea.** Of these, the members of **Ascidiacea,** commonly known as the ascidians, or sea squirts, are by far the most common and are the best known. Ascidians may be solitary, colonial, or compound. Each of the solitary and colonial forms has its own test, but among the compound forms many individuals may share the same test. In some of these compound ascidians each member has its own in-

current siphon, but the excurrent opening is common to the group.

The typical solitary ascidian (Fig. 24-6) is globose in form and is attached by its base to piles and stones. Lining the test or tunic is a membrane or **mantle.** On the outside are two projections: the **incurrent** and **excurrent siphons** (Fig. 24-6). Water enters the incurrent siphon and passes into the branchial sac (pharynx) through the mouth. On the midventral side of the branchial sac is a groove, the **endostyle,** which is ciliated and secretes mucus. Food material in the water is entangled by the mucus in this endostyle and carried into the esophagus and stomach. The intestine leads to the anus near the excurrent siphon. The water passes through the pharyngeal slits in the walls of the branchial sac into the atrial cavity. As the water passes through the slits, respiration occurs.

The circulatory system contains a ventral **heart** near the stomach and two large vessels, one connected to each end of the heart. The action of the heart is peculiar in that it drives the blood first in one direction and then in the other. This reversal of blood flow is found in no other animal. The excretory system is a

493

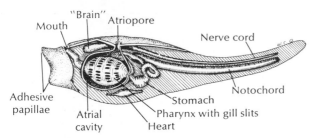

Fig. 24-7. Structure of tunicate (ascidian) tadpole larva. This larva is believed to closely resemble the ancestor of all vertebrates and shows all four principal chordate characteristics—notochord, dorsal nerve cord, pharyngeal gill slits, and postanal tail.

type of nephridium near the intestine. The nervous system is restricted to a nerve ganglion and a few nerves that lie on the dorsal side of the pharynx. A notochord is lacking. The animals are hermaphroditic, for both ovaries and testes are found in the same animal. The germ cells are carried out the excurrent siphon into the surrounding water, where cross-fertilization occurs.

Of the four chief characteristics of chordates, adult tunicates have only one, the pharyngeal gill slits. However, the larval form gives away the secret of their true relationship. The tadpole larva (Fig. 24-7) is an elongate, transparent form with all four chordate characteristics: a **notochord,** a hollow **dorsal nerve cord,** a propulsive **postanal tail,** and a large pharynx with endostyle and **gill slits.** The larva does not feed but swims about for some hours before fastening itself vertically by its adhesive papillae to some solid object. It then undergoes a retrograde metamorphosis to become the sessile adult.

The remaining two classes of the Urochordata—**Larvacea** and **Thaliacea**—are mostly small, transparent animals of the open sea. Some are small tadpolelike forms resembling the larval stage of ascidians. Others are spindle shaped or cylindric forms surrounded by delicate muscle bands. They are mostly carried along by the ocean currents and as such form a part of the plankton. Many are provided with luminous organs and emit a beautiful light at night.

SUBPHYLUM CEPHALOCHORDATA

The cephalochordates are the marine lancelets. These include *Amphioxus (Branchiostoma),* one of the classic animals in zoology. The amphioxus has the four distinctive characteristics of chordates in simple form; in this and in other ways it may be considered a blueprint of the phylum. It has a slender, laterally compressed body 2 to 3 inches long, with both ends pointed (Fig. 24-8).

Not only are the four chief characteristics of the chordates—dorsal nerve cord, notochord, pharyngeal gill slits, and postanal tail—well represented in amphioxus, but also the secondary characteristics, such as liver diverticulum, hepatic portal system, and the beginning of a ventral heart. The separation of the dorsal and ventral roots of the spinal nerves may indicate the early condition in the vertebrate ancestors. *Amphioxus* is often placed close in affinity to the higher chordates, the vertebrates. Many authorities place it near the primitive fish, ostracoderms, but whether it comes before or after these fish in the evolutionary line is not settled.

Despite its many distinctive chordate features, many regard the amphioxus as a highly specialized or degenerate member of the early chordates and believe that the overdeveloped notochord was developed in them as an adaptation for burrowing. The extension of the notochord into the tip of the snout is one of the reasons for the small development of the brain of the amphioxus. There are other objections to considering amphioxus as a generalized ancestral type of chordate, and some authorities therefore think it is a divergent side branch of some stage intermediate between the early filter-feeding prevertebrates and the vertebrates.

SUBPHYLUM VERTEBRATA

The third subphylum of the chordates is the large and eminently successful Vertebrata, the subject of the next five chapters of this book. The subphylum Vertebrata shares the basic chordate characteristics with the other two subphyla, but in addition it has a number of features that the others do not share. The characteristics that give the members of this group the name "Vertebrata" or "Craniata" are the presence of a braincase, or **cranium,** and a spinal column of vertebrae, which forms the chief skeletal axis of the body.

Characteristics

1. Chief diagnostic features of chordates—**notochord, dorsal nerve cord, pharyngeal gill slits,** and **postanal tail**—all present at some stage of the life cycle
2. **Integument** basically of two divisions, an outer **epidermis** of stratified epithelium from the ectoderm and an inner **dermis,** of

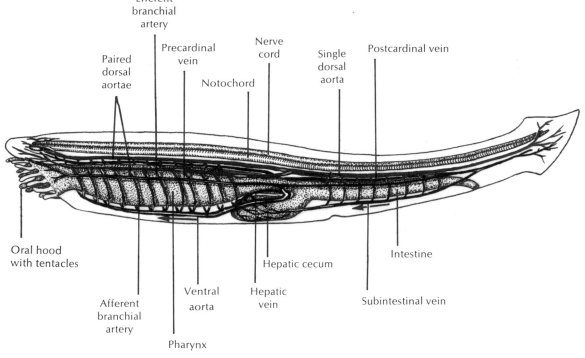

Efferent
branchial
artery

Precardinal
vein

Nerve
cord

Single
dorsal
aorta

Postcardinal vein

Paired
dorsal
aortae

Notochord

Oral hood
with tentacles

Intestine

Afferent
branchial
artery

Ventral
aorta

Hepatic
vein

Hepatic cecum

Subintestinal vein

Pharynx

Fig. 24-8. Amphioxus. This interesting bottom-dwelling cephalochordate possesses the four distinctive chordate characteristics (notochord, dorsal nerve cord, pharyngeal gill slits, and postanal tail) that once made it the prime candidate for our vertebrate ancestor. However, because it also bears many specialized and degenerate features, zoologists now consider it a divergent offshoot from the main line of chordate evolution. *Above,* Living specimen. *Below,* Structure of amphioxus showing scheme of circulation. (Photo by B. Tallmark.)

connective tissue derived from the mesoderm; many modifications of skin among the various classes, such as glands, scales, feathers, claws, horns, and hair

3. Notochord more or less replaced by the spinal column of vertebrae composed of cartilage or bone or both; distinctive **endoskeleton** consisting of vertebral column with the cranium, visceral arches, limb girdles, and two pairs of jointed appendages.

4. **Many muscles** attached to the skeleton to provide for movement

5. Complete **digestive system** ventral to the spinal column and provided with large digestive glands, liver, and pancreas

6. Circulatory system consisting of the **ventral heart** of two to four chambers; a closed blood vessel system of arteries, veins, and capillaries; blood fluid containing red blood corpuscles with hemoglobin and white corpuscles; paired aortic arches connecting the ventral and dorsal aortae and giving off branches to the gills among the aquatic vertebrates; in the terrestrial types modification of the aortic arch plan into pulmonary and systemic systems

7. Well-developed **coelom** largely filled with the visceral systems

8. **Excretory system** consisting of paired kidneys (mesonephric or metanephric types in adults) provided with ducts to drain the waste to the cloaca or anal region

9. Brain typically divided into five vesicles

10. Ten or twelve pairs of cranial nerves with both motor and sensory functions usually; a pair of spinal nerves for each primitive myotome; an autonomic nervous system in control of involuntary functions of internal organs

11. **Endocrine system** of ductless glands scattered through the body

12. Nearly always separate sexes; each sex containing paired gonads with ducts that discharge their products either into the cloaca or into special openings near the anus

13. **Body plan** consisting typically of **head, trunk,** and **postanal tail; neck** present in some, especially terrestrial forms; two pairs of appendages usually, although entirely absent in some; coelom divided into a pericardial space and a general body cavity; mammals with a thoracic cavity

REVIEW OF CONTRIBUTIONS OF THE CHORDATES

1. A **living endoskeleton** is characteristic of the entire phylum. Two endoskeletons are present in the group as a whole. One of these is the rodlike **notochord,** which is present in all members of the phylum at some time; the other is the **vertebral column,** which largely replaces the notochord in higher chordates.

2. The endoskeleton does not interfere with **continuous growth,** for it can increase in size with the rest of the body. There is therefore no necessity for shedding it, as is the case with the nonliving exoskeleton of the invertebrate phyla. Moreover, the endoskeleton allows for almost indefinite growth, so that many chordates are the largest of all animals.

3. The nature of the endoskeleton is such that it affords much surface for muscular attachment. Locomotory muscles are consequently well developed and powerful. Since the muscular and skeletal systems make up most of the bulk of animals, other bodily systems must also become specialized to serve the metabolic requirements of these two great systems.

4. A **postanal tail** is a new development in the animal kingdom and is present at some stage in most chordates. In fishes it is developed into a powerful propulsive device; in terrestrial vertebrates it serves a variety of functions in locomotion, grasping, and stability.

5. A **ventral heart** is a new characteristic, and a closed blood system is better developed for producing high pressures than it is in other phyla. Chordates have also developed a **hepatic portal system,** which is specialized for conveying food-laden blood from the digestive system to the liver.

6. **Perforated pharynx** (gill slits) are introduced for the first time. Gill slits or traces are present in the embryos of all chordates. Terrestrial chordates have developed lungs by modification of this same pharyngeal region.

7. A dorsal hollow nerve cord is universally present at some stage.

SELECTED REFERENCES

Barrington, E. J. W. 1965. The biology of Hemichordata and Protochordata. (Paperback.) San Francisco, W. H. Freeman & Co. *A synthesis of recent work on those deuterostomes most closely related to vertebrates; it discusses the possible homologies between the endostyle and the vertebrate thyroid gland.*

Berrill, N. J. 1955. The origin of vertebrates. New York, Oxford University Press. *The author stresses the tunicates as the basic stock from which other protochordates and vertebrates arose. He believes that such a sessile filter feeder was really the most primitive animal and was not a mere degenerate side branch of chordate evolution.*

Colbert, E. H. 1955. Evolution of the vertebrates. New York, John Wiley & Sons, Inc. *A clear and well-written presentation.*

Halstead, L. B. 1968. The pattern of vertebrate evolution. San Francisco, W. H. Freeman & Co. *In this thoughtful interpretation of fossil evidence the author considers physiologic and ecologic aspects of evolution.*

Romer, A. S. 1959. The vertebrate story. Chicago, University of Chicago Press. *A comprehensive background of the evolutionary trends and relationships of the various vertebrate groups leading up to that of man himself.*

Romer, A. S. 1966. Vertebrate paleontology, ed. 3. Chicago, University of Chicago Press. *An authoritative work by a distinguished paleontologist.*

Stahl, B. J. 1974. Vertebrate history: problems in evolution. New York, McGraw-Hill Book Co. *Vertebrate evolution and paleontology.*

Retracing an ancestral route to spawning grounds, two Pacific salmon, *Oncorhynchus* sp., leap rapids at Brook Falls, Alaska. The unerring return of salmon to their natal streams to spawn after spending 2 or more years at sea is one of the wonders of nature. (Photo by G. B. Kelez.)

25 FISHES

Fishes are the undisputed masters of the aquatic environment. Because fish live in a habitat that is basically hostile to man, we have not always found it easy to appreciate the incredible success of these vertebrates. Plato considered fish "senseless beings . . . which have received the most remote habitations as a punishment for their extreme ignorance." The average North American today is probably unconscious of and uninformed about fish unless he happens to be a sports fisherman or tropical fish enthusiast.

Nevertheless the world's fishes have enjoyed an adaptive radiation more spectacular than that of all the land vertebrates, with the possible exception of the mammals (the latter having succeeded in the air and in the sea as well as on land). Their numerous structural adaptations have produced a great variety of forms ranging from gracefully streamlined trout to grotesque creatures that dwell in the blackness of the ocean's abyssal depths. Considered either in numbers of species (more than 20,000 named species) or in numbers of individuals (countless billions), fishes at least equal, if not outnumber, the four terrestrial vertebrate classes combined.

Although fishes are the oldest vertebrate group, there is not the slightest evidence that, like their amphibian and reptile successors, they are declining from a period of earlier glory; certain groups of ancient fishes have become extinct, yet they have been replaced by successful, modern fishes. There are indeed more bony fishes today than ever before, and no other group threatens their domination of the seas.

Their success can be attributed to one thing: they are marvelously adapted to their dense medium. A trout or pike can hang motionless in the water, varying its neutral buoyancy by adding or removing air from the swim bladder, or dart forward or at angles, using its fins as brakes and tilting rudders. Fishes have excellent olfactory and visual senses and a unique lateral line system, which with its exquisite sensitivity to water currents and vibrations provides a "distance touch" in water. Their gills are the most effective respiratory devices in the animal kingdom for extracting oxygen from water. With highly developed organs for salt and water exchange, bony fishes are excellent osmotic regulators capable of fine tuning their body fluid composition in their chosen freshwater or seawater environment. Fishes have evolved complex behavioral mechanisms for dealing with emergencies, and many have evolved elaborate reproductive behavior con-cerned with courtship, nest building, and care of the young. These are only a few of many such examples. The adaptations evident in this varied phylogenetic assemblage, which includes four of the eight living vertebrate classes, are nearly as numerous as the number of species.

ANCESTRY OF MAJOR GROUPS OF FISHES

As described in the preceding chapter, the fishes are descended from an unknown common ancestor that may have arisen from a free-swimming larval tunicate. During the Cambrian, or perhaps even in the Pre-cambrian, the earliest fishlike vertebrates branched into the jawless Agnatha and the jawed Gnathostomata (Fig. 25-1). All vertebrates are descended from one or the other of these two great stems. The Agnatha include various extinct armored groups (ostracoderms in older terminology) and two remnant living groups: the hagfishes (Myxini) and lampreys (Petromyzontes). The ancestry of hagfishes and lampreys is disputed, but they are now so different from each other that they are placed in separate classes by many ichthyologists.

All the rest of the fishes are gnathostomes that have descended from one or more early jawed ancestors. The placoderms have often been suggested, but the fossil evidence is actually so fragmentary that it is impossible to pick out ancestral groups with certainty. Whatever the early lines of descent, ichthyologists now recognize several natural groups of living jawed fishes, which are depicted in Fig. 25-1.

The sharks, skates, and rays comprise one natural, compact group, the subclass Elasmobranchii of the class Chondrichthyes. These animals lost the heavy armor of the early placoderms and adopted cartilage instead of bone for the skeleton (a secondary degeneration), an active and predatory habit, and shark-like body form that has undergone only minor changes over the ages.

Obviously sharing a remote relationship to the elasmobranchs are the bizarre and strangely appealing chimaeras of the subclass Holocephali (class Chondrichthyes). They appeared in the Devonian (400 million years ago), perhaps independently derived from some placoderm.

The bony fishes (Osteichthyes) are the dominant fishes today. Three great stems of descent are recognized: first there are the ray-finned fishes (Actinopterygii), which radiated into the vast assemblage of

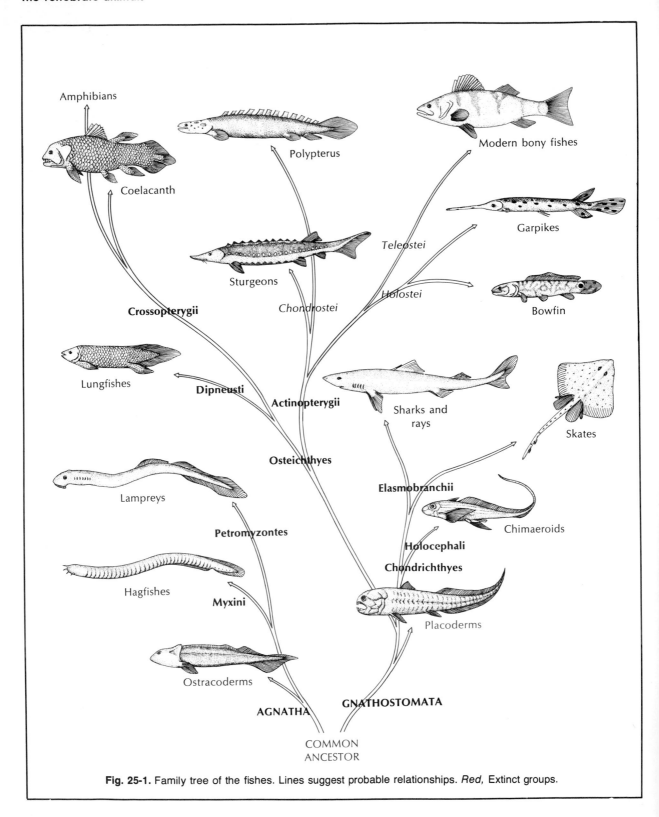

Fig. 25-1. Family tree of the fishes. Lines suggest probable relationships. *Red,* Extinct groups.

modern fishes that today occupy virtually all of the earth's aquatic environments; second are the lobe-finned fishes (Crossopterygii), from which the amphibians are descended; and third are the lungfishes (Dipneusti). Both the lobe-finned fishes and the lungfishes are relic groups. Only seven survivors remain (six species of lungfishes and one species of lobe-fin, the coelacanth)—meager evidence of important stocks that flourished in the Devonian. These survivors have remained relatively unchanged for 400 million years.

Classification of living fishes

The following broad classification is a composite of schemes by several contemporary ichthyologists. No one scheme is accepted by even the majority of ichthyologists. When we contemplate the incredible difficulty of ferreting out relationships among some 20,000 living species and a vast number of fossils of varying age, we can appreciate why fish classification has been and will continue to be undergoing continuous change.

Subphylum Vertebrata
  Superclass Agnatha (ag′na-tha) (Gr. *a*, not, + *gnathos*, jaw) **(Cyclostomata).** No jaws; cartilaginous skeleton; ventral fins absent; two semicircular canals; notochord persistent.
  Class Petromyzontes (pet′ro-my-zon′teez) (Gr. *petros*, stone, + *myzon*, sucking)—**lampreys.** Suctorial mouth with horny teeth; nasal sac not connected to mouth; seven pairs of gill pouches.
  Class Myxini (mik-sy′ny) (Gr. *myxa*, slime)—**hagfishes.** Terminal mouth with four pairs of tentacles; buccal funnel absent; nasal sac with duct to pharynx; 5 to 15 pairs of gill pouches; partially hermaphroditic.
  Superclass Gnathostomata (na′tho-sto′ma-ta) (Gr. *gnathos*, jaw, + *stoma*, mouth). Jaws present; usually paired limbs; three semicircular canals; notochord persistent or replaced by vertebral centra.
  Class Chondrichthyes (kon-drik′thee-eez) (Gr. *chondros*, cartilage, + *ichthys*, a fish)—**sharks, skates, rays,** and **chimaeras.** Streamlined body with heterocercal tail; cartilaginous skeleton; five to seven gills with separate openings, no operculum, no swim bladder.
  Subclass Elasmobranchii (e-laz′mo-bran′kee-i) (Gr. *elasmos*, a metal plate, + *branchia*, gills)—**sharks, skates,** and **rays.** Cartilaginous endoskeleton often calcified; placoid scales or no scales; five to seven gill arches and gills in separate clefts along pharynx.
  Subclass Holocephali (hol′o-sef′a-li) (Gr. *holos*, entire, + *kephale*, head)—**chimaeras** or **ghostfishes.** Gill slits covered with operculum; jaws with tooth plates; single nasal opening; without scales; accessory clasping organs in male; lateral line an open groove.
  Class Osteichthyes (os′te-ik′thee-eez) (Gr. *osteon*, bone, + *ichthys*, a fish) **(Teleostomi)—bony fishes.** Primitively fusiform body but variously modified; mostly ossified skeleton; single gill opening on each side covered with operculum; usually swim bladder or lung.
  Subclass Crossopterygii (cros-sop-ter-ij′ee-i) (Gr. *krossoi*, fringe or tassels, + *pteryx*, fin, wing)—**lobed-finned fishes.** Paired fins lobed with internal skeleton of basic tetrapod type; three-lobed diphycercal tail; skeleton with much cartilage; vestigial air bladder; hard gills with teeth; intestine with spiral valve; spiracle present.
  Subclass Dipneusti (dip-nyu′sti) (Gr. *di-*, two, + *pneusti-kos*, of breathing—**lungfishes.** All median fins fused to form diphycercal tail; fins lobed or of filaments; teeth of grinding plates; air bladder of single or paired lobes and specialized for breathing; intestine with spiral valve; spiracle absent.
  Subclass Actinopterygii (ak′ti-nop-te-rij′ee-i) (Gr. *aktis*, ray, + *pteryx*, fin, wing)—**ray-finned fishes.** Paired fins supported by dermal rays and without basal lobed portions; nasal sacs open only to outside.

JAWLESS FISHES—SUPERCLASS AGNATHA

The living members of the Agnatha are represented by some 60 species almost equally divided between two classes: Petromyzontes and Myxini (Fig. 25-2). They have in common the absence of jaws, internal ossification, scales, and paired fins, and both share porelike gill openings and an eellike body form. At the same time there are so many important differences, some of which are indicated in the following list, that they have been assigned to separate vertebrate classes.

Characteristics of the Agnatha

1. Slender, **eellike** body
2. Median fins but **no paired appendages**
3. **Fibrous** and **cartilaginous skeleton;** notochord persistent
4. Suckerlike oral disc with well-developed teeth in lampreys; mouth with two rows of eversible teeth in hagfish
5. Heart with one auricle and one ventricle; aortic arches in gill region
6. Seven pairs of gills in lampreys; 5 to 16 pairs of gills in hagfish
7. Mesonephric kidney in lampreys; pronephric kidney anteriorly and mesonephric kidney posteriorly in hagfish
8. Dorsal nerve cord with differentiated brain; 8 to 10 pairs of cranial nerves
9. Digestive system without stomach; intestine with spiral fold in lampreys; spiral fold absent in hagfish
10. Sense organs of taste, smell, hearing; eyes moderately developed in lampreys but highly degenerate in hagfish
11. External fertilization; single gonad without duct; separate sexes and long larval stage in lampreys; hermaphroditic and direct development with no larval stage in hagfish

501

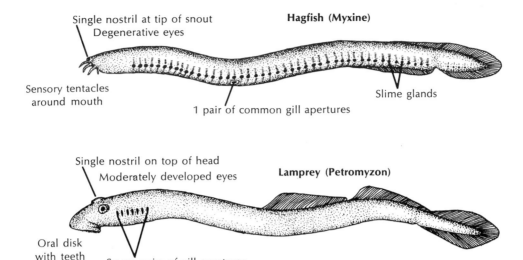

Fig. 25-2. Comparison of hagfish (class Myxini) and lamprey (class Petromyzontes), representatives of the superclass Agnatha.

Lampreys—class Petromyzontes

All lampreys of North and South America belong to the family Petromyzontidae. The destructive marine lamprey *Petromyzon marinus* is found on both sides of the Atlantic Ocean (America and Europe) and may attain a length of 1 m. Other genera also have a wide distribution in North America and Eurasia and are usually from 15 to 60 cm long. Of the 19 species of lampreys in North America, approximately half are of the nonparasitic, brook type; the others are parasitic.

All lampreys, marine as well as freshwater forms, spawn in the spring in North America in shallow gravel and sand in freshwater streams. The males begin nest building and are joined later by females. Using their oral discs to lift stones and pebbles and using vigorous body vibrations to sweep away light debris, they form an oval depression. At spawning, with the female attached to a rock to maintain position over the nest, the male attaches to the dorsal side of her head. As the eggs are shed into the nest, they are fertilized by the male. The sticky eggs adhere to pebbles in the nest and soon become covered with sand. The adults die soon after spawning.

The eggs hatch in approximately 2 weeks into small larvae (ammocoetes), which stay in the nest until they are approximately 1 cm long; they then burrow into the mud and sand and emerge at night to feed. The ammocoete period lasts from 3 to 7 years before the larva rapidly metamorphoses into an adult.

Parasitic lampreys either migrate to the sea, if marine, or else remain in fresh water, where they attach themselves by their suckerlike mouth to fish and with their sharp horny teeth rasp away the flesh and suck out the blood (Fig. 25-3). To promote the flow of blood, the lamprey injects an anticoagulant into the wound. When gorged, the lamprey releases its hold but leaves the fish with a wound that may prove fatal. The parasitic freshwater adults live a year or more before spawning and then die; the marine forms may live longer.

The nonparasitic lampreys do not feed after emerging as adults, for their alimentary canal degenerates; within a few months and after spawning, they die.

The invasion of the Great Lakes above Lake Ontario by the landlocked sea lamprey *Petromyzon marinus* in this century had a devastating effect on the fisheries. Lampreys first entered the Great Lakes after the Welland Canal around Niagara Falls was deepened between 1913 and 1918. Moving first through Lake Erie to Lakes Huron, Michigan, and Superior, sea lampreys caused the total collapse of a multimillion dollar lake trout fishery in the early 1950s. Other less valuable fish species were attacked and destroyed in turn. Sea lampreys are now declining because of both depletion of their food and control measures.

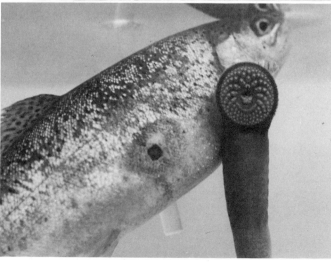

Fig. 25-3. Sea lampreys *(Petromyzon marinus)* attacking trout. **A,** Recently transformed sea lampreys, 15 to 18 cm long, attack 20 cm brook trout *(Salvelinus fontinalis)* in experimental aquarium. **B,** Head of 38 cm sea lamprey feeding on a rainbow trout *(Salmo gairdneri).* Note the single nostril on top of head and the eyes and gill aperatures. **C,** Lamprey detached from rainbow trout to show feeding wound that had penetrated body cavity and perforated gut. Trout died from wound. Note chitinous teeth on underside of lamprey head. (Courtesy United States Bureau of Sport Fisheries and Wildlife, Fish Control Laboratory, La Crosse, Wis.; **A,** photo by R. E. Lennon; **B** and **C,** photos by L. L. Marking.)

Hagfishes—class Myxini

The hagfishes are an entirely marine group that feed on dead or dying fishes, annelids, molluscs, and crustaceans. Thus they are neither parasitic, like lampreys, nor predaceous; rather they are scavengers. There are only 32 species of hagfishes, of which the best known in North America are the Atlantic hagfish *Myxine glutinosa* (Fig. 25-2) and the Pacific hagfish *Eptatretus (Bdellostoma) stouti*.

Hagfishes probably have fewer human admirers than any other group of fishes. The sports fisherman who catches one discovers that his hook is so deeply swallowed that retrieval is impossible. To cap the sportsman's misfortune, the animal secretes enormous quantities of slimy mucus from large and small mucous glands located all over the body and from special slime glands positioned along its sides (Fig. 25-2). A single hagfish is said to be capable of converting a bucket of water into a mass of whitish jelly in minutes. Their habit of biting into and entering the bodies of gill-netted fish has not endeared them to commercial fishermen either. Entering through the anus or gills, hagfishes eat out the contents of the body, leaving behind a loose sack of skin and bones. But, as fishing methods have passed from the use of drift nets and set lines to large and efficient otter trawls, hagfishes have ceased to be an important pest.

CARTILAGINOUS FISHES—CLASS CHONDRICHTHYES

There are more than 625 living species in the class Chondrichthyes, an ancient, compact, and highly developed group. Although a much smaller and less diverse assemblage than the bony fishes, their impressive combination of well-developed sense organs, powerful jaws and swimming musculature, and pre-daceous habits ensures them a secure and lasting niche in the aquatic community. One of their distinctive features is their cartilaginous skeleton, which must be considered degenerate instead of primitive. Although there is some calcification here and there, bone is entirely absent throughout the class.

Sharks, skates, and rays—subclass Elasmobranchii

With the exception of whales, sharks and their kin are the largest living vertebrates. The whale shark may reach 15 m in length. The dogfish sharks so widely used in zoologic laboratories rarely exceed 1 m. There are approximately 600 living species of elasmobranchs.

Characteristics of the elasmobranchs

1. Fusiform or spindle-shaped body
2. Ventral mouth; two olfactory sacs that do not break into the mouth cavity; jaws present
3. Skin with placoid scales and mucous glands; teeth of modified placoid scales
4. Entirely cartilaginous endoskeleton
5. Digestive system with a J-shaped stomach and intestine with a spiral valve
6. Circulatory system of several pairs of aortic arches; two-chambered heart
7. Respiration by means of five to seven pairs of gills with separate and exposed gill slits, no operculum
8. No swim bladder
9. Brain of two olfactory lobes, two cerebral hemispheres, two optic lobes, a cerebellum, and a medulla oblongata; ten pairs of cranial nerves
10. Separate sexes; oviparous or ovoviviparous; direct development; internal fertilization

Although sharks to most people have a sinister appearance and a fearsome reputation, they are at the same time among the most gracefully streamlined of

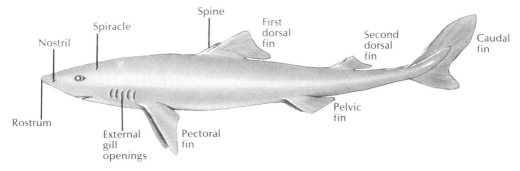

Fig. 25-4. Dogfish shark, *Squalus acanthias* (subclass Elasmobranchii).

all fishes (Fig. 25-4). Sharks are heavier than water and must always keep swimming forward to avoid sinking. The asymmetric **heterocercal tail**, in which the vertebral column turns upward and extends into the dorsal lobe of the tail, provides the necessary lift as it sweeps to and fro in the water, and the broad head and flat pectoral fins act as planes to provide head lift.

Sharks are well equipped for their predatory life. The tough leathery skin is covered with numerous knifelike **placoid scales** that are modified anteriorly to form replaceable rows of teeth in both jaws. Placoid scales in fact consist of dentine enclosed by an enamel-like substance, and they very much resemble the teeth of other vertebrates. Sharks have a keen sense of smell used to guide them to food. Vision is less acute than in most bony fishes, but a well-developed **lateral line system** serves as a "distance touch" in water for detecting and locating objects and moving animals (predators, prey, and social partners). It is composed of a canal system extending along the side of the body and over the head. The canal opens at intervals to the surface. Inside are special receptor organs **(neuromasts)** that are extremely sensitive to vibrations and currents in the water.

Skates and rays belong to a separate order (Batoidei) from the sharks (Selachii). Skates and rays are distinguished by their dorsoventrally flattened bodies and the much-enlarged pectoral fins that behave as wings in swimming. The gill openings are on the underside of the head, and the **spiracles** (on top of the head) are unusually large. Respiratory water is taken in through these spiracles to prevent clogging the gills, for their mouth is often buried in sand. The teeth are adapted for crushing the prey—mainly molluscs, crustaceans, and an occasional small fish.

In the stingrays and eagle rays (Fig. 25-5), the caudal and dorsal fins have disappeared and the tail is slender and whiplike. The stingray tail is armed with one or more saw-toothed spines that can inflict dangerous wounds. Electric rays have certain dorsal muscles modified into powerful electric organs, which can give severe shocks and stun their prey.

Chimaeras—subclass Holocephali

The members of the small group of chimaeras, approximately 25 species in all, are distinguished by such suggestive names as ratfish (Fig. 25-6), rabbitfish, spookfish, and ghostfish. They are remnants of an aberrant line that diverged from the placoderms at least 300 million years ago (Carboniferous or Devonian). Fossil chimaeras (ky-meer'uz) were first found in the Jurassic, reached their zenith in the

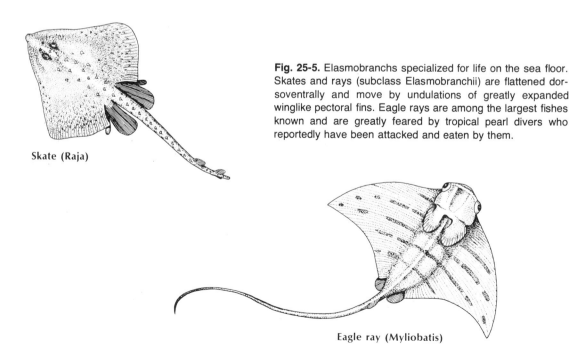

Skate (Raja)

Eagle ray (Myliobatis)

Fig. 25-5. Elasmobranchs specialized for life on the sea floor. Skates and rays (subclass Elasmobranchii) are flattened dorsoventrally and move by undulations of greatly expanded winglike pectoral fins. Eagle rays are among the largest fishes known and are greatly feared by tropical pearl divers who reportedly have been attacked and eaten by them.

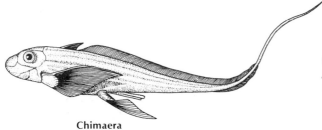

Chimaera

Fig. 25-6. Chimaera, or ratfish (subclass Holocephali), of North American west coast. This species is one of the most handsome of chimaeras, which tend toward bizarre appearances.

Cretaceous and early Tertiary (120 million to 50 million years ago), and have declined ever since. Anatomically they present an odd mixture of sharklike and bony fish–like features. Their food is a mixed diet of seaweed, molluscs, echinoderms, crustaceans, and fishes. Chimaeras are not commercial species and are seldom caught. Despite their grotesque shape, they are beautifully colored with a pearly iridescence and have vivid emerald-green eyes.

BONY FISHES—CLASS OSTEICHTHYES (TELEOSTOMI)

In no other major animal group do we see better examples of adaptive radiation than among the bony fishes. Their adaptations have fitted them for every aquatic habitat except the most completely inhospitable. Body form alone is indicative of this diversity. Some have fusiform (streamlined) bodies and other adaptations for reducing friction. Predaceous, pelagic fish have trim, elongate bodies and powerful tail fins and other mechanical advantages for swift pursuit. Sluggish bottom-feeding forms have flattened bodies for movement and concealment on the ocean floor. The elongate body of the eel is an adaptation for wriggling through mud and reeds and into holes and crevices. Some, such as pipefishes, are so whiplike that they are easily mistaken for filaments of marine algae waving in the current. Many other grotesque body forms are obviously cryptic or mimetic adaptations for concealment from predators or as predators. Such few examples cannot begin to express the amazing array of physiologic and anatomic specializations for defense and offense, food gathering, navigation, and reproduction in the diverse aquatic habitats to which bony fishes have adapted themselves.

The Osteichthyes are divided into three clearly distinct groups: Crossopterygii (lobe-finned fishes), Dipneusti (lung-fishes), and Actinopterygii (ray-finned fishes).

Characteristics

1. More or less bony skeleton, numerous vertebrae; usually homocercal tail
2. Skin with mucous glands and with embedded dermal scales of three types: ganoid, cycloid, or ctenoid; some without scales; no placoid scales
3. Both median and paired fins with fin rays of cartilage or bone
4. Terminal mouth with many teeth (some toothless); jaws present; olfactory sacs paired and may or may not open into mouth
5. Respiration by gills supported by bony gill arches and covered by a common operculum
6. Swim bladder often present with or without duct connected to pharynx
7. Circulation consisting of a two-chambered heart, arterial and venous systems, and four pairs of aortic arches
8. Nervous system of a brain with small olfactory lobes and cerebrum and large optic lobes and cerebellum; ten pairs of cranial nerves
9. Separate sexes; paired gonads; usually external fertilization; larval forms may differ greatly from adults

Bony fishes vary greatly in size. Some of the minnows are less than 2 cm long; other forms may exceed 3 m in length. The swordfish is one of the largest and may attain a length of 4 m. Most fishes, however, fall between 2 and 30 cm in length.

Classification

Class Osteichthyes (os-te-ik′thee-eez)—bony fishes. Three subclasses, 69 orders.

    Subclass Crossopterygii (cros-sop-te-rij′ee-i)—**lobe-finned fishes.** Four extinct orders, one living order (Coelacanthimorpha) containing one species, *Latimeria chalumnae.*

    Subclass Dipneusti (dip-nyu′sti)—**lungfishes.** Six extinct orders; two living orders, containing three genera: *Neoceratodus, Lepidosiren,* and *Protopterus.*

    Subclass Actinopterygii (ak′ti-nop-te-rij′ee-i)—**ray-finned fishes.** Three superorders and 59 orders.

        Superorder Chondrostei (kon-dros′tee-i) (Gr. *chrondros,* cartilage, + *osteon,* bone)—**primitive ray-finned fishes.** Ten extinct orders; two living orders containing the bichir *(Polypterus),* sturgeons, and paddlefishes.

        Superorder Holostei (ho-los′tee-i) (Gr. *holos,* entire, + *osteon,* bone)—**intermediate ray-finned fishes.** Four extinct orders; two living orders containing the bowfin *(Amia)* and the gars *(Lepidosteus).*

Superorder Teleostei (tel′e-os′tee-i) (Gr. *teleos,* complete + *osteon,* bone)—**climax bony fishes.** Body covered with thin scales without bony layer (cycloid or ctenoid) or scaleless; dermal and chondral parts of skull closely united; caudal fin mostly homocercal; terminal mouth; notochord a mere vestige; swim bladder mainly a hydrostatic organ and usually not opened to the esophagus; endoskeleton mostly bony. According to J. S. Nelson (1976), there are 31 living orders, 415 families, and approximately 18,000 living species, representing 96% of all living fishes. Seven of the larger orders are as follows

Anguilliformes—21 families; freshwater eels, moray eels, conger eels, snipe eels.

Salmoniformes—23 families; pikes, whitefishes, salmon, trout, smelts, deep-sea luminescent fishes.

Cypriniformes—25 families; suckers, minnows, carps, electric eels.

Siluriformes—30 families; catfishes.

Atheriniformes—15 families; flying fishes, medakas, killifishes, licebearers.

Scorpaeniformes—20 families; rockfishes, searobins, greenlings, sculpins, poachers.

Perciformes—146 families; barracudas, mullets, perches, darters, sunfishes, grunters, croakers, moorish idols, damsel fishes, viviparous perches, wrasses, parrot fishes, trumpeters, sand perches, stargazers, blennies, wolf fishes, eel pouts, mackerels, tunas, swordfishes, and many others.

Lobe-finned fishes—subclass Crossopterygii

The crossopterygians are represented today by a single known species, the famous coelacanth *Latimeria chalumnae* (Fig. 25-7). Since the last coelacanths (seal-a-canths) were believed to have become extinct 70 million years ago, the astonishment of the scientific world can be imagined when the remains of a coelacanth were found on a dredge off the coast of South Africa in 1938. An intensive search was begun in the Comoro Islands area near Madagascar where, it was learned, native Comoran fishermen occasionally caught them with hand lines at great depths. By the end of 1976, more than 80 specimens had been caught, many in excellent condition, although none have been kept alive after capture. The ''modern'' marine coelacanth is a descendant of the Devonian freshwater stock that reached their evolutionary peak in the Mesozoic and then disappeared—or so it was thought until 1938.

The lobe-finned fishes occupy an important position in vertebrate evolution because the amphibians—indeed all tetrapod vertebrates—arose from one or

Fig. 25-7. Coelacanth, *Latimeria chalumnae* (subclass Crossopterygii). This marine surviving relic of the crossopterygians that flourished some 350 million years ago has fleshy-based (''lobed'') fins with which its ancestors used to pull themselves across land from pond to pond. (Courtesy Vancouver Public Aquarium, British Columbia.)

more of their ancient members. The crossopterygians had lungs as well as gills, which would have been a decided advantage to survival during the Devonian, a capricious period of alternating droughts and floods. They used their strong lobed fins as four legs to scuttle from one disappearing swamp to another that offered more promise for a continuing aquatic existence.

Lungfishes—subclass Dipneusti

The lungfishes are another relic group of fishes represented today by only three surviving genera (Fig. 25-8). Like the crossopterygians to which they are related, the lungfishes were a distinct group during the swampy Devonian period when their lungs would have been a distinct asset for survival.

Of the three extant lungfishes, the least specialized is *Neoceratodus,* the living Australian lungfish, which may attain a length of 1.5 m. This lungfish is able to survive in stagnant, oxygen-poor water by coming to the surface and gulping air into its single lung, but it cannot live out of water.

The South American lungfish *Lepidosiren* and the African lungfish *Protopterus* are evolutionary side branches of the Dipneusti, and they can live out of water for long periods of time. *Protopterus* lives in African streams and rivers that run completely dry during the dry season, with their mud beds baked hard by the hot tropical sun. The fish burrows down at the approach of the dry season and secretes a copious slime that is mixed with mud to form a hard cocoon in which it estivates until the rains return (Fig. 25-9).

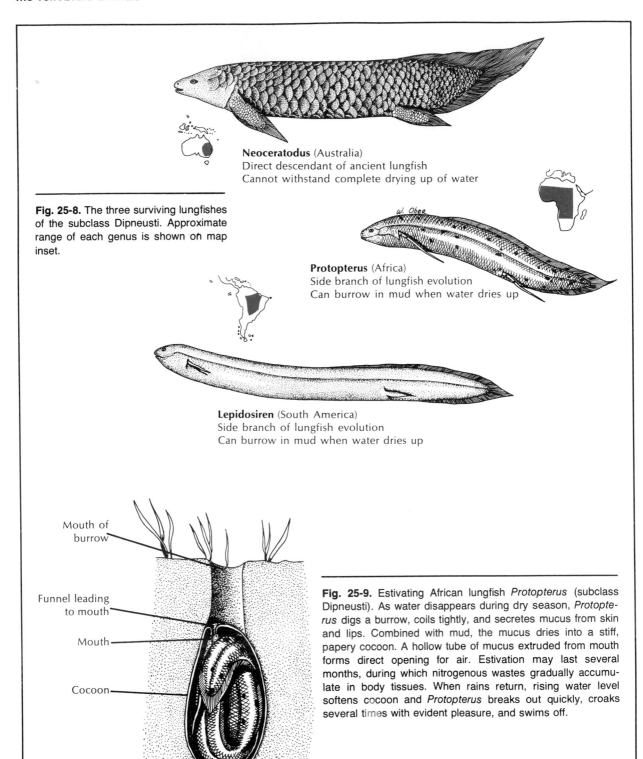

Neoceratodus (Australia)
Direct descendant of ancient lungfish
Cannot withstand complete drying up of water

Fig. 25-8. The three surviving lungfishes of the subclass Dipneusti. Approximate range of each genus is shown on map inset.

Protopterus (Africa)
Side branch of lungfish evolution
Can burrow in mud when water dries up

Lepidosiren (South America)
Side branch of lungfish evolution
Can burrow in mud when water dries up

Mouth of burrow

Funnel leading to mouth

Mouth

Cocoon

Fig. 25-9. Estivating African lungfish *Protopterus* (subclass Dipneusti). As water disappears during dry season, *Protopterus* digs a burrow, coils tightly, and secretes mucus from skin and lips. Combined with mud, the mucus dries into a stiff, papery cocoon. A hollow tube of mucus extruded from mouth forms direct opening for air. Estivation may last several months, during which nitrogenous wastes gradually accumulate in body tissues. When rains return, rising water level softens cocoon and *Protopterus* breaks out quickly, croaks several times with evident pleasure, and swims off.

Ray-finned fishes—subclass Actinopterygii

Ray-finned fishes are an enormous assemblage containing all of our familiar bony fishes. The actinopterygians had their beginnings in the Devonian freshwater lakes and streams. The ancestral forms were small, bony, heavily armored fishes with functional lungs as well as gills. In their evolution, the actinopterygians have passed through three stages.

The most primitive group is the Chondrostei, represented today by the freshwater and marine sturgeons and freshwater paddlefishes and the bichir, or reedfish, *Polypterus,* of African rivers (Fig. 25-1). *Polypterus* is an interesting relic with a lunglike swim bladder and many other primitive characteristics; it resembles an ancestral actinopterygian more than any other living descendant. There is no satisfactory explanation for the survival to the present of certain fish such as this and the coelacanth *Latimeria* when all of their kin perished millions of years ago.

A second actinopterygian group is the Holostei. There were several lines of descent within this group, which flourished during the Triassic and Jurassic. They declined toward the end of the Mesozoic as their successors, the teleosts, crowded them out. But they left two surviving lines, the bowfin, *Amia* (Fig. 25-1), of shallow, weedy waters of the Great Lakes and southeastern United States, and the gars of eastern North and Central America.

The third group is the Teleostei, the modern bony fishes (Fig. 25-10). Diversity appeared early in teleost

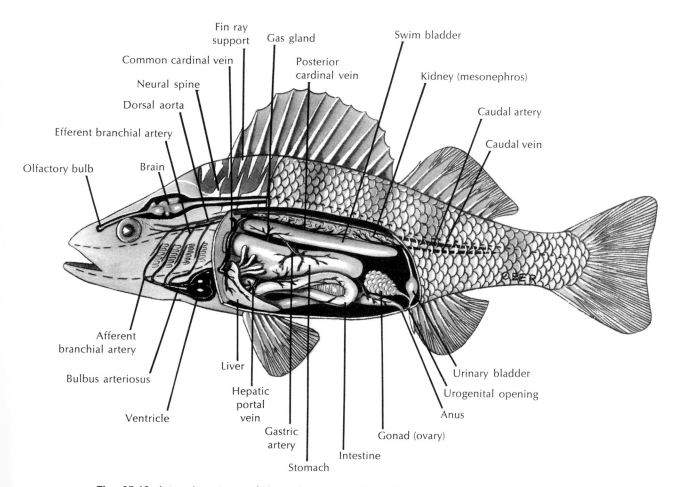

Fig. 25-10. Internal anatomy of the yellow perch, *Perca flavescens* (superorder Teleostei), a freshwater teleost.

evolution, foreshadowing the truly incredible variety of body forms among teleosts today. The skeleton of primitive fish was largely ossified, but this condition regressed to a partly cartilaginous state among many of the Chondrostei and Holostei. Teleosts, however, have an internal skeleton almost completely ossified like the primitive members. The dermal investing bones of the skull (dermatocranium) and the chondrocranium (endocranium) around the brain and sense organs form a closer union among the teleosts than they did in the primitive bony fish. Other evolutionary changes among the teleosts were a freeing of the upper jaw bones permitting a protrusible mouth, the movement of the pelvic fins forward to the head and thoracic region, the transformation of the lungs of primitive forms into air bladders (Figs. 25-10 and 25-12) with hydrostatic functions and without ducts, the changing of the heterocercal tail of primitive fish into a symmetric **homocercal** form, and the development of the thin cycloid and ctenoid scales from the thick ganoid type of early fish. Among other changes were the loss of the spiracles and the development of stout spines in the fins, especially in the pelvic, dorsal, and anal fins.

SOME STRUCTURAL AND FUNCTIONAL ADAPTATIONS OF FISHES
Locomotion in water

To the human eye, fishes appear capable of swimming at extremely high speeds. But our judgment is unconsciously tempered by our own experience of water as a highly resistant medium to move through. Most fishes, such as a trout or a minnow, can swim maximally approximately 10 body lengths per second, obviously an impressive performance by human standards. Yet when these speeds are translated into kilometers per hour it means that a 30 cm (1-foot) trout can swim only approximately 10.4 km (6.5 miles) per hour. The larger the fish, the faster it can swim. A 60 cm salmon can sprint 22.5 km per hour and a 1.2 m barracuda, the fastest fish measured, is capable of 43 km (27 miles) per hour. Fishes can swim this fast only for very brief periods during moments of stress; cruising speeds are much lower.

The propulsive mechanism of a fish is its trunk and tail musculature. The axial, locomotory musculature is composed of zigzag muscle bands (myotomes) that on the surface take the shape of a **W** lying on its side (Fig.

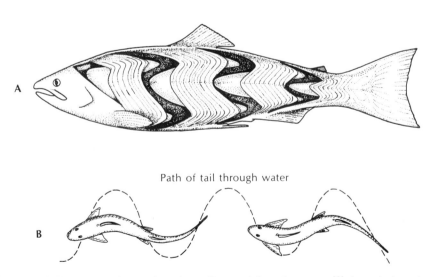

Path of tail through water

Fig. 25-11. A, Trunk musculature of a salmon. Segmental myotomes are W shaped when viewed from the surface. Musculature has been dissected away in four places to show internal anterior and posterior deflections of myotomes that improve muscular efficiency for swimming. **B,** Motion of swimming fish. Noncompressible water must be pushed aside by the forward motion of the head, driven by the snakelike stroke of the body. (**A** after Greene from Romer, A. 1970. The vertebrate body, ed. 4, Philadelphia, W. B. Saunders Co.; **B** modified from Marshall, P. T., and G. M. Hughes. 1967. The physiology of mammals and other vertebrates, New York, Cambridge University Press.)

25-11). Internally the muscle bands are deflected forward and backward in a complex fashion that apparently promotes efficiency of movement. The muscles are bound to broad sheets of tough connective tissue, which in turn tie to the highly flexible vertebral column.

The way fishes swim is seen best in a relatively primitive fish such as a shark. When swimming, the shark body assumes the form of a sine wave. Waves of contraction begin on one side of the body at the front and proceed to the tail. When this wave has moved some distance, another wave is initiated at the front on the opposite side of the body. The process continues, with waves of contraction moving posteriorly, alternating from one side to the other. In higher bony fishes, the sweeping movement of the tail assumes a greater role.

Swimming is possible only because the density and noncompressibility of water offers great purchase for forward thrust. As a medium for locomotion, water offers another advantage: since the density of water is only slightly less than that of protoplasm, aquatic animals are almost perfectly supported and need expend no energy overcoming the force of gravity. Consequently swimming is actually the most economic form of animal locomotion.

For example, the energetic cost per kilogram body weight of traveling 1 km is 0.39 kcal for a salmon (swimming), 1.45 for a gull (flying), and 5.43 for a ground squirrel (walking). However, the low energy cost of fish swimming is by no means fully understood. Relatively simple calculations show that a fish moves through water with only approximately one-tenth the drag of a rigid model of the fish's body. The amount of energy that is required to propel a submarine is many times greater than the amount of energy consumed by a whale of similar size and moving at the same speed.

Even when they are swimming at relatively high speeds, aquatic mammals and fishes create virtually no turbulence, a feat than man in his twentieth-century ingenuity is a long way from matching. The secret lies in the way aquatic animals bend their bodies and fins (or flukes) to swim and in the textural properties of the body surface. It has recently been shown, for example, that the slimy surface of a fish reduces water friction by at least 66%. Understanding the energetics of swimming remains part of the unfinished business of biology.

Neutral buoyancy and the swim bladder

All fishes are slightly heavier than water because their skeletons and other tissues contain heavy elements that are present only in trace amounts in natural waters. To keep from sinking, fishes must either carry some kind of flotation device or, as in the case of sharks, they must always keep moving forward in the water. Sharks are aided in their buoyancy problem by having very large livers containing a special fatty hydrocarbon called **squalene,** which has a density of only 0.86. The liver thus acts like a large sack of buoyant oil that helps to compensate for the shark's heavy body.

By far the most efficient flotation device is a gas-filled space. The **swim bladder** (or gas bladder as it is often called) serves this purpose in the bony fishes. It arose from the paired lungs of the primitive Devonian bony fishes (Fig. 25-12). Lungs were probably a ubiquitous feature of the Devonian freshwater bony fishes when, as we have seen, the alternating wet and dry climate probably made such an accessory respiratory structure essential for life. Swim bladders are present in all oceanic (pelagic) bony fishes but absent in most bottom dwellers such as flounders and sculpins.

By adjusting the volume of gas in the swim bladder, a fish can achieve neutral buoyancy and remain suspended indefinitely at any depth with no muscular effort. There are severe technical problems, however. If the fish descends to a greater depth, the swim bladder gas is compressed, so that the fish becomes heavier and tends to sink. Gas must be added to the bladder to establish a new equilibrium buoyancy. If the fish swims up, the gas in the bladder expands, making the fish lighter. Unless gas is removed, the fish rises with ever-increasing speed while the bladder continues to expand until it pops helplessly out of the water.

There are two ways fish adjust gas volume in the swim bladder. The less specialized fishes (trout, for example) have a **pneumatic duct** that connects the swim bladder to the esophagus (Fig. 25-12); these forms must come to the surface and gulp air to charge the bladder and obviously are restricted to relatively shallow depths. More specialized teleosts have lost the pneumatic duct (upper diagram in Fig. 25-12). Gas exchange depends on two highly specialized areas: a **gas gland,** which secretes gas into the bladder, and a **resorptive area,** or "oval," which can remove gas

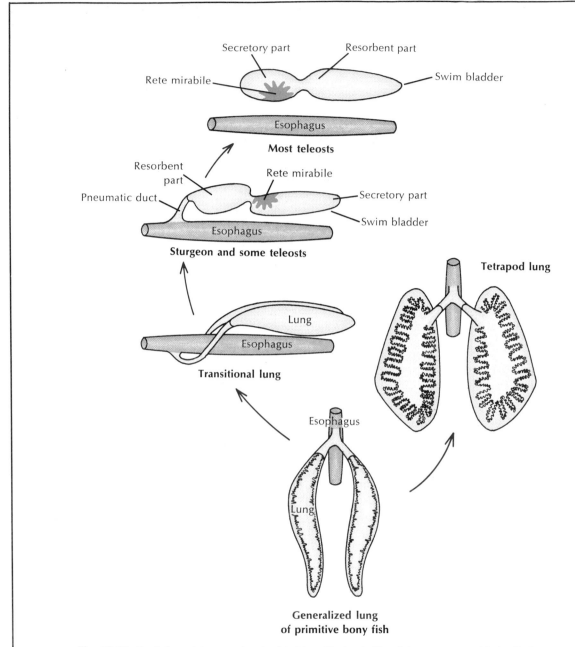

Fig. 25-12. Evolution of lung and swim bladder. Most primitive fishes were provided with lungs, adaptations for the oxygen-depleted environments that existed during the evolution of the Osteichthyes. Lung originated as diverticulum of the foregut. From this early, generalized lung two lines of evolution occurred. One led to swim bladder of modern teleost fish. Various transitional stages show that the swim bladder shifted to a dorsal position above the esophagus, becoming a buoyancy organ. The duct has been lost in most teleosts, and the swim bladder, with specialized gas secretion and reabsorption areas, is served by an independent blood supply. Second line of evolution has led to tetrapod lung found in land forms. There has been extensive internal folding, but no radical change in lung position.

from the bladder. The gas gland contains a remarkable network of blood vessels **(rete mirabile)** arranged so that a vast number of arteries and veins in a tight bundle run in opposite directions to each other. This is called **countercurrent flow,** an arrangement that makes possible a tremendous multiplication of gas concentration inside the swim bladder.

The amazing effectiveness of this device is exemplified by a fish living at a depth of 2,400 m (8,000 feet). To keep the bladder inflated, the gas inside (mostly oxygen, but also variable amounts of nitrogen, carbon dioxide, carbon monoxide, and argon) must have a pressure exceeding 240 atmospheres, much greater than the pressure in a fully charged steel gas cylinder. Yet the oxygen pressure in the fish's blood cannot exceed one-fifth atmosphere—equal to the oxygen pressure at the sea surface.

Respiration

Fish gills are composed of thin filaments covered with a thin epidermal membrane that is folded repeatedly into platelike **lamellae** (Fig. 25-13). These are richly supplied with blood vessels. The gills are located inside the pharyngeal cavity and covered with a movable flap, the **operculum.** This arrangement protects the delicate gill filaments, streamlines the body, and provides a pumping system for moving water through the mouth, across the gills, and out the operculum. Instead of opercular flaps as in bony fishes, the elasmobranchs have a series of **gill slits** out of which the water flows. In both elasmobranchs and bony fishes the branchial mechanism is arranged to pump water continuously and smoothly over the gills, even though to an observer it appears that fish breathing is pulsatile.

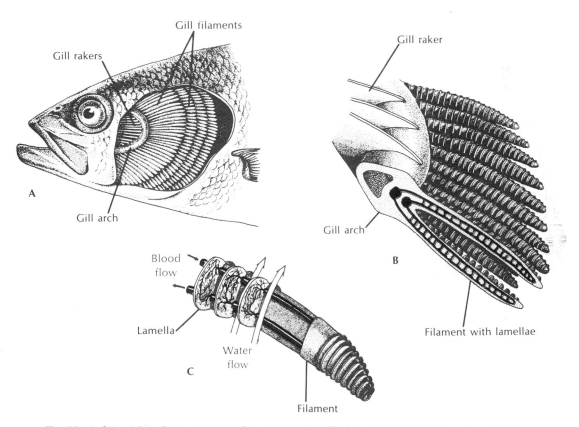

Fig. 25-13. Gills of fish. Bony, protective flap covering the gills (operculum) has been removed, **A,** to reveal branchial chamber containing the gills. Four gill arches are on each side, each bearing numerous filaments. A portion of gill arch, **B,** shows gill rakers that project forward to strain out food and debris and gill filaments that project to the rear. A single gill filament, **C,** is dissected to show the blood capillaries within the platelike lamellae. Direction of water flow *(large arrows)* is opposite the direction of blood flow.

The flow of water is opposite to the direction of blood flow (countercurrent flow), the best arrangement for extracting the greatest possible amount of oxygen from the water. Some bony fishes can remove as much as 85% of the oxygen from the water passing over their gills. Very active fishes, such as herring and mackerel, can obtain sufficient water for their high oxygen demands only by continually swimming forward to force water into the open mouth and across the gills. Such fishes are asphyxiated if placed in an aquarium that restricts free swimming movements, even though the water is saturated with oxygen.

Osmotic regulation

Fresh water is an extremely dilute medium with a salt concentration (0.001 to 0.005 gram moles per liter [M]) much below that of the blood of freshwater fishes (0.2 to 0.3 M). Water therefore tends to enter their bodies osmotically, and salt is lost by diffusion outward. Although the scaled and mucus-covered body surface is almost totally impermeable to water, water gain and salt loss do occur across the thin membranes of the gills.

Freshwater fishes are **hyperosmotic regulators** that have several defenses against these problems (Fig. 25-14). First, the excess water is pumped out by the **mesonephric kidney,** which is capable of forming a very dilute urine. Second, special **salt-absorbing cells** located in the gill epithelium are capable of actively moving salt ions, principally sodium and chloride, from the water to the blood. This, together with salt present in the fish's food, replaces diffusive salt loss. These mechanisms are so efficient that a freshwater fish devotes only a small part of its total energy expenditure to keeping itself in osmotic balance.

Marine bony fishes are **hypoosmotic regulators** that encounter a completely different set of problems. Having a much lower blood salt concentration (0.3 to 0.4 M) than the seawater around them (approximately 1 M), they tend to lose water and gain salt. The marine teleost fish literally risks drying out, much like a desert mammal deprived of water.

Again, marine bony fishes, like their freshwater counterparts, have evolved an appropriate set of defenses (Fig. 25-14). To compensate for water loss, the marine teleost drinks seawater. Although this behavior obviously brings needed water into the body, it is unfortunately accompanied by a great deal of unneeded salt. The latter is disposed of in two ways: (1) the major sea salts (sodium, chloride, and potassium) are pumped out of the body by special salt-secretory cells located in the gills and (2) divalent sea salts (magnesium, sulfate, and calcium) are excreted by the kidney.

Most bony fishes are restricted to either a freshwater or a seawater habitat. Fishes that can tolerate only very narrow ranges of salt concentration—most freshwater and marine fishes—are said to be **stenohaline** (Gr. *stenos,* narrow, + *hals,* salt). However, some 10% of all teleosts can pass back and forth with ease between both habitats. Examples of these **euryhaline fishes,** (Gr. *eurys,* broad, + *hals,* salt), which must have highly adaptable osmoregulatory mechanisms, are salmon, steelhead trout, many flounders and sculpins, killifish, sticklebacks, and eels. Those fishes that migrate from the sea to spawn in freshwater are **anadromous** (Gr. *ana,* up, + *dromos,* a running), such as salmon, shad, and marine lampreys. Freshwater forms that swim to the sea to spawn are **catadromous** (Gr. *kata,* down, + *dromos,* a running), such as the freshwater eel, *Anguilla.*

Coloration and concealment

Tropical coral reef fishes bear some of the most resplendent hues and strikingly brilliant color patterns in the animal kingdom. Viewed against the coral reef background of invertebrate and plant life that create a riot of color, the vivid markings and coloration of reef fishes attract relatively little attention. Their coloration is not always concealing, however, since tropical bony fishes tend toward vivid coloration even in areas of dull and somber backgrounds. Although conspicuous in these circumstances, they are protected by alertness and agility or by their poisonous flesh. In this instance the coloration is an advertisement, warning would-be predators that they should seek their meal elsewhere.

Outside of coral reefs and other littoral habitats of the tropical seas, fishes, like most other animals, characteristically bear colors and patterns that serve to conceal them from enemies. Freshwater fishes wear subdued shades of green, brown, or blue above, grading to silver and yellow-white below. This is **obliterative shading.** Seen from above against its normal background of water and stream bottom, the fish becomes almost invisible. Seen from beneath as it might be viewed by an aquatic predator, the pale belly of the fish blends with the water surface and sky above. The obliterative coloration is frequently en-

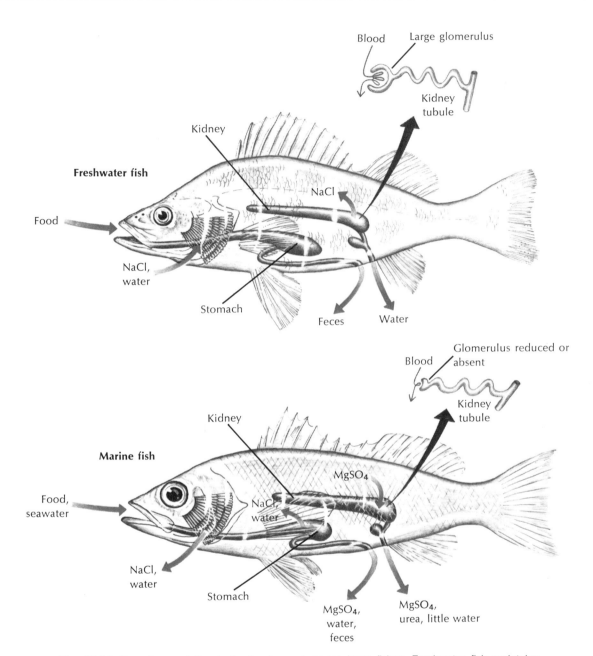

Fig. 25-14. Osmotic regulation in freshwater and marine bony fishes. Freshwater fish maintains osmotic and ionic balance in its dilute environment by actively absorbing sodium chloride across gills (some salt enters with food). To flush out excess water that constantly enters body, glomerular kidney produces a dilute urine by reabsorbing sodium chloride. Marine fish must drink seawater to replace water lost osmotically to its salty environment. Sodium chloride and water are absorbed from stomach. Excess sodium chloride is secreted outward by gills. Divalent sea salts, mostly magnesium sulfate, are eliminated with feces and secreted by tubular kidney. (Adapted from Webster, D., and M. Webster, 1974. Comparative vertebrate morphology. New York, Academic Press, Inc.)

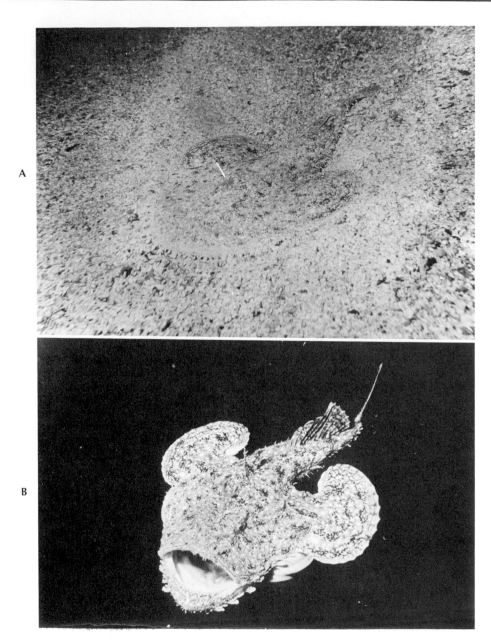

Fig. 25-15. A fish that fishes. **A,** Bedded down in the ocean bottom, beautifully concealed by protective coloration and fringed skin that breaks up the body outline, a goosefish, or angler, *Lophius piscatorius* (superorder Teleostei), awaits its meal. Above its head swings a modified dorsal fin spine on the end of which is a fleshy tentacle *(arrow)* that contracts and expands in a convincing wormlike manner. When a fish approaches the alluring bait, the huge mouth opens suddenly, creating a strong current that sweeps the prey inside; in a split second all is over. Prodded out of its resting place by a skin diver, **B,** the goosefish reveals its true appearance. Photographed off the Norwegian coast at 25 m depth. (Photos by T. Lundälv.)

hanced by blotches, spots, and bars—conflicting patterns that, like the camouflaged ships of the Second World War, tend to break up the outline of the body. Some fishes, such as many flatfish species, can change their color to harmonize with the patterns of their background.

Fish colors are chiefly the result of red, orange, yellow, and black pigments within **chromatophores** that can be blended to produce other shades. Fish also bear **guanine** in their skin, a purine compound that gives many fish their silvery appearance.

Another form of concealment is protective resemblance, or **mimicry.** There are many examples of protective body form that appear to turn their owners into fragmented seaweed, a leaf, floating debris, or a weed-covered rock on the bottom. The goosefish, or angler, with its obliterative coloration and numerous fringes and branched appendages of skin, not only becomes nearly invisible when bedded down on the ocean floor but sways a tempting bait above its huge mouth to attract prey (Fig. 25-15).

Migration

Eel. For centuries naturalists had been puzzled about the life history of the freshwater eel, *Anguilla* (angwil′la), a common and commercially important species of coastal streams of the North Atlantic. Each fall, large numbers of eels were seen swimming down the rivers toward the sea, but no adults ever returned. Each spring countless numbers of young eels, called ''elvers,'' each approximately the size of a wooden matchstick, appeared in the coastal rivers and began swimming upstream. Beyond the assumption that eels must spawn somewhere at sea, the location of their breeding grounds was totally unknown.

The first clue was provided by two Italian scientists, Grassi and Calandruccio, who in 1896 reported that elvers were not, in fact, larval eels; rather they were relatively advanced juveniles. The true larval eels, the Italians discovered, were tiny, leaf-shaped, completely transparent creatures that bore absolutely no resemblance to an eel. They had been called **leptocephali** by early naturalists who never suspected their true identity. In 1905 Johann Schmidt, supported by the Danish government, began a systematic study of eel biology, which he continued until his death in 1933. Through the cooperation of captains of commercial vessels plying the Atlantic, thousands of the leptocephali were caught in different areas of the Atlantic with the plankton nets that Schmidt supplied. By noting in what areas of the ocean larvae in different stages of development were captured, Schmidt and his colleagues eventually reconstructed the spawning migrations.

It is assumed that when the adult eels leave the coastal rivers of Europe and North America, they swim steadily and apparently at great depth for 1 to 2 months until they reach the Sargasso Sea, a vast area of warm oceanic water southeast of Bermuda (Fig. 25-16). Here, at depths of 1,000 feet or more, the eels spawn and die. The minute larvae then begin an incredible journey back to the coastal rivers of Europe. Drifting with the Gulf Stream and preyed upon constantly by numerous predators, they reach the middle of the Atlantic after 2 years. By the end of the third year they reach the coastal waters of Europe where the leptocephali metamorphose into elvers, with an unmistakable eellike body form. The males and females part company; the males remain in the brackish waters of coastal rivers and estuaries while the females continue up the rivers often penetrating hundreds of miles upstream. After 8 to 15 years of growth, the females, now 1 m or more in length, return to the sea to join the smaller males; both return to the ancestral breeding grounds thousands of miles away to complete the life cycle.

Schmidt found that the American eel *(Anguilla rostrata)* could be distinguished from the European eel *(A. anguilla)* because it had fewer vertebrae—an average of 107 in the American eel as compared to an average 114 in the European species. Since the American eel is much closer to the North American coastline, it requires only approximately 8 months to make the journey.

Homing salmon. The life history of salmon is nearly as remarkable as that of the eel and certainly has received far more popular attention. Salmon are usually **anadromous;** that is, they spend their adult lives at sea but return to fresh water to spawn. The Atlantic salmon *(Salmo salar)* and the Pacific salmon (five North American species of the genus *Oncorhynchus* [on-ko-rink′us]) have this practice, but there are important differences between the six species. The Atlantic salmon (as well as the related steelhead trout) make upstream spawning runs year after year. The five salmon species of the Pacific northwest (chinook, sockeye, coho, pink, and chum) each make a single spawning run, after which they die.

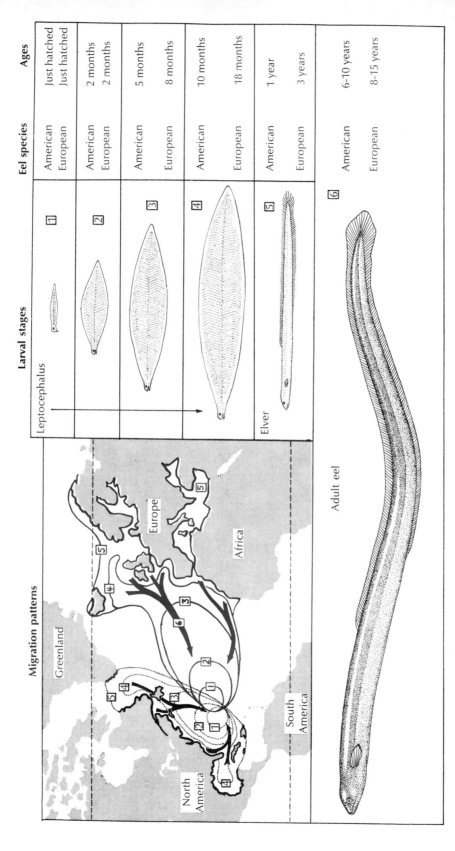

Fig. 25-16. Life histories of the European eel (*Anguilla anguilla*) and American eel (*Anguilla rostrata*). *Red*, Migration patterns of American species. *Black*, Migration patterns of European species. Boxed numbers refer to stages of development. Note that the American eel completes its larval metamorphosis and sea journey in 1 year. It requires nearly 3 years for the European eel to complete its much longer journey.

Larval stages

Leptocephalus

Elver

Adult eel

Eel species	Ages
American	Just hatched
European	Just hatched
American	2 months
European	2 months
American	5 months
European	8 months
American	10 months
European	18 months
American	1 year
European	3 years
American	6-10 years
European	8-15 years

Migration patterns

Greenland

Europe

Africa

North America

South America

The virtually infallible homing instinct of the Pacific species is legend: after migrating downstream as a young smolt, a sockeye salmon ranges many hundreds of miles over the North Pacific for 2 to 5 years, grows to 2 to 5 kg in weight, and then returns unerringly to spawn in its parent stream.

Many years ago Canadian biologists marked and released nearly a half million young sockeyes born in a tributary of British Columbia's Fraser River. Eleven thousand of these were recovered 4 years later in the same parent stream; not one was discovered to have strayed into any of the dozens of other Fraser River tributaries.

Experiments have shown that homing salmon are guided upstream by the characteristic odor of their parent stream. The salmon are apparently imprinted with the stream's odor while they are still unhatched embryos, since if the eggs are flown from the parent stream to another stream miles away, the adults still return to the parent stream after their residence at sea. The odor compound is a volatile organic substance, but its exact chemical nature remains unidentified; the embryonic salmon are probably conditioned to a mosaic of compounds released by the characteristic vegetation and soil in the water-shed of the parent stream.

It is not yet understood how salmon find their way to the mouth of the river from the trackless miles of the open ocean. It is impossible that they are capable of distinguishing the parent stream odor while still hundreds of miles at sea. Recent experiments suggest that adult salmon, like birds, are guided in the ocean by celestial cues (stars or azimuth position of the sun). To do this the salmon would require time-keeping abilities, certainly not an unreasonable possibility in view of the widespread presence of "biologic clocks" in many organisms from the simplest to the most advanced. But the crucial experiment remains to be performed.

SELECTED REFERENCES

Harden Jones, F. R. 1968. Fish migrations. London, Edward Arnold Ltd. *Scholarly review of fish migration and orientation.*

Hardisty, M. W., and I. C. Potter. 1971-1972. The biology of lampreys, 2 vols. New York, Academic Press, Inc. *Thorough presentation of systematics, life histories, ecology, behavior, physiology, and economic impact of this small but biologically interesting group.*

Hoar, W. S., and D. J. Randall (eds.). 1969-1972. Fish physiology, 6 vols. New York, Academic Press, Inc. *This series represents the most complete and authoritative treatise on the functional biology of fishes published. Technical and detailed.*

Lineaweaver, T. H., III, and R. H. Backus. 1970. The natural history of sharks. Philadelphia, J. B. Lippincott Co. *One of the best of many books dealing with these intriguing animals.*

Marshall, N. B. 1970. The life of fishes. New York, Universe Books. *Excellent general biology of fishes.*

Marshall, N. B. 1971. Explorations in the life of fishes. Cambridge, Mass., Harvard University Press. *More technical and selective than the author's 1970 book, dealing especially with deep-sea fishes and evolution.*

Nelson, J. S. 1976. Fishes of the world. New York, John Wiley & Sons, Inc. *A modern classification of all major groups of fishes.*

Netboy, A. 1974. The salmon: their fight for survival. Boston, Houghton Mifflin Co. *Thorough and readable treatise of the biology of the magnificent but embattled salmon.*

Norman, J. R. 1963. A history of fishes, ed. 2 (revised by P. H. Greenwood). New York, Hill & Wang. *A revision of a famous treatise that covers nearly all aspects of fish study.*

Rounsefell, G. A. 1975. Ecology, utilization, and management of marine fishes. St. Louis, The C. V. Mosby Co. *Comprehensive textbook of marine fishery science.*

Schultz, L. P., and E. M. Stern. 1948. The ways of fishes. New York, D. Van Nostrand Co. *Despite its age, this is still an excellent account of fish biology at the semipopular level. Recommended for the beginning student.*

SELECTED SCIENTIFIC AMERICAN ARTICLES

Brett, J. R. 1965. The swimming energetics of salmon. **213:**80-85 (Aug.). *Describes studies on the remarkable efficiency of swimming by fish.*

Carey, F. G. 1973. Fishes with warm bodies. **228:**36-44 (Feb.). *Some tuna and mackerel shark species employ a circulatory heat exchanger to conserve body heat and increase swimming power.*

Gilbert, P. W. 1962. The behavior of sharks. **207:**60-68 (July).

Gray, J. 1957. How fishes swim. **197:**48-54 (Aug.). *This British biologist's calculations of the seemingly impossible swimming efficiency of fishes have been referred to as "Gray's paradox."*

Grundfest, H. 1960. Electric fishes. **203:**115-124 (Oct.).

Hasler, A. D., and J. A. Larsen. 1955. The homing salmon. **193:**72-76 (Aug.). *Describes experiments that indicate that salmon locate parent stream by using sense of smell.*

Jensen, D. 1966. The hagfish. **214:**82-90 (Feb.).

Johansen, K. 1968. Air-breathing fishes. **219:**102-111 (Oct.). *Not all fishes are water breathers. Many remarkable alternatives for air breathing evolved in fishes; several are described in this article.*

Leggett, W. C. 1973. The migrations of the shad. **228:**92-98 (Mar.).

Lühling, K. H. 1963. The archer fish. **209:**100-108 (July). *This small fish of southeast Asia spouts a stream of water to down insects it sights on plants above the water.*

Millot, J. 1955. The coelacanth. **193:**34-39 (Dec.). *Its biology and evolutionary relationships are discussed.*

Ruud, J. T. 1965. The ice fish. **213:**108-114 (Nov.). *A family of transparent Antarctic fishes possesses neither red blood cells nor hemoglobin. The physiologic consequences are described.*

Shaw, E. 1962. The schooling of fishes. **206:**128-138 (June).

Red-backed salamander, *Plethodon cinereus,* encircles her hatching brood in their nest of forest humus. Unlike most amphibians, which after laying their eggs show no further interest in them, females of the large North American family Plethodontidae remain with their eggs during incubation and may even return to the same nest site year after year. (Drawn from photo by P A. Zahl.)

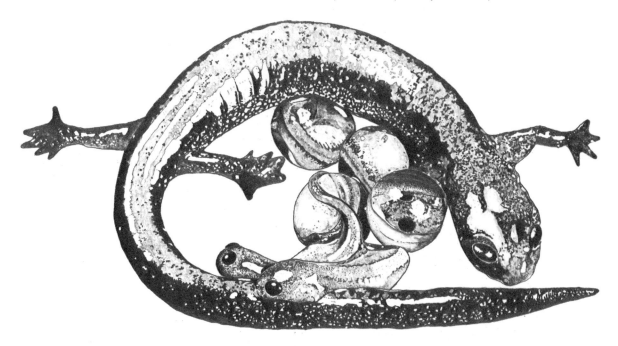

26 AMPHIBIANS

The chorus of frogs beside a pond on a spring evening heralds one of nature's dramatic events. Masses of frog eggs soon hatch into limbless, gill-breathing, fishlike tadpole larvae. Warmed by the late spring sun, they feed and grow. Then, almost imperceptibly, a remarkable transformation unfolds. Hind legs appear and gradually lengthen. The tail shortens. Larval teeth are lost, and the gills are replaced by lungs. The eyes develop lids. Forelegs emerge. In a matter of weeks the aquatic tadpole has completed its metamorphosis to adult frog.

The early members of the class Amphibia (am-fib′e-a) (Gr. *amphi,* both or double, + *bios,* life), of which our chorusing frogs are among the more vociferous modern descendants, originated not in weeks but over millions of years by a lengthy series of almost imperceptible alterations that gradually fitted the vertebrate body plan for life on land. The origin of land vertebrates is no less a remarkable feat for this fact—a feat that incidentally would have a poor chance of succeeding today because well-established competitors make it impossible for a poorly adapted transitional form to gain a foothold.

Even now after some 350 million years of evolution, the amphibians are not completely land adapted; they hover between aquatic and land environments. This double life is expressed in their name. Structurally they are between fishes and reptiles. Although adapted for a terrestrial existence, few can stray far from moist conditions. Many, however, have developed devices for keeping their eggs out of open water where the larvae would be exposed to enemies.

More than 2,500 species of amphibians are grouped into three living orders: the newts and salamanders (order Urodela), least specialized and most aquatic of all amphibians; the frogs and toads (order Salientia), largest and most successful group of amphibians and closest to the stock from which the higher tetrapods (animals with four legs) descended; and the highly specialized, secretive, earthworm-like tropical caecilians (order Gymnophiona).

MOVEMENT ONTO LAND

The movement from water to land is perhaps the most dramatic event in animal evolution since it involves the invasion of a habitat that in many respects is more hazardous for life. The origin of life was conceived in water, animals are mostly water in composition, and all cellular activities occur in water.

Nevertheless organisms eventually invaded the land, carrying their watery composition with them. To survive and maintain this fluid matrix, various structural, functional, and behavioral changes had to evolve. Considering that almost every system in the body required some modification, it is remarkable that all vertebrates are basically alike in fundamental structural and functional pattern: whether aquatic or terrestrial, vertebrates are obviously descendants of the same evolutionary branch.

Amphibians were not the first to move onto land. Insects made the transition earlier and plants much earlier still. The pulmonate snails were experimenting with land as a suitable place to live about the same time the early amphibians were. Yet of all these, the amphibian story is of particular interest because their descendants became the most successful and advanced animals on earth.

Physical contrast between aquatic and land habitats

Beyond the obvious difference in water content of aquatic and terrestrial habitats—water is wet and land is dry—several sharp differences between the two environments are of significance to animals attempting to move from water to land.

Greater oxygen content of air. Air contains at least 20 times more oxygen than water. Air has approximately 210 ml of oxygen per liter; water contains 3 to 9 ml per liter (depending upon temperature, presence of other solutes, and degree of saturation). Furthermore the diffusion rate of oxygen is low in water. Consequently aquatic animals must expend far more effort extracting oxygen from water than land animals expend removing oxygen from air.

Greater density of water. Water is approximately 1,000 times denser than air and approximately 100 times more viscous. Although water is a much more resistant medium to move through, its high density, approximately equal to that of animal protoplasm, buoys up the body. One of the major problems encountered by land animals was the need to develop strong limbs and remodel the skeleton to support their bodies in air.

Constancy of temperature in water. Natural bodies of water, containing a medium with tremendous thermal capacity, experience little fluctuation in temperature. The temperature of the oceans remains almost constant day after day. In contrast, both the range and the fluctuation in temperature are acute on land. Its harsh cycles of freezing, thawing, drying, and flooding, often in unpredictable sequence, present severe thermal problems to terrestrial animals.

Variety of land habitats. The variety of cover and shelter on land was a great inducement for its colonization. The rich offerings of terrestrial habitats include coniferous, temperate, and tropical forests, grasslands, deserts, mountains, oceanic islands, and polar regions. Even so, earth's hydrosphere (oceans, seas, lakes, rivers, and ice sheets), though offering a less diverse range of habitats, contains the greatest number and variety of living things on earth.

Opportunities for breeding on land. The provision of safe shelter for the protection of vulnerable eggs and young is much more readily accomplished on land than in water habitats.

ORIGIN AND RELATIONSHIPS OF AMPHIBIANS

The movement onto land required structural modifications of the fish body plan. Unlike a fish, which is supported and wetted by its medium and supplied with dissolved oxygen, a terrestrial animal must support its own weight, resist drying and rapid temperature change, and extract oxygen from air.

Appearance of lungs. The Devonian period, beginning some 400 million years ago, was a time of mild temperatures and alternating droughts and floods. During dry periods, pools and streams began to dry up, water became foul, and the dissolved oxygen disappeared. Only those fishes that were able to utilize the abundance of atmospheric oxygen could survive such conditions. Gills were unsuitable because in air the filaments collapsed into clumps that soon dried out.

Virtually all survivors of this period, including the lobe-finned fishes and the lungfishes, had a kind of lung developed as an outgrowth of the pharynx. It was relatively simple to enhance the efficiency of the air-filled cavity by improving its vascularity with a rich capillary network and by supplying it with arterial blood from the last (sixth) pair of aortic arches. Oxygenated blood was returned directly to the heart by a pulmonary vein to form a complete pulmonary circuit. Thus the double circulation characteristic of all tetrapods originated: a systemic circulation, serving the body, and a pulmonary circulation, supplying the lungs.

Limbs for travel on land. The evolution of limbs

was also a product of difficult times during the Devonian period. When pools dried up, fishes were forced to move to another pool that still contained water. Only the lobe-finned fishes (crossopterygians) were preadapted for the task. They had strong lobed fins used originally as swimming stabilizers, that could be adapted as paddles to lever their way across land in search of water. The pectoral fins were especially well developed, containing a series of skeletal elements in the fins and pectoral girdle that clearly foreshadowed the pentadactyl limb of tetrapods. We should note that the development first of strong fins and later of limbs did not happen so that fish could colonize land but rather so that they could find water and continue living like fish. Land travel was simply and paradoxically a means for survival in water. But the evolution of lungs and limbs were fortunate and essential specializations that preadapted vertebrates for life on land.

Earliest amphibians

All evidence points to the lobe-finned fishes as ancestors of the modern amphibians. The lobe-fins, abundant and successful in the Devonian, possessed lungs and strong, mobile fins. Their skull and tooth structure was similar to that of the earliest known amphibians, the Labyrinthodontia, a distinct salamander-like group of the late Devonian.

A representative of this group was a 350 million–year–old fossil called *Ichthyostega* (Fig. 26-1). *Ichthyostega* possessed several new adaptations that equipped it for life on land. It had jointed, pentadactyl limbs for crawling on land, a more advanced ear structure for picking up airborne sounds, a foreshortening of the skull, and a lengthening of the snout that announced improved olfactory powers for detecting dilute airborne odors. Yet *Ichthyostega* was still fish-like in retaining a fish tail complete with fin rays and in having opercular (gill) bones.

The capricious Devonian period was followed by the Carboniferous period, characterized by a warm, wet climate during which mosses and large ferns grew in profusion on a swampy landscape. Conditions were ideal for the amphibians. They radiated quickly into a great variety of species, feeding on the abundance of insects, insect larvae, and aquatic invertebrates available: this was the Age of Amphibians.

With water everywhere, however, there was little selective pressure to encourage movement onto land and many amphibians actually improved their ad-

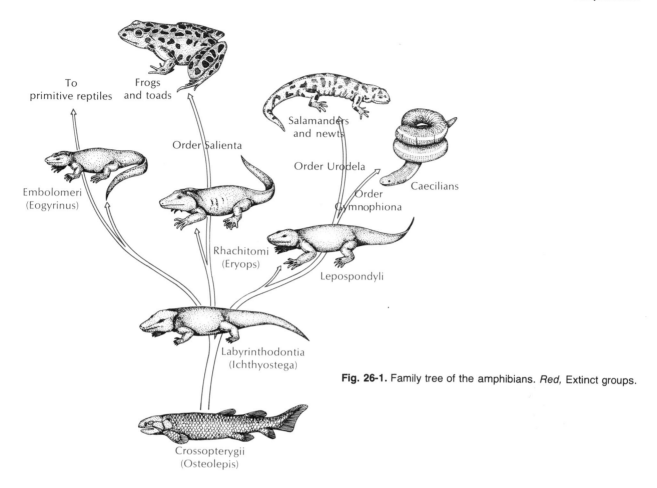

To
primitive reptiles

Frogs
and toads

Salamanders
and newts

Order Salienta

Order Urodela

Caecilians

Embolomeri
(Eogyrinus)

Order
Gymnophiona

Rhachitomi
(Eryops)

Lepospondyli

Labyrinthodontia
(Ichthyostega)

Fig. 26-1. Family tree of the amphibians. *Red,* Extinct groups.

Crossopterygii
(Osteolepis)

aptations for living in water. Their bodies became flatter for moving about in shallow water. Many of the urodeles (newts and salamanders), which may have descended from the lepospondyls (Fig. 26-1), developed weak limbs. The tail became better developed as a swimming organ. Even the anurans (frogs and toads), which are the most terrestrial of all amphibians, developed specialized hind limbs with webbed feet better suited for swimming than for movement on land. All groups of amphibians use their porous skin as an accessory breathing organ. This specialization was encouraged by the swampy surroundings of the Carboniferous but presented serious desiccation problems for life on land.

Amphibians' contribution to vertebrate evolution

Amphibians have met the problems of independent life on land only halfway. To be sure, they made

several important contributions to the transition that required the evolution of their own descendants, the reptiles, to complete. Of crucial importance were the change from gill to lung breathing and the development of limbs for locomotion on land. Amphibians also show strengthening changes in the skeleton so that the body can be supported on land. A start was also made toward shifting special sense priorities from the lateral line system of fish to the senses of smell and hearing. For this, both the olfactory epithelium and the ear required redesigning to improve sensitivities to airborne odors and sounds.

Despite these modifications, the amphibians are basically aquatic animals. They are ectothermic; that is, their body temperature is determined by and varies with the environmental temperature. Their skin is thin, moist, and unprotected from desiccation in air. An intact frog loses water nearly as rapidly as a skinless frog. Most important, the amphibians remain chained

Fig. 26-2. Female Surinam toad, carrying young on her back. As eggs are laid, male assists in positioning them on rough back skin of female. The skin swells, enclosing eggs. The approximately 60 young pass through the tadpole stage beneath the skin and emerge as small frogs. Surinam "toad," actually a frog, is found mainly in Amazon and Orinoco river systems of equatorial South America. (Courtesy American Museum of Natural History.)

to the aquatic environment by their mode of reproduction. Eggs are shed directly into the water or laid in moist surroundings and are externally fertilized (with very few exceptions). The larvae that hatch typically pass through an aquatic tadpole stage.

Many amphibians have developed ingenious devices for laying their eggs elsewhere to give their young protection and a better chance for life. They may lay eggs under logs or rocks, in the moist forest floor, in flooded tree holes, in pockets on the mother's back (Fig. 26-2), or in folds of the body wall. One species of Australian frog even broods its young in its stomach.

However, it remained for the reptiles to complete the conquest of land with the development of a shelled (amniotic) egg, which finally freed the vertebrates from a reproductive attachment to the aquatic environment. With the appearance of reptiles at the end of the Paleozoic, the halcyon era for the amphibians began to fade. The reptiles captured rule of both water and land and removed most amphibians from both environments. From the survivors have descended the three modern orders of amphibians.

Characteristics

1. Mostly bony skeleton; ribs present in some, absent in others
2. Body forms vary greatly from an elongated trunk with distinct head, neck, and tail to a compact, depressed body with fused head and trunk and no intervening neck

3. **Usually four limbs (tetrapod); webbed feet often present**
4. **Smooth and moist skin with many glands,** some of which may be poison glands; **pigment cells (chromatophores)** common; **no scales,** except concealed dermal ones in some
5. Mouth usually large with small teeth in upper or both jaws; **two nostrils open into anterior part of mouth cavity**
6. Respiration by gills, lungs, skin, and pharyngeal region either separately or in combination
7. **Circulation with three-chambered heart,** two auricles and one ventricle, and a double circulation through the heart
8. Separate sexes; metamorphosis usually present; **eggs with jellylike membrane coverings**

Brief classification

Order Gymnophiona (jim′no-fy′o-na) (Gr. *gymnos,* naked, + *ophioneos,* of a snake) **(Apoda)—caecilians.** Body wormlike; limbs and limb girdle absent; mesodermal scales may be present in skin; tail short or absent; tropical; one family, 17 genera, approximately 160 species

Order Urodela (yu′ro-dee′la) (Gr. *oura,* tail, + *dēlos,* visible) **(Caudata)—salamanders, newts.** Body with head, trunk, and tail; no scales; usually two pairs of equal limbs; eight families, 51 genera, approximately 300 species

Order Salientia (say′lee-ench′e-a) (L. *saliens,* leaping, + *-ia,* pl. suffix) **(Anura)—frogs, toads.** Head and trunk fused; no tail; no scales; two pairs of limbs; large mouth; lungs; 10 vertebrae, including urostyle; 13 families, 100 genera, approximately 2,000 species

STRUCTURE AND NATURAL HISTORY OF AMPHIBIAN ORDERS
Caecilians—order Gymnophiona (Apoda)

The little-known order Gymnophiona contains approximately 160 species of burrowing, wormlike creatures commonly called caecilians (Fig. 26-1). They are distributed in tropical forests of South America (their principal home), Africa, and southeast Asia. Blind or nearly blind, slender-bodied, and limbless, they feed on worms and small invertebrates underground. In their native tropical habitats they are seldom seen by man.

Salamanders and newts—order Urodela (Caudata)

As the name of the order suggests, the order Urodela consists of tailed amphibians, the salamanders and newts. This compact, natural group is the least specialized of all the amphibians. Although urodeles are found in almost all temperate and tropical regions of the world, most species occur in North America. Urodeles are typically small; most of the common North American salamanders are less than 6 inches

long. Some aquatic forms are considerably longer, and the carnivorous Japanese giant salamander may exceed 5 feet in length.

Urodeles have primitive limbs set at right angles to the body with forelimbs and hindlimbs of approximately equal size. In some the limbs are rudimentary. One group, the sirens, with minute forelimbs and no hindlimbs at all, is so different from other urodeles that some authorities place them in a completely separate order, Trachystomata.

Salamanders and newts prey on worms, small arthropods, and small molluscs. Most eat only things that are moving. Since their food is rich in proteins, they do not usually store in their bodies great quantities of fat or glycogen. Like all amphibians they are ectotherms and have a low metabolic rate.

Breeding behavior. Some urodeles are wholly aquatic throughout their life cycle, but most are terrestrial, living in moist places under stones and rotten logs, usually not far from water. They do not show as great a diversity of breeding habits as do frogs and toads. The eggs of most salamanders are fertilized internally, usually after the female picks up a packet of sperm **(spermatophore)** that previously has been deposited by the male on a leaf or stick. Aquatic species lay their eggs in clusters or stringy masses in the water. Terrestrial species deposit eggs in small grapelike clusters under logs or excavations in soft earth (Fig. 26-3). Unlike frogs and toads that hatch into fishlike tadpole larvae, the embryos of urodeles hatch from their eggs, resembling their parents. The larvae undergo metamorphosis in the course of development, but it is not nearly so revolutionary a change as is the metamorphosis of frog and toad tadpoles to the adult body form.

Respiration. All urodeles hatch with gills, but during development these are lost in all except the aquatic forms or in those that fail to undergo a complete metamorphosis. Gills would be useless for terrestrial salamanders since the filaments would collapse and dry out to become functionless. Lungs, the characteristic respiratory organ of terrestrial vertebrates, replace the larval gills in most adult amphibians.

Yet some salamanders have dispensed with lungs altogether and thus bear the distinction of being the only vertebrates to have neither lungs nor gills. Members of the large family Plethodontidae, a group containing most of the familiar North American sal-

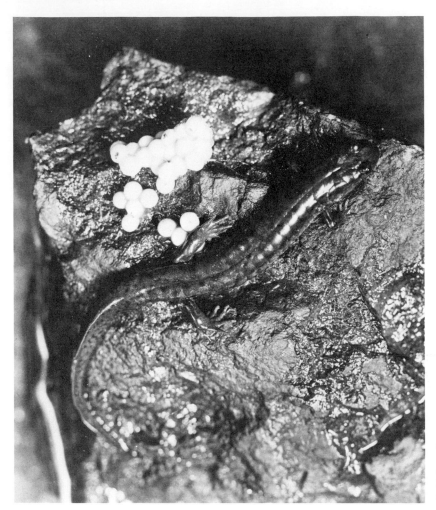

Fig. 26-3. A dusky salamander *(Desmognathus phoca)* broods her eggs. Many salamanders remain with their eggs throughout the incubation of about 2 months (Courtesy American Museum of Natural History.)

amanders (Fig. 26-4), are completely lungless, and some members of other urodele families exhibit reductions in lung development. In all amphibians the skin contains extensive vascular nets that serve in varying degrees for the respiratory exchange of oxygen and carbon dioxide. In lungless salamanders the efficiency of cutaneous respiration is increased by the penetration of a capillary network into the epidermis or by the thinning of the epidermis over superficial dermal capillaries. Cutaneous respiration is supplemented by the pumping of air in and out of the mouth where the respiratory gases exchange across the vascularized membranes of the buccal (mouth) cavity (buccopharyngeal breathing). Lungless plethodontid salamanders are believed to have originated in swift streams of the Appalachian mountains, where the

water is so cool and well oxygenated that cutaneous respiration alone was sufficient for life.

Paedogenesis and neoteny. Whereas most urodeles complete their development to the definitive adult body form by metamorphosis, there are some species that retain their gills and other larval characteristics even after becoming sexually mature. Some are **permanent larvae,** a genetically fixed condition in which the developing tissues fail to respond to the thyroid hormone that, in other amphibians, stimulates metamorphosis. This condition is called **paedogenesis** (Gr. *pais,* child, + *genesis,* descent).

Examples of *permanent larvae* are mud puppies of the genus *Necturus,* which live on bottoms of ponds and lakes and keep their external gills throughout life, and the congo eel *(Amphiuma means)* of the southeast-

Fig. 26-4. Common salamanders of the family Plethodontidae. All members of this large North American family of about 200 species lack lungs. **A,** Red-backed salamander, *Plethodon cinereus,* has dorsal reddish stripe with gray to black sides. **B,** Two-lined salamander, *Eurycea bislineata,* is yellow to brown with two dorsolateral black stripes. **C,** Long-tailed salamander, *Eurycea longicauda,* is yellow to orange with black spots that form vertical stripes on sides of tail. **D,** Slimy salamander, *Plethodon glutinosus,* is black with white spots. (Photos by F. M. Hickman.)

ern United States, which with its useless, rudimentary legs superficially resembles an eel more than an amphibian.

Other species of salamanders become sexually mature and breed in the larval state, but, unlike the paedogenetic forms, they may metamorphose to adults if environmental conditions change. This is called **neoteny** (Gr. *neos,* young, + *teinen,* to extend).

Examples are species of the genus *Ambystoma* and of the genus *Triturus.* The American axolotl, *Ambystoma tigrinum,* widely distributed over Mexico and the southwestern United States, remains in the aquatic, gill-breathing, and fully reproductive larval form unless the water begins to dry up; then it metamorphoses to an adult, loses its gills, develops lungs, and assumes the appearance of an ordinary salamander.

Axolotls can be made to metamorphose by treating them with the thyroid hormone, thyroxin. Thyroxin is essential for normal metamorphosis in all amphibians. Recent research suggests that for some reason the pituitary gland fails to become fully active in neotenous forms and does not release thyrotropin, which is required to stimulate the production of thyroxin by the thyroid gland.

Frogs and toads—order Salientia (Anura)

The more than 2,000 species of frogs and toads that comprise the order Salientia are the most familiar and most successful of amphibians. Frogs and toads are highly specialized for a jumping mode of locomotion, as suggested by the preferred name of the order,

Fig. 26-5. Some common North American frogs. **A,** Bullfrog, *Rana catesbeiana,* is largest of all American frogs. **B,** Green frog, *Rana clamitans,* is next to bullfrog in size. Body is usually green, especially around jaws; has dark bars on sides of legs. **C,** Leopard frog, *Rana pipiens,* has light-colored dorsolateral ridges and irregular spots. **D,** Spring peepers, *Hyla crucifer,* the darlings of warm spring nights, when their characteristic peeping is so often heard. They are small (2 to 3 cm) and light brown, with an "X" marked on back. (Photos by C. Alender.)

Salientia, meaning leaping. The alternate order name, Anura (Gr. *an-,* without, + *oura,* tail), refers to another obvious group characteristic, the absence of tails as adults (although all pass through a tailed larval stage during development).

The Salientia are further distinguished from the Urodela by their larvae and a dramatic metamorphosis during development. The eggs of most frogs hatch into a tadpole ("polliwog") stage, with a long, finned tail, both internal and external gills, no legs, specialized mouthparts for herbivorous feeding (salamander larvae, in distinction, are carnivorous), and a highly specialized internal anatomy. They look and act altogether differently from adult frogs. The metamorphosis of the frog tadpole to the adult frog is thus a striking transformation. Neoteny and paedogenesis are never exhibited in frogs and toads, as they are among salamanders.

The Salientia are an old group—fossil frogs are known from the Jurassic, 150 million years ago—and today they are a secure and successful group. Frogs and

toads occupy a great variety of habitats, despite their aquatic mode of reproduction and water-permeable skin, which prevent them from wandering too far afield from sources of water, and their ectothermy, which bars them from polar and subarctic habitats.

Notwithstanding their success as a distinct group, the frogs and toads are really a specialized side branch of amphibian evolution, and, despite their popularity for education purposes—approximately 20 million are used each year in the United States alone—they are not good representatives of the vertebrate body plan. The primitive and unspecialized salamander would be a much superior choice for the zoology laboratory were not frogs so readily available.

Frogs and toads are divided into 12 families. The best known frog families in North America are Ranidae, containing most of our familiar frogs, and Hylidae, the tree frogs (Fig. 26-5). True toads, belonging to the family Bufonidae, have short legs, stout bodies, and thick skins usually with prominent warts (Fig. 26-6). However, the term "toad" is used

Fig. 26-6. American toad, *Bufo americanus*. This principally nocturnal yet familiar amphibian feeds upon large numbers of insect pests as well as snails and earthworms. The warty skin contains numerous poison glands that produce a surprisingly poisonous milky fluid, providing the toad with excellent protection from a variety of potential predators. (Photo by L. L. Rue III.)

Fig. 26-7. *Gigantorana goliath* of West Africa, the world's largest frog. This specimen weighed 3.3 kg (approximately 7½ pounds). (Courtesy American Museum of Natural History.)

rather loosely to refer to more or less terrestrial members of several other families.

The largest anuran is the west African *Gigantorana goliath,* which is more than 30 cm long from tip of nose to anus (Fig. 26-7). This giant eats animals as big as rats and ducks. The smallest frog recorded is *Phyllobates limbatus,* which is only approximately 1 cm long. This tiny frog, which is more than covered by a dime, is found in Cuba. Our largest American frog is the bullfrog, *Rana catesbeiana* (Fig. 26-5, *A*), which reaches a head and body length of 20 cm.

Habitats and distribution. Probably the most abundant and successful of frogs are the 200 to 300 species of the genus *Rana,* found all over the temperate and tropical regions of the world except in New Zealand, the oceanic islands, and southern South America. They are usually found near water, although some, such as the wood frog *Rana sylvatica,* spend most of their time on damp forest floors, often some distance from the nearest water. The wood frog probably returns to pools only for breeding in early spring. The larger bullfrogs *R. catesbeiana* and green frogs *R. clamitans* (Fig. 26-5, *A* and *B*) are nearly always found in or near permanent water or swampy regions. The leopard frog *R. pipiens* (Fig. 26-5, *C*) has a wider variety of habitats and, with all its subspecies and phases, is perhaps the most widespread of all the North American frogs. This is the species most commonly used in biology laboratories and for classic electrophysiologic research. It has been found in some form in nearly every state, though sparingly represented along the extreme western part of the Pacific coast. It also extends far into northern Canada and as far south as Panama.

Within the range of any species of frogs, they are often restricted to certain habitats (for instance, to certain streams or pools) and may be absent or scarce in similar habitats of the range. The pickerel frog *(R. palustris)* is especially noteworthy this way, for it is known to be abundant only in certain localized regions.

Most of our larger frogs are solitary in their habits except during the breeding season. During the breeding period most of them, especially the males, are very

529

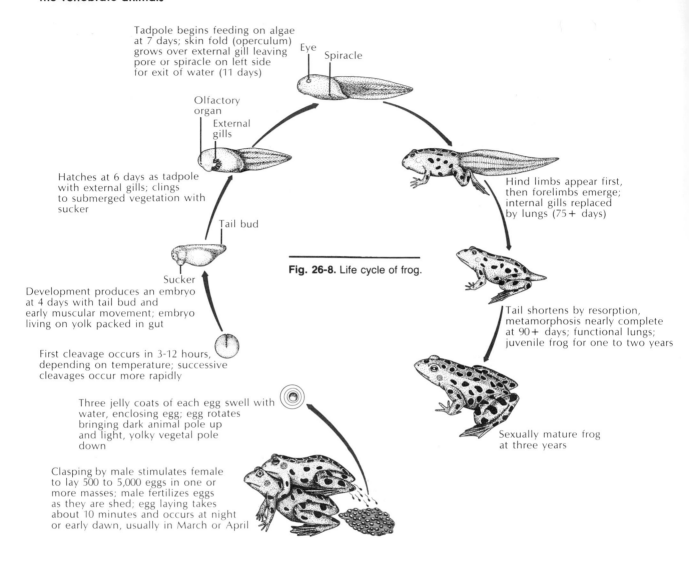

Tadpole begins feeding on algae at 7 days; skin fold (operculum) grows over external gill leaving pore or spiracle on left side for exit of water (11 days)

Eye   Spiracle

Olfactory organ
External gills

Hatches at 6 days as tadpole with external gills; clings to submerged vegetation with sucker

Tail bud

Hind limbs appear first, then forelimbs emerge; internal gills replaced by lungs (75+ days)

Fig. 26-8. Life cycle of frog.

Sucker

Development produces an embryo at 4 days with tail bud and early muscular movement; embryo living on yolk packed in gut

First cleavage occurs in 3-12 hours, depending on temperature; successive cleavages occur more rapidly

Tail shortens by resorption, metamorphosis nearly complete at 90+ days; functional lungs; juvenile frog for one to two years

Three jelly coats of each egg swell with water, enclosing egg; egg rotates bringing dark animal pole up and light, yolky vegetal pole down

Sexually mature frog at three years

Clasping by male stimulates female to lay 500 to 5,000 eggs in one or more masses; male fertilizes eggs as they are shed; egg laying takes about 10 minutes and occurs at night or early dawn, usually in March or April

noisy. Each male usually takes possession of a particular perch, where he may remain for a long time, trying to attract a female to that spot. At times frogs are mainly silent, and their presence is not detected until they are disturbed. When they enter the water, they dart about swiftly and reach the bottom of the pool, where they kick up a cloud of muddy water. In swimming, they hold the forelimbs near the body and kick backward with the webbed hindlimbs, which propel them forward. When they come to the surface to breathe, only the head and foreparts are exposed, and, as they usually take advantage of any protective vegetation, they are difficult to see.

During the winter months most frogs hibernate in the soft mud of the bottom of pools and streams. The wood frog hibernates under stones, logs, and stumps in the forest area. Naturally their life processes are at a very low ebb during their hibernation period, and such energy as they need is derived from the glycogen and fat stored in their bodies during the spring and summer months.

Adult frogs have numerous enemies, such as snakes, aquatic birds, turtles, raccoons, man, and many others; only a few tadpoles survive to maturity. Although usually defenseless, in the tropics and subtropics many frogs and toads are aggressive, jumping and biting at predators. Some defend themselves by feigning death. Most anurans can blow up their lungs so that they are difficult to swallow. When disturbed along the margin of a pond or brook, a frog often remains quite still; when it thinks it is detected, it jumps, not always into the water where enemies may be lurking but into grassy cover on the bank. When held in the hand a frog may cease its struggles for an instant to put its captor off guard and then leap violently, at the same time voiding its urine. Their best protection is their ability to leap and their use of poison glands. Bullfrogs in captivity do not hesitate to snap at tormenters and are capable of inflicting painful bites.

Reproduction. Because frogs and toads are ectotherms, they breed, feed, and grow only during the warmer seasons of the year. One of the first drives after the dormant period is breeding. In the spring males croak and call vociferously to attract females. When their eggs are mature, the females enter the water and are clasped by the males in a process called **amplexus.** As the female lays the eggs, the male discharges seminal fluid containing sperm over the eggs to fertilize them. The eggs are laid in large masses, usually anchored to vegetation.

Development of the fertilized egg (zygote) begins almost immediately (Fig. 26-8). By repeated division (cleavage) the egg is converted into a hollow ball of cells (blastula). This undergoes continued differentiation to form an embryo with a tail bud. At 6 to 9 days, depending on the temperature, a tadpole hatches from the protective jelly coats that had surrounded the original fertilized egg.

At the time of hatching, the tadpole has a distinct head and body with a compressed tail. The mouth is located on the ventral side of the head and is provided with horny jaws for scraping off vegetation from objects for food. Behind the mouth is a ventral adhesive disc for clinging to objects. In front of the mouth are two deep pits, which later develop into the nostrils. Swellings are found on each side of the head, and these later become external gills. There are finally three pairs of external gills, which are later replaced by three pairs of internal gills within the gill slits. On the left side of the neck region is an opening, the **spiracle,** through which water flows after entering the mouth and passing the internal gills. The hindlegs appear first, while the forelimbs are hidden for a time by the folds of the operculum. During metamorphosis the tail is resorbed, the intestine becomes much shorter, the mouth undergoes a transformation into the adult condition, lungs develop, and the gills are resorbed. The

leopard frog usually completes its metamorphosis within 3 months; the bullfrog takes 2 or 3 years to complete the process.

Migration of frogs and toads is correlated with their breeding habits. Males usually return to a pond or stream in advance of the females, which they then attract by their calls. Some salamanders are also known to have a strong homing instinct, returning year after year to the same pool for reproduction guided by olfactory cues. The initial stimulus for migration in many cases is attributable to a seasonal cycle in the gonads plus hormonal changes that increase their sensitivity to temperature and humidity changes.

SELECTED REFERENCES

Barbour, R. W. 1971. Amphibians and reptiles of Kentucky. Lexington, Ky., University of Kentucky Press. *Although primarily concerned with a state survey of amphibians and reptiles, this volume is additionally concerned with general distribution of species and contains a number of excellent photographs.*

Cochran, D. M., and C. J. Goin. 1970. The new field book of Reptiles and amphibians. New York, G. P. Putnam's Sons. *Probably the most useful field guide available for reptiles and amphibians. Most species descriptions are concise and informative.*

Goin, C. J., and O. B. Goin. 1971. Introduction to herpetology, ed. 2. San Francisco, W. H. Freeman & Co. Publishers. *A basic introductory text for the study of amphibians and reptiles.*

Mertens, R. 1960. The world of amphibians and reptiles. London, George G. Harrap. *(Translation) Excellent source of information, beautifully illustrated.*

Moore, J. (ed.). 1964. Physiology of the Amphibia. New York, Academic Press, Inc. *A valuable reference on the Amphibia for the advanced student.*

Noble, G. K. 1931. Biology of the Amphibia. New York, McGraw-Hill, Inc. *A dated but invaluable classic containing information to be found nowhere else.*

Oliver, J. A. 1955. The natural history of North American amphibians and reptiles. New York, D. Van Nostrand Co. *A good introduction to classification, distribution, and behavior of reptiles and amphibians.*

Porter, G. 1967. The world of the frog and the toad. Philadelphia, J. B. Lippincott Co. *Superbly illustrated natural history.*

Stahl, B. J. 1974. Vertebrate history: problems in evolution. New York, McGraw-Hill, Inc. *A contemporary treatment of vertebrate paleontology. Good account of amphibian beginnings.*

Twitty, V. C. 1966. Of scientists and salamanders. San Francisco, W. H. Freeman & Co., Publishers. *Delightfully written account of the author's personal study of the experimental biology of salamanders and newts.*

Wright, A. H., and A. A. Wright. 1949. Handbook of frogs of the United States and Canada, ed. 3. Ithaca, N.Y., Comstock Publishing Co. *A volume that has, with time, become a standard reference work for the study of frogs.*

Pushing his way across rough terrain, a giant tortoise, *Geochelone elephantopus,* of the Galápagos Islands leisurely searches for food. These huge, ponderous reptiles, some reaching 2 m in length and 275 kg (approximately 600 pounds) in weight, were once so numerous that the archipelago was named Islas Galápagos—in Spanish, the "tortoise islands."

27 REPTILES

The class Reptilia (rep-til'e-a) (L. *repere,* to creep) are the first truly terrestrial vertebrates. With some 7,000 species (approximately 300 species in the United States and Canada) occupying a great variety of aquatic and terrestrial habitats, they are clearly a successful group. Nevertherless, reptiles are perhaps remembered best for what they once were, rather than for what they presently are. The Age of Dinosaurs, which lasted 100 million years encompassing the Jurassic and Cretaceous periods of the Mesozoic, saw the appearance of a great radiation of reptiles, many of huge stature and awesome appearance, that completely dominated life on land. Then they suddenly declined.

Out of the dozen or so principal groups of reptiles that evolved, four remain today. The most successful of these are the lizards and snakes of the order Squamata. A second group is the crocodilians; having survived for 200 million years, they may finally be made extinct by man. To a third group belong the turtles of the order Testudines, an ancient group that has somehow survived and remained mostly unchanged from its early reptile ancestors. The last group is a relic stock represented today by a sole survivor, the tuatara of New Zealand.

Reptiles are easily distinguished from amphibians by several adaptations that permit them to live in arid regions and in the sea—habitats barred to amphibians by their reproductive requirements. Reptiles have a dry, scaly skin, almost free of glands, that resists desiccation. Most important, reptiles lay their eggs on land; amphibians must lay their eggs in fresh water or in moist places. This seemingly simple difference was, in fact, a remarkable evolutionary achievement that was to have a profound impact on subsequent vertebrate evolution. To abandon totally an aquatic life, there evolved a sophisticated internally fertilized egg containing a complete set of life support systems. This **shelled egg** could be laid on dry land. Within, the embryo floats and develops in an aquatic environment (amniotic fluid) enclosed within a membrane, the amnion. It is provided with a yolk sac containing its food supply; another membrane, the allantois, serves as a surface for gas exchange through the calcareous or parchmentlike shell; provision is made in it for storing toxic wastes that accumulate during development.

The early reptiles that developed this "land egg" must certainly have enjoyed an immediate advantage over the amphibians. They could hide their eggs in a protected situation away from water—and away from the numerous creatures that fed freely on the eggs provided by amphibians each spring. With the evolution of this ultimate adaptation, conquest of land by the vertebrates was possible.

ORIGIN AND ADAPTIVE RADIATION OF REPTILES

Biologists generally agree that reptiles arose from the labyrinthodont amphibians sometime before the Permian, which began approximately 280 million years ago. The oldest "stem reptiles" belonged to the order Cotylosaura (Fig. 27-1). Forming a transition between the labyrinthodont amphibians and stem reptiles was a lizardlike, partly aquatic animal, approximately 0.5 m long, called *Seymouria.* This creature possessed several distinct reptilian skeletal features yet had a lateral line system typical of amphibians with an aquatic larval stage. Thus *Seymouria* probably did not lay amniotic eggs as did the true terrestrial reptiles that followed. Nevertheless, *Seymouria* represents a nearly perfect transition form. From *Seymouria,* or a closely related form, arose the stem reptiles (cotylosaurs), showing more definite reptilian characteristics.

The adaptive radiation of reptiles, especially pronounced in the Triassic period (which followed the Permian), corresponded with the appearance of new ecologic niches. These were provided by the climatic and geologic changes that were taking place at that time, such as a variable climate from hot to cold, mountain building and terrain transformations, and a varied assortment of plant life.

The Mesozoic was the age of the great ruling reptiles. Then suddenly they disappeared near the close of the Cretaceous approximately 65 to 80 million years ago. What caused their demise? Many changes were occurring during the Cretaceous: modern flowering plants were spreading rapidly, as were the aggressive and intelligent mammals. In general, the modern fauna and flora as we know it today were becoming well established, and the dinosaurs were not sufficiently adaptable to survive. Their extinction probably resulted from the combined effect of climatic and ecologic factors, excessive specialization, and low reproductive potential. But this is speculation, and debate continues among paleontologists. Why did some reptiles survive against the fierce competition of the mammals? Turtles had their protective shells, snakes and lizards evolved in habitats of dense forests and rocks where they could meet the competition of

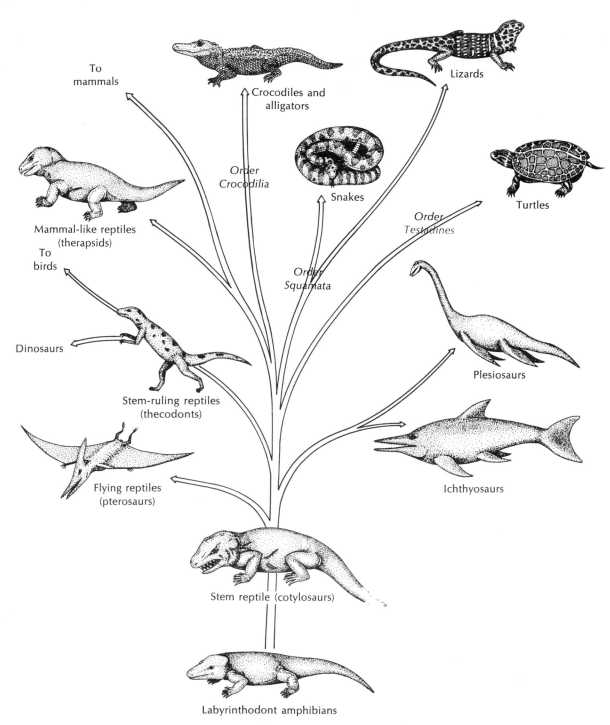

To mammals

Crocodiles and alligators

Lizards

Order Crocodilia

Snakes

Turtles

Mammal-like reptiles (therapsids)

Order Testudines

To birds

Dinosaurs

Order Squamata

Plesiosaurs

Stem-ruling reptiles (thecodonts)

Flying reptiles (pterosaurs)

Ichthyosaurs

Stem reptile (cotylosaurs)

Labyrinthodont amphibians

Fig. 27-1. For legend see opposite page.

any tetrapod, and crocodiles, because of their size, stealth, and aggressiveness, had few enemies in their aquatic habitats.

Characteristics

1. Body varied in shape, compact in some, elongated in others; **body covered with an exoskeleton of horny epidermal scales** with the addition sometimes of bony dermal plates; **integument with few glands**
2. **Paired limbs, usually with five toes,** and adapted for climbing, running, and paddling; absent in snakes
3. Well-ossified skeleton; ribs with sternum forming a complete thoracic basket; **skull with one occipital condyle**
4. Respiration by lungs; **no gills;** cloaca used for respiration by some; branchial arches in embryonic life
5. **Three-chambered heart; crocodiles with four-chambered heart;** usually one pair of aortic arches
6. **Metanephric (paired) kidney;** uric acid as main nitrogenous waste
7. Nervous system with the optic lobes on the dorsal side of brain; **12 pairs of cranial nerves** in addition to nervus terminalis
8. Separate sexes; **internal fertilization**
9. **Eggs covered with leathery or calcareous shells;** extraembryonic membranes **(amnion, chorion, yolk sac,** and **allantois)** present during embryonic life.

Brief classification

Order Squamata (squa-ma′ta) (L. *squamatus,* scaly + *-ata,* characterized by). Skin of horny epidermal scales or plates, which is shed; teeth attached to jaws; quadrate freely movable; vertebrae usually concave in front; anus consisting of a transverse slit. Snakes (3,000 species), lizards (3,800 species).

Order Testudines (tes-tu′din-eez) (L. *testudo,* tortoise) **(Chelonia).** Body in a bony case of dermal plates with dorsal carapace and ventral plastron; jaws without teeth but with horny sheaths; immovable quadrate; vertebrae and ribs fused to shell; anus consisting of a longitudinal slit. Turtles (250 species).

Order Crocodilia (croc′o-dil′e-a) (L. *crocodilus,* crocodile + *-ia,* pl. suffix) **(Loricata).** Four-chambered heart; vertebrae usually concave in front; forelimbs usually with five digits, hind limbs with four digits; immovable quadrate; anus consisting of longitudinal slit. Crocodiles and alligators (25 species).

Order Rhynchocephalia (rin′ko-se-fay′le-a) (Gr. *rhynchos,* snout, + *kephale,* head). Biconcave vertebrae; immovable quadrate; parietal eye fairly well developed and easily seen; anus consisting of transverse slit. *Sphenodon*—only species existing.

How reptiles show advancements over amphibians

Reptiles have tough, dry, scaly skin offering protection against desiccation and physical injury. The skin consists of a thin **epidermis,** shed periodically, and a much thicker well-developed **dermis.** The dermis is provided with **chromatophores,** the color-bearing cells that give many lizards and snakes their colorful hues. It is also the layer that, unfortunately for their bearers, is converted into alligator and snakeskin leather, so esteemed for expensive pocketbooks and shoes. The characteristic **scales** of reptiles are mostly derived from the epidermis and thus are not homologous to fish scales, which are bony, dermal structures. In some reptiles, such as alligators, the scales remain throughout life, growing gradually to replace wear. In others, such as snakes and lizards, new scales grow beneath the old, which are then shed at intervals. In snakes the old skin (epidermis and scales) is turned inside out when discarded; lizards split out of the old skin leaving it mostly intact and right side out, or it may slough off in pieces.

The shelled egg of reptiles contains food and protective membranes for supporting embryonic development on dry land. The great significance of this adaptation was described earlier in this chapter. The shelled egg is illustrated and described in some detail on p. 266. Amphibian eggs have a gelatinous covering and must be protected from drying. The appearance of the shelled egg marked a great division between amphibians and reptiles and, probably more than any

---

Fig. 27-1. Family tree of the reptiles. Transition from certain labyrinthodont amphibians to reptiles occurred in Carboniferous period to Mesozoic times. This transition was effected by development of a shelled egg, which made land existence possible, although this egg may well have developed before oldest reptiles had ventured far on land. Explosive adaptation by reptiles may have been due partly to variety of ecologic niches into which they could move. Fossil record shows that lines arising from stem reptiles led to ichthyosaurs, plesiosaurs, and stem-ruling reptiles. Some of these returned to the sea. Later radiations led to mammal-like reptiles, turtles, flying reptiles, birds, dinosaurs, etc. Of this great assemblage, the only reptiles now in existence belong to four orders (Testudines, Crocodilia, Squamata, and Rhynchocephalia). The order Rhynchocephalia, not shown in this diagram, is represented by only one living species, the tuatara *(Sphenodon)* of New Zealand. How are the mighty fallen!

other adaptation, contributed to the decline of amphibians and the ascendence of reptiles.

Reptiles have some form of copulatory organ, permitting internal fertilization. Internal fertilization is obviously a requirement for a shelled egg, since the sperm must reach the egg before the egg is enclosed. Sperm from the paired testes are carried by the vasa deferentia to the copulatory organ, which is an evagination of the cloacal wall. The female system consists of paired ovaries and oviducts. The glandular walls of the oviducts secrete albumin and shells for the large eggs.

Reptiles have a more efficient circulatory system and higher blood pressures than amphibians. In all reptiles the right atrium, which receives unoxygenated blood from the body, is completely partitioned from the left atrium, which receives oxygenated blood from the lungs. In the crocodilians there are two completely separated ventricles as well (Fig. 27-2); in other reptiles the ventricle is incompletely separated. The crocodilians are thus the first vertebrates with a four-chambered heart. Even in those reptiles with incomplete separation of the ventricles, flow patterns within the heart prevent admixture of pulmonary (oxygenated) and systemic (unoxygenated) blood; all reptiles therefore have two functionally separate circulations.

Reptile lungs are better developed than those of amphibians. Reptiles depend almost exclusively on the lungs for gas exchange, supplemented by pharyngeal membrane respiration in some of the aquatic turtles. The lungs have a larger respiratory surface in reptiles than amphibians, and air is *sucked* into the lungs, as in higher vertebrates, rather than *forced* in by mouth muscles, as in the amphibians. Cutaneous respiration, so important to most amphibians, has been completely abandoned by the reptiles.

The reptilian kidneys are of the advanced metanephros type with their own passageways (ureters) to the exterior. The kidneys are very efficient in producing small volumes of urine, thus conserving precious water. Nitrogenous wastes are excreted as uric acid, rather than urea or ammonia. Uric acid has a low solubility and precipitates out of solution readily; as a result the urine of many reptiles is a semisolid paste.

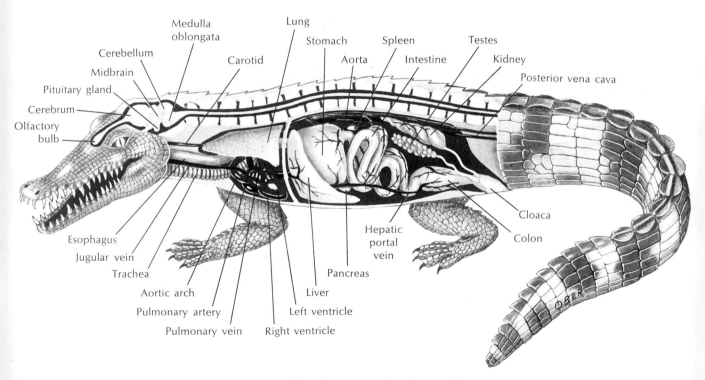

Fig. 27-2. Internal structures of male crocodile.

All reptiles, except the limbless members, have better body support than the amphibians and more efficiently designed limbs for travel on land. Many of the dinosaurs walked on powerful hindlimbs alone.

The reptilian nervous system is considerably more advanced than the amphibian. Although the reptile's brain is small, the **cerebrum** is increased in size relative to the rest of the brain. The crocodilians have the first true cerebral cortex (neopallium). Central nervous system connections are more advanced, permitting complex kinds of behavior unknown in the amphibians.

STRUCTURE AND NATURAL HISTORY OF REPTILIAN ORDERS
Lizards and snakes—order Squamata

The lizards and snakes of this order are the most recent products of reptile evolution and by all odds the most successful, comprising approximately 95% of all known living reptiles. The modern lizards began their adaptive radiation during the Cretaceous when the great dinosaurs were at the climax of their dominance of land. The immediate success of lizards was probably caused by a versatile and highly mobile jaw apparatus, which allowed them to capture large, struggling prey. The snakes appeared during the late Cretaceous probably from burrowing lizards, although the early fossil record of this group is poor. The flexible jaw apparatus of lizards was refined even further in snakes; in these the jaw and skull bones are so loosely connected that they can swallow prey several times their own diameter.

The order Squamata is divided into suborders Saura (Lacertilia), which includes the lizards, and Serpentes (Ophidia), which includes the snakes. In addition to the difference in jaw articulation and structure, lizards have movable eyelids, external ear openings, and legs (usually); snakes lack these characteristics.

Lizards—suborder Saura

The lizards are an extremely diversified group, including terrestrial, burrowing, aquatic, arboreal, and aerial members. Among the more familiar groups in this varied suborder are the **geckos,** small, agile, mostly nocturnal forms with adhesive toe pads that enable them to walk upside down and on vertical surfaces; the **iguanas,** New World lizards, often brightly colored with ornamental crests, frills, and throat fans, a group that includes the remarkable marine iguana of the Galápagos Islands (Fig. 27-3); the **skinks,** with elongate bodies and reduced limbs; and the **chameleons,** a group of arboreal lizards, mostly of Africa and Madagascar. These entertaining creatures catch insects with the sticky-tipped tongue that can be flicked accurately and rapidly to a distance greater than their own body length.

The radiation of lizards into so many different kinds of habitats has been accompanied by a great variety of specialized adaptations. *Draco,* a lizard inhabiting India, is able to volplane from tree to tree because of skin extensions on the side. A few lizards, such as the glass lizards, are limbless (Fig. 27-4).

Many lizards live in the world's hot and arid regions, aided by several adaptations for desert life. Since their skin lacks glands, water loss by this avenue is much reduced. They produce a semisolid urine with a high content of crystalline uric acid. This is an excellent adaptation for conserving water and is found in other groups living successfully in arid habitats (birds, insects, and pulmonate snails). Some, such as the Gila monster of the southwestern United States deserts, store fat in their tails, which they draw upon during drought to provide both energy and metabolic water (Fig. 27-5).

Especially interesting are the techniques desert lizards use to maintain a relatively constant body temperature, using what physiologists term "behavioral thermoregulation." Lizards, like other reptiles, are ectothermic; if a lizard is placed under constant temperature conditions in the laboratory, its body temperature soon becomes indistinguishable from that of its surroundings. But in its natural environment where surrounding temperatures vary widely, lizards modulate their body temperature by exploiting hour-to-hour changes in thermal flux from the sun. In the early morning they emerge from their burrows and bask in the sun with their bodies flattened to absorb heat. As the day warms they turn to face the sun, to reduce the body area exposed, and raise their bodies from the hot substrate. In the hottest part of the day they may retreat to their burrows. Later they emerge to bask as the sun sinks lower and the air temperature drops.

These behavioral patterns help to maintain a relatively steady body temperature of 36° to 39° C while the air temperature is varying between 29° and 44° C. Some lizards can tolerate intense midday heat without shelter. The desert iguana of the southwestern United States prefers a body temperature of 42° C when active

Fig. 27-3. Marine iguanas, *Amblyrhynchus cristatus,* of the Galápagos Islands bask in morning sunshine. This is the only marine lizard in the world. It has special salt-removing glands in the eye orbits and long claws that enable it to cling to the bottom while feeding on seaweed, its exclusive diet. It may dive to depths exceeding 10 m (35 feet) and remain submerged more than 30 minutes. (Photo by C. P. Hickman, Jr.)

Fig. 27-4. A glass lizard, *Ophisaurus* sp., of the southeastern United States. This legless lizard feels stiff and brittle to the touch and has an extremely long, fragile tail that readily fractures when the animal is struck or seized. Most specimens, such as this one, have only a partly regenerated tip to replace a much longer tail previously lost. Glass lizards can be readily distinguished from snakes by the deep, flexible groove running along each side of body (barely perceptible in photo). They feed on worms, insects, spiders, birds' eggs, and small reptiles. (Photo by L. L. Rue III.)

Fig. 27-5. Gila monster, *Heloderma suspectum*, of southwestern United States desert regions and the congeneric Mexican beaded lizard are the only venomous lizards known. These brightly colored, clumsy-looking lizards feed principally on birds' eggs, nestling birds and mammals, and insects. Unlike poisonous snakes, gila monster secrets venom from glands in lower jaw. Bite is painful to man but seldom fatal. (Courtesy American Museum of Natural History.)

and can tolerate a rise to 47° C, a temperature that is lethal to all birds and mammals and most other lizards. The term "cold-blooded" clearly does not apply to these animals!

Snakes—suborder Serpentes

Snakes are entirely limbless and lack both the pectoral and pelvic girdles (the latter persists as vestiges in pythons). The numerous vertebrae of snakes, shorter and wider than those of tetrapods, promote quick lateral undulations through grass and over rough terrain. The ribs increase rigidity of the vertebral column, providing more resistance to lateral stresses. The elevation of the neural spine gives the numerous muscles more leverage. Snakes also differ from lizards in having no movable eyelids (their eyes are permanently covered with a third eyelid) and no external ear. Snakes are, in fact, totally deaf, although they are

sensitive to low-frequency vibrations conducted through the ground.

Snakes employ a unique set of special senses to hunt down their prey. In addition to being deaf, most snakes have relatively poor vision, with the tree-living snakes of the tropical forest being a conspicuous exception. In fact, the latter possess excellent binocular vision that helps them track prey through the branches where scent trails would be impossible to follow. But most snakes live on the ground and rely on chemical senses to hunt food. In addition to the usual olfactory areas in the nose, which are not well developed, there are **Jacobson's organs,** a pair of pitlike organs in the roof of the mouth. These are lined with an olfactory epithelium and are richly innervated. The forked tongue, flicking through the air, picks up scent particles and conveys them to the mouth; the tongue is then drawn past Jacobson's organ or the tips of the

539

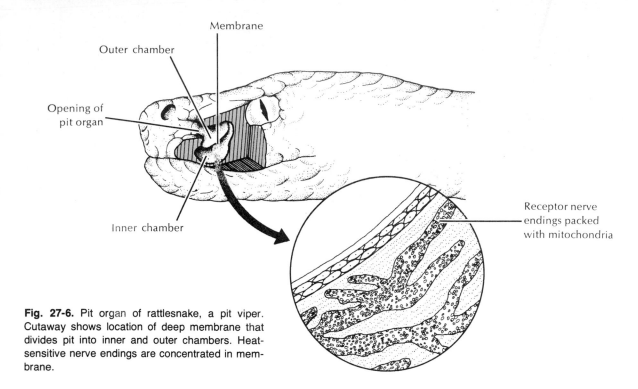

Fig. 27-6. Pit organ of rattlesnake, a pit viper. Cutaway shows location of deep membrane that divides pit into inner and outer chambers. Heat-sensitive nerve endings are concentrated in membrane.

forked tongue are inserted directly into the organs. Information is then transmitted to the brain where scents are identified.

Snakes of the subfamily Crotalinae within the family Viperidae are called **pit vipers** because of special heat-sensitive pits on their heads, between the nostrils and the eyes (Fig. 27-6). All of the best known North American poisonous snakes are pit vipers, such as the several species of rattlesnakes, the water moccasin and the copperhead (Fig. 27-7). The pits are supplied with a dense packing of free nerve endings from the fifth cranial nerve. They are exceedingly sensitive to radiant energy (long-wave infrared) and can distinguish temperature differences smaller than 0.2° C from a radiating surface. Pit vipers use the pits to track warm-blooded prey and to aim strikes, which they can make as effectively in total darkness as in daylight.

All pit vipers have a pair of teeth on the maxillary bones modified as fangs. These lie in a membrane sheath when the mouth is closed. When the viper strikes, a special muscle and bone lever system erects the fangs when the mouth opens. The fangs are driven into the prey by the thrust, and venom is injected into the wound along a groove in the fangs.

A pit viper immediately releases its prey after the bite and follows it until it is paralyzed or dies. Then the snake swallows it whole. Approximately 1,500 bites and 45 deaths from pit vipers are reported each year in the United States.

The tropical and subtropical countries are the homes of most species of snakes, both of the venomous and nonvenomous varieties. Even there, less than one third of them are venomous; the nonvenomous snakes kill their prey by constriction or by biting and swallowing. Their diet tends to be restricted, many feeding principally on rodents, whereas others feed on fishes, frogs, and insects. Some African, Indian, and Neotropical snakes have become specialized as egg eaters.

Poisonous snakes are usually divided into three groups based on the type of fangs. The vipers (family Viperidae) have tubular or grooved fangs at the front of the mouth; the group includes the American pit vipers previously mentioned and the Old World true vipers, which lack facial heat-sensing pits. Among the latter are the common European adder and the African puff adder. A second family of poisonous snakes (family Elapidae) has short, permanently erect fangs so that the venom must be injected by chewing. In this

Fig. 27-7. Copperhead snake, *Ancistrodon contortrix,* one of the best known of American pit vipers, is distributed through the east and central United States. Its distinctive brown to copper coloration blends with the dead-leaf ground cover of hardwood forests, making its detection by man difficult. Being a smaller snake than either the water moccasin or timber rattlesnake, its bite, though painful, is seldom fatal. As with any snakebite, the danger is much greater to children than to adults. Copperheads feed on frogs, small birds and mammals, and large insects. They are ovoviviparous, bearing six to nine young alive in each brood. (Photo by C. P. Hickman, Jr.)

Fig. 27-8. Milk snake, *Lampropeltis doliata,* with eggs. Soon after internal fertilization, these oviparous snakes lay elliptical eggs enclosed in tough, parchment-like shells in a hole in the ground. The name "milk snake" is derived from a ridiculous old wives' tale that it milks cows. It actually eats mice, lizards, and other small snakes and is common around farm buildings. (Photo by L. L. Rue III.)

Fig. 27-9. Timber rattlesnake, *Crotalus horridus horridus,* with newly born young. All pit vipers in the United States are ovoviviparous, giving birth to well-formed young fully capable of capturing their own food. No snake species feeds its young. Timber rattlesnakes once were common in eastern hardwood forests, now are restricted mostly to second-growth timbered terrain where their rodent food abounds. (Photo by L. L. Rue III.)

group are the cobras, mambas, coral snakes, and kraits. The highly poisonous sea snakes are placed in a third family (Hydrophiidae).

The saliva of all harmless snakes possesses limited toxic qualities, and it is logical that evolution should have stressed this toxic tendency. There are two types of snake venom. One type acts mainly on the nervous systems (neurotoxic), affecting the optic nerves (causing blindness) or the phrenic nerve of the diaphragm (causing paralysis of respiration). The other type is hemolytic; that is, it breaks down the red blood corpuscles and blood vessels and produces extensive extravasation of blood into the tissue spaces. Many venoms have both neurotoxic and hemolytic properties. The toxicity of a venom is determined by the minimal lethal dose on laboratory animals. By this standard the venoms of the Australian tiger snake and some of the sea snakes appear to be the most deadly of poisons drop for drop. However, several larger snakes are more dangerous. The aggressive king cobra, which may exceed 5.5 m in length, is the largest and probably the most dangerous of all poisonous snakes. In India, where snakes come in constant contact with people, cobra bites cause more than 10,000 deaths each year.

Most snakes are **oviparous** species that lay their shelled, elliptic eggs soon after fertilization (Fig. 27-8). They are laid beneath rotten logs, under rocks,

or in holes dug in the ground. Most of the remainder, including all the American pit vipers, except the tropical bushmaster, are **ovoviviparous,** giving birth to well-formed young (Fig. 27-9). A very few snakes are **viviparous;** a primitive placenta forms, permitting the exchange of materials between the embryonic and maternal bloodstreams.

Turtles — order Testudines (Chelonia)

The turtles are an ancient group that have plodded on from the Triassic to the present with very little change in their early basic morphology. They are enclosed in shells consisting of a dorsal **carapace** and a ventral **plastron.** Clumsy and unlikely as they appear to be within their protective shells, they are nonetheless a varied and successful group that seems able to accommodate to man's presence. The shell is so much a part of the animal that it is built in with the thoracic vertebrae and ribs. Into this shell the head and appendages can be retracted for protection. The jaws lack teeth but are covered with a sharp, horny cutting surface. On their toes are horny claws used for digging in the sand, where they lay their eggs. Some marine forms have paddle-shaped limbs used for swimming. Fertilization is internal by means of a cloacal penis on the ventral wall of the cloaca. All turtles are oviparous, and the eggs have firm, calcareous shells.

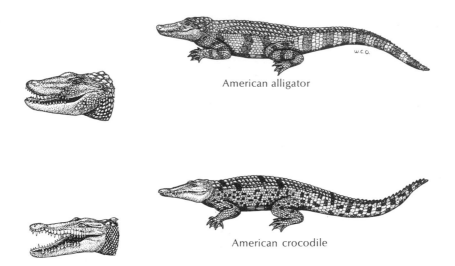

American alligator

American crocodile

Fig. 27-10. American crocodilians (order Crocodilia). The American alligator, *Alligator mississipiensis,* is found along the southeastern United States coast from central Texas to the Atlantic. The American crocodile, *Crocodylus acutus,* is now limited to the Everglades National Park in Florida. Both species are in danger of extinction. Enlargement of heads shows easily recognized differences between the two species. The crocodile has a more slender snout and the fourth tooth from the front of the lower jaw fits *outside* upper jaw and is visible when mouth is closed. It is not true, as commonly believed, that there is a difference in the way the jaws are hinged in the two species.

The terms "turtle," "tortoise," and "terrapin" are applied variously to different members of this order; they are all correctly called turtles. The term "tortoise" is frequently given to land turtles, especially the large forms.

The great marine turtles, buoyed by their aquatic environment, may reach 2 m in length and 725 kg in weight. One is the leatherback. The green turtle, so named because of its greenish body fat, may exceed 360 kg, although most individuals of this economically valuable and heavily exploited species seldom live long enough to reach anything approaching this size. Some land tortoises may weigh several hundred kilograms, such as the giant tortoises of the Galápagos Islands that so intrigued Darwin during his visit there in 1835. Most tortoises are rather slow moving; 1 hour of determined trudging carries a large Galápagos tortoise approximately 300 m. Their low metabolism probably explains their longevity, for some are believed to live more than 150 years.

Crocodiles and alligators—order Crocodilia

The modern crocodiles are the largest living reptiles. They are what remain of a once abundant group in the Jurassic and Cretaceous. Having managed to survive virtually unchanged for some 160 million years, the modern crocodilians face a forbidding, and perhaps short, future in man's world.

Crocodiles have relatively long slender snouts; alligators have short and broader snouts. With their powerful jaws and sharp teeth, they are formidable antagonists. The "man-eating" members of the group are found mainly in Africa and Asia. The estuarine crocodile *(Crocodylus porosus)* found in southern Asia grows to a great size and is very much feared. It is swift and aggressive and eats any bird or mammal it can drag from the shore to water where it is violently torn to pieces. Crocodiles are known to attack animals as large as cattle, deer, and men.

Alligators are usually less aggressive than crocodiles. They are almost unique among reptiles in being able to make definite sounds. The male alligator can give loud bellows in the mating season. Vocal sacs are found on each side of the throat and are inflated when he calls. In the United States, *Alligator mississipiensis* is the only species of alligator; *Crocodylus acutus,* restricted to extreme southern Florida, is the only species of crocodile (Fig. 27-10).

Alligators and crocodiles are oviparous. Usually

543

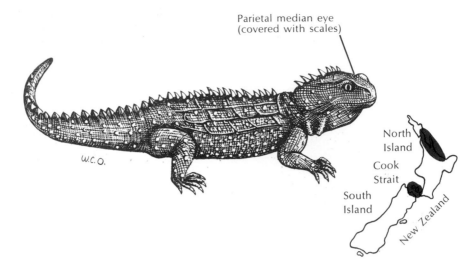

Parietal median eye
(covered with scales)

North
Island

Cook
Strait

South
Island

New Zealand

W.C.O.

Fig. 27-11. Tuatara, *Sphenodon punctatum,* the only living representative of order Rhynchocephalia. This "living fossil" reptile has well-developed parietal "eye" with retina and lens on top of head. Eye is covered with scales and is considered nonfunctional but may have been important sense organ in early reptiles. The tuatara is found only on small islets of Cook Strait and along the northern coast of North Island, New Zealand.

from 20 to 50 eggs are laid in a mass of dead vegetation. The eggs are approximately 8 cm long. The penis of the male is an outgrowth of the ventral wall of the cloaca.

The tuatara—order Rhynchocephalia

The order Rhynchocephalia is represented by a single living species, the tuatara *(Sphenodon punctatum)* of New Zealand (Fig. 27-11). This animal is the sole survivor of a group of primitive reptiles that otherwise became extinct 100 million years ago. The tuatara was once widespread on the North Island of New Zealand but is now restricted to islets of Cook Strait and off the northern coast of North Island where, under protection from the New Zealand government, it may recover.

It is a lizardlike form 66 cm long or less that lives in burrows often shared with petrels. They are slow-growing animals with a long life; one is recorded to have lived 77 years.

The tuatara has captured the interest of biologists because of its numerous primitive features that are almost identical to those of Mesozoic fossils 200 million years old. These include a primitive skull structure found in early Permian reptiles that were ancestors to the modern lizards. It also bears a well-developed parietal eye, complete with evidences of a

retina (Fig. 27-11), and a complete palate. It lacks a copulatory organ. A specialized feature is the teeth, which are fused to the edge of the jaws rather than being set in sockets. *Sphenodon* represents one of the slowest rates of evolution known among the vertebrates.

SELECTED REFERENCES

Barbour, R. W. 1971. Amphibians and reptiles of Kentucky. Lexington, Ky. University of Kentucky Press.

Bellairs, A. 1970. The life of reptiles, 2 vols. New York, Universe Books. *An accurate well-written treatise that is primarily concerned with anatomy and evolution. Perhaps a bit advanced for the beginning student.*

Carr, A. 1963. The reptiles. New York, Life Nature Library. *There is a wealth of reptile lore in this superbly illustrated volume.*

Ernst, C. H., and R. W. Barbour. 1972. Turtles of the United States. Lexington, Ky., University of Kentucky Press. *Authoritative monograph on identification, distribution, and detailed life histories of all turtles of the United States.*

Gans, C. (ed.). 1969. Biology of the Reptilia, 3 vols. New York, Academic Press, Inc. *A work for the advanced student or researcher. The series will eventually consist of morphologic, embryologic and physiologic, and ecologic and behavioral portions. Specialists contribute chapters on topics that correspond to their research interests.*

Gans, C. 1974. Biomechanics: an approach to vertebrate biology. Philadelphia, J. B. Lippincott Co. *Many examples of interesting functional adaptations, especially of snakes.*

Goin, C. J., and O. B. Goin. 1971. Introduction to herpetology, ed.

2. San Francisco, W. H. Freeman & Co., Publishers. *Popular, basic text.*

Klauber, L. M. 1956. Rattlesnakes, 2 vols. Berkeley and Los Angeles, University of California Press. *An excellent monograph on this group of reptiles.*

Schmidt, K. P., and R. F. Inger. 1957. Living reptiles of the world. Garden City, N.Y., Hanover House. *A magnificent volume with excellent illustrations of the various families of living reptiles.*

Smith, H. M. 1946. Handbook of lizards of the United States and Canada. Ithaca, N.Y., Comstock Publishing Co. *A standard reference for those interested in the natural history and taxonomy of lizards.*

Wright, A. H., and A. A. Wright. 1957. Handbook of snakes of the United States and Canada, 2 vols. Ithaca, N.Y., Comstock Publishing Co. *These two volumes are indispensable and are probably the best available on snakes of the United States and Canada.*

SELECTED SCIENTIFIC AMERICAN ARTICLES

Bogert, C. M. 1959. How reptiles regulate body temperature. **200:**105-120 (April). *Many reptiles, especially lizards, achieve a remarkable degree of temperature regulation by behavioral responses.*

Carr, A. 1965. The navigation of the green turtle. **212:**78-86 (May). *These marine turtles make regular migrations over vast ocean routes.*

Gamow, R. T., and J. E. Harris. 1973. The infrared receptors of snakes. **228:**94-100 (May). *The anatomy and physiology of the remarkable "sixth sense" of snakes is described.*

Gans, C. 1970. How snakes move. **222:**82-96 (June).

Minton, S. A., Jr. 1957. Snakebite. **196:**114-122 (Jan.). *A survey of the world's most poisonous snakes and their venoms.*

Pooley, A. C., and C. Gans. 1976. The Nile crocodile. **234:**114-124 (April). *The remarkably advanced social behavior of this large reptile is described.*

Riper, W. Van. 1953. How a rattlesnake strikes. **189:**100-102 (Oct.). *The rattlesnake strike is examined with high-speed photography.*

Revealing superb muscular coordination, a black skimmer passes with unerring precision across a lake, slicing a high-speed furrow on the surface to attract minnows. On a second pass, it will snap them up, its bill closing automatically upon contact. Birds have an especially well-developed cerebellum, which makes possible such perfect flight control. (Drawn from photo by F. K. Truslow.)

28 BIRDS

ORIGIN AND RELATIONSHIPS

> Characteristics
> Brief classification

ADAPTATIONS OF BIRD STRUCTURE AND FUNCTION FOR FLIGHT

FLIGHT

MIGRATION AND NAVIGATION

BIRD SOCIETY

BIRD POPULATIONS

Of the vertebrates, birds of the class Aves (ay′veez) (pl. of L. *avis,* bird) are the most studied, the most observable, the most melodious, and many think the most beautiful. With 8,600 species distributed over nearly the entire earth, birds far outnumber all other vertebrates except the fishes. Birds are found in forests and deserts, in mountains and prairies, and on all the oceans. Four species are known to have visited the North Pole, and one, a skua, was seen at the South Pole. Some birds live in total blackness in caves, finding their way about by echolocation, and others dive to depths greater than 45 m to prey on aquatic life.

The single unique feature that distinguishes birds from other animals is their feathers. If an animal has feathers, it is a bird; if it lacks feathers, it is not a bird. No other vertebrate group bears such an easily recognized and foolproof identification tag.

There is great uniformity of structure among birds. Despite approximately 130 million years of evolution, during which they proliferated and adapted themselves to specialized ways of life, we have no difficulty recognizing a bird as a bird. In addition to feathers, all birds have forelimbs modified into wings (although they may not be used for flight); all have hind limbs adapted for walking, swimming, or perching; all have horny beaks; and all lay eggs. Probably the reason for this great structural and functional uniformity is that birds evolved into flying machines. This fact greatly restricts diversity, so much more evident in other vertebrate classes. For example, birds do not begin to approach the diversity seen in their warm-blooded evolutionary peers, the mammals, a group that includes forms as unlike as a whale, a porcupine, a bat, and a giraffe.

Birds share with mammals the highest organ system development in the animal kingdom. But a bird's entire anatomy is designed around flight and its perfection. An airborne life for a large vertebrate is a highly demanding evolutionary challenge. A bird must, of course, have wings for support and propulsion. Bones must be light and hollow yet serve as a rigid airframe. The respiratory system must be highly efficient to meet the intense metabolic demands of flight and serve also as a thermo-regulatory device to maintain a constant body temperature. A bird must have a rapid and efficient digestion to process an energy-rich diet; it must have a high metabolic rate; and it must have a high-pressure circulatory system.

Above all, birds must have a finely tuned nervous system and acute senses, especially superb vision, to handle the complex problems of headfirst, high-velocity flight.

ORIGIN AND RELATIONSHIPS

Approximately 150 million years ago, a flying animal drowned and settled to the bottom of a tropical freshwater lake in what is now Bavaria, Germany. It was rapidly covered with a fine silt and eventually fossilized. There it remained until discovered in 1861 by a workman splitting slate in a limestone quarry. The fossil was approximately the size of a crow, with a skull not unlike that of modern birds except that the beaklike jaws bore bony teeth set in sockets like those of reptiles (Fig. 28-1). The skeleton was decidedly reptilian with a long bony tail, clawed fingers, and abdominal ribs. It might have been classified as a reptile except that it carried the unmistakable imprint of **feathers,** those marvels of biologic engineering that only birds possess. The finding was dramatic because it proved beyond reasonable doubt that birds had evolved from reptiles.

Archaeopteryx (ar-kee-op′ter-ix) meaning ''ancient wing,'' as the fossil was named, was an especially fortunate discovery because the fossil record of birds is disappointingly meager. The bones of birds are lightweight and quickly disintegrate, so that only under the most favorable conditions do they fossilize. Nevertheless, by 1952 more than 780 different fossil species had been recorded. Although most of these are relatively recent fossils, enough intermediate forms are known to provide a reasonable picture of bird evolution from the Jurassic, when *Archaeopteryx* lived, to recent times. By the close of the Cretaceous, approximately 63 million years ago, the characteristics of modern birds had been thoroughly molded. There remained only the emergence and proliferation of the modern orders of birds. Hundreds of thousands of bird species have appeared and nearly as many have disappeared, following *Archaeopteryx* to extinction. Only a minute fraction of these nameless species have been discovered as fossils.

Most paleontologists agree that the ancestors of both birds and dinosaurs were derived from a stem group of reptiles called thecodonts. (See Fig. 27-1.) Birds probably have a monophyletic origin (that is, evolved from a single ancestor). However, existing birds are divided into two groups: (1) **ratite** (rat′ite) (L. *ratitus,*

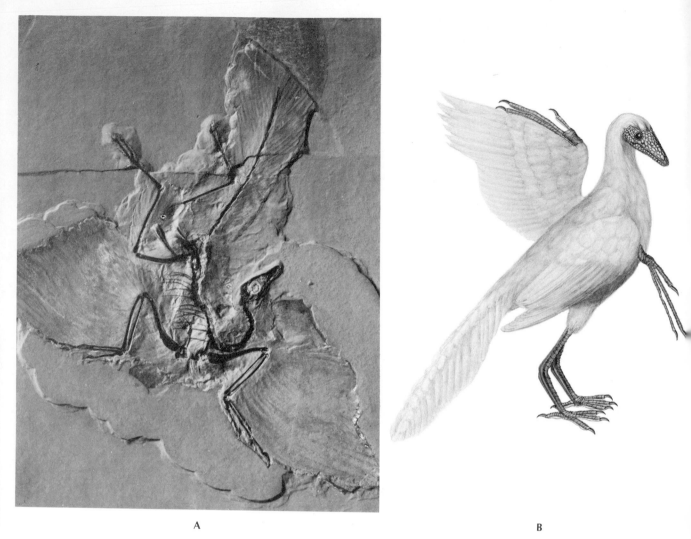

A

B

Fig. 28-1. *Archaeopteryx,* the 150-million–year–old ancestor of modern birds. **A,** Cast of the second and most nearly perfect fossil of *Archaeopteryx,* which was discovered in a Bavarian stone quarry. **B,** Reconstruction of *Archaeopteryx.* (**A** courtesy American Museum of Natural History.)

marked like a raft, from *ratis,* raft), the flightless ostrichlike birds that have a flat sternum with poorly developed pectoral muscles, and (2) **carinate** (L. *carina,* keel), the flying birds that have a keeled sternum upon which the powerful flight muscles insert. This division originated from the view that the flightless birds (ostrich, emu, kiwi, rhea, etc.) represented a separate line of descent that never attained flight. This idea is now completely rejected. The flightless birds are descended from a flying ancestor but have lost the use of their wings and then taken up a different mode

of life. Flightless forms are ground-living birds that can outrun predators, or they live where few carnivorous enemies are found.

It may seem paradoxical that birds with their agile, warm-blooded, colorful, and melodious way of life should have descended from lethargic, cold-blooded, and silent reptiles. Yet the numerous anatomic affinities of the two groups are abundant evidence of close kinship and led the great English zoologist Thomas Henry Huxley to call birds merely "glorified reptiles." This unflattering description has never

Fig. 28-2 Family tree of birds showing probable lines of descent and relationship. Thirteen of the most familiar of the 27 recognized living orders of birds are pictured. Two extinct orders, represented by *Archaeopteryx* and *Hesperornis*, are also pictured.

pleased bird lovers, who answer, "But how wondrously glorified."

Characteristics

1. Usually spindle-shaped body with four divisions; head, neck, trunk, and tail; **neck disproportionately long** for balancing and food gathering
2. Paired limbs, with the **forelimbs usually adapted for flying;** posterior pair variously adapted for perching, walking, and swimming; foot with four toes (chiefly)
3. Epidermal **exoskeleton of feathers** and **leg scales;** thin integument of epidermis and dermis; no sweat glands; **rudimentary pinna of ear**
4. **Fully ossified skeleton with air cavities;** jaws covered with **horny beaks;** small ribs; sternum well-developed with keel or reduced with no keel; **no teeth**
5. Well-developed nervous system with brain and 12 pairs of cranial nerves
6. Circulatory system of **four-chambered heart,** with the **right aortic arch persisting;** nucleated red blood cells
7. Respiration by slightly expansible lungs, with thin **air sacs** among the visceral organs and skeleton; **syrinx (voice box)** near junction of trachea and bronchi
8. Excretory system by metanephric kidney; ureters open into cloaca; **no bladder;** semisolid urine; uric acid as main nitrogenous waste
9. Separate sexes; **females with left ovary and oviduct only**
10. Internal fertilization; **eggs with much yolk and hard calcareous shells;** embryonic membranes in egg during development; **external incubation;** young, active at hatching (**precocial**) or helpless and naked (**altricial**)

Brief classification

The following classification includes most of the recognized 32 orders of birds. Excluded are 5 fossil orders and certain small living orders whose members are mostly unfamiliar to North Americans. Birds are the most thoroughly described class of animals. Probably only a very few species remain to be discovered and named.

Subclass Archaeornithes (ar′ke-or′ni-theez) (Gr. *archaios,* ancient, + *ornis, ornithes,* bird). Ancestral birds. *Archaeopteryx,* known from four fossil finds (Fig. 28-1).

Subclass Neornithes (ne-or′ni-theez) (Gr. *neos,* new, + *ornis,* bird). True, modern birds.

  Order Struthioniformes (stroo′thi-on-i-for′meez) (L. *struthio,* ostrich, + L. *forma,* form). Largest living birds; flightless; all African. **Ostriches.**

  Order Rheiformes (re′i-for′meez) (Greek mythology, *Rhea,* mother of Zeus, + form). Flightless birds of South America. **Rheas.**

  Order Casuariiformes (kazh′u-ar′ee-i-for′meez) (NL. *Casuarius,* type genus, + form). Flightless birds of Australia and New Guinea. **Emus, cassowaries.**

Order Sphenisciformes (sfe-nis′i-for′meez) (Gr. *spheniskos,* dim. of *sphen,* wedge, from the shortness of the wings, + form). Web-footed, flightless, marine swimmers; found from Antarctica north to Galapagos Islands. **Penguins.**

Order Gaviiformes (gay′vee-i-for′meez) (L. *gavia,* bird, probably sea mew, + form). Short legs, heavy body, good divers; North America and Eurasia. **Loons.**

Order Podicipediformes (pod′i-si-ped′i-for′meez) (L. *podex,* rump, + *pes, pedis,* foot). Short legs, lobate-webbed toes, good divers; worldwide. **Grebes.**

Order Procellariiformes (pro-sel-lar′ee-i-for′,meez) (L. *procella,* tempest, + form). Marine birds with tubular nostrils; worldwide. **Albatrosses, shearwaters, fulmars, petrels.**

Order Pelecaniformes (pel′e-can-i-for′meez) (Gr. *pelekan,* pelican, + form). Fish-eating birds with throat pouch; mostly colony nesters; worldwide, especially in tropics. **Pelicans, boobies, gannets, cormorants, frigate birds.**

Order Ciconiiformes (si-ko′nee-i-for′meez) (L. *ciconia,* stork, + form). Long-necked; long-legged waders; mostly colony nesters; worldwide. **Herons, storks, ibises, flamingos.**

Order Anseriformes (an′ser-i-for′meez) (L. *anser,* goose, + form). Broad bills with filtering ridge or "teeth" at margins; worldwide. **Ducks, geese, swans, screamers.**

Order Falconiformes (fal′ko-ni-for′meez) (NL. *falco,* falcon, + form). Diurnal birds of prey; keen vision; strong fliers; worldwide. **Eagles, hawks, vultures, falcons, condors, kites.**

Order Galliformes (gal′li-for′meez) (L. *gallus,* cock, + form). Henlike birds; vegetarians; with strong beaks and heavy feet; worldwide. **Grouse, ptarmigan, quails, pheasants, fowl, turkeys.**

Order Gruiformes (groo′i-for′meez) (L. *grus,* crane, + form). Diverse group of prairie and marsh dwellers; worldwide. **Cranes, rails, coots, gallinules, bustards.**

Order Charadriiformes (ka-rad′ree-i-for′meez) (NL. *Charadrius,* genus of plovers, + form). Large assemblage of shore birds; strong fliers; often colonial; worldwide. **Plovers, turnstones, lapwings, oyster catchers, snipe, sandpipers, avocets, phalaropes, skuas, gulls, terns, skimmers, auks, puffins.**

Order Strigiformes (strij′i-for′meez) (L. *strix,* screech owl, + form). Nocturnal predators with large eyes, powerful beak and feet, and silent flight; worldwide. **Owls.**

Order Caprimulgiformes (kap′ri-mul′ji-for′meez) (L. *caprimulgus,* goatsucker, + form). Twilight insect feeders with wide mouth; worldwide. **Goatsuckers, nighthawks, poorwills.**

Order Apodiformes (up-pod′i-for′meez) (Gr. *apous,* sandmartin; footless + form). Small birds with short legs and rapid wingbeat; worldwide. **Swifts, hummingbirds.**

Order Piciformes (pis′i-for′meez) (L. *picus,* woodpecker, + form). Highly specialized bills; toes, two front and two rear; nest in cavities; worldwide. **Woodpeckers, toucans, honeyguides.**

Order Passeriformes (pas′er-i-for′meez) (L. *passer,* sparrow, + form). Largest order of birds, containing 56 families and 60% of all birds; all songbirds; highly developed syrinx, thus excellent vocal powers; feet adapted for perching on stems or twigs; toes, three in front, one behind; young hatched blind, naked, and helpless; worldwide. **Perching songbirds.**

ADAPTATIONS OF BIRD STRUCTURE AND FUNCTION FOR FLIGHT

Just as an airplane must be designed and built according to rigid aerodynamic specifications if it is to fly, so too must birds meet stringent structural requirements if they are to stay airborne. All the special adaptations found in flying birds come down to two things: more power and less weight. Flight by man became possible when he developed an internal combustion engine and learned how to reduce the weight-to-power ratio to a critical point. Birds did this millions of years ago. But birds must do much more than fly. They must feed themselves and convert food into high-energy fuel; they must escape predators; they must be able to repair their own injuries; they must be able to air-condition themselves when overheated and heat themselves when too cool; and, perhaps most important of all, they must reproduce themselves.

Feathers

A feather is almost weightless yet possesses remarkable toughness and tensile strength. A typical **contour feather** consists of a hollow **quill,** or calamus, thrust into the skin and a **shaft,** or rachis, which is a continuation of the **quill** and bears numerous **barbs** (Fig. 28-3). The barbs are arranged in closely parallel fashion and spread diagonally outward from both sides of the central shaft to form a flat, expansive, webbed surface, the **vane.** There may be several hundred barbs in the vane.

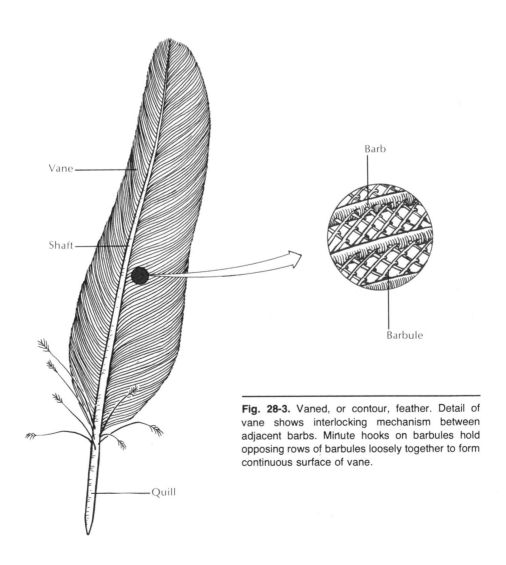

Vane

Shaft

Quill

Barb

Barbule

Fig. 28-3. Vaned, or contour, feather. Detail of vane shows interlocking mechanism between adjacent barbs. Minute hooks on barbules hold opposing rows of barbules loosely together to form continuous surface of vane.

If the feather is examined with a microscope, each barb appears to be a miniature replica of the feather, with numerous parallel filaments, called **barbules,** set in each side of the barb and spreading laterally from it. There may be 600 barbules on each side of a barb, which adds up to more than 1 million barbules for the feather. The barbules of one barb overlap the barbules of a neighboring barb in a herringbone pattern and are held together with great tenacity by tiny hooks. Should two adjoining barbs become separated—and considerable force is needed to pull the vane apart—they are instantly zipped together again by drawing the feather through the fingertips. The bird, of course, does it with its bill, and much of a bird's time is occupied with preening to keep its feathers in perfect condition.

Feathers are epidermal structures that evolved from the reptilian scale; a developing feather closely resembles a reptile scale when growth is just beginning. We can imagine that in its evolution the scale elongated and its edges frayed outward until it became the complex feather of birds. Strangely enough, though modern birds possess both scales (especially on their feet) and feathers, no intermediate stage between the two has been discovered on either fossil or living forms.

When fully grown, a feather is a dead structure. Tough as it is, it eventually frays and may even break from wear. All birds renew their feathers at regular intervals, usually in late summer after the nesting season. Many birds also undergo a second partial molt just before the mating season, to equip them with their breeding finery, so important for courtship display. The shedding of feathers is a highly ordered process and is scheduled to least disturb the demands for self-protection and food-gathering. Except in penguins, which molt all their short feathers at once and lose 50% of the body weight while waiting solemnly in one spot for new feathers to grow, feathers are discarded gradually thus avoiding the appearance of bare spots. Flight and tail feathers are lost in exact pairs, one from each side, so that balance is maintained (Fig. 28-13). Replacements emerge before the next pair is lost, and most birds can continue to fly unimpaired during the molting period; only ducks and geese are completely grounded during the molt.

Skeleton

One of the major adaptations that allow a bird to fly is its light skeleton (Fig. 28-4). Bones are phenome-nally light, delicate, and laced with air cavities, yet they are strong. The skeleton of a frigate bird with a 7-foot wingspan weighs only 114 g (4 ounces), less than the weight of all its feathers. A pigeon skull weighs only 0.21% of its body weight; the skull of a rat by comparison weighs 1.25%.

The bird skull is mostly fused into one piece. The braincase and orbits are large to accommodate a bulging cranium and the large eyes needed for quick motor coordination and superior vision. The anterior bones are elongated to form a beak. The lower mandible is a complex of several bones that hinge on two small movable bones, the quadrates. This provides a double-jointed action that permits the mouth to open widely. The upper jaw is usually fused to the forehead, but in some birds—parrots, for instance—the upper jaw is hinged also. This adaptation allows greater flexibility of the beak in food manipulation and provides insect-catching species with a wider gap for successful feeding on the wing.

The beaks of birds are strongly adapted to specialized food habits—from generalized types, such as the strong, pointed beaks of crows to grotesque, highly specialized ones in flamingoes, hornbills, and toucans. The beak of a woodpecker is a straight, hard, chisel-like device. Anchored to a tree trunk with its tail serving as a brace, the woodpecker delivers powerful, rapid blows to build nests or expose the burrows of wood-boring insects. It then uses its long, flexible, barbed tongue to seek out insects in their galleries. The woodpecker's skull is especially thick to absorb shock.

The bones of the pelvis are fused together and fused with the lumbar and most of the tail vertebrae to form a rigid plate of bone, the **synsacrum.** This light and stiff structure supports the legs and provides a rigid air-frame. To assist in this rigidity, the ribs are mostly fused with the vertebrae, pectoral girdle, and sternum. Except for the flightless birds, the sternum bears a large, thin keel that provides for the attachment of the powerful flight muscles. Of the body box, only the 8 to 24 (according to the species) vertebrae of the neck remain fully flexible.

The bones of the forelimbs have become highly modified for flight. They are hollow (for lightness) and reduced in number, and several are fused together. Despite these alterations, the bird wing is clearly a rearrangement of the basic vertebrate tetrapod limb from which it arose, and all the elements—upper arm,

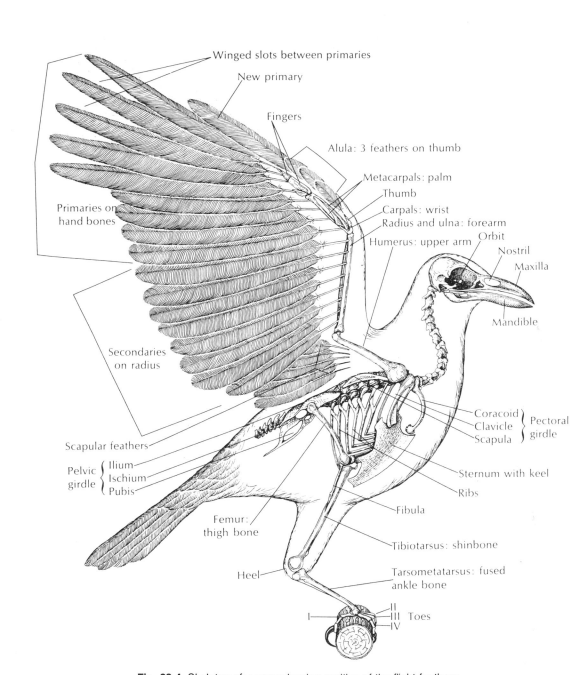

Winged slots between primaries

New primary

Fingers

Alula: 3 feathers on thumb

Metacarpals: palm

Thumb

Carpals: wrist

Radius and ulna: forearm

Humerus: upper arm

Orbit

Nostril

Maxilla

Mandible

Primaries on hand bones

Secondaries on radius

Coracoid
Clavicle
Scapula
} Pectoral girdle

Scapular feathers

Pelvic girdle {
Ilium
Ischium
Pubis

Sternum with keel

Ribs

Fibula

Femur: thigh bone

Tibiotarsus: shinbone

Heel

Tarsometatarsus: fused ankle bone

II
III Toes
I
IV

Fig. 28-4. Skeleton of a crow showing position of the flight feathers.

forearm, wrist, and fingers—are represented in modified form (Fig. 28-4). The birds' legs have undergone less pronounced modification than the wings since they are still designed principally for walking, as well as for perching and occasionally for swimming, as were those of their reptilian ancestors.

Muscular system

The locomotor muscles of the wings are relatively massive to meet the demands of flight. The largest of these is the **pectoralis,** which depresses the wings in flight. Its antagonist is the **supracoracoideus** muscle, which raises the wing. Surprisingly perhaps, this latter muscle is not located on the backbone (anyone who has been served the back of the chicken knows it offers little meat) but is positioned under the pectoralis on the breast. It is attached by a tendon to the upper side of the humerus of the wing so that it pulls from below by an ingenious rope-and-pulley arrangement. Both of these muscles are anchored to the keel. Thus with the main muscle mass low in the body, aerodynamic stability is improved.

The main leg muscle mass is located in the thigh, surrounding the femur, and a smaller mass lies over the tibiotarsus (shank, or "drumstick"). Strong but thin

tendons extend downward through sleevelike sheaths to the toes. Consequently the feet are nearly devoid of muscles, thus explaining the thin, delicate appearance of the bird leg. This arrangement places the main muscle mass near the bird's center of gravity and at the same time allows great agility to the slender, lightweight feet. Since the feet are made up mostly of bone, tendon, and tough, scaly skin, they are highly resistant to damage from freezing. When a bird perches on a branch, an ingenious toe-locking mechanism (Fig. 28-5) is activated, which prevents a bird from falling off its perch when asleep. The same mechanism causes the talons of a hawk or owl to automatically sink deeply into its victim as the legs bend under the impact of the strike. The powerful grip of a bird of prey was described by L. Brown*:

When an eagle grips in earnest, one's hand becomes numb, and it is quite impossible to tear it free, or to loosen the grip of the eagle's toes with the other hand. One just has to wait until the bird relents, and while waiting one has ample time to realize that an animal such as a rabbit would be quickly paralyzed, unable to draw breath, and perhaps pierced through and through by the talons in such a clutch.

*Brown, L. 1970. Eagles. New York, Arco Publishing Co., Inc.

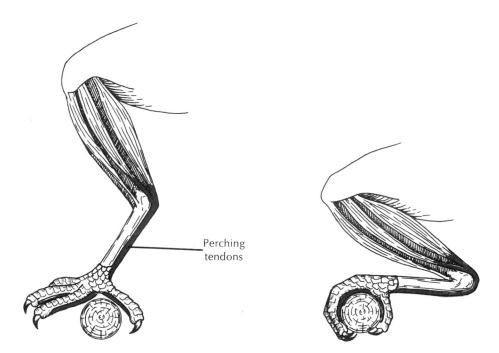

Fig. 28-5. Perching mechanism of bird. When bird settles on branch, tendons automatically tighten, closing toes around perch.

Perching tendons

Digestive system

Birds eat an energy-rich diet in large amounts and process it rapidly with efficient digestive equipment. A shrike can digest a mouse in 3 hours, and berries pass completely through the digestive tract of a thrush in just 30 minutes. Furthermore birds utilize a very high percentage of the food they eat. There are no teeth in the mouth, and the poorly developed salivary glands mainly secrete mucus for lubricating the food and the slender, horn-covered **tongue.** There are few taste buds. From the short **pharynx** a relatively long, muscular, elastic **esophagus** extends to the **stomach.** In many birds there is an enlargement **(crop)** at the lower end of the esophagus, which serves as a storage chamber.

In pigeons and some parrots the crop not only stores food but produces milk by the breakdown of epithelial cells of the lining. This "bird milk" is regurgitated by both male and female into the mouth of the young squabs. The milk has a much higher fat content than cow's milk.

The stomach proper consists of a **proventriculus,** which secretes gastric juice, and the muscular **gizzard,** which is lined with horny plates that serve as millstones for grinding the food. To assist in the grinding process, birds swallow coarse, gritty objects or pebbles, which lodge in the gizzard. Certain birds of prey, such as owls, form pellets of indigestible materials, for example, bones and fur, in the proventriculus and eject them through the mouth. At the junction of the intestine with the rectum there are paired **ceca,** which may be well developed in some birds. The terminal part of the digestive system is the **cloaca,** which also receives the genital ducts and ureters.

Circulatory system

The general plan of circulation in birds is not greatly different from that of mammals. The four-chambered heart is large with strong ventricular walls; thus birds share with mammals a complete separation of the respiratory and systemic circulations. The heartbeat is extremely fast, but as in mammals there is an inverse relationship between heart rate and body weight. For example, a turkey has a heart rate at rest of approximately 93 beats per minute, a chicken has 250 beats per minute, and a black-capped chickadee has a rate of 500 beats per minute when asleep, which may increase to a phenomenal 1,000 beats per minute during exercise. Blood pressure in birds is roughly

equivalent to that in mammals of similar size. Bird's blood contains nucleated, biconvex red corpuscles that are somewhat larger than those of mammals. The phagocytes, or mobile ameboid cells, of the blood are unusually active and efficient in birds in the repair of wounds and in destroying microbes.

The body temperature of birds is high, ranging between 40° and 42° C as compared to 36° to 39° C for mammals. Some thrushes operate at 43.5° C (110.5° F), a temperature well past the lethal limit for most mammals and only approximately 2.5° C below the upper lethal temperature for these birds. Small birds tend to have more variable body temperatures than do large birds; for example, that of a house wren may fluctuate 8 centigrade degrees during 24 hours.

Like their homothermic counterparts, the mammals, birds must also maintain relative constancy of their body temperature. If they begin to overheat, heat loss is accelerated by dilating blood vessels in the skin (to increase radiant heat loss) and by increasing the breathing rate (to increase evaporative cooling). Many species pant in extreme heat. In cold weather, birds ruffle their feathers to form a blanket of warm, insulative air next to the skin and the peripheral blood vessels vasoconstrict to reduce radiant heat loss. If these physical mechanisms are not sufficient to prevent a decrease in body temperature in very cold weather, a bird shivers. As in mammals, this muscular movement creates needed heat. It also increases food and oxygen consumption. A sparrow consumes twice as much oxygen and eats twice as much food at 0° C (32° F) as at 37° C (98° F). What person living in a northern climate has not marveled at the ability of the tiny chickadee, a minute furnace of cheerful activity, to survive direct exposure to the coldest winter weather?

Respiratory system

The respiratory system of birds differs radically from the lungs of reptiles and mammals and is marvelously adapted for meeting the high metabolic demands of flight. The lungs, which are relatively inexpansible because of their direct attachment to the body wall, are filled with numerous tiny **air capillaries** instead of alveoli of the mammalian type. Most unique, however, is the extensive system of nine interconnecting **air sacs** that are located in pairs in the thorax and abdomen and even extend by tiny tubes into the centers of the long bones (Fig. 28-6). The air sacs

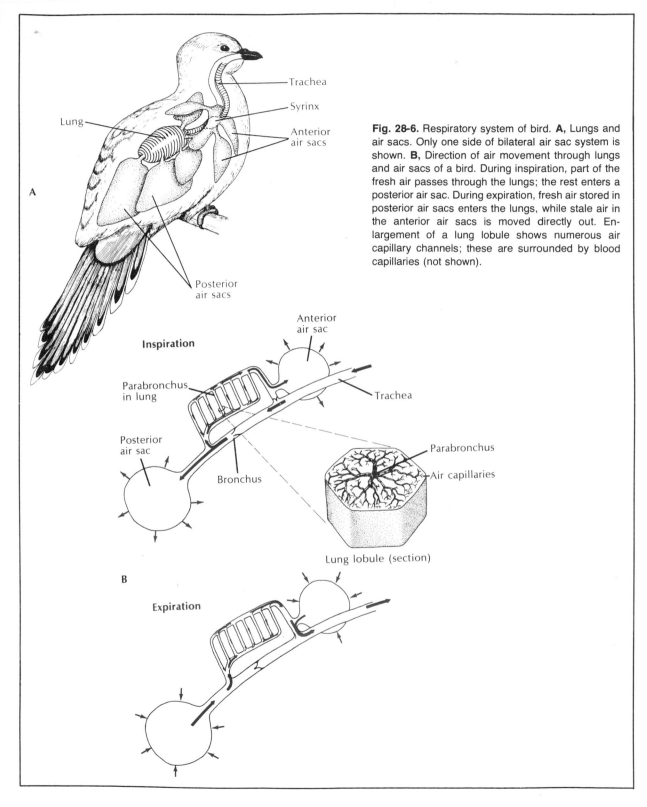

Fig. 28-6. Respiratory system of bird. **A,** Lungs and air sacs. Only one side of bilateral air sac system is shown. **B,** Direction of air movement through lungs and air sacs of a bird. During inspiration, part of the fresh air passes through the lungs; the rest enters a posterior air sac. During expiration, fresh air stored in posterior air sacs enters the lungs, while stale air in the anterior air sacs is moved directly out. Enlargement of a lung lobule shows numerous air capillary channels; these are surrounded by blood capillaries (not shown).

are connected to the lungs in such a way that perhaps 75% of the inspired air bypasses the lungs and flows directly into the air sacs, which serve as reservoirs for fresh air. On expiration, some of this fully oxygenated air is shunted through the lung, while the used air passes directly out. The advantage of such a system is obvious—the lungs receive fresh air during both inspiration and expiration. Rather than having the respiratory exchange surface deep within blind sacs, which, as in mammals, are difficult to ventilate, birds have a stream of fully oxygenated air passing continuously through a system of richly vascularized air capillaries (Fig. 28-6, *B*). Although many details of the bird's respiratory system are not yet understood, it is clearly the most efficient of all the vertebrates.

Excretory system

The relatively large paired metanephric kidneys are composed of many thousands of **nephrons,** each consisting of a renal corpuscle and a nephric tubule. As in other vertebrates, the urine is formed by glomerular filtration followed by the selective modification of the filtrate in the nephric tubule. However, the urine contains a high concentration of uric acid and creatine; when combined with fecal material in the cloaca, it becomes a white paste. There is a great advantage to excreting nitrogenous waste as uric acid instead of urea. Uric acid is so nearly insoluble that once excreted it precipitates out of solution and therefore does not contribute to the osmotic pressure of the urine. Birds thus excrete a "concentrated" urine with a relatively simple kidney. In the mammalian kidney, a highly sophisticated urine-concentrating device has evolved, enabling mammals to excrete their nitrogenous waste as soluble urea rather than insoluble uric acid.

Marine birds (also marine turtles) have evolved a unique solution for excreting the large loads of salt eaten with their food and in the seawater they drink. Sea water contains approximately 3% salt and is three times saltier than a bird's body fluids. Yet the bird kidney cannot concentrate salt in urine above approximately 0.3%. The problem is solved by special **salt glands,** one located above each eye. These glands are capable of excreting a highly concentrated solution of sodium chloride—as much as twice the concentration of seawater. The salt solution runs out the internal or external nostrils, giving gulls, petrels, and other sea birds a perpetual runny nose.

Nervous and sensory system

A bird's nervous and sensory system accurately reflects the complex problems of flight and a highly visible existence, in which it must gather food, mate, defend territory, incubate and rear young, and correctly distinguish friend from foe. The brain of a bird has well-developed **cerebral hemispheres, cerebellum,** and **midbrain tectum** (optic lobes). The **cerebral cortex**—the portion in mammals that becomes the chief coordinating center—is thin, unfissured, and poorly developed in birds. But the core of the cerebrum, the **corpus striatum,** has enlarged into the principal integrative center of the brain, controlling such activities as eating, singing, flying, and all the complex instinctive reproductive activities. Relatively intelligent birds, such as crows and parrots, have larger cerebral hemispheres than do less intelligent birds, such as chickens and pigeons. The **cerebellum** is a crucial coordinating center where muscle-position sense, equilibrium sense, and visual cues are all assembled and used to coordinate movement and balance. The **optic lobes,** laterally bulging structures of the midbrain, form a visual association apparatus comparable to the visual cortex of mammals.

Except in flightless birds and in ducks, the senses of smell and taste are poorly developed in birds. This deficiency, however, is more than compensated for by

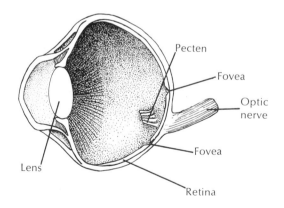

Fig. 28-7. Hawk eye has all the structural components of mammalian eye, plus a peculiar pleated structure, the pecten, believed to provide nourishment to the retina. The extraordinary keen vision of the hawk is attributed to the extreme density of cone cells in the foveae: 1.5 million per fovea compared to 0.2 million for man. Each hawk eye has two foveae as opposed to man's one, meaning that each hawk eye focuses on two objects simultaneously—the better to see its next meal!

good hearing and superb vision, the keenest in the animal kingdom. The organ of hearing, the **cochlea,** is much shorter than the coiled mammalian cochlea, yet birds can hear roughly the same range of sound frequencies as man. Actually the bird ear far surpasses man's in capacity to distinguish differences in intensities and to respond to rapid fluctuations in song.

The bird eye resembles that of other vertebrates in gross structure but is relatively larger, less spheric, and almost immobile; instead of turning their eyes, birds turn their heads with their long and flexible necks to scan the visual field. The light-sensitive **retina** (Fig. 28-7) is elaborately equipped with rods (for dim light vision) and cones (for color vision). Cones predominate in day birds, and rods are more numerous in nocturnal birds. The fovea, or region of keenest vision on the retina, is placed (in birds of prey and some others) in a deep pit, which makes it necessary for the bird to focus exactly on the source. Many birds moreover have two sensitive spots (foveas) on the retina (Fig. 28-7)—the central one for sharp monocular views and the posterior one for binocular vision. The visual acuity of a hawk is believed to be eight times that of man (enabling it to see clearly a crouching

Fig. 28-8. Screech owl, *Otus asio* (order Strigiformes) during brief interlude after successful forage. Owls possess eyes incredibly sensitive to light, enabling them to see prey in light one hundreth to one tenth the intensity required by man (0.000,000,73 footcandle). But this particular ultrasensitivity is traded off for relatively poor visual acuity, narrow visual field, and weak accommodation (ability to focus on near objects). (Photo by L. L. Rue III.)

rabbit 2 km away), and an owl's ability to see in dim light is more than 10 times that of the human eye (Fig. 28-8).

Reproductive system

The male reproductive organs, or gonads, are the testes. These are tiny bean-shaped bodies during most of the year, which undergo a great enlargement at the breeding season, as much as 300 times larger than the nonbreeding size. Before discharge, the millions of sperm are stored in a **seminal vesicle,** which, like the testes, enlarges greatly during the breeding season. Since most birds lack a penis, copulation is a matter of bringing the cloacal surfaces into contact, usually while male stands on the back of the female. Some swifts copulate in flight.

In the female, only the left ovary and oviduct develop; those on the right dwindle to vestigial structures. Eggs discharged from the ovary are picked up by the oviduct, which runs posteriorly to the cloaca. While the eggs are passing down the oviduct, **albumin,** or egg white, from special glands is added to them; farther down the oviduct, the shell membrane, shell, and shell pigments are also secreted about the egg.

Fertilization takes place in the upper oviduct several hours before the layers of albumin, shell membranes, and shell are added. Sperm remain alive in the female oviduct for many days after a single mating. Hen eggs show good fertility for 5 or 6 days after mating, but then fertility decreases rapidly. However, an occasional egg is fertile as long as 30 days after separation of the hen from the rooster.

Some birds, such as herring gulls, are determinate layers and lay only a fixed number of eggs (clutch) in a season. If any of the eggs of a set are removed, the deficit is not made up by additional laying. Indeterminate layers, however, continue to lay additional eggs for a long time if some of the first-laid eggs are continually removed (flickers, ducks, domestic poultry). Most birds are probably determinate layers. Many birds, such as songbirds, lay an egg a day until the clutch is completed; others stagger their egg laying and lay every other day or so.

FLIGHT

What prompted the evolution of flight in birds and the ability to rise free of earth-bound concerns, as almost every human being has dreamed of doing? The origin of flight was the result of complex adaptive pressures. The air was a relatively unexploited habitat stocked with flying insect food. Flight also offered escape from terrestrial predators and opportunity to travel rapidly and widely to establish new breeding areas and to benefit from year-round favorable climate by migrating north and south with the seasons.

Bird wing as a lift device

Bird flight, especially the familiar flapping flight of birds, is complex. Despite careful analysis by conventional aerodynamic techniques and high-speed pho-

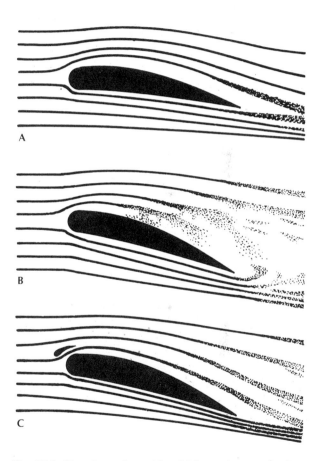

Fig. 28-9. Air patterns formed by airfoil, or wing, moving from right to left. **A,** Normal flight with low angle of attack. As air moves smoothly over wing, areas of low pressure on upper wing surface and of high pressure on lower wing surface create lift. **B,** Appearance of lift-destroying turbulence on upper wing surface when angle of attack becomes too great. Stalling occurs. **C,** Prevention of stalling by directing a layer of rapidly moving air over upper surface with a wing-slot. (From Welty, J. C. 1962. The life of birds, Philadelphia, W. B. Saunders Co.)

Fig. 28-10. Elliptic wing of the black-billed magpie, *Pica pica* (order Passeriformes), in slow flight. Note well-developed alula and separation (slotting) between primaries, typical of wings adapted for high maneuverability and low-speed flight. This member of the crow family is a familiar feature of the western North American landscape. (Photo by C. G. Hampson, University of Alberta.)

tography, there is much about it that is not understood. Nevertheless the bird wing is an airfoil that is subject to recognized laws of aerodynamics. It is adapted for high lift at low speeds, and, not surprisingly perhaps, it resembles the wings of early low-speed aircraft. The bird wing is streamlined in cross section, with a slightly concave lower surface (cambered) and with small tight-fitting feathers where the leading edge meets the air (Fig. 28-9). Air slips efficiently over the wing, creating lift with minimum drag. Some lift is produced by positive pressure against the undersurface of the wing. But on the upper side, where the airstream must travel farther and faster over the convex surface, a negative pressure is created that provides more than two thirds of the total lift.

The lift-to-drag ratio of an airfoil is determined by the angle of tilt (angle of attack) and the airspeed (Fig. 28-9). A wing carrying a given load can pass through the air at high speed and small angle of attack or at low speed and larger angle of attack. But as speed decreases, a point is reached at which the angle of attack

becomes too steep; turbulence appears on the upper surface, lift is destroyed, and stalling occurs. Stalling can be delayed or prevented by placing a **wing slot** along the leading edge so that a layer of rapidly moving air is directed across the upper wing surface. Wing slots were and still are used in aircraft traveling at low speed. In birds, two kinds of wing slots have developed: (1) the **alula,** or group of small feathers on the thumb (Fig. 28-4), which provides a midwing slot, and (2) **slotting between primary feathers,** which provides a wing-tip slot. In many songbirds, these together provide stall-preventing slots for nearly the entire outer (and aerodynamically most important) half of the wing.

Basic forms of bird wings

Bird wings vary in size and form because the successful exploitation of different habitats has imposed special aerodynamic requirements. The following four types of bird wings are easily recognized (Savile, 1957).

Fig. 28-11. High-speed wing of swallow-tailed gull, *Creagrus furcatus* (order Charadriiformes), of the Galápagos Islands. Sweepback, taper, and absence of wing slotting adapt gulls for high speed flight. This species is the only nocturnal gull in the world, feeding on fish and squid caught at night. (Photo by C. P. Hickman, Jr.)

Elliptic wings. Birds that must maneuver in forested habitats, such as sparrows, warblers, doves, woodpeckers, and magpies (Fig. 28-10), have elliptic wings. This type has a low **aspect ratio** (ratio of length to width). The outline of a sparrow wing is almost identical to that of the British Spitfire fighter plane of World War II fame—also a highly maneuverable flyer. Elliptic wings are highly slotted between the primary feathers (Fig. 28-10); this helps to prevent stalling during sharp turns, low-speed flight, and frequent landing and takeoff. Each separated primary behaves as a narrow wing with a high angle of attack, providing high lift at low speed. The high maneuverability of the elliptic wing is exemplified by the tiny chickadee, which, if frightened, can change course within 0.03 second.

High-speed wings. Birds that feed on the wing, such as swallows, hummingbirds, and swifts, or that make long migrations, such as plovers, sandpipers, terns, and gulls (Fig. 28-11), have wings that sweep back and taper to a slender tip. They are rather flat in section, have a moderately high aspect ratio, and lack the wing-tip slotting characteristic of the preceding group. Sweepback and wide separation of the wing tips reduce ''tip vortex,'' the drag-creating turbulence that tends to develop at wing tips. The fastest birds alive, such as falcons and sandpipers, clocked at 285 and 177

Fig. 28-12. Soaring wing of waved albatross, *Diomedea irrorata* (order Procellariiformes). The long, high-aspect wing, like that of a sailplane, has great aerodynamic efficiency and lift but is relatively weak and lacking in maneuverability. (Photo by C. P. Hickman, Jr.)

Fig. 28-13. High-lift wings of the osprey, *Pandion haliaetus* (order Falconiformes), landing on nest. Note alulas *(top arrows)* and new primary feathers *(side arrows).* Feathers are molted in sequence in exact pairs so that balance is maintained during flight. These fish-eating birds have suffered a severe population decline in the United States in recent years because of illegal hunting and poor nesting success. Pesticides concentrated in fish eaten by ospreys cause egg shells to be thin and burst during incubation. (Photo by B. Tallmark.)

km per hour (177 and 110 miles per hour), respectively, belong to this group.

Soaring wings. The oceanic soaring birds have high–aspect ratio wings resembling those of sailplanes. This group includes albatrosses (Fig. 28-12), frigate birds, and gannets. Such long, narrow wings lack wing slots and are adapted for high speed, high lift, and dynamic soaring. They have the highest aerodynamic efficiency of all wings but are less maneuverable than the wide, slotted wings of land soarers. Dynamic soarers have learned how to exploit the highly reliable sea winds, using adjacent air currents of different velocities.

High-lift wings. Vultures, hawks, eagles, owls, and ospreys (Fig. 28-13)—predators that carry heavy loads—have wings with slotting, alulas, and pronounced camber, all of which promote high lift at low speed. Many of these birds are land soarers, with broad, slotted wings that provide the sensitive response and maneuverability required for static soaring in the capricious air currents over land.

MIGRATION AND NAVIGATION

Perhaps it was inevitable that birds, having mastered the art of flight, would use this power to make the long and arduous seasonal migrations that have captured man's wonder and curiosity. The term **migration** refers to the regular, extensive, seasonal movements that birds make between their summer breeding regions and their wintering regions. The chief advantage seems obvious: it enables birds to live in an optimal climate all the time, where abundant and unfailing sources of food are available to sustain their intense metabolism. Migrations also provide optimal conditions for rearing young when demands for food are especially great. Broods are largest in the far North, where the long summer days and the abundance of insects combine to provide parents with ample food-gathering opportunity. Predators are relatively rare in the North, and the brief once-a-year appearance of vulnerable young birds does not encourage the build-up of predator populations. Migration also vastly increases the amount of space available for breeding and reduces aggressive

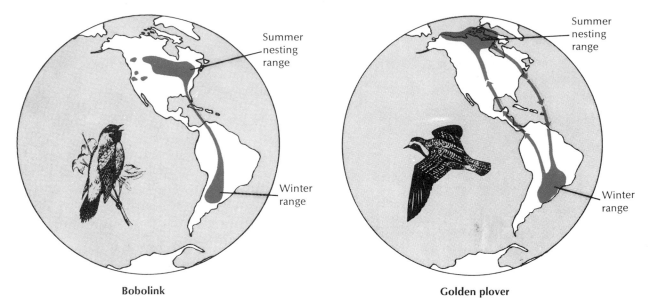

Bobolink   Golden plover

Fig. 28-14. Migrations of the bobolink and golden plover. The bobolink commutes 22,500 km (14,000 miles) each year between nesting sites in North America and its wintering range in Argentina, a phenomenal feat for such a small bird. Although the breeding range has extended to colonies in western areas, these birds take no shortcuts but adhere to the ancestral eastern seaboard route. The golden plover flies a loop migration, striking out across the Atlantic in its southward autumnal migration but returning in the spring via Central America and the Mississippi valley because ecologic conditions are more favorable at that time.

territorial behavior. Of course, migration favors homeostasis (constancy of the body's internal environment) by allowing birds to avoid climatic extremes.

Migration routes

Most migratory birds have well-established routes trending north and south. Since most birds (and other animals) live in the northern hemisphere where most of the earth's land mass is concentrated, most birds are south-in-winter and north-in-summer migrants. Of the 4,000 or more species of migrant birds (a little less than one half of the total bird species), most breed in the more northern latitudes of the hemisphere; the percentage of migrants in Canada is far higher than is the percentage of migrants in Mexico, for example. Some use different routes in the fall and spring (Fig. 28-14). Some, especially certain aquatic species, complete their migratory routes in a very short time. Others, however, make the trip in a leisurely manner, often stopping here and there to feed. Some of the warblers are known to take 50 to 60 days to migrate

from their winter quarters in Central America to their summer breeding areas in Canada.

Some species are known for their long-distance migrations. The arctic tern, greatest globe spanner of all, breeds north of the Arctic Circle and in winter is found in the Antarctic regions, 17,500 km away. This species is also known to take a circuitous route in migrations from North America, passing over to the coastlines of Europe and Africa and thence to their winter quarters. Other birds that breed in Alaska follow a more direct line down the Pacific coast to North and South America.

Many small songbirds also make great migration treks (Fig. 28-14). Africa is a favorite wintering ground for European birds, and many fly there from Central Asia as well.

Stimulus for migration

Man has known for centuries that the onset of the reproductive cycle of birds is closely related to season. Only within the last 50 years, however, has it been proved that the lengthening days of late winter and early spring stimulate the development of the gonads and accumulation of fat—both important internal changes that predispose birds to migrate northward. There is evidence that increasing day length stimulates the anterior lobe of the pituitary into activity. The release of pituitary gonadotropic hormone in turn sets in motion a complex series of physiologic and behavioral changes, resulting in gonadal growth, fat deposition, migration, courtship and mating behavior, and care of the young.

Direction finding in migration

Numerous experiments suggest that most birds navigate chiefly by sight. Birds recognize topographic landmarks and follow familiar migratory routes—a behavior assisted by flock migration, during which navigational resources and experience of older birds can be pooled. But in addition to visual navigation, birds make use of a variety of orientation cues at their disposal. Birds have an innate time sense, a built-in clock of great accuracy; they have an innate sense of direction; and very recent work adds much credence to an old, much debated theory that birds can detect and navigate by the earth's field of gravity. All of these resources are inborn and instinctive, although a bird's navigational abilities may improve with experience.

Recently experiments by German ornithologists G. Kramer and E. Sauer and American ornithologist S. Emlen have demonstrated convincingly that birds can navigate by celestial cues—the sun by day and the stars by night. Using special circular cages, Kramer concluded that birds possessed a built-in time sense that enabled them to maintain compass direction by re-

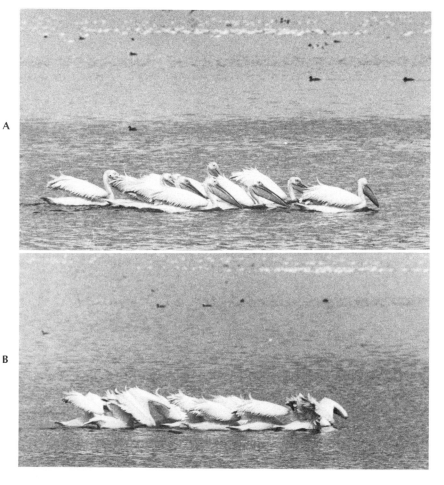

Fig. 28-15. Cooperative feeding behavior by the white pelican, *Pelecanus onocrotalus* (order Pelecaniformes). **A,** Pelicans on Lake Nakuru, East Africa, form a horseshoe to drive fish together. **B,** Then they plunge simultaneously to scoop up fish in their huge bills. Pelicans were attracted to lake in mid-1960s to feed on fish *(Tilapia grahami)* introduced to control malaria by eating mosquito larvae. The pictures were taken 2 seconds apart. (Photos by B. Tallmark.)

ferring to the sun, regardless of time of day. This is called **sun-azimuth orientation** (*azimuth,* compass-bearing of the sun). Sauer's and Emlen's ingenious planetarium experiments strongly suggest that some birds, probably many, learn to use the constellations to navigate at night.

Some of the remarkable feats of bird navigation still defy rational explanation. Most birds undoubtedly use a combination of environmental and innate cues to migrate. Migration is a rigorous undertaking; the target is often small, and natural selection relentlessly prunes off errors in migration, leaving only the best navigators to propagate the species.

BIRD SOCIETY

The adage says "birds of a feather flock together," and birds are indeed highly social creatures. Especially during the breeding seasons, sea birds gather, often in enormous colonies, to nest and rear young. Land birds, with some conspicuous exceptions (such as starlings and rooks), tend to be less gregarious than sea birds during breeding and seek isolation for rearing their brood. But these same species that covet separation from their kind during breeding may aggregate for migration or feeding. Togetherness offers advantages: mutual protection from enemies, greater ease in finding mates, less opportunity for individual straying during migration, and mass huddling for protection against low night temperatures during migration. Certain species may use highly organized cooperative behavior to feed, such as pelicans, as shown in Fig. 28-15. At no time are the highly organized social interactions of birds more evident than during the breeding season, as they stake out territorial claims, select mates, build nests, incubate and hatch their eggs, and rear their young.

Selection of territories and mates

The territory on which to raise a brood is selected in the spring by the male, who jealously guards it against all other males of the same species. The male sings a great deal to help him establish priority on his domain. Eventually he attracts a female, and the pair starts mating and nest building. The female apparently wanders from one territory to another until she settles down with a male.

How large a territory a pair takes over depends on location, abundance of food, natural barriers, etc. In the case of robins a house may serve as the dividing line between two adjacent domains, and each pair usually stays close to the lawn on its particular side of the house. When members of another species trespass, they are usually ignored; competition is greatest among the members of the same species. Song sparrows and mockingbirds, however, try to keep off members of other species as well as their own. Birds may also defend their territories against other species because of environmental limitations and changes, in which case competition for food or other factors between different species (usually closely related) may occur.

Nesting and care of young

To produce offspring, all birds lay eggs that must be incubated by one or both parents. The eggs of most songbirds require approximately 14 days for hatching; those of ducks and geese require at least twice that long. Most of the duties of incubation fall upon the female, although in many instances both parents share the task, and occasionally only the male performs this work.

Most birds build some form of nest in which to rear their young. Some birds simply lay their eggs on the bare ground or rocks and make no pretense of nest building (Fig. 28-16). Others build elaborate nests, such as the pendent nests constructed by orioles, the neat lichen-covered nests of hummingbirds (Fig. 28-17) and flycatchers, the chimney-shaped mud nests of cliff swallows, or the floating nest of the red-necked grebe. Most birds take considerable pains to conceal their nests from enemies. Woodpeckers, chickadees, bluebirds, and many others place their nests in tree hollows or other cavities; kingfishers excavate tunnels in the banks of streams for their nests; and birds of prey build high in lofty trees or on inaccessible cliffs. A few birds, such as the American cowbird, the European cuckoo, and some ducks, build no nests at all but simply lay their eggs in the nests of other birds. When the eggs hatch, the young are taken care of by their foster parents. Most of our songbirds lay from 3 to 6 eggs, but the number of eggs laid in a clutch varies from 1 or 2 (some hawks and pigeons) to 18 or 20 (quail).

Newly hatched birds are of two types: **precocial** or **altricial** (Fig. 28-18). The precocial young, such as quail, fowl, ducks, and most water birds, are covered with down when hatched and can run or swim as soon as their plumage is dry. The altricial ones, on the other hand, are naked and helpless at birth and remain in the

Fig. 28-16. Blue-footed booby, *Sula nebouxii* (order Pelecaniformes), of the Galápagos Islands with newly hatched chick and yet-unhatched egg in nest, a simple depression in the ground. Living in colonies of hundreds of individuals, these large birds depend on sea life for food, which they capture by plunge-diving. The young feed by taking food directly from the crop of the parent. (Photo by C. P. Hickman, Jr.)

nest for a week or more. The young of both types require care from the parents for some time after hatching. They must be fed, guarded, and protected against rain and sun. The parents of altricial species must carry food to their young almost constantly, for most young birds eat more than their weight each day. This enormous food consumption explains the rapid growth of the young and their quick exit from the nest.

BIRD POPULATIONS

Bird populations, like those of other animal groups, vary in size from year to year. Snowy owls, for example, are subject to population cycles that closely follow cycles in their food crop, mainly rodents. Voles, mice, and lemmings in the north have a fairly regular 4-year cycle of abundance; at population peaks, predator populations of foxes, weasels, and buzzards,

as well as snowy owls, increase because there is abundant food for rearing their young. After a crash in the rodent population, snowy owls move south, seeking alternate food supplies. They occasionally appear in large numbers in southern Canada and northern United States, where their total absence of fear of man makes them easy targets for thoughtless hunters.

Occasionally the activities of man bring about spectacular changes in bird distribution. Both starlings (Fig. 28-19) and house sparrows have been accidentally or deliberately introduced into numerous countries, where they have become the two most abundant bird species on earth with the exception of domestic fowl.

Man also is responsible for the extinction of many bird species. More than 80 species of birds have, since 1681, followed the last dodo to extinction. Many died naturally, victims of changes in their habitat or competition with better-adapted species. But several have been hunted to extinction, among them the passenger pigeon, which only a century ago darkened the skies over North America in incredible numbers estimated in the billions. Hunters kill millions of game birds annually, as well as many nongame birds that happen to make convenient targets. An even greater number of game birds die indirectly as the result of eating lead pellets (which they mistake for seeds) or from the crippling effects of embedded pellets. One survey in Wisconsin revealed that the average hunter required 36 shots to kill one goose. Of the Canada geese that survived the barrage to fly as far south as the Mississippi Valley, 44% contained embedded lead shot.

Man's most destructive effects on birds are usually unintentional. The draining of marshes—more than 99% of Iowa's once extensive marshland is now farmland—has destroyed waterfowl nesting. Deforestation has likewise had great impact on tree-nesting species. The vertical appendages of civilization, such as television towers, monuments, tall buildings, and electric transmission towers and lines take a fearful toll during bird migration in bad weather. Tree-spraying programs have virtually eradicated songbirds from certain areas. Most birds, through their impressive reproductive potential, can replace in numbers those that become victims of man's activities. Someone has calculated that a single pair of robins, producing two broods of four young a season, would leave 19,500,000 descendants in 10 years, should all survive at least that long. But, although some birds, such as robins, house sparrows, and starlings, thrive

Fig. 28-17. Ruby-throated hummingbird, *Archilochus colubris* (order Apodiformes), in nest built of plant down and spider webs and decorated on the outside with lichens. The female builds the nest, incubates the two pea-sized eggs, and rears the young with no assistance from the male. These frail-looking but pugnacious little birds make arduous seasonal migrations between Canada and Mexico. (Photo by L. L. Rue III.)

Altricial
One-day-old meadow lark

Precocial
One-day-old ruffed grouse

Fig. 28-18. Comparison of 1-day-old altricial and precocial young. The altricial meadowlark at left is born nearly naked, blind, and helpless. The precocial ruffed grouse at right is covered with down, alert, legs strong, and able to feed itself.

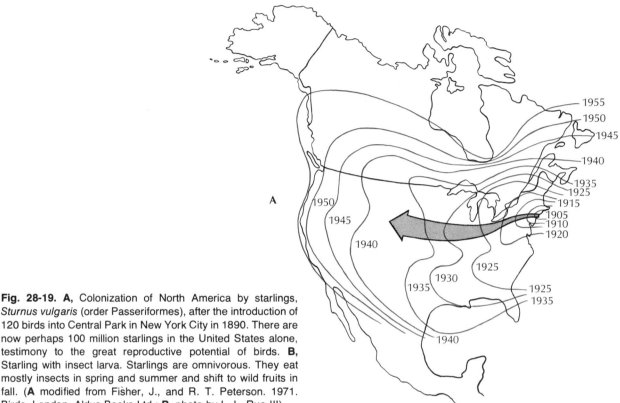

A

1955
1950
1945
1940
1935
1925
1915
1905
1910
1920

1950
1945
1940
1925
1930
1935
1925
1935
1940

Fig. 28-19. A, Colonization of North America by starlings, *Sturnus vulgaris* (order Passeriformes), after the introduction of 120 birds into Central Park in New York City in 1890. There are now perhaps 100 million starlings in the United States alone, testimony to the great reproductive potential of birds. **B,** Starling with insect larva. Starlings are omnivorous. They eat mostly insects in spring and summer and shift to wild fruits in fall. (**A** modified from Fisher, J., and R. T. Peterson. 1971. Birds. London, Aldus Books Ltd.; **B,** photo by L. L. Rue III).

on man's heavy-handed influence on his environment, most birds find the changes adverse, and to some species it is lethal.

SELECTED REFERENCES

Bent, A. C. 1919. Life histories of North American birds. United States National Museum Bulletins. *More than a score of monographs in this outstanding series provide a wealth of information on birds. Dover reprints of entire series are available.*

Fisher, J., and R. T. Peterson. 1971. An introduction to general ornithology. London, Aldus Books. *Beautifully illustrated general biology of birds by two famous ornithologists.*

Lorenz, K. Z. 1952. King Solomon's ring. New York, Thomas Y. Crowell Co., Inc. *A deservedly popular account of animal (especially bird) behavior by a great ethologist.*

Marshall, A. J. (ed.).1960-1961. Biology and comparative physiology of birds, 2 vols. New York, Academic Press, Inc. *A thorough, extensive, and technical treatment of bird physiology by different contributors.*

Peterson, R. T. 1947. Field guide to the birds (east of the Rockies), ed. 2. Boston, Houghton Mifflin Co. *This popular field guide has a better text than the Robbins guide but is not so conveniently arranged. There is a companion volume for western birds.*

Pettingill, O. S., Jr. 1970. Ornithology in laboratory and field. Minneapolis, Burgess Publishing Co. *Comprehensive ornithology text with an excellent section on field methods.*

Robbins, C. S., B. Brunn, H. Zim, and A. Singer. 1966. Birds of North America: a guide to field identification. New York, Golden Press. *Excellent full-color illustrations of all species with range maps on opposite page.*

Savile, D. B. O. 1957. Adaptive evolution in the avian wing. Evolution **11:**212-224.

Tinbergen, N. 1960. The herring gull's world New York, Basic Books, Inc., Publishers. *A superb account of bird behavior by a renowned ethologist. Excellent illustrations by the author.*

Welty, J. C. 1975. The life of birds, ed. 2. Philadelphia, W. B. Saunders Co. *Among the best of the ornithology texts; lucid style and excellent illustrations.*

B

Fig. 28-19, cont'd. For legend see opposite page.

SELECTED SCIENTIFIC AMERICAN ARTICLES

Cone, C. D. Jr. 1962. The soaring flight of birds. **206:**130-140 (April). *The nature of air currents and patterns of soaring flight are analysed.*

Eklund, C. R. 1964. The Antarctic skua. **210:**94-100 (Feb.). *Discusses the biology of this large, aggressive, and cold-adapted bird.*

Emlen, J. T., and R. L. Penny. 1966. The navigation of penguins. **215:**104-113 (Oct.). *Penguins depend on an innate biologic clock and the sun's direction to guide them across hundreds of miles of featureless Antarctic landscape.*

Emlen, S. T. 1975. The stellar-orientation system of a migratory bird. **233:**102-111 (Aug.). *Describes fascinating research with indigo buntings, revealing their ability to orient by the stars.*

Greenewalt, C. H. 1969. How birds sing. **221:**126-139 (Nov.). *The mechanism of bird song is quite different from that of musical instruments or the human voice.*

Keeton, W. T. 1974. The mystery of pigeon homing. **231:**96-107 (Dec.). *Birds use several compass systems in navigation.*

Nicoli, J. 1974. Mimicry in parasitic birds. **231:**92-98 (Oct.).

Peakall, D. B. 1970. Pesticides and the reproduction of birds. **222:**72-78 (April). *Pesticides threaten the survival of several species of birds of prey. The reasons are explained.*

Pennycuick, C. J. 1973. The soaring flight of vultures. **229:**102-109 (Dec.). *Patterns of soaring flight are studied with the aid of powered glider.*

Schmidt-Nielsen, K. 1959. Salt glands. **200:**109-116 (Jan.). *The anatomy and physiology of this special salt-excretory organ of marine birds are described.*

Schmidt-Nielsen, K. 1971. How birds breathe. **225:**72-79 (Dec.).

Stettner, L. J., and K. A. Matyniak. 1968. The brain of birds. **218:**64-76 (June).

Tickell, W. L. N. 1970. The great albatrosses. **223:**84-93 (Nov.). *Describes the reproductive behavior and movements of the largest of oceanic birds.*

Tucker, V. A. 1969. The energetics of bird flight. **220:**70-78 (May).

Welty, C. 1955. Birds as flying machines **192:**88-96 (March). *A description of the remarkable adaptations that fit birds for flight.*

A flying squirrel, *Glaucomys volans,* opens a nut, a favored item in its preferred plant diet of seeds, bark of twigs, lichens, fungi, fruits, and berries, as well as acorns and other nuts. The flying squirrel is nocturnal, unlike other squirrels. Incapable of true flight, it uses its furry gliding membranes to sail from tree to tree during its nightly travels.

29 MAMMALS

Mammals, with their highly developed nervous system and numerous ingenious adaptations, occupy almost every environment on earth that supports life. Although not a large group (4,500 species as compared to 8,600 species of birds, approximately 20,000 species of fishes, and 800,000 species of insects), the class Mammalia (mam-may′lee-a) (L. *mamma,* breast) is overall the most biologically successful group in the animal kingdom, with the possible exception of the insects. Many potentialities that dwell more or less latently in other vertebrates are highly developed in mammals. Mammals are exceedingly diverse in size, shape, form, and function. They range in size from the diminutive pigmy shrew, which has a body length of less than 4 cm and a weight of only a few grams, to the whales, which exceed 100 tons in weight.

Yet, despite their adaptability and in some instances because of it, mammals have been influenced by man's heavy-handed presence more than any other group of animals. He has domesticated numerous mammals for food and clothing, as beasts of burden, and as pets. He uses millions of mammals each year in biomedical research. He has introduced alien mammals into new habitats, occasionally with benign results but more frequently with unexpected disaster. Although history provides us with numerous warnings, we continue to overcrop valuable wild stocks of mammals. The whale industry threatens itself with total collapse by exterminating its own resource—a classic example of self-destruction in the modern world, in which competing segments of an industry are intent only on reaping all they can before bankruptcy enfolds them. In some cases destruction of a valuable mammalian resource has been deliberate, such as the officially sanctioned (and tragically successful) policy during the Indian wars of exterminating the bison to drive the plains Indians into starvation. Although commercial hunting by man has declined, the ever-increasing human population with the accompanying destruction of wild habitats has harassed and disfigured the mammalian fauna. We are becoming increasingly aware that our presence on this planet as the most powerful product of organic evolution makes us totally responsible for the character of our natural environment. Since man's welfare has been and continues to be closely related to that of the other mammals, it is clearly in our interest to preserve the natural environment of which all mammals, ourselves included, are a part. We need to remember that nature can do without man, but man cannot exist without nature.

ORIGIN AND RELATIONSHIPS

In the early Mesozoic, long before the great dinosaurs had reached the peak of their evolutionary success, a group of reptiles with mammallike characteristics appeared. These were the **therapsids** (Fig. 29-1). Their evolution and that of their descendants were accompanied by several structural changes that brought them ever closer to full mammalian status. The clumsy limbs of the reptile that stuck out laterally were replaced by straight legs held close to the body, which provided speed and efficiency for hunting. Since reptilian stability was sacrificed by raising the animal from the ground, the muscular coordination center of the brain, the cerebellum, took on a greatly expanded role. Among the many changes in the bony structure of the head was the separation of air and food passages in the mouth. This enabled the animal to breathe while holding prey in its mouth. It also made possible prolonged chewing and some predigestion of food. At some point the premammals acquired those two most characteristic of all mammalian identification tags: hair and mammary glands.

Most of the living mammals belong to the subclass Theria and have descended from a common ancestor of the Jurassic, some 150 million years ago. However, monotremes (subclass Prototheria), the egg-laying mammals of Australia, Tasmania, and New Guinea, are so different from the others and possess so many reptilian characters that they are believed to have descended from an entirely different mammallike reptile. The separation of Prototheria and Theria probably occurred approximately 50 million years earlier in the Triassic period. The geologic record during the following Jurassic and Cretaceous periods is fragmentary, in large part because the mammals of these periods were creatures the size of a rat or smaller, with fragile bones that fossilized only under the most ideal circumstances.

When the dinosaurs vanished near the beginning of the Cenozoic era, the mammals suddenly expanded. This is partly attributable to the numerous ecologic niches vacated by the reptiles, into which the mammals could move as their divergent adaptations fitted them. There were other reasons for their success. Mammals were agile, warm blooded, and insulated with hair; they had developed placental reproduction and suckled

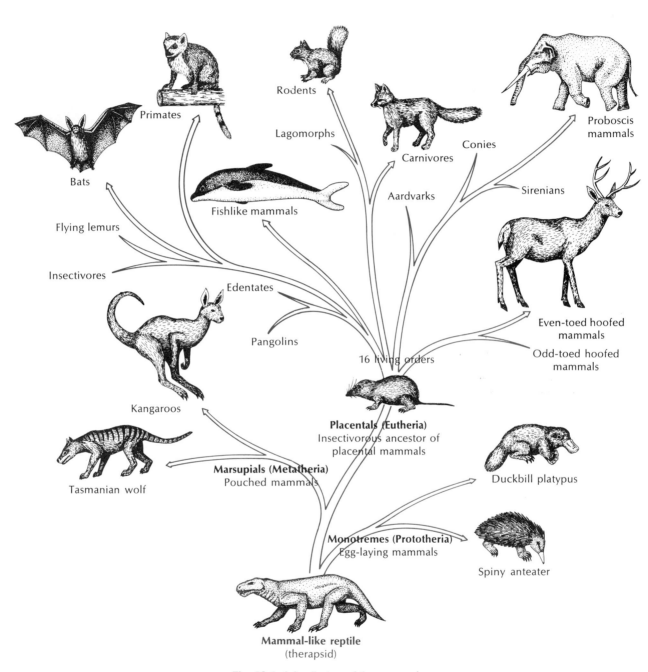

Bats

Primates

Rodents

Proboscis mammals

Lagomorphs

Carnivores

Conies

Aardvarks

Sirenians

Fishlike mammals

Flying lemurs

Insectivores

Edentates

Even-toed hoofed mammals

Odd-toed hoofed mammals

Pangolins

16 living orders

Kangaroos

Placentals (Eutheria)
Insectivorous ancestor of placental mammals

Duckbill platypus

Tasmanian wolf

Marsupials (Metatheria)
Pouched mammals

Monotremes (Prototheria)
Egg-laying mammals

Spiny anteater

Mammal-like reptile
(therapsid)

Fig. 29-1. A family tree of the mammals.

their young, thus dispensing with vulnerable eggs and nests; and they were more intelligent than any other animal alive. During the Eocene and Oligocene epochs of the Tertiary (55 to 30 million years ago), the mammals flourished and reached their peak. In terms of number of species, this was the golden age of mammals. Although they have declined somewhat in numbers since then, they are a secure group, dominating the land environment as thoroughly now as they did 50 million years ago.

Characteristics

Since mammals and birds both evolved from reptiles, we can expect to find and do find many structural similarities among the three groups. It is, in fact, much easier to point to numerous resemblances between the mammals and the reptiles than to point to characteristics that are unique and diagnostic for mammals. **Hair** is the most obvious mammalian characteristic, although it is vastly reduced in some (such as whales) and although reptilian scales, from which hair is derived, may persist (such as on tails of rats and beaver). A second unique characteristic of mammals is the method of nourishing their young with **milk-secreting glands;** reptiles have nothing remotely similar. Although less obvious, the mammalian skull contains several important differences in cranial and jaw structure and jaw articulation. Mammals have **diphyodont teeth** (milk teeth replaced by a permanent set of teeth) rather than reptilian **polyphyodont teeth** (successive sets of teeth). But the single most important factor contributing to the success of mammals is the remarkable development of the **neocerebrum,** permitting a level of adaptive behavior, learning, curiosity, and intellectual activity far beyond the capacity of any reptile.

We may summarize the mammalian characteristics as follows:

1. **Body covered with hair,** but reduced in some
2. **Integument with sweat, sebaceous,** and **mammary glands**
3. Mouth with teeth on both jaws
4. **Movable eyelids** and **fleshy external ears**
5. Four limbs (reduced or absent in some) adapted for many forms of locomotion
6. Circulatory system of a four-chambered heart, **persistent left aorta,** and **nonnucleated red blood corpuscles**
7. Respiratory system of lungs and a voice box
8. **Muscular partition (diaphragm) between thorax** and **abdomen**
9. Excretory system of metanephros kidneys and ureters that usually open into a bladder
10. Brain highly developed, especially **neocerebrum;** 12 pairs of cranial nerves
11. **Warm blooded (homeothermic)**
12. Separate sexes
13. Internal fertilization; **eggs developed in a uterus** with **placental attachment** (except in monotremes); **fetal membranes (amnion, chorion, allantois)**
14. Young nourished by **milk from mammary glands**

Classification

The classification given here recognizes 18 living orders of mammals, although some mammalogists remove the aquatic carnivores (seals and whales) from the order Carnivora and place them in a distinct order Pinnipedia. Fourteen extinct orders are not included in this classification.

Class Mammalia is divided into two subclasses as follows: subclass **Prototheria** includes the monotremes, or egg-laying mammals. Subclass **Theria** includes two infraclasses, the **Metatheria** with one order, the marsupials, and the **Eutheria** with the rest of the orders, all of which are placental mammals (Fig. 29-1).

Subclass Prototheria (pro′to-thir′e-a) (Gr. *prōtos*, first, + *thēr*, wild animal)—**egg-laying mammals.**

Order Monotremata (mon′o-tre′mah-tah) (Gr. *monos*, single, + *trēma*, hole—**egg-laying mammals: duck-billed platypus** and **spiny anteater.** Only oviparous mammals; name of group referring to single opening from cloaca serving reproductive, digestive, and excretory systems; restricted to Australian region.

Subclass Theria (thir′e-a) (Gr. *thēr*, wild animal).

Infraclass Metatheria (met′a-thir′e-a) (Gr. *meta*, after, + *thēr*, wild animal)—**marsupial mammals.**

Order Marsupialia (mar-su′pe-ay′le-a) (Gr. *marsypion*, little pouch)—**pouched mammals,** for example, **opossums, kangaroos, koala.** Primitive mammals characterized in most by an abdominal pouch, the **marsupium,** in which young are reared; usually no typical placenta; mostly Australian with representatives in the Americas.

Infraclass Eutheria (yu-thir′e-a) (Gr. *eu*, good, + *thēr*, wild animal—**placental mammals.**

Order Insectivora (in-sec-tiv′o-ra) (L. *insectum*, an insect, + *vorare*, to devour)—**insect-eating mammals,** for example, **shrews, hedgehogs, moles.** Most primitive of placental mammals; includes the smallest mammals (shrews); worldwide except Australia.

Order Chiroptera (ky-rop′ter-a) (Gr. *cheir*, hand, + *pteron*, wing—**bats.** All flying mammals with forelimbs modified into wings; use of echolocation by most bats; mostly nocturnal; worldwide.

Order Dermoptera (der-mop′ter-a) (Gr. *derma*, skin, + *pteron*, wing)—**flying lemurs.** Relatives of the true bats; they do not fly but glide like flying squirrels; nocturnal; restricted to Malay Peninsula and East Indies.

Order Carnivora (car-niv′o-ra) (L. *carn,* flesh, + *vorare,* to devour)—**flesh-eating mammals,** for example, **dogs, wolves, cats, bears, weasels.** Some of the most intelligent and strongest of animals; all with predatory habits; teeth especially adapted for tearing flesh; in most, canines used for killing their prey; animals divided among two suborders: Fissipedia, whose feet contain toes, and Pinnipedia, whose limbs are modified for aquatic life; **suborder Fissipedia,** worldwide except in the Australian and antarctic regions and divided into certain familiar families, among which are **Canidae,** the dog family, **Felidae,** the cat family, **Ursidae,** the bear family, and **Mustelidae,** the fur-bearing family; **suborder Pinnipedia,** the sea lions, seals, sea elephants, and walruses.

Order Tubulidentata (tu′byu-li-den-ta′ta) (L. *tubulus,* tube, + *dens,* tooth)—**aardvarks.** Aardvark, the Dutch name for earth pig, a peculiar piglike anteater found in Africa.

Order Rodentia (ro-den′che-a) (L. *rodere,* to gnaw)—**gnawing mammals,** for example, **squirrels, rats, woodchucks.** Most numerous of all mammals both in numbers and species; dentition with two upper and two lower chisellike incisors that grow continually and are adapted for gnawing; of the more than 30 families, some of the best known in North America are **Sciuridae** (squirrels and woodchucks), **Cricetidae** (hamsters, deer mice, gerbils, voles, lemmings), **Muridae** (rats and house mice), **Castoridae** (beavers), **Erethizontidae** (porcupines), and **Geomyidae** (pocket gophers); worldwide.

Order Pholidota (fol′i-do′ta) (Gr. *pholis,* horny scale)—**pangolins.** An odd group of only eight species; bodies covered with overlapping horny scales; found in tropical Asia and Africa.

Order Lagomorpha (lag′o-mor′fa) (Gr. *lagōs,* hare, + *morphē,* form)—**rabbits, hares, pikas.** Dentition resembling that of rodents but with four upper incisors rather than two as in rodents; worldwide.

Order Edentata (ee′den-ta′ta) (L. *edentatus,* toothless)—**toothless mammals,** for example, **sloths, anteaters, armadillos.** Either toothless or with degenerate peglike teeth; most representatives in South America; the nine-banded armadillo in the southern United States.

Order Cetacea (see-tay′she-a) (L. *cetus,* whale)—**fishlike mammals,** for example, **whales, dolphins, porpoises.** Anterior limbs modified into broad flippers; posterior limbs absent; nostrils represented by a single or double blowhole on top of the head; teeth usually absent but when present all alike and lacking enamel; **suborder Odontoceti** made up of toothed members, represented by the sperm whales, porpoises, and dolphins; **suborder Mysticeti,** containing the whalebone whales, and possessing a peculiar straining device of whalebone (baleen), instead of teeth, attached to the palate and used to filter plankton (microscopic animals) out of the water.

Order Proboscidea (pro′ba-sid′e-a) (Gr. *proboskis,* elephant's trunk, from *pro,* before, + *boskein,* to feed)—**proboscis mammals: elephants.** Once contained the largest herbivores of the Cenozoic; only two genera remaining: the Indian elephant, with relatively small ears, and the African elephant, with large ears.

Order Hyracoidea (hy′ra-coi′de-a) (Gr. *hyrax,* shrew)—**hyraxes,** for example, **conies.** Rabbitlike but short-eared herbivores, restricted to Africa and Syria.

Order Sirenia (sy-ree′nee-a) (Gr. *seirēn,* sea nymph)—**manatees** and **dugongs.** Large, clumsy, aquatic animals known as sea cows; no hind limbs; forelimbs modified into swimming flippers; only two genera living at present: *Trichechus,* the manatee found in the rivers of Florida, West Indies, Brazil, and Africa; and *Halicore,* the dugong of India and Australia.

Order Perissodactyla (pe-ris′so-dak′ti-la) (Gr. *perissos,* odd, + *dactylos,* toe)—**odd-toed hoofed mammals.** Mammals with an odd number (one or three) of toes and with well-developed hooves; includes horses, zebras, tapirs, and rhinoceroses; all herbivorous; both Perissodactyla and Artiodactyla often referred to as ungulates, or hoofed mammals, with teeth adapted for chewing; native distribution in Africa, Asia, Central and South America.

Order Artiodactyla (ar′te-o-dak′ti-la) (Gr. *artios,* even, + *daktylos,* toe)—**even-toed hoofed mammals.** Even-toed ungulates, including swine, camels, deer, hippopotamuses, antelopes, cattle, pronghorn, sheep, and goats; each toe sheathed in a cornified hoof; many, such as the cow, deer, and sheep, with horns; many ruminants, that is, herbivores with partitioned stomachs; worldwide.

Order Primates (pry-may′teez) (L. *prima,* first)—**highest mammals,** for example, **lemurs, monkeys, apes, man.** First in the animal kingdom in brain development with especially large cerebral hemispheres; five digits (usually provided with flat nails) on both forelimbs and hind limbs; group singularly lacking in claws, scales, horns, and hoofs; two suborders.

Suborder Prosimii (pro-sem′ee-i) (Gr. *pro,* before, + *simia,* ape)—**lemurs, tree shrews, tarsiers, lorises, pottos.** Primitive, arboreal, mostly nocturnal, primates restricted to the tropics of the Old World.

Suborder Anthropoidea (an′thro-poi′de-a) (Gr. *anthropos,* man). Consists of monkeys, gibbons, apes, and man; three superfamilies.

Superfamily Ceboidea (se-boi′de-a) (Gr. *kēbos,* long-tailed monkey). New World monkeys, characterized by the broad flat nasal septum, nonopposable thumbs, prehensile tails, and absence of ischial (hip) callosities and cheek pouches; includes capuchin monkey *(Cebus)* of the organ grinder; spider monkey *(Ateles);* and howler monkey *(Alouatta).*

Superfamily Cercopithecoidea (sur′ko-pith′e-koi′de-a) (Gr. *kerkos,* tail, + *pithēkos,* monkey)—**(Catarrhinii).** Old World monkeys, with external nares (nostrils) close together, opposable thumbs, and calloused tuberosities on buttocks; tails not prehensile; includes the savage mandrill *(Cynocephalus),* the rhesus monkey *(Macacus)* widely used in biologic investigation, and the proboscis monkey *(Nasalis).*

Superfamily Hominoidea (hom′i-noi′de-a) (L. *homo, hominis,* man). Members (anthropoid apes and man) lacking tail and cheek pouches but with large braincases; Pongidae family includes the higher apes, gibbon *(Hylobates),* orangutan *(Simia),* chimpanzee *(Pan),* and gorilla *(Gorilla);* Hominidae family represented by a single living species *(Homo sapiens),* modern man.

STRUCTURAL AND FUNCTIONAL ADAPTATIONS OF MAMMALS
Integument and its derivatives

The mammalian skin and its modifications especially distinguish mammals as a group. As the interface between the animal and its environment, the skin is strongly molded by the animal's way of life. In general the skin is thicker in mammals than in other classes of vertebrates, although as in all vertebrates it is made up of **epidermis** and **dermis** (Fig. 7-1, p. 130) Among the mammals the dermis becomes much thicker than the epidermis. The epidermis is relatively thin where it is well protected by hair, but in places subject to much contact and use, such as the palms or soles, its outer layers become thick and cornified with keratin.

Hair

Mammals characteristically have two kinds of hair forming the **pelage** (fur coat): (1) dense and soft **underhair** for insulation and (2) coarse and longer **guard hair** for protection against wear and to provide coloration. The underhair traps a layer of insulating air; in aquatic animals, such as the fur seal, otter, and beaver, it is so dense that it is almost impossible to wet it. In water the guard hairs wet and mat down over the underhair, forming a protective blanket (Fig. 29-2). A quick shake when the animal emerges flings off the water and leaves the outer guard hair almost dry.

When a hair reaches a certain length, it stops growing. Normally it remains in the follicle until a new growth starts, whereupon it falls out. In man, hair is shed and replaced throughout life. But in most mammals, there are periodic molts of the entire coat.

In the simplest cases, such as in foxes and seals, the coat is shed once each year during the summer months. Most mammals have two annual molts, one in the spring and one in the fall. The summer coat is always much thinner than the winter coat and is usually a different color. Several of the northern mustelid carnivores, such as the weasel, have white winter coats and colored summer coats. It was once believed that the white winter pelage of arctic animals served to conserve body heat by reducing radiation loss, but recent research has shown that dark and white pelages radiate heat equally well. The winter white of arctic animals is simply camouflage in a land of snow. The varying hare of North America has three annual molts: the white winter coat is replaced by a brownish gray

Fig. 29-2. American beaver, *Castor canadensis* (order Rodentia, family Castoridae), cutting up a trembling aspen tree. This second largest rodent has a heavy waterproof pelage consisting of long, tough guard hairs overlying the thick, silky underhair so valued in the fur trade. Other adaptations for an aquatic life are nostrils and ear canals provided with valves, webbed hind feet, flexible forefeet for gripping branches, a flat and broad paddle-like tail for swimming, diving, and signaling, and a mouth that closes behind the incisors to permit gnawing underwater. (Photo by L. L. Rue III.)

summer coat, and this is replaced in autumn by a grayer coat, which is soon shed to reveal the winter white coat beneath (Fig. 29-3).

Outside of the arctic, most mammals wear somber colors that are protective. Often the species is marked with "salt-and-pepper" coloration or a disruptive pattern that helps to make it inconspicuous in its natural surroundings. Examples are the leopard's spots, the stripes of the tiger, and the spots of fawns. Other mammals, for example, skunks, advertise their presence with conspicuous warning coloration.

An interesting aspect of color is seen in the pair of rump patches of the pronghorn antelope, which are composed of long white hairs erected by special muscles. When alarmed, the animal can flash these patches in a manner visible for a long distance. They may be used as a warning signal to other members of the herd. The well-known "flag" of the white-tailed deer serves a similar purpose.

Fig. 29-3. Snowshoe, or varying, hare, *Lepus americanus* (order Lagomorpha), in brown summer coat, **A,** and white winter coat, **B.** In winter, extra hair growth on the hind feet broadens the animal's support in snow. Snowshoe hares are common residents of the taiga (northern coniferous forests) and are an important food for lynxes, foxes, and other carnivores. Population fluctuations of hares and their predators are closely related. (Photos by L. L. Rue III.)

The hair of mammals has become modified to serve many purposes. The bristles of hogs, vibrissae on the snouts of most mammals, and spines of porcupines and their kin are examples.

Vibrissae, commonly called "whiskers," are really sensory hairs that provide an additional special sense to many mammals. The bulb at the base of each follicle is provided with a large sensory nerve. The slightest movement of a vibrissa generates impulses in the nerve endings that travel to a special sensory area in the brain. The vibrissae are especially long in nocturnal and burrowing animals. In seals they apparently serve as a "distance touch" sensitive to pressure waves and turbulence in the water caused by objects or passing fishes. Vision is of little use to seals hunting in turbid water, where they are frequently found; investigators have noted that blind seals remain just as fat and healthy as normal seals.

Porcupines, hedgehogs, echidnas, and a few other mammals have developed an effective and dangerous spiny armor; the spines of the common North American porcupine break off at the bases when struck and, aided by backward-pointing hooks on the tips, work deeply into their victims. To assist slow learners, like dogs, in understanding what they are dealing with, porcupines rattle the spines and prominently display the white markings on the quills toward their tormentors.

Horns and antlers

Three kinds of horns or hornlike substances are found in mammals (Fig. 29-4). **True horns** found in ruminants, for example, sheep and cattle, are hollow sheaths of keratinized epidermis that embrace a core of bone arising from the skull. Horns are not normally shed, are not branched (although they may be greatly curved), and are found in both sexes. The horns of North American pronghorn antelope are unique in that they are shed each year after the breeding season. But, unlike the shedding of deer antlers, the new horn replaces the old by growing up inside and pushing off the outer sheath.

Antlers of the deer family are entirely bone when mature. During their annual growth, antlers develop beneath a covering of highly vascular soft skin called **"velvet."** When growth of the antlers is complete just

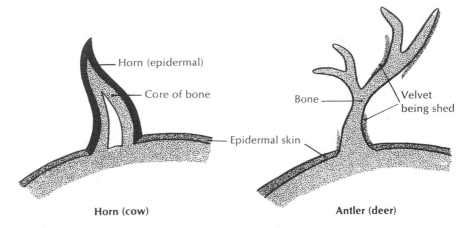

Horn (epidermal)

Core of bone

Bone

Velvet
being shed

Epidermal skin

Horn (cow)

Antler (deer)

Fig. 29-4. Chief differences between horns and antlers. Bone, a dermal (mesoderm) derivative, forms basic part of each type, and when epidermal velvet with its hair is shed, bone forms all of antlers. Horns do not branch and are not shed. Antlers are shed annually (in winter) when zone of constriction below burr appears near skull.

A

B

Fig. 29-5. White-tailed deer, *Odocoileus virginianus* (order Artiodactyla, family Cervidae), with antlers in velvet, **A,** and shedding velvet, **B.** Antlers begin growth in the spring as projections of frontal bone, enclosed within a soft, nourishing skin, the "velvet." When growth is complete in late summer, vessels at the bases of the antlers constrict and the velvet dies and peels off. After the breeding season in early winter the antlers are shed. Antlers become larger and more elaborate with each successive growth and shedding. (Photos by L. L. Rue III.)

prior to the breeding season, the blood vessels constrict and the stag tears off the velvet by rubbing the antlers against trees (Fig. 29-5). The antlers are dropped after the breeding season. New buds appear a few months later to herald the next set of antlers. For several years each new pair of antlers is larger and more elaborate than the previous set. The annual growth of antlers places a strain on the mineral metabolism since during the growing season a large moose or elk must accumulate 50 or more pounds of calcium salts from its vegetable diet.

The **rhinoceros horn** is the third kind of horn. Hairlike horny fibers arise from dermal papillae and are cemented together to form a single horn.

Glands

Of all vertebrates, mammals have the greatest variety of integumentary glands. Whatever the type of gland, they all appear to fall into one of three classes: eccrine, apocrine, and holocrine.

Eccrine glands (sweat glands) are found only in hairless regions (food pads, etc.) in most mammals, although in some apes, man, and a few others such as the horse they are scattered all over the body. Sweat glands are used mainly to regulate the body temperature. Although common in the horse and man, they are greatly reduced on the carnivores (cats) and are entirely lacking in shrews, whales, and others. Dogs are now known to have sweat glands all over the body. In human beings, racial differences are pronounced. Blacks, who have more than whites, can withstand warmer weather.

Apocrine glands are larger than eccrine glands and have longer and more winding ducts. Their secretory coil is in the subdermis. They always open into the follicle of a hair or where a hair has been. Blacks have more apocrine glands than do whites, and women have twice as many as men. They develop approximately at sexual puberty and are restricted (in the human species) to the axillae (arm pits), mons pubis, breasts, external auditory canals, prepuce, scrotum, and a few other places. Their secretion is not watery, like ordinary sweat (eccrine gland), but is a milky, whitish or yellow secretion that dries on the skin to form a plasticlike film. Apocrine glands are not involved in heat regulation, but their activity is known to be correlated with certain aspects of the sex cycle, among other possible functions.

The most common apocrine glands include the **scent glands,** found in all terrestrial mammals. Their location and functions vary greatly. Some are defensive in nature; others convey information to members of the same species; still others are involved in the mating process. These glands are often located in the preorbital, metatarsal, and interdigital regions (deer); preputial region on the penis (muskrats, beavers, canine family, etc.); base of tail (wolves and foxes); and anal region (skunks, minks, weasels). These last, the most odoriferous of all glands, open by ducts into the anus; their secretions can be discharged forcefully for several feet. During the mating season many mammals give off strong scents for attracting the opposite sex. Man also is endowed with scent glands. But civilization has taught us to dislike our own scent, a concern that has stimulated a lucrative deodorant industry to produce an endless output of soaps and odor-masking compounds.

Mammary glands, which provide the name for mammals, are probably modified apocrine glands, although recent studies suggest that they may have been derived from sebaceous glands. Whatever their evolutionary origin, they occur on all female mammals and in a rudimentary form on all male mammals. They develop by the thickening of the epidermis to form a milk line along each side of the abdomen in the embryo. On certain parts of these lines the mammae appear, while the intervening parts of the ridge disappear.

Milk varies in composition; in marine mammals (whales and seals) and arctic mammals (polar bears and caribou), where rapid growth of young is important for species survival, the milk may contain 30% to 40% fat, as compared to 4% fat in domestic cow milk. In the human female, the mammary glands begin at puberty to increase in size because of fat accumulation and reach their maximum development in approximately the twentieth year. The breasts (or mammae) undergo additional development during pregnancy. In other mammals, the breasts are swollen only periodically when they are distended with milk during pregnancy and subsequent nursing of the young.

The third type of gland **(holocrine)** is one in which the entire cell is discharged in the secretory process and must be renewed for further secretion. Most of them open into hair follicles, but some are free and open directly onto the surface. The **sebaceous gland** is the most common example. In most mammals sebaceous glands are found all over the body; in man they are most numerous in the scalp, forehead, and face. Sebaceous glands that open into the hair follicles keep the skin and hair soft and glossy.

Food and feeding

Mammals have exploited an enormous variety of food sources; some mammals require highly specialized diets, whereas others are opportunistic feeders that thrive on diversified diets. In all, food habits and physical structure are inextricably linked together. A mammal's adaptations for attack and defense and its

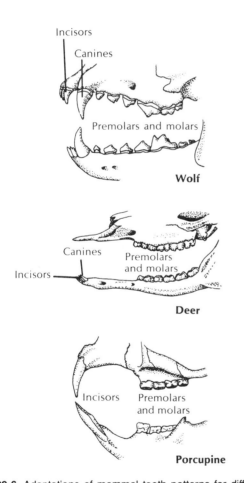

Fig. 29-6. Adaptations of mammal tooth patterns for different kinds of diet. Sharp canines of wolf are designed for stabbing, and premolars and molars are for cutting rather than grinding. Browsing deer has predominantly grinding teeth; lower incisors and canines bite against a horny pad in the upper jaw. Porcupine has no canines; self-sharpening incisors are used for gnawing. (Modified from Carrington, R. 1968. The mammals, New York, Life Nature Library.)

specialization for finding, capturing, reducing, swallowing, and digesting food all determine a mammal's shape and habits.

Teeth, perhaps more than any other single physical characteristic, reveal the life-style of a mammal (Fig. 29-6). It has been claimed that, if all mammals except man were extinct and represented only by fossil teeth, we could still construct a classification as correct as the one we have now, which is based on all anatomic features. All mammals have teeth, except certain whales, monotremes, and anteaters, and their modifications are correlated with what the mammal eats.

Typically, mammals have a **diphyodont** dentition, that is, two sets of teeth: a set of deciduous, or milk, teeth that are replaced by a set of permanent teeth. In any given species, mammalian teeth are modified to perform specialized tasks such as cutting, nipping, gnawing, seizing, tearing, grinding, and chewing. Teeth differentiated in this manner in the individual are called **heterodont,** in contrast to the uniform, **homodont** dentition characteristic of lower vertebrates.

Usually four types of teeth are recognized. **Incisors,** with simple crowns and slightly sharp edges, are mainly for snipping or biting; **canines,** with long conic crowns, are specialized for piercing; **premolars,** with compressed crowns and one or two cusps, are suited for shearing and slicing; and **molars,** with large bodies and variable cusp arrangement, are for crushing and mastication. Molars always belong to the permanent set.

On the basis of food habits, animals may be divided into herbivores, carnivores, omnivores, and insectivores.

Herbivorous animals that feed upon grasses and other vegetation form two main groups: **browsers** or **grazers,** such as the ungulates (horses, swine, deer, antelope, cattle, sheep, and goats), and the **gnawers** and **nibblers,** such as the rodents and rabbits. In herbivores the canines are suppressed, whereas the molars are broad, and high crowned and bear enamel ridges for grinding. Rodents have chisel-shaped incisors that grow throughout life and must be worn away to keep pace with their continual growth.

Herbivorous mammals have a number of interesting adaptations for dealing with their massive diet of plant food. **Cellulose,** the structural carbohydrate of plants, is a potentially nutritious foodstuff, comprised of long chains of glucose. However, the glucose molecules in cellulose are linked by a type of chemical bond that

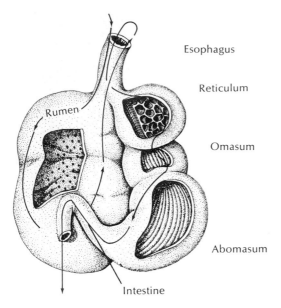

Esophagus

Reticulum

Rumen

Omasum

Abomasum

Intestine

Fig. 29-7. Ruminant's stomach. Food passes first to rumen (sometimes via reticulum) and then is returned to mouth for chewing (chewing the cud, or rumination) *(black arrow).* After reswallowing, food returns to the rumen or passes directly to reticulum, omasum, and abomasum for final digestion *(red arrow).* See text for further explanation.

few enzymes can attack. No vertebrates synthesize cellulose-splitting enzymes. Instead, the herbivorous vertebrates harbor a microflora of anaerobic bacteria in huge fermentation chambers in the gut. These bacteria break down the cellulose, releasing a variety of fatty acids, sugars, and starches that the host animal can absorb and utilize.

In some herbivores, such as horse and rabbit, the gut has a capacious sidepocket, or diverticulum, called a **cecum,** which serves as a fermentation chamber and absorptive area. Hares and rabbits often eat their fecal pellets, giving the food a second pass through the fermenting action of the intestinal bacteria.

The **ruminants** (cattle, sheep, antelope, deer, giraffe, and other hooved mammals) have a huge **four-chambered stomach** (Fig. 29-7). When a ruminant feeds, grass passes down the esophagus to the **rumen,** where it is broken down by the rich microflora and then formed into small balls of cud. At its leisure the ruminant returns the cud to its mouth where it is deliberately chewed at length to crush the fiber. Swallowed again, the food returns to the rumen where it is digested by the cellulolytic bacteria. The pulp

passes to the **reticulum,** then to the **omasum,** and finally to the **abomasum** ("true" stomach), where proteolytic enzymes are secreted and normal digestion takes place.

Herbivores in general have large and long digestive tracts and must eat a large amount of plant food to survive. A large African elephant weighing 6 tons must consume between 135 to 150 kg (300 to 400 pounds) of rough fodder each day to obtain sufficient nourishment for life.

Carnivorous mammals feed mainly on herbivores. This group includes foxes, weasels, cats, dogs, wolverines, fishers, lions, and tigers. Carnivores are well equipped with biting and piercing teeth and powerful clawed limbs for killing their prey. Since their protein diet is much more easily digested than is the woody food of herbivores, their digestive tract is shorter and the cecum small or absent. Carnivores eat separate meals and have much more leisure time for play and exploration.

In general, carnivores lead more active—and by man's standards more interesting—lives than do the herbivores. Since a carnivore must find and catch its prey, there is a premium on intelligence; many carnivores, the cats, for example, are noted for their stealth and cunning in hunting prey. Although evolution seems to have favored the carnivores, their success has led to a selection of herbivores capable of either defending themselves or of detecting and escaping carnivores. Thus for the herbivores, there has been a premium on keen senses and agility. Some herbivores, however, survive by virtue of their sheer size, for example, elephants, or by defensive group behavior, for example, muskoxen (Fig. 29-8).

Man has changed the rules in the carnivore-herbivore contest. Carnivores, despite their intelligence, have suffered much from man's presence and have been virtually exterminated in some areas. Herbivores, on the other hand, especially the rodents with their potent reproductive ability, have consistently defeated man's most ingenious efforts to banish them from his environment. The problem of rodent pests in agriculture has been intensified; man has removed carnivores, which served as the herbivores' natural population control, but has not been able to devise a suitable substitute.

Omnivorous mammals live on both plant food and animals. Examples are pigs, raccoons, rats, bears, man, and most other primates. Many carnivorous

Fig. 29-8. Muskoxen, *Ovibos moschatus* (order Artiodactyla, family Bovidae), of arctic Canada forming a defensive phalanx against predators. Fully exposed to the coldest blizzards the north can offer, these large herbivores depend on their excellent insulation, large size (some weigh nearly 400 kg [approximately 900 pounds]), and defensive huddling for protection from cold as well as against wolves. Muskoxen were nearly exterminated during the last century by excessive hunting but are now recovering in numbers under complete legal protection. The long, flowing guard hair covers fine, dense underhair that can be spun into high-quality wool. (Photo by L. L. Rue III.)

forms also eat fruits, berries, and grasses when hard pressed (Fig. 29-9). The fox, which usually feeds upon mice, small rodents, and birds, eats frozen apples, beechnuts, and corn when its normal sources are scarce.

Insectivorous mammals are those that subsist chiefly on insects and grubs. Examples are moles, shrews, anteaters, and some bats. The insectivorous category is not a well-distinguished one, however, because many omnivores, carnivores, and even some herbivores eat insects on occasion.

For most mammals, searching for food and eating occupy most of their active life. Seasonal changes in food supplies are considerable in temperate zones. Living may be easy in the summer when food is abundant, but in winter many carnivores must range far and wide to eke out a narrow existence (Fig. 29-10). Some migrate to regions where food is more abundant. Others hibernate and sleep the winter months away.

But there are many provident mammals that build up food stores during periods of plenty. This habit is most pronounced in many of our rodents, such as squirrels, chipmunks, gophers, and certain mice. All

Fig. 29-9. Alaskan brown bear, *Ursus arctos* (order Carnivora), with a large salmon captured during the spawning run. All bears are opportunistic omnivores that take advantage of any seasonally abundant resource. They are highly dependent on vegetable food but also kill small mammals, scavange, and rob all bees' nests they can find and enter. Bears are the least carnivorous of the carnivores. (Photo by L. L. Rue III.)

Fig. 29-10. Red fox, *Vulpes fulva* (order Carnivora), eating a cottontail rabbit. Foxes are opportunistic feeders that eat small rodents, rabbits, insects, and wild fruits and berries in proportion to their abundance. During the berry season, blueberries, raspberries, and wild cherries may comprise 100% of the diet. In winter, mice, voles, and rabbits form a high proportion of the diet, as well as winter-starved deer or deer shot and lost by hunters. (Photo by L. L. Rue III.)

tree squirrels—red, fox, and gray—collect nuts, conifer seeds, and fungi and bury these in caches for winter use. Often each item is hidden in a different place (scatter hoarding) and scent marked to assist relocation in the future. The chipmunk is one of the greatest providers, for it spends the autumn months collecting nuts and seeds. Some of its caches may exceed a bushel.

Body temperature regulation

Mammals share with birds the ability to maintain their body temperature well above that of their surroundings. They are **homeothermic,** or "warm-blooded," animals, as distinguished from the invertebrates and lower vertebrates, which are "cold-blooded" **(poikilothermic)** animals with variable body temperatures. The terms "warm-blooded" and "cold-blooded" are hopelessly subjective and nonspecific but are so firmly entrenched in our vocabulary that most biologists find it easier to accept the usage than to try to change people.

Homeothermy allows mammals to stabilize their internal temperature so that biochemical processes and nervous function can proceed at steady high levels of activity. Thus they can remain active in winter and

exploit habitats unavailable to the poikilotherms. Most mammals have body temperatures between 36° and 38° C (somewhat lower than those of birds, which range between 40° and 42° C). This constant temperature is maintained by a delicate balance between heat production and heat loss—not a simple matter when mammals are constantly alternating between periods of rest and bursts of activity.

Heat is produced by the animal's metabolism, which includes the oxidation of foodstuffs, basal cellular metabolism, and muscular contraction. Heat is lost by radiation and conduction to a cooler environment and by the evaporation of water. The mammal can control both processes of heat production and heat loss within rather wide limits. If it becomes too cool, it can increase heat production by increasing muscular activity (exercise or shivering) and by decreasing heat loss by increasing its insulation. If the mammal becomes too warm, it decreases heat production and increases heat loss. We will examine these processes in the following examples.

Adaptations for hot environments

Despite the harsh conditions of deserts—intense heat during the day, cold at night, and scarcity of water,

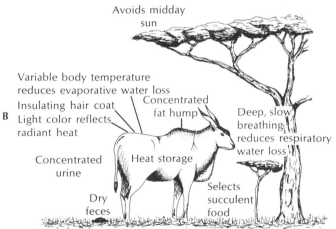

Fig. 29-11. **A,** The common eland, *Taurotragus oryx* (order Artiodactyla), inhabitant of the arid, open savanna of central Africa. It is one of 72 species of African antelopes that occupy a variety of habitats that include open savanna, bush savanna, marshes, and flooded grassland. Special food preferences reduce competition between different species. **B,** Drawing shows physiologic and behavioral adaptations of the eland for maintaining a constant body temperature in a hot environment. See text for explanation. (**A,** Photo by C. G. Hampson, University of Alberta; **B** from Tallmark, B. 1972. Fauna och Flora [Stockholm] **67:**163-175; after Taylor, C. P. 1969. Sci. Amer. **220:**88-95.)

vegetation, and cover—many kinds of animals live there successfully. The smaller desert mammals are mostly fossorial (fitted for digging burrows) and nocturnal. The lower temperature and higher humidity of burrows help to reduce water loss by evaporation. Water loss is replaced by free water in their food or by drinking water if it is available. Water is also formed in the cells by the metabolic oxidation of foods. This gain from **oxidation water,** as it is called, can be very significant since water is not always available for drinking. In fact, some desert mammals, such as the kangaroo rats and ground squirrels of American deserts, can, if necessary, derive all the water they need from their dry food, thus drinking no water at all. Such animals produce a highly concentrated urine and form nearly solid feces.

The large desert ungulates obviously cannot escape the desert heat by living in burrows. Animals such as camels and desert antelopes (gazelle, oryx, and eland) possess a number of adaptations for coping with heat and dehydration. Those of the eland are shown in Fig. 29-11. The mechanisms for controlling water loss and preventing overheating are closely linked together. The glossy, pallid color of the fur reflects direct sunlight, and the fur itself is an excellent insulation that works to keep heat out. Heat is lost by convection and conduction from the underside of the eland where the pelage is very thin. Fat tissue of the eland, an essential food reserve, is concentrated in a single hump on the back, instead of being uniformly distributed under the skin where it would impair heat loss by radiation. The eland avoids evaporative water loss—the only device an animal has for cooling itself when the environmental temperature is higher than that of the body—by permitting its body temperature to decrease during the cool night and then increase slowly during the day as the body stores heat. Only when the body temperature reaches 41° C must the eland prevent further rise through **evaporative cooling** by sweating and panting. Water is also conserved by means of concentrated urine and dry feces. All of these adaptations are also found developed to a similar or even greater degree in camels, the most perfectly adapted of all large desert mammals.

Adaptations for cold environments

In cold environments mammals use two major mechanisms to maintain homeothermy: (1) **decreased conductance,** that is, reduction of heat loss by increas-

ing the effectiveness of the insulation and (2) **increased heat production.**

The excellent insulation of the thick pelage of arctic animals is familiar. In all mammals living in cold regions of the earth fur thickness increases in winter, sometimes by as much as 50%. As described earlier, the thick underhair is the major insulating layer, whereas the longer and more visible guard hair serves as protection against wear and for protective coloration.

But the body extremities (legs, tail, ears, nose) of arctic mammals cannot be insulated as well as can the thorax. To prevent these parts from becoming major avenues of heat loss, they are allowed to cool to low temperatures, often approaching the freezing point. As warm arterial blood passes into a leg, for example, heat is shunted directly from artery to vein and carried back to the core of the body. This device prevents the loss of valuable body heat through the poorly insulated distal regions of the leg. A consequence of this **peripheral heat exchange system** is that the legs and feet must operate at low temperatures. The temperatures of the feet of the arctic fox and barren-ground caribou are just above the freezing point; in fact, the temperature may be below 0° C in the footpads and hooves. To keep feet supple and flexible at such low temperatures, fats in the extremities have very low melting points, perhaps 30° C lower than ordinary body fats.

In severely cold conditions all mammals can produce more heat by **augmented muscular activity** through exercise or shivering. We are all familiar with the effectiveness of both activities. A man can increase his heat production as much as eighteenfold by violent shivering when maximally stressed by cold. Another source of heat is the increased oxidation of foodstuffs, especially brown fat stores. This mechanism is called **nonshivering thermogenesis.**

Small mammals the size of lemmings, voles, and mice meet the challenge of cold environments in a different way. Small mammals are not as well insulated as large mammals because there is an obvious practical limit to how much pelage a mouse, for example, can carry before it becomes an immobile bundle of fur. Consequently these forms have successfully exploited the excellent insulating qualities of snow by living under it in runways on the forest floor, where incidentally their food is also located. In this **subnivean environment** the temperature seldom drops

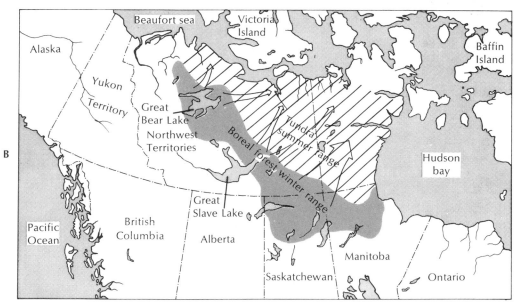

Fig. 29-12. Barren-ground caribou, *Ranger tarandus groenlandicus,* of Canada. **A,** Caribou moving southwest in autumn toward forest winter range. Two males in foreground in autumn pelage and velvet-covered antlers. Adult females in background. Cows and younger animals normally lead the herd; males follow. Caribou feed while moving, grazing on low ground vegetation, principally though not exclusively lichens. **B,** Summer and winter ranges of the barren-ground caribou of Canada. The principal spring migration routes are indicated by arrows; routes vary considerably from year to year. Caribou in Alaska have similar migrations. (**A** courtesy D. Thomas, Canadian Wildlife Service, Ottawa; **B** adapted from Kelsall, J. P. 1968. The migratory barren-ground caribou of Canada. Ottawa, Canadian Wildlife Service, Queens Printers.)

below −5° C even though the air temperature above may fall to −50° C. The snow insulation decreases thermal conductance from small mammals in the same way that pelage does for large mammals. Living beneath the snow is really a type of avoidance response to cold.

Two additional ways in which mammals may survive low temperatures are migration and hibernation.

Migration

Migration is a much more difficult undertaking for mammals than for birds; not surprisingly, few mammals make regular seasonal migrations, preferring instead to center their activities in a defined and limited home range. Nevertheless, there are some striking examples of mammalian migrations. More migrators are found in North America than on any other continent.

An example is the barren-ground caribou of Canada and Alaska, which undertakes direct and purposeful mass migrations spanning 160 to 1,100 km (100 to 700 miles) twice annually (Fig. 29-12). From winter ranges in the boreal forests (taiga) they migrate rapidly in late winter and spring to calving ranges on the barren grounds (tundra). The calves are born in mid-June. Harassed by warble and nostril flies that bore into their flesh, they move southward in July and August, feeding little along the way. In September they reach the forest, feeding there almost continuously on low ground vegetation. Mating (rut) occurs in October.

The caribou have suffered a drastic decline in numbers. In primitive times there were several million of them; by 1958 they numbered less than 200,000. The decline has been caused by man's excessive hunting, poor calf crops, and destruction of the vulnerable forested wintering areas caused by fires accidentally started by man during recent exploration and exploitation activities in the North. However, the population has increased slowly since 1958 under protection (Kelsall, 1968).

The plains bison, before its deliberate near extinction by the white man, made huge circular migrations to separate summer and winter ranges.

The longest mammal migrations of all are made by the oceanic seals and whales. One of the most remarkable migrations is that of the fur seal, which breeds on the Pribilof Islands approximately 300 km off the coast of Alaska and north of the Aleutian Islands. From wintering grounds off southern Califor-

nia the females journey 4,800 km (3,000 miles) across open ocean, arriving at the Pribilofs in the spring where they congregate in enormous numbers. The young are born within a few hours or days after arrival of the cows. Then the bulls, having already arrived and established territories, collect harems of cows, which they guard with vigilance. After the calves have been nursed for approximately 3 months, cows and juveniles leave for their long migration southward. The bulls do not follow but remain in the Gulf of Alaska during the winter.

Although we might expect the only winged mammals, bats, to use their gift to migrate, few of them do. Most spend the winter in hibernation. The four species of American bats that do migrate, the red bat, the silver-haired bat, the hoary bat, and the Brazilian free-tailed bat, spend their summers in the northern or western states and their winters in the South.

Hibernation

Many small and medium-sized mammals in northern temperate regions solve the problem of winter scarcity of food and low temperature by entering a prolonged and controlled state of dormancy. True hibernators, such as ground squirrels, jumping mice, marmots, and woodchucks (Fig. 29-13), prepare for hibernation by building up large amounts of body fat. Entry into hibernation is gradual. After a series of "test drops" during which body temperature decreases a few degrees and then returns to normal, the animal cools to within a degree or less of the ambient temperature. Metabolism decreases to a fraction of normal. In the ground squirrel, for example, the respiratory rate decreases from a normal rate of 200 per minute to 4 or 5 per minute, and the heart rate from 150 to 5 beats per minute. During arousal, the hibernator both shivers violently and employs nonshivering thermogenesis to produce heat.

Some mammals, such as bears, badgers, raccoons, and opossums, enter a state of prolonged sleep in winter with little or no decrease in body temperature. This is not true hibernation. Bears of the northern forest den-up for several months. Heart rate may decrease from 40 to 10 beats per minute, but body temperature remains normal and the bear is awakened if sufficiently disturbed. One intrepid but reckless biologist learned how lightly a bear sleeps when he crawled into a den and attempted to measure the bear's rectal temperature with a thermometer!

Fig. 29-13. Hibernating woodchuck, *Marmota monax* (order Rodentia), in den exposed by roadbuilding work sleeps on, unaware of the intrusion. Woodchucks begin hibernation in late September while the weather is still warm and may sleep 6 months. The animal is rigid and decidedly cold to the touch. Breathing is imperceptible, as slow as one breath every 5 minutes. Although it appears to be dead, it will awaken if den temperature drops dangerously low. (Photo by L. L. Rue III.)

Flight and echolocation

Bats, the only group of flying mammals, are nocturnal insectivores and thus occupy a niche left vacant by birds (Fig. 29-14). Their outstanding success is attributed to two things: flight and the capacity to navigate by echolocation. Together these adaptations enable bats to fly and avoid obstacles in absolute darkness, to locate and catch insects with precision, and to find their way deep into caves (another habitat largely ignored by both mammals and birds) where they sleep during the daytime hours.

When they are in flight, bats emit short pulses 5 to 10 milliseconds in duration in a narrow directed beam from the mouth. Each pulse is frequency modulated; that is, it is highest at the beginning, as much as 100,000 Hertz (Hz, cycles per second) and decreases to perhaps 30,000 Hz at the end. Sounds of this frequency are ultrasonic to the human ear, which has an upper limit of approximately 20,000 Hz. The pulses are produced at a rate of 30 to 40 per second, increasing to perhaps 50 per second as the bat nears an object. Since the transmission-to-reception time decreases as the bat approaches an object, it can increase the pulse frequency to obtain more information about the object. The pulse length is also shortened as it nears the object.

The external ears of bats are large, like hearing trumpets, and shaped variously in different species. Less is known about the bat's inner ear, but it obviously is capable of receiving the ultrasonic sounds emitted. Bat navigation is so refined that biologists believe that the bat builds up a mental image of its

Fig. 29-14. Red bat, *Lasiurus borealis,* in flight with four young. This species is unusual in giving birth to three or four young; most bats have one or two. The mother carries the young until their combined weight may exceed her own weight; then they are left in the roost, usually a tree. Mortality among the young is rather high. These medium-sized bats are distributed over all the eastern and southern United States. They are strong fliers, migrating northward in spring and southward in the fall. (Photo by L. L. Rue III.)

surroundings from echo scanning that is virtually as complete as the visual image from eyes of diurnal animals.

For reasons not fully understood, all bats are nocturnal, even the fruit-eating bats that use vision and olfaction instead of sonar to find their food. The tropics have many kinds of bats, including the famed vampire bat. This species is provided with razor-sharp incisors used to shave away the epidermis to expose underlying capillaries. After infusing an anticoagulant to keep the blood flowing, it laps up its meal and stores it in a specially modified stomach. It is said that sleeping dogs can hear an approaching vampire's sonar and thus awaken and escape.

Reproduction

Fertilization is always internal in mammals. All mammals except monotremes, which lay eggs, are **viviparous:** the embryo develops in a uterus and is nourished by an intimate connection between embryo and mother, the **placenta.** Most mammals have definite mating seasons, usually in the winter or spring and timed to coincide with the most favorable time of the year for rearing the young after birth. Many male mammals are capable of fertile copulation at any time, but the female mating function is restricted to a periodic cycle, known as the **estrous cycle.** The female receives the male only during a relatively brief period known as **estrus,** or heat (Fig. 29-15).

The estrous cycle is divided into stages marked by characteristic changes in the ovary, uterus, and vagina. **Proestrus,** or period of preparation, when new ovarian follicles grow is followed by **estrus,** when mating occurs. Almost simultaneously the ovarian follicles burst, releasing the eggs **(ovulation),** which are fertilized. Implantation of the fertilized egg and **pregnancy** follow. However, should mating and fertilization not occur, estrus is followed by **metestrus,** a period of repair. This stage is followed by **diestrus,** during which the uterus becomes small and anemic. The cycle then repeats itself, beginning with proestrus.

How often females are in heat varies greatly among the different mammals. Those animals that have only a

Fig. 29-15. Mating elk, *Cervus canadensis* (order Artiodactyla). Male North American elk, more correctly called wapiti, begin bugling in August to herald the beginning of the rutting season. Elk are polygamous and males attempt to round up and mate with as many cows as they can defend against rival males. The rut ends in late October and calves are born the following June. (Photo by L. L. Rue III.)

single estrus during the breeding season are called **monestrus;** those that have a recurrence of estrus during the breeding season are called **polyestrus.** Dogs, foxes, and bats belong to the first group; field mice and squirrels are all polyestrus, as are many mammals living in the more tropical regions of the earth. The Old World monkeys and humans have a somewhat different cycle in which the postovulation period is terminated by **menstruation,** during which the lining of the uterus (endometrium) collapses and is discharged with some blood. This is called a **menstrual cycle** and is described in Chapter 13.

Gestation, or period of pregnancy, also varies greatly among the mammals. Mice and rats have a gestation period of approximately 21 days; rabbits and hares, 30 to 36 days; cats and dogs, 60 days; cows, 280 days; and elephants, 22 months. The marsupials (opossum) have a very short gestation period of 13 days; at the end of that time the tiny young leave the vaginal orifice and make their way to the marsupial pouch where they attach themselves to nipples. Here they remain for more than 2 months before emerging.

The number of young produced by mammals in a season depends on many factors. Usually the larger the animal, the smaller the number of young in a litter. One of the greatest factors involved is the number of enemies a species has. Small rodents, which serve as prey for so many carnivores, produce as a rule more than one litter of several young each season. Field mice are known to produce as many as 17 litters of four to nine young in a year. Most carnivores have but one litter of three to five young per year. Large mammals, such as elephants and horses, have only one young.

Territory and home range

Virtually all mammals have territories—areas from which individuals of the *same* species are excluded. In fact, many wild mammals, like many people, are basically unfriendly to their own kind, especially so to their own sex during the breeding season. If the mammal dwells in a burrow or den, this area forms the center of its territory. If it has no fixed address, the territory is marked out, usually with the highly developed scent glands described earlier in this chapter. Territories vary greatly in size, of course, depending on the size of the animal and its feeding habits. The

grizzly bear has a territory of several square miles, which it guards zealously against all other grizzlies.

Mammals usually use natural features of their surroundings in staking their claims. These are marked with secretions from the scent glands or by urinating or defecating. When an intruder knowingly enters another's marked territory, it is immediately placed at a psychologic disadvantage. Should a challenge follow, the intruder almost invariably breaks off the encounter in a submissive display characteristic for the species.

An interesting exception to the territorial nature of most mammals is the prairie dog, which lives in large, friendly communities called prairie-dog "towns" (Fig. 29-16). When a new litter has been reared, the adults relinquish the old home to the young and move to the edge of the community to establish a new home. Such a practice is totally antithetic to the behavior of most mammals, which drive off the young when they are self-sufficient.

The **home range** of a mammal is a much larger

Fig. 29-16. Family of prairie dogs, *Cynomys ludovicianus* (order Rodentia). These highly social prairie dwellers are plant eaters that comprise an important source of food to many animals. They live in elaborate tunnel systems so closely interwoven that they form "towns" of as many as 1,000 individuals. Towns are subdivided into wards, in turn divided into coteries, the basic family unit, containing one or two adult males, several females, and their litters. Although prairie dogs display ownership of burrows with territorial calls, they are friendly with inhabitants of adjacent burrows. The name "prairie dogs" derives from the sharp, doglike bark they make when danger threatens. Western cattle and sheep ranches have nearly eradicated prairie dogs in some areas by mass poisoning programs with disastrous results. The tunnel systems, which served as a natural sponge to prevent flash floods, filled in and serious erosion followed. (Photo by L. L. Rue III.)

foraging area surrounding a defended territory. Home ranges are not defended in the same way as is a territory; home ranges may, in fact, overlap, producing a neutral zone used by the owners of several territories for seeking food.

Mammal populations

A population of animals includes all the members of a species that interbreed and share a particular space (see Chapter 1). All mammals live in communities, each composed of numerous populations of different animal and plant species. Each species is affected by the activities of other species and by the changes, especially climatic, that occur. Thus populations are always changing in size. Populations of small mammals are lowest before the breeding season and greatest just after the addition of the new members. Beyond these expected changes in population size, animal populations may fluctuate from other causes.

Irregular fluctuations are commonly produced by variations in climate, such as unusually cold, hot, or dry weather, or by natural catastrophes, such as fires, hailstorms, and hurricanes. These are **density-independent** causes since they affect a population whether it is crowded or dispersed. However, the most spectacular fluctuations are **density dependent;** that is, they are correlated with population crowding. Cycles of abundance are common among many rodent species.

The population peaks and mass migrations of the Scandinavian and arctic North American lemmings are well known. Lemmings breed all year round, although more in the summer than in winter. The gestation period is only 21 days; young born at the beginning of the summer are weaned in 14 days and are themselves capable of reproducing by the end of the summer. Lemmings experience a 4-year cycle in abundance. Their numbers gradually increase from a density of 1 animal in 10 acres to 40 to 70 per acre at the peak of the fourth year. At this density, having devastated the vegetation by tunneling and grazing, they begin long, mass migrations to find new undamaged habitats for food and space. They swim across streams and small lakes as they go but cannot distinguish these from large lakes, rivers, and the sea, in which they drown. Since lemmings are the main diet of many carnivorous mammals and birds, any change in lemming population density affects all their predators as well.

The varying hare (snowshoe rabbit) of North America shows 10-year cycles in abundance. The well-known fecundity of rabbits enables them to produce litters of three or four young as many as five times per year. The density may increase to 4,000 hares competing for food in each square mile of northern forest. Predators (owls, minks, foxes, and especially lynxes) also increase (Fig. 29-17). Then the population crashes precipitously, for reasons that have long been a puzzle to scientists. Rabbits die in great numbers, not from lack of food or from an epidemic disease (as was once believed) but evidently from some density-dependent psychogenic cause. As crowding increases, hares

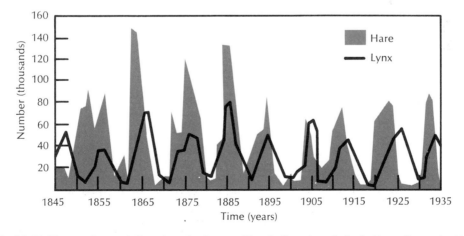

Fig. 29-17. Changes in population of varying hare and lynx in Canada as indicated by pelts received by the Hudson's Bay Company. The abundance of lynx (predator) follows that of the hare (prey). (After Odum, E. P. 1959. Fundamentals of ecology. Philadelphia, W. B. Saunders Co.)

become more aggressive, show signs of fear and defense, and stop breeding. The entire population reveals symptoms of pituitary–adrenal gland exhaustion, an endocrine imbalance called "shock disease," which results in death. There is much about these dramatic crashes that is not understood. Whatever the causes, population crashes that follow superabundance, though harsh, are clearly advantageous to the species, for the vegetation is allowed to recover, thus providing the survivors with a much better chance for successful breeding.

SELECTED REFERENCES

Burt, W. H., and R. P. Grossenheider. 1963. A field guide to the mammals, ed. 2. Boston, Houghton Mifflin, Co. *Full color illustrations and range maps.*

Carrington, R. 1975. The mammals. New York, Life Nature Library, Div. of Time, Inc. *Well-written, beautifully illustrated semi-popular treatment.*

Davis, E. E., and F. B. Golley. 1964. Principles of mammalogy. New York, Reinhold Publishing Corp. *This introductory text deals with classification, adaptations, evolution, distribution, populations, and behavior of mammals.*

Gunderson, H. L. 1976. Mammalogy. New York, McGraw-Hill, Inc. *Comprehensive text emphasizing adaptations.*

Hall, E. R., and K. R. Kelson. 1959. The mammals of North America, 2 vols. New York, The Ronald Press Co. *Full descriptions of species and subspecies with distribution maps. Taxonomic keys, records, and revealing line drawings of skull characteristics are included in this authoritative work.*

Kelsall, J. D. 1968. The migratory barren-ground caribou of Canada. Ottawa, Canadian Wildlife Service, Queen's Printers.

Matthews, L. H. 1969. The life of mammals, 2 vol. New York, Universe Books. *The first volume of this excellent set treats the origins, structure, and adaptations of mammals; the second volume reviews the orders and families of mammals.*

Richards, S. A. 1973. Temperature regulation. London, Wykeham Publications. *Clearly written primer on the subject.*

Sadleir, R. M. F. S. 1969. The ecology of reproduction in wild and domestic mammals. London, Methuen & Co. *Wealth of information on mammalian reproductive physiology.*

Sanderson, I. T. 1955. Living mammals of the world. New York, Garden City Books (Hanover House). *A beautiful and informative work of many photographs and concise text material. It is a delight to any zoologist regardless of his specialized interest.*

Scheffer, V. B. 1958. Seals, sea lions, and walruses. A review of the Pinnipedia. Stanford, Calif., Stanford University Press. *Treats the group's evolution, characteristics, and classification. Many fine photographs and a good bibliography are included.*

Seton, E. T. 1925-1928. Lives of game animals, 4 vols. New York, Doubleday, Doran & Co. *A classic.*

Vaughn, T. A. 1972. Mammalogy. Philadelphia, W. B. Saunders Co. *This basic text takes a systematic approach to the subject.*

Walker, E. P. 1968. Mammals of the world, 3 vols. Baltimore, The Johns Hopkins University Press. *The only single compendium of information on all known and living mammalian genera. A valuable reference work.*

Young, J. Z. 1957. The life of mammals. New York, Oxford University Press. *This is a well-known work about mammals, especially their anatomy, histology, physiology, and embryology. Classification is not treated.*

SELECTED SCIENTIFIC AMERICAN ARTICLES

Bartholomew, G. A., and J. W. Hudson. 1961. Desert ground squirrels. **205**:107-116 (Nov.). *Two species living in California's Mojave Desert have developed interesting adaptations for survival in desert heat and aridity.*

Flyger, V., and M. R. Townsend. 1968. The migration of polar bears. **218**:108-116 (Feb.). *The habitat, biology, and wide migrations of these large arctic carnivores are described.*

Griffin, D. R. 1958. More about bat "radar." **199**:40-44 (July).

Irving, L. 1966. Adaptations to cold. **214**:94-101 (Jan.). *The homeothermic birds and mammals have evolved several adaptations that permit them to survive in cold environments.*

King, J. A. 1959. The social behavior of prairie dogs. **201**:128-140 (Oct.).

Kooyman, G. L. 1969. The Weddell seal. **221**:100-106 (Aug.). *This Antarctic seal swims for miles under shelf ice on one breath of air, returning unerringly to its breathing hole.*

McVay, S. 1966. The last of the great whales. **215**:13-21 (Aug.). *Most of the 12 commercially hunted whale species have been nearly exterminated. The indifference and unrestricted fishing that has characterized the whaling industry are recounted in this article.*

Modell, W. 1969. Horns and antlers. **220**:114-122 (April). *Their differences, growth, and structure are described.*

Montagna, W. 1965. The skin. **212**:56-66 (Feb.). *The structure and diverse functions of mammalian skin are described.*

Mrosovsky, N. 1968. The adjustable brain of hibernators. **218**:110-118 (March). *Hibernation is preceded by a remarkable series of changes and a resetting of the hypothalamic thermostat.*

Mykytowycz, R. 1968. Territorial marking by rabbits. **218**:116-126 (May). *Describes the use of odor-producing glands by Australian colonial rabbits to mark territories.*

Pruitt, W. O., Jr. 1960. Animals in the snow. **202**:60-68 (Jan.). *Describes the adaptations of the many homeotherms that live where snow persists for more than half the year.*

Schmidt-Nielsen, K., and B. Schmidt-Nielsen. 1953. The desert rat. **189**:73-78 (July). *The kangaroo rat possesses several adaptations that enable it to live in hot, arid regions, eating only dry food and drinking no water at all.*

Scholander, P. F. 1963. The master switch of life. **209**:92-106 (Dec.). *Diving vertebrates are obviously specialized for making prolonged dives without breathing. One of the most important adaptations is the capacity to grossly redistribute the circulation.*

Taylor, C. R. 1969. The eland and the oryx. **220**:88-95 (Jan.). *These large African antelopes thrive in desert or near-desert regions without drinking water. Their adaptations for heat and aridity are described.*

GLOSSARY

aboral (ab-o′rəl) (L. *ab*, from, + *os*, mouth). A region opposite the mouth.

acanthor (ə-kan′thor) (Gr. *akantha*, spine or thorn, + *-or*). First larval form of acanthocephalans in the intermediate host.

acclimatization (ə-klī′mə-də-zā′shən) (L. *ad*, to, + Gr. *klima*, climate). Gradual physiologic adaptation in response to relatively long-lasting environmental changes.

acetabulum (as′ə-tab′ū-lum) (L. a little saucer for vinegar). True sucker, especially in flukes and leeches. The socket in the hip bone that receives the thigh bone.

acinus, pl. **acini** (as′ə-nəs, as′ə-ni) (L. grape). A small lobe of a compound gland or a saclike cavity at the termination of a passage.

acoelomate (ā-sēl′ə-māt′) (Gr. *a*, not, + *koilōma*, cavity). Without a coelom, as in flatworms and proboscis worms.

acontium (ə-kän′chē-əm), pl. **acontia** (Gr. *akontion*, dart). Thread-bearing nematocysts located on mesentery of sea anemone.

actin (Gr. *aktis*, ray). A protein in the contractile tissue that forms the thin myofilaments of striated muscle.

adaptation (L. *adaptatus*, fitted). Adjustment to environment by an organism so that it becomes more fit for existence.

adductor (ə-duk′tər) (L. *ad*, to, + *ducere*, to lead). A muscle that draws a part toward a median axis, or a muscle that draws the two valves of a mollusc shell together.

adenine (ad′nēn, ad′ə-nēn) (Gr. *adēn*, gland, + *-ine*, suffix). A purine base; component of nucleotides and nucleic acids.

adenosine (ə-den′ə-sēn) **(di-, tri-) phosphate** (ADP and ATP). A nucleotide composed of adenine, ribose, sugar, and two (ADP) or three (ATP) phosphate units; ATP is an energy-rich compound that, with ADP, serves as a phosphate bond–energy transfer system in cells.

adipose (ad′ə-pōs) (L. *adeps*, fat). Fatty tissue; fatty.

adrenaline (ə-dren′ə-lən) (L. *ad*, to, + *renalis*, pertaining to kidneys). A hormone produced by the adrenal, or suprarenal, gland; epinephrine.

adsorption (ad-sorp′shən) (L. *ad*, to, + *sorbere*, to suck in). The adhesion of molecules to solid bodies.

aerobic (a-rō′bik) (Gr. *aēr*, air, + *bios*, life). Oxygen-dependent form of respiration.

afferent (af′ə-rənt) (L. *ad*, to, + *ferre*, to bear). Adj. Leading or bearing toward some organ, for example, nerves conducting impulses toward the brain or blood vessels carrying blood toward an organ; opposed to efferent.

aggression (L. *aggredi*, attack). A primary instinct usually associated with emotional states; an offensive behavior action.

agonistic behavior (Gr. *agōnistēs*, combatant). Any behavior related to aggression, such as threat, attack, fighting, as well as defensive responses to aggression.

alate (ā′lāt) (L. *alatus*, wing). Winged.

allantois (ə-lan′tois) (Gr. *allas*, sausage, + *eidos*, form). One of the extraembryonic membranes of the amniotes that functions in respiration and excretion in birds and reptiles and plays an important role in the development of the placenta in most mammals.

allele (ə-lēl′) (Gr. *allēlōn*, of one another). One of a pair, or series, of genes that are alternative to each other in heredity and are situated at the same locus in homologous chromosomes.

allograft (a′lō-graft) (Gr. *allos*, other, + graft). A piece of tissue or an organ transferred from one individual to another individual of the same species, not identical twins; homograft.

allometry (ə-lom′ə-trē) (Gr. *allos*, other, + *metry*, measure). Relative growth of a part in relation to the whole organism.

allopatric (Gr. *allos*, other, + *patra*, native land). In separate and mutually exclusive geographic regions.

alpha helix (Gr. *alpha*, first, + L. *helix*, spiral). Literally the first spiral arrangement of the genetic DNA molecule; regular coiled arrangement of polypeptide chain in proteins; secondary structure of proteins.

altricial (al-tri′shəl) (L. *alcatrices*, nourishers). Referring to young animals (especially birds) having the young hatched in an immature, dependent condition.

alula (al′yə-lə) (L. dim. of *ala*, wing). The first digit or thumb of a bird's wing, much reduced in size.

alveolus (al-vē′ə-ləs) (L. dim. of *alveus*, cavity, hollow). A small cavity or pit, such as a microscopic air sac of the lungs, terminal part of an alveolar gland, or bony socket of a tooth.

ambulacra (am′byə-lak′rə) (L. *ambulare*, to walk). In echinoderms, radiating grooves where podia of water-vascular system project to outside.

amebocyte (ə-mē′bə-sīt) (Gr. *amoibē*, change, + *kytos*, hollow vessel). Any free body cell capable of movement by pseudopodia; certain types of blood cells and tissue cells.

ameboid (ə-mē′boid) (Gr. *amoibē*, change, + *-oid*, like). Amebal-ike in putting forth pseudopodia.

amictic (ə-mik′tic) (Gr. *a*, without, + *miktos*, mixed or blended). Pertaining to female rotifers, which produce only diploid eggs that cannot be fertilized, or to the eggs produced by such females.

amino acid (ə-mē′nō) (amine, an organic compound). An organic acid with an amino radical (NH_2). Makes up the structure of proteins.

amitosis (ā′mī-tō′səs) (Gr. *a*, not, + *mitos*, thread). A form of cell division in which mitotic nuclear changes do not occur; cleavage without separation of daughter chromosomes.

amnion (am′nē-än) (Gr. *amnion*, membrane around the fetus). The innermost of the extraembryonic membranes forming a fluid-filled sac around the embryo in amniotes.

amniote (am′nē-ōt′). Having an amnion; as a noun, an animal that

---

bat / āpe / ärmadillo / herring / fēmale / finch / līce / crocodile / crōw / duck / ūnicorn / ə indicates unaccented vowel sound ''uh'' as in mammal, fishes, cardinal, heron, vulture / stress as in bi-ol′o-gy, bi′o-log′i-cal

develops an amnion in embryonic life, that is, reptiles, birds, and mammals.

amphiblastula (am'fə-blas'chə-lə) (Gr. *amphi,* on both sides, + *blastos,* germ, + L. *-ula,* small). Free-swimming larval stage of certain marine sponges; blastula-like, but with only the cells of the animal pole flagellated; those of the vegetal pole unflagellated.

amphid (am'fəd) (Gr. *amphidea,* anything that is bound around). One of a pair of anterior sense organs in certain nematodes.

amplexus (am-plek'səs) (L. embrace). The copulatory embrace of frogs or toads.

amylase (am'ə-lās') (L. *amylum,* starch, + *ase,* suffix meaning enzyme). An enzyme that breaks down starch into smaller units.

anadromous (an-ad'rə-məs) (Gr. *anadromos,* running upward). Refers to those fish that migrate up streams from the sea to spawn.

anaerobic (an'ə-rō'bik) (Gr. *an,* not, + *aēr,* air, + *bios,* life). Not dependent on oxygen for respiration.

analogy (L. *analogus,* ratio). Similarity of function but not in origin.

anastomosis (ə-nas'tə-mō'səs) (Gr. *ana,* again, + *stoma,* mouth). A union of two or more arteries, veins, or fibers to form a branching network.

androgen (an'drə-jən) (Gr. *anēr, andros,* man, + *genēs,* born). Any of a group of vertebrate male sex hormones.

Angstrom (after Ångström, Swedish physicist). A unit of one tenmillionth of a millimeter (one ten-thousandth of a micron); it is represented by the symbol Å.

anhydrase (an-hī'drās) (Gr. *an,* not, + *hydōr,* water, + *ase,* enzyme suffix). An enzyme involved in the removal of water from a compound. Carbonic anhydrase promotes the conversion of carbonic acid into water and carbon dioxide.

anlage (än'lä-gə) (Ger. *anlage,* laying-out, foundation). Rudimentary form; primordium.

antenna (L. *antenna, antemna,* sail yard). A sensory appendage on the head of arthropods, or the first pair of the two such pairs of structures in Crustacea.

anterior (L. comparative of *ante,* before). The head end of an organism, or (as adj.) toward that end.

anticodon (an'tī-kō'don). A sequence of three nucleotides in transfer RNA that is complementary to a codon in messenger RNA.

aperture (ap'ər-chər) (L. *apertura,* from *aperire,* to uncover). An opening; the opening into the first whorl of a gastropod shell.

apical (ap'ə-kl) (L. *apex,* tip). Pertaining to the tip or apex.

apopyle (ap'-ə-pīl) (Gr. *apo,* away from, + *pylē,* gate). In sponges, opening of the radial canal into the spongocoel.

appendicular (L. *ad,* to, + *pendare,* to hang). Pertaining to appendages; pertaining to vermiform appendix.

arboreal (är-bōr'ē-al) (L. *arbor,* tree). Living in trees.

archenteron (ärk-en'tə-rän') (Gr. *archē,* beginning, + *enteron,* gut). The main cavity of an embryo in the gastrula stage; it is lined with endoderm, and represents the future digestive cavity.

archeocytes (ärk'ē-ō-sītes) (Gr. *archaios,* beginning, + *kytos,* hollow vessel). Ameboid cells of varied function in sponges.

artiodactyl (är'ti-o-dak'təl) (Gr. *artios,* even, + *daktylos,* toe). One of an order or suborder of mammals with two or four digits on each foot.

ascon (Gr. *askos,* bladder). Simplest form of sponges, with canals leading directly from the outside to the interior.

asexual. Without distinct sexual organs; not involving gametes.

assimilation (L. *assimilatio,* bringing into conformity). Absorption and building up of digested nutriments into complex organic protoplasmic materials.

atoke (ā'-tōk) (Gr. *a,* without, + *tokos,* offspring). Anterior part of a marine polychaete, as distinct from the posterior part (epitoke) during the breeding season.

ATP. In biochemistry, an ester of adenosine and triphosphoric acid. Adenosine triphosphate.

atrium (ā'trē-əm) (L.). One of the chambers of the heart; the tympanic cavity of the ear; the large cavity containing the pharynx in tunicates and cephalochordates.

auricle (aw'ri-kl) (L. *auricula,* dim. of *auris,* ear). One of the less muscular chambers of the heart; atrium; the external ear, or pinna; any earlike lobe or process.

autosome (aw'tō-sōm) (Gr. *autos,* self, + *sōma,* body). Any chromosome that is not a sex chromosome.

autotomy (aw-täd'ə-mē) (Gr. *autos, self,* + *tomos,* a cutting). The automatic breaking off of a part of the body.

autotroph (aw'tō-trōf) (Gr. *autos,* self, + *trophos,* feeder). An organism that makes its organic nutrients from inorganic raw materials.

autotrophic nutrition (Gr. *autos,* self, + *trophia,* denoting nutrition). Nutrition characterized by the ability to utilize simple inorganic substances for the synthesis of more complex organic compounds, as in green plants and some bacteria and true fungi.

avicularium (L. *avicula,* small bird, + *aria,* like or connected with). Modified zooid that is attached to the surface of the major zooid in Ectoprocta and resembles a bird's beak.

axial (L. *axis,* axle). Relating to the axis, or stem; on or along the axis.

axolotl (ak'sə-lot'l) (Nahuatl *atl,* water, + *xolotl,* doll, servant, spirit). Larval stage of any of several species of the genus *Ambystoma* (such as *A. tigrinum),* exhibiting neotenic reproduction.

axopodium (ak'sə-pō'di-um) (L. *axis,* an axis, + Gr. *podion,* small foot). Long, slender, more or less permanent pseudopodium found in certain sarcodine protozoans. (Also **axopod**).

basal body. Specialized granule in a cell, found basal to a flagellum or cilium, often consisting of a centriole or centrosome; a blepharoplast.

benthos (ben'thäs) (Gr. depth of the sea). Those organisms that live along the bottom of seas and lakes; adj., **benthic.**

biogenesis (bī'ō-jen'ə-səs) (Gr. *bios,* life, + *genesis,* birth). The doctrine that life originates only from preexisting life.

bioluminescence. Method of light production by living organisms in which usually certain proteins (luciferins), in the presence of oxygen and an enzyme (luciferase), are converted to oxyluciferins with the liberation of light.

biomass (Gr. *bios,* life, + *maza,* lump or mass). The weight of total living organisms or of a species population per unit of area.

biome (bī'ōm) (Gr. *bios,* life, + *ōma,* abstract group suffix). Complex of plant and animal communities characterized by climatic and soil conditions; the largest ecologic unit.

biosphere (Gr. *bios,* life, + *sphaira,* globe). That part of earth containing living organisms.

bipinnaria (L. *bi-,* double, + *pinna,* wing, + *-aria,* like or

595

connected with). Free-swimming, ciliated, bilateral larva of the asteroid echinoderms; develops into the brachiolaria larva.

blastocoel (blas′tō-sēl′) (Gr. *blastos*, germ, + *koilos*, hollow). Cavity of blastula.

blastomere (Gr. *blastos*, germ, + *meros*, part). An early cleavage cell.

blastopore (Gr. *blastos*, germ, + *poros*, passage, pore). External opening of the archenteron in the gastrula.

blastula (Gr. *blastos*, germ, + L. *-ula*, diminutive). Early embryologic stage of many animals; consists of a hollow mass of cells.

blepharoplast (blə-fā′rə-plast) (Gr. *blepharon*, eyelid, + *plastos*, formed). Specialized granule in a cell, connected with a basal granule and the base of a flagellum or cilium; often consists of a centriole or centrosome.

brachial (brak′ē-əl) (L. *brachium*, forearm). Referring to the forearm.

brachiolaria (brak′ē-ō-lār′ē-ə) (L. *brachiola*, little arm, + *-aria*, pertaining to). This asteroid larva develops from the bipinnaria larva and has three preoral holdfast processes.

branchial (brank′ē-əl) (Gr. *branchia*, gills). Referring to gills.

buccal (buk′əl) (L. *bucca*, cheek). Referring to the mouth cavity.

buffer Any substance or chemical compound that tends to keep pH constant when acids or bases are added.

carapace (kar′ə-pās) (F., fr. Sp. *carapacho*). Shieldlike plate covering the cephalothorax of certain crustaceans; dorsal part of the shell of a turtle.

carbohydrate (L. *carbo*, charcoal, + Gr. *hydōr*, water). Compounds of carbon, hydrogen, and oxygen; aldehyde or ketone derivatives of polyhydric alcohols, with hydrogen and oxygen atoms attached in a 2 to 1 ratio.

carboxyl (kär-bäk′səl) (carbon + oxygen + -yl, chemical radical suffix). The acid group of organic molecules—COOH.

carnivore (kar′nə-vōr′) (L. *carnivorus*, flesh-eating). One of the flesh-eating mammals of the order Carnivora.

carotene (kär′ə-tēn) (L. *carota*, carrot, + -ene, unsaturated straight-chain hydrocarbons). A red, orange, or yellow pigment belonging to the group of carotenoids; precursor of vitamin A.

cartilage (L. *cartilago*; akin to L. *cratis*, wickerwork). A translucent elastic tissue that makes up most of the skeleton of embryos and very young vertebrates; in higher forms much of it is converted into bone.

caste (kast) (L. *castus*, pure, separated). One of the polymorphic forms within an insect society, each caste having its specific duties, as queen, worker, soldier, etc.

catadromous (kə-tad′rə-məs) (Gr. *kata*, down, + *dromos*, a running). Refers to those fish that migrate from freshwater to the ocean to spawn.

catalyst (kad′ə-ləst) (Gr. *kata*, down, + *lysis*, a loosening). A substance that accelerates a chemical reaction but does not become a part of the end product.

cecum, caecum (sē′kəm) (L. *caecus*, blind). A blind pouch at the beginning of the large intestine; any similar pouch.

cellulose (sel′ū-lōs) (L. *cella*, small room). Chief polysaccharide constituent of the cell wall of green plants and some fungi; an

insoluble carbohydrate $(C_6H_{10}O_5)_x$ that is converted into glucose by hydrolysis.

centriole (sen′trē-ol) (Gr. *kēntron*, center of a circle, + L. *-ola*, small). A minute cytoplasmic organelle usually found in the centrosome and considered to be the active division center of the cell; organizes spindle fibers during mitosis and meiosis.

centrolecithal (sen′tro-les′ə-thəl) (Gr. *kentron*, center, + *lekithos*, yolk, + Eng. -al, adj.). Pertaining to an insect egg with the yolk concentrated in the center.

centromere (sen′trə-mir) (Gr. *kentron*, center, + *meros*, part). A small body or constriction on the chromosome to which a spindle fiber attaches during mitosis or meiosis.

cephalization (sef′ə-li-zā′shən) (Gr. *kephale*, head). The process by which specialization, particularly of the sensory organs and appendages, became localized in the head end of animals.

cephalothorax (sef′ə-lä-thō′raks) (Gr. *kephale*, head, + thorax). A body division found in many Arachnida and higher Crustacea, in which the head and some or all of the thoracic segments are fused.

cercaria (ser-kār′ē-ə) (Gr. *kerkos*, tail, + L. *-aria*, like or connected with). Tadpolelike larva of trematodes (fluke).

chelicera (kə-lis′ə-rə), pl. **chelicerae** (Gr. *chēlē*, claw, + *keras*, horn). One of a pair of pincerlike head appendages on the members of the subphylum Chelicerata.

chelipeds (kēl′ə-peds) (L., clamfoot). First pair of legs in most decapod crustaceans; specialized for seizing and crushing.

chiasma (kī-az′mə), pl. **chiasmata** (Gr. *chiasma*, cross). An intersection or crossing, as of nerves; an exchange of partners in meiosis.

chitin (kī′tən) (Fr. *chitine*, fr. Gr. *chitōn*, tunic). A horny substance that forms part of the cuticle of arthropods and is found sparingly in certain other invertebrates; a nitrogenous polysaccharide insoluble in water, alcohol, dilute acids, and digestive juices of most animals.

chloragogue cells (klōr′ə-gog) (Gr. *chlōros*, light green, + *agōgos*, a leading, a guide). Modified peritoneal cells, greenish or brownish, clustered around the digestive tract of certain annelids; apparently they aid in elimination of nitrogenous wastes and in food transport.

chlorocruorin (klō′rō-kroo′ə-rən) (Gr. *chlōros*, light green, + L. *cruor*, blood). A greenish iron-containing respiratory pigment dissolved in the blood plasma of certain marine polychaetes.

chlorophyll (klō′ro-fil) (Gr. *chlōros*, light green, + *phyllōn*, leaf). Green pigment found in plants and in some animals; necessary for photosynthesis.

choanocyte (kō-an′ō-sīt) (Gr. *choanē*, funnel, + *kytos*, hollow vessel). One of the flagellate collar cells that line cavities and canals of sponges.

cholinergic (kōl′-i-nər′jik) (Gr. *chōle*, bile, + *ergon*, work). Type of nerve fiber that releases acetylcholine from axon terminal.

chorion (kō′rē-on) (Gr. *chorion*, skin). The outer of the double membrane that surrounds the embryo of reptiles, birds, and mammals; in mammals it helps form the placenta.

chromatid (krō′mə-tid) (Gr. *chromato-*, fr. *chrōma*, color, + L. *-id*, feminine stem for particle of specified kind). A half chromosome between early prophase and metaphase in mitosis; a half chromosome between synapsis and second metaphase in meiosis. At the anaphase stage each chromatid is known as a daughter chromosome.

---

bat / āpe / ärmadillo / herring / fēmale / finch / līce / crocodile / crōw / duck / ūnicorn / ə indicates unaccented vowel sound ′′uh′′ as in mammal, fishes, cardinal, heron, vulture / stress as in bi-ol′o-gy, bi′o-log′i-cal

chromatin (krō′mə-tin) (Gr. *chrōma*, color). The nucleoprotein material of a chromosome; regarded as the physical basis of heredity.

chromatophore (krō-mat′ə-fōr) (Gr. *chrōma*, color, + *phorein*, to bear). Pigment cell, usually in the dermis, in which usually the pigment can be dispersed or concentrated. (Sometimes used to refer to chloroplast or chromoplast.)

chromomere (krō′mō-mir) (Gr. *chrōma*, color, + *meros*, part). One of the chromatin granules of characteristic size on the chromosome; may be identical with a gene or a cluster of genes.

chromonema (krō-mə-nē′mə) (Gr. *chrōma*, color, + *nema*, thread). A convoluted thread in prophase of mitosis or the central thread in a chromosome.

chromosome (krō′mə-sōm) (Gr. *chrōma*, color, + *sōma*, body). One of the DNA bodies that arises from the nuclear network during mitosis, splits longitudinally, and carries the linear sequence of genes.

chrysalis (kris′ə-lis) (L. from Gr. *chrysos*, gold). The pupal stage of a butterfly.

cilium, pl. cilia (sil′i-əm) (L. *cilium*, eyelid). A hairlike, vibratile organelle process found on many animal cells. Cilia may be used in moving particles along the cell surface or, in ciliate protozoans, for locomotion.

cinclides (sing′klid-əs), sing. **cinclis** (sing′kləs) (Gr. *kinklis*, latticed gate or partition). Small pores in the external body wall of sea anemones for extrusion of acontia.

circadian (sər′kə-dē′-ən) (L. *circa*, around, + *dies*, day). Occurring at a period of approximately 24 hours.

cirrus (sir′əs) (L. curl). A hairlike tuft on an insect appendage; locomotor organelle of fused cilia; male copulatory organ of some invertebrates.

cleavage (OE. *cleofan*, to cut). Process of nuclear and cell division in animal zygote.

climax (klī′maks) (Gr. *klimax*, ladder). A state of dynamic equilibrium; a culmination of the succession in the life forms of a community.

cline (klīn) (Gr. *klinein*, slope, bend). A pattern of gradual genetic change in a population according to its geographic range.

clitellum (klī-tel′əm) (L. *clitellae*, packsaddle). Thickened saddle-like portion of certain midbody segments of many oligochaetes and leeches.

cloaca (klō-ā′kə) (L. *cloaca*, sewer). Posterior chamber of digestive tract in many vertebrates, receiving feces and urogenital products. In certain invertebrates, a terminal portion of digestive tract that serves also as respiratory, excretory, or reproductive duct.

clone (klōn) (Gr. *klōn*, twig). All descendants derived by asexual reproduction from a single individual.

cnidoblast (nī′dō-blast) (Gr. *knidē*, nettle, + *blastos*, germ). Modified interstitial cell that holds the nematocyst.

cnidocil (nī′dō-sil) (Gr. *knidē*, nettle, + L. *cilium*, hair). Triggerlike spine on nematocyst.

coacervate (kō′ə-sər′vət) (L. *coacervatus*, to heap up). An aggregate of colloidal droplets held together by electrostatic forces.

cochlea (kōk′lēə) (L. snail, fr. Gr. *kochlos*, a shellfish). A tubular cavity of the inner ear containing the essential organs of hearing; occurs in crocodiles, birds, and mammals; spirally coiled in mammals.

codon (kō′dän) (L. code, + on). A sequence of three adjacent nucleotides that code for one amino acid.

coelenteron (sē-len′tər-on) (Gr. *koilos*, hollow, + *enteron*, intestine). Internal cavity of a coelenterate; gastrovascular cavity; archenteron.

coelom (sē′lōm) (Gr. *koilōma*, cavity). The body cavity in triploblastic animals, lined with mesodermal peritoneum.

coelomocyte (sē-lō′mə-cīt) (Gr. *koilōma*, cavity, + *kytos*, hollow vessel). Another name for amebocyte; primitive or undifferentiated cell of the coelom and the water-vascular system.

coelomoduct (sē-lō′mə-dukt) (Gr. *koilos*, hollow, + L. *ductus*, a leading). A duct that carries gametes or excretory products (or both) from the coelom to the exterior.

coenecium, coenoecium (sə-nēs(h)′ē-um) (Gr. *koinos*, common, + *oikion*, house). The common secreted investment of an ectoproct colony; may be chitinous, gelatinous, or calcareous.

coenzyme (kō-en′zīm) (L. prefix *co-*, with, + Gr. *enzymos*, leavened, from *en*, in, + *zymē*, leaven). A required substance in the activation of an enzyme; a prosthetic or nonprotein constituent of an enzyme.

collenchyme (käl′ən-kīm) (Gr. *kolla*, glue, + *enchyma*, infusion). A gelatinous mesenchyme containing undifferentiated cells; found in coelenterates and ctenophores.

colloblast (käl′ə-blast) (Gr. *kolla*, glue, + *blastos*, germ). A glue-secreting cell on the tentacles of ctenophores.

colloid (kä′loid) (Gr. *kolla*, glue, + *eidos*, form). A two-phase system in which particles of one phase are dispersed in the second phase.

comb plate. One of the plates of fused cilia that are arranged in rows for ctenophore locomotion.

commensalism (kə-men′səl-iz′əm) (L. *cum*, together with, + *mensa*, table). A symbiotic relationship in which one individual benefits and the other is unharmed.

community (L. *communitas*, community, fellowship). An assemblage of organisms that are associated in a common environment and interact with each other in a self-sustaining and self-regulating relation.

conjugation (kon′jū-gā′shun) (L. *conjugare*, to yoke together). Temporary union of two ciliate protozoans while they are exchanging chromatin material and undergoing nuclear phenomena resulting in binary fission.

conspecific (*con*, together, + *species*). A member of the same species.

contractile vacuole. A clear fluid-filled cell vacuole in protozoans and a few lower metazoans; takes up water and releases it to the outside in a cyclic manner, for osmoregulation and some excretion.

copulation (kop′ū-lā′shun) (F. fr. L. *copulare*, to couple). Sexual union to facilitate the reception of sperm by the female.

corium (kō′re-um) (L.). The deep layer of the skin; dermis.

cornea (kor′nē-ə) (L. *corneus*, horny). The outer transparent coat of the eye.

cortex (kor′teks) (L. bark). The outer layer of a structure.

crista (kris′ta), pl. **cristae** (L. *crista*, crest). A crest or ridge on a body organ or organelle.

cryptobiotic (Gr. *kryptos*, hidden, + *biōticus*, pertaining to life). Living in concealment; refers to insects and other animals that live in secluded situations, such as underground or in wood; also tardigrades and some nematodes, rotifers, and others that survive harsh environmental conditions by assuming for a time a state of very low metabolism.

ctenoid scales (ten′oid) (Gr. *kteis, ktenos,* comb). Thin, overlapping dermal scales of the more advanced fishes; posterior margins are serrate, or comblike.

cuticle (kū′ti-kəl) (L. *cutis,* skin). A protective, noncellular, organic layer secreted by the external epithelium (hypodermis) of many invertebrates. In higher animals, the term refers to the epidermis or outer skin.

cycloid scales (sī′-kloid) (Gr. *klyos,* circle). Thin, overlapping dermal scales of the more primitive fishes; posterior margins are smooth.

cydippid larva (sī-dip′pid) (Gr. *kydippe,* mythological Athenian maiden). Free-swimming type of most ctenophores; superficially similar to the adult.

cystid (sis′tid) (Gr. *kystis,* bladder). In an ectoproct, the dead secreted outer parts plus the adherent underlying living layers.

cytochrome (sī′tō-krōm) (Gr. *kytos,* hollow vessel, + *chrōma,* color). One of several iron-containing pigments that serve as hydrogen carriers in aerobic respiration.

cytokinesis (sī′tō-kin-ē′sis) (Gr. *kytos,* hollow, + *kinesis,* movement). Changes that occur in the cytoplasm during cell division.

cytopharynx (Gr. *kytos,* hollow vessel, + *pharynx,* throat). Short tubular gullet in ciliate protozoans.

cytoplasm (sī′tō-plazm) (Gr. *kytos,* hollow, + *plasma,* mold). The living matter of the cell, excluding the nucleus.

cytosome (sī′tə-sōm) (Gr. *kytos,* hollow vessel, + *sōma,* body). The cell body inside the plasma membrane.

cytostome (sī′tə-stōm) (Gr. *kytos,* hollow vessel, + *stoma,* mouth). The cell mouth in ciliate protozoans.

deme (dēm) (Gr. populace). A local population of closely related animals.

demography (də-mäg′grə-fē) (Gr. *demos,* people, + *graphy*). The properties of the rate of growth and the age structure of populations.

deoxyribose (dē-ok′sē-rī′bōs) (*deoxy,* loss of oxygen, + *ribose,* pentose sugar). A 5-carbon sugar having 1 oxygen atom less than ribose; a component of deoxyribose nucleic acid (DNA).

dermal (Gr. *derma,* skin). Pertaining to the skin; cutaneous.

dermis. The inner, sensitive mesodermal layer of skin; corium.

desmosome (des′mə-sōm) (Gr. *desmos,* bond + *soma,* body). Buttonlike plaque serving as an intercellular connection.

determinate cleavage. The type of cleavage, usually spiral, in which the fate of the early blastomeres can be foretold; mosaic cleavage.

detritus (də-trī′tus) (L. that which is rubbed or worn away). Any fine particulate debris of organic or inorganic origin.

diapause (dī′ə-pawz) (Gr. *diapausis,* pause). A period of arrested development in the life cycle of insects and certain other animals; physiologic activity is very low and they are highly resistant to unfavorable external conditions.

diffusion (L. *diffusus,* dispersion). The movement of particles or molecules from area of high concentration of the particles or molecules to area of lower concentration.

digitigrade (dij′ə-də-grād′) (L. *digitus,* finger, toe, + *gradus,* step,

degree). Walking on the digits with the posterior part of the foot raised; compare plantigrade.

dihybrid (Gr. *dis,* twice, + L. *hibrida,* mixed offspring). A hybrid whose parents differ in two distinct characters; an offspring having different alleles at two different loci, for example, AaBb.

dimorphism (dī-mor′fizm) (Gr. *di,* two, + *morphē,* form). Existence within a species of two distinct forms according to color, sex, size, organ structure, etc. Occurrence of two kinds of zooids in a colonial organism.

dioecious (dī-ē′shəs) (Gr. *di-,* two, + *oikos,* house). Having male and female organs in separate individuals.

diphycercal (dif′i-ser′kəl) (Gr. *diphyēs,* two-fold, + *kerkos,* tail). A tail that tapers to a point, as in lungfishes; vertical column extends to tip without upturning.

dipleurula (dī-ploor′ū-lə) (Gr. *di-,* two, + *pleura,* rib, side, + L. *-ula,* small). A hypothetical, simple ancestral form of echinoderms; elongated and bilaterally symmetric, with three pairs of coelomic sacs; collective term used for various types of bilateral echinoderm larvae.

diploid (dip′loid) (Gr. *diploos,* double, + *eidos,* form). Having the somatic (double, or 2n) number of chromosomes or twice the number characteristic of a gamete of a given species.

disulfide. A compound of two atoms of sulfur combined with a chemical element or radical.

DNA. Deoxyribose nucleic acid.

dominance hierarchy. A social ranking, formed through agonistic behavior, in which individuals are associated with each other so that some have greater access to resources than do others.

dorsal (dor′səl) (L. *dorsum,* back). Toward the back, or upper surface, of an animal.

DPN. Abbreviation for diphosphopyridine nucleotide, a hydrogen carrier in respiration; now called NAD, nicotinamide adenine dinucleotide.

drive. A state of activity directed toward satisfying a specific need.

dyad (dī′əd) (Gr. *dyas,* two). One of the groups of two chromosomes formed by the division of a tetrad.

ecdysis (ek′də-sis) (Gr. *ekdysis,* a stripping, escape). Shedding of outer cuticular layer; molting, as in insects or crustaceans.

ecologic niche. The role of an organism in a community; its unique way of life and relationship to other biotic and abiotic factors.

ecology (Gr. *oikos,* house, + *logos,* discourse). That part of biology that deals with the relationship between organisms and their surroundings.

ecosystem (ek′ō-sis-təm) (eco[logy], from Gr. *oikos,* house, + system). An ecologic unit consisting of both the biotic communities and the nonliving (abiotic) environment, which interact to produce a stable system.

ecotone (ek′ō-tōn) (eco[logy], from Gr. *oikos,* home, + *tonos,* stress). The transition zone between two adjacent communities.

ectoderm (ek′tō-derm) (Gr. *ektos,* outside, + *derma,* skin). Outer layer of cells of an early embryo (gastrula stage); one of the germ layers, also sometimes used to include tissues derived from ectoderm.

ectoplasm (ec′tō-plazm) (Gr. *ektos,* outside, + *plasma,* form). The cortex of a cell or that part of cytoplasm just under the cell surface; contrasts with endoplasm.

ectothermic (ek′tō-therm′ic) (Gr. *ectos,* external, + *thermē,* heat).

bat / āpe / ärmadillo / herring / fēmale / finch / līce / crocodile / crōw / duck / ūnicorn / ə indicates unaccented vowel sound ''uh'' as in mammal, fishes, cardinal, heron, vulture / stress as in bi-ol′o-gy, bi′o-log′i-cal

Having a variable body temperature derived from heat acquired from the environment; contrasts with endothermic.

effector (L. *efficere,* bring to pass). An organ, tissue, or cell that becomes active in response to stimulation.

efferent (ef′ə-rənt) (L. *ex,* out, + *ferre,* to bear). Adj. Leading or conveying away from some organ, for example, nerve impulses conducted away from the brain or blood conveyed away from an organ; opposed to afferent.

emigrate (L. *emigrare,* to move out). To move *from* one area to another to take up residence. (Opposite of immigrate.)

emulsion (ə-məl′shən) (L. *emulsus,* milked out). A colloidal system in which both phases are liquids.

endemic (en-dem′ik) (Gr, *en-,* in, + *demos,* populace). Peculiar to a certain region or country; native to a restricted area; not introduced.

endergonic (en-dər-gän′ik) (Gr. *endon,* within, + *ergon,* work). Used in reference to a chemical reaction that requires energy.

endocrine (en′də-krən) (Gr. *endon,* within, + *krinein,* to separate). Refers to a gland that is without a duct and that releases its product directly into the blood or lymph.

endocytosis (en′dō-sī-tō′sis) (Gr. *endon,* within, + *kytos,* hollow vessel). The engulfment of matter by phagocytosis and of macromolecules by pinocytosis.

endoderm (en′də-dərm) (Gr. *endon,* within, + *derma,* skin). Innermost germ layer of an embryo, forming the primitive gut; also may refer to tissues derived from endoderm.

endometrium (en′də-mē′trē-əm) (Gr. *endon,* within, + mētra, womb). The mucous membrane lining the uterus.

endomixis (en′də-mik′səs) (Gr. *endon,* within, + *mixis,* a mixing). Reorganization of the nuclear material in the protozoans.

endoplasm (en′də-pla-zəm) (Gr. *endon,* within, + *plasma,* mold or form). That portion of cytoplasm that immediately surrounds the nucleus.

endoplasmic reticulum The cytoplasmic double membrane with ribosomes (rough) or without ribosomes (smooth).

endoskeleton (Gr. *endon,* within, + *skeletos,* hard. An internal skeleton or supporting framework, as opposed to exoskeleton.

endostyle (en′də-stīl) (Gr. *endon,* within, + *stylos,* a pillar). Ciliated groove(s) in the floor of the pharynx of tunicates, cephalochordates, and larval cyclostomes, used for accumulating and moving food particles to the stomach.

endothermic (en′də-therm′ic) (Gr. *endon,* within, + thermē, heat). Having a body temperature determined by heat derived from the animal's own oxidative metabolism; contrasts with ectothermic.

enterocoel (en′tər-ō-sēl′) (Gr. *enteron,* gut, + *koilos,* hollow). A type of coelom that is formed by the outpouching of a mesodermal sac from the endoderm of the primitive gut.

enterocoelic mesoderm formation. Embryonic formation of mesoderm by a pouchlike outfolding from the archenteron, which then expands and obliterates the blastocoel, thus forming a large cavity, the coelom, lined with mesoderm.

enterocoelomate (en′ter-ō-sēl′ō-māte) (Gr. *enteron,* gut, + *koilōma,* cavity, + Eng. *-ate,* state of). An animal having an enterocoel, such as an echinoderm or a vertebrate.

enteron (en′tə-rän) (Gr. *enteron,* intestine). The digestive cavity.

entozoic (en-tə-zō′ic) (Gr. *entos,* within, + zoon, animal). Living within another animal; internally parasitic (chiefly parasitic worms).

enzyme (en′zīm) (Gr. *enzymos,* leavened, from *en,* in, + *zyme,* leaven). A protein substance, produced by living cells, that is capable of speeding up specific chemical transformations, such as hydrolysis, oxidation, or reduction, but is unaltered itself in the process; a biologic catalyst.

ephyra (ef′ə-rə) (Gr. *Ephyra,* Greek city). Refers to castle-like appearance. Stage in development of Scyphozoa.

epidermis (ep′ə-dər′məs) (Gr. *epi,* upon, + *derma,* skin). The outer, nonvascular layer of skin of ectodermal origin; in invertebrates, a single layer of ectodermal epithelium.

epididymis (ep′ə-did′ə-məs) (Gr. *epi,* over, + *didymos,* testicle). That part of the sperm duct that is coiled and lying near the testis.

epigenesis (ep′ə-jen′ə-sis) (Gr. *epi,* over, + *genesis,* birth). The embryologic (and generally accepted) view that an embryo is a new creation that develops and differentiates step by step from an initial stage; the progressive production of new parts that were nonexistent as such in the original zygote.

epigenetics (ep′ə-jə-net′iks) (Gr. *epi,* over, + *genesis,* birth). Study of those mechanisms by which the genes produce phenotypic effects.

epithelium (ep′i-thē′lē-um) (Gr. *epi,* upon, + *thele,* nipple). A cellular tissue covering a free surface or lining a tube or cavity.

epitoke (ep′i-tōk) (Gr. *epitokos,* fruitful). Posterior part of a marine polychaete when swollen with developing gonads during the breeding season.

estrus (es′trəs) (Gr. *oistros,* a gadfly). Heat or rut, especially of the female during ovulation.

ethology (e-thäl′-ə-jē) (*ethos,* character, + *logy*). The study of animal behavior in natural environments.

eukaryotic, eucaryotic (ū′-ka-rē-ot′ik) (*eu* good, true, + Gr. *karyon,* nut, kernel). Containing a visible nucleus or nuclei.

euryhaline (yū′-rə-hā′-līn) (Gr. *eurys,* broad, + *hals,* salt). Able to tolerate wide ranges of saltwater concentrations.

eurytopic (yū-rə-täp′ik) (Gr. *eurys,* broad, + *topos,* place). Refers to an organism with a wide distribution range.

evagination (ē-vaj′ə-nā′shən) (L. *e,* out, + *vagina,* sheath). An outpocketing from a hollow structure.

evolution (L. *evolvere,* to unfold). Organic evolution is any genetic change in organisms, or more strictly a change in gene frequency, from generation to generation.

exergonic (ek′sər-gän′ik) (Gr. *exo,* outside of, + *ergon,* work). An energy-yielding reaction.

exocrine (ek′sə-krən) (Gr. *exo,* outside, + *krinein,* to separate). That type of gland that releases its secretion through a duct; opposed to endocrine.

exoskeleton (ek′sō-skel′ə-tən) (Gr. *exo,* without, + *skeletos,* hard). A hard supporting structure secreted by ectoderm or epidermis; external, as opposed to endoskeleton.

exteroceptor (ek′stər-ō-sep′tər) (L. *exter,* outward, + *capere,* to take). A sense organ excited by stimuli from the external world.

FAD. Abbreviation for flavine adenine dinucleotide, a hydrogen acceptor in the respiratory chain.

fatty acid. Any of a series of saturated acids having the general formula $C_nH_{2n}O_2$; occurs in natural fats of animals and plants.

fiber, fibril (L. *fibra,* thread). These two terms are often confused. Fiber is a fiberlike cell or a strand of protoplasmic material produced or secreted by a cell and lying outside the cell. Fibril is a strand of protoplasm produced by a cell and lying within the cell.

filter feeding. Any feeding process by which food (usually in fine particles) is filtered from water in which it is suspended.

fission (L. *fissio,* a splitting). Asexual reproduction by a division of the body into two or more parts.

flagellum (flə-jel′əm) (L. *a whip*). Whiplike organelle of locomotion.

flame bulb. Specialized hollow excretory structure of one or several small cells containing a tuft of cilia (the ''flame'') and situated at the end of a minute tubule; connected tubules ultimately open to the outside. (See **solenocyte, protonephridium.**)

fluke (O.E. *flōc,* flatfish). A member of class Trematoda.

food vacuole. A digestive organelle in the cell.

fovea (fō′vē-ə) (L. a small pit). A small pit or depression; especially the fovea centralis, a small rodless pit in the retina of some vertebrates, a point of acute vision.

free energy. The energy available for doing work in a chemical system.

gamete (ga′mēt, gə-mēt′) (Gr. *gamos,* marriage). A mature haploid sex cell, either male or female.

ganoid scales (ga′noid) (Gr. *ganos,* brightness). Thick, bony, rhombic scales of some primitive bony fishes; not overlapping.

gastrodermis (gas′tro-dər′mis) (Gr. *gastēr,* stomach, + *derma,* skin). Lining of the digestive cavity of coelenterates.

gastrovascular cavity (Gr. *gaster,* stomach, + L. *vasculum,* small vessel). Body cavity in certain lower invertebrates that functions in both digestion and circulation and has a single opening serving as both mouth and anus.

gastrula (gas′trə-lə) (Gr. *gastēr,* stomach, + L. *ula,* dim.). Embryonic stage, usually cap or sac shaped, with walls of two layers of cells surrounding a cavity (archenteron) with one opening (blastopore).

gel (jel) (from gelatin, from L. *gelare,* to freeze). That state of a colloidal system in which the solid particles form the continuous phase and the fluid medium the discontinuous phase.

gemmule (je′mūl) (L. *gemma,* bud, + *ula,* dim.). Asexual, cystlike reproductive unit in freshwater sponges; formed in summer or autumn and capable of wintering over.

gene (Gr. *genos,* descent). The part of a chromosome that is the hereditary determiner and is transmitted from one generation to another. It occupies a fixed chromosomal locus and can best be defined only in a physiologic or operational sense.

gene pool. A collection of all of the alleles of all of the genes in a population.

genetic drift. Change in gene frequencies by chance processes in the evolutionary process of animals. In small populations, one allele may drift to fixation, becoming the only representative of that gene locus.

genome (jē′nōm) (Gr. *genos,* offspring, + L. *-oma,* abstract group). The total number of genes in a haploid set of chromosomes.

genotype (jēn′ō-tīp) (Gr. *genos,* offspring, + *typos,* form). The genetic constitution, expressed and latent, of an organism; the total set of genes present in the cells of an organism; opposed to phenotype.

genus (jē-nus), pl. **genera** (L. *genus,* race). A taxonomic rank between family and species.

germ layer. In the animal embryo, one of three basic layers (ectoderm, endoderm, mesoderm) from which the various organs and tissues arise in the multicellular animal.

germ plasm. The germ cells of an organism, as distinct from the somatoplasm; the hereditary material (genes) of the germ cells.

gestation (je-stā′shən) (L. *gestare,* to bear). The period in which offspring are carried in the uterus.

glochidium (glō-kid′e-əm) (Gr. *glochis,* point, + *-idion,* diminutive). Bivalved larval stage of freshwater mussels.

glomerulus (glä-mer′u-ləs) (L. *glomus,* ball). A ball-shaped network of capillaries projecting into a renal corpuscle in kidney. Also, a small spongy mass of tissue in proboscis of hemichordates, presumed to have excretory function. Also, a concentration of nerve fibers situated in the olfactory bulb.

glycogen (glī′kə-jən) (Gr. *glykys,* sweet, + *genēs,* produced). A polysaccharide constituting the principal form in which carbohydrate is stored in plants and animals; animal starch.

glycolysis (glī-kol′i-sis) (Gr. *glykys,* sweet, + *lyein,* to loosen). Enzymatic breakdown of glucose (especially) or glycogen into phosphate derivatives with release of energy.

gnathobase (nāth′ə-bās′) (Gr. *gnathos,* jaw, + base). A median basic process on certain appendages in some arthropods, usually for biting or crushing food.

Golgi complex (gōl′jē) (after Golgi, Italian histologist). An organelle in cells that serves as a collecting and packaging center for secretory products.

gonad (gō′nad) (Gr. *gonos,* a primary sex gland). A sex gland (ovary in the female and testis in the male).

gonangium (gō-nan′jē-əm) (Gr. *gonos,* seed, + *angeion,* dim. of vessel). Reproductive zooid of hydroid colony (Cnidaria).

gonoduct (Gr. *gonos,* offspring, + duct). Duct leading from a gonad to the exterior.

gonophore (gän′ə-fōr) (Gr. *gonos,* offspring, + *phorein,* to bear). A reproducing zooid of a hydroid colony representing the medusa stage, but differing from a medusa in remaining attached.

gonopore (gän′ə-pōr) (Gr. *gonos,* offspring, seed, + *poros,* an opening). A genital pore found in many invertebrates.

gregarious (L. *grex,* herd). Living in groups or flocks.

guanine (gwä′nēn) (Sp. fr. Quechura *huanu,* dung). A white crystalline purine base, $C_5H_5N_5O$, occurring in various animal tissues and in guano and other animal excrements.

gynandromorph (ji-nan′drə-mawrf) (Gr. *gyn,* female, + *andr,* male, + *morph,* form). A bisexual form with the characteristics of both sexes; bisexual mosaic.

habitat (L. *habitare,* to dwell). The place where an organism normally lives or where individuals of a population live.

habituation. A kind of learning in which continued exposure to the same stimulus produces diminishing responses.

halter (hal′tər), pl. **halteres** (hal-ti′rēz) (Gr. *halter,* leap). In Diptera, small club-shaped structure on each side of the metathorax representing the hindwings; believed to be sense organs for balancing; also called balancer.

haploid (Gr. *haploos,* single). The reduced, or *n* number of chromosomes, typical of gametes, as opposed to the diploid, or 2 *n,* number found in somatic cells. In certain lower phyla, some mature animals have a haploid number of chromosomes.

hectocotylus (hek-tə-kät′ə-ləs) (Gr. *hekaton,* hundred, + *kotylē,*

bat / āpe / ärmadillo / herring / fēmale / fīnch / līce / crocodile / crōw / duck / ūnicorn / ə indicates unaccented vowel sound ''uh'' as in mammal, fishes, cardinal, heron, vulture / stress as in bi-ol′o-gy, bi′o-log′i-cal

cup). Specialized, and sometimes autonomous, arm that serves as a male copulatory organ in cephalopods.

hemal system (hē′məl) (Gr. *haima*, blood). System of small vessels in echinoderms, presumably functioning in circulation.

hemerythrin (hē′mə-rith′rin) (Gr. *haima*, blood, + *erythros*, red). A red iron-containing respiratory pigment found in the blood of some polychaetes, sipunculids, priapulids, and brachiopods.

hemoglobin (Gr. *haima*, blood, + L. *globulus*, globule). An iron-containing respiratory pigment occurring in vertebrate red blood cells and in blood plasma of many invertebrates; a compound of an iron porphyrin heme and a protein globin.

hepatic (hə-pat′ic) (Gr. *hēpatikos*, of the liver). Pertaining to the liver.

herbivore ((h)ərb′ə-vōr′) (L. *herba*, green crop, + *vorare*, to devour). Any organism subsisting on plants. Adj. **herbivorous.**

hermaphrodite (hə(r)-maf′rə-dīt) (Gr. *hermaphroditos*, containing both sexes; from greek mythology, *Hermaphroditos*, son of Hermes and Aphrodite). An organism with both male and female functional reproductive organs. **Hermaphroditism** may refer to an aberration in unisexual animals; **monoecism** implies that this is the normal condition for the species.

heterocercal (het′ər-o-sər′kəl) (Gr. *heteros*, other, + *kerkos*, tail). In some fishes, a tail with the upper lobe larger than the lower, and the end of the vertebral column somewhat upturned in the upper lobe, for example, as in sharks.

heterotroph (hət′ə-rō-träf) (Gr. *heteros*, another, + *trophos*, feeder). An organism that obtains both organic and inorganic raw materials from the environment in order to live; includes most animals and those plants that do not carry on photosynthesis.

heterozygote (het′ə-rō-zī′gōt) (Gr. *heteros*, another, + *zygōtos*, yoked). An organism in which the pair of alleles for a trait is composed of different genes (usually dominant and recessive); derived from a zygote formed by the union of gametes of dissimilar genetic constitution.

hibernation (L. *hibernus*, wintery). The act of passing the winter in a resting state.

histone (hi′stōn) (Gr. *histos*, tissue). Any of several simple proteins found in cell nuclei and complexed at one time or another with DNA. Histones yield a high proportion of basic amino acids upon hydrolysis

holoblastic cleavage (Gr. *holos*, whole, + *blastos*, germ). Complete and approximately equal division of cells in early embryo. Found in mammals, *Amphioxus,* and many aquatic invertebrates that have eggs with a small amount of yolk.

holometabolism (hō′lō-mə-ta′bə-liz-əm) (Gr. *holo*, complete, + *metabolē*, change). Complete metamorphosis during development.

holophytic nutrition (hōl′ō-fit′ik) (Gr. *holo*, whole, + *phyt*, plant). Occurs in green plants and certain protozoans and involves synthesis of carbohydrates from carbon dioxide and water in the presence of light, chlorophyll and certain enzymes.

holozoic nutrition (Gr. *holos*, whole, + *zoikos*, of animals). That type of nutrition that involves ingestion of liquid or solid organic food particles.

homeostasis (hō′mē-ō-stā′sis) (G. *homeo*, similar, + *stasis*, state or standing). Maintenance of an internal steady state by means of self-regulation.

homeothermic (hō-mē-ō-thər′mik) (Gr. *homeo*, alike, + *thermē*, heat). Having a nearly uniform body temperature, regulated independently of the environmental temperature; "warm-blooded."

home range. The area over which an animal ranges in its activities. Unlike territories, home ranges are not defended.

hominid (häm′ə-nid) (L. *homo, homonis,* man). A member of the family Hominidae, now represented by one living species, *Homo sapiens.*

homocercal (hō′mə-ser′kal) (Gr. *homos,* the same, common, + *kerkos,* tail). A tail with the upper and lower lobes symmetric and the vertebral column ending near the middle of the base, as in most teleost fishes.

homograft. See **allograft.**

homology (hō-mäl′ə-jē) (Gr. *homologos,* agreeing). Similarity of parts or organs of different organisms caused by similar embryonic origin and evolutionary development from a corresponding part in some remote ancestor. Also, correspondence in structure of different parts of the same individual. May also refer to a matching pair of chromosomes. Adj. **homologous.**

homozygote (hō-mə-zī′gōt) (Gr. *homos,* same, + *zygotos,* yoked). An organism in which the pair of alleles for a trait is composed of the same genes (either dominant or recessive but not both). Adj. **homozygous.**

honeydew. A sweet secretion produced by aphids, leafhoppers, and phyllas that is the principal food of some ants; also eaten by bees and wasps.

humoral (hū′mər-əl) (L. *humor,* a fluid). Pertaining to an endocrine secretion.

hyaline (hī′ə-lən) (Gr. *hyalos,* glass). Adj., glassy, translucent. Noun, a clear, glassy, structureless material occurring, for example, in cartilage, vitreous body, mucin, and glycogen.

hydranth (hī′dranth) (Gr. *hydōr,* water, + *anthos,* flower). Nutritive zooid of hydroid colony.

hydroid. The polyp form of coelenterate (cnidarian) as distinguished from the medusa form. Any coelenterate of the class Hydrozoa.

hydrolysis (Gr. *hydōr,* water, + *lysis,* a loosening). The decomposition of a chemical compound by the addition of water; the splitting of a molecule into its groupings so that the split products acquire hydrogen and hydroxyl groups.

hydrostatic skeleton. A mass of fluid or plastic parenchyma enclosed within a muscular wall to provide the support necessary for antagonistic muscle action; for example, parenchyma in acoelomates and perivisceral fluids in pseudocoelomates serve as hydrostatic skeletons.

hydroxyl (hydrogen + oxygen, + yl). Containing an OH⁻ group, a negatively charged ion formed by alkalies in water.

hypertonic (Gr. *hyper,* over, + *tonos,* tension). Refers to a solution whose osmotic pressure is greater than that of another solution with which it is compared; contains a greater concentration of dissolved particles and gains water through a semipermeable membrane from a solution containing particles. Hyperosmotic. Opposite of hypotonic.

hypertrophy (hī-pər′trə-fē) (Gr. *hyper,* over, + *trophē,* nourishment). Abnormal increase in size of a part or organ.

hypodermis (hī′pə-dər′mis) (Gr. *hypo,* under, + L. *dermis,* skin). The cellular layer lying beneath and secreting the cuticle of annelids, arthropods, and certain other invertebrates.

hypothalamus (hī-pō-thal′ə-mis) (Gr. *hypo,* under, + *thalamos,* inner chamber). A ventral part of the forebrain beneath the thalamus; one of the centers of the autonomic nervous system.

hypotonic (Gr. *hypo*, under, + *tonos*, tension). Refers to a solution whose osmotic pressure is less than that of another solution with which it is compared or taken as standard; contains a lesser concentration of dissolved particles and loses water during osmosis. Hypoosmotic. Contrast with hypertonic

imago (ə mā′gō). The adult and sexually mature insect.

indeterminate cleavage. A type of early cleavage in which the fate of the blastomeres is not predetermined as to tissues or organs, for example, in echinoderms and vertebrates.

inquiline (in′kwə-līn) (L. *inquilinus*, lodger). A relation in which a socially parasitic species lives in the abode of its host insect.

instar (inz′tär) (L. *instar*, form). Stage in the life of an insect or other arthropod between molts.

instinct (L. *instinctus*, impelled). Genetically programmed behavior; unlearned, stereotyped behavior.

integument (ən-teg′ū-mənt) (L. *integumentum*, covering). An external covering or enveloping layer.

interstitial (in-tər-sti′shəl). (L. *inter*, among, + *sistere*, to stand). Situated in the interstices or spaces between body cells or organs.

introvert (L. *intro*, inward, + *vertere*, to turn). The anterior narrow portion that can be withdrawn (introverted) into the trunk of a sipunculid worm.

invagination (in-vaj′ə-nā′shən) (L. *in*, in, + *vagina*, sheath). An infolding of a layer of tissue to form a saclike structure.

irritability (L. *irritare*, to provoke). A general property of all organisms involving the ability to respond to stimuli or changes in the environment.

isolecithal (ī′sə-les′ə-thəl) (Gr. *isos*, equal, + *lekithos*, yolk, + -*al*). Pertaining to a zygote (or ovum) with yolk evenly distributed. Homolecithal.

isotonic (Gr. *isotonos*, *isos*, equal, + *tonikos*, tension). Said of solutions having the same or equal osmotic pressure; isosmotic.

isotope (Gr. *isos*, equal, + *topos*, place). One of several different forms (species) of a chemical element, differing from each other in atomic mass but not in atomic number.

keratin (ker′ə-tən) (Gr. *kera*, horn, + -*in*, suffix of proteins). A scleroprotein found in epidermal tissues and modified into hard structures such as horns, hair, and nails.

kinetosome (kin-et′ə-sōm) (Gr. *kinētos*, moving, + *sōma*, body). The self-duplicating granule at the base of the flagellum or cilium; similar to centriole.

kinin (kī′nin) (Gr. *kinein*, to move, + -*in*, suffix of hormones). A type of local hormone that is released near its site of origin; also called parahormone or tissue hormone.

labium (lā′bē-əm) (L. a lip). The lower lip of the insect formed by fusion of the second pair of maxillae.

labrum (lā′brəm) (L. a lip). The upper lip of insects and crustaceans situated above or in front of the mandibles; also refers to the outer lip of a gastropod shell.

labyrinthodont (lab′ə-rin′thə-dänt) (Gr. *labyrinthos*, labyrinth, + *odous*, *odontos*, tooth). A group of fossil stem amphibians from which most amphibians later arose. They date from the late Paleozoic.

lacteal (lak′tē-əl) (L. *lacteus*, of milk). Refers to one of the lymph vessels in the villus of the intestine. Adj., relating to milk.

lacuna (lə-kū′nə), pl. **lacunae** (L. pit, cavity). A sinus; a space between cells; a cavity in cartilage or bone.

lagena (lə-jē′nə) (L. large flask). Portion of the primitive ear in which sound is translated into nerve impulses; evolutionary beginning of cochlea.

lamella (lə-mel′ə) (L. dim. of *lamina*, plate.). One of the two plates forming a gill in a bivalve mollusc. One of the thin layers of bone laid concentrically around a Haversian canal. Any thin, platelike structure.

larva (lär′və), pl. **larvae** (L. a ghost). An immature stage that is quite different from the adult.

lek (lek) (Sw. play, game). An area where animals assemble for communal courtship display and mating.

lemniscus (lem-nis′kəs) (L. ribbon). One of a pair of internal projections of the epidermis from the neck region of Acanthocephala, which functions in fluid control in the protrusion and invagination of the proboscis.

leukocyte (lū′kə-sīt′) (Gr. *leukos*, white, + *kytos*, hollow vessel). A common type of white blood cell with beaded nucleus.

lipase (lī′pās) (Gr. *lipos*, fat, + -*ase*, enzyme suffix). An enzyme that accelerates the hydrolysis or synthesis of fats.

lipid, lipoid (li′pid) (Gr. *lipos*, fat). Certain fatlike substances that often contain other groups such as phosphoric acid; lipids combine with proteins and carbohydrates to form principal structured components of cells.

littoral (lit′ə-rəl) (L. *litoralis*, seashore). Adj., Pertaining to the shore. Noun, that portion of the sea floor from the shore to the continental shelf; in lakes, the shallow part from the shore to the lakeward limit of aquatic plants.

locus (lō′kəs), pl. **loci** (lō-cī) (L. place). Position of a gene in a chromosome.

lophophore (lōf′ə-fōr) (Gr. *lophos*, crest, + *phoros*, bearing). Tentacle-bearing ridge or arm that is an extension of the coelomic cavity in lophophorate animals (ectoprocts, brachiopods, and phoronids).

lumen (lū′mən) (L. light). The cavity of a tube or organ.

lysosome (lī′sə-sōm) (Gr. *lysis*, loosing, + *soma*, body). Intracellular organelle consisting of a membrane enclosing several digestive enzymes that are released when the lysosome ruptures.

macronucleus (ma′krō-nū-klē-əs) (Gr. *makros*, long, large, + *nucleus*, kernel). The larger of the two kinds of nuclei in ciliate protozoa; controls all cell functions except reproduction.

madreporite (ma′drə-pōr′īt) (Fr., *madrépore*, reef-building coral, + -*ite*, suffix for some body parts). Sievelike structure that is the intake for the water-vascular system of echinoderms.

malacostracan (mal′ə-käs′trə-kən) (Gr. *malako*, soft, + *ostracon*, shell). Any member of the crustacean subclass Malacostraca, which includes both aquatic and terrestrial forms of crabs, lobsters, shrimps, pill bugs, sand fleas, and others.

Malpighian tubules (mal-pig′ē-ən) (Marcello Malpighi, Italian anatomist, 1628-1694). Blind tubules opening into the hindgut of nearly all insects and some myriapods and arachnids, and functioning primarily as excretory organs.

mantle. Soft extension of the body wall in certain invertebrates, for example, brachiopods and molluscs, in which it usually secretes a shell; thin body wall of tunicates.

---

bat / āpe / ärmadillo / herring / fēmale / finch / līce / crocodile / crōw / duck / ūnicorn / ə indicates unaccented vowel sound ''uh'' as in mammal, fishes, cardinal, heron, vulture / stress as in bi-ol′o-gy, bi′o-log′i-cal

marsupial (mär-sū′pē-əl) (Gr. *marsypion*, little pouch). One of the pouched mammals of the subclass Metatheria.

matrix (mā′triks) (L. *mater*, mother). The intercellular substance of a tissue, or that part of a tissue into which an organ or process is set.

maturation (L. *maturus*, ripe). The process of ripening; the final stages in the preparation of gametes for fertilization.

maxilla (mak-sil′ə) (L. diminutive of *mala*, jaw). One of the upper jawbones in vertebrates; one of the head appendages in arthropods.

maxilliped (mak-sil′ə-ped) (L. *maxilla*, jaw, + *-ped-*, foot). One of the three pairs of head appendages located just posterior to the maxilla in crustaceans.

medulla (mə-dul′ə) (L. marrow). The inner portion of an organ in contrast to the cortex or outer portion; hindbrain.

medusa (mə-dū-sə) (Greek mythology, female monster with snake-entwined hair). A jellyfish, or the free-swimming stage in the life cycle of coelenterates.

meiosis (mī-ō′səs) (Gr. from *meioun*, to make small). The nuclear changes that occur in the last two divisions in the formation of the mature egg or sperm, by means of which the chromosomes are reduced from the diploid to the haploid number.

melanin (mel′ə-nin) (Gr. *melas*, black). Black or dark brown pigment found in plant or animal structures.

menstruation (men′stroo-ā′shən) (L. *menstrua*, the menses, from *mensis*, month). The discharge of blood and uterine tissue from the vagina at the end of a menstrual cycle.

meroblastic (mer-ə-blas′tik) (Gr. *meros*, part, + *blastos*, germ). Pertaining to cleavage occurring in zygotes having a large amount of yolk at the vegetal pole; cleavage restricted to a small area on the surface of the egg.

mesenchyme (me′zn-kīm) (Gr. *mesos*, middle, + enchyma, infusion). Embryonic connective tissue; irregular or amebocytic cells often embedded in gelatinous matrix.

mesoderm (me′zə-dərm) (Gr. *mesos*, middle, + *derma*, skin). The third germ layer, formed in the gastrula between the ectoderm and endoderm; gives rise to connective tissues, muscle, urogenital and vascular systems, and the peritoneum.

mesoglea (mez′ō-glē′ə) (Gr. *mesos*, middle, + *gloios*, glutinous substance). The layer of jellylike or cement material between the epidermis and gastrodermis in coelenterates and ctenophores; also may refer to jellylike matrix between epithelial layers in sponges.

mesonephros (me-zō-nef′rōs) (Gr. *mesos*, middle, + *nephros*, kidney). The middle of three pairs of embryonic renal organs in vertebrates. Functional kidney of fishes and amphibians; its collecting duct is a Wolffian duct. Adj. **mesonephric.**

messenger RNA (m-RNA). A form of ribonucleic acid that carries genetic information from the gene to the ribosome, where it determines the order of amino acids as a polypeptide is formed.

metabolism (Gr. *metabolē*, change). A group of processes that includes nutrition, production of energy (respiration), and synthesis of protoplasm; the sum of the constructive (anabolism) and destructive (catabolism) processes.

metacercaria (mə′tə-sər-ka′rē-ə) (Gr. *meta*, after, + *kerkos*, tail, + L. *aria*, connected with). Fluke larva (cercaria) that has lost its tail and become encysted in an intermediate aquatic host.

metamere (met′ə-mər) (Gr. *meta*, after, + *meros*, part). A repeated unit of structure; a somite, or segment.

metamerism (mə-ta′mə-ri′zəm) (Gr. *meta*, after, + *meros*, part). Condition of being made up of serially repeated parts (metameres); serial segmentation.

metamorphosis (Gr. *meta*, after, + *morphē*, form, + *-osis*, state of). Sharp change in form during postembryonic development; for example, tadpole to frog or larval insect to adult.

metanephridium (me′tə-nə-fri′di-əm) (Gr. *meta*, after, + *nephros*, kidney). A type of tubular nephridium with the inner open end draining the coelom and the outer open end discharging to the exterior.

metanephros (me′tə-ne′fräs) (Gr. *meta*, between, after, + *nephros*, kidney). Embryonic renal organs of vertebrates arising behind the mesonephros; the functional kidney of reptiles, birds, and mammals. It is drained from a ureter.

micron (μ) (mī′krän) (Gr. neuter of *mikros*, small). One one-thousandth of a millimeter; about 1/25,000 of an inch. Now largely replace by micrometer (μm).

micronucleus. A small nucleus found in ciliate protozoa; controls the reproductive functions of these organisms.

microtubule (Gr. *mikros*, small, + L. *tubule*, pipe). A long, tubular cytoskeletal element with an outside diameter of 200 to 270 Å. Microtubules influence cell shape and play important roles during cell division.

microvillus (Gr. *mikros*, small, + L. *villus*, shaggy hair). Narrow, cylindric cytoplasmic projection from epithelial cells; microvilli form the brush border of several types of epithelial cells.

mictic (mik′tik) (Gr. *miktos*, mixed or blended). Pertaining to haploid egg of rotifers or the females that lay such eggs.

miracidium (mīr′ə-sid′ē-əm) (Gr. *meirakidion*, youthful person). A minute ciliated larval stage in the life of flukes.

mitochondrian (mīd′ə-kän′drē-ən) (Gr. *mitos*, a thread, + *chondrion*, diminutive of *chondros*, corn, grain). An organelle in the cell in which aerobic metabolism takes place.

mitosis (mī-tō′səs) (Gr. *mitos*, thread, + *-osis*, state of). Cell division in which there is an equal qualitative and quantitative division of the chromosomal material between the two resulting nuclei; ordinary cell division (indirect).

monoecious (mə-nē′shəs) (Gr. *monos*, single, + *oikos*, house). Having both male and female gonads in the same organism; hermaphroditic.

monohybrid (Gr. *monos*, single, + L. *hybrida*, mongrel). A hybrid offspring of parents different in one specified character.

monomer (mä′nə-mər) (Gr. *monos*, single, + *meros*, part). A molecule of simple structure, but capable of linking with others to form polymers.

monophyletic (mä′nə-phī-le′tik) (Gr. *mono*, single, + *phyletikos*, pertaining to a phylum). Referring to a taxon whose units all evolved from a single parent stock. Compare with polyphyletic, diphyletic.

monosaccharide (mä′nə-sa′kə-rīd) (Gr. *monos*, one, + *sakcharon*, sugar, from Sanskrit *sarkarā*, gravel, sugar). A simple sugar that cannot be decomposed into smaller sugar molecules; the most common are pentoses (such as fructose) and hexoses (such as glucose).

morphogenesis (mor′fə-je′nə-səs) (Gr. *morphē*, form, + *genesis*, origin). Development of the architectural features of organisms; formation and differentiation of tissues and organs.

morphology (Gr. *morphē*, form, + L. *logia*, study, from Gr. *logos*, word). The science of structure. Includes cytology, or the study of cell structure; histology, or the study of tissue structure; and anatomy, or the study of gross structure.

morula (mär'u-lə) (L. *morum*, mulberry, + *ula*, diminutive). Solid group of cells in early stage of segmentation.

mosaic cleavage. Type characterised by independent differentiation of each part of the embryo; determinative cleavage.

Müller's larva. Free-swimming ciliated larva that resembles a modified ctenophore; characteristic of certain marine polyclad turbellarians.

mutation (myū-tā'shən) (L. *mutare*, to change). A stable and abrupt change of a gene; the heritable modification of a character.

mutualism (myū'chə-wə-li'zəm) (L. *mutuus*, lent, borrowed, reciprocal). A type of symbiosis in which two different species derive benefit from their association.

myofibril (Gr. *mys*, muscle, mouse, + L. dim. of *fibra*, fiber). A contractile filament within muscle or muscle fiber.

myoneme (mī'ə-nēm') (Gr. *mys*, mouse, + *nēma*, thread). Long contractile fibril in certain protozoans.

myosin (mī'ə-sin) (Gr. *mys,* muscle, mouse). A large protein of contractile tissue that forms the thick myofilaments of striated muscle. During contraction it combines with actin to form actomyosin.

myotome (mī'ə-tōm') (Gr. *mys,* muscle + *tomos,* cutting). A voluntary muscle segment in cephalochordates and vertebrates; that part of a somite destined to form muscles; the muscle group innervated by a single spinal nerve.

nacre (nā'kər) (F. mother-of-pearl). Innermost lustrous layer of mollusc shell, secreted by mantle epithelium. Adj. **nacreous.**

NAD. Abbreviation of nicotinamide adenine dinucleotide; see **DPN.**

naiad (nā'əd) (Gr. *naias,* a water nymph). An aquatic, gill-breathing nymph.

nares (na'rēz), sing. **naris** (L. nostrils). Openings into the nasal cavity, both internally and externally, in the head of a vertebrate.

natural selection. A nonrandom reproduction of genotypes that results in the survival of those best adapted to their environment and elimination of those less well adapted; leads to evolutionary change.

nauplius (naw'plē-əs) (L. a kind of shellfish). A free-swimming microscopic larval stage of certain crustaceans, with three pairs of appendages (antennules, antennae and mandibles) and median eye. Characteristic of ostracods, decapods, barnacles, and some others.

nekton (nek'tən) (Gr. neuter of *nēktos,* swimming). Term for actively swimming organisms, essentially independent of wave and current action. Compare with plankton.

nematocyst (ne-mad'ə-sist') (Gr. *nēma,* thread, + *kystis,* bladder). Stinging organoid of coelenterates.

neoteny (nē'ə-tē'nē, nē-ot'ə-nē) (Gr. *neos,* new, + *teinein,* to extend). The attainment of sexual maturity in the larval condition. Also the retention of larval characters into adulthood.

nephridium (nə-frid'ē-əm) (Gr. *nephridios,* of the kidney). One of the segmentally arranged, paired excretory tubules of many

invertebrates, notably the annelids. In a broad sense, any tubule specialized for excretion and/or osmoregulation; with an external opening and with or without an internal opening.

nephron (ne'frän) (Gr. *nephros,* kidney). Functional unit of kidney structure of reptiles, birds and mammals, consisting of a Bowman's capsule body, its glomerulus, and the attached uriniferous tubule.

neurosecretory cell (nu'rō-sə-krēd'ə-rē). Any cell (neuron) of the nervous system that produces a hormone.

nitrogen fixation (Gr. *nitron,* soda, + *gen,* producing). Oxidation of ammonia to nitrites and of nitrites to nitrates, as by action of bacteria.

notochord (nōd'ə-kord') (Gr. *nōtos,* back, + *chorda,* cord). An elongated cellular cord, enclosed in a sheath, which forms the primitive axial skeleton of chordate embryos and adult cephalochordates.

nucleic acid (nu-klē'ik) (L. *nucleus,* kernel). One of a class of molecules composed of joined nucleotides; chief types are deoxyribonucleic acid (DNA), found in cell nuclei (chromosomes), and ribonucleic acid (RNA), found both in cell nuclei (chromosomes and nucleoli) and in cytoplasm ribosomes.

nucleolus (nu-klē'ə-ləs) (dim. of nucleus). A deeply staining body within the nucleus of a cell and containing RNA.

nucleoplasm (nu'klē-ə-plazm') (L. *nucleus,* kernel, + Gr. *plasma,* mold). Protoplasm of nucleus, as distinguished from cytoplasm.

nucleoprotein. A molecule composed of nucleic acid and protein; occurs in the nucleus and cytoplasm of all cells.

nucleotide (nu'klē-ə-tīd). A molecule consisting of phosphate, 5-carbon sugar (ribose or deoxyribose), and a purine or a pyrimidine; the purines are adenine and guanine, and the pyrimidines are cytosine, thymine, and uracil.

nuptial flight (nəp'shəl). The mating flight of insects, especially that of the queen with male or males.

nymph (L. *nympha,* nymph, bride). An immature stage (following hatching) of a hemimetabolic insect that lacks a pupal stage.

ocellus (ō-sel'əs) (L. dim. of *oculus,* eye). A simple eye or eyespot in many types of invertebrates.

ommatidium (ä'mə-tid'ē-əm) (Gr. *omma,* eye, + *idium,* small). One of the optical units of the compound eye of arthropods and molluscs.

ontogeny (än-tä'jə-nē) (Gr. *ontos,* being, + *-geneia,* act of being born, from *genēs,* born). The course of development of an individual from egg to senescence.

ooecium (ō-ēs'ē-əm) (Gr. *ōion,* egg, + *oikos,* house, + L. *-ium,* from Ger. -ion, diminutive). Brood pouch; compartment for developing embryos in ectoproct.

oogonium (ō'ə-gōn'ē-əm) (Gr. *ōion,* egg, + *gonos,* offspring). A cell that, by continued division, gives rise to oocytes; an ovum in a primary follicle immediately before the beginning of maturation.

operculum (ō-per'kyə-ləm) (L. cover). The gill cover in bony fishes; horny plate in some snails.

operon (äp'ə-rän). A genetic unit consisting of a cluster of genes that are under the control of an operator and a repressor.

ophthalmic (äf-thal'mik) (Gr. *ophthalmos,* an eye). Pertaining to the eye.

organelle (Gr. *organon,* tool, organ, + L. *ella,* diminutive). Specialized part of a cell; literally, a small organ.

bat / āpe / ärmadillo / herring / fēmale / finch / līce / crocodile / crōw / duck / ūnicorn / ə indicates unaccented vowel sound "uh" as in mammal, fishes, cardinal, heron, vulture / stress as in bi-ol'o-gy, bi'o-log'i-cal

osmoregulation. Maintenance of proper internal salt and water concentrations in a cell or in the body of a living organism; active regulation of internal osmotic pressure.

osmosis (oz-mō´sis) (Gr. *ōsmos*, act of pushing, impulse). The flow of solvent (usually water) through a semipermeable membrane.

osphradium (äs-frā´dē-əm) (Gr. *osphradion*, small bouquet, dim. of *osphra*, smell). A sense organ in aquatic snails and bivalves that tests incoming water.

ossicles (L. *ossiculum*, small bone). Small separate pieces of echinoderm endoskeleton; tiny bones of middle ear of vertebrates.

ostium (L. door). Opening.

otolith (ōd´əl-ith´) (Gr. *ous, otos*, ear, + *lithos*, stone). Calcareous concretions in the membranous labyrinth of the inner ear of lower vertebrates, or in the auditory organ of certain invertebrates.

oviparity (ō´və-pa´rəd-ē) (L. *ovum*, egg, + *parere*, to bring forth). Reproduction in which eggs are released by the female; development of offspring occurs outside the maternal body. Adj. **oviparous** (ō-vip´ə-rəs).

ovipositor (ō´və-päz´əd-ər) (L. *ovum*, egg, + *parere*, to bring forth). In many female insects a structure at the posterior end of the abdomen for laying eggs.

ovoviviparity (ō´vo-vī-və-par´əd-ē) (L. *ovum*, egg, + *vivere*, to live, + *parere*, to bring forth). Reproduction in which eggs develop within the maternal body without additional nourishment from the parent and hatch within the parent or immediately after laying. Adj., **ovoviviparous** (ō´vo-vī-vip´ə-rəs).

oxidation (äk´sə-dā´shən) (Fr. *oxider*, to oxidize, fr. Gr. *oxys*, sharp, + *-ation*). The loss of an electron by an atom or molecule; addition of oxygen chemically to a substance.

oxidative phosphorylation (äk´sə-dād´iv fäs´fər-i-lā´shən). The conversion of inorganic phosphate to energy-rich phosphate of ATP.

paedogenesis (pē-dō-jen´ə-sis) (Gr. pais, child, + *genēs*, born). Reproduction by immature or larval animals.

pair bond. An affiliation between an adult male and an adult female for reproduction. Characteristic of monogamous species.

papilla (pə-pil´ə) (L. nipple). A small nipplelike projection. A vascular process that nourishes the root of a hair, feather, or developing tooth.

parabiosis (pa´rə-bī-ō´sis) (Gr. *para*, beside, + *biosis*, mode of life). The fusion of two individuals, resulting in mutual physiologic intimacy.

parapodium (pa´rə-pō´dē-əm) (Gr. *para*, subsidiary, + *pous, podos*, foot). One of the paired flat, lateral processes on each side of most segments in polychaete annelids; variously modified for locomotion, respiration or feeding.

parasitism (par´ə-sīd´izəm) (Gk. *parasitos*, from *para*, beside, + *sitos*, food). The condition of an organism living in or on another organism at whose expense the parasite is maintained; destructive symbiosis.

parasympathetic (par´ə-sim-pə-thed´ik) (Gr. *para*, beside, + *sympathes*, sympathetic, from *syn*, with, + *pathos*, feeling). One of the subdivisions of the autonomic nervous system, whose fibers originate in the brain and in anterior and posterior parts of the spinal cord.

parenchyma (pə-ren´kə-mə) (Gr. anything poured in beside). In lower animals, a spongy mass of vacuolated mesenchyme cells filling spaces between viscera, muscles, or epithelia; the specialized tissue of an organ as distinguished from the supporting connective tissue.

parthenogenesis (pär´thə-nō-gen´ə-sis) (Gr. *parthenos*, virgin, + L., fr. Gr. *genesis*, origin). Unisexual reproduction involving the production of young by females not fertilized by males; common in rotifers, cladocerans, aphids, bees, ants, and wasps. A parthenogenetic egg may be diploid or haploid.

pathogenic (path´ə-jen´ik) (Gr. *pathos*, disease, + *gennan*, to produce). Producing or capable of producing disease.

peck order. A hierarchy of social privilege in a flock of birds.

pecten (L. comb). A pigmented, vascular, and comblike process that projects into the vitreous humor from the retina at point of entrance of the optic nerve in the eyes of all birds and many reptiles.

pectoral (pek´tə-rəl) (L. *pectoralis*, from *pectus*, the breast). Of or pertaining to the breast or chest; to the pectoral girdle; or to a pair of horny shields of the plastron of certain turtles.

pedicel (ped´ə-sel) (L. *pediculus*, little foot). A small or short stalk or stem. In insects, the second segment of an antenna or the waist of an ant.

pedicellaria (ped´ə-sə-lar´ē-ə) (L. *pediculus*, little foot, + *-aria*, like or connected with). One of many minute pincerlike organs on surface of certain echinoderms.

pedipalps (ped´ə-palps´) (L. *pes, pedis*, foot, + *palpus*, stroking, caress). Second pair of appendages of arachnids.

pedogenesis. See **paedogenesis.**

peduncle (pē´dən-kəl) (L. *pedunculus* dim. of *pes*, foot). A stalk; a band of white matter joining different parts of the brain.

pelagic (pə-laj´ik) (Gr. *pelagos*, the open sea). Pertaining to the open ocean.

pellicle (pel´ə-kəl) (L. *pellicula*, dim. of *pellis*, skin). Thin, translucent, secreted envelope covering many protozoans.

pentadactyl (pen-tə-dak´təl) (Gr. *pente*, five, + *daktylos*, finger). With five digits, or five fingerlike parts, to the hand or foot.

peptidase (pep´tə-dās) (Gr. *peptein*, to digest, + *-ase*, enzyme suffix). An enzyme that breaks down simple peptides, releasing amino acids.

peptide bond. A bond that binds amino acids together into a polypeptide chain, formed by removing an OH from the carboxy group of one amino acid and an H from the amino group of another to form an amide group −CO−NH−.

periostracum (pe-rē-äs´trə-kəm) (Gr. *peri*, around, + *ostrakon*, shell). Outer horny layer of mollusc shell.

periproct (per´ə-präkt) (Gr. *peri*, around, + *prōktos*, anus). Region of aboral plates around the anus of echinoids.

perisarc (per´ə-särk) (Gr. *peri*, around, + *sarx*, flesh). Sheath covering the stalk and branches of a hydroid.

perissodactyl (pə-ris´ə-dak´təl) (Gr. *perissos*, odd, + *dactylos*, finger, toe). Pertaining to an order of ungulate mammals with an odd number of digits.

peristalsis (per´ə-stal´səs) (Gr. *peristaltikos*, compressing around). The series of alternate relaxations and contractions that serve to force food through the alimentary canal.

peristomium (per´ə-stō´mē-əm) (Gr. *peri*, around, + *stoma*, mouth). Foremost true segment of annelid; it bears the mouth.

peritoneum (per´ə-tə-nē´əm) (Gr. *peritonaios*, stretched across). The membrane that lines the coelom and is reflected over the coelomic viscera.

605

pH (*p*otential of *h*ydrogen). A symbol of the relative concentration of hydrogen ions in a solution; pH values are from 0 to 14, and the lower the value, the more acid or hydrogen ions in the solution. Equal to the negative logarithm of the hydrogen ion concentration.

phagocyte (fag′ə-sīt) (Gr. *phagein*, to eat, + *kytos*, hollow vessel). Any cell that engulfs and devours microorganisms or other foreign particles.

phagocytosis (fag′ə-sī-tō′səs) (Gr. *phagein*, to eat, + *kytos*, hollow). The engulfment of a foreign particle by a phagocyte.

pharynx (far′inks), pl. **pharynges** (Gr). The part of the digestive tract between the mouth cavity and the esophagus that, in vertebrates, is common to both digestive and respiratory tracts. In cephalochordates the gill slits open from it.

phasmid (faz′mid) (Gr. *phasma*, apparition, phantom, + -id). One of a pair of glands or sensory structures found in the posterior end of certain nematodes.

phenotype (fē′nə-tīp′) (Gr. *phainein*, to show). The visible characters of an organism; opposed to genotype of the hereditary constitution.

pheromone (fer′ə-mōn) (Gk. *phorein*, to carry). Chemical scent released by one organism that influences the behavior of another organism; ectohormone.

phosphagen (fäs′fə-jən) (phosphate + gen). A term for creatine-phosphate and arginine-phosphate, which store and may be sources of high-energy phosphate bonds.

phosphatide (fäs′fə-tīd′) (phosphate + -ide). A lipid with phosphorus, such as lecithin. A complex phosphoric ester lipid, such a lecithin, found in all cells. Phospholipid.

phosphorylation (fäs′fə-rə-lā′shən). The addition of a phosphate group, such as H_2PO_3, to a compound.

photosynthesis (fōd′ō-sin′thə-sis) (Gr. *phōs*, light, + *synthesis*, action or putting together). The synthesis of carbohydrates from carbon dioxide and water in chlorophyll-containing cells exposed to light.

phototropism (fō-tä′trō-piz′m) (Gr. *phōs*, light, + *trope*, turn). A tropism in which light is the orienting stimulus. An involuntary tendency for an organism to turn toward (positive) or away from (negative) light.

phylogeny (fī-läj′ə-nē) (Gr. *phylon*, tribe, race, + *geneia*, origin). The origin and development of any taxon, or the evolutionary history of its development.

phylum (fī′ləm), pl. **phyla** (NL. Gr. *phylon*, race, tribe). A chief category of taxonomic classification into which are grouped organisms of common descent that share a fundamental pattern of organization.

pilidium (pī-lid′ē-əm) (Gr. *pilidion*, dim. of *pilos*, felt cap). Free-swimming hat-shaped larva of nemertine worms.

pinna (pin′ə) (L. feather, sharp point). The external ear. Also a feather, wing, or fin or similar part.

pinocytosis (pin′o-cī-tō′sis, pīn′o-cī-to′sis) (Gr. *pinein*, to drink, + *kytos*, hollow vessel, + *-osis*, condition). Taking up of fluid by living cells; cell-drinking.

placenta (plə-sen′tə) (L. flat cake, Gr. *plakous*, from Gr. *plax*, *plakos*, anything flat and broad). The vascular structure, em-bryonic and maternal, through which the embryo and fetus are nourished while in the uterus.

placoid scale (pla′koid) (Gr. *plax*, *plakos*, tablet, plate). Type of scale found in cartilaginous fishes, with basal plate of dentine embedded in the skin and a backward-pointing spine tipped with enamel.

plankton (plank′tən) (Gr. neuter of *planktos*, wandering). The passively floating animal and plant life of a body of water. Compare with nekton.

plantigrade (plan′tə-grād′) (L. *planta*, sole, + *gradus*, step, degree). Pertaining to animals that walk on the whole surface of the foot (for example, man and bear). Compare with digitigrade.

planula (plan′yə-lə) (NL. dim. fr. L. *planus*, flat). Free-swimming, ciliated larval type of coelenterates; usually flattened and ovoid, with an outer layer of ectodermal cells and an inner mass of endodermal cells.

plasma membrane (plaz′mə) (Gr. *plasma*, a form, mold). A living, external, limiting, protoplasmic structure that functions to regulate exchange of nutrients across the cell surface.

pleopod (plē′ə-päd′) (Gr. *plein*, to sail, + *pous*, *podos*, foot). One of the swimming appendages on the abdomen of a crustacean.

pleura (plu′rə) (Gr. side, rib). The membrane that lines each half of the thorax and covers the lungs.

plexus (plek′səs) (L. network, braid). A network, especially of nerves or of blood vessels.

pluteus (plū′dē-əs), pl. **plutei** (L. *pluteus*, movable shed, reading desk). Echinoid larva with elongated processes like the supports of a desk; originally called "painter's easel larva."

poikilothermic (poi-ki′lə-thər′mik) (Gr. *poikilos*, variable, + thermal). Pertaining to animals whose body temperature is variable and fluctuates with that of the environment; cold-blooded. Compare with ectothermic.

polarization (L. *polaris*, polar, + Gr. -iz-, make). The arrangement of positive electric charges on one side of a surface membrane and negative electric charges on the other side (in nerves and muscles).

polymer (pä′lə-mər) (Gr. *polys*, many, + *meros*, part). A chemical compound composed of repeated structural units called monomers.

polymerization (pə-lim′ər-ə-zā′shən). The process of forming a polymer or polymeric compound.

polymorphism (pä′lē-mor′fi-zəm) (Gr. *polys*, many, + *morphe*, form). The presence in a species of more than one type of individual.

polynucleotide (poly + nucleotide). A nucleotide of many mononucleotides combined.

polyp (päl′əp) (Fr. *polype*, octopus, from L. *polypus*, many-footed). The sessile stage in the life cycle of coelenterates.

polypeptide (pä-lē-pep′tīd) (Gr. *polys*, many, + *peptein*, to digest). A molecule consisting of many joined amino acids, not as complex as a protein.

polyphyletic (pä′lē-fī-led′ik) (Gr. *polys*, many, + *phylon*, tribe). Derived from more than one ancestral source; opposed to monophyletic.

polyploid (pä′lə-ploid′) (Gr. *polys*, many, + *ploidy*, number of chromosomes). Characterized by a chromosome number that is greater than two full sets of homologous chromosomes.

polysaccharide (pä′lē-sak′ə-rid -rīd) (Gr. *polys*, many, + *sakcharon*, sugar, from Sanskrit śarkarā, gravel, sugar). A car-

bat / āpe / ärmadillo / herring / fēmale / finch / līce / crocodile / crōw / duck / ūnicorn / ə indicates unaccented vowel sound "uh" as in mammal, fishes, cardinal, heron, vulture / stress as in bi-ol′o-gy, bi′o-log′i-cal

606

bohydrate composed of many monosaccharide units, such as glycogen, starch, and cellulose.

polysome (polyribosome) (Gr. *polys*, many, + *soma*, body). Two or more ribosomes connected by a molecule of messenger RNA.

population (L. *populus*, people). A group of organisms of the same species inhabiting a specified geographic locality.

portal system (L. *porta*, gate). System of large veins beginning and ending with a bed of capillaries; for example, hepatic portal and renal portal systems in vertebrates.

preadaptation. The possession of a condition not adapted to the ancestral environment but that predisposes an organism for survival in some other environment.

prebiotic synthesis. The chemical synthesis which occurred before the emergence of life.

precocial (prē-kō′shəl) (L. *praecoquere*, to ripen beforehand). Referring (especially) to birds whose young are covered with down and are able to run about when newly hatched.

predaceous, predacious (prē-dā′shəs) (L. *praeda*, prey). Living by killing and consuming other animals, predatory.

predator (pred′ə-tər) (L. *praeda*, plunder). An organism that preys upon other organisms for its food.

prehensile (prē-hen′səl) (L. *prehendere*, to seize). Adapted for grasping.

primate (prī′māt) (L. *primus*, first). Any mammal of the order Primates, which includes the tarsiers, lemurs, marmosets, monkeys, apes, and man.

primitive (L. primus, first). Primordial; ancient; little evolved; said of species closely approximating their early ancestral types.

proboscis (prō-bäs′əs) (Gr. *pro*, before, + *boskein*, feed). A snout or trunk; tubular sucking or feeding organ with mouth at the end as in planarians, leeches, and insects; the sensory and defensive organ at the anterior end of certain invertebrates.

producers (L. *producere*, to bring forth). Organisms, such as plants, able to produce their own food from inorganic substances.

progesterone (prō-jes′tə-rōn′) (L. *pro*, before, + *gestare*, to carry). Hormone secreted by the corpus luteum and the placenta; prepares the uterus for the fertilized egg and maintains the capacity of the uterus to hold the embryo and fetus.

proglottid (prō-gläd′əd) (Gr. *proglōttis*, tongue tip, from *pro*, before, + *glōtta*, tongue, + *-id*, suffix). A segment of a tapeworm.

prokaryotic, procaryotic (pro-kar′ē-ot′ik) (Gr. *pro-*, before, + *karyon*, kernel, nut). Not having a visible nucleus or nuclei. Prokaryotic cells were more primitive than eukaryotic cells and persist today in the bacteria and blue-green algae.

pronephros (prō-nef′rəs) (Gr. *pro-*, before, + *nephros*, kidney). Most anterior of three pairs of embryonic renal organs of vertebrates; functional only in adult hagfishes and larval fishes and amphibians; vestigial in mammalian embryos. Adj. **pronephric.**

proprioceptor (prō′prē-ə-sep′tər) (L. *proprius*, own, particular, + receptor). Sensory receptor located deep within the tissues, especially muscles, tendons, and joints, that is responsive to changes in muscle stretch, body position, and movement.

prosimian (prō-sim′ē-ən) (Gr. *pro*, before + L. *simia*, ape). Any member of a group of primitive, arboreal primates: lemurs, tarsiers, lorises, etc.

prosoma (prō-sōm′ə) (Gr. *pro*, before, + *sōma*, body). Anterior part of an invertebrate in which primitive segmentation is not visible; fused head and thorax of arthropod; cephalothorax.

prosopyle (präs′ə-pīl) (Gr. *prosō*, forward, + *pylē*, gate). Connection between the incurrent and radial canal in some sponges.

prostaglandins (prös′tə-glan′dəns). A family of fatty acid tissue hormones, originally discovered in semen, known to have powerful effects on smooth muscle, nerves, circulation, and reproductive organs.

prostomium (prō-stō′mē-əm) (Gr. *pro*, before, + *stoma*, mouth). In most annelids and some molluscs, that part of the head located in front of the mouth.

protease (prō′tē-ās) (Gr. *protein* + *-ase*, enzyme). An enzyme that digests proteins; includes proteinases and peptidases.

protein (prō′tēn, prō′tē-ən) (Gr. *protein* fr. *proteios*, primary). A macromolecule of carbon, hydrogen, oxygen, and nitrogen and sometimes sulfur and phosphorus; composed of chains of amino acids joined by peptide bonds; present in all cells.

prothrombin (pro-thräm′bən) (Gr. *pro*, before, + *thrombos*, clot). A constituent of blood plasma that is changed to thrombin by a catalytic sequence that includes thromboplastin, calcium, and plasma globulins; involved in blood clotting.

protist (prō′tist) (*protos*, first). A member of the kingdom Protista, generally considered to include the unicellular eukaryotic organisms (protozoans and eukaryotic algae).

protonephridium (prō′tō-nə-frid′ē-əm) (Gr. *protos*, first, + *nephros*, kidney). Primitive osmoregulatory or excretory organ consisting of tubule with terminating flame bulb or solenocyte; the unit of a flame bulb system.

protoplasm (prō′tə-plazm) (Gr. *protos*, first + *plasma*, form). Organized living substance; cytoplasm and karyoplasm of the cell.

Protostomia (prō′də-stō′mē-ə) (Gr. *protos*, first, + *stoma*, mouth). A group of higher phyla in which cleavage is determinate, mesoderm and coelom are formed by proliferation of mesodermal bands (schizocoelic formation), and the mouth is derived from the blastopore. Includes the Annelida, Arthropoda, Mollusca, and a number of minor phyla. Compare with Deuterostomia.

proventriculus (pro′ven-trik′yə-ləs) (L. *pro-*, before, + *ventriculum*, ventricle). In birds, the glandular stomach between the crop and gizzard. In insects, a muscular dilation of foregut armed internally with chitinous teeth.

proximal (L. *proximus*, nearest). Situated toward or near the point of attachment; opposite of **distal,** distant.

psammolittoral (sam′ə-lid′ə-rəl) (Gr. *psammos*, sand, + L. *litoralis*, seashore). Pertaining to the intertidal areas of sandy beaches, or the intertidal biota of such regions.

psammon (sa′män) (Gr. *psammos*, sand). Microfauna and microflora inhabiting the interstices between grains of sand of sandy beaches; the psammolittoral biota.

pseudocoel (sū′dō-sēl) (Gr. *pseudēs*, false, + *koilōma*, cavity). A body cavity not lined with peritoneum and not a part of the blood or digestive systems.

pseudopodium (sū′də-pō′dē-əm) (Gr. *pseudēs*, false, + *podion*, small foot, + *eidos*, form). A temporary cytoplasmic protrusion extended out from a protozoan or ameboid cell, and serving for locomotion or for taking up food.

puff. The pattern of swelling of specific bands or gene loci on giant chromosomes during the larval and imaginal stages of flies.

pupa (pyū′pə) (L. girl, doll, puppet). Inactive quiescent stage of the holometabolous insects. It follows the larval and precedes the adult stage.

purine (pyū′rēn) (L. *purus*, pure, + *urina*, urine). Organic base with carbon and nitrogen atoms in two interlocking rings. The parent substance of adenine, guanine, and other naturally occurring bases.

pyrimidine (pī-rim′ə-dēn) (alter. of pyridine, fr. Gr. *pyr*, fire, + -*id*, adjective suffix, + *ine*). An organic base composed of a single ring of carbon and nitrogen atoms; parent substance of several bases found in nucleic acids.

queen. In entomology, the single fully-developed female in a colony of social insects such as bees, ants, and termites, distinguished from workers, unproductive females, and soldiers.

radial cleavage. Type in which early cleavage planes are symmetric to the polar axis, each blastomere of one tier lying directly above the corresponding blastomere of the next layer; indeterminate cleavage.

radula (ra′jə-lə) (L. scraper). Rasping tongue of certain molluscs.

recombinant DNA. DNA technique in which restriction enzymes are used to divide DNA molecules at particular sequences, thus allowing a hybrid molecule to be formed by joining a genetic segment from one organism to a similarly cut segment from another organism; a technique for joining DNA molecules of unrelated organisms.

redia (rē′dē-ə), pl. **rediae** (rē′dē-ē) (from Redi, Italian biologist). A larval stage in the life cycle of flukes; it is produced by a sporocyst larva, and in turn gives rise to many cercariae.

releaser (L. *relaxare*, to unloose). Simple stimulus that elicits an innate behavior pattern.

replication (L. *replicatio*, a folding back). In genetics, the duplication of one or more DNA molecules from the preexisting molecule.

respiration (L. *respiratio*, breathing). Gaseous interchange between an organism and its surrounding medium. In the cell, the release of energy by the oxidation of food molecules.

rete mirabile (rē′tē mə-rab′ə-lē) (L. wonderful net). A network of small blood vessels so arranged that the incoming blood runs countercurrent to the outgoing blood and thus makes possible efficient exchange between the two bloodstreams. Such a mechanism serves to maintain the high concentration of gases in the fish swim bladder.

rhabdoid (rab′doid) (Gr. *rhabdos*, rod). Rodlike structures in the cells of the epidermis or underlying parenchyma in certain turbellarians. They are discharged in mucous secretions.

rheoreceptor (rē′ə-rē-cep′tər) (Gr. *rheos*, a flowing, + receptor). A sensory organ of aquatic animals that responds to water current.

rhopalium (rō-pā′lē-əm) (NL. fr. Gr. *rhopalon*, a club). One of the marginal, club-shaped sense organs of certain jellyfishes; tentaculocyst.

rhynchocoel (ring′kō-sēl) (Gr. *rhynchos*, snout, + *koilos*, hollow). In nemertines, the dorsal tubular cavity that contains the inverted proboscis. It has no opening to the outside.

ribosome (rī′bō-sōm). A small organelle composed of protein and ribonucleic acid. May be free in the cytoplasm or attached to the membranes of the endoplasmic reticulum; thought to function in protein synthesis.

ritualization. In ethology, the evolutionary modification, usually intensification, of a behavior pattern to serve communication.

RNA. Ribonucleic acid, of which there are several different kinds, such as messenger RNA, ribosomal RNA, and transfer RNA.

rostrum (räs′trəm) (L. ship's beak). A snoutlike projection on the head.

sagittal (saj′ə-dəl) (L. *sagitta*, arrow). Pertaining to the median anteroposterior plane that divides a bilaterally symmetric organism into right and left halves.

saprobe (sa′prōb) (Gr. *sapros*, rotten, + *bios*, life). Any organism living on dead or decaying plant or animal life.

saprophagous (sə-präf′ə-gəs) (Gr. *sapros*, rotten, + -*phagos*, from *phagein*, to eat). Feeding on decaying matter; saprobic; saprozoic.

saprophyte (sap′rə-fīt) (Gr. *sapros*, rotten, + phyton, plant). A plant living on dead or decaying organic matter.

saprozoic nutrition (sap-rə-zō′ik) (Gr. *sapros*, rotten, + *zoion*, animal). Animal nutrition by absorption of dissolved salts and simple organic nutrients from surrounding medium; also refers to feeding on decaying matter.

sarcolemma (sär′kə-lem′ə) (Gr. *sarx*, flesh, + *lemma*, rind). The thin noncellular sheath that encloses a striated muscle fiber.

schizocoel (skiz′ō-sēl) (Gr. *schizo*, fr. *schizein*, to split, + *koilōma*, cavity). A body cavity formed by the splitting of embryonic mesoderm. Noun **schizocoelomate** refers to an animal with a schizocoel, such as an arthropod or mollusc. Adj. **schizocoelous.**

schizocoelous mesoderm formation. (skiz′ō-sēl-ləs) Embryonic formation of mesoderm as cords of cells between ectoderm and endoderm; splitting of these cords results in the coelomic space.

schizogony (skə-zä′gə-nē) (Gr. *schizein*, to split, + *gonos*, seed). Multiple asexual fission.

sclerite (skle′rīt) (Gr. *sklēros*, hard). A hard chitinous or calcareous plate or spicule; one of the plates making up the exoskeleton of arthropods, especially insects.

scleroblast (skler′ō-blast) (Gr. *sklēros*, hard, + *blastos*, germ). An amoebocyte specialized to secrete a spicule, found in sponges.

sclerotization (skli′rə-tə-zā′shən). Process of hardening of the cuticle of arthropods by the formation of stabilizing cross-linkages in the protein of chitin.

scolex (skō′leks) (Gr. *skōlēx*, worm, grub). The holdfast, or so-called head, of a tapeworm; it bears suckers and, in some, hooks, and from it new proglottids are budded off.

scrotum (skrō′təm) (L. bag). The pouch that contains the testes in most mammals.

scyphistoma (sī-fis′tə-mə) (Gr. *skyphos*, cup, + *stoma*, mouth). A stage in the development of scyphozoan jellyfish just after the larva becomes attached.

seminiferous (sem-ə-nif′-rəs) (L. *semen*, semen, + *ferre*, to bear). Pertains to the tubules that produce or carry semen in the testes.

semipermeable (L. *semi*, half, + *permeabilis*, capable of being passed through). Permeable to small particles, such as water and certain inorganic ions, but not to larger molecules.

septum, pl. **septa** (L. fence). A wall between two cavities.

serosa (sə-rō′sə) (NL. from L. *serum*, serum). The outer embryonic membrane of birds and reptiles. Chorion.

---

bat / āpe / ärmadillo / herring / fēmale / finch / līce / crocodile / crōw / duck / ūnicorn / ə indicates unaccented vowel sound "uh" as in mammal, fishes, cardinal, heron, vulture / stress as in bi-ol′o-gy, bi′o-log′i-cal

serotonin (sir′ə-tōn′ən) (L. *serum,* serum). A phenolic amine, found in the serum of clotted blood and in many other tissues, that possesses several poorly understood metabolic, vascular, and neural functions; 5-hydroxytryptamine.

serum (sir′əm) (L. whey, serum). The liquid that separates from the blood after coagulation; blood plasma from which fibrinogen has been removed. Also, the clear portion of a biological fluid separated from its particulate elements.

sessile (ses′əl) (L. *sessilis,* low, dwarf). Attached at the base; sedentary; fixed to one spot, not able to move about.

seta, pl. **setae** (sēd′ə, sē′tē) (L. bristle). A needlelike chitinous structure of the integument of annelids and related forms.

siliceous (sə-li′shəs) (L. *silex,* flint). Containing silica.

simian (sim′ē-ən) (L. *simia,* ape). Pertaining to monkeys or apes.

sinus (sī′nəs) (L. curve) A cavity or space in tissues or in bone.

siphonoglyph (sī-fän′ə-glif′) (Gr. *siphōn,* reed, tube, siphon, + *glyphē,* carving). Ciliated furrow in the gullet of sea anemones.

solenocyte (sō-len′ə-sīt) (Gr. *solēn,* pipe, + *kytos,* hollow vessel). Special type of flame bulb in which the bulb bears a flagellum instead of a tuft of cilia. See **flame bulb, protonephridium.**

soma (sō′mə) (Gr. body). The whole of an organism except the germ cells (germ plasm).

somatic (sō-mad′ik) (Gr. *sōma,* body). Refers to the body, for example, somatic cells in contrast to germ cells.

somatoplasm (sō′mə-də-pla′zəm) (Gr. *sōma,* body, + *plasma,* anything formed). The living matter that makes up the mass of the body as distinguished from germ plasm, which makes up the reproductive cells. The protoplasm of body cells.

somite (sō′mīt) (Gr. *sōma,* body). One of the blocklike masses of mesoderm arranged segmentally (metamerically) in a longitudinal series beside the neural tube of the embryo; metamere.

speciation (spē′sē-ā′shən) (L. *species,* kind). The evolutionary process by which new species arise; the process by which variations become fixed.

species (spē′shez, spē′sēz) sing. and pl. (L. *species,* particular kind). A group of interbreeding individuals of common ancestry that are reproductively isolated from all other such groups; a taxonomic unit ranking below a genus and designated by a binomial consisting of its genus and the species name.

spermatheca (spər′mə-thē′kə) (Gr. *sperma,* seed, + *thēkē,* a case). A sac in the female reproductive organs for the reception and storage of sperm.

spermatogonium (spər′mad-ə-gō′nē-əm) (Gr. *sperma,* seed, + *gonē,* offspring). Precursor of mature male reproductive cell; gives rise directly to a spermatocyte.

spermatophore (spər-mad′ə-for′) (Gr *sperma, spermatos,* seed, + *phorein,* to bear). Capsule or packet enclosing sperm, produced by males of several invertebrate groups and a few vertebrates.

sphincter (sfingk′tər) (Gr. *sphinkter,* band, sphincter, from *sphingein,* to bind tight). A ring-shaped muscle capable of closing a tubular opening by constriction.

spicule (spi′kyūl) (L. dim. of *spica,* point). One of the minute calcareous or siliceous skeletal bodies found in sponges, radiolarians, soft corals, and sea cucumbers.

spiracle (spi′rə-kəl) (L. *spiraculum,* fr. *spirare,* to breathe). External opening of a trachea in arthropods. One of a pair of openings on the head of elasmobranchs for passage of water. Exhalent aperture of tadpole gill chamber.

spiral cleavage. A type of early embryonic cleavage in which cleavage planes are diagonal to the polar axis and unequal cells are produced by the alternate clockwise and counterclockwise cleavage around the axis of polarity; determinate cleavage.

spongin (spən′jin) (L. *spongiai,* sponge). Fibrous, scleroprotein material making up the skeletal network of horny sponges.

spongocoel (spən′jō-sēl′) (Gr. *spongos,* sponge, + *koilos,* hollow). Central cavity in sponges.

sporocyst (spō′rə-sist) (Gr. *sporos,* seed, + *kystis,* pouch). A larval stage in the life cycle of flukes; it originates from a miracidium.

sporozoite (spō′rə-zō′it) (Gr. *sporos,* seed, + *zōon,* animal, + *-ite,* suffix for body part). A stage in the life history of many sporozoan Protozoa; released from spores.

statoblast (stad′ə-blast) (Gr. *statos,* standing, fixed, + *blastos,* germ). Biconvex capsule containing germinative cells and produced by most freshwater ectoprocts by asexual budding. Under favorable conditions it germinates to give rise to new zooid.

statocyst (Gr. *statos,* standing, + *kystis,* bladder). Sense organ of equilibrium; a fluid-filled cellular cyst containing one or more granules (statoliths) used to sense direction of gravity.

statolith (Gr. *statos,* standing, + *lithos,* stone). Small calcareous body resting on tufts of cilia in the statocyst.

stenohaline (sten-ə-hā′līn, -lən) (Gr. *stenos,* narrow, + *hals,* salt). Pertaining to aquatic organisms that have restricted tolerance to changes in environmental saltwater concentration.

stenotopic (sten-ə-tä′pik) (Gr. *stenos,* narrow, + *topos,* place). Refers to an organism with a narrow range of adaptability to environmental change; having a restricted geographical distribution.

stereogastrula (ste′rē-ə-gas′trə-lə) (Gr. *stereos,* solid, + *gastēr,* stomach, + L. *-ula,* diminutive). A solid type of gastrula, such as the planula of coelenterates.

sterol (ste′rōl) **steroid** (ste′roid) (Gr. *stereos,* solid, + *-ol* [L. *oleum,* oil]). One of a class of organic compounds containing a molecular skeleton of four fused carbon rings; it includes cholesterol, sex hormones, adrenocortical hormones, and vitamin D.

stigma (Gr. *stigma,* mark, tattoo mark). Eyespot in certain protozoans. Spiracle of certain terrestrial arthropods.

stolon (stō′lən) (L. *stolō, stolonis,* a shoot, or sucker of a plant). A rootlike extension of the body wall that gives rise to buds that may develop into new zooids, thus forming a compound animal in which the zooids remain united by the stolon. Found in some colonial anthozoans, hydrozoans, ectoprocts, and ascidians.

stoma (stō′mə) (Gr. mouth). A mouthlike opening.

stomochord (stō′mə-kord) (Gr. *stoma,* mouth, + L. *chorda,* cord). Anterior evagination of the dorsal wall of the buccal cavity into the proboscis of hemichordates; the buccal diverticulum.

strobila (strō′bə-lə) (Gr. *strobilē,* lint plug like a pine cone [*strobilos*]). A stage in the development of the scyphozoan jellyfish.

sycon (Gr. *sykon,* fig). A type of canal system in certain sponges. Sometimes called syconoid.

symbiosis (sim′bī-ōs′əs, sim′bē-ōs′əs) (Gr. *syn,* with, + *bios,* life). The living together of two different species in an intimate relationship; includes mutualism, commensalism, and parasitism.

sympatry (sim′pə-trē) (Gr. *syn,* with, + *patra,* native land). Having the same or overlapping regions of distribution. Adj. **sympatric.**

synapse (si′naps, si-naps′) (Gr. *synapsis,* contact, union). The place at which a nerve impulse passes between neuron processes, typically from an axon of one nerve cell to a dendrite of another nerve cell.

syncytium (sin-sish′e-əm) (Gr. *syn,* with, + *kytos,* hollow vessel). A mass of protoplasm containing many nuclei and not divided into cells.

syrinx (sir′inks) (Gr. shepherd's pipe). The vocal organ of birds located at the base of the trachea.

tactile (tak′til) (L. *tactilis,* able to be touched, from *tangere,* to touch). Pertaining to touch.

tagma, pl. **tagmata** (Gr. *tagma,* arrangement, order, row). A compound body section of an arthropod resulting from embryonic fusion of two or more segments; for example head, thorax, abdomen.

tagmatization, tagmosis. Organization of the arthropod body into tagmata.

taiga (tī′gä) (Russ.). Habitat zone characterized by large tracts of coniferous forests, long, cold winters, and short summers; most typical in Canada and Siberia.

taxis (taks′əs) (Gr. *taxis,* arrangement). An orientation movement by a (usually) simple organism in response to an environmental stimulus.

tegument (teg′yə-ment) (L. *tegumentum,* fr. *tegere,* to cover). An integument; external covering in cestodes and trematodes, formerly thought to be a cuticle.

telencephalon (tel′en-sef′ə-lon) (Gr. *telos,* end, + *encephalon,* brain). The most anterior vesicle of the brain; the anteriormost subdivision of the prosencephalon that becomes the cerebrum and associated structures.

teleology (tel′ē-äl′ə-jē) (Gr. *telos,* end, + L. *-logia,* study of, from Gr. *logos,* word). The philosphic view that natural events are goal directed and are preordained; contrasts with scientific view of mechanical determinism.

telolecithal (te-lō-les′ə-thəl) (Gr. *telos,* end, + *lekithos,* yolk, + -al). Having the yolk concentrated at one end of an egg.

telson (tel′sən). Posterior projection of the last body segment in many crustaceans.

tentaculocyst (ten-tak′u-lō-sist) (L. *tentaculum,* feeler, + Gr. *kystis,* pouch). One of the sense organs along the margin of medusae; a rhopalium.

territory (L. *territorium,* fr. *terra,* earth). A restricted area preempted by an animal or pair of animals, usually for breeding purposes, and guarded from other individuals of the same species.

test. A shell or hardened outer covering.

tetrad (te′trad) (Gr. *tetras,* four). Group of four chromatids formed by synapsis and resulting from the splitting of paired homologous chromosomes.

tetrapods (te′trə-päds) (Gr. *tetras,* four + *pous,* foot). Four-footed vertebrates; the group includes, amphibians, reptiles, birds, and mammals.

therapsid (thə-rap′sid) (Gr. *theraps,* an attendant). Extinct Mesozoic mammal-like reptile from which true mammals evolved.

tornaria (tor-na′rē-ə) (L. *tornare,* to turn). A free-swimming larva of enteropneusts that rotates as it swims; resembles somewhat the bipinnaria larva of echinoderms.

torsion (LL. *torsio,* fr. L. *torquere,* to twist). A twisting phenomenon in gastropod development that alters the position of the visceral and pallial organs by 180°.

trachea (trā′kē-ə) (ML. windpipe, trachea, from Gr. [*artēria*] *tracheia,* rough [artery]). The windpipe; any of the air tubes of insects.

transcription. Formation of messenger RNA from the coded DNA.

transfer RNA (t-RNA). A form of RNA of about 70 nucleotides, which are adapter molecules in the synthesis of proteins. A specific amino acid molecule is carried by transfer RNA to a ribosome–messenger RNA complex for incorporation into a polypeptide.

translation (L. *translation,* a transferring). The process in which the genetic information present in messenger RNA is used to direct the order of specific amino acids during protein synthesis.

trichocyst (trik′ə-sist) (Gr. *thrix,* hair, + *kystis,* bladder). Saclike protrusible organelle in the ectoplasm of ciliates, which discharges as a threadlike weapon of defense.

triploblastic (trip′lō-blas′tik) (Gr. *triploos,* triple, + *blastos,* germ). Pertaining to metazoans in which the embryo has three primary germ layers—ectoderm, mesoderm, and endoderm.

trochophore (trōk′ə-fōr) (Gr. *trochos,* wheel, + *pherein,* to bear). A free-swimming ciliated marine larva characteristic of most molluscs and certain bryozoans, brachiopods, and marine worms; an ovoid or pyriform body with preoral circlet of cilia and sometimes a secondary circlet behind the mouth.

trophallaxis (trōf′ə-lak′səs) (Gr. *trophē,* food, + *allaxis,* barter. exchange). Exchange of food between young and adults, especially among those of certain social insects.

trophoblast (trōf′ə-blast) (Gr. *trephein,* to nourish, + *blastos,* germ.) Outer ectodermal nutritive layer of blastodermic vesicle; in mammals it is part of the chorion and attaches to the uterine wall.

trophozoite (trōf′ə-zō′īt) (Gr. *trophē,* food, + *zōon,* animal). Adult stage in the life cycle of a sporozoan in which it is actively absorbing nourishment from the host.

tube feet. Numerous small, muscular, fluid-filled tubes projecting from body of echinoderms; part of water-vascular system; used in locomotion, clinging, food handling, and respiration.

tundra (tən′drə) (Russ. from Lapp *tundar,* hill). Terrestrial habitat zone, between taiga in south and polar region in north; characterized by absence of trees, short growing season, and mostly frozen soil during much of the year.

tunic (L. *tunica,* tunic, coat). In tunicates, a cuticular, cellulose-containing covering of the body secreted by the underlying body wall.

turbellarian (tər′bə-lar′ə-an) (L. *turbellae,* a stir or tumult). Free-living flatworm of phylum Platyhelminthes.

typhlosole (tif′lə-sōl′) (Gr. *typhlos,* blind, + *sōlēn,* channel, pipe). A longitudinal fold projecting into the intestine in certain invertebrates such as the earthworm.

umbilical (L. *umbilicus,* navel). Refers to the navel, or umbilical cord.

umbo (əm′bō), pl. **umbones** (əm-bō′nēz) (L. boss of a shield). One of the prominences on either side of the hinge region in a bivalve mollusc shell; the "beak" of a brachiopod shell.

---

bat / āpe / ärmadillo / herring / fēmale / finch / līce / crocodile / crōw / duck / ūnicorn / ə indicates unaccented vowel sound "uh" as in mammal, fishes, cardinal, heron, vulture / stress as in bi-ol′o-gy, bi′o-log′i-cal

ungulate (ən′gyə-lət) (L. *ungula,* hoof). Hoofed; any hoofed mammal.

urethra (yə-rē′thrə) (Gr. *ourethra,* urethra). The tube from the urinary bladder to the exterior in both sexes.

uriniferous tubule (yu′rə-nif′rəs) (L. *urina,* urine, + *ferre,* to bear). One of the tubules in the kidney extending from a Malpighian body to the collecting tubule.

utricle (yū′trə-kəl) (L. *utriculus,* little bag). That part of the inner ear containing the receptors for dynamic body balance; the semicircular canals lead from and to the utricle.

vacuole (vak′yə-wōl′) (L. *vacuus,* empty, + Fr. *-ole,* diminutive). A membrane-bounded, fluid-filled space in a cell.

valve. One of the two shells of a typical bivalve mollusc or brachiopod.

vector (L. a bearer, carrier, fr. *vehere, vectum,* to carry). An animal, usually an insect, that carries and transmits pathogenic microorganisms from one host to another host.

veliger (vēl′ə-jər, vel-) (L. sail-bearing). Larval form of certain molluscs; develops from the trochophore and has the beginning of a foot, mantle, shell, etc.

velum (vē′ləm) (L. *veil,* sail). A membrane on the subumbrella surface of jellyfish of class Hydrozoa.

vestige (ves′tij) (L. *vestigium,* footprint). A rudimentary organ that may have been well developed in some ancestor or in the embryo.

villus (vil′əs), pl. **villi** (L. tuft of hair). A small fingerlike, vascular process on the wall of the small intestine; one of the branching, vascular processes on the embryonic portion of the placenta.

virus (vī′rəs) (L. slimy liquid, poison). A submicroscopic noncellular particle composed of a nucleoprotein core and a protein shell; parasitic and will grow and reproduce in a host cell.

viscera (vis′ər-ə), sing. **viscus** (L., pl. of *viscus,* internal organ). Internal organs in the body cavity.

vitalism (L. *vita,* life). The view that natural processes are controlled by supernatural forces and cannot be explained through the laws of physics and chemistry alone; contrasts with mechanism.

vitamin (L. *vita,* life, + *amine,* from former supposed chemical origin). An organic substance required in small amounts as a catalyst for normal metabolic function; must be supplied in the diet or by intestinal flora because the organism cannot synthesize it.

vitelline membrane (və-tel′ən, vī′təl-ən) (L. *vitellus,* yolk of an egg). The noncellular membrane that encloses the egg cell.

viviparity (vī′və-par′ə-dē) (L. *vivus,* alive, + *parere,* to bring forth). Reproduction in which eggs develop within the female body, with nutritional aid of maternal parent as in therian mammals, many reptiles, and some fishes; offspring are born as juveniles. Adj. **viviparous** (vī-vip′ə-rəs).

water-vascular system. System of fluid-filled closed tubes and ducts peculiar to echinoderms; used to move tentacles and tube feet that serve variously for clinging, food handling, locomotion, and respiration.

zoecium, zooecium (zō-ē′shē-əm) (Gr. *zōon,* animal, + *oikos,* house). Cuticular sheath or shell of Ectoprocta.

zoochlorella (zō′ə-klōr-el′ə) (Gr. *zōion,* life, + *Chlorella*). Any of various minute green algae (usually *Chlorella*) that live symbiotically within the cytoplasm of some protozoans and other invertebrates.

zooid (zō-oid) (Gr. *zōion,* life). An individual member of a colony of animals, such as colonial cnidarians and ectoprocts.

zygote (Gr. *zygōtos,* yoked). The fertilized egg.

INDEX

ORIGIN OF LIFE AND GEOLOGIC TIME TABLE

Millions of years ago	Bacteria	Blue-green algae	Protista	Fungi	Higher plants	Animals	Oxygen in atmosphere	Stages in evolution of life
Present day								
1000								Appearance of multicellular organisms
2000								
								Appearance of oxygen in atmosphere
								Photosynthesis
3000								Anaerobic metabolism
								First living systems
								Biopolymers: proteins, polysaccharides, nucleic acids
4000								Chemical evolution
								Early reducing atmosphere of methane, ammonia, water, nitrogen, carbon dioxide
								Origin of solar system
5000								